Springer Proceedings in Mathematics & Statistics

Volume 314

Springer Proceedings in Mathematics & Statistics

This book series features volumes composed of selected contributions from workshops and conferences in all areas of current research in mathematics and statistics, including operation research and optimization. In addition to an overall evaluation of the interest, scientific quality, and timeliness of each proposal at the hands of the publisher, individual contributions are all refereed to the high quality standards of leading journals in the field. Thus, this series provides the research community with well-edited, authoritative reports on developments in the most exciting areas of mathematical and statistical research today.

More information about this series at http://www.springer.com/series/10533

José Ignacio Burgos Gil · Kurusch Ebrahimi-Fard ·
Herbert Gangl
Editors

Periods in Quantum Field Theory and Arithmetic

ICMAT, Madrid, Spain,
September 15 – December 19, 2014

Springer

Editors
José Ignacio Burgos Gil
Institute of Mathematical Sciences (ICMAT)
Spanish National Research Council (CSIC)
Madrid, Spain

Kurusch Ebrahimi-Fard
Department of Mathematical Sciences
Norwegian University of Science
and Technology
Trondheim, Norway

Herbert Gangl
Department of Mathematical Sciences
Durham University
Durham, UK

ISSN 2194-1009 ISSN 2194-1017 (electronic)
Springer Proceedings in Mathematics & Statistics
ISBN 978-3-030-37030-5 ISBN 978-3-030-37031-2 (eBook)
https://doi.org/10.1007/978-3-030-37031-2

Mathematics Subject Classification (2010): 11M32, 17B81, 20E08, 11G09

This Springer imprint is published by the registered company Springer Nature Switzerland AG
The registered company address is: Gewerbestrasse 11, 6330 Cham, Switzerland

Preface

Theoretical Physics and Mathematics have a long tradition of cross-fertilization with questions, insights and tools having been transferred from either one to the other. A particularly active area over the past few decades has been the relationship between amplitudes in perturbative quantum field theories (pQFT) on the physical side and a certain class of periods and motives on the mathematical side.

Feynman graphs and their associated Feynman integrals play the quintessential role in the perturbative approach to QFT. The calculation of Feynman integrals is crucial for accurate and precise theoretical predictions, and these have been corroborated by experiments probing the realm of elementary particles to spectacularly high accuracy. The computation associated to a particular physical process may a priori involve thousands or even millions of such integrals, though, and in quite a few instances physicists have found ways to invoke structure and symmetry that cut down on that number drastically.

From the point of view of pure mathematics, more precisely, from an arithmetic–geometric perspective, it is profitable to interpret a typical Feynman integral as an integral of a rational algebraic differential form over a domain defined by algebraic equations and inequalities. Such an integral goes under the name *period* in mathematics, and each such period is associated to—or rather a shadow of—a mathematical object called a *motive*. Motives have a very rich structure that can be used to obtain relations between Feynman integrals, dramatically simplifying many computations.

For instance, at low loop order, Feynman integrals tend to produce periods of the simplest motives, the so-called mixed Tate motives (MTM's), more specifically those MTM's which are defined over a field (in the mathematical sense). The periods of MTM's over a number field of the complex numbers are multiple zeta values and values of multiple polylogarithms.

Multiple zeta values (MZV's) and multiple polylogarithms (MPL's) are very active and rather recent research topics in modern mathematics. In fact, a systematic study only started in the early 1990s with the (independent) seminal works of Hoffman and Zagier, although the prehistory can be traced back to Euler in the eighteenth century. In QFT, some of the first instances can be attributed to the work of the theoretical physicists t'Hooft–Veltman (involving dilogarithms, 1972) and to Broadhurst (first conjecturally new irreducible MZV in weight 8, mid-80s). Research on MZV's, MPL's and related problems has rapidly grown in the past two decades, and involves several areas of advanced mathematics: number theory, algebraic K-theory, hyperbolic geometry, combinatorics, arithmetic and algebraic geometry (motives), Lie group theory (Grothendieck–Teichmüeller theory, Drinfel'd associators, Kashiwara–Vergne Conjecture, Broadhurst–Kreimer Conjecture), and as indicated above, show deep connections with high-energy physics.

We would like to mention also that the theory of MZV's and their relatives has a strong combinatorial flavour. The common algebraic structures of these numbers, e.g. shuffle and quasi-shuffle algebras, Lie algebras and Lie series, as well as Hopf algebras, call for a careful study of algebraic and advanced combinatorics in the context of underlying mathematical theory.

When studying Feynman integrals at higher loop order, periods associated to other motives appear, like elliptic polylogarithms. This insight triggered further interest among mathematicians and physicists alike as it revealed rich and fruitful contact points between these communities.

In the year 2014, the Instituto de Ciencias Matemáticas—ICMAT in Madrid hosted a special semester devoted to the study of multiple zeta values and connections to high-energy physics. The present volume is a compilation of papers that arose during two conferences that took place as part of the special semester.

Let us explain quickly the content of the present volume. The reader can find here a survey article by Todorov on the relationship between number theory and QFT as well as a survey by Panzer concerning open problems on periods associated with Feynman integrals.

The review by Stieberger unfolds and discusses periods appearing in superstring amplitudes. Similarly, Schlotterer exposes the links between superstring amplitudes and both classical and elliptic MZV's. The latter ones are being investigated and related to elliptic associators in the paper by Matthes on the mathematical side. Moreover, elliptic polylogarithms play a crucial role in studying the massive sunrise integral in the paper by Adams, Bogner and Weinzierl on the physical side. Further relations between periods and high-energy physics are explained in Vergu's article, which surveys the role of cluster algebras in scattering amplitudes and in particular in $\mathcal{N} = 4$ supersymmetric Yang–Mills theory.

There are competing approaches suggesting good definitions of q-analogues of multiple zeta values satisfying good properties. One version is based on Bachmann and Kühn's bi-brackets, which in turn arose from the study of multiple Eisenstein series. In the present volume, the authors give a conjectural dimension formula for q-MZV's, while Bachmann further provides a review of the subject. In a different direction, Zhao uniformizes different approaches to q-MZV's by generalizing the respective double shuffle relations and 'duality relations', while Singer extends both MZV's and q-MZV's to non-positive integers in a way that respects the shuffle relations using Rota–Baxter algebras. The latter contribution treats these q-analogues in connection with the theory of renormalization, a topic which is also at the heart of the papers by Nikolov who treats it from an operadic point of view, invoking vertex algebras, and by Rejzner who investigates algebraic structures in the so-called Epstein–Glaser renormalization.

The language of moulds developed by Ecalle in the context of his resurgence theory aims at studying local properties of dynamical systems. It is well known that Ecalle provides an alternative presentation of multiple zeta values. It is discussed in the papers by Malvenuto and Patras, who view the shuffle and quasi-shuffle properties in the context of generalized bialgebras, and the one by Salerno and Schneps who use the mould language to give a simple and natural proof of Racinet's theorem showing that formal MZV's, subject to double shuffle relations and modulo products, form a Lie coalgebra.

Other combinatorial structures related to multiple zeta values and polylogarithms are the subject of the papers by Chapoton, by Ebrahimi-Fard, Gray and Manchon, by Manchon and by Foissy and Patras.

Furusho investigates the action of the Grothendieck–Teichmüller group, which, in turn, is closely connected with MZV's, on a vector space of tangles with an application to pro-algebraic knots.

A distribution formula for complex and l-adic polylogarithms is the main result of the paper by Nakamura and Wojtkowiak.

It turns out that many of the above-mentioned objects have a close relationship also to string amplitudes, as reviewed by Stieberger and Schlotterer.

Finally, the paper by Zudilin gives a new proof of the identity

$$\zeta(\{2,1\}^{\ell}) = \zeta(\{3\}^{\ell})$$

using generating series.

As a closing remark, we would like to extend our sincere gratitude to the ICMAT for its hospitality and to its staff, whose professionalism made the organization of the semester possible, as well as to all the participants who contributed to its success.

Last but not least, this volume would not have been possible without the commitment of all the referees to which we extend our sincere gratitude.

Madrid, Spain — José Ignacio Burgos Gil
Trondheim, Norway — Kurusch Ebrahimi-Fard
Durham, UK — Herbert Gangl

Contents

Perturbative Quantum Field Theory Meets Number Theory

Ivan Todorov

Dedicated to the memory of Raymond Stora, mentor and friend.

Abstract Feynman amplitudes are being expressed in terms of a well structured family of special functions and a denumerable set of numbers—*periods*, studied by algebraic geometers and number theorists. The periods appear as residues of the poles of regularized primitively divergent multidimensional integrals. In low orders of perturbation theory (up to six loops in the massless φ^4 theory) the family of hyperlogarithms and multiple zeta values (MZVs) serves the job. The (formal) hyperlogarithms form a double shuffle differential graded Hopf algebra. Its subalgebra of single valued multiple polylogarithms describes a large class of euclidean Feynman amplitudes. As the grading of the double shuffle algebra of MZVs is only conjectural, mathematicians are introducing an abstract graded Hopf algebra of *motivic zeta values* whose weight spaces have dimensions majorizing (hopefully equal to) the dimensions of the corresponding spaces of real MZVs. The present expository notes provide an updated version of 2014's lectures on this subject presented by the author to a mixed audience of mathematicians and theoretical physicists in Sofia and in Madrid.

Keywords Residue · Transcendental · Polylogarithm · Shuffle · Stuffle product · Formal multizeta values · Single-valued hyperlogarithm

I. Todorov (✉)
Institut des Hautes Etudes Scientifiques, 91440 Bures-sur-Yvette, France
e-mail: ivbortodorov@gmail.com

Institute for Nuclear Research and Nuclear Energy, Tsarigradsko Chaussee 72, 1784 Sofia, Bulgaria

J. I. Burgos Gil et al. (eds.), *Periods in Quantum Field Theory and Arithmetic*, Springer Proceedings in Mathematics & Statistics 314,
https://doi.org/10.1007/978-3-030-37031-2_1

1 Introduction

In the period preceding the start of the Large Hadron Collider (LHC) at CERN the "theoretical theorists" indulged into physically inspired speculations. That produced (occasionally) interesting mathematical insights but the contact of the resulting activity with real physics, as much as it existed at all, mainly came through its impact on quantum field theory (QFT). When LHC began working at full swing the major part of the theory which does have true applications in particle physics turned out to be good old perturbative QFT—as it used to be over sixty years ago with quantum electrodynamics. There is a difference, however. Half a century ago the dominating view still was that QFT is "plagued with divergences" and that renormalization merely "hides the difficulties under the rug". In the words of Freeman Dyson [42] perturbative QFT was an issue for divorce between mathematics and physics. The work of Stueckelberg, Bogolubov, Epstein and Glaser, Stora and others gradually made it clear (in the period 1950–1980, although it took quite a bit longer to get generally acknowledged) that perturbative renormalization can be neatly formulated as a problem of extension of distributions, originally defined for non-coinciding arguments in position space. A parallel development, due to Stueckelberg and Petermann, Gell-Mann and Low, Bogolubov and Shirkov (see [12]), culminating in the work of Kenneth Wilson, the renormalization group became a tool to study QFT—well beyond keeping track of renormalization ambiguities. (The authors of [44] have felt the need, even in 2012—the year of the final confirmation of the Standard Model through the discovery of the Higgs boson—to appeal to fellow theorists "to stop worrying [about divergences] and love QFT".) It was however a newer development, pioneered by David Broadhurst that led to an unlikely confluence between particle physics and number theory (see e.g. [15] and references to earlier work cited there). In a nutshell, renormalization consists in subtracting a pole term whose residue is an interesting number—a *period* in the sense of [51]—associated with the corresponding Feynman amplitude, independent of the ambiguities inherent to the renormalization procedure. These numbers also appear in the renormalization group beta-function [48, 68] and, somewhat mysteriously, in the successive approximation of such an all important physical quantity as the anomalous magnetic moment of the electron (see [69] as well as Eq. (36) below). More generally, for rational ratios of invariants and masses, euclidean Feynman amplitudes are periods [11]. Theorists are trying to reduce the evaluation of Feynman amplitudes to an expansion with rational coefficients in a basis of transcendental functions and numbers (see [1, 40] and references therein). Thanks to the rich algebraic structure of the resulting class of functions, this development did not make mathematically minded theorists redundant—substituted by computer programmers.

The present lecture provides an introductory survey of the double shuffle and Hopf algebra of (formal) hyperlogarithms and of the associated multiple zeta values and illustrates their applications to QFT on simple examples of evaluating massless Feynman amplitudes in the position space picture. We note by passing that this picture is advantageous for exhibiting the causal factorization principle of Epstein-Glaser

[43, 63] and it allows an extension to a curved space-time (see [31, 36, 41, 50]). It is also preferable from computational point of view when dealing with off-shell massless amplitudes (see [37, 70, 73–75]). On the other hand, on shell scattering amplitudes are studied for good reasons in momentum space. Moreover, the pioneering work of Bloch, Kreimer and others [6, 7, 9, 28, 30] that displayed the link between Feynman amplitudes, algebraic geometry and number theory is using the "graph polynomial" in the Schwinger (or Feynman) momentum space α-representation. The equivalence of the definition of *quantum periods* in the different pictures is established in [69].

We begin in Sect. 2 with a brief introduction to position space renormalization highlighting the role of "Feynman periods". We point out in Sect. 2.3 that (primitive) 4-point functions in the φ^4 theory are conformally invariant and can be expressed as functions of a complex variable z (that appears subsequently as the argument of multiple polylogarithms). Section 3 is devoted to the double shuffle algebra of hyperlogarithms including the Knizhnik-Zamolodchikov equation for their generating function $L(z)$. The definition of monodromy of $L(z)$ (31) involves the "Drinfeld associator"—the generating series of multiple zeta values (MZVs) whose formal and motivic generalizations are surveyed in Sect. 4. We give, in particular, a pedestrian summary of Brown's derivation of the Hilbert-Poincaré series of the dimensions of weight spaces of motivic zeta values and formulate the more refined Broadhurst-Kreimer's conjecture. The Hopf algebra of MZVs is extended at the end of Sect. 4 to a comodule structure of a quotient Hopf algebra of multiple polylogarithms. In Sect. 5 we review Brown's theory [17, 18] of single valued hyperlogarithms and end up with a couple of illustrative applications. An appendix is devoted to a brief historical survey, including a glimpse into the life and work of Leonhard Euler with whom originates to a large extent the theory of MZVs and polylogarithms.

2 Residues of Primitively Divergent Amplitudes

2.1 Periods in Position Space Renormalization

A position space Feynman integrand $G(\vec{x})$ in a massless QFT is a rational homogeneous function of $\vec{x} \in \mathbb{R}^N$. If G corresponds to a connected graph with $V(\geq 2)$ vertices then, in a four-dimensional (4D) space-time, $N = 4(V-1)$. The integrand is *convergent* if it is locally integrable everywhere so that it defines a homogeneous distribution on $\mathbb{R}^N$. G is *superficially divergent* if it gives rise to a homogeneous density in $\mathbb{R}^N$ of non-positive degree:

$$G(\lambda\vec{x})\, d^N\lambda x = \lambda^{-\kappa} G(\vec{x})\, d^N x\,, \quad \kappa \geq 0\,, \quad \vec{x} \in \mathbb{R}^N \quad (\lambda > 0)\,; \tag{1}$$

κ is called the (superficial) *degree of divergence*. In a scalar QFT with massless propagators a connected graph with a set $\mathscr{L}$ of internal lines gives rise to a Feynman amplitude that is a multiple of the product

$$G(\vec{x}) = \prod_{(i,j)\in\mathcal{L}} \frac{1}{x_{ij}^2}, \; x_{ij} = x_i - x_j, \; x^2 = \sum_\alpha x^\alpha x_\alpha. \tag{2}$$

If G is superficially divergent (i.e. if $\kappa = 2L - N \geq 0$ where L is the number of lines in $\mathcal{L}$) then it is *divergent*—that is, it does not admit a homogeneous extension as a distribution on $\mathbb{R}^N$. (For more general spin-tensor fields whose propagators have polynomial numerators a superficially divergent amplitude may, in fact, turn out to be convergent—see Sect. 5.2 of [63].) A divergent amplitude is *primitively divergent* if it defines a homogeneous distribution away from the *small diagonal* ($x_i = x_j$ for all i, j). The following proposition (Theorem 2.3 of [62]) serves as a definition of both the *residue* $Res\,G$ and of a *renormalized* (primitively divergent) *amplitude* $G^\rho(\vec{x})$.

Proposition 1 *If $G(\vec{x})$ (2) is primitively divergent then for any smooth norm $\rho(\vec{x})$ on $\mathbb{R}^N$ one has*

$$[\rho(\vec{x})]^\varepsilon G(\vec{x}) - \frac{1}{\varepsilon}(Res\;G)(\vec{x}) = G^\rho(\vec{x}) + O(\varepsilon). \tag{3}$$

Here $Res\;G$ is a distribution with support at the origin. Its calculation is reduced to the case $\kappa = 0$ of a logarithmically divergent graph by using the identity

$$(Res\;G)(\vec{x}) = \frac{(-1)^\kappa}{\kappa!}\,\partial_{i_1}...\partial_{i_\kappa} Res\;(x^{i_1}...x^{i_\kappa}G)(\vec{x}) \tag{4}$$

where summation is assumed (from 1 to N) over the repeated indices $i_1, ..., i_\kappa$. If G is homogeneous of degree $-N$ then

$$(Res\;G)(\vec{x}) = res\;(G)\,\delta(\vec{x}) \quad (for\;\partial_i(x^i G) = 0)\,. \tag{5}$$

Here $\delta(\vec{x})$ is the N-dimensional Dirac delta function while the numerical residue $res\;G$ is given by an integral over the hypersurface $\Sigma_\rho = \{\vec{x}|\,\rho(\vec{x}) = 1\}$:

$$res\;G = \frac{1}{\pi^{N/2}}\int_{\Sigma_\rho} G(\vec{x})\sum_{i=1}^N (-1)^{i-1}x^i dx^1\wedge...\widehat{dx^i}...\wedge dx^N, \tag{6}$$

(a hat over an argument meaning, as usual, that this argument is omitted). The residue $res\;G$ is independent of the (transverse to the dilation) surface Σ_ρ since the form in the integrand is closed in the projective space $\mathbb{P}^{N-1}$.

We note that N is even, in fact, divisible by 4, so that $\mathbb{P}^{N-1}$ is orientable.

The functional $res\,G$ is a *period* according to the definition of [51, 61]. Such residues are often called "Feynman" or "quantum" periods in the present context (see e.g. [69]). The same numbers appear in the expansion of the renormalization group beta function (see [48, 68]).

The convention of accompanying the 4D volume d^4x by a π^{-2} ($2\pi^2$ being the volume of the unit sphere $\mathbb{S}^3$ in four dimensions), reflected in the prefactor, goes back at least to Broadhurst and is adopted in [28, 69]; it yields rational residues for one- and two-loop graphs. For graphs with three or higher *number of loops* ℓ ($= h_1$, the *first Betti number* of the graph) one encounters, in general, multiple zeta values of overall weight not exceeding $2\ell - 3$ (cf. [15, 69, 70]). If we denote by L and V the numbers of internal lines and vertices of a connected graph then $\ell = L - V + 1 (= V - 1$ for a connected 4-point graph in the φ^4 theory). With the above choice of the 4D volume form the only residues at three, four and five loops (in the φ^4 theory) are integer multiples of $\zeta(3)$, $\zeta(5)$ and $\zeta(7)$, respectively. The first double zeta value, $\zeta(3, 5)$, appears at six loops (with a rational coefficient) (see the census in [69]). All *known* residues were (up to 2013) rational linear combinations of multiple zeta values (MZVs) [15, 69]. A seven loop graph was recently demonstrated [14, 64] to involve *multiple Deligne values*—i.e., values of *hyperlogarithms* at sixth roots of unity.

Remark 1 The definition of a period is deceptively simple: a complex number is a period if its real and imaginary parts can be written as absolutely convergent integrals of rational functions with rational coefficients in domains given by polynomial inequalities with rational coefficients. The set $\mathbb{P}$ of all periods would not change if we replace everywhere in the definition "rational" by "algebraic". If we denote by $\bar{\mathbb{Q}}$ the *field of algebraic numbers* (the inverse of a non-zero algebraic number being also algebraic) then we would have the inclusions

$$\mathbb{Q} \subset \bar{\mathbb{Q}} \subset \mathbb{P} \subset \mathbb{C}. \tag{7}$$

The periods form a ring (they can be added and multiplied) but the inverse of a period needs not be a period. Feynman amplitudes in an arbitrary (relativistic, local) QFT can be normalized in such a way that the only numerical coefficient to powers of coupling constants and ratios of dimensional parameters that appear are periods [11]. The set of all periods is still countable although it contains infinitely many transcendent numbers. A useful criterion for transcendence is given by the *Hermite-Lindemann theorem: if z is a non-zero complex number then either z or e^z is transcendental*. It follows that $e (= e^1)$ is transcendental and so is π as $e^{i\pi} = -1$ and i is algebraic. Furthermore, the natural logarithm of an algebraic number different from 0 and 1 is transcendental. Examples of periods include the transcendentals

$$\pi = \iint_{x^2+y^2 \le 1} dxdy, \quad \ln n = \int_1^n \frac{dx}{x}, \quad n = 2, 3, \ldots, \tag{8}$$

as well as the values of iterated integrals, to be introduced in Sect. 3, at algebraic arguments. They include both the classical MZVs as well as the above mentioned multiple Deligne values. The basis e of natural logarithms, the Euler constant $\gamma = -\Gamma'(1)$, as well as $\ln(\ln n), \ln(\ln(\ln n)), \ldots$, and $1/\pi$ are believed (but not proven) not to be periods.

2.2 Vacuum Completion of 4-Point Graphs in φ^4

In the important special case of the φ^4 theory (in four space-time dimensions) the definition of residue admits an elegant generalization which also simplifies its practical calculation. Following Schnetz [69, 70] we associate to each 4-point graph Γ of the φ^4 theory a *completed vacuum graph* $\bar{\Gamma}$, obtained from Γ by joining all four external lines in a new vertex "at infinity". An n-vertex *4-regular vacuum graph*—having four edges incident with each vertex and no tadpole loops—gives rise to n 4-point graphs (with $(n-1)$ vertices each) corresponding to the n possible choices of the vertex at infinity. The introduction of such completed graphs is justified by the following result (see Proposition 2.6 and Theorem 2.7 of [69] as well as Sect. 3.1 of [75]).

Theorem 1 *A 4-regular vacuum graph $\bar{\Gamma}$ with at least three vertices is said to be completed primitive if the only way to split it by a four edge cut is by splitting off one vertex. A 4-point Feynman amplitude corresponding to a connected 4-regular graph Γ is primitively divergent iff its completion $\bar{\Gamma}$ is completed primitive. All 4-point graphs with the same primitive completion have the same residue.*

The period of a completed primitive graph $\bar{\Gamma}$ is equal to the residue of each 4-point graph $\Gamma = \bar{\Gamma} - v$ (obtained from $\bar{\Gamma}$ by cutting off an arbitrary vertex v). The resulting common period can be evaluated from $\bar{\Gamma}$ by choosing arbitrarily three vertices $\{0, e\ (s.t.\, e^2 = 1), \infty\}$, setting all propagators corresponding to edges of the type (x_i, ∞) equal to 1 and integrating over the remaining $n-2$ vertices of Γ $(n = V(\Gamma))$:

$$Per(\bar{\Gamma}) \equiv res(\Gamma) = \int \Gamma(e, x_2, ..., x_{n-1}, 0) \prod_{i=2}^{n-1} \frac{d^4 x_i}{\pi^2}\,. \tag{9}$$

The proof uses the conformal invariance of residues in the φ^4-theory.

There are infinitely many primitively divergent 4-point graphs (while there is a single primitive 2-point graph—corresponding to the self-energy amplitude $(x_{12}^2)^{-3}$). A remarkable sequence of ℓ-loop graphs ($\ell \geq 3$) with four external lines, the *zig-zag graphs*, can be characterized by their n-point vacuum completions $\bar{\Gamma}_n$, $n = \ell + 2$ as follows. $\bar{\Gamma}_n$ admits a closed *Hamiltonian cycle* that passes through all vertices in consecutive order such that each vertex i is also connected with $i \pm 2\ (mod\, n)$. These graphs were conjectured by Broadhurst and Kreimer [15] in 1995 and proven by Brown and Schnetz [29] to have residues

$$\begin{aligned} Per(\bar{\Gamma}_{\ell+2}) &= \frac{4 - 4^{3-\ell}}{\ell} \binom{2\ell - 2}{\ell - 1} \zeta(2\ell - 3) \quad \text{for } \ell = 3, 5, ...\,; \\ &= \frac{4}{\ell} \binom{2\ell - 2}{\ell - 1} \zeta(2\ell - 3) \quad \text{for } \ell = 4, 6, ...\end{aligned} \tag{10}$$

We note that the periods for $\ell = 3, 4$ also belong to the with ℓ spokes series and are given by $\binom{2\ell-2}{\ell-1}\zeta(2\ell-3)$ (cf. (75) below).

2.3 Primitive Conformal Amplitudes

Each primitively divergent Feynman amplitude $G(x_1, ..., x_4)$ defines a conformally covariant (locally integrable) function away from the small diagonal $x_1 = ... = x_4$. On the other hand, every four points, $x_1, ..., x_4$, can be confined by a conformal transformation to a projective 2-plane (for instance by sending a point to infinity and another to the origin). Then we can represent each *euclidean* point x_i by a complex number z_i so that

$$x_{ij}^2 = |z_{ij}|^2 = (z_i - z_j)(\bar{z}_i - \bar{z}_j). \tag{11}$$

To make the correspondence between 4-vectors x and complex numbers z explicit we fix a unit vector e and let n be a variable unit vector parametrizing a 2-sphere orthogonal to e. Then any euclidean 4-vector x can be written (in spherical coordinates) in the form:

$$x = r(cos\rho\, e + sin\rho\, n)\,,\ e^2 = 1 = n^2\,,\ en = 0\,,\ r \geq 0\,,\ 0 \leq \rho \leq \pi\,. \tag{12}$$

In these coordinates the 4D volume element takes the form

$$d^4x = r^3 dr sin^2\rho\, d\rho\, d^2n\,, \quad \int_{\mathbb{S}^2} d^2n = 4\pi\,. \tag{13}$$

We associate with the vector x (12) a complex number z such that:

$$z = re^{i\rho} \to x^2 (= r^2) = z\bar{z}\,, \quad (x-e)^2 = |1-z|^2 = (1-z)(1-\bar{z}) \tag{14}$$

$$\int_{n\in\mathbb{S}^2} \frac{d^4x}{\pi^2} = |z-\bar{z}|^2 \frac{d^2z}{\pi}\,, \quad \int_{\mathbb{S}^2} \delta(x) d^4x = \delta(z) d^2z\,. \tag{15}$$

For a graph with four distinct external vertices in the φ^4 theory the amplitude (integrated over the internal vertices) has scale dimension 12 (in mass or inverse length units) and can be written in the form:

$$G(x_1, ..., x_4) = \frac{g(u,v)}{\prod_{i<j} x_{ij}^2} = \frac{F(z)}{\prod_{i<j} |z_{ij}|^2} \tag{16}$$

where the indices run in the range $1 \leq i < j \leq 4$, the (positive real) variables u, v, and (the complex) z are conformally invariant crossratios:

$$u = \frac{x_{12}^2 x_{34}^2}{x_{13}^2 x_{24}^2} = z\bar{z}\,, \quad v = \frac{x_{14}^2 x_{23}^2}{x_{13}^2 x_{24}^2} = |1-z|^2\,, \quad z = \frac{z_{12} z_{34}}{z_{13} z_{24}}\,. \tag{17}$$

The crossratios z and $\bar{z}$ are the simplest realizations of the argument z of the hyperlogarithmic functions introduced in the next section. They also appear (as a consequence of the so called *dual conformal invariance* [38, 39]) in the expressions of momentum space integrals like

$$T(p_1^2, p_2^2, p_3^2) = \int \frac{d^4k}{\pi^2 k^2 (p_1 + k)^2 (k - p_3)^2} = \frac{F(z)}{p_3^2} \tag{18}$$

where $p_1 + p_2 + p_3 = 0$, $\frac{p_1^2}{p_3^2} = z\bar{z}$, $\frac{p_2^2}{p_3^2} = |1-z|^2$ (see Eqs. (5–9) of [40]).

3 Double Shuffle Algebra of Hyperlogarithms

The story of polylogarithms begins with the dilogarithmic function (see the inspired and inspiring lecture [81] as well as the brief historical survey in the Appendix). Here we shall start instead with the modern general notion of a hyperlogarithm [17, 19] whose physical applications are surveyed in [40, 64].

Let $\sigma_0 = 0, \sigma_1, ..., \sigma_N$ be distinct complex numbers corresponding to an alphabet $X = \{e_0, ..., e_N\}$. Let X^* be the set of words w in this alphabet including the empty word $\emptyset$. The hyperlogarithm $L_w(z)$ is an iterated integral [19, 33] defined recursively in any dense simply connected open subset U of the punctured complex plane $D = \mathbb{C} \setminus \Sigma,\ \Sigma = \{\sigma_0, ..., \sigma_N\}$ by the differential equations[1]

$$\frac{d}{dz} L_{w\sigma}(z) = \frac{L_w(z)}{z-\sigma}\,, \quad \sigma \in \Sigma\,, \quad L_\emptyset = 1\,, \tag{19}$$

and the initial condition

$$L_w(0) = 0 \quad \text{for} \quad w \neq 0^n (= 0 \ldots 0), \quad L_{0^n}(z) = \frac{(\ln z)^n}{n!}, \ L_\emptyset = 1. \tag{20}$$

Denoting by σ^n a word of n consecutive σ's we find, for $\sigma \neq 0$,

$$L_{\sigma^n}(z) = \frac{(\ln(1 - \frac{z}{\sigma}))^n}{n!}. \tag{21}$$

There is a correspondence between iterated integrals and multiple power series:

[1]We use following [20, 70] concatenation to the right. Other authors, [14, 40], use the opposite convention. This also concerns the definition of coproduct (63) (65) below.

$$(-1)^d L_{\sigma_1 0^{n_1-1}\ldots\sigma_d 0^{n_d-1}}(z) = Li_{n_1,\ldots,n_d}(\frac{\sigma_2}{\sigma_1}, \ldots, \frac{\sigma_d}{\sigma_{d-1}}, \frac{z}{\sigma_d}) \tag{22}$$

where $Li_{n_1,\ldots,n_d}$ is given by the d-fold series

$$Li_{n_1,\ldots,n_d}(z_1, \ldots, z_d) = \sum_{1\le k_1<\ldots<k_d} \frac{z_1^{k_1}\ldots z_d^{k_d}}{k_1^{n_1}\ldots k_d^{n_d}}. \tag{23}$$

More generally, we have

$$(-1)^d L_{0^{n_0}\sigma_1 0^{n_1-1}\ldots\sigma_d 0^{n_d-1}}(z) =$$

$$\sum_{\substack{k_0\ge 0\, k_i\ge n_i, 1\le i\le d\\ k_0+\ldots+k_d=n_0+\ldots+n_d}} (-1)^{k_0+n_0} \prod_{i=1}^{d} \binom{k_i-1}{n_i-1} L_{0^{k_0}}(z)\, L_{k_1-k_r}(\frac{\sigma_2}{\sigma_1}, \ldots, \frac{\sigma_d}{\sigma_{d-1}}, \frac{z}{\sigma_d}). \tag{24}$$

In particular, $L_{01}(z) = Li_2(z) - \ln z Li_1(z) = Li_2(z) + \ln z \ln(1-z)$. The number of letters $|w| = n_0 + \ldots + n_d$ of a word w defines its *weight*, while the number d of non zero letters is its *depth*. We observe that the product $L_w L_{w'}$ of two hyperlogarithms of weights $|w|$, $|w'|$ and depths d, d' can be expanded in hyperlogarithms of weight $|w| + |w'|$ and depth $d + d'$ (as the product of simplices can be expanded into a sum of higher dimensional simplices). This observation can be formalized as follows. The set X^* of words can be equipped with a commutative *shuffle product* $w ⧢ w'$ defined recursively by

$$\emptyset ⧢ w = w (= w ⧢ \emptyset), \quad au ⧢ bv = a(u ⧢ bv) + b(au ⧢ v) \tag{25}$$

where u, v, w are (arbitrary) words while a, b are letters (note that the empty word $\emptyset$ is *not* a letter). We denote by

$$\mathscr{O}_\Sigma = \mathbb{C}\big[z, \Big(\frac{1}{z-\sigma_i}\Big)_{i=1,\ldots,N}\big] \tag{26}$$

the ring of regular functions on D. Extending by $\mathscr{O}_\Sigma$ linearity the correspondence $w \to L_w$ one proves that it defines a homomorphism of shuffle algebras $\mathscr{O}_\Sigma \otimes \mathbb{C}(X) \to \mathscr{L}_\Sigma$ where $\mathscr{L}_\Sigma$ is the $\mathscr{O}_\Sigma$ span of $L_w, w \in X^*$. The commutativity of the shuffle product is made obvious by the identity

$$L_{u⧢v} = L_u L_v (= L_v L_u). \tag{27}$$

It is easy to verify, in particular, that the dilogarithm $Li_2(z) (= -L_{10}(z))$ given by (23) for $d = 1, n_1 = 2$ disappears from the shuffle product:

$$L_{0⧢1}(z) = L_{01}(z) + L_{10}(z) = L_0(z) L_1(z). \tag{28}$$

If the shuffle relations are suggested by the expansion of products of iterated integrals, the product of series expansions of type (23) suggests the (also commutative) *stuffle product*. Rather than giving a cumbersome general definition we shall illustrate the rule on the simple example of the product of depth one and depth two factors (cf. [40]):

$$\begin{aligned} Li_{n_1,n_2}(z_1, z_2)Li_{n_3}(z_3) &= Li_{n_1,n_2,n_3}(z_1, z_2, z_3) + \\ Li_{n_1,n_3,n_2}(z_1, z_3, z_2) &+ Li_{n_3,n_1,n_2}(z_3, z_1, z_2) + \\ Li_{n_1,n_2+n_3}(z_1, z_2z_3) &+ Li_{n_1+n_3,n_2}(z_1z_3, z_2). \end{aligned} \tag{29}$$

We observe the that the multiple polylogarithms of one variable (with $z_1 = ... = z_{d-1} = 1$ considered in [18, 70] span a shuffle but not a stuffle algebra. As seen from the above example the stuffle product also respects the weight but (in contrast to the shuffle product) only filters the depth (the depth of each term in the right hand side does not exceed the sum of depths of the factors in the left hand side (which is three in Eq. (29)).

It is convenient to rewrite the definition of hyperlogarithms in terms of a formal series $L(z)$ with values in the (free) tensor algebra $\mathbb{C}(X)$ (the complex vector space generated by all words in X^*) which satisfies the *Knizhnik-Zamoldchikov equation*:

$$L(z) := \sum_w L_w(z)w, \quad \frac{d}{dz} L(z) = L(z) \sum_{i=0}^{N} \frac{e_i}{z - \sigma_i}. \tag{30}$$

One assigns weight -1 to e_σ, so that $L(z)$ carries weight zero. If the index of the hyperlogarithm L_w is expressed by its (potential) singularities σ_i the word w which multiplies it in the series (30) should be written in terms of the corresponding (non-commuting) symbols e_i (thus justifying the apparent doubling of notation). In the special case when the alphabet X consists of just two letters e_0, e_1 corresponding to $\sigma_0 = 0, \sigma_1 = 1,\ L(z)$ is the generating function of the *classical multipolylogarithms* while its value at $z = 1, Z := L(1)$ is the generating function of MZVs. In these notations the *monodromy* of L around the points 0 and 1 is given by

$$\mathcal{M}_0\, L(z) = e^{2\pi i e_0}\, L(z)\,, \quad \mathcal{M}_1\, L(z) = Z\, e^{2\pi i e_1}\, Z^{-1} L(z),\ Z = \sum_w \zeta_w w, \tag{31}$$

so that $\mathcal{M}_0 L_{0^n}(z) = L_{0^n}(z) + 2\pi i\, L_{0^{(n-1)}}(z)$, $\mathcal{M}_1 Li_n(z) = Li_n(z) - 2\pi i\, L_{0^{(n-1)}}(z)$. The first relation (31) follows from the fact that $L(z)$ is the unique solution of the Knizhnik-Zamolodchikov equation obeying the "initial" condition

$$L(z) = e^{e_0 \ln z} h_0(z),\ h_0(0) = 1, \tag{32}$$

$h_0(z)$ being a formal power series in the words in X^* that is holomorphic in z in the neighborhood of $z = 0$. The second relation (31) is implied by the fact that there exists a counterpart $h_1(z)$ of h_0, holomorphic around $z = 1$ and satisfying $h_1(1) = 1$

such that

$$L(z) = Z\, e^{e_1 \ln(1-z)}\, h_1(z) \tag{33}$$

(see [70] or Appendix C of [73]). This construction can be viewed as a special case (corresponding to $N = 1$) of a monodromy representation of the fundamental group of the punctured plane D (studied in Sect. 6 of [17]).

The next simplest case, $N = 2$, including the square roots of unity ± 1, has been considered by physicists [65] under the name *harmonic polylogarithms*. The value of the function

$$L_{-10^{n-1}}(z) = \sum_{k=1}^{\infty} (-1)^{k-1} \frac{z^k}{k^n} = -Li_n(-z) \tag{34}$$

at $z = 1$ is the *Euler phi function* [4] (alias, Dirichlet eta function)

$$\phi(n) = \sum_{k=1}^{\infty} (-1)^{k-1} \frac{1}{k^n} = (1 - 2^{1-2^{1-n}})\zeta(n), \quad \phi(1) = \ln(2), \tag{35}$$

a special case of the Dirichlet *L-functions* [71]. It is remarkable that the anomalous magnetic moment of the electron, the most precisely measured quantity in physics, is expressed in terms of values of the Dirichlet eta function at integer points:

$$\frac{g-2}{2} = \frac{1}{2}\frac{\alpha}{\pi} + (\phi(3) - 6\phi(1)\phi(2) + \frac{197}{2^4 3^2})(\frac{\alpha}{\pi})^2 + \ldots \tag{36}$$

(see [69] where also the next (α^3-)contribution is expressed in terms of (multiple) eta values). The same weight three combination, $\phi(3) - 6\phi(1)\phi(2)$, appears in the second order of the Lamb shift calculation (see [57]).

Remark 2 The repeated application of the recursive differential equations (19) leads to $dL_\sigma(z) = \frac{dz}{z-\sigma}$ $(d1 = 0)$. Brown [17] calls such differential equations *unipotent* and proves that the double shuffle algebra L_Σ is a *differential graded* (by the weight) *algebra*.

The weight of consecutive terms in the expansion of $L(z)$ (33) is the sum of the weights of hyperlogarithms and the zeta factors. It is thus natural to begin the study of multiple polylogarithms with the algebra of MZVs.

4 Formal Multizeta Values

4.1 Shuffle Regularized MZVs

We now turn to the alphabet X of two letters e_0, e_1 corresponding to $\sigma_0 = 0, \sigma_1 = 1$ and restrict the multiple polylogarithm (23) to a single variable:

$$Li_{n_1,\dots,n_d}(z) = \sum_{1 \le k_1 < \dots < k_d} \frac{z^{k_d}}{k_1^{n_1} \dots k_d^{n_d}}.$$

The MZV $\zeta(n_1, \dots, n_d)$ is then defined as its value at 1 whenever the corresponding series converges. Using also (22) we can write:

$$(-1)^d \zeta_{10^{n_1-1}\dots 10^{n_d-1}} = \zeta(n_1, \dots, n_d) = \sum_{1 \le k_1 < \dots < k_d} \frac{1}{k_1^{n_1} \dots k_d^{n_d}} \text{ for } n_d > 1. \qquad (37)$$

The convergent MZVs of a given weight satisfy a number of shuffle and stuffle identities. Looking for instance at the shuffle (sh) and the stuffle (st) products of two $-\zeta_{10} = \zeta(2)$ we find:

$$sh : \zeta_{10}^2 = 4\zeta_{1100} + 2\zeta_{1010} (= 4\zeta(1,3) + 2\zeta(2,2)); \;\; st : \zeta(2)^2 = 2\zeta(2,2) + \zeta(4);$$
$$\text{hence } \zeta(4) = 4\zeta(1,3) = \zeta(2)^2 - 2\zeta(2,2). \qquad (38)$$

There are no non-zero convergent words of weight 1 and hence no shuffle or stuffle relations of weight 3. On the other hand, already Euler has discovered the relation: $\zeta(1, 2) = \zeta(3)$. Thus shuffle and stuffle relations among convergent words do not exhaust all known relations among MZVs of a given weight. Introducing the divergent zeta values which correspond to $n_d = 1$ we observe that they cancel in the difference between the shuffle and stuffle products $u \sqcup\!\sqcup v - u * v$ of divergent words. For instance, at weight 3 we have

$$\zeta((1) \sqcup\!\sqcup (2)) = 2\zeta(1,2) + \zeta(2,1); \;\; \zeta((1) * (2)) = \zeta(1,2) + \zeta(3) + \zeta(2,1). \qquad (39)$$

Extending the homomorphism $w \to \zeta(w)$ as a homomorphism of both the shuffle and the stuffle algebras to divergent words, assuming, in particular, that $\zeta((1) \sqcup\!\sqcup (2)) = \zeta((1) * (2)) = \zeta(1)\zeta(2)$ and taking the difference of the two equations (39) we observe that all divergent zeta's cancel and we recover Euler's identity above.

Remark 3 In fact, the shuffle product is naturally defined (as we did in (25)) in the two-letter alphabet $\{0, 1\}$ used as lower indices, while the stuffle product has a simple formulation in the infinite alphabet of all positive integers, appearing (in parentheses) as arguments of zeta. Equation (37) provides the translation between the two:

$$\vec{n} = (n_1, \dots, n_d) \leftrightarrow (-1)^d, \rho(\vec{n}) \text{ for } \rho(\vec{n}) = 10^{n_1-1} \dots 10^{n_d-1}. \qquad (40)$$

Using this correspondence one obtains, in particular, the first relation (39).

It is useful to introduce shuffle regularized MZVs using the following result (see Lemma 2.2 of [20]).

Proposition 2 *There is a unique way to define a set of real numbers* $I(a_0; a_1, \dots, a_n; a_{n+1})$ *for any* $a_i \in \{0, 1\}$, *such that*

$$(i)\ I(0;1,a_2,....a_{n-1},0;1)=(-1)^d\zeta(n_1,...,n_d)\ for\ \rho(\vec{n})=(1,a_2,...,a_{n-1},0);$$
$$(ii)\ I(a_0;a_1;a_2)=0,\ I(a_0,a_1)=1\ for\ all\ a_0,a_1,a_2\in\{0,1\};$$
$$(iii)\ I(a_0;a_1,...,a_r;a_{n+1})I(a_0;a_{r+1},...,a_{r+s})=$$
$$\sum_{\sigma\in\Sigma(r,s)} I(a_0;a_{\sigma(1)},...,a_{\sigma(r+s)};a_{n+1})\ (r+s=n);$$
$$(iv)\ I(a;a_1,...,a_n;a)=0\ for\ n>0,a\in\{0,1\}$$
$$(v)\ I(a_0;a_1,...,a_n;a_{n+1})=(-1)^n I(a_{n+1};a_n,...,a_1;a_0);$$
$$(vi)\ I(a_0;a_1,...,a_n;a_{n+1})=I(1-a_{n+1};1-a_n,...,1-a_1;1-a_0). \tag{41}$$

Here $\Sigma(r,s)$ is the set of permutations of the indices $(1,...,n)$ preserving the order of the first r and the last s among them; Eq. (v) is the reverse of path formula, while (vi) expresses functoriality with respect to the map $t\to 1-t$. Equation $\zeta(n_1,...,n_d)=(-1)^d I(0;\rho(\vec{n});1)$ then defines the *shuffle regularized zeta values for all* $n_d\geq 1$. Condition (ii) implies, in particular, $\zeta(1)=0$.

In fact, it suffices to add a condition involving multiplication by the divergent word (1),

$$\zeta((1)\sqcup\!\sqcup w-(1)*w)=0 \text{ for all convergent words } w, \tag{42}$$

to the shuffle and stuffle relations among convergent words in order to obtain all known relations among MZVs of a given weight. For $w=(n), n\geq 2$ (a word of depth 1), Eq. (42) gives

$$\zeta((1)\sqcup\!\sqcup(n)-(1)*(n))=\sum_{i=1}^{n-1}\zeta(i,n+1-i)-\zeta(n+1)=0 \tag{43}$$

(a relation known to Euler). The discovery (and the proof) that

$$\zeta(2n)=-\frac{B_{2n}}{2(2n)!}(2\pi i)^{2n},\ B_2=\frac{1}{6},\ B_4=-\frac{1}{30},\ B_6=\frac{1}{42},\ (-1)^{n-1}B_{2n}\in\mathbb{Q}_{>0}, \tag{44}$$

where B_n are the (Jacob) Bernoulli numbers, was among the first that made Euler famous (see Appendix). Nothing is known about the transcendence of $\zeta(n)$ (or of $\frac{\zeta(n)}{\pi^n}$) for odd n. We introduce following Leila Schneps [67] the notion of the $\mathbb{Q}$-algebra $\mathscr{F}\mathscr{Z}$ of *formal MZVs* ζ^f which satisfy the relations:

$$\zeta^f(1)=0,\ \zeta^f(u)\zeta^f(v)=\zeta^f(u\sqcup\!\sqcup v)=\zeta^f(u*v),\ \zeta^f((1)\sqcup\!\sqcup w-(1)*w)=0. \tag{45}$$

The algebra $\mathscr{F}\mathscr{Z}=\bigoplus_n\mathscr{F}\mathscr{Z}_n$ is *weight graded* and

$$\mathscr{F}\mathscr{Z}_0=\mathbb{Q},\mathscr{F}\mathscr{Z}_1=\{0\},\mathscr{F}\mathscr{Z}_2=\langle\zeta(2)\rangle,\mathscr{F}\mathscr{Z}_3=\langle\zeta(3)\rangle,\mathscr{F}\mathscr{Z}_4=\langle\zeta(4)\rangle,$$
$$\mathscr{F}\mathscr{Z}_5=\langle\zeta(5),\zeta(2)\zeta(3)\rangle,\mathscr{F}\mathscr{Z}_6=\langle\zeta(2)^3,\zeta(3)^2\rangle,$$
$$\mathscr{F}\mathscr{Z}_7=\langle\zeta(7),\zeta(2)\zeta(5),\zeta(2)^2\zeta(3)\rangle, \tag{46}$$

where $< x, y, ... >$ is the $\mathbb{Q}$ vector space spanned by $x, y, ...$ (and we have replaced ζ^f by ζ in the right hand side for short). Clearly, there is a surjection $\zeta^f \to \zeta$ of $\mathcal{FZ}$ onto $\mathcal{Z}$. **The main conjecture in the theory of MZVs** is that *this surjection is an isomorphism of graded algebras*. This is a strong conjecture. If true it would imply that there is no linear relation among MZVs of different weights over the rationals. Actually, a less obvious statement is valid: such an isomorphism would imply that all MZVs are transcendental. Indeed, if a non-zero multiple zeta value is algebraic, then expanding out its minimal polynomial according to the shuffle relation $\zeta(u)\zeta(v) = \zeta(u \sqcup\!\sqcup v)$ (starting with $\zeta^2(w)$) would give a linear combination of multiple zetas in different weights equal to zero, contradicting the weight grading. In fact, we only know that there are infinitely many linearly independent over $\mathbb{Q}$ odd zeta values (Ball and Rivoal, 2001) and that $\zeta(3)$ is irrational (Apéry, 1978). From now on, we shall follow the physicists' practice to treat this conjecture as true and to omit the f's in the notation for (formal) MZVs.

Examples: E1. In order to see that the space $\mathcal{Z}_4$ of weight four zeta values is 1-dimensional we should add to Eqs. (38) the relation (43) for $n = 3$ and its depth three counterpart:

$$\zeta((1) \sqcup\!\sqcup (1, 2) - (1) * (1, 2)) = \zeta(1, 1, 2) - \zeta(1, 3) - \zeta(2, 2)(= \zeta(1, 1, 2) - \zeta(4)) = 0. \tag{47}$$

This allows to express all zeta values of weight four as (positive) integer multiples of $\zeta(1, 3)$ (see Eq. (B.8) of [73]).

E2. The shuffle and the stuffle products corresponding to $\zeta(2)\zeta(3)$ give two relations which combined with (43) for $n = 4$ allow to express the three double zeta values of weight five in terms of simple ones:

$$\zeta(1, 4) = 2\zeta(5) - \zeta(2)\zeta(3),\ \zeta(2, 3) = 3\zeta(2)\zeta(3) - \frac{11}{2}\zeta(5),$$
$$\zeta(3, 2) = \frac{9}{2}\zeta(5) - 2\zeta(2)\zeta(3). \tag{48}$$

In general, the number of convergent words of weight n and depth d in the alphabet $\{0, 1\}$ is $\binom{n-2}{d}$, so their number at weight n is 2^{n-2}. As it follows from Eq. (46) the number of relations also grows fast: there are six relations among the eight MZVs at weight five; 14 such relations at weight six, 29, at weight seven. One first needs a double zeta value, say $\zeta(3, 5)$, in order to write a basis (of four elements) at weight eight (there being 60 relations among the 2^6 elements of $\mathcal{FZ}_8$). Taking the identities among (formal) zeta values into account we can write the generating series Z of MZV (also called *Drinfeld's associator*) in terms of multiple commutators of e_0, e_1:

$$Z = 1 + \zeta(2)[e_0, e_1] + \zeta(3)[[e_0, e_1], e_0 + e_1] + ... \tag{49}$$

It is natural to ask what is the dimension d_n of the space $\mathcal{FZ}_n$ of (formal) MZVs of any given weight n and then to construct a basis of independent elements. These problems have only been solved for the so called *motivic MZV*. Here is a simple-minded substitute of their abstract construction.

4.2 Hopf Algebra of Motivic Zeta Values

Consider the *concatenation algebra*

$$\mathscr{C} = \mathbb{Q}\langle f_3, f_5, ...\rangle, \tag{50}$$

the free algebra over $\mathbb{Q}$ on the countable alphabet $\{f_3, f_5, ...\}$ (see Example 21 in [77]). *The algebra of motivic zeta values* is identified (non-canonically) with the algebra

$$\mathscr{C}[f_2] = \mathscr{C} \otimes_{\mathbb{Q}} \mathbb{Q}[f_2], \tag{51}$$

which plays an important role in the theory of mixed Tate motives (see Sect. 3 of [20]). The algebra $\mathscr{C}[f_2]$ is graded by the weight (the sum of indices of f_i) and it is straightforward to compute the dimension d_n of the weight spaces$\mathscr{C}[f_2]_n$ for any n. Indeed, the generating (or *Hibert-Poincaré*) series for the dimensions $d_n^{\mathscr{C}}$ of the weight n subspace of $\mathscr{C}$ is given by

$$\sum_{n\geq 0} d_n^{\mathscr{C}} t^n = \frac{1}{1-t^3-t^5-...} = \frac{1-t^2}{1-t^2-t^3} \tag{52}$$

while the corresponding series of the second factor $\mathbb{Q}[f_2]$ in (51) is $(1-t^2)^{-1}$. Multiplying the two we obtain the dimensions d_n of the weight spaces conjectured by Don Zagier:

$$\sum_{n\geq 0} d_n t^n = \frac{1}{1-t^2-t^3}, \; d_0 = 1, d_1 = 0, d_2 = 1, d_{n+2} = d_n + d_{n-1}. \tag{53}$$

Here is a wonderful more detailed conjecture advanced by Broadhurst and Kreimer [15] (1997); its motivic version is still occupying mathematicians.

Let $\mathscr{Z}_n^r$ be the linear span of $\zeta(n_1, ..., n_k)$, $n_1 + ... + n_k = n$, $k \leq r$; we define $d_{n,r}$ as the dimension of the quotient space $\mathscr{Z}_n^r / \mathscr{Z}_n^{r-1}$. Broadhurst and Kreimer have advanced the following conjecture for the generating series of $d_{n,r}$ (based on experience with MZVs appearing in Feynman amplitudes):

$$D(X, Y) = \frac{1 + \mathscr{E}(X)Y}{1 - \mathscr{O}(X)Y + \mathscr{S}(X)Y^2(1-Y^2)} = \sum d_{n,r} X^n Y^r. \tag{54}$$

Here $\mathscr{E}(X)$ and $\mathscr{O}(X)$ generate series of even and odd powers of X,

$$\mathscr{E}(X) = \frac{X^2}{1-X^2} = X^2 + X^4 + ..., \ \mathscr{O}(X) = \frac{X^3}{1-X^2} = X^3 + X^5 + ..., \tag{55}$$

while $\mathscr{S}(X)$ is the generating series for the dimensions of the spaces of *cuspidal modular forms* (see for background the physicists' oriented survey [80]):

$$\mathscr{S}(X) = \frac{X^{12}}{(1-X^4)(1-X^6)}. \tag{56}$$

Setting in (54) $Y = 1$ we recover the Zagier conjecture (53) (with $d_n = \sum_r d_{n,r}$), proven for motivic MZVs. The ansatz (54) can presently only be derived in the motivic case under additional assumptions (cf. [32]).

The concatenation algebra $\mathscr{C}$, identified with the quotient

$$\mathscr{C} = \mathscr{C}[f_2]/\mathbb{Q}[f_2], \tag{57}$$

can be equipped with a *Hopf algebra structure* (with f_i as primitive elements) with the *deconcatenation coproduct* $\Delta : \mathscr{C} \to \mathscr{C} \otimes \mathscr{C}$ given by

$$\Delta(f_{i_1}...f_{i_r}) = 1 \otimes f_{i_1}...f_{i_r} + f_{i_1}...f_{i_r} \otimes 1 + \sum_{k=1}^{r-1} f_{i_1}...f_{i_k} \otimes f_{i_{k+1}}...f_{i_r}. \tag{58}$$

This coproduct can be extended to the trivial comodule $\mathscr{C}[f_2]$ (51) by setting

$$\Delta : \mathscr{C}[f_2] \to \mathscr{C} \otimes \mathscr{C}[f_2], \ \Delta(f_2) = 1 \otimes f_2 \tag{59}$$

(and assuming that f_2 commutes with f_{odd}). Remarkably, there appear to be a one-to-one (albeit non-canonical) correspondence between the bases of the weight spaces $\mathscr{Z}_n$ and $\mathscr{C}[f_2]_n$ as displayed in the following list ([20], 3.4)

$$\begin{aligned}
&\langle\zeta(2)\rangle \leftrightarrow \langle f_2\rangle; \langle\zeta(3)\rangle \leftrightarrow \langle f_3\rangle; \langle\zeta(2)^2\rangle \leftrightarrow \langle f_2^2\rangle;\\
&\langle\zeta(5), \zeta(2)\zeta(3)\rangle \leftrightarrow \langle f_5, f_2 f_3 (= f_3 f_2)\rangle; \langle\zeta(2)^3, \zeta(3)^2\rangle \leftrightarrow \langle f_2^3, f_3 \sqcup\!\sqcup f_3\rangle;\\
&\langle\zeta(7), \zeta(2)\zeta(5), \zeta(2)^2\zeta(3)\rangle \leftrightarrow \langle f_7, f_2 f_5, f_2^2 f_3\rangle;\\
&\langle\zeta(2)^4, \zeta(2)\zeta(3)^2, \zeta(3)\zeta(5), \zeta(3,5)\rangle \leftrightarrow \langle f_2^4, f_3 \sqcup\!\sqcup f_3 f_2, f_3 \sqcup\!\sqcup f_5, f_5 f_3\rangle.
\end{aligned} \tag{60}$$

There is a counterpart of Proposition 2 defining *motivic iterated integrals* whose Hopf algebra,[2] [47], is non-canonically isomorphic to $\mathscr{C}[f_2]$. It allows to define a surjective *period map* $\mathscr{C}[f_2] \to \mathscr{Z}$ onto the algebra of real MZVs ([20] Theorem 3.5). Since, on the other hand, $\mathscr{C}[f_2]$ satisfies the defining relations of the formal zeta values we have the surjections $\mathscr{F}\mathscr{Z} \to \mathscr{C}[f_2] \to \mathscr{Z}$. Our main conjecture would then mean that the two (surjective) maps are also injective and thus define isomorphisms

[2]Brown's definition which we follow differs from Goncharov's (adopted in [32]) in that the motivic $\zeta^m(2)$ is non-zero.

of graded algebras.. If true it would imply that the (infinite sequence of) numbers $\pi, \zeta(3), \zeta(5), \ldots$ are transcendental algebraically independent over the rationals (cf. [77]). It would also fix the dimension of the weight spaces $\mathscr{Z}_n$ to be equal to d_n (53). Presently, we only know that this is true for $n = 0, 1, 2, 3, 4$; in general, the above cited results prove that

$$dim\mathscr{Z}_n \leq d_n \;, \quad dim\mathscr{Z}_n = d_n \quad \text{for} \quad n \leq 4 \,. \tag{61}$$

Remark 4 The validity of the above sharpened conjecture would imply, in particular, that $\zeta(2n+1)$ are primitive elements of the Hopf algebra of MZVs:

$$\Delta(\zeta(2n+1)) = \zeta(2n+1) \otimes 1 + 1 \otimes \zeta(2n+1). \tag{62}$$

Eq. (44) precludes the possibility of extending this property to even zeta values. Indeed, it implies the relation $\zeta(2n) = b_n \zeta(2)^n$, $b_n = \frac{(24)^n |B_{2n}|}{2(2n)!}$ which is only compatible with the one-sided coproduct $\Delta\zeta(2) = 1 \otimes \zeta(2)$.

If for weights $n \leq 7$ one can express all MZVs in terms of (products of) simple zeta values (of depth one) the last equation (60) shows that for $n \geq 8$ this is no longer possible. Brown [21] has established that the *Hoffman elements* $\zeta(n_1, \ldots, n_d)$ with $n_i \in \{2, 3\}$ form a basis of motivic zeta values for all n (see also [35, 77]).

The coproduct for MZV, described in Remark 4 extends to hyperlogarithms and can be formulated in terms of the regularized iterated integrals of Proposition 2—see Theorem 3.8 of [20] and Sect. 5.3 of [40]. Here we shall just reproduce the special case of the coproduct of a classical polylogarithm:

$$\Delta Li_n(z) = Li_n(z) \otimes 1 + \sum_{k=0|}^{n-1} \frac{(\ln z)^k}{k!} \otimes Li_{n-k}(z). \tag{63}$$

According to Remark 4, specializing to $z = 1$ in (63) for even n leads to a contradiction unless we factor the algebra of hyperlogarithms by $\zeta(2)$ or, better, by $\ln(-1) = i\pi (= \sqrt{-6\zeta(2)})$ setting

$$\mathscr{H} := \mathscr{L}_\Sigma / i\pi \mathscr{L}_\Sigma \quad \text{so that} \quad \mathscr{L}_\Sigma = \mathscr{H}[i\pi] \,. \tag{64}$$

The coaction Δ is then defined on the comodule $\mathscr{L}_\Sigma$ as follows:

$$\Delta : \mathscr{L}_\Sigma \to \mathscr{H} \otimes \mathscr{L}_\Sigma \,, \quad \Delta(i\pi) = 1 \otimes i\pi \,. \tag{65}$$

The asymmetry of the coproduct is also reflected in its relation to differentiation and to the discontinuity $disc_\sigma = \mathscr{M}_\sigma - 1$:

$$\Delta(\frac{\partial}{\partial z} F) = (\frac{\partial}{\partial z} \otimes id)\Delta F \,, \quad \Delta(disc_\sigma F) = (id \otimes disc_\sigma)\Delta F \,. \tag{66}$$

We leave it to the reader to verify that e.g. for $F = Li_2(z)$ both sides of (66) give the same result. This allows us, in particular, to consider $\mathcal{L}_\Sigma$ as a *differential graded Hopf algebra*.

5 Single-Valued Hyperlogarithms. Applications

Knowing the action of the monodromy M_{σ_i} around each singular point of a hyperlogarithm one can construct single valued hyperlogarithms in the tensor product of $\mathcal{L}_\Sigma$ with its complex conjugate [17]. We shall survey this construction for classical multiple polylogarithms $\mathcal{L}_\Sigma = \mathcal{L}_c$, defined as $\mathcal{O}$-linear combination of $L_w(z)$ for w, words in the "Morse alphabet" $X = \{e_0, e_1\} \leftrightarrow \{0, 1\}$, where $\mathcal{O} = \mathbb{C}[z, \frac{1}{z}, \frac{1}{z-1}]$. This case is spelled out in [18, 70]. The tensor product $\bar{\mathcal{L}}_c \otimes \mathcal{L}_c$ contains functions of $(\bar{z}, z)$ transforming under arbitrary representations of the monodromy group (see Theorem 7.4 of [17]) including the trivial one,—i.e. the *single-valued multiple polylogarithms* (SVMPs). We introduce an $\bar{\mathcal{O}}\mathcal{O}$ basis of homogeneous SVMPs $P_w(z)$ and will denote by

$$P_X(z) = \sum_{w \in X^*} P_w(z) w \tag{67}$$

its generating series. Their significance stems from the fact that a large class of euclidean Feynman amplitudes are single valued. The following theorem is a special case of Theorem 8.1 proven by Brown [17] (coinciding with Theorem 2.5 of [70]).

Theorem 2 *There exists a unique family of single-valued functions $P_w(z), w \in X^*, z \in \mathbb{C} \setminus \{0, 1\}$ such that their generating function (67) satisfies the following Knizhnik-Zamolodchikov equations and initial condition:*

$$\begin{aligned}
&\partial P_X(z) = P_X(z)(\frac{e_0}{z} + \frac{e_1}{z-1}) \,, \ \partial := \frac{\partial}{\partial z} \,, \ \bar{\partial} := \frac{\partial}{\partial \bar{z}} , \\
&\bar{\partial} P_X(z) = (\frac{e_0}{\bar{z}} + \frac{e_1'}{1-\bar{z}}) P_X(z) \,, \ Z_{-e_0,-e_1'} e_1' Z^{-1}_{-e_0,-e_1'} = Z_{e_0,e_1} e_1 Z^{-1}_{e_0,e_1} \,, \\
&P_X(z) \sim e^{e_0 \ln(z\bar{z})} \ \ for \ \ z \sim 0 \,.
\end{aligned} \tag{68}$$

The functions $P_w(z)$ are linearly independent over $\bar{\mathcal{O}}\mathcal{O}$ and satisfy the shuffle relations. Every element of their linear span has a primitive with respect to $\frac{\partial}{\partial z}$, and every single valued function $F(z) \in \bar{\mathcal{L}}_c \mathcal{L}_c$ can be written as a unique $\bar{\mathcal{O}}\mathcal{O}$-linear combination of $P_w(z)$.

The equation for e_1' is dictated by the expression for the monodromy of $L_w(z)$ (31) around $z = 1$ and can be solved recursively in terms of elements of the Lie algebra over the *ring of zeta integers* $\mathbb{Z}[\mathcal{Z}]$, generated by e_0, e_1 and their multiple commutators (see Lemma 2.6 of [70]). The result is:

$$e_1' = e_1 + 2\zeta(3)[[[e_0, e_1], e_1]\,,\; e_0 + e_1] + \zeta(5)(...) + ...\,, \tag{69}$$

where the parenthesis multiplying $\zeta(5)$ consists of eight bracket words of degree six in $\{e_0, e_1\}$. It follows, in particular, that $e_1' = e_1$ for words of weight not exceeding three or depth not exceeding one.

We proceed to constructing some simple examples of basic SVMPs. For words involving (repeatedly) a single letter we have

$$P_{0^n}(z) = \frac{(\ln \bar{z}z)^n}{n!}\; (P_w(0) = 0 \text{ for } w \neq 0^n, w \neq \emptyset),\; P_{1^n}(z) = \frac{(\ln|1-z|^2)^n}{n!}. \tag{70}$$

The depth-one-weight-two SVMPs, which satisfy the differential equations

$$\partial P_{01} = \frac{P_0}{z-1}\,,\quad \bar{\partial} P_{01} = \frac{P_1}{\bar{z}}\; (P_{01}(0) = 0 = P_{10}(0))\,,$$
$$\partial P_{10} = \frac{P_1}{z}\,,\quad \bar{\partial} P_{10} = \frac{P_0}{\bar{z}-1}\,, \tag{71}$$

are given by

$$P_{01} = L_{10}(\bar{z}) + L_{01}(z) + L_0(\bar{z})L_1(z) = Li_2(z) - Li_2(\bar{z}) + \ln \bar{z}z \ln(1-z),$$
$$P_{10} = L_{01}(\bar{z}) + L_{10}(z) + L_1(\bar{z})L_0(z) = Li_2(\bar{z}) - Li_2(z) + \ln \bar{z}z \ln(1-\bar{z})\,. \tag{72}$$

They obey the shuffle relation $P_{01} + P_{10} = P_0 P_1$ so that the only new weight two function is their difference,

$$P_{01} - P_{10} = 2(Li_2(z) - Li_2(\bar{z}) + \ln \bar{z}z \ln \frac{1-z}{1-\bar{z}}) = 4i D(z)\,, \tag{73}$$

proportional to the *Bloch-Wigner dilogarithm* (see [5] as well as the stimulating survey [81]), $D(z) = Im(Li_2(z) + \ln(1-z)\ln|z|)$. One can also write down depth-one SVMPs of arbitrary weight encountered in the expression $F_n(z)$ for the graphical function associated with the *wheel diagram with* $(n+1)$ *spokes*, first computed by Broadhurst in 1985 (for a modern treatment and references to earlier work—see [70]):

$$F_n(z) = (-1)^n \frac{P_{0^{n-1}10^n}(z) - P_{0^n10^{n-1}}(z)}{z-\bar{z}} =$$
$$= \sum_{k=0}^{n} (-1)^{n-k} \binom{n+k}{n} P_{0^{n-k}}(z) \frac{Li_{n+k}(z) - Li_{n+k}(\bar{z})}{z-\bar{z}}\,. \tag{74}$$

The period of the wheel amplitude is given by the limit of this expression for $z \to 1$

$$F_n(1) = \binom{2n}{n} Li_{2n-1}(1) = \binom{2n}{n} \zeta(2n-1). \tag{75}$$

Just like MZVs appear as values at $z = 1$ of multiple polylogarithms the values at one of SVMPs define *single-valued periods* [22] which find applications in QFT (and in superstring theory—in the hands of Stephan Stieberger). Their generating function is

$$Z^{sv} = P_{e_0,e_1}(1) = 1 + 2\zeta(3)[e_0, [e_1, e_0]] + 2\zeta(5)(...) + ... \Rightarrow \zeta^{sv}(2) = 0. \quad (76)$$

The structure of a graded Hopf algebra of the family of hyperlogarithms allows to read off there symmetry properties from the simpler properties of ordinary logarithms, as illustrated in Example 25 of [40] which begins with a derivation of the inversion formula for the dilog: $Li_2\left(\frac{1}{x}\right) = i\pi \ln x - Li_2(x) - \frac{1}{2}\ln^2 x + 2\zeta(2)$. Remarkably, the SVMPs satisfy simpler symmetry relations under the permutation group $\mathscr{S}_3$ of Möbius transformations of $\mathbb{P}^1(\mathbb{C})\backslash\{0, 1, \infty\}$ that interchange the singular points (see Sect. 2.6 of [70]). $\mathscr{S}_3$ is generated by two involutive transformations, $s_1 : z \to 1 - z$, $s_2 : z \to \frac{1}{z}$ such that $s_1 s_2 : z \to \frac{z-1}{z}$, $(s_1 s_2)^3 = 1$. The formal power series $P_{e_0,e_1}(z)$ satisfies simple symmetry relations under s_1 and s_2 (cf. Lemma 2.17 of [70]):

$$P_{e_0,e_1}(1-z) = P_{e_0,e_1}(1)P_{e_1,e_0}(z), \ P_{e_0,e_1}(\frac{1}{z}) = P_{e_0,-e_0-e_1}(1)P_{-e_0-e_1,e_1}(z). \quad (77)$$

According to (76) the first factor in the right hand side of (77) does not contribute to the transformation law of SVMPs of weight one and two; s_1 just permutes the indices 0 and 1 while $P_0(\frac{1}{z}) = -P_0(z)$, $P_1(\frac{1}{z}) = P_1(z) - P_0(z)$ and

$$P_{01}(\frac{1}{z}) = P_{00}(z) - P_{01}(z), \ P_{10}(\frac{1}{z}) = P_{00}(z) - P_{10}(z) \Rightarrow D(\frac{1}{z}) = -D(z) \quad (78)$$

where D(z) is the Bloch-Wigner dilogarithm (73).

Finally we shall demonstrate as a simple illustration of the theory how one can calculate—without really integrating—the integral

$$I(x_1, x_2, x_3, x_4) = \int \frac{d^4x}{\pi^2} \prod_{i=1}^{4} \frac{1}{(x - x_i)^2} = \frac{f(u, v)}{x_{13}^2 x_{24}^2}, \quad (79)$$

where u, v are the crossratios (17). Using the conformal invariance of $f(u, v)$ we can set $x_1 \to \infty$, $x_2 = e$ $(e^2 = 1)$, $x_4 = 0$; $x_3^2 = \bar{z}z$, $(x_3 - e)^2 = |1 - z|^2$ (cf. Sect. 2.3). Applying to the result the 4-dimensional Laplacian with respect to x_3 which acts on $F(z) = f(u, v)$ as $\frac{1}{4}\Delta_3 F(z) = \frac{1}{z-\bar{z}}\bar{\partial}\partial[(z - \bar{z})F(z)]$, and using the fact that the massless scalar propagator is the Green function of $-\Delta$ we obtain:

$$\bar{\partial}\partial[(z-\bar{z})F(z)] = \frac{\bar{z}-z}{\bar{z}z|1-z|^2} = \frac{1}{\bar{z}(z-1)} - \frac{1}{z(\bar{z}-1)}$$
$$\Rightarrow F(z) = \frac{P_{01}(z) - P_{10}(z)}{z-\bar{z}}. \tag{80}$$

Thus $F(z)$ is given by (74) for $n = 1$, $(z - \bar{z})F(z)$ being the only odd with respect to complex conjugation SVMP of weight two. We note that the odd denominator $z - \bar{z}$ also comes from the Jacobian J of the change of integration variables $\{x^\alpha\} \to \{D_i = (x - x_i)^2\}$, $\alpha, i = 1, ..., 4$ in (79):

$$I(x_1, ..., x_4) = \frac{1}{\pi^2}\int \frac{1}{J}\prod_{i=1}^{4}\frac{dD_i}{D_i},\ J = \det\left(\frac{\partial D_i}{\partial x^\alpha}\right). \tag{81}$$

Indeed, at the singularity $D_i = 0$ we have

$$J|_{D_i=0} = 4x_{13}^2 x_{24}^2\sqrt{2(u+v+uv) - 1 - u^2 - v^2} = 4x_{13}^2 x_{24}^2\sqrt{-(z-\bar{z})^2}. \tag{82}$$

(More about the "d(log) forms and generalized unitarity cuts" the reader will find in Sect. 6 of [49].) Integrals of the type of (79) have been calculated long ago by more conventional methods [76]. For an application of the present techniques to a (previously unknown) 3-loop correlator—see [37].

6 Outlook

Multidimensional Feynman integrals give rise to a family of functions and numbers with the structure of a differential graded double shuffle Hopf algebra. It is displayed most readily for conformally invariant position space amplitudes in a massless QFT.

The dimensions of weight spaces of MZVs (which exhaust the Feynman periods up to six loops in the massless φ^4 theory) do not exceed—and are conjectured to coincide with—their motivic counterparts studied by Francis Brown [20]. Values of hyperlogarithms at sixth roots of unity first appear at seven loops. For the two-loop sunrise integral with massive propagators one encounters multiple elliptic polylogarithms [2, 8, 10] (studied previously in [27]).

The interplay between algebraic geometry, number theory and perturbative QFT is a young and vigorous subject and our survey is far from complete. We have not touched upon the application of cluster algebras to multileg on-shell Feynman amplitudes—see [46] for a remarkable first step in this direction. As hyperlogarithms and associated numbers do not suffice for expressing massive and higher order Feynman amplitudes, mathematicians are exploring their generalizations [5, 24, 25] and physicists are closely following this development [2, 12]. For the connections of MZVs with other parts of mathematics (including the Grothendieck-Teichmüller Lie algebra, mixed Tate motives and modular forms)—see [67].

Acknowledgements It is a pleasure to thank Francis Brown and Herbert Gangl for enlightening discussions at different stages of this work and Kurusch Ebrahimi-Fard for his invitation to the 2014 ICMAT Research Trimester. I thank IHES for its hospitality during the completion of these notes (January, 2016). The author's work has been supported in part by Grant DFNI T02/6 of the Bulgarian National Science Foundation.

Appendix. Historical Notes

Leonhard Euler (1707-1783)

Just as Archimedes (287–212) dominated the mathematics of the 3d century BC so, 2000 years later, Euler is dominating the mathematics of 18th century. Born in the family of the Protestant minister of the parish church in Riehen, a suburb of the free (Swiss) city of Basel (see [45] where pictures of the church and of the parish residence are reproduced), he entered the University of Basel at the age of 13 to study theology. His mathematics professor, Leibniz's student Johann Bernoulli (1667–1748—who had inherited the Basel chair of his brother Jacob, 1654–1705), offered to give private consultations at his home to the diligent boy on Saturday afternoons. At the age of twenty Euler accepted a call to the Academy of Sciences of St. Petersburg (founded a few years earlier by the czar Peter I, the Great) where two of Johann's sons, Daniel and Niklaus Bernoulli, were already active. Contrary to most other foreign members he mastered quickly the Russian language, both in writing and speaking. It is during this very active first St. Petersburg period that Euler first became interested (around 1729—along with major work on mechanics, music theory, and naval architecture) in the "Basel problem"—the problem of finding what we would now call (after Riemann) $\zeta(2)(=\sum_1^\infty 1/n^2)$. (It is actually a problem which a young professor in Bologna, Pietro Mengoli, successor of the great Cavalieri, posed in 1644/1650 [4] and which excited the brothers-rivals Jacob and Johann Bernoulli in Basel.) Euler started by devising efficient approximation for calculating the (slowly convergent) series for $\zeta(2)$. As Weil [79] puts it "as with most of the questions that ever attracted his attention, he never abandoned it, soon making a number of fundamental contributions ...". In 1731 the 24-year-old Euler introduced the "Euler-Mascheroni constant" (see [54]):

$$\gamma = \lim_{n\to\infty}(\sum_{k=1}^{n}\frac{1}{k} - \ln n) = \sum_{n=2}^{\infty}(-1)^n\frac{\zeta(n)}{n}(= 0.5772...) . \qquad (83)$$

He then discovered the so-called Euler-MacLaurin formula and introduced for the first time the Bernoulli numbers into the subject [79]. Next came, in 1735, his sensational discovery of the formula $\zeta(2) = \frac{\pi^2}{6}$, based on a bold application of the theory of algebraic equations to the transcendental equation $1 - \sin x = 0$. This was soon followed by the calculation of $\zeta(m)$ for $m = 4, 6$, etc. In the same period Euler calculated $\zeta(3)$ up to ten significant digits and convinced himself that it is not a rational multiple of π^3 (with a small denominator) and found on the way the identity $\zeta(1,2) = \zeta(3)$ [35]. Peeling off consecutive prime factors from $\zeta(s)$, starting with two (see [45]),

$$(1 - 2^{-s})\zeta(s) = 1 + 3^{-s} + 5^{-s} + \dots ,$$

Euler discovered in 1737 the fabulous product formula,

$$\prod_p (1 - p^{-s})\zeta(s) = 1 \,. \tag{84}$$

It was during the subsequent *Berlin period* (1741-1766), invited by Frederick II, that Euler conjectured, in 1749, the functional equation for the zeta function that became, 110 years later, the basis of Riemann's great 1859 paper, [79]. Euler's work on number theory was done, as Fermat's a century earlier, against a background of contempt towards the field by the majority of mathematicians. He was not deterred. As he once observed "one may see how closely and wonderfully infinitesimal analysis is related not only to ordinary analysis but even to the theory of numbers, however repugnant the latter may seem to that higher kind of calculus" (see [78] Chapter III Sect. V). Euler's "defense of Christianity" of 1747, as Weil ([78] Chapter III, Sect. II) puts it, "did nothing to ingratiate its author with the would be philosopher-king Frederick." Disgusted by the superficial (but fashionable at the court of Frederick) anticlerical Voltaire, Euler took the opportunity offered to him by Empress Catherine II (the Great) to return to St. Petersburg where he spent the last (most productive!) period of his life (1766–1783).

Polylogarithms and multiple zeta values

The study of polylogarithms has started with the dilogarithm function. Its integral representation (that serves as an analytic continuation of the series)

$$Li_2(z) \; (= \sum_{n=1}^{\infty} \frac{z^n}{n^2}) = - \int_0^z \frac{\ln(1-t)}{t} dt \tag{85}$$

first appears in 1696 in a letter of Leibniz (1646–1716) to Johann Bernoulli. As already noted, Euler started playing with the corresponding series around 1729. According to Maximon [59] (who is taking care to establish the priority of British mathematicians) the first study of the properties of the integral (85) for complex z belongs to John Landen (1719–1790) whose memoir appears in the *Phil. Trans. R. Soc. Lond.* in 1760, albeit most authors credit for this Euler (who wrote on the subject later, in 1768). The first comprehensive study of the dilog was given in the book of Spence of 1809 (see [3, 59]) whose results are usually attributed to later work of Abel. The name *Euler dilogarithm* was only introduced by Hill (1828). The two-variable, five-term relation

$$Li_2(x) + Li_2(y) + Li_2(\frac{1-x}{1-xy}) + Li_2(1-xy) + Li_2(\frac{1-y}{1-xy}) =$$
$$\frac{\pi^2}{6} - \ln x \ln(1-x) - \ln y \ln(1-y) + \ln(\frac{1-x}{1-xy}) \ln(\frac{1-y}{1-xy}) \tag{86}$$

was discovered and rediscovered by Spence (1809), Abel (1827), Hill (1828), Kummer (1840), Schaeffer (1846) (see [81]). The oriented (odd under permutations of the vertices) volume of the ideal tetrahedron in hyperbolic space is expressed in terms of the Bloch-Wigner function (73): $\tilde{D}(z_1, ..., z_4) := D(\frac{z_{12}z_{34}}{z_{13}z_{24}})$. It has been found by Lobachevsky in 1836 (for a review see [60]). An early (1955) reference, in which Clausen's (1832) dilogarithm appears in a fourth order calculation in QFT, is [52]. Two years later A. Petermann and (independently) C.A. Sommerfield uncovered $\zeta(3)$ in a calculation of the electron magnetic moment (see the lively review in [72] where important later work by Laporta and Remiddi [56] is also surveyed—see also [74]). It reappeared in another calculation in perturbative quantum electrodynamics in the mid 1960's [66]. The modern notations and a survey of dilogarithmic identities and their polylogarithmic generalizations are given by an electrical engineer [58] (for a nice informative review of his book—see [3]). The *harmonic polylogarithms* (with singularities at the three roots of the equation $x^3 = x$) are surveyed in [65]. David Broadhurst was a pioneer in the systematic study of MZV in QFT (see [15] and references to earlier work cited there as well as his popular talk [13] in which he shares his enthusiasm with beautiful numbers—like $\zeta(3)$—appearing in various branches of physics). The saga of the anomalous electron magnetic moment $(g-2)$ calculation continues to this day (for a review of a leading author in the field—see [53]; a later four-loop analytic calculation is reported in [55]). The resurgence of polylogarithms in pure mathematics, anticipated by 19th century work of Kummer and Poincaré and a 20 century contribution by Lappo-Danilevsky, was prepared by the work of Chen [19, 33] on iterated path integrals. The coproduct of hyperlogarithms was written down by Goncharov [47] as a planar decorated version of Connes-Kreimer's Hopf algebra of rooted trees [34]. One of a number of recent conferences dedicated to this topic had the telling title *Polylogarithms as a Bridge between Number Theory and Particle Physics* (see the notes [82] which contain a historical survey with a bibliography of some 394 entries). Recent developments and perspectives are surveyed in Francis Brown's lecture [23] at the 2014 International Congress of Mathematicians and [26] as well as in the lectures ([6, 32]) in these proceedings.

References

1. Abreu, S., Britto, R., Duhr, C., Gardi, E.: From multiple unitarity cuts to the coproduct of Feynman integrals. arXiv:1401.3546v2 [hep-th]
2. Adams, L., Bogner, C., Weinzierl, S.: The two-loop sunrise graph in two space-time dimensions with arbitrary masses in terms of elliptic dilogarithms. J. Math. Phys. **55**, 102301 (2014). arXiv:1405.5640 [hep-ph]; see also arXiv:1504.03255, arXiv:1512.05630 [hep-ph]
3. Askey, R.: Polylogarithms and associated functions, by Leonard Lewin. Bull. Amer. Math. Soc. **6**(2), 248–251 (1982)
4. Ayoub, R.: Euler and the zeta function. Amer. Math. Monthly **81**, 1067–1086 (1974)
5. Bloch, S.: Applications of the dilogarithm function in algebraic K-theory and algebraic geometry, In: Proceedings of the Internat. Symposium on Algebraic Geometry. Kinokuniya, Tokyo (1978)
6. Bloch, S.: Feynman amplitudes in mathematics and physics, August 2014 lectures at ICMAT, Madrid. arXiv:1509.00361 [math.AG]
7. Bloch, S., Esnault, H., Kreimer, D.: On motives and graph polynomials. Commun. Math. Phys. **267**, 181–225 (2006). [math/0510011]
8. Bloch, S., Kerr, M., Vanhove, P.: A Feynman integral via higher normal functions. arXiv:1406.2664v3 [hep-th]
9. Bloch, S., Kreimer, D.: Mixed Hodge structures and renormalization in physics. Commun. Number Theory Phys. **2**, 637–718 (2008). arXiv:0804.4399 [hep-th]; Feynman amplitudes and Landau singularities for 1-loop graphs, arXiv:1007.0338 [hep-th]
10. Bloch, S., Vanhove, P.: The elliptic dilogarithm for the sunset graph. J. Number Theory **148**, 328–364 (2015). arXiv:1309.5865 [hep-th]
11. Bogner, C., Weinzierl, S.: Periods and Feynman integrals. J. Math. Phys. **50**, 042302 (2009). arXiv:0711.4863v2 [hep-th]
12. Bogoliubov, N.N., Shirkov, D.V.: Introduction to the Theory of Quantized Fields, 3d edn. Wiley (1980) (first Russian edition, 1957)
13. Broadhurst, D.J.: Feynman's sunshine numbers. arXiv:1004.4238 [physics.pop-ph]
14. Broadhurst, D.J.: Multiple Deligne values: a data mine with empirically tamed denominators. arXiv:1409.7204 [hep-th]
15. Broadhurst, D.J., Kreimer, D.: Knots and numbers in ϕ^4 to 7 loops and beyond. Int. J. Mod. Phys. **6C** 519–524 (1995); Association of multiple zeta values with positive knots via Feynman diagrams up to 9 loops. Phys. Lett. **B393**, 403–412 (1997). [hep-th/9609128]
16. Broadhurst, D.J., Schnetz, O.: Algebraic geometry informs perturbative quantum field theory. Proc. Sci. **211**, 078 (2014). arXiv:1409.5570
17. Brown, F.: Single-valued hyperlogarithms and unipotent differential equations. IHES notes (2004)
18. Brown, F.: Single valued multiple polylogarithms in one variable. C.R. Acad. Sci. Paris Ser. I 338, 522–532 (2004)
19. Brown, F.: Iterated integrals in quantum field theory. In: Cardona, A. et al. (eds.) Geometric and Topological Methods for Quantum Field Theory, Proceedings of the 2009 Villa de Leyva Summer School, pp.188–240. Cambridge Univ. Press (2013)
20. Brown, F.: On the decomposition of motivic multiple zeta values. Adv. Stud. Pure Math. **63**, 31–58 (2012). arXiv:1102.1310v2 [math.NT]
21. Brown, F.: Mixed Tate motives over $\mathbb{Z}$. Annals of math. **175**, 949–976 (2012). arXiv:1102.1312 [math.AG]
22. Brown, F.: Single-valued periods and multiple zeta values. arXiv:1309.5309 [math.NT]
23. Brown, F.: Motivic periods and $\mathbb{P}^1 \setminus \{0, 1, \infty\}$. In: Jang, S.Y. et al. (eds.) Proc. ICM, Invited Lectures **II**, 295–318. Seoul (2014). arXiv:1407.5165 [math.NT]; see also arXiv:1512.09265 [math-ph]
24. Brown, F.: Multiple modular values for $SL(2, \mathbb{Z})$. arXiv:1407.5167
25. Brown, F.: Zeta elements of depth 3 and the fundamental Lie algebra of a punctured elliptic curve. arXiv:1504.04737 [math.NT]

26. Brown, F.: Periods and Feynman amplitudes, Talk at the ICMP, Santiago de Chile. arXiv:1512.09265 [math-ph]; – Notes on motivic periods, arXiv:1512.06409v2 [math-ph], arXiv:1512.06410 [math.NT]
27. Brown, F., Levin, A.: Multiple elliptic polylogarithms. arXiv:1110.6917v2 [math.NT]
28. Brown, F., Schnetz, O.: A K3 in ϕ^4. Duke Math. Jour. **161**(10), 1817–1862 (2012). arXiv:1006.4064v5 [math.AG]
29. Brown, F., Schnetz, O.: Proof of the zig-zag conjecture. arXiv:1208.1890v2 [math.NT]
30. Brown, F., Schnetz, O.: Modular forms in quantum field theory. arXiv:1304.5342v2 [math.AG]
31. Brunetti, R., Fredenhagen, K.: Microlocal analysis and interacting quantum field theories: renormalization on physical backgrounds. Commun. Math. Phys. **208**, 623–661 (2000). [math-ph/990328]
32. Carr, S., Gangl, H., Schneps, L.: On the Broadhurst-Kreimer generating series for multiple zeta values. In: Proceedings of the Madrid-ICMAT conference on Multizetas (2015)
33. Chen, K.T.: Iterated path integrals. Bull. Amer. Math. Soc. **83**, 831–879 (1977)
34. Connes, A., Kreimer, D.: Renormalization in quantum field theory and the Riemann-Hilbert problem I, II. Commun. Math. Phys. **210**, 249–273 (2000), **216**, 215–241 (2001). [hep-th/9912092, hep-th/0003188]; Insertion and elimination: the doubly infinite algebra of Feynman graphs. Ann. Inst. Henri Poincaré **3**, 411–433 (2002). [hep-th/0201157]
35. Deligne, P.: Multizetas d'aprés Francis Brown. Séminaire Bourbaki 64ème année, 1048
36. de Medeiros, P., Hollands, S.: Superconformal quantum field theory in curved spacetime. arXiv:1305.5191 [gr-qc]
37. Drummond, J., Duhr, C., Eden, B., Heslop, P., Pennington, J., Smirnov, V.A.: Leading singularities and off shell conformal amplitudes. JHEP **1308**, 133 (2013). arXiv:1303.6909v2 [hep-th]
38. Drummond, J.M., Henn, J., Korchemsky, G.P., Sokatchev, E.: Dual superconformal symmetry of scattering amplitudes in N=4 super-Yang-Mills theory. arXiv:0807.1095v2 [hep-th]
39. Drummond, J.M., Henn, J., Smirnov, V.A., Sokatchev, E.: Magic identities for conformal four-point integrals. JHEP **0701**, 064 (2007). arXiv:hep-th/0607160
40. Duhr, C.: Mathematical aspects of scattering amplitudes. arXiv:1411.7538 [hep-ph]
41. Dütsch, M., Fredenhagen, K.: Causal perturbation theory in terms of retarded products, and a proof of the action Ward identity. Rev. Math. Phys. **16**(10), 1291–1348 (2004). arXiv:hep-th/0403213v3
42. Dyson, F.J.: Missed opportunities. Bull. Amer. Math. Soc. **78**(5), 635–652 (1972)
43. Epstein, H., Glaser, V.: The role of locality in perturbation theory. Ann. Inst. H. Poincaré **A19**(3), 211–295 (1973)
44. Flory, M., Helling, R.C., Sluka, C.: How I learned to stop worrying and love QFT. arXiv:1201.2714
45. Gautschi, W.: Leonhard Euler: his life, the man, and his works. SIAM Rev. **50**(1), 3–33 (2008)
46. Golden, J.K., Goncharov, A.B., Spradlin, M., Vergu, C., Volovich, A.: Motivic amplitudes and cluster coordinates. arXiv:1305.1617 [hep-th]; Golden, J.K., Spradlin, M.: The differential of all two-loop MHV amplitudes in N=4 Yang Mills theory. arXiv:1306.1833 [hep-th]
47. Goncharov, A.: Galois symmetry of fundamental groupoids and noncommutative geometry. Duke Math. J. **128**(2), 209–284 (2005). [math/0208144v4]
48. Gracia-Bondia, J.M., Gutierrez-Garro, H., Varilly, J.C.: Improved Epstein-Glaser renormalization in x-space. III Versus differential renormalization. Nucl. Phys. **B886**, 824–826 (2014). arXiv:1403.1785v3
49. Henn, J.M.: Lectures on differential equations for Feynman integrals. J. Phys. A. arXiv:1412.2296v3 [hep-ph]
50. Hollands, S., Wald, R.M.: Existence of local covariant time ordered products of quantum fields in curved spacetime. Commun. Math. Phys. **231**, 309–345 (2002). [gr-qc/0111108]
51. Kontsevich, M., Zagier, D.: Periods. In: Engquist, B., Schmid, W. (eds.) Mathematics - 20101 and beyond, pp. 771–808. Springer, Berlin (2001)
52. Källen, G., Sabry, A.: Forth order vacuum polarization. Dan. Mat. Fys. Med. **29**(17), 1–20 (1955)

53. Kinoshita, T.: Tenth-order QED contribution to the electron $g-2$ and high precision test of quantum electrodynamics. In: Proceedings of the Conference in Honor of the 90th Birthday of Freeman Dyson, pp.148–172. World Scientific (2014)
54. Lagarias, J.C.: Euler's constant: Euler's work and modern developments. Bull. Amer. Math. Soc. **50**(4), 527–628 (2013)
55. Laporta, S.: High precision calculation of the 4-loop contribution to the electron $g-2$ in QED. arXiv:1704.06996 [hep-ph]
56. Laporta, S., Remiddi, E.: The analytical value of the electron $g-2$ at order α^3 in QED. Phys. Lett. **B379**, 283–291 (1996). arXiv:hep-ph/9602417
57. Lautrup, B.E., Peterman, A., de Rafael, E.: Recent developments in the comparison between theory and experiment in quantum electrodynamics. Phys. Rep. **3**(4), 193–260 (1972)
58. Lewin, L.: Polylogarithms and Associated Functions, North Holland, Amsterdam (1981); Structural Properties of Polylogarithms, Mathematical Surveys and Monographs, vol. 37. AMS, Providence, R.I. (1991)
59. Maximon, L.C.: The dilogarithm function for complex argument. Proc. Roy. Soc. Lond. A **459**, 2807–2819 (2003)
60. Milnor, J.W.: Hyperbolic geometry: the first 150 years. Bull. Amer. Math. Soc. **5**(1) (1982)
61. Müller-Stach, S.: What is a period?. Not. AMS (2014). arXiv:1407.2388 [math.NT]
62. Nikolov, N.M., Stora, R., Todorov, I.: Euclidean configuration space renormalization, residues and dilation anomaly. In: Dobrev, V.K. (eds.) Lie Theory and Its Applications in Physics (LT9), pp. 127–147. Springer, Japan, Tokyo (2013). CERN-TH-PH/2012-076, LAPTH-Conf-016/12
63. Nikolov, N.M., Stora, R., Todorov, I.: Renormalization of massless Feynman amplitudes as an extension problem for associate homogeneous distributions. Rev. Math. Phys. **26**(4), 1430002 (65 pages) (2014). CERN-TH-PH/2013-107; arXiv:1307.6854 [hep-th]
64. Panzer, E.: Feynman integrals via hyperlogarithms. Proc. Sci. **211**, 049 (2014). arXiv:1407.0074 [hep-ph]; Feynman integrals and hyperlogarithms, 220 pp. Ph.D. thesis. arXiv:1506.07243 [math-ph]
65. Remiddi, E., Vermaseren, J.A.M.: Harmonic polylogarithms. Int. J. Mod. Phys. **A15**, 725–754 (2000). arXiv:hep-ph/9905237
66. Rosner, J.: Sixth order contribution to Z_3 in finite quantum electrodynamics. Phys. Rev. Letters **17**(23), 1190–1192 (1966)
67. Schneps, L.: Survey of the theory of multiple zeta values (2011)
68. Schnetz, O.: Natural renormalization. J. Math. Phys. **38**, 738-758 (1997). arXiv:9610025
69. Schnetz, O.: Quantum periods: a census of ϕ^4 transcendentals. Commun. Number Theory Phys. **4**(1), 1–48 (2010). arXiv:0801.2856v2
70. Schnetz, O.: Graphical functions and single-valued multiple polylogarithms. Commun. Number Theory Phys. **8**(4), 589-685 (2014). arXiv:1302.6445v2 [math.NT]
71. Steuding, J.: An Introduction to the theory of L-functions. Course Giv. Madr **06** (2005)
72. Styer, D.: Calculation of the anomalous magnetic moment of the electron (2012) (available electronically)
73. Todorov, I.: Polylogarithms and multizeta values in massless Feynman amplitudes, In: Dobrev, V. (eds.) Lie Theory and Its Applications in Physics (LT10). Springer Proceedings in Mathematics and Statistics, vol. 111, pp. 155–176. Springer, Tokyo (2014). Bures-sur-Yvette, IHES/P/14/10
74. Todorov, I.: Hyperlogarithms and periods in Feynman amplitudes, Chapter 10. In: Dobrev, V.K. (eds.) Springer Proceedings in Mathematics and Statistics, International Workshop on Lie Theory and Its Applications in Physics (LT-11), vol. 191, pp. 151–167 June 2015, Varna, Bulgaria. Springer, Tokyo-Heidelberg (2016). arXiv:1611.09323 [math-ph]
75. Todorov, I.: Renormalization of position space amplitudes in a massless QFT, PEPAN 48(2), 227–236 (2017) (Special Issue); (see also CERN-PH-TH-2015-016)
76. Ussyukina, N.I., Davyddychev, A.I.: An approach to the evaluation of 3- and 4-point ladder diagrams. Phys. Letters B **298**, 363–370 (1993)
77. Waldschmidt, M.: Lectures on multiple zeta values. IMSc, Chennai (2011)

78. Weil, A.: Number Theory-An Approach through history from Hammurapi to Legendre. Birkhäuser, Basel 1983 (2007)
79. Weil, A.: Prehistory of the zeta-function. Number Theory, Trace Formula and Discrete Groups, pp. 1–9. Academic Press, N.Y (1989)
80. Zagier, D.: Introduction to modular forms. In: From Number Theory to Physics, pp. 238–291. Springer, Berlin (1992)(Les Houches,1989)
81. Zagier, D.: The dilogarithm function. In: Frontiers in Number Theory, Physics and Geometry II, pp. 3–65.Springer, Berlin (2006)
82. Zhao, J.: Multiple Polylogarithms. In: Notes for the Workshop Polylogarithms as a Bridge between Number Theory and Particle Physics, Durham, July 3–13 2013

Some Open Problems on Feynman Periods

Erik Panzer

Abstract Feynman integrals of quantum field theories that contain non-scalar particles go beyond the well-studied leading period associated to primitive Feynman graphs. It is therefore necessary to study the space of periods spanned by all convergent Feynman integrals for a given graph. Even when the leading period is known, this total space of periods is not understood and carries non-trivial structures. After reviewing the leading period, we consider all convergent integrals of a graph and related open questions.

Keywords Feynman integrals · Periods · Multiple zeta values

1 The Leading Feynman Period

Let G be a connected graph and $\mathscr{T}(G)$ the set of its spanning trees. We associate a variable α_e to each edge of G and define its *graph polynomial* [11, 31] as

$$\psi_G := \sum_{T \in \mathscr{T}(G)} \prod_{e \notin T} \alpha_e. \tag{1}$$

The inverse of this polynomial defines the integrand of logarithmically divergent Feynman amplitudes in a scalar quantum field theory. Concretely, let N denote the number of edges in G and note that ψ_G is homogeneous of degree $h_1(G) = \dim H_1(G)$, the number of independent cycles in G. This is also known as the *loop number* of the graph. If $N = 2h_1(G)$ and

$$\mathscr{P}(G) := \int_0^\infty \cdots \int_0^\infty \frac{\mathrm{d}\alpha_1 \cdots \mathrm{d}\alpha_{N-1}}{\psi_G^2\big|_{\alpha_N=1}} > 0 \tag{2}$$

E. Panzer (✉)
All Souls College, OX1 4AL Oxford, UK
e-mail: erik.panzer@all-souls.ox.ac.uk

J. I. Burgos Gil et al. (eds.), *Periods in Quantum Field Theory and Arithmetic*, Springer Proceedings in Mathematics & Statistics 314,
https://doi.org/10.1007/978-3-030-37031-2_2

converges, then G is called *primitive log. div.* and the number $\mathscr{P}(G)$ is independent of the choice of the last edge. We will call these numbers *leading periods*. Their physical significance stems from the fact that they provide renormalization scheme independent contributions to the β-function of a four-dimensional theory.[1] Typical examples are

$\mathscr{P}\Big($ 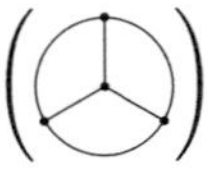 $\Big) = 6\zeta(3)$, $\mathscr{P}\Big($ 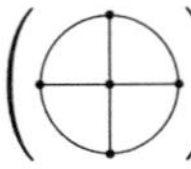 $\Big) = 20\zeta(5)$ and

$\mathscr{P}\Big($ 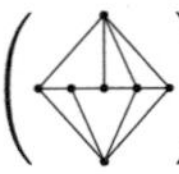 $\Big) = \frac{1063}{9}\zeta(9) + 8\zeta^3(3).$

Already thirty years ago, the periods for all primitive log. div. graphs with $h_1(G) \leq 6$ loops were evaluated in terms of Riemann zeta values $\zeta(n) = \sum_{k=1}^{\infty} k^{-n}$, their products and only one further number [13]. It took ten years [14] to identify this number,

$\mathscr{P}\Big($ 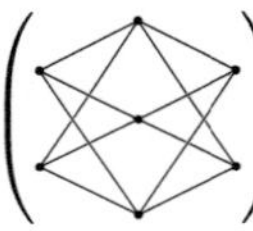 $\Big) = \frac{16704}{5}\zeta(8) - \frac{6912}{5}\zeta(3,5) - 2592\zeta(3)\zeta(5),$ (3)

because it involves the *multiple* zeta value $\zeta(3,5) = \sum_{1 \leq k < r} 1/(k^3 r^5)$ which, conjecturally, cannot be expressed as a polynomial in Riemann zeta values with rational coefficients. More generally, multiple zeta values (MZV) are defined by

$$\zeta(n_1, \ldots, n_d) = \sum_{1 \leq k_1 < \cdots < k_d} \frac{1}{k_1^{n_1} \cdots k_d^{n_d}} \qquad (n_d \geq 2) \tag{4}$$

and they are of great interest in their own right. Many more occurrences of MZV in Feynman integrals have been found [39], which triggered intensive research by mathematicians and physicists. By now, Feynman integrals in general are well known to link methods from calculus, combinatorics, algebraic geometry and number theory (as illustrated in this proceedings volume; see also [5, 6, 16, 17]).

In these notes we will continue to consider only logarithmically divergent integrals, which means that they are determined by the polynomial ψ_G alone and do not depend on physical data like masses or momenta of elementary particles. For primitive log. div. graphs G, the leading periods (2) are now known for all G with $h_1(G) \leq 7$ loops [38]. In particular we now know the first explicit examples where $\mathscr{P}(G)$ is (conjecturally) not an MZV, but expressible as a linear combination of multiple polylogarithms (MPL)

---

[1]The β-function determines the running of the coupling constant under the evolution by the renormalization group and is a fundamental property of a quantum field theory [30].

$$\mathrm{Li}_{n_1,\ldots,n_d}(z) := \sum_{1\le k_1<\cdots<k_d} \frac{z^{k_d}}{k_1^{n_1}\cdots k_d^{n_d}} \tag{5}$$

at second or sixth roots of unity z [36, 38]. We also know many graphs for which there are strong indications to believe that they cannot be evaluated in terms of MPL at any algebraic arguments [21–23, 40].

While there are still many open questions and unresolved conjectures for the leading periods, we refer to [38] for a thorough discussion. Instead, we want to look at generalizations of the periods $\mathscr{P}(G)$ from (2) which are particularly important for gauge theories.

2 Generalized Feynman Periods

Feynman integrals are essential to the perturbative formulation of quantum field theory. Namely, observables like cross-sections of scattering processes are expressed as series over a (typically infinite) set of Feynman graphs. To each of these graphs, there is an associated Feynman integral which gives a contribution to the cross-section. We refer to [30] for a detailed introduction to these concepts and to [41] for a thorough discussion of Feynman integrals (we will work exclusively in the manifoldly named Schwinger-, Feynman- or α-representation).

What is important for these notes is that different types of particles give rise to different Feynman integrals. The leading period from Eq. (2) is only a very special case of a Feynman integral, namely under the assumptions that

1. only scalar particles partake in the interaction,
2. the integral is logarithmically divergent,
3. there are no subdivergences and
4. the dimension of space-time is four.

The first constraint excludes all particles of the standard model (gauge bosons and fermions) with the sole exception of a scalar Higgs boson. However, there are well-known techniques which allow to reduce such Feynman integrals to linear combinations of scalar integrals [42]. The integrals arising after this procedure take the form (up to simple prefactors which are irrelevant to the following discussion)

$$I_{\boldsymbol{\nu}}(G) := \int_0^\infty \cdots \int_0^\infty \frac{\alpha_1^{\nu_1-1}\mathrm{d}\alpha_1\cdots\alpha_{N-1}^{\nu_{N-1}}\mathrm{d}\alpha_{N-1}}{\psi_G^{d/2}\Big|_{\alpha_N=1}} \tag{6}$$

and are indexed by a vector $\boldsymbol{\nu} = (\nu_e)_{e=1\ldots N} \in \mathbb{N}^N$ which encodes a monomial $\prod_e \alpha_e^{\nu_e-1}$ multiplying the integrand. The exponent $d/2$ in Eq. (6) is not fixed to 2; instead it can also take higher integer values subject to the condition

$$\sum_{e=1}^{N} \nu_e = \frac{d}{2} h_1(G) \tag{7}$$

which encodes that the integrand is homogeneous of degree zero (in particular, d is determined by $\boldsymbol{\nu}$). This ensures that $I_{\boldsymbol{\nu}}(G)$ is independent of the choice of the N'th edge, corresponding to a logarithmic divergence. As a consequence, our integrals are independent of any kinematic data of the interacting particles like masses and momenta. Nevertheless these periods play a crucial role, because they compute massless propagators [35] and determine the renormalization group functions (β-functions and anomalous dimensions) in minimal subtraction schemes via standard techniques [32].

2.1 Convergence

The convergence of a generalized period can be inferred from a simple power-counting procedure [45]. In the integral representation (6), this amounts to considering the growth of the integrand as $\alpha_e \to 0$ for some subset $\gamma \ni e$ of edges.[2]

Lemma 1 *The generalized period $I_{\boldsymbol{\nu}}(G)$ for $\boldsymbol{\nu} \in \mathbb{N}^N$ converges precisely when*

$$\sum_{e \in \gamma} \nu_e > \frac{d}{2} h_1(\gamma) \tag{8}$$

holds for all non-empty proper subgraphs $\gamma \subsetneq G$. We call such indices $\boldsymbol{\nu}$ convergent in d dimensions.

Example 1 For the wheel with 3 spokes graph $\mathrm{W}_3 =$ , the convergence conditions from triangle subgraphs γ are (with respect to the edge labels in Table 1)

$$\nu_1 + \nu_2 + \nu_3,\ \nu_1 + \nu_5 + \nu_6,\ \nu_2 + \nu_4 + \nu_6,\ \nu_3 + \nu_4 + \nu_5 > \frac{d}{2}. \tag{9}$$

The 2-loop subgraphs $\gamma = G \setminus e$ yield the constraint $d < (\sum_{i=1}^{6} \nu_i) - \nu_e$, which is equivalent to $\nu_e < d/2$ via Eq. (7). Together with Eq. (9) and $\nu_e > 0$ (from $\gamma = \{e\}$), these conditions are also sufficient for the convergence of $I_{\boldsymbol{\nu}}(\mathrm{W}_3)$.[3] The vector $\nu = (1, 1, 1, 2, 2, 2)$ is not convergent, because it gives $d/2 = 3$ and thus violates the first triangle condition in Eq. (9). Examples of convergent periods are given in Table 1.

[2] Note that due to the projective nature of the integral (6), this region is equivalently described by $\alpha_e \to \infty$ for all $e \notin \gamma$, which we thus do not have to consider separately.

[3] More generally, for an arbitrary graph G, the only independent constraints on convergence are those arising from 2-connected subgraphs γ. Such graphs are usually called 1PI in physics. In the case of $G = \mathrm{W}_3$, the 2-connected subgraphs are precisely the edges, triangles and the graphs $G \setminus e$.

Table 1 Some periods of the wheel with 3 spokes graph W_3, including the leading period $\mathscr{P}(\mathrm{W}_3) = 6\zeta(3)$ from Eq. (2)

$\mathrm{W}_3 =$ 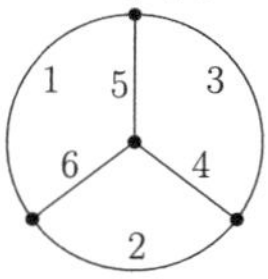

$\prod_e \alpha_e^{\nu_e - 1}$	1	$\alpha_1\alpha_2\alpha_3$	$\alpha_1\alpha_2\alpha_4$	$\alpha_1^2\alpha_2^2\alpha_3\alpha_4$	$\alpha_1\alpha_2\alpha_3\alpha_4\alpha_5\alpha_6$
d	4	6	6	8	8
$I_{\vec{\nu}}(\mathrm{W}_3)$	$6\zeta(3)$	$\frac{1}{2}$	$\frac{1}{2}\zeta(3) - \frac{1}{6}$	$\frac{1}{24}$	$\frac{1}{10}\zeta(3) - \frac{7}{60}$

Remark 1 One can check that a graph must be 2-connected in order to have any convergent periods (otherwise, Eq. (8) has no solutions).[4] We will therefore only consider 2-connected graphs.

Remark 2 Recall that d is a function of $\boldsymbol{\nu}$ by Eq. (7). Condition (8) is equivalent to

$$h_1(G)\sum_{e\in\gamma}\nu_e > h_1(\gamma)\sum_{e\in G}\nu_e. \tag{10}$$

Definition 1 Given a graph G, we denote by $\widehat{\mathscr{P}}(G)$ the $\mathbb{Q}$-vector space generated by the convergent generalized Feynman periods of G in even dimensions:

$$\widehat{\mathscr{P}}(G) := \operatorname{lin}_{\mathbb{Q}}\left\{I_{\boldsymbol{\nu}}(G)\colon\ \boldsymbol{\nu} \text{ is convergent in an even dimension}\right\}. \tag{11}$$

Remark 3 Equivalently, $\widehat{\mathscr{P}}(G)$ is the set of all convergent integrals of the form

$$I_P(G) := \int_0^\infty \mathrm{d}\alpha_1 \cdots \int_0^\infty \mathrm{d}\alpha_{N-1}\, \frac{P}{\psi_G^k}\bigg|_{\alpha_N=1} \tag{12}$$

where $k \in \mathbb{N}$ and $P \in \mathbb{Q}[\boldsymbol{\alpha}]$ is a homogeneous polynomial of degree $kh_1(G) - N$. Since we can multiply with $1 = \frac{\psi_G}{\psi_G}$, every period in dimension d is a linear combination of periods in dimension $d+2$. Also, from cohomology theory it is clear that $\widehat{\mathscr{P}}(G)$ is a finite dimensional vector space over $\mathbb{Q}$ (see Sect. 2.4.1).

Example 2 The simplest graph to consider is the bubble $G =$  with $\psi_G = \alpha_1 + \alpha_2$ and $d/2 = \nu_1 + \nu_2$. It has only rational periods, $\widehat{\mathscr{P}}($$) = \mathbb{Q}$, because

$I_{\vec{\nu}}($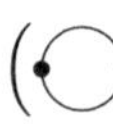$)$ $$= \int_0^\infty \frac{\alpha_1^{\nu_1-1}\mathrm{d}\alpha_1}{(\alpha_1+1)^{\nu_1+\nu_2}} = \frac{\Gamma(\nu_1)\Gamma(\nu_2)}{\Gamma(\nu_1+\nu_2)} = \frac{(\nu_1-1)!(\nu_2-1)!}{(\nu_1+\nu_2-1)!} \in \mathbb{Q}. \tag{13}$$

---

[4] G is 2-connected if it is connected and remains connected even after deletion of an arbitrary vertex.

2.2 General Properties and Relations

This section is a summary of very general results, while the next section will be more specific to particular examples of graphs.

At first, let us recall the series-parallel operations on graphs (depicted in Fig. 1):

(S) replace two sequential edges (joined at a two-valent vertex) by a single edge,
(P) replace two parallel edges by a single edge.

A 2-connected graph is called *series-parallel* if it can be reduced to the bubble (equivalently, to a single edge) by a sequence of the operations (S) and (P). The following well-known result (see [16, 41]) follows from integrations of Euler's beta function similar to Eq. (13).

Lemma 2 *If G_1 can be obtained from G_2 by series-parallel operations, then we have $\widehat{\mathscr{P}}(G_1) = \widehat{\mathscr{P}}(G_2)$. In particular, if G is series-parallel, then $\widehat{\mathscr{P}}(G) = \mathbb{Q}$.*

Due to this result, it suffices to study graphs without any parallel or sequential edges. The next well-known simplification arises when G has a 2-cut, that is, there exist 2 vertices v and w such that $G \setminus \{v, w\}$ is disconnected. In such a situation we can partition the edges of G into two subgraphs G_1' and G_2' such that G_1' and G_2' intersect only at v and w (see Fig. 2). In this situation, we call G a *2-vertex-join* of $G_1 := G_1' \,\dot{\cup}\, \{v, w\}$ and $G_2 := G_2' \,\dot{\cup}\, \{v, w\}$ and write $G = G_1 \mathbin{:} G_2$. The following factorization is immediate in the momentum-space representation of Feynman integrals; a derivation directly in Schwinger parameters is given in [16, Proposition 40].

Lemma 3 *If $G = G_1 \mathbin{:} G_2$ is a 2-vertex join, then $\widehat{\mathscr{P}}(G) = \widehat{\mathscr{P}}(G_1) \cdot \widehat{\mathscr{P}}(G_2)$.*

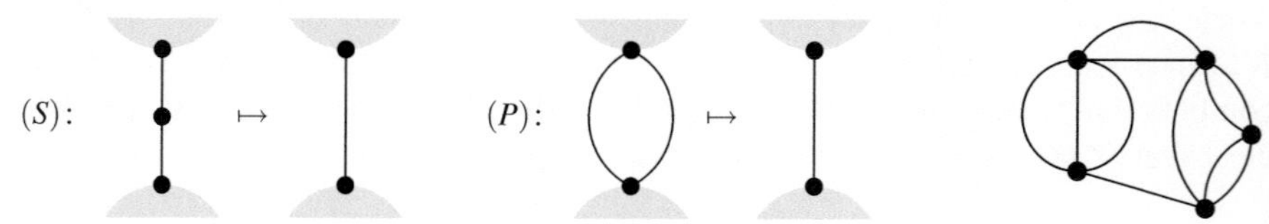

Fig. 1 The series and parallel operations (S) and (P) acting on a graph (the grey areas indicate that only a part of the actual graph is shown). On the right is an example of a series-parallel graph

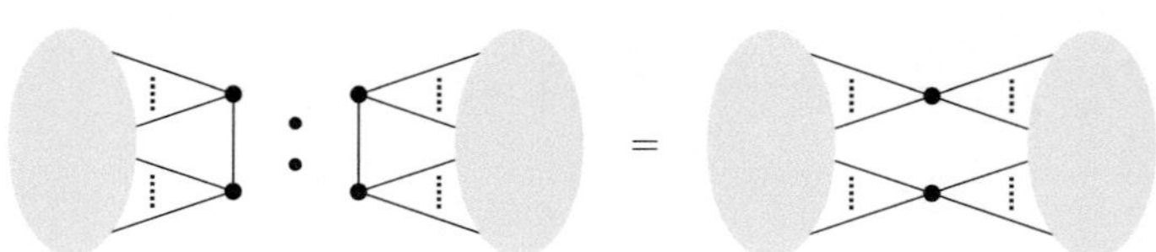

Fig. 2 The 2-vertex join of two graphs identifies a pair of vertices and deletes the edge between them

Example 3 The self-join of the wheel with 3 spokes gives the unique graph

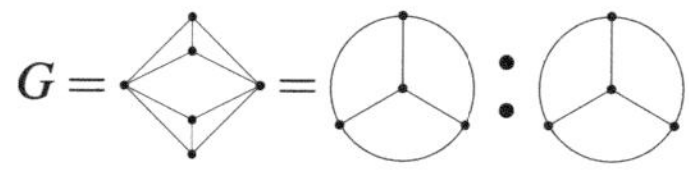

(14)

with periods $\widehat{\mathscr{P}}(G) = \{p \cdot q \colon p, q \in \widehat{\mathscr{P}}(\mathrm{W}_3)\}$. From Table 1 we know $\mathbb{Q} + \mathbb{Q}\zeta(3) \subseteq \widehat{\mathscr{P}}(\mathrm{W}_3)$ and conclude that $\mathbb{Q} + \mathbb{Q}\zeta(3) + \mathbb{Q}\zeta^2(3) \subseteq \widehat{\mathscr{P}}(G)$. For instance, the leading period of G is $\mathscr{P}(G) = 36\zeta^2(3) = \mathscr{P}(\mathrm{W}_3)^2$.

Remark 4 The reduction of series-parallel graphs in Lemma 2 is actually a special case of the factorization Lemma 3 (by cutting the graph at the endpoints of the parallel or sequential edges).

We can now restrict our attention to 3-connected graphs (these are precisely the graphs *without* any 2-cuts). Recall that γ is called a *minor* of G if γ can be obtained from G by a sequence of deletions and contractions of edges. A proof of the following is given in [16, Proposition 37].

Theorem 1 (Minor monotonicity) *If γ is a minor of G, then $\widehat{\mathscr{P}}(\gamma) \subseteq \widehat{\mathscr{P}}(G)$.*

Example 4 Consider the family of wheel graphs depicted in Fig. 3. Their leading periods are known from [12]:

$$\mathscr{P}(\mathrm{W}_n) = \binom{2n-2}{n-1}\zeta(2n-3). \tag{15}$$

It is easy to see that each W_m with $m \leq n$ occurs as a minor of W_n. Hence their periods must appear as periods of W_n:

$$\mathbb{Q} + \mathbb{Q}\zeta(3) + \cdots + \mathbb{Q}\zeta(2n-3) \subseteq \widehat{\mathscr{P}}(\mathrm{W}_n)\,. \tag{16}$$

One can furthermore check that actually all minors of a wheel are either series-parallel or equivalent to another wheel (under series-parallel operations).

Example 5 The situation is quite different for another famous family of graphs, the zig-zags (depicted in Fig. 4). Their leading periods are also Riemann zetas [24],

$$\mathscr{P}(\mathrm{Z}_n) = 4\frac{(2n-2)!}{n!(n-1)!}\left(1 - \frac{1-(-1)^n}{2^{2n-3}}\right)\zeta(2n-3), \tag{17}$$

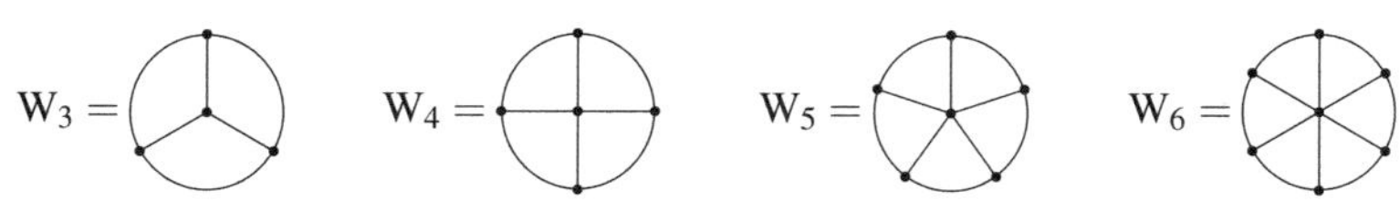

Fig. 3 The wheel graphs with three, four, five and six loops

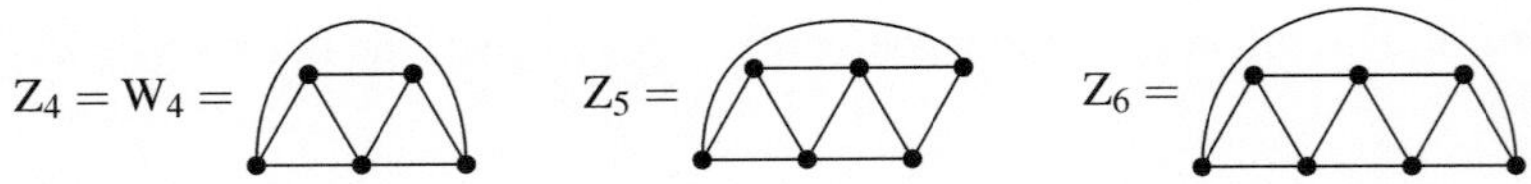

Fig. 4 The zig-zag graphs with four, five and six loops

and they contain the smaller zig-zags as minors. But they have more minors, including products. For example, we find $W_3 \mathbin{:} W_3$ in Z_6 (delete the middle edge of the baseline at the bottom and contract the outer arc) and hence deduce $\zeta^2(3) \in \widehat{\mathscr{P}}(Z_6)$. The same reasoning shows that $\widehat{\mathscr{P}}(Z_n)$ must contain many products of MZV when n gets large. We do not expect such products for the wheels (Conjecture 1 below).

Remark 5 The reasoning above does not explain all products. For example, $W_3 \mathbin{:} W_3$ is not a minor of Z_5, but still we find $\zeta^2(3) \in \widehat{\mathscr{P}}(Z_5)$. It shows up, for instance, in

$$I_\nu(Z_5) = -\frac{5}{3} + \frac{161}{6}\zeta(3) + \frac{70}{3}\zeta(5) + \zeta^2(3) - \frac{441}{8}\zeta(7) \tag{18}$$

where we set $\nu_e = 2$ for the thick edges in 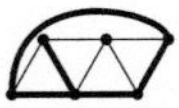 and $\nu_e = 1$ otherwise.

Because a minor is a quotient of a subgraph, Theorem 1 is a special case of

Theorem 2 *If γ is a subgraph of G, then $\widehat{\mathscr{P}}(\gamma) \cdot \widehat{\mathscr{P}}(G/\gamma) \subseteq \widehat{\mathscr{P}}(G)$.*

Proof It is well-known that for a subgraph γ of G, the graph polynomial of G factorizes to leading order in the subgraph variables [5]. Concretely, if we substitute $\alpha_e = t\tilde{\alpha}_e$ for all $e \in \gamma$, then (recall that the degree of ψ is the loop number)

$$\psi_G(\boldsymbol{\alpha}, \tilde{\boldsymbol{\alpha}}, t) = \psi_\gamma(\tilde{\boldsymbol{\alpha}})\psi_{G/\gamma}(\boldsymbol{\alpha})t^{h_1(\gamma)} + \mathscr{O}\left(t^{h_1(\gamma)+1}\right).$$

Let us label the edges of γ with $1, \ldots, N_\gamma$ and those of G/γ with $(N_\gamma + 1) \ldots N_G$. Consider a pair of convergent periods $I_{(\nu_1,\ldots,\nu_{N_\gamma})}(\gamma)$ and $I_{(\nu_{N_\gamma+1},\ldots,\nu_{N_G})}(G/\gamma)$. We may assume that they lie in the same dimension d (see Remark 3). We claim that the period

$$\left(\prod_{e\in G}\int_0^\infty \alpha_e^{\nu_e-1}\mathrm{d}\alpha_e\right)\frac{d}{2}\frac{\left[\sum_{e\in\gamma}\alpha_e\partial_{\alpha_e} - h_1(\gamma)\right]\psi_G}{\psi_G^{d/2+1}}\delta(1-\alpha_{N_G}) \in \widehat{\mathscr{P}}(G)$$

is equal to the product $I_{\ldots}(\gamma)I_{\ldots}(G/\gamma)$ of our given pair of periods. To see this, we multiply with $1 = \int_0^\infty \mathrm{d}t\,\delta(\alpha_{N_\gamma} - t)$ and change variables $\alpha_e = t\tilde{\alpha}_e$ for all $e \in \gamma$ as above. The integrand then becomes

$$\left(\prod_{e\in\gamma}\tilde{\alpha}_e^{\nu_e-1}\mathrm{d}\tilde{\alpha}_e\right)\delta(1-\tilde{\alpha}_{N_\gamma})\left(\prod_{e\in G/\gamma}\alpha_e^{\nu_e-1}\mathrm{d}\alpha_e\right)\delta(1-\alpha_{N_G})(-\partial_t)\frac{t^{h_1(\gamma)d/2}}{\psi_G^{d/2}}$$

such that the integration of t becomes trivial. It produces precisely the product of the integrands of the sought-after periods of γ and G/γ, because the factorization of ψ_G mentioned above provides

$$\lim_{t\to 0} \frac{t^{h_1(\gamma)d/2}}{\psi_G^{d/2}} = \frac{1}{[\psi_\gamma(\tilde{\alpha})]^{d/2}} \frac{1}{[\psi_{G/\gamma}(\alpha)]^{d/2}}. \qquad \square$$

Example 6 Inserting W_3 into a vertex of another copy of W_3 yields a graph which, according to Theorem 2, has $\zeta^2(3)$ as a period. Note that the leading period itself [35],

$\mathcal{P}\Bigg($ 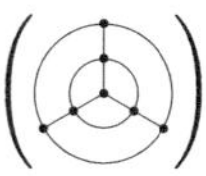 $\Bigg) = 72\zeta^2(3) - \frac{189}{2}\zeta(7),$ (19)

is a combination of $\zeta^2(3)$ and $\zeta(7)$. Consequently, we know that $\zeta(7)$ on its own must also be a period of this graph.

Remark 6 This product structure in terms of sub-quotients is essential to the motivic theory of Feynman periods. In fact, a motivic version of Theorem 2 is true; see [19, Section 7.4].

2.3 Families of Graphs with Polylogarithmic Periods

In general it is extremely hard to get a handle on all periods of a Feynman graph, because in most cases it is unknown what kind of numbers to expect. We will restrict here to very special cases where the periods can be expressed as MPL from Eq. (5). At the moment it is unknown how to decide if an arbitrary given graph belongs to this class. However, there are sufficient criteria which cover many cases of interest. In this section we summarize results from the integration of the integrals (6) with hyperlogarithms and refer to [9, 10, 15, 16, 36, 37] for a discussion of this method.

Definition 2 (*from* [16]) A graph G has *vertex-width* 3 if its edges can be ordered in such a way that the subgraphs formed by $\{e_1, \ldots, e_k\}$ and $\{e_{k+1}, \ldots, e_N\}$ have at most 3 vertices in common (for all $1 < k < N$).

Equivalently, G has vertex-width 3 if it can be constructed from the triangle by a sequence of the operations shown in Fig. 5 (the three white vertices mark the intersection of the subgraphs in Definition 2). Note that all wheels and zig-zags can be obtained this way, hence they are covered by the following result from [16, 36]:

Theorem 3 *If G has vertex-width 3, then all of its periods are rational linear combinations of MZV.*

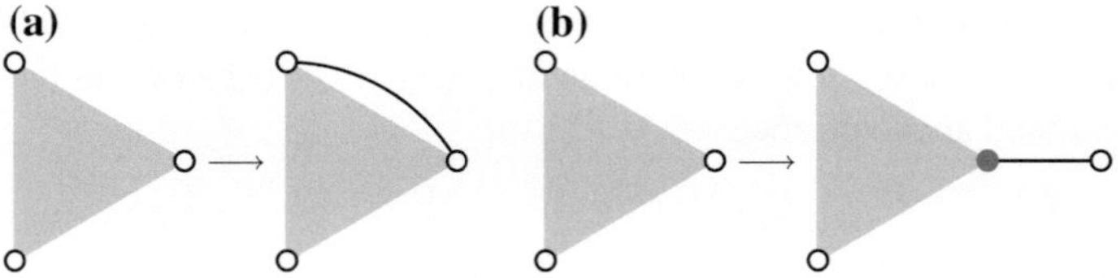

Fig. 5 The allowed steps to construct a graph of vertex-width 3 are: **a** add an edge between two of the three marked vertices, **b** attach a new vertex with a single (new) edge to one of the marked vertices (the new vertex becomes now marked, and the attachment point is not marked anymore)

Furthermore, only $\zeta(n_1, \ldots, n_r)$ of *weight* $n_1 + \ldots + n_r \leq N - 3$ can appear. Using the known relations among MZV, this implies for example that

$$\widehat{\mathscr{P}}(\mathrm{W}_4) \subseteq \mathbb{Q} + \mathbb{Q}\zeta(2) + \mathbb{Q}\zeta(3) + \mathbb{Q}\zeta(4) + \mathbb{Q}\zeta(5) + \mathbb{Q}\zeta(2)\zeta(3).$$

In fact, $\zeta(2)\zeta(3)$ can be excluded, because for graphs with $N = 2h_1(G)$ the weight $N - 3$ part of the periods is one-dimensional [21]. In this case it is spanned by the leading period $\mathscr{P}(\mathrm{W}_4) = 20\zeta(5)$. Comparing this with the lower bound of Eq. (16), we miss the even zeta values. We make

Conjecture 1 *Wheel graphs* W_n *(*$n \geq 3$*) have only odd Riemann zetas as periods:*

$$\widehat{\mathscr{P}}(\mathrm{W}_n) = \mathbb{Q} + \mathbb{Q}\zeta(3) + \cdots + \mathbb{Q}\zeta(2n-3). \tag{20}$$

This conjecture is supported by explicit computation of periods of W_3, W_4 and W_5 using the methods of [36, 37]. It would also follow from several conjectures about *motivic* Feynman periods as explained in [19, Example 9.7]. Notice how far Theorem 3 still is from Eq. (20): We not only have to exclude *multiple* zeta values, but also all even Riemann zeta values $\zeta(2k)$.

Remark 7 Integrals of rational functions which evaluate to linear combinations of only odd Riemann zeta values are known from work on the irrationality of zeta values [3, 27]. This topic is nicely summarized in [18]. However, it seems unlikely that these integrals can be related to the Feynman integrals of the wheels in a straightforward way.

Note that Eq. (20) is false for the zig-zags Z_n due to the presence of products as demonstrated in Example 5 and Eq. (18). It is still striking that no even zeta values seem to appear in their periods, see Eq. (18). In fact, while $\zeta(12)$ is known to appear in periods [39], the even zeta values with lower weight have not been observed as periods of *any* graph.

Conjecture 2 $\zeta(2)$ *is not a period of any graph:* $\zeta(2) \notin \widehat{\mathscr{P}}(G)$.

Also this conjecture is supported by the motivic approach to Feynman periods [19]. Furthermore, the cosmic Galois group would imply that together with $\zeta(2)$, also all

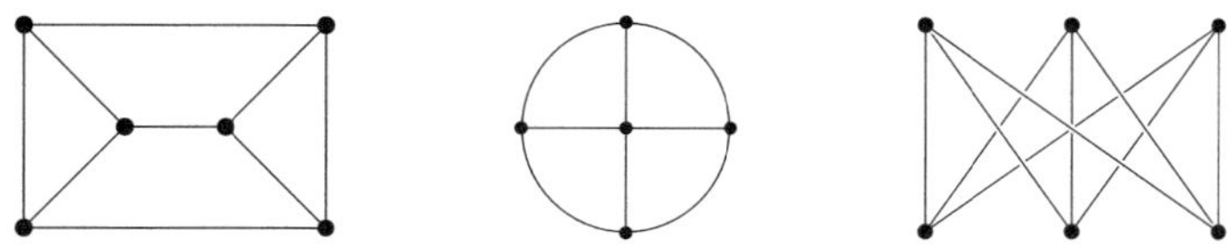

Fig. 6 All 3-connected graphs with four loops

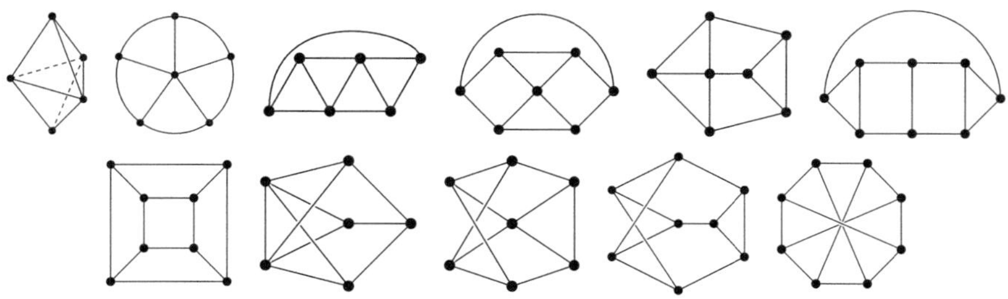

Fig. 7 All 3-connected graphs with five loops

products with odd zeta values (like $\zeta(2)\zeta(3)$) would be excluded as periods of graphs. For a discussion of these ideas we refer to [19, 20], and physical interpretations of the absence of $\zeta(2)$ are given in [2, 33].

Inside the full space of 3-connected graphs, those with vertex-width 3 form a tiny subset. In particular note that graphs with vertex-width 3 are always planar. The first non-planar graph occurs at 4 loops (see Fig. 6), where the other two graphs have vertex-width 3. At 5 loops however there are also planar graphs which do not have vertex-width 3, like the cube in the bottom left of Fig. 7. The following was proved in [35].

Theorem 4 *The periods of all planar graphs with ≤ 5 loops, except for the cube, are MZV. For the cube and the non-planar graphs with ≤ 5 loops, all periods are rational linear combinations of MPL at $z = -1$ (so-called* alternating sums*).*

The interesting observation is that all graphs with ≤ 5 loops seem to have only MZV as periods [1, 34, 35]; that is, all periods of the supposedly more complicated graphs (the non-planar ones and the cube) turn out to be MZV.

Conjecture 3 *All periods of graphs with ≤ 5 loops are multiple zeta values.*

The phenomenon that suitable linear combinations of alternating sums can combine to MZV is well-known. Such alternating sums are also called *honorary MZV*; one example is

$$\mathrm{Li}_{1,2,3}(-1) + \frac{7}{2}\,\mathrm{Li}_{1,5}(-1) = \frac{17}{64}\zeta^2(3) - \frac{1}{15}\zeta^3(2). \tag{21}$$

Very recently a basis of MZV in terms of such honorary alternating sums was constructed [29], using tools of motivic periods [28]. At the moment, however, we can only use these to show that a given individual computed period is an MZV (for this purpose it is also sufficient to look up the corresponding entry in the *datamine* [7]). The difficulty is to understand why this miracle has to happen for *all* periods of the graphs with ≤ 5 loops.

2.4 Further Remarks

Finally we briefly mention further approaches and ideas which might help to understand the spaces of periods of Feynman graphs.

2.4.1 Integration by Parts and Master Integrals

The periods of a graph fulfill a myriad of relations, because $\widehat{\mathscr{P}}(G)$ is finite-dimensional (over $\mathbb{Q}$). This comes about because the de Rham cohomology of the complement of the graph hypersurface $\{\boldsymbol{\alpha}\colon\ \psi_G(\boldsymbol{\alpha}) = 0\}$ is finite dimensional. This cohomology was first studied in [5] and has been further explored since then [21, 26]. One might hope that at least for some interesting family of graphs, an algorithm to construct a finite generating set of cohomology classes might be devised. It would then suffice to compute each of the corresponding periods, which is manageable at a large scale because powerful integration algorithms have been implemented [8, 37].

Similar period relations coming from integration by parts (IBP) in momentum space have been studied systematically since [25] and have been exceedingly successful; currently they form an essential part of almost all perturbative calculations of loop corrections. However, the application to the high loop numbers we are interested in is extremely difficult. Only very recently it has been achieved at the 5-loop level [43]. Since this approach works in dimensional regularization (see Sect. 2.4.2), this alone does not yet solve our problem, because one needs to control the order of poles in ε.

For the wheels W_n the IBP identities were solved explicitly in [13] and yield

Lemma 4 *All periods of wheels are polynomials in Riemann zeta values:*

$$\widehat{\mathscr{P}}(\mathrm{W}_n) \subseteq \mathbb{Q}[\zeta(k)\colon k \geq 2]. \tag{22}$$

This follows because IBP identities express every period of a wheel as a linear combination of products of coefficients of the ε-expansion of the Γ function,

$$\Gamma(1-\varepsilon) = \exp\left[\varepsilon\gamma_E + \sum_{n\geq 2}\frac{\zeta(n)}{n}\varepsilon^n\right], \tag{23}$$

and one can show that the Euler-Mascheroni constant γ_E cancels in the end. Hence, MZV like $\zeta(3,5)$ are excluded from $\widehat{\mathscr{P}}(\mathrm{W}_n)$ and Lemma 4 gives a much better bound than the result of Theorem 3, but it is still a long way from Conjecture 1.

2.4.2 Logarithmic Periods and Regularization

A further generalization of the period integrals $I_{\boldsymbol{\nu}}(G)$ from Eq. (6) are the so called *logarithmic periods*

$$\int_0^\infty \cdots \int_0^\infty \mathrm{d}\alpha_1 \cdots \mathrm{d}\alpha_{N-1} \left. \frac{P(\alpha_1, \ldots, \alpha_N, \log(\alpha_1), \ldots, \log(\alpha_N), \log(\psi_G))}{\psi_G^{d/2}} \right|_{\alpha_N=1} \tag{24}$$

where the polynomial P is a homogeneous function of $\boldsymbol{\alpha}$. These integrals arise as coefficients if one expands $I_{\boldsymbol{\nu}}(G)$ as a function of the ν_e and the dimension d. In physics, this *dimensional regularization* is a popular method to regularize divergent integrals (which appear abundantly before renormalization) and also essential to some renormalization schemes like minimal subtraction [32].

The vector space of logarithmic periods of a graph is infinite dimensional: Already the bubble  generates all Riemann zeta values and products from Eq. (13) via Eq. (23). But still there are interesting structures and several relations from Sect. 2.2 also hold for logarithmic periods [16]. In particular, it is known that all logarithmic periods of graphs with vertex-width 3 or loop order ≤ 5 are MZV or alternating sums and Conjecture 3 is expected to hold also for the logarithmic periods.

However, as soon as logarithmic periods are considered, even zeta values do appear (so Conjecture 2 does not hold for the spaces of logarithmic periods).

2.4.3 Motivic Feynman Periods

Recently, the theory of motivic periods [20] was applied to Feynman integrals and appears very promising to understand some of the observed phenomena, as we already mentioned in several places above. Unfortunately, an adequate discussion of these ideas would be too lengthy to fit in here. Instead we advocate the comprehensive notes [19].

Acknowledgements I wish to thank the organizers of the *Research Trimester on Multiple Zeta Values, Multiple Polylogarithms, and Quantum Field Theory* at ICMAT, Madrid (September 15–December 19, 2014) for their generous invitation and all participants for creating a very stimulating atmosphere. In particular I want to thank Kurusch Ebrahimi-Fard, Dominique Manchon and Wadim Zudilin. Furthermore I am indebted to Francis Brown for numerous discussions on topics of these notes. Figures were generated with `JaxoDraw` [4] and `Axodraw` [44].

References

1. Baikov, P.A., Chetyrkin, K.G.: Four loop massless propagators: an algebraic evaluation of all master integrals. Nucl. Phys. B **837**, 186–220 (2010). https://doi.org/10.1016/j.nuclphysb.2010.05.004, arXiv:1004.1153 [hep-ph]

2. Baikov, P.A., Chetyrkin, K.G.: No-π theorem for Euclidean massless correlators. In: Loops and Legs in Quantum Field Theory 2018, St. Goar, Germany, April 29–May 04, vol. LL2018, p. 008 (2018). https://doi.org/10.22323/1.303.0008, arXiv:1808.00237 [hep-ph]
3. Ball, K., Rivoal, T.: Irrationalité d'une infinité de valeurs de la fonction zêta aux entiers impairs. Invent. Math. **146**(1), 193–207 (2001). https://doi.org/10.1007/s002220100168
4. Binosi, D., Theußl, L.: JaxoDraw: a graphical user interface for drawing Feynman diagrams. Comput. Phys. Commun. **161**, 76–86(2004). https://doi.org/10.1016/j.cpc.2004.05.001, arXiv:hep-ph/0309015
5. Bloch, S., Esnault, H., Kreimer, D.: On motives associated to graph polynomials. Commun. Math. Phys. **267**(1), 181–225 (2006). https://doi.org/10.1007/s00220-006-0040-2, arXiv:math/0510011
6. Bloch, S., Kreimer, D.: Mixed Hodge structures and renormalization in physics. Commun. Number Theory Phys. **2**(4), 637–718 (2008). https://doi.org/10.4310/CNTP.2008.v2.n4.a1, arXiv:0804.4399 [hep-th]
7. Blümlein, J., Broadhurst, D.J., Vermaseren, J.A.M.: The Multiple Zeta Value data mine. Comput. Phys. Commun. **181**, 582–625 (2010). https://doi.org/10.1016/j.cpc.2009.11.007, arXiv:0907.2557 [math-ph]
8. Bogner, C.: MPL-a program for computations with iterated integrals on moduli spaces of curves of genus zero. Comput. Phys. Commun. **203**, 339–353 (2016). https://doi.org/10.1016/j.cpc.2016.02.033, arXiv:1510.04562 [physics.comp-ph]
9. Bogner, C., Brown, F.C.S.: Symbolic integration and multiple polylogarithms. Proc. Sci. **LL2012**, 053 (2012). http://pos.sissa.it/cgi-bin/reader/conf.cgi?confid=151, arXiv:1209.6524 [hep-ph]
10. Bogner, C., Brown, F.C.S.: Feynman integrals and iterated integrals on moduli spaces of curves of genus zero. Commun. Number Theory Phys. **9**(1), 189–238 (2015). https://doi.org/10.4310/CNTP.2015.v9.n1.a3, arXiv:1408.1862 [hep-th]
11. Bogner, C., Weinzierl, S.: Feynman graph polynomials. Int. J. Mod. Phys. A **25**, 2585–2618 (2010). https://doi.org/10.1142/S0217751X10049438, arXiv:1002.3458 [hep-ph]
12. Broadhurst, D.J.: Evaluation of a class of Feynman diagrams for all numbers of loops and dimensions. Phys. Lett. B **164**(4–6), 356–360 (1985). https://doi.org/10.1016/0370-2693(85)90340-5
13. Broadhurst, D.J.: Massless scalar Feynman diagrams: five loops and beyond. Tech. Rep. OUT-4102-18, Open University, Milton Keynes (1985). http://cds.cern.ch/record/164890, arXiv:1604.08027 [hep-th]
14. Broadhurst, D.J., Kreimer, D.: Knots and numbers in ϕ^4 theory to 7 loops and beyond. Int. J. Mod. Phys. C **6**, 519–524 (1995). https://doi.org/10.1142/S012918319500037X arXiv:hep-ph/9504352
15. Brown, F.C.S.: The massless higher-loop two-point function. Commun. Math. Phys. **287**, 925–958 (2009). https://doi.org/10.1007/s00220-009-0740-5, arXiv:0804.1660 [math.AG]
16. Brown, F.C.S.: On the periods of some Feynman integrals. preprint (2009). http://arxiv.org/abs/0910.0114 arXiv:0910.0114 [math.AG]
17. Brown, F.C.S.: Multiple zeta values and periods: from moduli spaces to Feynman integrals, in Combinatorics and Physics. In: Ebrahimi-Fard, K., Marcolli, M., van Suijlekom, W.D. (eds.): Contemp. Math. 27–52. Proceedings of the mini-workshop on Renormalization (December 15–16, 2006) and the conference on Combinatorics and Physics (March 19–23, 2007), both at Max-Planck-Institut für Mathematik. American Mathematical Society, Bonn, Germany (May 2011). https://doi.org/10.1090/conm/539
18. Brown, F.C.S.: Irrationality proofs for zeta values, moduli spaces and dinner parties. Mosc. J. Combinatorics Number Theory **6**(2–3), 102–165 (2016). http://mjcnt.phystech.edu/en/article.php?id=115 arXiv:1412.6508 [math.NT]
19. Brown, F.C.S.: Feynman amplitudes, coaction principle, and cosmic Galois group. Commun. Number Theory Phys. **11**(3), 453–556 (2017). https://doi.org/10.4310/CNTP.2017.v11.n3.a1, arXiv:1512.06409 [math-ph]. based on lectures (links: 1, 2, 3 and 4), given at the IHÉS, May 2015

20. Brown, F.C.S.: Notes on motivic periods. Commun. Num. Theor. Phys. **11**(3), 557–655 (2017). https://doi.org/10.4310/CNTP.2017.v11.n3.a2, arXiv:1512.06410 [math.NT]. based on lectures (links to recordings: 1, 2, 3 and 4), given at the IHÉS, May 2015
21. Brown, F.C.S., Doryn, D.: Framings for graph hypersurfaces. preprint (2013). http://arxiv.org/abs/1301.3056, arXiv:1301.3056 [math.AG]
22. Brown, F.C.S., Schnetz, O.: A K3 in ϕ^4. Duke Math. J **161**, 1817–1862 (2012). https://doi.org/10.1215/00127094-1644201, arXiv:1006.4064 [math.AG]
23. Brown, F.C.S., Schnetz, O.: Modular forms in quantum field theory. Commun. Num. Theor. Phys. **7**(2), 293–325 (2013). https://doi.org/10.4310/CNTP.2013.v7.n2.a3, arXiv:1304.5342 [math.AG]
24. Brown, F.C.S., Schnetz, O.: Single-valued multiple polylogarithms and a proof of the zig-zag conjecture. J. Number Theory **148**, 478–506 (2015). https://doi.org/10.1016/j.jnt.2014.09.007, arXiv:1208.1890 [math.NT]
25. Chetyrkin, K.G., Tkachov, F.V.: Integration by parts: the algorithm to calculate β-functions in 4 loops. Nucl. Phys. B **192**, 159–204 (1981). https://doi.org/10.1016/0550-3213(81)90199-1
26. Doryn, D.: Cohomology of graph hypersurfaces associated to certain Feynman graphs. Commun. Num. Theor. Phys. **4**(2), 365–415 (2010). https://doi.org/10.4310/CNTP.2010.v4.n2.a3, arXiv:0811.0402 [math.AG]
27. Dupont, C.: Odd zeta motive and linear forms in odd zeta values. Compos. Math. **154** (2018)(2), 342–379. https://doi.org/10.1112/S0010437X17007588, arXiv:1601.00950 [math.AG]
28. Glanois, C.: Motivic unipotent fundamental groupoid of $\mathbb{G}_m \setminus \mu_N$ for $N = 2, 3, 4, 6, 8$ and Galois descents. J. Number Theory **160**, 334–384 (2016). https://doi.org/10.1016/j.jnt.2015.08.003, arXiv:1411.4947 [math.NT]
29. Glanois, C.: Unramified Euler sums and Hoffman $\star$ basis. preprint (2016). arXiv:1603.05178 [math.NT]
30. Itzykson, C., Zuber, J.-B.: Quantum Field Theory. Dover Publications, Inc., (2006). first published by McGraw-Hill in 1980
31. Kirchhoff, G.: Ueber die Auflösung der Gleichungen, auf welche man bei der Untersuchung der linearen Vertheilung galvanischer Ströme geführt wird. Annalen der Physik und Chemie **72**(12), 497–508 (1847). https://doi.org/10.1002/andp.18471481202
32. Kleinert, H., Schulte-Frohlinde, V.: Critical Properties of φ^4-theories. World Scientific (2001)
33. Kotikov, A.V., Teber, S.: On the Landau-Khalatnikov-Fradkin transformation and the mystery of even ζ-values in Euclidean massless correlators. preprint (2019). arXiv:1906.10930 [hep-th]
34. Lee, R.N., Smirnov, A.V., Smirnov, V.A.: Master integrals for four-loop massless propagators up to weight twelve. Nucl. Phys. B **856**, 95–110 (2012). https://doi.org/10.1016/j.nuclphysb.2011.11.005 arXiv:1108.0732 [hep-th]
35. Panzer, E.: On the analytic computation of massless propagators in dimensional regularization. Nucl. Phys. B **874**, 567–593 (2013). https://doi.org/10.1016/j.nuclphysb.2013.05.025, arXiv:1305.2161 [hep-th]
36. Panzer, E.: Feynman integrals and hyperlogarithms. Ph.D. thesis, Humboldt-Universität zu Berlin (2014). https://doi.org/10.18452/17157, arXiv:1506.07243 [math-ph]
37. Panzer, E.: Algorithms for the symbolic integration of hyperlogarithms with applications to Feynman integrals. Comput. Phys. Commun. **188**, 148–166 (2015). https://doi.org/10.1016/j.cpc.2014.10.019, arXiv:1403.3385 [hep-th]. maintained and available at https://bitbucket.org/PanzerErik/hyperint
38. Panzer, E., Schnetz, O.: The Galois coaction on ϕ^4 periods. Commun. Num. Theor. Phys. **11**(3), 657–705 (2017). https://doi.org/10.4310/CNTP.2017.v11.n3.a3, arXiv:1603.04289 [hep-th]
39. Schnetz, O.: Quantum periods: a Census of ϕ^4-transcendentals. Commun.Num. Theor. Phys. **4**(1), 1–47 (2010). arXiv:0801.2856 [hep-th]
40. Schnetz, O.: Quantum field theory over $\mathbb{F}_q$. Electron. J. Combin. **18**, P102 (2011). arXiv:0909.0905 [math.CO]
41. Smirnov, V.A.: Analytic Tools for Feynman integrals, Springer Tracts in Modern Physics, vol. 250. Springer, Berlin, Heidelberg (2012). https://doi.org/10.1007/978-3-642-34886-0

42. Tarasov, O.V.: Connection between Feynman integrals having different values of the space-time dimension. Phys. Rev. D **54**, 6479–6490 (1996). https://doi.org/10.1103/PhysRevD.54.6479, arXiv:hep-th/9606018
43. Ueda, T., Ruijl, B., Vermaseren, J.A.M.: Calculating four-loop massless propagators with Forcer. In: 17th International workshop on Advanced Computing and Analysis Techniques in physics research (ACAT 2016) Valparaiso, Chile, January 18–22, 2016, April 2016. arXiv:1604.08767 [hep-ph]
44. Vermaseren, J.A.M.: Axodraw. Comput. Phys. Commun. **83**, 45–58 (1994). https://doi.org/10.1016/0010-4655(94)90034-5
45. Weinberg, S.: High-energy behavior in quantum field theory. Phys. Rev. **118**, 838–849 (1960). https://doi.org/10.1103/PhysRev.118.838

Periods and Superstring Amplitudes

S. Stieberger

Abstract Scattering amplitudes which describe the interaction of physical states play an important role in determining physical observables. In string theory the physical states are given by vibrations of open and closed strings and their interactions are described (at the leading order in perturbation theory) by a world–heet given by the topology of a disk or sphere, respectively. Formally, for scattering of N strings this leads to $N-3$–dimensional iterated real integrals along the compactified real axis or $N-3$–dimensional complex sphere integrals, respectively. As a consequence the physical observables are described by periods on $\mathscr{M}_{0,N}$–the moduli space of Riemann spheres of N ordered marked points. The mathematical structure of these string amplitudes share many recent advances in arithmetic algebraic geometry and number theory like multiple zeta values, single–valued multiple zeta values, Drinfeld, Deligne associators, Hopf algebra and Lie algebra structures related to Grothendiecks Galois theory. We review these results, with emphasis on a beautiful link between generalized hypergeometric functions describing the real iterated integrals on $\mathscr{M}_{0,N}(\mathbf{R})$ and the decomposition of motivic multiple zeta values. Furthermore, a relation expressing complex integrals on $\mathscr{M}_{0,N}(\mathbf{C})$ as single–valued projection of iterated real integrals on $\mathscr{M}_{0,N}(\mathbf{R})$ is exhibited.

Keywords Periods · Multiple zeta values · Single-valued multiple zeta values · String amplitudes

1 Introduction

During the last years a great deal of work has been addressed to the problem of revealing and understanding the hidden mathematical structures of scattering amplitudes in both field–and string theory. Particular emphasis on their underlying geometric

S. Stieberger (✉)
Max–Planck–Institut für Physik, Werner–Heisenberg–Institut, 80805 München, Germany
e-mail: stephan.stieberger@mpp.mpg.de

J. I. Burgos Gil et al. (eds.), *Periods in Quantum Field Theory and Arithmetic*, Springer Proceedings in Mathematics & Statistics 314,
https://doi.org/10.1007/978-3-030-37031-2_3

structures seems to be especially fruitful and might eventually yield an alternative[1] way of constructing perturbative amplitudes by methods residing in arithmetic algebraic geometry . In such a framework physical quantities are given by periods (or more generally by L–functions) typically describing the volume of some polytope or integrals of a discriminantal configuration (a configuration of multivariate hyperplanes). The mathematical quantities which occur in string amplitude computations are periods which relate to fundamental objects in number theory and algebraic geometry. A period is a complex number whose real and imaginary parts are given by absolutely convergent integrals of rational functions with rational coefficients over domains in $\mathbf{R}^n$ described by polynomial inequalities with rational coefficients. More generally, periods are values of integrals of algebraic differential forms over certain chains in algebraic varieties [35]. Example in quantum field theory the coefficients of the Laurent series in the parameter $\varepsilon = \frac{1}{2}(4-D)$ of dimensionally regulated Feynman integrals are numerical periods in the Euclidian region with all ratios of invariants and masses having rational values [10]. Furthermore, the power series expansion in the inverse string tension α' of tree–level superstring amplitudes yields iterated integrals [12, 39, 43], which are periods of the moduli space $\mathscr{M}_{0,N}$ of genus zero curves with N ordered marked points [30] and integrate to $\mathbf{Q}$–linear combinations of multiple zeta values (MZVs) [15, 48]. Similar considerations [18] are expected to hold at higher genus in string perturbation theory, cf. [22] for some recent investigations at one–loop. At any rate, the analytic dependence on the inverse string tension α' of string amplitudes furnishes an extensive and rich mathematical structure, which is suited to exhibit and study modern developments in number theory and arithmetic algebraic geometry.

The forms and chains entering the definition of periods may depend on parameters (moduli). As a consequence the periods satisfy linear differential equations with algebraic coefficients. This type of differential equations is known as generalized Picard–Fuchs equations or Gauss–Manin systems. A subclass of the latter describes the A–hypergeometric system[2] or Gelfand–Kapranov–Zelevinsky (GKZ) system relevant to tree–level string scattering. One notorious example of periods are multivariate (multidimensional) or generalized hypergeometric functions.[3] In the non–resonant case the solutions of the GKZ system can be represented by generalized Euler integrals [26], which appear as world–sheet integrals in superstring tree–level amplitudes and integrate to multiple Gaussian hypergeometric functions [39]. Other occurrences of periods as physical quantities are string compactifications on Calabi–Yau manifolds. According to Batyrev the period integrals of Calabi–Yau

[1]In field–theory with $\mathscr{N} = 4$ supersymmetry such methods have recently been pioneered by using tools in algebraic geometry [1, 2] and arithmetic algebraic geometry [27, 31].

[2]The initial data for a GKZ–system is an integer matrix $A \in \mathbf{Z}^{r\times n}$ together with a parameter vector $\gamma \in \mathbf{C}^r$. For a given matrix A the structure of the GKZ–system depends on the properties of the vector γ defining non–resonant and resonant systems. Example a non–resonant system of A–hypergeometric equations is irreducible [5].

[3]More precisely, at an algebraic value of their argument their value is $\frac{1}{\pi}\wp$, with $\wp$ being the set of periods.

toric varieties are also governed by GKZ systems. Therefore, the GKZ system is ubiquitous to functions describing physical effects in string theory as periods.

2 Periods on $\mathcal{M}_{0,N}$

The object of interest is the moduli space $\mathcal{M}_{0,N}$ of Riemann spheres (genus zero curves) of $N \geq 4$ ordered marked points modulo the action of $PSL(2,\mathbf{C})$ on those points. The connected manifold $\mathcal{M}_{0,N}$ is described by the set of N–tuples of distinct points $(z_1,\ldots,z_N)$ modulo the action of $PSL(2,\mathbf{C})$ on those points. As a consequence with the choice

$$z_1 = 0, \quad z_{N-1} = 1, \quad z_N = \infty \tag{1}$$

there is a unique representative

$$(z_1,\ldots,z_N) = (0,t_1,\ldots,t_{N-3},1,\infty) \tag{2}$$

of each equivalence class of $\mathcal{M}_{0,N}$

$$\mathcal{M}_{0,N} \simeq \{\ (t_1,\ldots,t_{N-3}) \in \left(\mathbf{P}^1\backslash\{0,1,\infty\}\right)^{N-3}\ \mid t_i \neq t_j \text{ for all } i \neq j\ \}\ , \tag{3}$$

and the dimension of $\mathcal{M}_{0,N}(\mathbf{C})$ is $N-3$. On the other hand, the real part of (3) describing the space of points

$$\mathcal{M}_{0,N}(\mathbf{R}) := \{(0,t_1,\ldots,t_{N-3},1,\infty) \mid t_i \in \mathbf{R}\} \tag{4}$$

is not connected. Up to dihedral permutation each of its $\frac{1}{2}(N-1)!$ connected components (open cells γ)

$$\gamma = (z_1,z_2,\ldots,z_N) \tag{5}$$

is completely described by the (real) ordering of the N marked points

$$z_1 < z_2 < \ldots < z_N\ , \tag{6}$$

with:

$$\bigcup_{i=1}^{N}\{z_i\} = \{0,t_1,\ldots,t_{N-3},1,\infty\}\ . \tag{7}$$

In the compactification $\overline{\mathcal{M}}_{0,N}(\mathbf{R})$ the components γ become closed cells. Each cell corresponds to a triangulation of a regular polygon with N sides. The number of triangulations is given by $C_{N-2} = \frac{2^{N-2}(2N-5)!!}{(N-1)!}$ (with C_N the Catalan number). In total an underlying K_{N-1} associahedron (Stasheff polytope) can naturally be associated with each vertex describing one triangulation [15]. The standard cell of $\mathcal{M}_{0,N}$ is denoted by δ and given by the set of real marked points $(z_1, z_2, \ldots, z_N) = (0, t_1, t_2, \ldots, t_{N-3}, 1, \infty)$ on $\mathcal{M}_{0,N}$ subject to the (canonical) ordering (6), i.e.:

$$\delta = \{\, t_l \in \mathbf{R} \mid 0 < t_1 < t_2 < \ldots < t_{N-3} < 1 \,\} . \tag{8}$$

A period on $\mathcal{M}_{0,N}$ is defined to be a convergent integral [30]

$$\int_\delta \omega \tag{9}$$

over the standard cell (8) in $\mathcal{M}_{0,N}(\mathbf{R})$ and $\omega \in H^{N-3}(\mathcal{M}_{0,N})$ a regular algebraic $(N-3)$–form, which converges on δ and has no poles along $\overline{\delta}$. Every period on $\mathcal{M}_{0,N}$ is a $\mathbf{Q}$–linear combination of MZVs [15]. Furthermore, every MZV can be written as (9).

To each cell γ a unique $(N-3)$–form can be associated [16]

$$\omega_\gamma = \prod_{i=2}^{N} (z_i - z_{i-1})^{-1}\, dt_1 \wedge \ldots \wedge dt_{N-3} , \tag{10}$$

subject to (7) with $z_l = \infty$ dismissed in the product. The form (10) is unique up to scalar multiplication, holomorphic on the interior of γ and has simple poles on the boundary of that cell. To a cell (6) in $\mathcal{M}_{0,N}(\mathbf{R})$ modulo rotations an oriented N–gon (N–sided polygons) may be associated by labelling clockwise its sides with the marked points $(z_1, z_2, \ldots, z_N)$. Example according to (6) the polygon with the cyclically labelled sides $\gamma = (0, 1, t_1, t_3, \infty, t_2)$ is identified with the cell $0 < 1 < t_1 < t_3 < \infty < t_2$ in $\mathcal{M}_{0,6}(\mathbf{R})$ and the corresponding cell form is:

$$\omega_\gamma = \pm \frac{dt_1 dt_2 dt_3}{(-t_2)\,(t_3 - t_1)\,(t_1 - 1)} .$$

The cell form (10) refers to the ordering (6). A cyclic structure γ corresponds to the cyclic ordering $(\gamma(1), \gamma(2), \ldots, \gamma(N))$ of the elements $\{1, 2, \ldots, N\}$ and refers to the standard N–gon $(1, 2, \ldots, N)$ modulo rotations. There is a unique ordering σ of the N marked points (2) as

$$z_{\sigma(1)} < z_{\sigma(2)} < \ldots < z_{\sigma(N)} . \tag{11}$$

with $\sigma(N) = N$ and compatible with the cyclic structure γ. The cell–form corresponding to γ is defined as [16]

$$\omega_\gamma = \prod_{i=2}^{N-1} \left(z_{\sigma(i)} - z_{\sigma(i-1)}\right)^{-1} dt_1 \wedge \ldots \wedge dt_{N-3} . \tag{12}$$

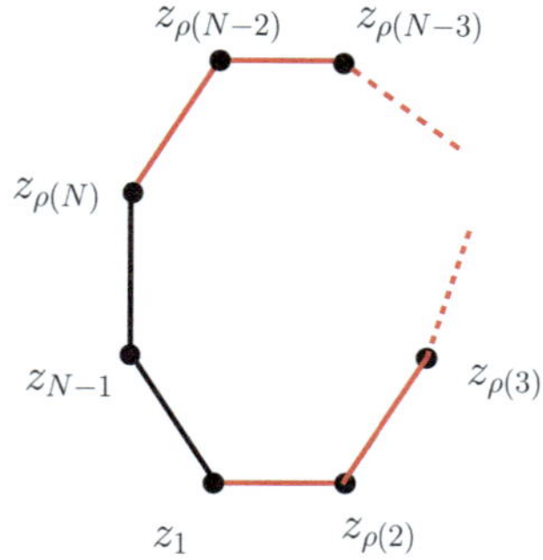

Fig. 1 N–gon describing the 01 cyclic structure $\gamma = (0, 1, \rho)$

For example the cyclic structure $(2, 5, 1, 6, 4, 3)$ the unique ordering σ compatible with the latter and with $\sigma(6) = 6$ is the ordering $(4, 3, 2, 5, 1, 6)$, i.e. $\gamma = (t_3, t_2, t_1, 1, 0, \infty)$.

In the following, we consider orderings (6) (01 cyclic structure γ) of the set $\bigcup_{i=1}^{N}\{z_i\} = \{0, t_1, \ldots, t_{N-3}, 1, \infty\}$ with the elements $z_1 = 0$ and $z_{N-1} = 1$ being consecutive, i.e. $\gamma = (0, 1, \rho)$ with $\rho \in S_{N-2}$ some ordering of the $N - 2$ points $\{t_1, \ldots, t_{N-3}, \infty\}$. The corresponding cell–function is given by

$$\omega_\rho = z_{\rho(2)}^{-1} \prod_{i=3}^{N-2} \left(z_{\rho(i)} - z_{\rho(i-1)}\right)^{-1} dt_1 \wedge \ldots \wedge dt_{N-3} \quad , \quad \rho \in S_{N-2} \,, \tag{13}$$

it is called 01 cell–function [16] and its associated N–gon, in which the edge referring to 0 appears next to that referring to 1, is depicted in Fig. 1. The $(N-2)!$ 01 cell–functions (13) generate the top–dimensional cohomology group $H^{N-3}(\mathcal{M}_{0,N})$ of $\mathcal{M}_{0,N}$ by constituting a basis of $H^{N-3}(\mathcal{M}_{0,N}, \mathbf{Q})$, i.e. [16]:

$$\dim H^{N-3}(\mathcal{M}_{0,N}, \mathbf{Q}) = (N-2)! \,. \tag{14}$$

As a consequence the cohomology group $H^{N-3}(\mathcal{M}_{0,N})$ is canonically isomorphic to the subspace of polygons having the vertex (edge) 0 adjacent to edge 1 [16].

Generically, in terms of cells a period (9) on $\mathcal{M}_{0,N}$ may be defined as the integral [16]

$$\int_\beta \omega_\gamma \tag{15}$$

over the cell β in $\mathcal{M}_{0,N}(\mathbf{R})$ and the cell–form ω_γ with the pair (β, γ) referring to some polygon pair. Therefore, generically the cell–forms (10) integrated over cells (5) give rise to periods on $\mathcal{M}_{0,N}$, which are $\mathbf{Q}$–linear combinations of MZV. By changing variables the period integral (15) can be brought into an integral over the standard cell δ parameterized in (8). To obtain a convergent integral (15) in [16] certain linear combinations of 01 cell–forms (13) (called insertion forms) have been constructed with the properties of having no poles along the boundary of the standard cell δ and converging on the closure $\overline{\delta}$. Example in the case of $\mathcal{M}_{0,5}$ the cell–form ω_γ corresponding to the cell $\gamma = (0, 1, t_1, \infty, t_2)$ can be integrated over the compact standard cell $\overline{\delta}$ defined in (8)

$$\int_{\overline{\delta}} \omega_\gamma = \int_{0\leq t_1\leq t_2\leq 1} \frac{dt_1 dt_2}{(1-t_1)\, t_2} = \zeta_2 \, , \tag{16}$$

with the period ζ_2 following from the general definition for the Riemann zeta function:

$$\zeta_a = \sum_{k=1}^{\infty} k^{-a} \quad , \quad a \in \mathbf{N},\ a \geq 2 \, . \tag{17}$$

3 Volume Form and Period Matrix on $\mathcal{M}_{0,N}$

For a regular algebraic $(N-3)$–form ω_δ on $\mathcal{M}_{0,N}$ conditions exist for the integral (9) over the standard cell δ to converge. The set of all regular $(N-3)$–forms can be written in terms of the canonical cyclically invariant form [15]:

$$\omega_\delta = \frac{dt_1 \wedge \ldots \wedge dt_{N-3}}{t_2\, (t_3 - t_1)\, (t_4 - t_2) \cdot \ldots \cdot (t_{N-3} - t_{N-5})\, (1 - t_{N-4})} \, . \tag{18}$$

(Up to multiplication by $\mathbf{Q}^+$) this form is the canonical volume form on $\mathcal{M}_{0,N}(\mathbf{R})$ without zeros or poles along the standard cell (8). An algebraic volume form Ω on $\mathcal{M}_{0,N}(\mathbf{R})$ may be supplemented by the $PSL(2,\mathbf{C})$ invariant factor $\prod_{i<j}^{N-1} |z_i - z_j|^{s_{ij}}$ (subject to (1) and with some conditions on the parameter s_{ij}, which turn into physical conditions, cf. (100)) as

$$\Omega = \frac{dt_1 \wedge \ldots \wedge dt_{N-3}}{t_2\, (t_3 - t_1)\, (t_4 - t_2) \cdot \ldots \cdot (t_{N-3} - t_{N-5})\, (1 - t_{N-4})} \left(\prod_{i<j}^{N-1} |z_i - z_j|^{s_{ij}} \right) , \tag{19}$$

with $s_{ij} \in \mathbf{Z}$. The form (19) gives rise to the family of periods $\int_{\overline{\delta}} \Omega$ of $\mathcal{M}_{0,N}$

$$I_\delta(a,b,c) = \int_{\overline{\delta}} dt_1 \cdot \ldots \cdot dt_{N-3} \prod_{i=1}^{N-3} t_i^{a_i}\, (1-t_i)^{b_i} \prod_{1\leq i<j\leq N-3} (t_i - t_j)^{c_{ij}} \, , \tag{20}$$

for suitable choices of integers $a_i, b_i, c_{ij} \in \mathbf{Z}$ such that the integral converges. The latter refers to the compactified standard cell δ defined in (8). It has been shown by Brown and Terasoma, that integrals of the form (20) yield linear combinations of MZVs with rational coefficients. In cubical coordinates $x_1 = \frac{t_1}{t_2},\ x_2 = \frac{t_2}{t_3}, \ldots, x_{N-4} = \frac{t_{N-4}}{t_{N-3}},\ x_{N-3} = t_{N-3}$ parameterizing the integration region (8) as $t_k = \prod_{l=k}^{N-3} x_l,\ k = 1, \ldots, N-3$ with $0 < x_i < 1$, the integral (20) becomes

$$I_\delta(a',b',c')=\left(\prod_{i=1}^{N-3}\int_0^1 dx_i\right)\prod_{j=1}^{N-3}x_j^{a'_j}\,(1-x_j)^{b'_j}\prod_{l=j+1}^{N-3}\left(1\ -\ \prod_{k=j}^{l}x_k\right)^{c'_{jl}}\ , \tag{21}$$

with some integers $a'_i, b'_i, c'_{ij}\in\mathbf{Z}$.

Moreover, the form (19) can be generalized to the family of real period integrals $\int_{\overline{\delta}}\Omega$ on $\overline{\delta}$, with $s_{ij}\in\mathbf{R}$. Then, Taylor expanding (19) w.r.t. s_{ij} at integral points $s_{ij}\in\mathbf{Z}^+$ yields coefficients representing period integrals of the form (20). Similar observations have been made in [39, 43] when computing α'–expansions of string amplitudes which can be described by integrals of the type (21). In this setup the additional $PSL(2,\mathbf{C})$ invariant factor $\prod_{i<j}^{N-1}|z_i-z_j|^{\alpha' s_{ij}}$ represents the so–called Koba–Nielsen factor with the parameter α' being the inverse string tension and the kinematic invariants s_{ij} specified in (100).

Similarly to (19) in the following let us consider all the $(N-2)!$ 01 cell–forms (13) supplemented by the $PSL(2,\mathbf{C})$ invariant factor $\prod_{i<j}^{N-1}|z_i-z_j|^{\alpha' s_{ij}}$ and integrated over the standard cell δ in $\mathcal{M}_{0,N}(\mathbf{R})$, i.e.:

$$\int_\delta\left(\prod_{i<j}^{N-1}|z_i-z_j|^{\alpha' s_{ij}}\right)\omega_\rho=\int_\delta\frac{dt_1\wedge\ldots\wedge dt_{N-3}}{z_{\rho(2)}\prod\limits_{i=3}^{N-2}\left(z_{\rho(i)}-z_{\rho(i-1)}\right)}\left(\prod_{i<j}^{N-1}|z_i-z_j|^{\alpha' s_{ij}}\right)\ ,\ \ \rho\in S_{N-2}\,. \tag{22}$$

Integration by part allows to express the $(N-2)!$ integrals (22) in terms of a basis of $(N-3)!$ integrals, i.e.:

$$\dim H^{N-3}(\mathcal{M}_{0,N},\mathbf{R})=(N-3)!\ . \tag{23}$$

For a given cell π in $\mathcal{M}_{0,N}(\mathbf{R})$ we can choose the 01 cell–form ω_γ with $\gamma=(0,1,\infty,\rho),\ \rho\in S_{N-3}$ and the following basis (subject to (1)) [12]

$$Z_\pi^\rho:=Z_\pi(1,\rho(2,3,\ldots,N-2),N,N-1) \tag{24}$$

$$=\int_\pi\left(\prod_{i=2}^{N-2}dz_i\right)\frac{\prod\limits_{i<j}^{N-1}|z_{ij}|^{\alpha' s_{ij}}}{z_{1\rho(2)}z_{\rho(2),\rho(3)}\cdots z_{\rho(N-3),\rho(N-2)}}\ ,\quad \pi,\rho\in S_{N-3}\ ,$$

with

$$z_{ij}:=z_i-z_j\ , \tag{25}$$

and ρ describing some ordering of the $N-3$ points $\bigcup_{i=2}^{N-2}\{z_i\}=\{t_1,\ldots,t_{N-3}\}$ along the N–gon depicted in Fig. 2. The iterated integrals (24) represent generalized Euler (Selberg) integral and integrate to multiple Gaussian hypergeometric functions [39]. Furthermore, the integrals (24) can also be systematized within the framework of Aomoto-Gelfand hypergeometric functions or GKZ structures [26].

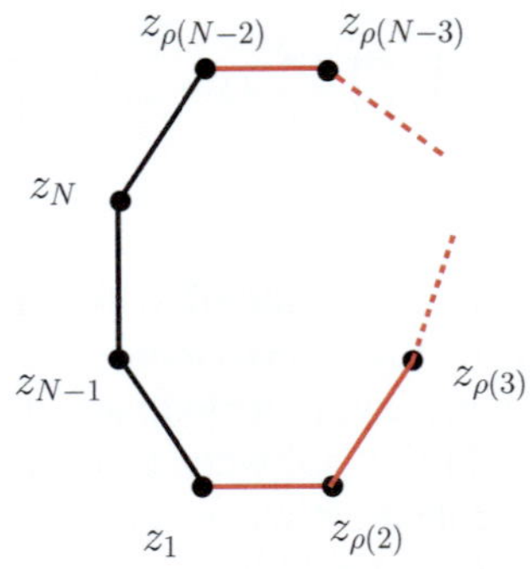

Fig. 2 N–gon describing the cyclic structure $\gamma = (0, 1, \infty, \rho)$

The integrals (24) can be Taylor expanded w.r.t. α' around the point $\alpha' = 0$, e.g.:

$$\begin{aligned}&\int_{\delta}\left(\prod_{l=2}^{3} dz_l\right)\frac{\prod\limits_{i<j}^{4}|z_{ij}|^{\alpha' s_{ij}}}{z_{12}\, z_{23}\, z_{41}}\\ &= \alpha'^{-2}\left(\frac{1}{s_{12}s_{45}}+\frac{1}{s_{23}s_{45}}\right)+\zeta_2\left(1-\frac{s_{34}}{s_{12}}-\frac{s_{12}}{s_{45}}-\frac{s_{23}}{s_{45}}-\frac{s_{51}}{s_{23}}\right)+\mathcal{O}(\alpha')\ .\end{aligned} \tag{26}$$

Techniques for computing α' expansions for the type of integrals (24) have been exhibited in [39, 43], systematized in [12], and pursued in [40]. In fact, the lowest order contribution of (24) in the Taylor expansion around the point $\alpha' = 0$ is given by

$$Z|_{\alpha'^{3-N}} = (-1)^{N-3}\ S^{-1}\ , \tag{27}$$

with the kernel[4] [3, 7, 34]

$$S[\rho|\sigma] := S[\,\rho(2,\ldots,N-2)\,|\,\sigma(2,\ldots,N-2)\,] = \alpha'^{N-3}\prod_{j=2}^{N-2}\left(s_{1,j_\rho}+\sum_{k=2}^{j-1}\theta(j_\rho,k_\rho)\, s_{j_\rho,k_\rho}\right), \tag{28}$$

with $j_\rho = \rho(j)$ and $\theta(j_\rho, k_\rho) = 1$ if the ordering of the legs j_ρ, k_ρ is the same in both orderings $\rho(2,\ldots,N-2)$ and $\sigma(2,\ldots,N-2)$, and zero otherwise. The matrix elements $S[\rho|\sigma]$ are polynomials of the order $N-3$ in the parameters (100).

A natural object to define is the $(N-3)! \times (N-3)!$–matrix

$$F_{\pi\sigma} = (-1)^{N-3}\sum_{\rho\in S_{N-3}} Z_\pi(\rho)\ S[\rho|\sigma]\ , \tag{29}$$

[4]The matrix S with entries $S_{\rho,\sigma} = S[\rho|\sigma]$ is defined as a $(N-3)! \times (N-3)!$ matrix with its rows and columns corresponding to the orderings $\rho \equiv \{\rho(2),\ldots,\rho(N-2)\}$ and $\sigma \equiv \{\sigma(2),\ldots,\sigma(N-2)\}$, respectively. The matrix S is symmetric, i.e. $S^t = S$.

which according to (27) satisfies:

$$F|_{\alpha'^{3-N}} = 1 \ . \tag{30}$$

The matrix F has rank

$$\mathrm{rk}(F) = (N-3)! \tag{31}$$

and represents the period matrix of $\mathcal{M}_{0,N}$ [32].

In [41] it has been observed, that F can be written in the following way[5]

$$F = P\ Q\ :\exp\left\{\sum_{n\geq 1} \zeta_{2n+1}\ M_{2n+1}\right\}: , \tag{32}$$

with the Riemann zeta–functions (17). This decomposition is guided by its organization w.r.t. multiple zeta values (MZVs) $\zeta_{n_1,\ldots,n_r}$ as

$$\begin{aligned} M_{2n+1} &= F|_{\zeta_{2n+1}} \ , \\ P_{2n} &= F|_{\zeta_2^n} \ , \end{aligned} \tag{33}$$

with:

$$P = 1 + \sum_{n\geq 1} \zeta_2^n\ P_{2n} \ , \tag{34}$$

$$\begin{aligned} Q &= 1 + \sum_{n\geq 8} Q_n = 1 + \frac{1}{5}\,\zeta_{3,5}\,[M_5, M_3] + \left\{\frac{3}{14}\,\zeta_5^2 + \frac{1}{14}\,\zeta_{3,7}\right\}\,[M_7, M_3] \\ &\quad + \left\{9\,\zeta_2\,\zeta_9 + \frac{6}{25}\,\zeta_2^2\,\zeta_7 - \frac{4}{35}\,\zeta_2^3\,\zeta_5 + \frac{1}{5}\,\zeta_{3,3,5}\right\}\,[M_3, [M_5, M_3]] + \cdots \ . \end{aligned} \tag{35}$$

MZVs are generalizations of single zeta functions (17)

$$\zeta_{n_1,\ldots,n_r} := \zeta(n_1,\ldots,n_r) = \sum_{0<k_1<\ldots<k_r} \prod_{l=1}^{r} k_l^{-n_l} \ , \quad n_l \in \mathbf{N}^+ \ , \ n_r \geq 2 \ , \tag{36}$$

with r specifying its depth and $w = \sum_{l=1}^{r} n_l$ denoting its weight. Hence, all the information is kept in the matrices P and M and the particular form of Q. The entries of the matrices M_{2n+1} are polynomials in s_{ij} of degree $2n+1$ (and hence of the order α'^{2n+1}), while the entries of the matrices P_{2n} are polynomials in s_{ij} of degree $2n$ (and hence of the order α'^{2n}). Example for $N=5$ we have

---

[5]The ordering colons $:\ldots:$ are defined such that matrices with larger subscript multiply matrices with smaller subscript from the left, i.e. $:M_i\ M_j: := \begin{cases} M_i\ M_j \ , \ i \geq j \ , \\ M_j\ M_i \ , \ i < j \ . \end{cases}$ The generalization to iterated matrix products $:M_{i_1} M_{i_2} \ldots M_{i_p}:$ is straightforward.

$$P = \alpha'^2 \begin{pmatrix} -s_{34}s_{45} + s_{12}\,(s_{34} - s_{51}) & s_{13}\,s_{24} \\ s_{12}\,s_{34} & (s_{12} + s_{23})\,(s_{23} + s_{34}) - s_{45}s_{51} \end{pmatrix}, \tag{37}$$

and

$$M_3 = \alpha'^3 \begin{pmatrix} m_{11} & m_{12} \\ m_{21} & m_{22} \end{pmatrix}, \tag{38}$$

with:

$$\begin{aligned} m_{11} &= s_{34}\,[\,-s_{12}\,(s_{12} + 2s_{23} + s_{34}) + s_{34}s_{45} + s_{45}^2\,] + s_{12}s_{51}\,(s_{12} + s_{51})\,, \\ m_{12} &= -s_{13}\,s_{24}\,(s_{12} + s_{23} + s_{34} + s_{45} + s_{51})\,, \\ m_{21} &= s_{12}\,s_{34}\,[\,s_{12} + s_{23} + s_{34} - 2\,(s_{45} + s_{51})\,]\,, \\ m_{22} &= (s_{23} + s_{34})\,[\,(s_{12} + s_{23})(s_{12} + s_{34}) - 2\,s_{12}s_{45}\,] \\ &\quad -[\,2s_{12}s_{34} - s_{45}^2 + 2s_{23}\,(s_{34} + s_{45})\,]s_{51} + s_{45}s_{51}^2\,. \end{aligned} \tag{39}$$

As we shall see in Sect. 5 the form (32) is bolstered by the algebraic structure of motivic MZVs. The form (32) exactly appears in F. Browns decomposition of motivic MZVs [17]. In Sect. 6 we shall demonstrate, that the period matrix F has also a physical meaning describing scattering amplitudes of open and closed strings.

4 Motivic and Single–valued Multiple Zeta Values

MZVs (36) can be represented as period integrals. With the iterated integrals of the following form

$$I_\gamma(a_0; a_1, \ldots, a_n; a_{n+1}) = \int_{\Delta_{n,\gamma}} \frac{dz_1}{z_1 - a_1} \ldots \frac{dz_n}{z_n - a_n}\,, \tag{40}$$

with γ a path in $M = \mathbf{C}/\{a_1, \ldots, a_n\}$ with endpoints $\gamma(0) = a_0 \in M,\ \gamma(1) = a_{n+1} \in M$ and $\Delta_{n,\gamma}$ a simplex consisting of all ordered n–tuples of points $(z_1, \ldots, z_n)$ on γ and for the map

$$\rho(n_1, \ldots, n_r) = 10^{n_1-1} \ldots 10^{n_r-1}\,, \tag{41}$$

with $n_r \geq 2$ Kontsevich observed that

$$\begin{aligned} \zeta_{n_1,\ldots,n_r} &= (-1)^r\, I_\gamma(0; \rho(n_1 \ldots n_r); 1) \\ &= (-1)^r \int\limits_{0\leq t_1 \leq \ldots \leq t_n \leq 1} \frac{dt_1}{t_1 - a_1} \ldots \frac{dt_n}{t_n - a_n}\,, \end{aligned} \tag{42}$$

with the sequence of numbers $(a_1, \ldots, a_n)$ given by $(1, 0^{n_1-1}, \ldots, 1, 0^{n_r-1})$. Note, that the integral (42) defines a period. Furthermore, the numbers (36) arise as coefficients of the Drinfeld associator $Z(e_0, e_1)$ [24]. The latter is a function in terms of the generators e_0 and e_1 of a free Lie algebra g and is given by the non–commutative generating series of (shuffle–regularized) MZVs [36]

$$Z(e_0, e_1) = \sum_{w \in \{e_0, e_1\}^\times} \zeta(w)\, w\,, \tag{43}$$

with the symbol $w \in \{e_0, e_1\}^\times$ denoting a non–commutative word $w_1 w_2 \ldots$ in the letters $w_i \in \{e_0, e_1\}$. Furthermore, we have the shuffle product $\zeta(w_1)\zeta(w_2) = \zeta(w_1 ⧢ w_2)$ and $\zeta(e_0) = 0 = \zeta(e_1)$ and $\zeta(e_1 e_0^{n_1-1} \ldots e_1 e_0^{n_r-1}) = \zeta_{n_1,\ldots,n_r}$. Explicitly, (43) becomes:

$$\begin{aligned} Z(e_0, e_1) = &\sum_{w \in \{e_0, e_1\}^\times} \zeta(w)\, w = 1 + \zeta_2\, [e_0, e_1] + \zeta_3\, (\, [e_0, [e_0, e_1]] - [e_1, [e_0, e_1]\,]\,) \\ &+ \zeta_4 \left([e_0, [e_0, [e_0, e_1]]] - \frac{1}{4}\, [e_1, [e_0, [e_0, e_1]]] + [e_1, [e_1, [e_0, e_1]]] + \frac{5}{4}\, [e_0, e_1]^2 \right) \\ &+ \zeta_2\, \zeta_3 \left(([e_0, [e_0, e_1]] - [e_1, [e_0, e_1]])\, [e_0, e_1] + [e_0, [e_1, [e_0, [e_0, e_1]]]] \right. \\ &\left. - [e_0, [e_1, [e_1, [e_0, e_1]]]] \right) + \zeta_5 \left([e_0, [e_0, [e_0, [e_0, e_1]]]] \right. \\ &\left. - \frac{1}{2}\, [e_0, [e_0, [e_1, [e_0, e_1]]]] - \frac{3}{2}\, [e_1, [e_0, [e_0, [e_0, e_1]]]] + (e_0 \leftrightarrow e_1) \right) + \cdots\,. \end{aligned} \tag{44}$$

The set of integral linear combinations of MZVs (36) is a ring, since the product of any two values can be expressed by a (positive) integer linear combination of the other MZVs [50]. There are many relations over $\mathbf{Q}$ among MZVs. We define the (commutative) $\mathbf{Q}$–algebra $\mathcal{Z}$ spanned by all MZVs over $\mathbf{Q}$. The latter is the (conjecturally direct) sum over the $\mathbf{Q}$–vector spaces $\mathcal{Z}_N$ spanned by the set of MZVs (36) of total weight $w = N$, with $n_r \geq 2$, i.e. $\mathcal{Z} = \bigoplus_{k \geq 0} \mathcal{Z}_k$. For a given weight $w \in \mathbf{N}$ the dimension $\dim_{\mathbf{Q}}(\mathcal{Z}_N)$ of the space $\mathcal{Z}_N$ is conjecturally given by $\dim_{\mathbf{Q}}(\mathcal{Z}_N) = d_N$, with $d_N = d_{N-2} + d_{N-3},\ N \geq 3$ and $d_0 = 1,\ d_1 = 0,\ d_2 = 1$ [50]. Starting at weight $w = 8$ MZVs of depth greater than one $r > 1$ appear in the basis. By applying stuffle, shuffle, doubling, generalized doubling relations and duality it is possible to reduce the MZVs of a given weight to a minimal set. Strictly speaking this is explicitly proven only up to weight 26 [8]. For $D_{w,r}$ being the number of independent MZVs at weight $w > 2$ and depth r, which cannot be reduced to primitive MZVs of smaller depth and their products, it is believed, that $D_{8,2} = 1,\ D_{10,2} = 1,\ D_{11,3} = 1,\ D_{12,2} = 1$ and $D_{12,4} = 1$ [11]. For $Z = \frac{\mathcal{Z}_{>0}}{\mathcal{Z}_{>0} \cdot \mathcal{Z}_{>0}}$ with $\mathcal{Z}_{>0} = \oplus_{w>0} \mathcal{Z}_w$ the graded space of irreducible MZVs we have: $dim(Z_w) \equiv \sum_r D_{w,r} = 1, 0, 1, 0, 1, 1, 1, 1, 2, 2, 3, 3, 4, 5$ for $w = 3, \ldots, 16$, respectively [8, 11].

An important question is how to decompose a MZV of a certain weight w in terms of a given basis of the same weight w. Example for the decomposition

$$\zeta_{4,3,3} = \frac{4336}{1925}\,\zeta_2^5 + \frac{1}{5}\,\zeta_2^2\zeta_3^2 + 10\,\zeta_2\zeta_3\zeta_5 - \frac{49}{2}\,\zeta_5^2 - 18\,\zeta_3\zeta_7 - 4\,\zeta_2\zeta_{3,5} + \zeta_{3,7} \quad (45)$$

we wish to find a method to determine the rational coefficients. Clearly, this question cannot be answered within the space of MZV $\mathscr{Z}$ as we do not know how to construct a basis of MZVs for any weight. Eventually, we seek to answer the above question within the space $\mathscr{H}$ of motivic MZVs with the latter serving as some auxiliary objects for which we assume certain properties [17]. For this purpose the actual MZVs (36) are replaced by symbols (or motivic MZVs), which are elements of a certain algebra. We lift the ordinary MZVs ζ to their motivic versions[6] ζ^{m} with the surjective projection (period map) [19, 31]:

$$\mathrm{per} : \zeta^{\mathrm{m}} \longrightarrow \zeta\ . \quad (46)$$

Furthermore, the standard relations among MZV (like shuffle and stuffle relations) are supposed to hold for the motivic MZVs ζ^{m}. In particular, $\mathscr{H}$ is a graded Hopf algebra[7] $\mathscr{H}$ with a coproduct Δ, i.e.

$$\mathscr{H} = \bigoplus_{n\geq 0} \mathscr{H}_n\ , \quad (47)$$

and for each weight n the Zagier conjecture is assumed to be true, i.e. $\dim_{\mathbf{Q}}(\mathscr{H}_n) = d_n$. To explicitly describe the structure of the space $\mathscr{H}$ one introduces the (trivial) algebra–comodule:

$$\mathscr{U} = \mathbf{Q}\langle f_3, f_5, \ldots\rangle \otimes_{\mathbf{Q}} \mathbf{Q}[f_2]\ . \quad (48)$$

The multiplication on

$$\mathscr{U}' = \mathscr{U} \big/ f_2\mathscr{U} = \mathbf{Q}\langle f_3, f_5, \ldots\rangle \quad (49)$$

is given by the shuffle product $\sqcup\!\sqcup$

$$f_{i_1}\ldots f_{i_r} \sqcup\!\sqcup f_{i_{r+1}}\ldots f_{i_{r+s}} = \sum_{\sigma\in\Sigma(r,s)} f_{i_{\sigma(1)}}\ldots f_{i_{\sigma(r+s)}}\ , \quad (50)$$

[6]In [19, 31] motivic MZVs ζ^{m} are defined as elements of a certain graded algebra $\mathscr{H}$ equipped with a period homomorphism (46).

[7]A Hopf algebra is an algebra $\mathscr{A}$ with multiplication $\mu : \mathscr{A}\otimes\mathscr{A} \to \mathscr{A}$, i.e. $\mu(x_1\otimes x_2) = x_1\cdot x_2$ and associativity. At the same time it is also a coalgebra with coproduct $\Delta : \mathscr{A} \to \mathscr{A}\otimes\mathscr{A}$ and coassociativity such that the product and coproduct are compatible: $\Delta(x_1\cdot x_2) = \Delta(x_1)\cdot\Delta(x_2)$, with $x_1, x_2 \in \mathscr{A}$.

$\Sigma(r,s)=\{\sigma\in\Sigma(r+s)\,|\,\sigma^{-1}(1)<\ldots<\sigma^{-1}(r)\cap\sigma^{-1}(r+1)<\ldots<\sigma^{-1}(r+s)\}$. The Hopf–algebra $\mathscr{U}'$ is isomorphic to the space of non–commutative polynomials in f_{2i+1}. The element f_2 commutes with all f_{2r+1}. Again, there is a grading $\mathscr{U}_k$ on $\mathscr{U}$, with $\dim(\mathscr{U}_k)=d_k$. Then, there exists a morphism ϕ of graded algebra–comodules

$$\phi:\ \mathscr{H}\longrightarrow\mathscr{U}\ , \tag{51}$$

normalized[8] by:

$$\phi\big(\zeta_n^{\mathfrak{m}}\big)=f_n\quad,\quad n\geq 2\ . \tag{52}$$

Furthermore, (51) respects the shuffle multiplication rule (50):

$$\phi(x_1x_2)=\phi(x_1)\sqcup\!\sqcup\,\phi(x_2)\quad,\quad x_1,x_2\in\mathscr{H}\ . \tag{53}$$

The map (51) is defined recursively from lower weight and sends every motivic MZV $\xi\in\mathscr{H}_{N+1}$ of weight $N+1$ to a non–commutative polynomial in the f_i. The latter is given as series expansion up to weight $N+1$ w.r.t. the basis $\{f_{2r+1}\}$

$$\phi(\xi)=c_{N+1}\ f_{N+1}+\sum_{3\leq 2r+1\leq N} f_{2r+1}\ \xi_{2r+1}\in\mathscr{U}_{N+1}\ , \tag{54}$$

with the coefficients $\xi_{2r+1}\in\mathscr{U}_{N-2r}$ being of smaller weight than ξ and computed from the coproduct as follows. The derivation $D_r:\mathscr{H}_m\to\mathscr{A}_r\otimes\mathscr{H}_{m-r}$, with $\mathscr{A}=\mathscr{H}/\zeta_2\mathscr{H}$ takes only a subset of the full coproduct, namely the weight $(r,m-r)$–part. Hence, $D_{2r+1}\xi$ gives rise to a weight $(2r+1,N-2r)$–part $x_{2r+1}\otimes y_{N-2r}\in\mathscr{A}_{2r+1}\otimes\mathscr{H}_{N-2r}$ and $\xi_{2r+1}:=c^{\phi}_{2r+1}(x_{2r+1})\cdot\phi(y_{N-2r})$ The operator $c^{\phi}_{2r+1}(x_{2r+1})$, with $x_{2r+1}\in\mathscr{A}_{2r+1}$ determines the rational coefficient of f_{2r+1} in the monomial $\phi(x_{2r+1})\in\mathscr{U}_{2r+1}$. Note, that the right hand side of ξ_{2r+1} only involves elements from $\mathscr{H}_{\leq N}$ for which ϕ has already been determined. On the other hand, the coefficient c_{N+1} cannot be determined by this method unless we specify[9] a basis B and compute ϕ for this basis giving rise to the basis dependent map ϕ^B. Example for the basis $B=\{\zeta_2^{\mathfrak{m}}\zeta_3^{\mathfrak{m}},\zeta_5^{\mathfrak{m}}\}$ we have $\phi^B(\zeta_2^{\mathfrak{m}}\zeta_3^{\mathfrak{m}})=f_2f_3$ and $\phi^B(\zeta_5^{\mathfrak{m}})=f_5$, while $\phi^B(\zeta_{2,3}^{\mathfrak{m}})=3f_3f_2+cf_5$ with c undetermined.

---

[8] Note, that there is no canonical choice of ϕ and the latter depends on the choice of motivic generators of $\mathscr{H}$.

[9] The choice of ϕ describes for each weight $2r+1$ the motivic derivation operators ∂^{ϕ}_{2r+1} acting on the space of motivic MZVs $\partial^{\phi}_{2r+1}:\mathscr{H}\to\mathscr{H}$ [17] as:

$$\partial^{\phi}_{2r+1}=(c^{\phi}_{2r+1}\otimes id)\circ D_{2r+1}\ , \tag{55}$$

with the coefficient function c^{ϕ}_{2r+1}.

To illustrate the procedure for computing the map (51) and determining the decomposition let us consider the case of weight 10. First, we introduce a basis of motivic MZVs

$$B_{10} = \{\, \zeta^{\rm m}_{3,7},\ \zeta^{\rm m}_3\, \zeta^{\rm m}_7,\ (\zeta^{\rm m}_5)^2,\ \zeta^{\rm m}_{3,5}\, \zeta^{\rm m}_2,\ \zeta^{\rm m}_3\, \zeta^{\rm m}_5\, \zeta^{\rm m}_2,\ (\zeta^{\rm m}_3)^2\, (\zeta^{\rm m}_2)^2,\ (\zeta^{\rm m}_2)^5 \,\}\,, \tag{56}$$

with $\dim(B_{10}) = d_{10}$. Then for each basis element we compute (51):

$$\begin{aligned}
\phi^B\left(\zeta^{\rm m}_{3,7}\right) &= -14\ f_7 f_3 - 6\ f_5 f_5,\quad \phi^B\left(\zeta^{\rm m}_3 \zeta^{\rm m}_7\right) = f_3 \sqcup\!\sqcup f_7\,,\\
\phi^B\left((\zeta^{\rm m}_5)^2\right) &= f_5 \sqcup\!\sqcup f_5,\quad \phi^B\left(\zeta^{\rm m}_{3,5} \zeta^{\rm m}_2\right) = -5\ f_5 f_3 f_2\,,\\
\phi^B\left(\zeta^{\rm m}_3 \zeta^{\rm m}_5 \zeta^{\rm m}_2\right) &= f_3 \sqcup\!\sqcup f_5 f_2,\quad \phi^B\left((\zeta^{\rm m}_3)^2 (\zeta^{\rm m}_2)^2\right) = f_3 \sqcup\!\sqcup f_3 f_2^2\,,\\
\phi^B\left((\zeta^{\rm m}_2)^5\right) &= f_2^5\,.
\end{aligned} \tag{57}$$

The above construction allows to assign a **Q**–linear combination of monomials to every element $\zeta^{\rm m}_{n_1,\ldots,n_r}$. The map (51) sends every motivic MZV of weight less or equal to N to a non–commutative polynomial in the f_i's. Inverting the map ϕ gives the decomposition of $\zeta^{\rm m}_{n_1,\ldots,n_r}$ w.r.t. the basis B_n of weight n, with $n = \sum_{l=1}^r n_l$. We construct operators acting on $\phi(\xi) \in \mathcal{U}$ to detect elements in $\mathcal{U}$ and to decompose any motivic MZV ξ into a candidate basis B. The derivation operators $\partial_{2n+1} : \mathcal{U} \to \mathcal{U}$ are defined as [17]:

$$\partial_{2n+1}(f_{i_1} \ldots f_{i_r}) = \begin{cases} f_{i_2} \ldots f_{i_r},\ i_1 = 2n+1\,,\\ 0\,, \qquad\qquad \text{otherwise}\,, \end{cases} \tag{58}$$

with $\partial_{2n+1} f_2 = 0$. Furthermore, we have the product rule for the shuffle product:

$$\partial_{2n+1}(a \sqcup\!\sqcup b) = \partial_{2n+1} a \sqcup\!\sqcup b + a \sqcup\!\sqcup \partial_{2n+1} b \quad , \quad a, b \in \mathcal{U}'\,. \tag{59}$$

Finally, c_2^n takes the coefficient of f_2^n. By first determining the map (51) for a given basis B_n we then can construct the motivic decomposition operator ξ_n such that it acts trivially on this basis. This is established for the weight ten basis (57) in the following.

With the differential operator (58) we may consider the following operator

$$\begin{aligned}
\xi_{10} = a_0\, (\zeta^{\rm m}_2)^5 &+ a_1\, (\zeta^{\rm m}_2)^2\, (\zeta^{\rm m}_3)^2 + a_2\, \zeta^{\rm m}_2\, \zeta^{\rm m}_3\, \zeta^{\rm m}_5 + a_3\, (\zeta^{\rm m}_5)^2\\
&+ a_4\, \zeta^{\rm m}_2\, \zeta^{\rm m}_{3,5} + a_5\, \zeta^{\rm m}_3\, \zeta^{\rm m}_7 + a_6\, \zeta^{\rm m}_{3,7}
\end{aligned} \tag{60}$$

with the operators

$$\begin{aligned}
a_1 &= \tfrac{1}{2}\, c_2^2\, \partial_3^2,\ a_2 = c_2\, \partial_5 \partial_3,\ a_3 = \tfrac{1}{2}\, \partial_5^2 + \tfrac{3}{14}\, [\partial_7, \partial_3]\\
a_4 &= \tfrac{1}{5}\, c_2\, [\partial_5, \partial_3],\ a_5 = \partial_7 \partial_3,\ a_6 = \tfrac{1}{14}\, [\partial_7, \partial_3]
\end{aligned} \tag{61}$$

acting on $\phi^B(\xi_{10})$. Clearly, for the basis (57) we exactly verify (60) to a be a decomposition operator acting trivially on the basis elements.

Let us now discuss a special class of MZVs (36) identified as single–valued MZVs (SVMZVs)

$$\zeta_{sv}(n_1, \ldots, n_r) \in \mathbf{R} \tag{62}$$

originating from single–valued multiple polylogarithms (SVMPs) at unity [14]. The latter are generalization of the Bloch–Wigner dilogarithm:

$$D(z) = \Im\{\mathscr{L}i_2(z) + \ln|z|\ \ln(1-z)\}\ . \tag{63}$$

Thus, e.g.:

$$\zeta_{sv}(2) = D(1) = 0\ . \tag{64}$$

SVMZVs represent a subset of the MZVs (36) and they satisfy the same double shuffle and associator relations than the usual MZVs and many more relations [20]. SVMZVs have recently been studied by Brown in [20] from a mathematical point of view. They have been identified as the coefficients in an infinite series expansion of the Deligne associator [23] in two non–commutative variables. The latter is defined through the equation [20]

$$W(e_0, e_1) = Z(-e_0, -e_1')^{-1}\ Z(e_0, e_1)\ , \tag{65}$$

with the Drinfeld associator (44) and $e_1' = We_1W^{-1}$. The equation (65) can systematically be worked out at each weight yielding [46]:

$$\begin{aligned} W(e_0, e_1) &= 1 + 2\,\zeta_3\,([e_0,[e_0,e_1]] - [e_1,[e_0,e_1]) + 2\,\zeta_5\,\Big([e_0,[e_0,[e_0,[e_0,e_1]]]] \\ &\quad - \tfrac{1}{2}\,[e_0,[e_0,[e_1,[e_0,e_1]]]] - \tfrac{3}{2}\,[e_1,[e_0,[e_0,[e_0,e_1]]]] + (e_0 \leftrightarrow e_1)\Big) + \cdots\ . \end{aligned} \tag{66}$$

Strictly speaking, the numbers (62) are established in the Hopf algebra (47) of motivic MZVs ζ^{m}. In analogy to the motivic version of the Drinfeld associator (44)

$$Z^{\mathrm{m}}(e_0, e_1) = \sum_{w \in \{e_0, e_1\}^\times} \zeta^{\mathrm{m}}(w)\, w \tag{67}$$

in Ref. [20] Brown has defined the motivic single–valued associator as a generating series

$$W^{\mathrm{m}}(e_0, e_1) = \sum_{w \in \{e_0, e_1\}^\times} \zeta^{\mathrm{m}}_{sv}(w)\, w\ , \tag{68}$$

whose period map (46) gives the Deligne associator (65). Hence, for the motivic MZVs there is a map from the motivic MZVs to SVMZVs furnished by the following homomorphism

$$\mathrm{sv} : \mathscr{H} \to \mathscr{H}^{\mathrm{sv}}\ , \tag{69}$$

with:

$$\mathrm{sv}:\zeta^{\mathfrak{m}}_{n_1,\ldots,n_r}\mapsto\zeta^{\mathfrak{m}}_{\mathrm{sv}}(n_1,\ldots,n_r)\ . \tag{70}$$

In the algebra $\mathcal{H}$ the homomorphism (69) together with

$$\zeta^{\mathfrak{m}}_{\mathrm{sv}}(2)=0 \tag{71}$$

can be constructed [20]. The motivic SVMZVs $\zeta^{\mathfrak{m}}_{\mathrm{sv}}(n_1,\ldots,n_r)$ generate the subalgebra $\mathcal{H}^{\mathrm{sv}}$ of the Hopf algebra $\mathcal{H}$ and satisfy all motivic relations between MZVs.

In practice, the map sv is constructed recursively in the (trivial) algebra–comodule (48) with the first factor given by (49) and generated by all non–commutative words in the letters f_{2i+1}. We have $\mathcal{H}\simeq\mathcal{U}$, in particular $\zeta^{\mathfrak{m}}_{2i+1}\simeq f_{2i+1}$. The homomorphism

$$\mathrm{sv}:\mathcal{U}'\longrightarrow\mathcal{U}'\ , \tag{72}$$

with

$$w\longmapsto\sum_{uv=w}u\sqcup\!\sqcup\,\tilde{v}\ , \tag{73}$$

and

$$\mathrm{sv}(f_2)=0 \tag{74}$$

maps the algebra of non–commutative words $w\in\mathcal{U}$ to the smaller subalgebra $\mathcal{U}^{\mathrm{sv}}$, which describes the space of SVMZVs [20]. In Eq. (73) the word $\tilde{v}$ is the reversal of the word v and $\sqcup\!\sqcup$ is the shuffle product. For more details we refer the reader to the original reference [20] and subsequent applications in [46]. With (73) the image of sv can be computed very easily, e.g.:

$$\mathrm{sv}(f_{2i+1})=2f_{2i+1}\ . \tag{75}$$

Eventually, the period map (46) implies the homomorphism

$$\mathrm{sv}:\zeta_{n_1,\ldots,n_r}\mapsto\zeta_{\mathrm{sv}}(n_1,\ldots,n_r)\ , \tag{76}$$

and with (73) we find the following examples (cf. Ref. [46] for more examples):

$$\mathrm{sv}(\zeta_2)=\zeta_{\mathrm{sv}}(2)=0\ , \tag{77}$$

$$\mathrm{sv}(\zeta_{2n+1})=\zeta_{\mathrm{sv}}(2n+1)=2\,\zeta_{2n+1}\ ,\quad n\geq 1\ , \tag{78}$$

$$\mathrm{sv}(\zeta_{3,5})=-10\,\zeta_3\,\zeta_5\quad,\quad\mathrm{sv}(\zeta_{3,7})=-28\,\zeta_3\,\zeta_7-12\,\zeta_5^2\ , \tag{79}$$

$$\mathrm{sv}(\zeta_{3,3,5})=2\,\zeta_{3,3,5}-5\,\zeta_3^2\,\zeta_5+90\,\zeta_2\,\zeta_9+\frac{12}{5}\,\zeta_2^2\,\zeta_7-\frac{8}{7}\,\zeta_2^3\,\zeta_5^2\ ,\ldots\ . \tag{80}$$

5 Motivic Period Matrix $\mathbf{F}^{\mathrm{m}}$

The motivic version F^{m} of the period matrix (32) is given by passing from the MZVs $\zeta \in \mathscr{Z}$ to their motivic versions $\zeta^{\mathrm{m}} \in \mathscr{H}$ as

$$F^{\mathrm{m}} = P^{\mathrm{m}}\ Q^{\mathrm{m}}\ : \exp\left\{\sum_{n\geq 1} \zeta^{\mathrm{m}}_{2n+1}\ M_{2n+1}\right\} : , \tag{81}$$

with

$$P^{\mathrm{m}} = P|_{\zeta_2 \to \zeta_2^{\mathrm{m}}} \quad , \quad Q^{\mathrm{m}} = Q|_{\zeta_{n_1,\ldots,n_r} \to \zeta^{\mathrm{m}}_{n_1,\ldots,n_r}} , \tag{82}$$

and the matrices P, M and Q defined in (33) and (35), respectively. Extracting e.g. the weight $w = 10$ part of (81)

$$\begin{aligned}
F^{\mathrm{m}}\Big|_{\zeta_3^{\mathrm{m}}\zeta_7^{\mathrm{m}}} &= M_7\ M_3 ,\\
F^{\mathrm{m}}\Big|_{\zeta_{3,7}^{\mathrm{m}}} &= \frac{1}{14}\ [M_7, M_3] ,\\
F^{\mathrm{m}}\Big|_{(\zeta_5^{\mathrm{m}})^2} &= \frac{1}{2}\ M_5^2 + \frac{3}{14}\ [M_7, M_3] ,\\
F^{\mathrm{m}}\Big|_{\zeta_2^{\mathrm{m}}\zeta_3^{\mathrm{m}}\zeta_5^{\mathrm{m}}} &= P_2\ M_5\ M_3 ,\\
F^{\mathrm{m}}\Big|_{\zeta_2^{\mathrm{m}}\zeta_{3,5}^{\mathrm{m}}} &= \frac{1}{5}\ P_2\ \ [M_5, M_3] ,\\
F^{\mathrm{m}}\Big|_{(\zeta_2^{\mathrm{m}})^2(\zeta_3^{\mathrm{m}})^2} &= \frac{1}{2}\ P_4\ \ M_3^2 ,\\
F^{\mathrm{m}}\Big|_{(\zeta_2^{\mathrm{m}})^5} &= P_{10} ,
\end{aligned} \tag{83}$$

and comparing with the motivic decomposition operators (61) yields a striking exact match in the coefficients and commutator structures by identifying the motivic derivation operators with the matrices (33) as:

$$\begin{aligned}
\partial_{2n+1} &\simeq M_{2n+1} \quad , \quad n \geq 1 ,\\
c_2^k &\simeq P_{2k} \quad , \quad k \geq 1 .
\end{aligned} \tag{84}$$

This agreement has been shown to exist up to the weight $w = 16$ in [41] and extended through weight $w = 22$ in [12]. Hence, at least up to the latter weight the decomposition of motivic MZVs w.r.t. to a basis of MZVs encapsulates the α'–expansion of the motivic period matrix written in terms of the same basis elements (81).

In the following we shall demonstrate, that the isomorphism (51) encapsulates all the relevant information of the α'–expansion of the motivic period matrix (81) without further specifying the latter explicitly in terms of motivic MZVs ζ^{m}. In the sequel we shall apply the isomorphism ϕ to F^{m}. The action (51) of ϕ on the motivic MZVs is explained in the previous section. The first hint of a simplification under ϕ

occurs by considering the weight $w=8$ contribution to $F^{\rm m}$, where the commutator term $[M_5, M_3]$ from $Q_8^{\rm m}$ together with the prefactor $\frac{1}{5}\zeta_{3,5}^{\rm m}$ conspires[10] into (with $\phi^B(\zeta_{3,5}^{\rm m}) = -5 f_5 f_3$):

$$\phi^B\left(\zeta_3^{\rm m}\zeta_5^{\rm m}\, M_5M_3 + Q_8^{\rm m}\right) = f_3 f_5\, M_5M_3 \;+\; f_5 f_3\, M_3M_5\ . \tag{85}$$

The right hand side obviously treats the objects f_3, M_3 and f_5, M_5 in a democratic way. The effect of the map ϕ is, that in the Hopf algebra $\mathcal{U}$, every non–commutative word of odd letters f_{2k+1} multiplies the associated reverse product of matrices M_{2k+1}. Powers f_2^k of the commuting generator f_2 are accompanied by P_{2k}, which multiplies all the operators M_{2k+1} from the left. Most notably, in contrast to the representation in terms of motivic MZVs, the numerical factors become unity, i.e. all the rational numbers in (35) drop out. Our explicit results confirm, that the beautiful structure with the combination of operators $M_{i_p}\ldots M_{i_2}M_{i_1}$ accompanying the word $f_{i_1}f_{i_2}\ldots f_{i_p}$, continues to hold through at least weight $w = 16$. To this end, we obtain the following striking and short form for the motivic period matrix $F^{\rm m}$ [41]:

$$\phi^B(F^{\rm m}) = \left(\sum_{k=0}^{\infty} f_2^k\ P_{2k}\right)\left(\sum_{p=0}^{\infty}\ \sum_{\substack{i_1,\ldots,i_p\\ \in 2\mathbb{N}^+ +1}} f_{i_1}f_{i_2}\ldots f_{i_p}\ M_{i_p}\ldots M_{i_2}M_{i_1}\right)\ . \tag{86}$$

In (86) the sum over the combinations $f_{i_1}f_{i_2}\ldots f_{i_p}M_{i_p}\ldots M_{i_2}M_{i_1}$ includes *all* possible non–commutative words $f_{i_1}f_{i_2}\ldots f_{i_p}$ with coefficients $M_{i_p}\ldots M_{i_2}M_{i_1}$ graded by their length p. Matrices P_{2k} associated with the powers f_2^k always act by left multiplication. The commutative nature of f_2 w.r.t. the odd generators f_{2k+1} ties in with the fact that in the matrix ordering the matrices P_{2k} have the well–defined place left of all matrices M_{2k+1}. Alternatively, we may write (86) in terms of a geometric series:

$$\phi^B(F^{\rm m}) = \left(\sum_{k=0}^{\infty} f_2^k\ P_{2k}\right)\left(1-\sum_{k=1}^{\infty} f_{2k+1}\ M_{2k+1}\right)^{-1}\ . \tag{87}$$

Thus, under the map ϕ the motivic period matrix $F^{\rm m}$ takes a very simple structure $\phi^B(F^{\rm m})$ in terms of the Hopf–algebra.

After replacing in (86) the matrices (33) by the operators as in (84) the operator (86) becomes the canonical element in $\mathcal{U}\otimes\mathcal{U}^*$, which maps any non–commutative word in $\mathcal{U}$ to itself. In this representation (86) gives rise to a group like action on $\mathcal{U}$. Hence, the operators ∂_{2n+1} and c_2^k are dual to the letters f_{2n+1} and f_2^k and have the matrix representations M_{2n+1} and P_{2k}, respectively. By mapping the motivic MZVs $\zeta^{\rm m}$ of the period matrix $F^{\rm m}$ to elements $\phi^B(\zeta^{\rm m})$ of the Hopf algebra $\mathcal{U}$ the map ϕ endows $F^{\rm m}$ with its motivic structure: it maps the latter into a very short and intriguing form in terms of the non–commutative Hopf algebra $\mathcal{U}$. In particular,

[10] Note the useful relation $\phi^B(Q_8^{\rm m}) = f_5 f_3\,[M_3, M_5]$ for $Q_8^{\rm m} = \frac{1}{5}\zeta_{3,5}^{\rm m}\,[M_5, M_3]$.

the various relations among different MZVs become simple algebraic identities in the Hopf algebra $\mathscr{U}$. Moreover, in this representation the final result (86) for period matrix does not depend on the choice of a specific set[11] of MZVs as basis elements. In fact, this feature follows from the basis–independent statement in terms of the motivic coaction (subject to matrix multiplication) [25]

$$\Delta F^{\mathfrak{m}} = F^{a} \otimes F^{\mathfrak{m}} , \tag{88}$$

with the superscripts a and m referring to the algebras $\mathscr{A}$ and $\mathscr{H}$, respectively. Furthermore with [13]

$$\partial_{2n+1} F^{\mathfrak{m}} = F^{\mathfrak{m}} \, M_{2n+1} \tag{89}$$

one can explicitly prove (86).

It has been pointed out in [21] that the simplification occurring in (86) can be interpreted as a compatibility between the motivic period matrix and the action of the Galois group of periods. Let us introduce the free graded Lie algebra $\mathscr{F}$ over $\mathbf{Q}$, which is freely generated by the symbols τ_{2n+1} of degree $2n+1$. Ihara has studied this algebra to relate the Galois Lie algebra $\mathscr{G}$ of the Galois group G to the more tractable object $\mathscr{F}$ [33]. The dimension $\dim(\mathscr{F}_m)$ can explicitly given by [49]

$$\dim(\mathscr{F}_m) = \sum_{d|m} \frac{1}{d}\, \mu(d) \sum_{\lceil \frac{m}{3d} \rceil \leq n \leq \lfloor \frac{m-d}{2d} \rfloor} \frac{1}{n} \binom{n}{\frac{m}{d} - 2n} , \tag{90}$$

with the Möbius function μ. Alternatively, we have [33]

$$\dim(\mathscr{F}_m) = \frac{1}{m} \sum_{d|m} \mu\left(\frac{m}{d}\right) \left(\sum_{i=1}^{3} \alpha_i^d - 1 - (-1)^d \right) , \tag{91}$$

with α_i being the three roots of the cubic equation $\alpha^3 - \alpha - 1 = 0$. The graded space of irreducible (primitive) MZVs $Z = \frac{\mathscr{Z}_{>0}}{\mathscr{Z}_{>0} \cdot \mathscr{Z}_{>0}}$ with $\mathscr{Z}_{>0} = \oplus_{w>0} \mathscr{Z}_w$ is isomorphic to the dual of $\mathscr{F}$, i.e. $\dim(Z_m) = \dim(\mathscr{F}_m)$ [28, 29]. This property relates linearly independent elements $\mathscr{F}$ in the α'–expansion of (32) or (81) to primitive MZVs. The linearly independent algebra elements of $\mathscr{F}$ and irreducible (primitive) MZVs (in lines of [8]) at each weight m are displayed in the Table 1 through weight $m = 22$.

[11]For instance instead of taking a basis containing the depth one elements $\zeta^{\mathfrak{m}}_{2n+1}$ one also could choose the set of Lyndon words in the Hoffman elements $\zeta^{\mathfrak{m}}_{n_1,\ldots,n_r}$, with $n_i = 2, 3$ and define the corresponding matrices (33).

Table 1 Linearly independent elements in $\mathscr{L}_m$ and primitive MZVs for $m = 1, \ldots, 22$

m	$\dim(\mathscr{F}_m)$	linearly independent elements at α'^{m}	irreducible MZVs
1	0	–	–
2	0	–	–
3	1	τ_3	ζ_3
4	0	–	–
5	1	τ_5	ζ_5
6	0	–	–
7	1	τ_7	ζ_7
8	1	$[\tau_5, \tau_3]$	$\zeta_{3,5}$
9	1	τ_9	ζ_9
10	1	$[\tau_7, \tau_3]$	$\zeta_{3,7}$
11	2	$\tau_{11},\ [\tau_3, [\tau_5, \tau_3]]$	$\zeta_{11},\ \zeta_{3,3,5}$
12	2	$[\tau_9, \tau_3],\ [\tau_7, \tau_5]$	$\zeta_{3,9}\ ,\zeta_{1,1,4,6}$
13	3	$\tau_{13},\ [\tau_3, [\tau_7, \tau_3]],\ [\tau_5, [\tau_5, \tau_3]]$	$\zeta_{13},\ \zeta_{3,3,7},\ \zeta_{3,5,5}$
14	3	$[\tau_{11}, \tau_3],\ [\tau_9, \tau_5],\ [\tau_3, [\tau_3, [\tau_5, \tau_3]]]$	$\zeta_{3,11},\ \zeta_{5,9},\ \zeta_{3,3,3,5}$
15	4	$\tau_{15},\ [\tau_3, [\tau_9, \tau_3]],\ [\tau_5, [\tau_7, \tau_3]],$ $[\tau_7, [\tau_5, \tau_3]]$	$\zeta_{15},\ \zeta_{5,3,7},\ \zeta_{3,3,9},\ \zeta_{1,1,3,4,6}$
16	5	$[\tau_{13}, \tau_3],\ [\tau_{11}, \tau_5],\ [\tau_9, \tau_7],$	$\zeta_{3,13},\ \zeta_{5,11},\ \zeta_{1,1,6,8},$
		$[\tau_3, [\tau_5, [\tau_5, \tau_3]]],\ [\tau_3, [\tau_3, [\tau_7, \tau_3]]]$	$\zeta_{3,3,3,7},\ \zeta_{3,3,5,5}$
17	7	$\tau_{17},\ [\tau_3, [\tau_3, [\tau_3,$ $[\tau_5, \tau_3]]]],\ [\tau_7, [\tau_7, \tau_3]],\ [\tau_5, [\tau_7, \tau_5]],$	$\zeta_{17},\ \zeta_{3,3,3,3,5},\ \zeta_{1,1,3,6,6},\ \zeta_{5,5,7},$
		$[\tau_3, [\tau_{11}, \tau_3]],\ [\tau_9,$ $[\tau_5, \tau_3]],\ [\tau_5, [\tau_9, \tau_3]]$	$\zeta_{3,3,11},\ \zeta_{5,3,9},\ \zeta_{3,5,9}$
18	8	$[\tau_{15}, \tau_3],\ [\tau_{13}, \tau_5],\ [\tau_{11}, \tau_7],$	$\zeta_{3,15},\ \zeta_{5,13},\ \zeta_{1,1,6,10},$
		$[\tau_5, [\tau_5, [\tau_5, \tau_3]]],$ $[\tau_3, [\tau_3, [\tau_7, \tau_5]]],$	$\zeta_{3,5,5,5},\ \zeta_{5,3,3,7},$
		$[\tau_5, [\tau_3, [\tau_7, \tau_3]]],\ [\tau_3, [\tau_3, [\tau_9, \tau_3]]],$ $[\tau_3, [\tau_5, [\tau_7, \tau_3]]]$	$\zeta_{3,3,3,9}, \zeta_{3,5,3,7},\ \zeta_{1,1,3,3,4,6}$
19	11	$\tau_{19},\ [\tau_3, [\tau_{13}, \tau_3]],$ $[\tau_7, [\tau_9, \tau_3]],\ [\tau_9, [\tau_7, \tau_3]],$	$\zeta_{19},\ \zeta_{3,3,13},\ \zeta_{7,3,9},\ \zeta_{1,1,3,6,8},$
		$[\tau_5, [\tau_{11}, \tau_3]],\ [\tau_{11}, [\tau_5, \tau_3]],$ $[\tau_5, [\tau_9, \tau_5]],\ [\tau_7, [\tau_7, \tau_5]],$	$\zeta_{5,3,11},\ \zeta_{3,5,11},\ \zeta_{5,5,9},$ $\zeta_{1,1,5,6,6},$
		$[\tau_3, [\tau_3, [\tau_5, [\tau_5, \tau_3]]]],\ [\tau_5, [\tau_3, [\tau_3,$ $[\tau_5, \tau_3]]]],$	$\zeta_{3,3,5,3,5}, \zeta_{3,3,3,5,5},$
		$[\tau_3, [\tau_3, [\tau_3, [\tau_7, \tau_3]]]]$	$\zeta_{3,3,3,3,7}$
20	13	$[\tau_{17}, \tau_3],\ [\tau_{15}, \tau_5],\ [\tau_{13}, \tau_7],\ [\tau_{11}, \tau_9],$	$\zeta_{7,13},\ \zeta_{5,15},\ \zeta_{3,17},\ \zeta_{1,1,8,10},$
		$[\tau_3, [\tau_3, [\tau_3, [\tau_3, [\tau_5, \tau_3]]]]],$	$\zeta_{3,3,3,3,3,5},$
		$[\tau_5, [\tau_5, [\tau_7, \tau_3]]],\ [\tau_3, [\tau_5, [\tau_7, \tau_5]]],$ $[\tau_3, [\tau_3, [\tau_9, \tau_5]]],$	$\zeta_{5,5,3,7},\ \zeta_{3,5,5,7},\ \zeta_{5,3,3,9},$
		$[\tau_3, [\tau_7, [\tau_7, \tau_3]]],\ [\tau_3, [\tau_5, [\tau_9, \tau_3]]],$ $[\tau_3, [\tau_3, [\tau_{11}, \tau_3]]],$	$\zeta_{3,3,7,7},\ \zeta_{3,5,3,9},\ \zeta_{3,3,3,11},$
		$[\tau_5, [\tau_3, [\tau_7, \tau_5]]],\ [\tau_5, [\tau_3, [\tau_9, \tau_3]]]$	$\zeta_{1,1,3,3,4,8},\ \zeta_{1,1,5,3,4,6},$

(continued)

Table 1 (continued)

m	$\dim(\mathscr{F}_m)$	linearly independent elements at α'^m	irreducible MZVs
21	17	$\tau_{21},\ [\tau_9,[\tau_9,\tau_3]],$ $[\tau_7,[\tau_{11},\tau_3]],\ [\tau_7,[\tau_9,\tau_5]],$	$\zeta_{21},\ \zeta_{3,9,9},\ \zeta_{1,1,3,6,10},\ \zeta_{7,5,9},$
		$[\tau_5,[\tau_{13},\tau_3]],\ [\tau_5,[\tau_{11},\tau_5]],\ [\tau_5,[\tau_9,\tau_7]],$	$\zeta_{5,3,13},\ \zeta_{1,1,5,4,10},\ \zeta_{1,1,5,6,8},$
		$[\tau_3,[\tau_{15},\tau_3]],\ [\tau_3,[\tau_{13},\tau_5]],\ [\tau_3,[\tau_{11},\tau_7]],$	$\zeta_{3,3,15},\ \zeta_{3,5,13},\ \zeta_{7,3,11},$
		$[\tau_7,[\tau_3,[\tau_3,[\tau_5,\tau_3]]]],\ [\tau_5,[\tau_3,[\tau_5,[\tau_5,\tau_3]]]],$	$\zeta_{3,5,3,3,7},\ \zeta_{3,5,3,5,5},$
		$[\tau_5,[\tau_3,[\tau_3,[\tau_7,\tau_3]]]],\ [\tau_3,[\tau_5,[\tau_5,[\tau_5,\tau_3]]]],$	$\zeta_{5,3,3,3,7},\ \zeta_{3,3,5,5,5},$
		$[\tau_3,[\tau_3,[\tau_3,[\tau_7,\tau_5]]]],\ [\tau_3,[\tau_5,[\tau_3,[\tau_7,\tau_3]]]],$	$\zeta_{3,3,5,3,7},\ \zeta_{1,1,3,3,3,4,6},$
		$[\tau_3,[\tau_3,[\tau_3,[\tau_9,\tau_3]]]]$	$\zeta_{3,3,3,3,9}$
22	21	$[\tau_{19},\tau_3],\ [\tau_{17},\tau_5],\ [\tau_{15},\tau_7],\ [\tau_{13},\tau_9]$	$\zeta_{3,19},\ \zeta_{5,17},\ \zeta_{7,15},\ \zeta_{1,1,8,12}$
		$[\tau_3,[\tau_3,[\tau_{13},\tau_3]]]$	$\zeta_{3,3,3,13}$
		$[\tau_{11},[\tau_3,[\tau_5,\tau_3]]],\ [\tau_3,[\tau_5,[\tau_{11},\tau_3]]],$ $[\tau_3,[\tau_{11},[\tau_5,\tau_3]]]$	$\zeta_{5,3,3,11},\ \zeta_{3,5,3,11},\ \zeta_{3,3,5,11}$
		$[\tau_9,[\tau_3,[\tau_7,\tau_3]]],\ [\tau_7,[\tau_3,[\tau_9,\tau_3]]],$ $[\tau_3,[\tau_9,[\tau_7,\tau_3]]]$	$\zeta_{3,7,3,9},\ \zeta_{1,1,7,3,2,8},\ \zeta_{1,1,3,3,6,8}$
		$[\tau_7,[\tau_5,[\tau_7,\tau_3]]],\ [\tau_7,[\tau_7,[\tau_5,\tau_3]]]],$ $[\tau_5,[\tau_7,[\tau_7,\tau_3]]]]$	$\zeta_{5,3,7,7},\ \zeta_{3,7,5,7},\ \zeta_{1,1,3,5,6,6}$
		$[\tau_5,[\tau_9,[\tau_5,\tau_3]]],\ [\tau_5,[\tau_5,[\tau_9,[\tau_3]]]]],$ $[\tau_3,[\tau_5,[\tau_9,\tau_5]]]$	$\zeta_{5,3,5,9},\ \zeta_{3,5,5,9},\ \zeta_{5,5,3,9}$
		$[\tau_5,[\tau_5,[\tau_7,\tau_5]]]$	$\zeta_{1,1,5,5,4,6}$
		$[\tau_3,[\tau_3,[\tau_3,[\tau_5,[\tau_5,\tau_3]]]]],$ $[\tau_3,[\tau_5,[\tau_3,[\tau_3,[\tau_5,\tau_3]]]]]$	$\zeta_{3,3,3,5,3,5},\ \zeta_{3,3,3,3,5,5}$
		$[\tau_3,[\tau_3,[\tau_3,[\tau_3,[\tau_7,\tau_3]]]]]$	$\zeta_{3,3,3,3,3,7}$

The generators M_{2n+1} defined in (33) are represented as $(N-3)! \times (N-3)!$–matrices and enter the commutator structure

$$[M_{n_2},[M_{n_3},\ldots,[M_{n_r},M_{n_1}]]\ldots] \tag{92}$$

in the expansion of (32) or (81). These structures can be related to a graded Lie algebra over $\mathbf{Q}$

$$\mathscr{L} = \bigoplus_{r\geq 1} \mathscr{L}_r\ , \tag{93}$$

which is generated by the symbols M_{2n+1} with the Lie bracket $(M_i, M_j) \mapsto [M_i, M_j]$. The grading is defined by assigning M_{2n+1} the degree $2n+1$. More precisely, the algebra $\mathscr{L}$ is generated by the following elements:

$$M_3,\ M_5,\ M_7,\ [M_5,M_3],\ M_9,\ [M_7,M_3],\ M_{11},\ [M_3[M_5,M_3]],\ [M_9,M_3],\ [M_7,M_5],\ldots\,. \tag{94}$$

However, this Lie algebra $\mathscr{L}$ is not free for generic matrix representations M_{2i+1} referring to any $N \geq 5$. Hence, generically $\mathscr{L} \not\simeq \mathscr{F}$. In fact, for $N = 5$ at weight $w = 18$ we find the relation $[M_3,[M_5,[M_7,M_3]]] = [M_5,[M_3,[M_7,M_3]]]$ leading to $\dim(\mathscr{L}_{18}) = 7$ in contrast to $\dim(\mathscr{F}_{18}) = 8$.

For a given multiplicity N the generators M_{2n+1}, which are represented as $(N-3)! \times (N-3)!$–matrices, are related to their transposed M_l^t by a similarity (conjugacy) transformation S_0

$$S_0^{-1}\ M_l^t\ S_0 = M_l\ , \tag{95}$$

i.e. M_l and M_l^t are similar (conjugate) to each other. The matrix S_0 is symmetric and has been introduced in [41]. The relation (95) implies, that the matrices M_l are conjugate to symmetric matrices. An immediate consequence is the set of relations

$$S_0\ \mathscr{Q}_{(r)} + (-1)^r\ \mathscr{Q}_{(r)}^t S_0 = 0\ , \tag{96}$$

for any nested commutator of generic depth r

$$\mathscr{Q}_{(r)} = [M_{n_2},[M_{n_3},\ldots,[M_{n_r},M_{n_1}]]\ldots]\ ,\qquad r \geq 2\ .$$

As a consequence any commutator $\mathscr{Q}_{2r}$ is similar to an anti–symmetric matrix and any commutator $\mathscr{Q}_{2r+1}$ is similar to a symmetric and traceless matrix. Depending on the multiplicity N the relations (96) impose constraints on the number of independent generators at a given weight m given in the Table 1. Example for $N = 5$ the constraints (96) imply[12]:

$$r_1 + r_2 \in 2\mathbf{Z}^+\ :\quad [\ \mathscr{Q}_{(r_1)},\tilde{\mathscr{Q}}_{(r_2)}\ ] = 0\ , \tag{97}$$

$$r_1 + r_2 \in 2\mathbf{Z}^+ + 1\ :\quad \{\ \mathscr{Q}_{(r_1)},\tilde{\mathscr{Q}}_{(r_2)}\ \} = 0\ . \tag{98}$$

As a consequence, for $N = 5$ the number of independent elements at a given weight m does not agree with the formulae (90) nor (91) starting at weight $w = 18$. The actual number of independent commutator structures at weight w is depicted in Table 2.

Therefore, for $N = 5$ an other algebra $\mathscr{L}^{(5)}$ rather than $\mathscr{F}$ is relevant for describing the expansion of (32) or (81). For $N \geq 6$ we expect the mismatch $\dim(\mathscr{F}_m) \neq \dim(\mathscr{L}_m)$ to show up at higher weights m. This way, for each $N \geq 5$ we obtain a

[12]The relation (96) implies, that any commutator $\mathscr{Q}_{(2)}$ is similar to an anti–symmetric matrix, and hence (97) implies $[\,[M_a,M_b],[M_c,M_d]\,] = 0$, which in turn as a result of the Jacobi relation yields the following identity: $[M_a,[M_b,[M_c,M_d]]] - [M_b,[M_a,[M_c,M_d]]] = [\,[M_a,M_b],[M_c,M_d]\,] = 0$. Furthermore, (96) implies that the commutator $\mathscr{Q}_{(3)}$ is similar to a symmetric and traceless matrix. As a consequence from (98), we obtain the following anti–commutation relation: $\{\,[M_a,M_b],[M_c,[M_d,M_e]\,\} = 0$. Relations for $N = 5$ between different matrices M_{2i+1} have also been discussed in [9].

Table 2 Linearly independent elements in $\mathscr{F}$, $\mathscr{L}_m^{(5)}$ and primitive MZVs for $m = 18, \ldots, 23$ for $N = 5$

m	$\dim(\mathscr{F}_m)$	$\dim(\mathscr{L}_m^{(5)})$	irreducible MZVs
18	8	7	7
19	11	11	11
20	13	11	11
21	17	16	16
22	21	16	16
23	28	25	25

different algebra $\mathscr{L}^{(N)}$, which is not free. However, we speculate that for N large enough, the matrices M_{2k+1} should give rise to the free Lie algebra $\mathscr{F}$, i.e.:

$$\lim_{N\to\infty} \mathscr{L}^{(N)} \simeq \mathscr{F}\ . \tag{99}$$

6 Open and Closed Superstring Amplitudes

The world–sheet describing the tree–level scattering of N open strings is depicted in Fig. 3. Asymptotic scattering of strings yields the string S–matrix defined by the emission and absorption of strings at space–time infinity, i.e. the open strings are incoming and outgoing at infinity. In this case the world–sheet can conformally be mapped to the half–sphere with the emission and absorption of strings taking place at the boundary through some vertex operators. Source boundaries representing the emission and absorption of strings at infinity become points accounting for the vertex operator insertions along the boundary of the half–sphere (disk). After projection onto the upper half plane $\mathbf{C}^+$ the strings are created at the N positions $z_i,\ i = 1, \ldots, N$ along the (compactified) real axis $\mathbf{RP}^1$. By this there appears a natural ordering $\Pi \in S_N$ of open string vertex operator insertions z_i along the boundary of the disk given by $z_{\Pi(1)} < \ldots < z_{\Pi(N)}$. To conclude, the topology of the string world–sheet describing tree–level scattering of open strings is a disk or upper half plane $\mathbf{C}^+$. On the other hand, the tree–level scattering of closed strings is characterized by a complex sphere $\mathbf{P}^1$ with vertex operator insertions on it.

At the N positions z_i massless strings carrying the external four–momenta $k_i,\ i = 1, \ldots, N$ and other quantum numbers are created, subject to momentum conservation $k_1 + \cdots + k_N = 0$. Due to conformal invariance one has to integrate over all vertex operator positions z_i in any amplitude computation. Therefore, for a given ordering Π open string amplitudes $\mathscr{A}^o(\Pi)$ are expressed by integrals along the boundary of the world–sheet disk (real projective line) as iterated (real) integrals on $\mathbf{RP}^1$ giving rise to multi–dimensional integrals on the space $\mathscr{M}_{0,N}(\mathbf{R})$ defined in (4). The N external four–momenta k_i constitute the kinematic invariants of the scattering process:

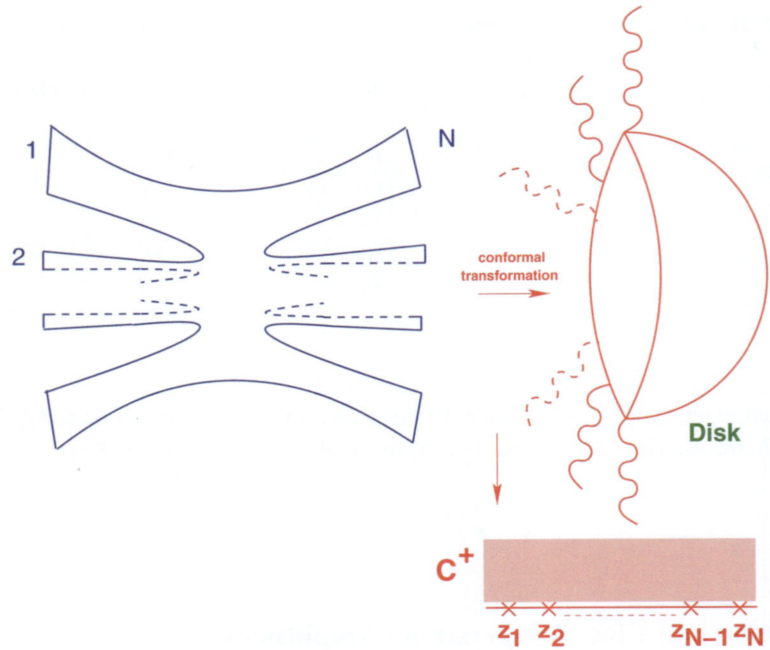

Fig. 3 World–sheet describing the scattering of N open strings

$$s_{ij} = (k_i + k_j)^2 = 2k_ik_j \ . \tag{100}$$

Out of (100) there are $\frac{1}{2}N(N-3)$ independent kinematic invariants involving N external momenta $k_i,\ i = 1,\ldots,N$. Any amplitude analytically depends on those independent kinematic invariants s_{ij}.

A priori there are $N!$ orderings Π of the vertex operator positions z_i along the boundary. However, string world–sheet symmetries like cyclicity, reflection and parity give relations between different orderings. In fact, by using monodromy properties on the world–sheet further relations are found and any superstring amplitude $\mathscr{A}^o(\Pi)$ of a given ordering Π can be expressed in terms of a minimal basis of $(N-3)!$ amplitudes [6, 44]:

$$\mathscr{A}^o(\sigma) := \mathscr{A}^o(1,\sigma(2,\ldots,N-2),N-1,N) \quad , \quad \sigma \in S_{N-3} \ . \tag{101}$$

The amplitudes (101) are functions of the string tension α'. Power series expansion in α' yields iterated integrals (20) multiplied by some polynomials in the parameters (100).

On the other hand, closed string amplitudes are given by integrals over the complex world–sheet sphere $\mathbf{P}^1$ as iterated integrals integrated independently on all choices of paths. While in the α'–expansion of open superstring tree–level amplitudes generically the whole space of MZVs (36) enters [39, 41, 45], closed superstring tree–level amplitudes exhibit only a subset of MZVs appearing in their α'–expansion [41, 45]. This subclass can be identified [46] as the single–valued multiple zeta values (SVMZVs) (62).

The open superstring N–gluon tree–level amplitude $\mathfrak{A}_N^o$ in type I superstring theory decomposes into a sum

$$\mathfrak{A}_N^o = (g_{YM}^o)^{N-2} \sum_{\Pi \in S_N/\mathbf{Z}_2} \mathrm{Tr}(T^{a_{\Pi(1)}} \ldots T^{a_{\Pi(N)}})\, \mathscr{A}^o(\Pi(1), \ldots, \Pi(N)) \tag{102}$$

over color ordered subamplitudes $\mathscr{A}^o(\Pi(1), \ldots, \Pi(N))$ supplemented by a group trace over matrices T^a in the fundamental representation. Above, the YM coupling is denoted by g_{YM}^o, which in type I superstring theory is given by $g_{YM}^o \sim e^{\Phi/2}$ with the dilaton field Φ. The sum runs over all permutations S_N of labels $i = 1, \ldots, N$ modulo cyclic permutations $\mathbf{Z}_2$, which preserve the group trace. The $\alpha' \to 0$ limit of the open superstring amplitude (102) matches the N–gluon scattering amplitude of super Yang–Mills (SYM):

$$\mathscr{A}^o(\Pi(1), \ldots, \Pi(N))\big|_{\alpha'=0} = A(\Pi(1), \ldots, \Pi(N))\ . \tag{103}$$

As a consequence from (101) also in SYM one has a minimal basis of $(N-3)!$ independent partial subamplitudes [4]:

$$A(\sigma) := A(1, \sigma(2, \ldots, N-2), N-1, N) \quad , \quad \sigma \in S_{N-3}\ . \tag{104}$$

Hence, for the open superstring amplitude we may consider a vector $\mathscr{A}^o$ with its entries $\mathscr{A}_\sigma^o = \mathscr{A}^o(\sigma)$ describing the $(N-3)!$ independent open N–point superstring subamplitudes (101), while for SYM we have another vector A with entries $A_\sigma = A(\sigma)$:

$$\begin{aligned} \mathscr{A}^o &= (N-3)! \text{ dimensional vector encompassing} \\ &\qquad \textbf{all} \text{ independent superstring subamplitudes } \mathscr{A}_\sigma^o = \mathscr{A}^o(\sigma),\ \sigma \in S_{N-3}\ , \\ A &= (N-3)! \text{ dimensional vector encompassing} \\ &\qquad \textbf{all} \text{ independent SYM subamplitudes } A_\sigma = A(\sigma),\ \sigma \in S_{N-3}\ . \end{aligned}$$

The two linear independent $(N-3)!$–dimensional vectors $\mathscr{A}$ and A are related by a non–singular matrix of rank $(N-3)!$. An educated guess is the following relation

$$\mathscr{A}^o = F\ A\ , \tag{105}$$

with the period matrix F given in (29). Note, that with (30) the Ansatz (105) matches the condition (103). In components the relation (105) reads:

$$\mathscr{A}^o(\pi) = \sum_{\sigma \in S_{N-3}} F_{\pi\sigma}\ A(\sigma) \quad , \quad \pi \in S_{N-3}\ . \tag{106}$$

In fact, an explicit string computation proves the relation (105) [37, 38].

Let us now move on to the scattering of closed strings. In heterotic string vacua gluons are described by massless closed strings. Therefore, we shall consider the closed superstring N–gluon tree–level amplitude $\mathfrak{A}_N^c$ in heterotic superstring theory. The string world–sheet describing the tree–level scattering of N closed strings has the topology of a complex sphere with N insertions of vertex operators. The closed string has holomorphic and anti–holomorphic fields. The anti–holomorphic part is similar to the open string case and describes the space–time (or superstring) part. On the other hand, the holomorphic part accounts for the gauge degrees of freedom through current insertions on the world–sheet. As in the open string case (102), the single trace part decomposes into the sum

$$\mathfrak{A}^c_{N,\,\text{s.t.}} = (g_{YM}^{\text{HET}})^{N-2} \sum_{\Pi \in S_N/\mathbf{Z}_2} \text{tr}(T^{a_{\Pi(1)}} \ldots T^{a_{\Pi(N)}})\ \mathscr{A}^c(\Pi(1), \ldots, \Pi(N)) \qquad (107)$$

over partial subamplitudes $\mathscr{A}^c(\Pi)$ times a group trace over matrices T^a in the vector representation. In the $\alpha' \to 0$ limit the latter match the N–gluon scattering subamplitudes of SYM

$$\mathscr{A}^c(\Pi(1), \ldots, \Pi(N))\Big|_{\alpha'=0} = A(\Pi(1), \ldots, \Pi(N))\ , \qquad (108)$$

similarly to open string case (103). Again, the partial subamplitudes $\mathscr{A}^c(\Pi)$ can be expressed in terms of a minimal basis of $(N-3)!$ elements $\mathscr{A}^c(\rho),\ \rho \in S_{N-3}$. The latter have been computed in [47] and are given by

$$\mathscr{A}^c(\rho) = (-1)^{N-3} \sum_{\sigma \in S_{N-3}} \sum_{\overline{\rho} \in S_{N-3}} J[\rho\,|\,\overline{\rho}]\ S[\overline{\rho}|\sigma]\ A(\sigma)\ ,\quad \rho \in S_{N-3}\ , \qquad (109)$$

with the complex sphere integral[13]

$$J[\rho\,|\,\overline{\rho}] := V_{\text{CKG}}^{-1} \left(\prod_{j=1}^{N} \int_{z_j \in \mathbf{C}} d^2 z_j \right) \prod_{i<j}^{N} |z_{ij}|^{2\alpha' s_{ij}}\ \frac{1}{z_{1\rho(2)} z_{\rho(2),\rho(3)} \cdots z_{\rho(N-2),N-1} z_{N-1,N} z_{N,1}}$$
$$\times\ \frac{1}{\overline{z}_{1\overline{\rho}(2)} \overline{z}_{\overline{\rho}(2),\overline{\rho}(3)} \cdots \overline{z}_{\overline{\rho}(N-3),\overline{\rho}(N-2)} \overline{z}_{\overline{\rho}(N-2),N} \overline{z}_{N,N-1}\ \overline{z}_{N-1,1}}\ , \qquad (110)$$

the kernel S introduced in (28) and the SYM amplitudes (104). In (110) the rational function comprising the dependence on holomorphic and anti–holomorphic vertex operator positions shows some pattern depicted in Fig. 4. Based on the results [46] the following (matrix) identity has been established in [47]

$$J = \text{sv}(Z)\ , \qquad (111)$$

[13] The factor V_{CKG} accounts for the volume of the conformal Killing group of the sphere after choosing the conformal gauge. It will be canceled by fixing three vertex positions according to (1) and introducing the respective c–ghost factor $|z_{1,N-1} z_{1,N} z_{N-1,N}|^2$.

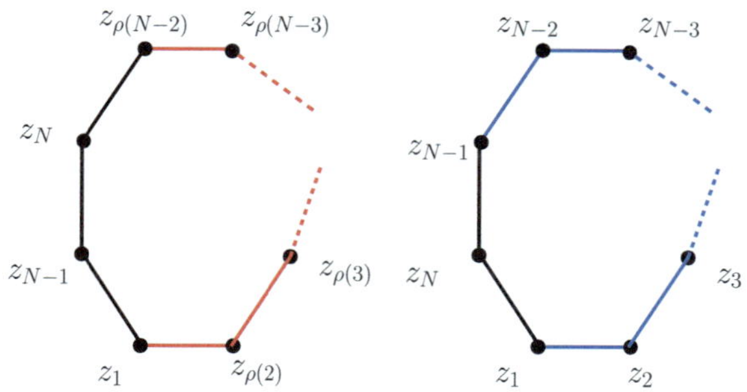

Fig. 4 N–gons describing the cyclic structures of holomorphic and anti–holomorphic cell forms

relating the complex integral (110) to the real iterated integral (24). The holomorphic part of (110) simply turns into the corresponding integral ordering of (24). As a consequence of (111) we find the following relation between the closed (109) and open (106) superstring gluon amplitude [47]:

$$\mathscr{A}^c(\rho) = \text{sv}\left(\mathscr{A}^o(\rho)\right) \quad , \quad \rho \in S_{N-3} \, . \tag{112}$$

To conclude, the single trace heterotic gauge amplitudes $\mathscr{A}^c(\rho)$ referring to the color ordering ρ are simply obtained from the relevant open string gauge amplitudes $\mathscr{A}^o(\rho)$ by imposing the projection sv introduced in (73). Therefore, the α'–expansion of the heterotic amplitude $\mathscr{A}^c(\rho)$ can be obtained from that of the open superstring amplitude $\mathscr{A}^o(\rho)$ by simply replacing MZVs by their corresponding SVMZVs according to the rules introduced in (73). The relation (112) between the heterotic gauge amplitude $\mathscr{A}^c$ and the type I gauge amplitude A^o establishes a non–trivial relation between closed string and open string amplitudes: the α'–expansion of the closed superstring amplitude can be cast into the same algebraic form as the open superstring amplitude: the closed superstring amplitude is essentially the single–valued (sv) version of the open superstring amplitude.

Also closed string amplitudes other than the heterotic (single–trace) gauge amplitudes (109) can be expressed as single–valued image of some open string amplitudes. From (111) the closed string analog of (27) follows:

$$J|_{\alpha'=0} = (-1)^{N-3} \, S^{-1} \, . \tag{113}$$

Hence, the set of complex world–sheet sphere integrals (110) are the closed string analogs of the open string world–sheet disk integrals (24) and serve as building blocks to construct any closed string amplitude. After applying partial integrations to remove double poles, which are responsible for spurious tachyonic poles, further performing partial fraction decompositions and partial integration relations all closed superstring amplitudes can be expressed in terms of the basis (110), which in turn through (111) can be related to the basis of open string amplitudes (24). As a consequence the α'–dependence of any closed string amplitude is given by that of the underlying open string amplitudes. This non–trivial connection between open and closed string amplitudes at the string tree–level points into a deeper connection between gauge and gravity amplitudes than what is implied by Kawai–Lewellen–Tye relations [34].

7 Complex Versus Iterated Integrals

Perturbative open and closed string amplitudes seem to be rather different due to their underlying different world–sheet topologies with or without boundaries, respectively. On the other hand, mathematical methods entering their computations reveal some unexpected connections. As we have seen in the previous section a new relation (111) between open (24) and closed (110) string world–sheet integrals holds.

Open string world–sheet disk integrals (24) are described as real iterated integrals on the space $\mathscr{M}_{0,N}(\mathbf{R})$ defined in (4), while closed string world–sheet sphere integrals (110) are given by integrals on the space $\mathscr{M}_{0,N}(\mathbf{C})$ defined in (3). The latter integrals, which can be considered as iterated integrals on $\mathbf{P}^1$ integrated independently on all choices of paths, are more involved than the real iterated integrals appearing in open string amplitudes. The observation (111) that complex integrals can be expressed as real iterated integrals subject to the projection sv has exhibited non–trivial relations between open and closed string amplitudes (112). In this section we shall elaborate on these connections at the level of the world–sheet integrals.

The simplest example of (111) arises for $N = 4$ yielding the relation

$$\int_{\mathbf{C}} d^2z\ \frac{|z|^{2s}\ |1-z|^{2u}}{z\ (1-z)\ \overline{z}} = \text{sv}\left(\int_0^1 dx\ x^{s-1}\ (1-x)^u\right)\ , \tag{114}$$

with $s, u \in \mathbf{R}$ such that both integrals converge. While the integral on the l.h.s. of (114) describes a four–point closed string amplitude the integral on the r.h.s. describes a four–point open string amplitude. Hence, the meaning of (114) w.r.t. to the corresponding closed versus open string world–sheet diagram describing four–point scattering (112) can be depicted as Fig. 5.

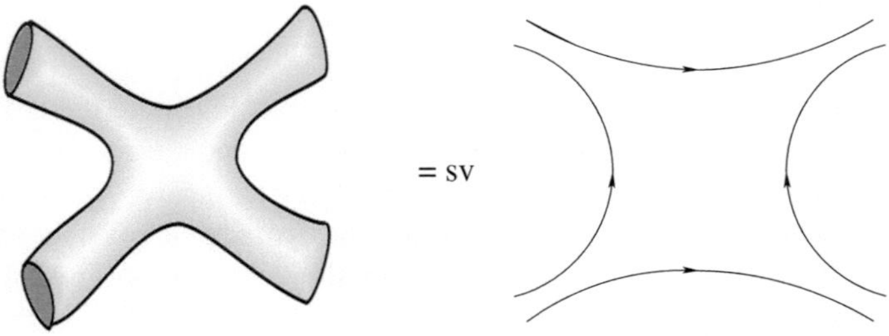

Fig. 5 Relation between closed and open string world–sheet diagram describing four–point scattering

After performing the integrations the relation (114) becomes (with $s+t+u=0$):

$$\frac{\Gamma(s)\,\Gamma(u)\,\Gamma(t)}{\Gamma(-s)\,\Gamma(-u)\,\Gamma(-t)}=\mathrm{sv}\left(\frac{\Gamma(1+s)\,\Gamma(1+u)}{\Gamma(1+s+u)}\right)\,. \tag{115}$$

Essentially, this equality (when acting on $[e_0, e_1]$) represents the relation between the Deligne (66) and Drinfeld (44) associators in the explicit representation of the limit $\mathrm{mod}(g')^2$ with $(g')^2=[g,g]^2, s=-\mathrm{ad}_{e_1},\ u=\mathrm{ad}_{e_0}$ and $\mathrm{ad}_x y=[x,y]$, i.e. dropping all quadratic commutator terms [25, 46]. Note, that applying Kawai–Lewellen–Tye (KLT) relations [34] to the complex integral of (114) rather yields

$$\int_{\mathbf{C}} d^2z\,\frac{|z|^{2s}\,|1-z|^{2u}}{z\,(1-z)\,\overline{z}}=\sin(\pi u)\left(\int_0^1 dx\,x^{s-1}\,(1-x)^{u-1}\right)\left(\int_1^\infty dx\,x^{t-1}\,(1-x)^{u}\right), \tag{116}$$

expressing the latter in terms of a square of real iterated integrals instead of a single real iterated integral as in (114). In fact, any direct computation of this complex integral by means of a Mellin representation or Gegenbauer decomposition ends up at (116).

Similar (114) explicit and direct correspondences (111) between the complex sphere integrals J and the real disk integrals Z can be made for $N\geq 5$. In order to familiarize with the matrix notation let us explicitly write the case (111) for $N=5$ (with (1)):

$$\begin{pmatrix}\displaystyle\int\limits_{z_2,z_3\in\mathbf{C}} d^2z_2\,d^2z_3\,\frac{\prod_{i<j}^4|z_{ij}|^{2\alpha' s_{ij}}}{z_{12}z_{23}z_{34}\,\overline{z}_{12}\overline{z}_{23}} & \displaystyle\int\limits_{z_2,z_3\in\mathbf{C}} d^2z_2\,d^2z_3\,\frac{\prod_{i<j}^4|z_{ij}|^{2\alpha' s_{ij}}}{z_{12}z_{23}z_{34}\,\overline{z}_{13}\overline{z}_{32}}\\ \displaystyle\int\limits_{z_2,z_3\in\mathbf{C}} d^2z_2\,d^2z_3\,\frac{\prod_{i<j}^4|z_{ij}|^{2\alpha' s_{ij}}}{z_{13}z_{32}z_{24}\,\overline{z}_{12}\overline{z}_{23}} & \displaystyle\int\limits_{z_2,z_3\in\mathbf{C}} d^2z_2\,d^2z_3\,\frac{\prod_{i<j}^4|z_{ij}|^{2\alpha' s_{ij}}}{z_{13}z_{32}z_{24}\,\overline{z}_{13}\overline{z}_{32}}\end{pmatrix}$$

$$=\mathrm{sv}\begin{pmatrix}\displaystyle\int\limits_{0<z_2<z_3<1} dz_2\,dz_3\,\frac{\prod_{i<j}^4|z_{ij}|^{\alpha' s_{ij}}}{z_{12}z_{23}} & \displaystyle\int\limits_{0<z_2<z_3<1} dz_2\,dz_3\,\frac{\prod_{i<j}^4|z_{ij}|^{\alpha' s_{ij}}}{z_{13}z_{32}}\\ \displaystyle\int\limits_{0<z_3<z_2<1} dz_2\,dz_3\,\frac{\prod_{i<j}^4|z_{ij}|^{\alpha' s_{ij}}}{z_{12}z_{23}} & \displaystyle\int\limits_{0<z_3<z_2<1} dz_2\,dz_3\,\frac{\prod_{i<j}^4|z_{ij}|^{\alpha' s_{ij}}}{z_{13}z_{32}}\end{pmatrix} \tag{117}$$

In (117) we explicitly see how the presence of the holomorphic gauge insertion in the complex integrals results in the projection onto real integrals involving only the right–moving part. Similar matrix relations can be extracted from (111) beyond $N>5$.

From (111) it follows that the α'–expansion of the closed string amplitude can be obtained from that of the open superstring amplitude by simply replacing MZVs by their corresponding SVMZVs according to the rules introduced in (76). Hence, closed string amplitudes use only the smaller subspace of SVMZVs. From a physical point of view SVMZVs appear in the computation of graphical functions (positive functions on the punctured complex plane) for certain Feynman amplitudes [42]. In supersymmetric Yang–Mills theory a large class of Feynman integrals in four space–time dimensions lives in the subspace of SVMZVs or SVMPs. As pointed out by Brown in [20], this fact opens the interesting possibility to replace general amplitudes with their single–valued versions (defined in (76) by the map sv), which should lead to considerable simplifications. In string theory this simplification occurs by replacing open superstring amplitudes by their single–valued versions describing closed superstring amplitudes.

Acknowledgements We wish to thank the organizers (especially José Burgos, Kurush Ebrahimi-Fard, and Herbert Gangl) of the workshop *Research Trimester on Multiple Zeta Values, Multiple Polylogarithms, and Quantum Field Theory* and the conference *Multiple Zeta Values, Modular Forms and Elliptic Motives II* at ICMAT, Madrid for inviting me to present the work exhibited in this publication and creating a stimulating atmosphere.

References

1. Arkani-Hamed, N., Bourjaily, J.L., Cachazo, F., Goncharov, A.B., Postnikov, A., Trnka, J.: Scattering amplitudes and the positive grassmannian. arXiv:1212.5605 [hep-th]
2. Arkani-Hamed, N., Trnka, J.: The Amplituhedron. JHEP **1410**, 30 (2014). arXiv:1312.2007 [hep-th]
3. Bern, Z., Dixon, L.J., Perelstein, M., Rozowsky, J.S.: Multileg one loop gravity amplitudes from gauge theory. Nucl. Phys. B **546**, 423 (1999). [hep-th/9811140]
4. Bern, Z., Carrasco, J.J.M., Johansson, H.: New relations for gauge-theory amplitudes. Phys. Rev. D **78**, 085011 (2008). arXiv:0805.3993 [hep-ph]
5. Beukers, F.: Algebraic A-hypergeometric functions. Invent. Math. **180**, 589–610 (2010)
6. Bjerrum-Bohr, N.E.J., Damgaard, P.H., Vanhove, P.: Minimal basis for gauge theory amplitudes. Phys. Rev. Lett. **103**, 161602 (2009). arXiv:0907.1425 [hep-th]
7. Bjerrum-Bohr, N.E.J., Damgaard, P.H., Sondergaard, T., Vanhove, P.: The momentum kernel of gauge and gravity theories. JHEP **1101**, 001 (2011). arXiv:1010.3933 [hep-th]
8. Blümlein, J., Broadhurst, D.J., Vermaseren, J.A.M.: The multiple zeta value data mine. Comput. Phys. Commun. **181**, 582 (2010). arXiv:0907.2557 [math-ph]
9. Boels, R.H.: On the field theory expansion of superstring five point amplitudes. Nucl. Phys. B **876**, 215 (2013). arXiv:1304.7918 [hep-th]
10. Bogner, C., Weinzierl, S.: Periods and Feynman integrals. J. Math. Phys. **50**, 042302 (2009). arXiv:0711.4863 [hep-th]
11. Broadhurst, D.J., Kreimer, D.: Association of multiple zeta values with positive knots via Feynman diagrams up to 9 loops. Phys. Lett. B **393**, 403 (1997). [hep-th/9609128]
12. Broedel, J., Schlotterer, O., Stieberger, S.: Polylogarithms, multiple zeta values and superstring amplitudes. Fortsch. Phys. **61**, 812 (2013). arXiv:1304.7267 [hep-th]
13. Broedel, J., Schlotterer, O., Stieberger, S., Terasoma, T.: Notes on Lie Algebra structure of Superstring Amplitudes. unpublished
14. Brown, F.: Single-valued multiple polylogarithms in one variable. C.R. Acad. Sci. Paris, Ser. I **338**, 527–532 (2004)

15. Brown, F.: Multiple zeta values and periods of moduli spaces $M_{0,n}(\mathbf{R})$. Ann. Sci. Ec. Norm. Sup. **42**, 371 (2009). arXiv:math/0606419 [math.AG]
16. Brown, F.C.S., Carr, S., Schneps, L.: Algebra of cell-zeta values. Compositio Math. **146**, 731–771 (2010)
17. Brown, F.: On the decomposition of motivic multiple zeta values. Galois-Teichmüller Theory Arithmetic Geometry Adv. Stud. Pure Math. **63**, 31–58 (2012). arXiv:1102.1310 [math.NT]
18. Brown, F.C.S., Levin, A.: Multiple Elliptic Polylogarithms. arXiv:1110.6917 [math.NT]
19. Brown, F.: Mixed tate motives over Z. Ann. Math. **175**, 949–976 (2012)
20. Brown, F.: Single-valued Motivic periods and multiple zeta values. SIGMA **2**, e25 (2014) arXiv:1309.5309 [math.NT]
21. Brown, F.: Periods and Feynman amplitudes. arXiv:1512.09265 [math-ph]
22. Brown, F.: A class of non-holomorphic modular forms I. arXiv:1707.01230 [math.NT]
23. Deligne, P.: Le groupe fondamental de la droite projective moins trois points. In: Galois groups over **Q**, Springer, MSRI publications **16** (1989), 72-297; "Periods for the fundamental group," Arizona Winter School (2002)
24. Drinfeld, V.G.: On quasitriangular quasi-Hopf algebras and on a group that is closely connected with $Gal(\overline{Q}, Q)$. Alg. Anal. **2**, 149: English translation: leningrad Math. J. **2**(1991), 829–860 (1990)
25. Drummond, J.M., Ragoucy, E.: Superstring amplitudes and the associator. JHEP **1308**, 135 (2013). arXiv:1301.0794 [hep-th]
26. Gelfand, I.M., Kapranov, M.M., Zelevinsky, A.V.: Generalized Euler integrals and A-hypergeometric functions. Adv. Math. **84**, 255–271 (1990)
27. Golden, J., Goncharov, A.B., Spradlin, M., Vergu, C., Volovich, A.: Motivic amplitudes and cluster coordinates. JHEP **1401**, 091 (2014). arXiv:1305.1617 [hep-th]
28. Goncharov, A.B.: Multiple zeta-values, Galois groups, and geometry of modular varieties. arXiv:math/0005069v2 [math.AG]
29. Goncharov, A.B.: Multiple polylogarithms and mixed Tate motives. arXiv:math/ 0103059v4 [math.AG]
30. Goncharov, A., Manin, Y.: Multiple ζ-motives and moduli spaces $\overline{\mathscr{M}}_{0,n}$. Compos. Math. **140**, 1–14 (2004). arXiv:math/0204102
31. Goncharov, A.B.: Galois symmetries of fundamental groupoids and noncommutative geometry. Duke Math. J. **128**, 209–284 (2005). arXiv:math/0208144v4 [math.AG]
32. Goncharov, A.B.: Private communication
33. Ihara, Y.: Some arithmetic aspects of Galois actions in the pro-p fundamental group of $P^1 - \{0, 1, \infty\}$. In: Arithmetic Fundamental Groups and Noncommutative Algebra. Proceedings of Symposia in Pure Mathematics **70** (2002)
34. Kawai, H., Lewellen, D.C., Tye, S.H.H.: A relation between tree amplitudes of closed and open strings. Nucl. Phys. B **269**, 1 (1986)
35. Kontsevich, M., Zagier, D.: Periods. In: Engquist, B., Schmid, W. (eds.) Mathematics unlimited—2001 and beyond, Berlin, pp. 771–808. Springer, New York (2001)
36. Le, T.Q.T., Murakami, J.: Kontsevich's integral for the Kauffman polynomial. Nagoya Math. J. **142**, 39–65 (1996)
37. Mafra, C.R., Schlotterer, O., Stieberger, S.: Complete N-point superstring disk amplitude I. Pure spinor computation. Nucl. Phys. B **873**, 419 (2013). arXiv:1106.2645 [hep-th]
38. Mafra, C.R., Schlotterer, O., Stieberger, S.: Complete N-point superstring disk amplitude II. Amplitude and hypergeometric function structure. Nucl. Phys. B **873**, 461 (2013). arXiv:1106.2646 [hep-th]
39. Oprisa, D., Stieberger, S.: Six gluon open superstring disk amplitude, multiple hypergeometric series and Euler-Zagier sums. [hep-th/0509042]
40. Puhlfürst, G., Stieberger, S.: Differential equations, associators, and recurrences for amplitudes. Nucl. Phys. B **902**, 186 (2016) arXiv:1507.01582 [hep-th]. A feynman integral and its recurrences and associators. Nucl. Phys. B **906**, 168 (2016). arXiv:1511.03630 [hep-th]
41. Schlotterer, O., Stieberger, S.: Motivic multiple zeta values and superstring amplitudes. J. Phys. A **46**, 475401 (2013). arXiv:1205.1516 [hep-th]

42. Schnetz, O.: Graphical functions and single-valued multiple polylogarithms. Commun. Num. Theor. Phys. **08**, 589 (2014). arXiv:1302.6445 [math.NT]
43. Stieberger, S., Taylor, T.R.: Multi-gluon scattering in open superstring theory. Phys. Rev. D **74**, 126007 (2006). [hep-th/0609175]
44. Stieberger, S.: Open & Closed vs. Pure Open String Disk Amplitudes. arXiv:0907.2211 [hep-th]
45. Stieberger, S.: Constraints on tree-level higher order gravitational couplings in superstring theory. Phys. Rev. Lett. **106**, 111601 (2011). arXiv:0910.0180 [hep-th]
46. Stieberger, S.: Closed superstring amplitudes, single-valued multiple zeta values and the Deligne associator. J. Phys. A **47**, 155401 (2014). arXiv:1310.3259 [hep-th]
47. Stieberger, S., Taylor, T.R.: Closed string amplitudes as single-valued open string amplitudes. Nucl. Phys. B **881**, 269 (2014). arXiv:1401.1218 [hep-th]
48. Terasoma, T.: Selberg integrals and multiple zeta values. Compos. Math. **133**, 1–24 (2002)
49. Tsunogai, H.: On ranks of the stable derivation algebra and Deligne's problem. Proc. Japan Acad. Ser. A **73**, 29–31 (1997)
50. Zagier, D.: Values of zeta functions and their applications. In: Joseph, A., et al. (eds.) First European Congress of Mathematics (Paris, 1992), vol. II, pp. 497–512. Basel, Birkhäuser (1994)

The Number Theory of Superstring Amplitudes

Oliver Schlotterer

Abstract The following article is intended as a survey of recent results at the interface of number theory and superstring theory. We review the expansion of scattering amplitudes—central observables in field and string theory—in the inverse string tension where elegant patterns of multiple zeta values occur. More specifically, the Drinfeld associator and the Hopf algebra structure of motivic multiple zeta values are shown to govern tree-level amplitudes of the open superstring. Partial results on the quantum corrections are discussed where elliptic analogues of multiple zeta values play a central rôle.

Keywords Scattering amplitudes · Superstring theory · Multiple zeta values · Hopf algebras

1 Introduction

Around 1970, string theory was born out of an attempt to describe pion scattering, see [1] for a recent historic account. Even though the rôle of string theory has changed a lot over the past 45 years—most notably from a model of hadrons and mesons to a candidate framework for quantum gravity—its scattering amplitudes have been of constant interest. On the one hand, they provide fertile testing grounds for string dualities [2] or possible phenomenological signatures of string theory [3–5] in connection with a low string scale [6, 7]. On the other hand, string amplitudes are a prominent tool to obtain a novel viewpoint on interacting quantum field theories and perturbative gravity which arise in the limit where strings shrink to point particles. In many instances, the hidden simplicity of and relations between gauge-theory and gravity amplitudes are invisible to conventional methods (Lagrangians or Feynman diagrams) but follow naturally from string theory, see for instance [8–11].

O. Schlotterer (✉)
Max-Planck-Institut für Gravitationsphysik, Albert-Einstein-Institut, 14476 Potsdam, Germany
e-mail: oliver.schlotterer@physics.uu.se

Department of Physics and Astronomy, Uppsala University, 75108 Uppsala, Sweden

J. I. Burgos Gil et al. (eds.), *Periods in Quantum Field Theory and Arithmetic*, Springer Proceedings in Mathematics & Statistics 314,
https://doi.org/10.1007/978-3-030-37031-2_4

In this work, we review recent encounters of string amplitudes with modern topics in number theory. In the "tree-level" approximation, open-string amplitudes depend on the strings' fundamental length scale through iterated integrals in the unit interval and therefore involve multiple zeta values (MZVs). As we will see, the rich mathematical patterns of the MZVs' appearance can be understood from the Drinfeld associator [12, 13] and the Hopf algebra structure of motivic MZVs [14]. We also report on tree-level amplitudes of the closed-string [14, 15] as well as recent results [16] on the leading quantum corrections, "one-loop amplitudes". In the open-string sector, the latter are governed by iterated integrals on a genus-one surface and thus involve elliptic analogues of MZVs as studied by Enriquez [17, 18].

1.1 The Disk Amplitude

Tree-level scattering amplitudes of open superstring states are given by iterated integrals along the boundary of a disk. The integrand is a correlation function of vertex operators which insert the degrees of freedom of the external states onto a worldsheet of disk topology. Using the pure spinor formulation of the superstring [19], the correlator has been evaluated recently for any number of massless external legs [20],

$$A(1,2,\ldots,n;\alpha') = \sum_{\sigma\in S_{n-3}} F^{\sigma}(s_{ij}) A_{\rm YM}(1,\sigma(2,3,\ldots,n{-}2),n{-}1,n)\,, \tag{1}$$

where the labels $1, 2, \ldots, n$ on the left hand side refer to the polarizations and momenta of the external gauge bosons or their supersymmetry partners. Their ordering specifies a cyclic arrangement of punctures along the disk boundary, and the additional argument α' denotes the inverse string tension or the squared string length scale. On the right hand side, $A_{\rm YM}(1,\sigma(2,3,\ldots,n{-}2),n{-}1,n)$ are partial tree amplitudes in the super Yang-Mills theory obtained in the point particle limit $\alpha' \to 0$ [8]. They encode sums of Feynman diagrams obtained in degeneration limits of the disk worldsheet (see Fig. 1) and also depend on the external states in a cyclic ordering which is governed by $(n-3)!$ permutations $\sigma \in S_{n-3}$.

The objects of central interest to this work are the integrals $F^{\sigma}(s_{ij})$ in (1), we will report on the results of [13, 14] on their expansion in α'. In a parametrization of the disk boundary through real coordinates $z_j \in \mathbb{R}$ with $z_{ij} \equiv z_i - z_j$ [20],

$$F^{\sigma}(s_{ij}) \equiv (-1)^{n-3} \int\limits_{0\leq z_2\leq z_3\leq \ldots \leq z_{n-2}\leq 1} \mathrm{d}z_2\, \mathrm{d}z_3 \,\ldots\, \mathrm{d}z_{n-2} \left(\prod_{i<j}^{n-1} |z_{ij}|^{s_{ij}} \right) \sigma\left\{ \prod_{k=2}^{n-2} \sum_{m=1}^{k-1} \frac{s_{mk}}{z_{mk}} \right\}. \tag{2}$$

We have fixed the SL(2) symmetry on the disk by choosing $z_1 = 0$, $z_{n-1} = 1$ and $z_n = \infty$. The permutation $\sigma \in S_{n-3}$ is understood to act on the labels $2, 3, \ldots, n-2$ in the curly bracket while leaving $\sigma(1) = 1$. The integrals in (2) carry the entire α'-dependence of the disk amplitude through dimensionless combinations

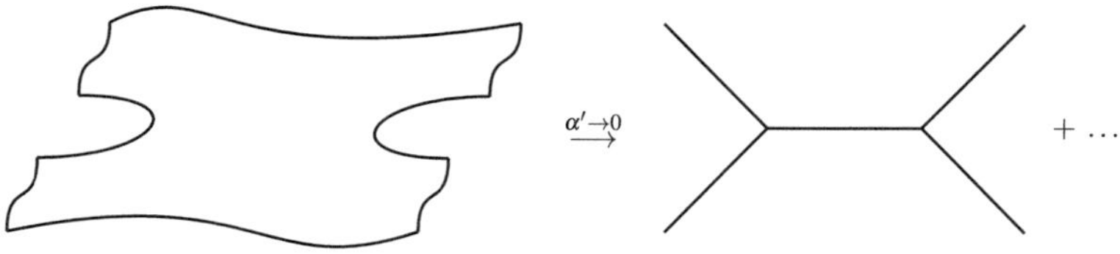

Fig. 1 The disk worldsheet describing open-string scattering at tree level degenerates to Feynman diagrams in the point-particle or field-theory limit $\alpha' \to 0$, where the ellipsis refers to further representatives of Feynman diagrams

$$s_{ij} \equiv \alpha'(k_i + k_j)^2 \tag{3}$$

of the external momenta k_i which are vectors of the D-dimensional Lorentz group. Momentum conservation $\sum_{i=1}^n k_i = 0$ and the on-shell condition $(k_i)^2 = 0$ for massless particles leave $\frac{n}{2}(n-3)$ independent Mandelstam variables s_{ij}. As we will demonstrate, the integrals in (2) reduce as follows in the field-theory limit $\alpha' \to 0$,

$$\lim_{\alpha' \to 0} F^\sigma(s_{ij}) = \begin{cases} 1 : \ \sigma(2,3,\ldots,n-2) = 2,3,\ldots,n-2 \\ 0 : \ \text{otherwise} \end{cases}, \tag{4}$$

i.e. their Taylor expansion w.r.t. s_{ij} in (3) encodes the string-corrections to super Yang-Mills theory. The expansion w.r.t. s_{ij} and thereby α' turns out to exhibit uniform transcendentality[1]: The w'th order in α' is accompanied by MZVs of transcendence weight w.

In the following sections, we will describe two organizing principles underlying the α'-expansion of the $F^\sigma(s_{ij})$. More specifically,

- A matrix representation of the Drinfeld associator generates the Taylor expansion in s_{ij} in a recursive manner w.r.t. the multiplicity n [13], see Sect. 2.
- Motivic MZVs and their Hopf algebra structure allow to extract the complete information on $F^\sigma(s_{ij})$ from its coefficients along with primitive MZVs ζ_w [14], see Sect. 3.

In Sect. 4, we conclude with a brief discussion of generalizations to closed strings or quantum corrections and raise open questions.

2 The α'-Expansion from the Drinfeld Associator

In this section, we review the recursion in [13] to obtain the α'-expansion of the integrals in (2) from the Drinfeld associator [21, 22]. This is achieved by establishing a Knizhnik-Zamolodchikov (KZ) equation for a deformation of the integrals in question through an auxiliary worldsheet puncture z_0. Certain boundary values of

[1]The terminology here and in later places relies on the commonly trusted conjectures on the transcendentality of MZVs.

the deformed integrals as $z_0 \to 0$ and $z_0 \to 1$ are found to yield the original $F^\sigma(s_{ij})$ at multiplicity $n-1$ and n, respectively. Recalling that the superscript σ denotes permutations of the legs $2, 3, \dots, n-2$, one can write the resulting recursion as [13]

$$F^{\sigma_i} = \sum_{j=1}^{(n-3)!} \big[\Phi(e_0, e_1)\big]_{ij} \, F^{\sigma_j}\big|_{k_{n-1}=0} \, , \tag{5}$$

where the kinematic regime $k_{n-1} = 0$ on the right hand side gives rise to $(n-1)$-point integrals,

$$F^{\sigma(23\dots n-2)}\big|_{k_{n-1}=0} = \begin{cases} F^{\sigma(23\dots n-3)} & \text{if } \sigma(n-2) = n-2 \\ 0 & \text{otherwise} \end{cases} . \tag{6}$$

The expressions for and derivation of the matrices e_0 and e_1 will be discussed in the subsequent.

2.1 Background on MZVs and the Drinfeld Associator

Before setting up the construction of the integrals $F^\sigma(s_{ij})$, we shall review the convention for MZVs and selected properties of the Drinfeld associator. MZVs of transcendental weight $w \in \mathbb{N}_0$ can be defined through iterated integrals labelled by a word in the two-letter alphabet $v_j \in \{0, 1\}$,

$$\zeta_{\{v_1 v_2 \dots v_w\}} \equiv (-1)^{\sum_{j=1}^w v_j} \int\limits_{0 \le z_1 \le z_2 \le \dots \le z_w \le 1} \frac{\mathrm{d}z_1}{z_1 - v_1} \frac{\mathrm{d}z_2}{z_2 - v_2} \cdots \frac{\mathrm{d}z_w}{z_w - v_w} \, , \tag{7}$$

where $v_1 = 1$ and $v_w = 0$ ensure convergence. Divergent integrals arising for $v_1 = 0$ or $v_w = 1$ can be addressed using the shuffle regularization prescription [23],

$$\zeta_{\{0\}} = \zeta_{\{1\}} = 0 \, , \qquad \zeta_{\{v\}} \cdot \zeta_{\{u\}} = \zeta_{\{v \sqcup\!\sqcup u\}} \, , \tag{8}$$

with the standard shuffle product $\sqcup\!\sqcup$ on words $v = v_1 v_2 \dots$ and $u \equiv u_1 u_2 \dots$. The representation of MZVs as nested sums can be recovered from the above integrals via

$$\zeta_{n_1, n_2, \dots, n_r} \equiv \sum_{0 < k_1 < k_2 < \dots < k_r}^{\infty} k_1^{-n_1} k_2^{-n_2} \dots k_r^{-n_r} = \zeta_{\{\underbrace{10\dots0}_{n_1}\underbrace{10\dots0}_{n_2} \cdots\cdots \underbrace{10\dots0}_{n_r}\}} \, , \tag{9}$$

such that for example $\zeta_{\{10\}} = -\zeta_{\{01\}} = \zeta_2$.

The Drinfeld associator governs the universal monodromy of the KZ equation[2] with $z_0 \in \mathbb{C}\backslash\{0,1\}$ and Lie-algebra generators e_0, e_1:

$$\frac{\mathrm{d}\hat{\mathbf{F}}(z_0)}{\mathrm{d}z_0} = \left(\frac{e_0}{z_0} + \frac{e_1}{1-z_0}\right)\hat{\mathbf{F}}(z_0)\,. \tag{10}$$

The solution $\hat{\mathbf{F}}(z_0)$ of the KZ equation lives in the vector space the representation of e_0 and e_1 acts upon. This general setup will later on be specialized to $(n-2)!$-component realizations of $\hat{\mathbf{F}}(z_0)$ closely related to the disk integrals F^σ.

Given the singularities of the differential operator in (10) as $z_0 \to 0$ and $z_0 \to 1$, non-analytic behaviour as $z_0^{e_0}$ and $(1-z_0)^{-e_1}$ has to be compensated when considering boundary values,

$$C_0 \equiv \lim_{z_0\to 0} z_0^{-e_0}\hat{\mathbf{F}}(z_0)\,, \quad C_1 \equiv \lim_{z_0\to 1}(1-z_0)^{e_1}\hat{\mathbf{F}}(z_0)\,. \tag{11}$$

As a defining property of the Drinfeld associator, it relates the regularized boundary values in (11) via [21, 22]

$$C_1 = \Phi(e_0, e_1)\, C_0\,. \tag{12}$$

At the same time, the Drinfeld associator in (12) can be written as a generating series of MZVs. In terms of their integral representation (7), we have [24]

$$\begin{aligned}\Phi(e_0,e_1) &= \sum_{\nu\in\{0,1\}^\times} e_{\nu_1}e_{\nu_2}\ldots e_{\nu_j}\ldots\zeta_{\{\ldots\nu_j\ldots\nu_2\nu_1\}} \\ &= 1 + \zeta_2[e_0,e_1] + \zeta_3[e_0-e_1,[e_0,e_1]] + \cdots\,.\end{aligned} \tag{13}$$

Hence, the Drinfeld associator plays a two-fold rôle as a generating series for MZVs in (13) and the universal monodromy of the KZ equation as in (12). Like this, it will be shown to hold the key to the recursion in (5) for disk integrals.

2.2 Deforming the Disk Integrals

In order to relate the disk integrals (2) to the Drinfeld associator, we will follow the lines of [25] and study a deformation that satisfies the KZ equation (10). In addition to an additional disk puncture $z_0 \in [0,1]$, auxiliary Mandelstam invariants $s_{02},\ldots,s_{0,n-2} \in \mathbb{R}$ as well as an integer parameter $\nu = 1,2,\ldots,n-2$ are introduced in

---

[2] The sign convention for e_1 varies in the literature.

$$\hat{F}_\nu^\sigma(s_{ij}, s_{0k}, z_0) \equiv (-1)^{n-3} \int_{0\leq z_2 \leq z_3 \leq \ldots \leq z_{n-2} \leq z_0} \mathrm{d}z_2\, \mathrm{d}z_3 \ldots \mathrm{d}z_{n-2} \left(\prod_{i<j}^{n-1} |z_{ij}|^{s_{ij}} \right) \tag{14}$$
$$\times \left(\prod_{k=2}^{n-2} |z_{0k}|^{s_{0k}} \right) \sigma \left\{ \prod_{l=2}^{\nu} \sum_{m=1}^{l-1} \frac{s_{ml}}{z_{ml}} \prod_{p=\nu+1}^{n-2} \sum_{q=p+1}^{n-1} \frac{s_{pq}}{z_{pq}} \right\} .$$

The integration domain for $z_2, \ldots, z_{n-2}$ reduces to the original one in (2) if $z_0 \to 1$ and sends all integration variables to zero if $z_0 \to 0$. As a consequence of the extra Mandelstam invariants s_{0k}, different values of $\nu = 1, 2, \ldots, n-2$ yield inequivalent integrals[3] such that the $(n-3)!$ permutations $\sigma \in S_{n-3}$ together with the range of ν yield a total of $(n-2)!$ functions in (14). It will be convenient to combine these objects to an $(n-2)!$-component vector whose entries are ordered as $\hat{\mathbf{F}} = (\hat{F}_{n-2}, \hat{F}_{n-3}, \ldots, \hat{F}_1)$.

The $(n-2)!$ components in (14) exceeding the number of $(n-3)!$ desired integrals in (2) are required to ensure that the deformed vector $\hat{\mathbf{F}}$ satisfies the KZ equation (10). Clearly, the variables e_0, e_1 therein become $(n-2)! \times (n-2)!$ matrices, and it will be illustrated by the later examples that their entries are linear in the Mandelstam variables s_{ij} as well as their auxiliary counterparts s_{0k}. Hence, the regularized boundary values (11) of $\hat{\mathbf{F}}$ will be related as in (12) by a $(n-2)! \times (n-2)!$ matrix representation of the Drinfeld associator. As is explained in more detail in [13], the components in (14) give rise to regularized boundary values

$$C_0\big|_{s_{0k}=0} = (F^\sigma\big|_{k_{n-1}=0}, \mathbf{0}_{(n-3)(n-3)!})^t , \qquad C_1\big|_{s_{0k}=0} = (F^\sigma, \ldots)^t \tag{15}$$

upon setting the auxiliary Mandelstam invariants s_{0k} to zero. The $(n-3)(n-3)!$ components of C_1 in the ellipsis do not need to be evaluated. In (15) and many subsequent equations, the dependence on s_{ij} is suppressed. With the regularized boundary values in (15), the relation (12) becomes

$$\begin{pmatrix} F^\sigma \\ \vdots \end{pmatrix} = \big[\Phi(e_0, e_1)\big]_{(n-2)! \times (n-2)!} \begin{pmatrix} F^\sigma\big|_{k_{n-1}=0} \\ \mathbf{0}_{(n-3)(n-3)!} \end{pmatrix} \tag{16}$$

---

[3] In the original disk integrals (2), rearranging the curly bracket of the integrand as

$$\prod_{l=2}^{n-2} \sum_{m=1}^{l-1} \frac{s_{ml}}{z_{ml}} \to \prod_{l=2}^{\nu} \sum_{m=1}^{l-1} \frac{s_{ml}}{z_{ml}} \prod_{p=\nu+1}^{n-2} \sum_{q=p+1}^{n-1} \frac{s_{pq}}{z_{pq}}$$

amounts to adding total derivatives w.r.t. $z_2, \ldots, z_{n-2}$ which vanish in presence of the Koba-Nielsen factor $\prod_{i<j}^{n-1} |z_{ij}|^{s_{ij}}$. Tentative boundary contributions at $z_j = z_{j\pm1}$ are manifestly suppressed by $|z_j - z_{j\pm1}|^{s_{j,j\pm1}}$ for positive real part of $s_{j,j\pm1}$ which propagates to generic complex values by analytic continuation.

upon taking $s_{0k} \to 0$, and the zeros in the vector on the right hand side reduce the recursion (16) to the form given in (5). From the linearity of e_0 and e_1 in s_{ij} (and therefore α'), two central properties of $F^\sigma(s_{ij})$ stated above can be easily verified:

- The $\alpha' \to 0$ limit of the disk integrals in (4) follows from the fact that the only contribution of the associator to this order is $\Phi(e_0, e_1) = 1 + \mathcal{O}(\alpha')$.
- Uniform transcendentality follows from the expansion (13) of the associator where MZVs of weight w are accompanied by w powers of e_0, e_1 and, by their linearity in s_{ij}, w powers of α'.

2.3 *Four- and Five-Point Examples*

In this subsection, we firstly illustrate the recursion (5) by examples with $n = 4, 5$ external states and secondly explain the mechanisms leading to a KZ equation for the functions in (14) as well as the explicit form of e_0, e_1 at various multiplicities. As a convenient shorthand, we introduce

$$X_{ij} \equiv \frac{s_{ij}}{z_{ij}} \ . \tag{17}$$

<u>$n=4$</u> points:

Here, the auxiliary vector made of (14) has two components

$$\begin{pmatrix} \hat{F}_2^{(2)} \\ \hat{F}_1^{(2)} \end{pmatrix} = \int_0^{z_0} \mathrm{d}z_2 \, |z_{12}|^{s_{12}} |z_{23}|^{s_{23}} z_{02}^{s_{02}} \begin{pmatrix} X_{21} \\ X_{32} \end{pmatrix} , \tag{18}$$

where the derivative w.r.t. z_0 introduces a factor of $\frac{s_{02}}{z_{02}}$ into the integrand.[4] Given the $SL(2)$-fixing $(z_1, z_3, z_4) = (0, 1, \infty)$, the extra dependence on z_0 can be rearranged into factors of $\frac{1}{z_{01}} = \frac{1}{z_0}$ and $\frac{1}{z_{03}} = \frac{1}{z_0 - 1}$ via partial fraction $(z_{12}z_{02})^{-1} = (z_{12}z_{01})^{-1} - (z_{01}z_{02})^{-1}$ and integration by parts:

$$\begin{aligned} 0 &= - \int_0^{z_0} \mathrm{d}z_2 \, \frac{\mathrm{d}}{\mathrm{d}z_2} \, |z_{12}|^{s_{12}} |z_{23}|^{s_{23}} z_{02}^{s_{02}} \\ &= \int_0^{z_0} \mathrm{d}z_2 \, |z_{12}|^{s_{12}} |z_{23}|^{s_{23}} z_{02}^{s_{02}} \left(\frac{s_{02}}{z_{02}} + \frac{s_{12}}{z_{12}} - \frac{s_{23}}{z_{23}} \right) . \end{aligned} \tag{19}$$

---

[4]The derivative w.r.t. z_0 directly acts at the level of the integrand since the boundary contribution from the z_0-dependence in the upper limit is suppressed as $\lim_{z_{n-2} \to z_0} (z_0 - z_{n-2})^{s_{0,n-2}} = 0$. As before, the limit is obvious if $s_{0,n-2}$ has a positive real part and otherwise follows from analytic continuation.

These manipulations lead to

$$\frac{\mathrm{d}}{\mathrm{d}z_0}\,\hat{F}_2^{(2)} = \frac{1}{z_0}\big[(s_{12}+s_{02})\hat{F}_2^{(2)} - s_{12}\hat{F}_1^{(2)}\big] \tag{20}$$

$$\frac{\mathrm{d}}{\mathrm{d}z_0}\,\hat{F}_1^{(2)} = \frac{1}{1-z_0}\big[s_{23}\hat{F}_2^{(2)} - (s_{23}+s_{02})\hat{F}_1^{(2)}\big]\,, \tag{21}$$

which allow to read off the following 2×2 matrix representations for e_0, e_1 upon setting $s_{02}\to 0$:

$$e_0 = \begin{pmatrix} s_{12} & -s_{12} \\ 0 & 0 \end{pmatrix}, \qquad e_1 = \begin{pmatrix} 0 & 0 \\ s_{23} & -s_{23} \end{pmatrix}. \tag{22}$$

Given the regularized boundary values (15), the main result (5) specializes to

$$\begin{pmatrix} F^{(2)} \\ \vdots \end{pmatrix} = \big[\Phi(e_0,e_1)\big]_{2\times 2}\begin{pmatrix} 1 \\ 0 \end{pmatrix}. \tag{23}$$

Note that the explicit form of the matrices (22) renders any nested commutator $\mathrm{ad}_0^k\mathrm{ad}_1^l[e_0,e_1]$ with $k,l\in\mathbb{N}_0$ and $\mathrm{ad}_i x\equiv[e_i,x]$ proportional to the nilpotent matrix $\left(\begin{smallmatrix} 1 & -1 \\ 1 & -1\end{smallmatrix}\right)$. As a consequence, the MZVs in $\big[\Phi(e_0,e_1)\big]_{2\times 2}$ can be expressed in terms of primitives ζ_w and are consistent with

$$F^{(2)} = \frac{\Gamma(1+s_{12})\Gamma(1+s_{23})}{\Gamma(1+s_{12}+s_{23})} = \exp\Big(\sum_{n=2}^{\infty}\frac{\zeta_n}{n}(-1)^n\big[s_{12}^n+s_{23}^n-(s_{12}+s_{23})^n\big]\Big)\,, \tag{24}$$

see [12] for a connection with a quotient of the associator. While the expression in (24) is more suitable to manifest the MZV-content of the four-point amplitude as compared to (23), the construction of the F^σ from the associator becomes significantly more rewarding at $n\geq 5$.

$n=5$ points:

At five-points, the auxiliary vector built from (14) has six components,

$$\begin{pmatrix} \hat{F}_3^{(23)} \\ \hat{F}_3^{(32)} \\ \hat{F}_2^{(23)} \\ \hat{F}_2^{(32)} \\ \hat{F}_1^{(23)} \\ \hat{F}_1^{(32)} \end{pmatrix} = \int_0^{z_0}\mathrm{d}z_3\int_0^{z_3}\mathrm{d}z_2\prod_{i<j}^{4}|z_{ij}|^{s_{ij}}\,z_{02}^{s_{02}}z_{03}^{s_{03}}\begin{pmatrix} X_{12}(X_{13}+X_{23}) \\ X_{13}(X_{12}+X_{32}) \\ X_{12}X_{34} \\ X_{13}X_{24} \\ (X_{23}+X_{24})X_{34} \\ (X_{32}+X_{34})X_{24} \end{pmatrix}. \tag{25}$$

Following the methods from the $n=4$ case, the z_0-derivatives can be cast into the form (10) using a sequence of partial fraction relations and integrations by parts. After setting $s_{0k}\to 0$, we can read off the resulting 6×6 matrix representation (with the shorthand $s_{ijk}\equiv s_{ij}+s_{ik}+s_{jk}$):

$$e_0 = \begin{pmatrix} s_{123} & 0 & -s_{13}-s_{23} & -s_{12} & -s_{12} & s_{12} \\ 0 & s_{123} & -s_{13} & -s_{12}-s_{23} & s_{13} & -s_{13} \\ 0 & 0 & s_{12} & 0 & -s_{12} & 0 \\ 0 & 0 & 0 & s_{13} & 0 & -s_{13} \\ 0 & 0 & 0 & 0 & 0 & 0 \\ 0 & 0 & 0 & 0 & 0 & 0 \end{pmatrix} \tag{26}$$

$$e_1 = \begin{pmatrix} 0 & 0 & 0 & 0 & 0 & 0 \\ 0 & 0 & 0 & 0 & 0 & 0 \\ s_{34} & 0 & -s_{34} & 0 & 0 & 0 \\ 0 & s_{24} & 0 & -s_{24} & 0 & 0 \\ s_{34} & -s_{34} & s_{23}+s_{24} & s_{34} & -s_{234} & 0 \\ -s_{24} & s_{24} & s_{24} & s_{23}+s_{34} & 0 & -s_{234} \end{pmatrix} . \tag{27}$$

The regularized boundary values in (15) then imply the following associator construction for the functions F^{σ} in the five-point amplitude:

$$\begin{pmatrix} F^{(23)} \\ F^{(32)} \\ \vdots \end{pmatrix} = \left[\Phi(e_0, e_1)\right]_{6\times 6} \begin{pmatrix} F^{(2)} \\ 0 \\ \mathbf{0}_4 \end{pmatrix} \tag{28}$$

Note that the five-point α'-expansion in (28) can also be obtained from the representation of $F^{(23)}$ and $F^{(32)}$ in terms of the hypergeometric functions ${}_3F_2$ [26–30].

2.4 *Higher Multiplicity*

The techniques to establish the KZ equation of $\hat{\mathbf{F}}(z_0)$ and to determine the matrices e_0, e_1 are universal to any value of n. Expressions for e_0, e_1 are straightforward to compute and additionally take a suggestive form; the resulting instances up to $n = 9$ can be downloaded from the website [31]. While the results for $n = 6, 7$ reproduce the α'-expansions in [27, 28, 32] as well as [33] to the orders tested, the associator-based method firstly makes high multiplicities $n > 7$ accessible. Even though the setup in [33] based on polylogarithms does not impose any limitations on n, its growing manual effort (e.g. in the treatment of poles) suggests to preferably rely on the Drinfeld associator at large multiplicities.

3 Motivic MZVs and the α'-Expansion

The main result (5) of the previous section together with the expressions for e_0 and e_1 in (22), (26), (27) as well as [31] make the s_{ij}-dependence of the disk integrals fully explicit. The MZVs originate from the Drinfeld associator as in (13) and carry redundancies in view of the relations over $\mathbb{Q}$ among the iterated integrals $\zeta_{\{v\}}$ with $v \in \{0, 1\}^{\times}$. In this section, we investigate the structure of the α'-expansion once the MZVs are reduced to their conjectural bases over $\mathbb{Q}$. In a conjectural model for MZVs using non-commutative generators $f_3, f_5, f_7, \ldots$ and a commutative variable f_2 [34], the end result for F^{σ} is captured by the neat expression [14]

$$\left(\sum_{k=0}^{\infty} f_2^k P_{2k} \right) \sum_{n=0}^{\infty} \Big(f_3 M_3 + f_5 M_5 + f_7 M_7 + \cdots \Big)^n , \tag{29}$$

where M_w and P_w are $(n-3)! \times (n-3)!$ matrices to be specified below. Most importantly, the coefficients P_{2k} and M_{2i+1} of the primitives f_2^k and f_{2i+1} completely determine the α'-dependence along with compositions such as $f_2 f_{2i+1}$ and $f_{2i+1} f_{2j+1}$.

3.1 Matrix-Valued Approach to Disk Amplitudes

In order to see the aforementioned relations between the coefficients of various basis MZVs over $\mathbb{Q}$, it is convenient to promote the disk integrals in (2) to a $(n-3)! \times (n-3)!$ matrix

$$F_{\tau}{}^{\sigma}(s_{ij}) \equiv (-1)^{n-3} \int\limits_{0 \leq z_{\tau(2)} \leq z_{\tau(3)} \leq \cdots \leq z_{\tau(n-2)} \leq 1} \mathrm{d}z_2\, \mathrm{d}z_3 \cdots \mathrm{d}z_{n-2} \left(\prod_{i<j}^{n-1} |z_{ij}|^{s_{ij}} \right) \sigma \Big\{ \prod_{k=2}^{n-2} \sum_{m=1}^{k-1} \frac{s_{mk}}{z_{mk}} \Big\} . \tag{30}$$

The additional index τ refers to permutations in S_{n-3} of the integration variables $2, 3, \ldots, n-2$ and distinguishes different integration domains $0 \leq z_{\tau(2)} \leq z_{\tau(3)} \leq \ldots \leq z_{\tau(n-2)} \leq 1$. The matrix of disk integrals in (30) allows to simultaneously address an $(n-3)!$ family of different tree-level subamplitudes,

$$A(1, \tau(2, 3, \ldots, n-2), n-1, n; \alpha') = \sum_{\sigma \in S_{n-3}} F_{\tau}{}^{\sigma}(s_{ij}) A_{\mathrm{YM}}(1, \sigma(2, 3, \ldots, n-2), n-1, n) . \tag{31}$$

They furnish a basis of arbitrary string subamplitudes $A(\pi(1, 2, \ldots, n); \alpha')$ with $\pi \in S_n$ [10, 11] in the same way as the $A_{\mathrm{YM}}(\ldots)$ on the right hand side are a basis of field-theory subamplitudes [35].

In principle, it suffices to know a single line of (30) with fixed τ since the remaining entries of the matrix can be generated by relabeling of the s_{ij} and corresponding changes in σ and τ. The description of disk integrals through a square matrix $F(s_{ij})$ as in (30) is useful in view of matrix multiplication. Let P_w and M_w denote $(n-3)! \times (n-3)!$ matrices whose entries are degree w polynomials in Mandelstam invariants with rational coefficients, then a reduction of MZVs to their conjectural $\mathbb{Q}$-bases at weight $w \leq 8$ yields

$$\begin{aligned} F(s_{ij}) &= 1_{(n-3)!\times(n-3)!} + \zeta_2 P_2 + \zeta_3 M_3 + \zeta_2^2 P_4 + \zeta_2\zeta_3 P_2 M_3 + \zeta_5 M_5 \\ &+ \zeta_2^3 P_6 + \frac{1}{2}\zeta_3^2 M_3 M_3 + \zeta_7 M_7 + \zeta_2\zeta_5 P_2 M_5 + \zeta_2^2\zeta_3 P_4 M_3 \\ &+ \zeta_2^4 P_8 + \zeta_3\zeta_5 M_5 M_3 + \frac{1}{2}\zeta_2\zeta_3^2 P_2 M_3 M_3 + \frac{1}{5}\zeta_{3,5}[M_5, M_3] + \mathcal{O}(\alpha'^9)\,. \end{aligned} \tag{32}$$

Remarkably, the matrix product $P_2 M_3$ along with the weight-five product $\zeta_2\zeta_3$ is determined by the coefficients P_2 and M_3 of ζ_2 and ζ_3, respectively. The different parental letters P_w, M_w for matrices of even and odd order w in α' goes back to the different nature of the associated primitives: At even weight, $\zeta_{2n} \in \mathbb{Q}\pi^{2n}$ can be reduced to powers of $\zeta_2 = \frac{\pi^2}{6}$ with rational prefactors while no relations among ζ_{2n+1} of different odd weight[5] and powers of π are known or expected. Also, only a single left-multiplicative matrix factor of P_w is seen in each term of the expansion in (32) and its generalization to higher weight.

The depth-two MZVs $\zeta_{3,5}$ in the last line of in (32) is accompanied by a matrix commutator $[M_5, M_3] = M_5 M_3 - M_3 M_5$, but its rational prefactor $\frac{1}{5}$ is less intuitive than the lower-weight counterparts. The even more dramatic proliferation of rational prefactors at weight eleven,

$$\begin{aligned} F(s_{ij})\big|_{(\alpha')^{11}} &= \zeta_{11} M_{11} + \frac{1}{2}\zeta_3^2\zeta_5 M_5 M_3^2 + \frac{1}{6}\zeta_2\zeta_3^3 P_2 M_3^3 + \zeta_2\zeta_9 P_2 M_9 \\ &+ \zeta_2^2\zeta_7 P_4 M_7 + \zeta_2^3\zeta_5 P_6 M_5 + \zeta_2^4\zeta_3 P_8 M_3 + \frac{1}{5}\zeta_{3,5}\zeta_3[M_5, M_3]M_3 \\ &+ \left(9\zeta_2\zeta_9 + \frac{6}{25}\zeta_2^2\zeta_7 - \frac{4}{35}\zeta_2^3\zeta_5 + \frac{1}{5}\zeta_{3,3,5}\right)[M_3, [M_5, M_3]]\,, \end{aligned} \tag{33}$$

calls for a systematic understanding of how the matrix commutators enter at generic weight, see [14] for the analogous expressions at weight $w \leq 16$. The required mathematical framework will be introduced in the following subsection.

3.2 *Motivic MZVs*

The basis MZVs over $\mathbb{Q}$ in the α'-expansion (32) and (33) have been chosen as in [38], following the guiding principle of preferring short and simple representatives. An alternative handle on the choice of basis can be obtained by switching to a conjecturally equivalent language for MZVs: a Hopf algebra comodule, which is composed from words

$$f_{2i_1+1}\ldots f_{2i_r+1}\, f_2^k, \quad \text{with} \quad r, k \geq 0 \quad \text{and} \quad i_1, \ldots, i_r \geq 1 \tag{34}$$

[5] Also, none of the odd ζ-values has been proven to be transcendental so far: the only known facts are the irrationality of ζ_3 as well as the existence of an infinite number of odd irrational ζ's [36, 37].

and graded by their weight $w = 2(i_1 + \cdots + i_r) + r + 2k$. The non-commutative generators f_{2i+1} of odd weight by themselves furnish a Hopf algebra, and the additional commutative variable f_2 extend it to a Hopf algebra comodule [34]. At each weight, the enumeration of all non-commutative words of the form in (34) yields the same basis dimension over $\mathbb{Q}$ as conjectured for MZVs of the same weight [39].

The mapping of MZVs to non-commutative words in (34) is slightly involved and relies on (commonly trusted) conjectures such as the exclusion of additional algebraic relations between MZVs beyond the known double-shuffle identities. In order to circumvent the currently intractable challenges to prove the outstanding conjectures, one lifts MZVs ζ to so-called motivic MZVs ζ^{m} whose more elaborate definition [34, 40–42] will not be reviewed in the subsequent. As a key property of motivic MZVs, they obey the same shuffle and stuffle product formulæ as the MZVs, e.g. (8) carries over to $\zeta^{\mathrm{m}}_{\{v\}}\zeta^{\mathrm{m}}_{\{u\}} = \zeta^{\mathrm{m}}_{\{v \sqcup\mathchoice{\mkern-7mu}{\mkern-7mu}{\mkern-3.2mu}{\mkern-3.8mu}\sqcup u\}}$. The transition from MZVs to their motivic counterparts, $\zeta_{n_1,\ldots,n_r} \to \zeta^{\mathrm{m}}_{n_1,\ldots,n_r}$, has the advantage that many of the desirable, but currently unproven facts about MZVs are in fact proven for motivic MZVs. In particular, the commutative algebra of motivic MZVs is graded by definition, and the motivic coaction, first written down by Goncharov [40] and further studied by Brown [34, 41, 43], is well-defined.

In the framework of motivic MZVs, one can construct an isomorphism ϕ of graded algebra comodules which map any $\zeta^{\mathrm{m}}_{n_1,\ldots,n_r}$ to non-commutative words of the form (34), see [43] for a thorough description. Once the normalization is fixed as

$$\phi(\zeta^{\mathrm{m}}_w) = f_w \ , \qquad f_{2k} \equiv \frac{\zeta_{2k}}{(\zeta_2)^k} \, f_2^k \ , \tag{35}$$

the map ϕ can be largely determined by demanding compatibility with the algebraic structures:

$$\phi(\zeta^{\mathrm{m}}_{n_1,\ldots,n_r}\zeta^{\mathrm{m}}_{m_1,\ldots,m_r}) = \phi(\zeta^{\mathrm{m}}_{n_1,\ldots,n_r}) \sqcup\mathchoice{\mkern-7mu}{\mkern-7mu}{\mkern-3.2mu}{\mkern-3.8mu}\sqcup \phi(\zeta^{\mathrm{m}}_{m_1,\ldots,m_r}) \tag{36}$$

$$\Delta\phi(\zeta^{\mathrm{m}}_{n_1,\ldots,n_r}) = \phi(\Delta\zeta^{\mathrm{m}}_{n_1,\ldots,n_r}) \ . \tag{37}$$

While the motivic coaction on the right hand side of (37) [40] can become combinatorically involved at higher weights, the coaction on the non-commutative words from (34) is given by simple deconcatenation

$$\Delta(f_2^k f_{i_1} f_{i_2} \ldots f_{i_r}) = \sum_{j=0}^{r} (f_2^k f_{i_1} f_{i_2} \ldots f_{i_j}) \otimes (f_{i_{j+1}} \ldots f_{i_r}) \ , \qquad i_j \in 2\mathbb{N}+1 \ . \tag{38}$$

In combination with (37), this largely determines the ϕ-image of higher-depth MZVs such as

$$\phi(\zeta^{\mathfrak{m}}_{3,5}) = -5 f_3 f_5\,, \qquad \phi(\zeta^{\mathfrak{m}}_{3,7}) = -14 f_3 f_7 - 6 f_5 f_5 \tag{39}$$

$$\phi(\zeta^{\mathfrak{m}}_{3,3,5}) = -5\, f_3 f_3 f_5 + \frac{4}{7}\, f_5 f_2^3 - \frac{6}{5}\, f_7 f_2^2 - 45\, f_9 f_2\,, \tag{40}$$

see [43] for an efficient algorithm based on an infinitesimal version of the coaction.

However, the insensitivity of the coaction constraint (37) to primitives introduces an ambiguity of adding rational multiples of f_2^4, f_2^5 and f_{11} to the right hand sides of (39) and (40). The absence of such primitives in the above expressions reflects a specific choice of the isomorphism ϕ. It is convenient to tailor the ϕ-map to the choice of $\mathbb{Q}$-basis for motivic MZVs at weight w by suppressing f_w in the ϕ-images of all basis elements except for (35). The ϕ-images at weights $w \leq 16$ displayed in [14] rely on reference bases of motivic MZVs over $\mathbb{Q}$ as in [38].

3.3 Cleaning up the α'-Expansion

The language of non-commutative words as in (34) turns out to reveal the pattern of MZVs in the α'-expansions in (32) and (33). Upon passing to a motivic version of the matrix $F(s_{ij})$ in (30),

$$F^{\mathfrak{m}}(s_{ij}) \equiv F(s_{ij})\big|_{\zeta_{n_1,\ldots,n_r} \to \zeta^{\mathfrak{m}}_{n_1,\ldots,n_r}}\,, \tag{41}$$

the above expansions (with weights $w = 9, 10$ restored) translate into the following ϕ-image under (39) and (40):

$$\begin{aligned}
\phi(F^{\mathfrak{m}}(s_{ij})) = \ & (1_{(n-3)!\times(n-3)!} + f_2 P_2 + f_2^2 P_4 + f_2^3 P_6 + f_2^4 P_8 + f_2^5 P_{10}) \\
\times\ & (1_{(n-3)!\times(n-3)!} + f_3 M_3 + f_5 M_5 + f_3 f_3 M_3^2 + f_7 M_7 + f_3 f_5 M_3 M_5 + f_5 f_3 M_5 M_3 \\
& + f_9 M_9 + f_3 f_3 f_3 M_3^3 + f_5 f_5 M_5^2 + f_3 f_7 M_3 M_7 + f_7 f_3 M_7 M_3 \\
& + f_{11} M_{11} + f_3 f_3 f_5 M_3 M_3 M_5 + f_3 f_5 f_3 M_3 M_5 M_3 + f_5 f_3 f_3 M_5 M_3 M_3) + \mathcal{O}(\alpha'^{12})\,.
\end{aligned} \tag{42}$$

The coefficients P_{2k} of the commutative variables f_2^k build up a left-multiplicative matrix factor and can be cleanly disentangled from the odd-weight contributions involving f_{2i+1} and M_{2i+1}. Within the odd-weight sector, the democratic appearance of any non-commutative word in $f_{2i+1} M_{2i+1}$ with unit coefficient motivates the following generalization to arbitrary weight [14]:

$$\phi(F^{\mathfrak{m}}(s_{ij})) = \left(\sum_{k=0}^{\infty} f_2^k P_{2k} \right) \sum_{p=0}^{\infty} \sum_{\substack{i_1,i_2,\ldots,i_p \\ \in 2\mathbb{N}+1}} f_{i_1} f_{i_2} \ldots f_{i_p} M_{i_1} M_{i_2} \ldots M_{i_p}\,. \tag{43}$$

This all-order expression remains a conjecture beyond weights $\leq 21, 9, 7$ at multiplicity $n = 5, 6, 7$ where explicit checks have been performed in [33]. It is tempting to rewrite the right-multiplicative factor made of $f_{2i+1}M_{2i+1}$ as a formal geometric series,

$$\sum_{p=0}^{\infty} \sum_{\substack{i_1,i_2,\ldots,i_p \\ \in 2\mathbb{N}+1}} f_{i_1} f_{i_2} \ldots f_{i_p} M_{i_1} M_{i_2} \ldots M_{i_p} = \frac{1}{1 - (f_3 M_3 + f_5 M_5 + f_7 M_7 + \cdots)} , \tag{44}$$

whose equivalence to (43) and (29) relies on the expansion $(1-x)^{-1} = \sum_{n=0}^{\infty} x^n$ under certain assumptions on the infinite series $x \equiv f_3 M_3 + f_5 M_5 + f_7 M_7 + \cdots$. Given the disappearance of the exotic rational prefactors $\frac{6}{25}$ or $-\frac{4}{35}$ in (33), the striking simplicity of (43) is manifested by the language based on words in f_2, f_{2i+1} where the coaction (38) takes a more intuitive form as compared to $\Delta\zeta^{\mathfrak{m}}_{n_1,\ldots,n_r}$. Hence, the understanding of the pattern in the α'-expansion can ultimately be attributed to the Hopf algebra structure of motivic MZVs.

Even though the coefficients of $f_{n_1+n_2+\cdots+n_p}$ in a given $\phi(\zeta^{\mathfrak{m}}_{n_1,n_2,\ldots,n_p})$ and thereby P_w, M_w depend on the choice of basis MZVs, the form of the end result (43) is universal. Explicit expressions for the matrices P_w, M_w at various weights w and multiplicities n are available for download at [31], they are associated with the MZV bases of [38]. At multiplicity $n = 4$, they become scalars such that the vanishing of any commutator $[M_i, M_j]$ ensures the absence of depth ≥ 2 MZVs in the four-point amplitude. The closed-form expressions

$$M_{2i+1}\big|_{n=4} = -\frac{1}{2i+1}\big[s_{12}^{2i+1} + s_{23}^{2i+1} + s_{13}^{2i+1}\big] , \qquad P_{2k}\big|_{n=4} = \frac{\zeta_{2k}}{2k(\zeta_2)^k}\big[s_{12}^{2k} + s_{23}^{2k} - s_{13}^{2k}\big] \tag{45}$$

with $s_{13} = -s_{12} - s_{23}$ can be inferred from the representation of $F^{(2)}$ in (24).

We emphasize that the complete information on $F^{\mathfrak{m}}(s_{ij})$ in contained in (43) since ϕ can be inverted to recover motivic MZVs from f_w. More importantly, only one matrix P_w, M_w along with f_w needs to be specified at each weight: The matrix-multiplicative pattern in (43) determines the coefficients of any other word in f_2 and f_{2i+1} of the same weight from matrices seen at lower weight. Given that the conjectural number [39] of linearly independent weight-w MZVs over $\mathbb{Q}$ grows with the order of $\left(\frac{4}{3}\right)^w$, this amounts to an enormous compression of information.

As firstly pointed out in [12], the form of the α'-expansion in (43) implies a simple expression for the motivic coaction,

$$\Delta F^{\mathfrak{m}}(s_{ij}) = F^{\mathfrak{m}}(s_{ij}) \otimes F^{\mathfrak{m}}(s_{ij})\big|_{\zeta_2^{\mathfrak{m}}=0} , \tag{46}$$

where matrix multiplication is understood between the two sides of the tensor product. This resembles the coaction of the motivic Drinfeld associator $\Phi^{\mathfrak{m}}(e_0, e_1) \equiv \Phi(e_0, e_1)\big|_{\zeta_{n_1,\ldots,n_r} \to \zeta^{\mathfrak{m}}_{n_1,\ldots,n_r}}$ [12]

$$\Delta\Phi^{\mathfrak{m}}(e_0, e_1) = \Phi^{\mathfrak{m}}(e_0, e_1) \overset{\triangleleft}{\otimes} \Phi^{\mathfrak{m}}(e_0, e_1)\big|_{\zeta_2^{\mathfrak{m}}=0} , \tag{47}$$

where the operation $\triangleleft$ on top of the tensor product denotes the Ihara action among the words in e_0 and e_1 on the two sides. The expansion of the Drinfeld associator in a conjectural basis of MZVs over $\mathbb{Q}$ takes a form analogous to (43) where matrix multiplication among P_w and M_w is replaced by Ihara products among elementary words [12].

4 Further Directions and Open Questions

In the previous sections, we have described the mathematical structure of tree-level amplitudes (1) among any number of massless open-superstring states. The string-corrections to the corresponding gauge-theory amplitudes are governed by the disk integrals in (2) whose α'-expansion exhibits elegant patterns of MZVs. As elaborated in Sect. 2, the Drinfeld associator generates the dependence on dimensionless kinematic invariants $\alpha' k_i \cdot k_j$ in a recursive fashion, see in particular (5). Once the resulting MZVs are cast into their (conjectural) basis over $\mathbb{Q}$, their coefficients are related by matrix multiplication as displayed in (29). The systematics discussed in Sect. 3 only become fully apparent if the MZVs are translated into a language based on non-commutative words. This dictionary is guided by the Hopf algebra structure, most notably by the coaction, and its mathematical validity relies on the use of motivic MZVs.

A couple of natural follow-up questions have already been addressed in the literature, so we shall conclude with a sketch of the subsequent developments before pointing out open problems.

4.1 *The Closed String at Genus Zero and Single-Valued MZVs*

Tree-level scattering of closed strings is described by worldsheets of sphere topology. The integrations over vertex operator positions can be deformed in a way described in [9] such that closed-string tree amplitudes are composed from squares of open-string subamplitudes. This so-called "KLT-formula" [9] relies on the fact that the closed-string spectrum is contained in the tensor product of open-string excitations. At the massless level, for instance, closed-string excitations furnish a supersymmetry multiplet containing the graviton which arises from doubling gauge-boson supermultiplets in the open-string sector.

Once the $(n-3)!$-element basis of open-string subamplitudes [10, 11] is organized as in (31), the n-point closed-string tree amplitude $\mathcal{M}_n$ takes the form

$$\mathcal{M}_n(\alpha') = \sum_{\tau,\sigma,\rho,\pi \in S_{n-3}} \tilde{A}_{\mathrm{YM}}(1,\tau,n-1,n) F_{\rho}{}^{\tau}(s_{ij}) \mathcal{S}_{\alpha'}^{\rho,\pi}(s_{ij}) F_{\pi}{}^{\sigma}(s_{ij}) A_{\mathrm{YM}}(1,\sigma,n-1,n)\,, \tag{48}$$

by the KLT-formula [9]. We use shorthands $\tilde{A}_{\rm YM}(1,\tau,n-1,n)$ and $A_{\rm YM}(1,\sigma,n-1,n)$ for the two independent gauge-theory factors $\tilde{A}_{\rm YM}(1,\tau(2,\ldots,n-2),n-1,n)$ and $A_{\rm YM}(1,\sigma(2,\ldots,n-2),n-1,n)$, respectively. The entries of the $(n-3)!\times(n-3)!$ matrix $\mathscr{S}^{\rho,\pi}_{\alpha'}(s_{ij})$ are degree $(n-3)!$ polynomials in $\sin(\pi s_{ij})$, see [9, 44, 45] for more details and various representations. Their α'-expansion is based on $\sin(\pi s_{ij})=\pi s_{ij}\sum_{n=1}^{\infty}\frac{(-1)^n(\pi s_{ij})^{2n}}{(2n+1)!}$ and clearly interferes with the "even-zeta" sector represented by f_2 and P_{2k} in the ϕ-image (43) of disk integrals. In supergravity amplitudes obtained from the field-theory limit $\alpha'\to 0$ of (48), the sine functions are reduced to their argument, leaving behind

$$S_0^{\rho,\pi}(s_{ij})\equiv\mathscr{S}^{\rho,\pi}_{\alpha'}(s_{ij})\big|_{\sin(\pi s_{ij})\to\pi s_{ij}}\ . \tag{49}$$

It turns out that the properties of the matrices P_{2k} and $\mathscr{S}^{\rho,\pi}_{\alpha'}(s_{ij})$ lead to the striking cancellation of f_2 in the ϕ-image of the closed-string amplitude (48) [14],

$$\left(\sum_{k=0}^{\infty}f_2^k P^t_{2k}\right)\mathscr{S}_{\alpha'}(s_{ij})\left(\sum_{l=0}^{\infty}f_2^l P_{2l}\right)=S_0(s_{ij})\ , \tag{50}$$

which is tested to very high orders in α' (21, 9 and 7 at $n=5$, 6 and 7) but remains conjectural beyond that. Another observational identity on the same footing concerns the matrices M_{2i+1} [14],

$$M^t_{2i+1}S_0(s_{ij})=S_0(s_{ij})M_{2i+1}\ , \tag{51}$$

which leads to additional cancellations among MZVs in the "odd-zeta" sector represented by the f_{2i+1} in the open-string α'-expansion. Taking both of (50) and (51) into account, the motivic version of the closed-string amplitude (48) defined in analogy to (41) can be simplified to [14]

$$\phi(\mathscr{M}_n^{\rm m}(\alpha'))=\tilde{A}_{\rm YM}S_0(s_{ij})\sum_{p=0}^{\infty}\sum_{\substack{i_1,i_2,\ldots,i_p\\ \in 2\mathbb{N}+1}}M_{i_1}M_{i_2}\ldots M_{i_p}\sum_{j=0}^{p}f_{i_1}f_{i_2}\ldots f_{i_j}\sqcup\!\sqcup\, f_{i_p}\ldots f_{i_{j+1}}A_{\rm YM}\ . \tag{52}$$

In all of (50) to (52), we have suppressed the S_{n-3}-"indices" present in (49) since the pattern of their summation is clear from the relative ordering of the matrices and vectors.

As pointed out in [15], the arrangement of the odd-weight variables f_{2i+1} in (52) implements the single-valued projection of MZVs [42, 46],

$${\rm sv}:\ f_{i_1}f_{i_2}\ldots f_{i_p}\to\sum_{j=0}^{p}f_{i_1}f_{i_2}\ldots f_{i_j}\sqcup\!\sqcup\, f_{i_p}\ldots f_{i_{j+1}}\ . \tag{53}$$

On these grounds, the α'-expansion in the representation (52) of the closed-string amplitude can be traced back to the single-valued version of the open-string amplitude (43) [15]

$$\mathcal{M}_n(\alpha') = \tilde{A}_{\rm YM} S_0(s_{ij}) {\rm sv}[A(\alpha')] \,, \tag{54}$$

where the $(n-3)!$ components of the vector $A(\alpha')$ on the right hand side are spelt out in (31). As detailed in [47], analogous statements hold for tree-level amplitudes of the heterotic string.

At the level of the associators, the single-valued projection (53) maps the Drinfeld associator to the Deligne associator [42] which therefore captures the structure of the closed-string amplitude [15]. It would be of central importance to find the closed-string counterpart of the recursive associator construction in Sect. 2 [13]. The emergence of the single-valued projection in (52) and (54) could be rigorously proven from a direct derivation of the closed-string integrals from the Deligne associator and would not rely on the empirical properties of the matrices P_w and M_w in (50) and (51) which remain conjectural beyond certain orders.

4.2 *The Open String at Genus One and Elliptic MZVs*

Apart from their implications for the closed string, the above results on open-string tree amplitudes call for a generalization to their quantum corrections and thereby to Riemann surfaces of higher genus. At the one-loop order of superstring perturbation theory, the worldsheet topologies relevant to open-string scattering are cylinder and Moebius-strip diagrams. For appropriate choice of the gauge group, these topologies conspire in a way to cancel infinities in the amplitudes considered in this section, and infinity cancellation in more general situations additionally involves the Klein-bottle topology [48]. Even though the cylindrical topology allows for insertions of vertex operators on both boundaries (see [49, 50] for the implications on anomaly cancellations), we shall now report on recent studies [16] of the "planar" cylinder where the iterated integration is performed on a single boundary.

4.2.1 Definition and Properties of Elliptic MZVs

The mathematical framework for worldsheet integrals in planar one-loop amplitudes of the open superstring in known under the name of elliptic MZVs (eMZVs) [17, 18]. In the same way as MZVs can be defined as the expansion coefficients of the Drinfeld associator, see (13), eMZVs are defined [17] as the expansion coefficients of the elliptic Knizhnik-Zamolodchikov-Bernard (KZB) associator [18] which governs the regularized monodromy of the universal elliptic KZB equation. This definition identifies eMZVs as iterated integrals on an elliptic curve $\frac{\mathbb{C}}{\mathbb{Z}+\tau\mathbb{Z}}$ with $\Im(\tau) > 0$, in agreement with the approach via elliptic polylogarithms [51, 52]. The two homology cycles of the elliptic curve parametrized through the paths from $[0, 1]$ and $[0, \tau]$ give rise to two types of eMZVs, namely A-elliptic and B-elliptic MZVs. They descend from the two components $(A(\tau), B(\tau))$ of the elliptic KZB associator describing the monodromies of the elliptic KZB equation w.r.t. the paths $[0, 1]$ and $[0, \tau]$ and are related through the modular transformation $\tau \to -\frac{1}{\tau}$.

We will focus on A-elliptic MZVs associated with the homology cycle $[0, 1] \subset \mathbb{R}$ and, given that modular transformations restore the information on B-elliptic MZVs, refer to the former as eMZVs for simplicity. In this context, the definition of MZVs in (7) via iterated integrals generalizes to

$$\omega(n_1, n_2, \ldots, n_r; \tau) \equiv \int_{0\leq z_1\leq z_2\leq \ldots \leq z_r \leq 1} \mathrm{d}z_1\, f^{(n_1)}(z_1, \tau)\, \mathrm{d}z_2\, f^{(n_2)}(z_2, \tau)\, \ldots\, \mathrm{d}z_r\, f^{(n_r)}(z_r, \tau) \tag{55}$$

with $n_j \in \mathbb{N}_0$ for $j = 1, 2, \ldots, r$. Instead of a two-letter alphabet $\{\frac{\mathrm{d}z}{z}, \frac{\mathrm{d}z}{1-z}\}$ of differential forms seen at genus zero, eMZVs in (55) exhibit an infinity of doubly-periodic functions $f^{(n)}$ which can be defined from their generating series

$$\exp\left(2\pi i \alpha \frac{\Im(z)}{\Im(\tau)}\right) \frac{\theta'(0, \tau)\theta(z + \alpha, \tau)}{\theta(z, \tau)\theta(\alpha, \tau)} = \sum_{n=0}^{\infty} \alpha^{n-1} f^{(n)}(z, \tau)\,, \tag{56}$$

for instance $f^{(0)}(z, \tau) = 1$ and $f^{(1)}(z, \tau) = \frac{\partial}{\partial z} \ln \theta(z, \tau) + 2\pi i \frac{\Im(z)}{\Im(\tau)}$. The non-negative integers r and $w = n_1 + n_2 + \cdots + n_r$ in (55) are referred to as the length and the weight of an eMZV. The tick along with $\theta'(0, \tau)$ in (56) denotes a derivative of the odd Jacobi θ function w.r.t. its first argument z. Performing the integrals in the definition of eMZVs (55) yields a Fourier series in $q \equiv e^{2\pi i \tau}$ whose coefficients are MZVs along with integer powers of $2\pi i$ [17, 18].

By their definition (55) as iterated integrals, eMZVs satisfy shuffle relations

$$\omega(n_1, n_2, \ldots, n_r; \tau)\omega(k_1, k_2, \ldots, k_s; \tau) = \omega\big((n_1, n_2, \ldots, n_r) \sqcup\!\sqcup (k_1, k_2, \ldots, k_s); \tau\big)\,, \tag{57}$$

and the parity properties $f^{(n)}(-z, \tau) = (-1)^n f^{(n)}(z, \tau)$ following from $\theta(-z, \tau) = -\theta(z, \tau)$ and (56) imply the reflection identities

$$\omega(n_1, n_2, \ldots, n_{r-1}, n_r; \tau) = (-1)^{n_1+n_2+\cdots+n_r} \omega(n_r, n_{r-1}, \ldots, n_2, n_1; \tau)\,. \tag{58}$$

The combination of (57) and (58) is particularly constraining if the length r and the weight w are both even or odd, i.e. if $r + w$ is even. In these cases, shuffle- and reflection identities can be used to express any such eMZV in terms of products of lower-length eMZVs [53].

As a higher-genus analogue of partial-fraction relations $\frac{1}{z(z-1)} = \frac{1}{z-1} - \frac{1}{z}$ among the genus-zero forms in MZVs (7), the doubly-periodic functions $f^{(n)}$ obey Fay relations [16, 52]

$$
\begin{aligned}
f^{(n_1)}(z-x,\tau)f^{(n_2)}(z,\tau) &= \sum_{j=0}^{n_2}\binom{n_1-1+j}{j} f^{(n_2-j)}(x,\tau)f^{(n_1+j)}(z-x,\tau)\\
&\quad + \sum_{j=0}^{n_1}\binom{n_2-1+j}{j}(-1)^{n_1+j} f^{(n_1-j)}(x,\tau)f^{(n_2+j)}(z,\tau)\\
&\quad - (-1)^{n_1} f^{(n_1+n_2)}(x,\tau)\,, \qquad (59)
\end{aligned}
$$

which play a central rôle in deriving the subsequent α'-expansion of open-string one-loop amplitudes. Together with the shuffle- and reflection identities (57) and (58), the Fay relations were observed to generate all identities between eMZVs across a wide range of weights and lengths [53].

4.2.2 Elliptic MZVs in Open-String Amplitudes

The simplest non-vanishing one-loop amplitude of the open superstring involves four external massless states [8]. In the aforementioned planar cylinder topology, the four-point amplitude

$$
A^{\text{1-loop}}(1,2,3,4;\alpha') = s_{12}s_{23}A_{\text{YM}}(1,2,3,4)\int_0^\infty \mathrm{d}t\; I_{1234}(\tau=it,s_{ij}) \qquad (60)
$$

is governed by the following iterated integral,

$$
I_{1234}(\tau,s_{ij}) \equiv \int_0^1 \mathrm{d}z_4 \int_0^{z_4}\mathrm{d}z_3\int_0^{z_3}\mathrm{d}z_2 \prod_{i<j}^4 e^{s_{ij}G(z_i-z_j,\tau)}\,, \qquad (61)
$$

with $z_1=0$ and $\Re(\tau)=0$. The Mandelstam variables s_{ij} are defined in (3), and the bosonic Green function $G(z_i-z_j,\tau)$ satisfies

$$
\frac{\partial}{\partial z}G(z,\tau) = f^{(1)}(z,\tau)\,, \qquad G(z,\tau) = \int_0^z \mathrm{d}x\; f^{(1)}(x,\tau)\,, \qquad (62)
$$

reflecting a regularization prescription for its zero mode that amounts to $G(0,\tau)\to 0$. Like this, the integral in (61) can be related to eMZVs in (55) and expanded at fixed values of τ [16],

$$
\begin{aligned}
I_{1234}(\tau,s_{ij}) &= \omega(0,0,0;\tau) \;-\; 2\omega(0,1,0,0;\tau)\,(s_{12}+s_{23})\\
&\quad + 2\omega(0,1,1,0,0;\tau)\,\big(s_{12}^2+s_{23}^2\big) \;-\; 2\omega(0,1,0,1,0;\tau)\,s_{12}s_{23} \qquad (63)\\
&\quad + \beta_5(\tau)\,(s_{12}^3+2s_{12}s_{23}(s_{12}+s_{23})+s_{23}^3) \;+\; \beta_{2,3}(\tau)\,s_{12}s_{23}(s_{12}+s_{23}) \;+\; \mathcal{O}(\alpha'^4)\,,
\end{aligned}
$$

with shorthands

$$\begin{aligned}\beta_5(\tau) &= \frac{4}{3}\big[\omega(0,0,1,0,0,2;\tau) + \omega(0,1,1,0,1,0;\tau)\\ &\qquad - \omega(2,0,1,0,0,0;\tau) - \zeta_2\omega(0,1,0,0;\tau)\big] \\ \beta_{2,3}(\tau) &= \frac{\zeta_3}{12} + \frac{8\zeta_2}{3}\,\omega(0,1,0,0;\tau) - \frac{5}{18}\,\omega(0,3,0,0;\tau)\,.\end{aligned} \tag{64}$$

At the third order in α', the Fay identities (59) are crucial to express the iterated integrals over three powers of the Green function (62) in terms of eMZVs. In an equivalent parametrization of the cylinder boundary via $z \in [0, \tau]$ instead of $z \in [0, 1]$ as chosen in (61), the A-elliptic MZVs in (63) are traded for their B-elliptic analogues. In contrast to A-elliptic MZVs, however, B-elliptic MZVs are not periodic w.r.t. $\tau \to \tau+1$ and do not have Fourier expansion such as

$$\omega(0,1,0,0;\tau) = \frac{\zeta_3}{4\pi^2} + \frac{3}{2\pi^2}\sum_{m,n=1}^{\infty}\frac{1}{m^3}q^{mn} \tag{65}$$

for the A-elliptic MZV along with first α'-correction in (63). The Fourier expansion of the cylinder integral admitted by the parametrization in (61) has been exploited to check [16] that (63) reproduces the expected tadpole divergence [54]. The latter arises from the integration region $t \to \infty$ in (60) and eventually cancels upon combination with the Moebius-strip diagram [48].

The polarization-dependence of the four-point amplitude (60) is represented by $A_{\text{YM}}(1,2,3,4)$ and thereby follows the organization principle (1) of tree-level n-point amplitudes in terms of an $(n-3)!$-basis of subamplitudes $A_{\text{YM}}(\ldots)$ [10, 11, 35]. Similarly, the five-point one-loop amplitude can be written as [16, 55]

$$A^{\text{1-loop}}(1,2,3,4,5;\alpha') = \int_0^\infty \mathrm{d}t \sum_{\sigma\in S_2} I_{1\sigma(23)45}(\tau = it, s_{ij})A_{\text{YM}}(1,\sigma(2,3),4,5)\,, \tag{66}$$

see Sect. 5.1 of [16] for more details on the integrals I_{12345} and I_{13245}. At higher multiplicity $n \geq 6$, a gauge invariant sector of open-string one-loop amplitudes has been reduced to field-theory subamplitudes as well [55]. However, the cancellation mechanism of the hexagon anomaly [49, 50] requires additional kinematic structures in $(n \geq 6)$-point amplitudes,[6] so it remains an open problem to identify a suitable generalization of gauge-theory tree amplitudes to carry the polarization dependence of the string amplitude.

[6] In the pure spinor framework [19], kinematic building blocks suitable to describe the anomaly sector have been constructed in [56], see [57] for their appearance in the integrand of ten-dimensional field-theory amplitudes.

4.2.3 Bases of Elliptic MZVs over $\mathbb{Q}$

Starting from the third subleading order in α', the increasing length and complexity of the eMZV-coefficients (64) calls for a systematic study of relations among eMZVs over $\mathbb{Q}$ and guiding principles to select a suitable basis. This has been done in [53], also see [58] for a particularly elaborate treatment of the length-two case. The number of independent eMZVs at given weight and length is bounded by their differential equation

$$\begin{aligned}
2\pi i \frac{\mathrm{d}}{\mathrm{d}\tau}\omega(n_1,\ldots,n_r;\tau) &= n_1 \mathrm{G}_{n_1+1}(\tau)\omega(n_2,\ldots,n_r;\tau) - n_r \mathrm{G}_{n_r+1}(\tau)\omega(n_1,\ldots,n_{r-1};\tau) \\
&+\sum_{i=2}^{r}\Big\{(-1)^{n_i}(n_{i-1}+n_i)\mathrm{G}_{n_{i-1}+n_i+1}(\tau)\omega(n_1,\ldots,n_{i-2},0,n_{i+1},\ldots,n_r;\tau) \\
&-\sum_{k=0}^{n_{i-1}+1}(n_{i-1}-k)\binom{n_i+k-1}{k}\mathrm{G}_{n_{i-1}-k+1}(\tau)\omega(n_1,\ldots,n_{i-2},k+n_i,n_{i+1},\ldots,n_r;\tau) \\
&+\sum_{k=0}^{n_i+1}(n_i-k)\binom{n_{i-1}+k-1}{k}\mathrm{G}_{n_i-k+1}(\tau)\omega(n_1,\ldots,n_{i-2},k+n_{i-1},n_{i+1},\ldots,n_r;\tau)\Big\}
\end{aligned} \tag{67}$$

with $\mathrm{G}_n(\tau)$ denoting holomorphic Eisenstein series

$$\mathrm{G}_n(\tau) = \begin{cases} \displaystyle\sum_{\substack{k,m\in\mathbb{Z}\\(k,m)\neq(0,0)}} \frac{1}{(k+\tau m)^n} & : \ n>0 \\ -1 & : \ n=0\,. \end{cases} \tag{68}$$

As a consequence of (67), eMZVs can be expressed in terms of iterated integrals over Eisenstein series, special cases of iterated Shimura integrals [59, 60]. In this picture, the iterated integration is carried out over the argument τ, and the counting of (shuffle-independent) iterated Eisenstein integrals sets an upper bound on the numbers of independent eMZVs.

On top of that, selection rules on the admissible Eisenstein integrals within eMZVs are encoded in an algebra of derivations [61–64] which appear in the differential equation of the elliptic KZB associator [17, 18], the generating series of eMZVs. In view of the central rôle of the Drinfeld associator for tree-level amplitudes seen in Sect. 2, the elliptic associator is expected to carry essential information on one-loop open-string amplitudes including the α'-expansion (63).

A careful bookkeeping of eMZV relations within the above framework leads to the numbers $N(r,w)$ of indecomposable eMZVs[7] of length r and weight w as shown in Table 1 [53]. The data in the table is compatible with the all-weight formulæ [53]

---

[7] A set of indecomposable eMZVs of weight w and length r is a minimal set of eMZVs such that any other eMZV of the same weight and length can be expressed as a linear combination of elements from this set as well as products of eMZVs with strictly positive weights and eMZVs of lengths smaller than r or weight lower than w. The coefficients are understood to comprise MZVs (including rational numbers) and integer powers of $2\pi i$.

Table 1 Numbers $N(r, w)$ of indecomposable eMZVs at length r and weight w

$r \backslash w$	1	2	3	4	5	6	7	8	9	10	11	12	13	14	15	16	17	18	19	20	21	22	23
2	1		1		1		1		1		1		1		1		1		1		1		1
3		1		1		1		2		2		2		3		3		3		4		4	
4	1		1		2		3		4		5		7		8		10		12		14		16
5		1		2		4		6		9		13		17		23		30		37		47	
6	1		2		4		8		13		22		31		45		?		?		?		?
7		1		4		8		16		29		48		?		?		?		?		?	

$$N(2,w) = 1\,, \qquad N(3,w) = \left\lceil \frac{1}{6} w \right\rceil\,, \qquad N(4,w) = \left\lfloor \frac{1}{2} + \frac{1}{48}(w+5)^2 \right\rfloor\,, \tag{69}$$

which only hold for odd values of $r + w$ and remain conjectural at $r = 4$.

Across a variety of lengths and weights, the decomposition of eMZVs in terms of such bases can be downloaded from [65], this website also contains new relations in the derivation algebra.

4.3 *The Closed String at Higher Genus*

Closed-string amplitudes at one loop originate from a worldsheet of torus topology. Again, the simplest non-vanishing superstring amplitude involves four massless external states [8], and the study of its α'-expansion has a rich history as well as strong motivation from S-duality of type-IIB superstring theory [66–68]. The α'-dependence stems from the worldsheet integral in the second line of

$$\begin{aligned} \mathcal{M}_4^{\text{1-loop}}(\alpha') &= s_{12}^2 s_{23}^2 A_{\text{YM}}(1,2,3,4) \tilde{A}_{\text{YM}}(1,2,3,4) \\ &\quad \times \int_{\mathcal{F}} \frac{\mathrm{d}^2\tau}{(\Im(\tau))^5} \int_{(\mathcal{T}_\tau)^3} \mathrm{d}^2 z_2 \, \mathrm{d}^2 z_3 \, \mathrm{d}^2 z_4 \prod_{i<j}^4 e^{s_{ij} G(z_i - z_j, \tau)}\,, \end{aligned} \tag{70}$$

analogous to (61) for the open string. The integration domain $\mathcal{T}_\tau$ is specified by the complex parametrization of the torus through a parallelogram with corners $0, 1, \tau + 1, \tau$. The Green function in the exponent is defined in (62) and ensures modular invariance of the τ-integrand in (70) with $\mathcal{F}$ denoting the fundamental domain.

The integration over τ leads to branch cuts in the dependence of the closed-string amplitude (70) on the Mandelstam variables s_{ij}, as required by unitarity. A procedure to reconcile the associated logarithmic dependence on s_{ij} with the naive Taylor expansion of the integral (70) has been described in [69], see [70] for recent updates. The discontinuity structure of the open-string one-loop amplitude follows the same principles and can be traced back to the integration over t in (60).

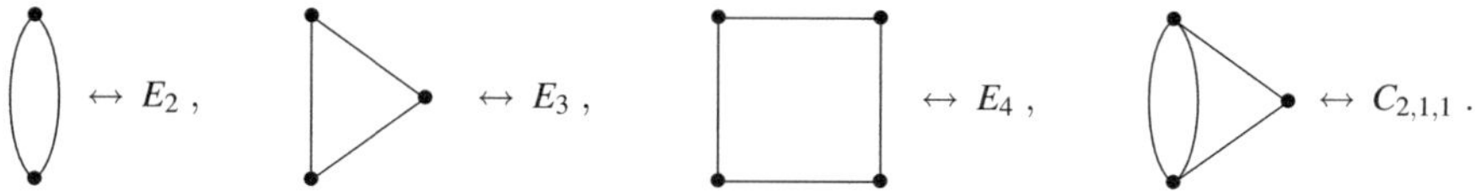

Fig. 2 Graphical organization of several sample contributions to (70): Vertices represent the punctures z_i, $i = 1, 2, 3, 4$ and edges between the vertices for z_i and z_i are associated with a factor of $G(z_i - z_j, \tau)$. The integrals over z_2, z_3, z_4 become elementary in a Fourier expansion of the Green functions and yield the modular invariant lattice sums in (71) and (72)

The systematic α'-expansion of the integrals arising from Taylor expanding $e^{s_{ij} G(z_i - z_j, \tau)}$ in (70) has been initiated in [71] and pursued in [69, 70]. In a representation of Green functions $G(z_i - z_j, \tau)$ as an edge between vertices i and j, intuitive graphical methods have been developed in these references, see [72, 73] for an extension to the five-point one-loop amplitude. Since the zero mode of the Green function decouples from (70), only one-particle irreducible graphs contribute to the α'-expansion. The simplest class of such graphs have the topology of an n-gon, see Fig. 2, and the integration over z_2, z_3, z_4 in (70) gives rise to non-holomorphic Eisenstein series

$$E_n(\tau) \equiv \sum_{\substack{k,m \in \mathbb{Z} \\ (k,m) \neq (0,0)}} \frac{(\Im(\tau))^n}{\pi^n \, |k + m\tau|^{2n}} \,, \qquad n \in \mathbb{N} \,, \quad n \geq 2 \,. \tag{71}$$

Beyond that, an infinite family of modular invariants has been classified and investigated in [70] (also see [74]), starting with the function

$$C_{2,1,1}(\tau) \equiv \sum_{\substack{k_1,k_2,m_1,m_2 \in \mathbb{Z} \\ (k_1,m_1),(k_2,m_2) \neq (0,0) \\ (k_1+k_2,m_1+m_2) \neq (0,0)}} \frac{(\Im(\tau))^4}{\pi^4 \, |k_1 + m_1\tau|^2 \, |k_2 + m_2\tau|^2 \, |k_1 + k_2 + (m_1 + m_2)\tau|^4} \tag{72}$$

associated with the two-loop graph depicted in Fig. 2.

The central rôle played by Laplace eigenvalue equations in the discussions of [70] such as

$$(\Delta - n(n-1))E_n(\tau) = 0 \,, \qquad (\Delta - 2)C_{2,1,1}(\tau) = 9E_4(\tau) - E_2^2(\tau) \tag{73}$$

with $\Delta = 4(\Im(\tau))^2 \frac{\partial}{\partial \tau} \frac{\partial}{\partial \bar{\tau}}$ bears similarities to the methods of [17, 18, 53] to compute eMZVs from their differential equation in τ. It appears promising to investigate the parallels in the expansion of the open-string integral (61) and its closed-string counterpart (70) and to exploit cross-fertilizations of the methods used. Ultimately, it is tempting to hope for a one-loop generalization of the single-valued projection which was seen in (52) and (54) to map tree-level amplitudes of the open string to those of the closed string [15].

Certainly, the above structures deserve an investigation on higher-genus surfaces on the long run. While higher-genus generalizations of eMZVs have not yet appeared in the literature, the α'-expansion of the two-loop closed-string amplitude has been pushed beyond the leading order [75, 76] and led to a connection with mathematics literature on the so-called Zhang-Kawazumi invariant [77, 78]

$$\varphi(\Omega) \equiv \int_{\Sigma^2} \mathscr{G}(z,w,\Omega) \sum_{\substack{I,J,K,L\\=1,2}} \big[2(\Im\Omega)^{-1}_{IL}(\Im\Omega)^{-1}_{JK} - (\Im\Omega)^{-1}_{IJ}(\Im\Omega)^{-1}_{KL}\big] \\ \times \omega_I(z)\overline{\omega_J(z)}\omega_K(w)\overline{\omega_L(w)}\,. \tag{74}$$

The arguments z, w of the genus-two Green function $\mathscr{G}(z,w,\Omega)$ are integrated over a genus-two Riemann surface with 2×2 period matrix Ω, and $\{\omega_I(z) : I = 1, 2\}$ is a canonically normalized basis of holomorphic one forms. The Zhang-Kawazumi invariant φ in (74) can be viewed as the simplest two-loop analogue of the non-holomorphic Eisenstein series (71) and the modular invariants for more involved graph topologies in the one-loop α'-expansion.

It is not unlikely that string-theory questions at higher order in loops and α' encourage and even inspire the development of new mathematical structures.

Acknowledgements I am very grateful to Johannes Broedel, Carlos Mafra, Nils Matthes, Stephan Stieberger and Tomohide Terasoma for collaboration on the projects on which this article is based. Moreover, I would like to thank Johannes Broedel, Nils Matthes and Federico Zerbini for valuable comments on the draft. I am indebted to the organizers of the conference "Numbers and Physics" in Madrid in September 2014 which strongly shaped the research directions leading to [16, 53] and possibly further results. I also acknowledge financial support by the European Research Council Advanced Grant No. 247252 of Michael Green.

References

1. Di Vecchia, P.: The Birth of string theory. Lect. Notes Phys. **737**, 59 (2008). arXiv:0704.0101 [hep-th]
2. Vafa, C.: Lectures on strings and dualities. [hep-th/9702201]
3. Lüst, D., Stieberger, S., Taylor, T.R.: The LHC string Hunter's companion. Nucl. Phys. B **808**, 1 (2009). arXiv:0807.3333 [hep-th]
4. Lüst, D., Schlotterer, O., Stieberger, S., Taylor, T.R.: The LHC string Hunter's companion (II): five-particle amplitudes and universal properties. Nucl. Phys. B **828**, 139 (2010). arXiv:0908.0409 [hep-th]
5. Feng, W.Z., Lüst, D., Schlotterer, O., Stieberger, S., Taylor, T.R.: Direct production of lightest regge resonances. Nucl. Phys. B **843**, 570 (2011). arXiv:1007.5254 [hep-th]
6. Arkani-Hamed, N., Dimopoulos, S., Dvali, G.R.: The Hierarchy problem and new dimensions at a millimeter. Phys. Lett. B **429**, 263 (1998). [hep-ph/9803315]
7. Antoniadis, I., Arkani-Hamed, N., Dimopoulos, S., Dvali, G.R.: New dimensions at a millimeter to a Fermi and superstrings at a TeV. Phys. Lett. B **436**, 257 (1998). [hep-ph/9804398]
8. Green, M.B., Schwarz, J.H., Brink, L.: N = 4 Yang-Mills and N = 8 Supergravity as Limits of String Theories. Nucl. Phys. B **198**, 474 (1982)

9. Kawai, H., Lewellen, D.C., Tye, S.H.H.: A relation between tree amplitudes of closed and open strings. Nucl. Phys. B **269**, 1 (1986)
10. Bjerrum-Bohr, N.E.J., Damgaard, P.H., Vanhove, P.: Minimal basis for gauge theory amplitudes. Phys. Rev. Lett. **103**, 161602 (2009). arXiv:0907.1425 [hep-th]
11. Stieberger, S.: Open & Closed vs. Pure Open String Disk Amplitudes. arXiv:0907.2211 [hep-th]
12. Drummond, J.M., Ragoucy, E.: Superstring amplitudes and the associator. JHEP **1308**, 135 (2013). arXiv:1301.0794 [hep-th]
13. Broedel, J., Schlotterer, O., Stieberger, S., Terasoma, T.: All order α'-expansion of superstring trees from the Drinfeld associator. Phys. Rev. D **89**(6), 066014 (2014) arXiv:1304.7304 [hep-th]
14. Schlotterer, O., Stieberger, S.: Motivic multiple zeta values and superstring amplitudes. J. Phys. A **46**, 475401 (2013). arXiv:1205.1516 [hep-th]
15. Stieberger, S.: Closed superstring amplitudes, single-valued multiple zeta values and the Deligne associator. J. Phys. A **47**, 155401 (2014). arXiv:1310.3259 [hep-th]
16. Broedel, J., Mafra, C.R., Matthes, N., Schlotterer, O.: Elliptic multiple zeta values and one-loop superstring amplitudes. JHEP **1507**, 112 (2015). arXiv:1412.5535 [hep-th]
17. Enriquez, B.: Analogues elliptiques des nombres multizétas. Bull. Soc. Math. France **144**, 395–427 (2016). arxiv:1301.3042
18. Enriquez, B.: Elliptic associators. Selecta Math. (N.S.) **20**, 491 (2014)
19. Berkovits, N.: Super Poincare covariant quantization of the superstring. JHEP **0004**, 018 (2000). [hep-th/0001035]
20. Mafra, C.R., Schlotterer, O., Stieberger, S.: Complete N-point superstring disk amplitude I. Pure Spinor computation. Nucl. Phys. B **873**, 419 (2013) arXiv:1106.2645 [hep-th]
21. Drinfeld, V.G.: Quasi Hopf algebras. Leningrad Math. J **1**, 1419 (1989)
22. Drinfeld, V.G.: On quasitriangular quasi-Hopf algebras and a group that is closely connected with Gal($\bar{\mathbb{Q}}/\mathbb{Q}$). Leningrad Math. J **2**(4), 829(1991)
23. Racinet, G.: Doubles mélanges des polylogarithmes multiples aux racines de l?unité. Publ. Math. Inst. Hautes Etudes Sci. **185** (2002)
24. Le, T., Murakami, J.: Kontsevich's integral for the Kauffman polynomial. Nagoya Math J. **142**, 93 (1996)
25. Terasoma, T.: Selberg Integrals and Multiple Zeta Values. Compositio Mathematica **133**, 1 (2002)
26. Barreiro, L.A., Medina, R.: 5-field terms in the open superstring effective action. JHEP **0503**, 055 (2005). [hep-th/0503182]
27. Oprisa, D., Stieberger, S.: Six gluon open superstring disk amplitude, multiple hypergeometric series and Euler-Zagier sums. hep-th/0509042
28. Stieberger, S., Taylor, T.R.: Multi-gluon scattering in open superstring theory. Phys. Rev. D **74**, 126007 (2006). [hep-th/0609175]
29. Boels, R.H.: On the field theory expansion of superstring five point amplitudes. Nucl. Phys. B **876**, 215 (2013). arXiv:1304.7918 [hep-th]
30. Puhlfuerst, G., Stieberger, S.: Differential equations, associators, and recurrences for amplitudes. Nucl. Phys. B **902**, 186 (2016). arXiv:1507.01582 [hep-th]
31. Broedel, J., Schlotterer, O., Stieberger, S. http://mzv.mpp.mpg.de
32. Mafra, C.R., Schlotterer, O., Stieberger, S.: Complete N-point superstring disk amplitude II. Amplitude and Hypergeometric Function Structure. Nucl. Phys. B **873**, 461 (2013) arXiv:1106.2646 [hep-th]
33. Broedel, J., Schlotterer, O., Stieberger, S.: Polylogarithms, multiple zeta values and superstring amplitudes. Fortsch. Phys. **61**, 812 (2013). arXiv:1304.7267 [hep-th]
34. Brown, F.: Mixed Tate motives over $\mathbb{Z}$. Ann. Math. **175**, 949 (2012)
35. Bern, Z., Carrasco, J.J.M., Johansson, H.: New relations for gauge-theory amplitudes. Phys. Rev. D **78**, 085011 (2008). arXiv:0805.3993 [hep-ph]
36. Apéry, R.: Irrationalité de $\zeta(2)$ et $\zeta(3)$. Astérisque **61**, 11 (1979)
37. Ball, K., Rivoal, T.: Irrationalité d'une infinité de valeurs de la fonction zeta aux entiers impairs. Invent. Math. **146**, 193 (2001)

38. Blumlein, J., Broadhurst, D.J., Vermaseren, J.A.M.: The multiple zeta value data mine. Comput. Phys. Commun. **181**, 582 (2010). arXiv:0907.2557 [math-ph]
39. Zagier, D.: Values of zeta functions and their applications. In: First European Congress of Mathematics, Vol. II (Paris, 1992) Birkhaeuser, Basel, p. 497 (1994)
40. Goncharov, A.B.: Galois symmetries of fundamental groupoids and noncommutative geometry. Duke Math. J. **128**, 209 (2005). [math/0208144]
41. Brown, F.: Motivic Periods and the Projective Line minus Three Points. ICM14 arXiv:1304.7267
42. Brown, F.: Single-valued motivic periods and multiple zeta values. SIGMA **2**, e25 (2014). arXiv:1309.5309 [math.NT]
43. Brown, F.: On the decomposition of motivic multiple zeta values. In: Galois-Teichmüller theory and arithmetic geometry, Adv. Stud. Pure Math. **63**, 31–58 (2012) arXiv:1102.1310
44. Bern, Z., Dixon, L.J., Perelstein, M., Rozowsky, J.S.: Multileg one loop gravity amplitudes from gauge theory. Nucl. Phys. B **546**, 423 (1999). [hep-th/9811140]
45. Bjerrum-Bohr, N.E.J., Damgaard, P.H., Sondergaard, T., Vanhove, P.: The momentum kernel of gauge and gravity theories. JHEP **1101**, 001 (2011). arXiv:1010.3933 [hep-th]
46. Schnetz, O.: Graphical functions and single-valued multiple polylogarithms. Commun. Num. Theor. Phys. **08**, 589 (2014). arXiv:1302.6445 [math.NT]
47. Stieberger, S., Taylor, T.R.: Closed string amplitudes as single-valued open string amplitudes. Nucl. Phys. B **881**, 269 (2014). arXiv:1401.1218 [hep-th]
48. Green, M.B., Schwarz, J.H.: Infinity cancellations in $SO(32)$ superstring theory. Phys. Lett. B **151**, 21 (1985)
49. Green, M.B., Schwarz, J.H.: Anomaly cancellation in supersymmetric $D = 10$ gauge theory and superstring theory. Phys. Lett. B **149**, 117 (1984)
50. Green, M.B., Schwarz, J.H.: The Hexagon Gauge anomaly in Type I superstring theory. Nucl. Phys. B **255**, 93 (1985)
51. Levin, A., Racinet, G.: Towards multiple elliptic polylogarithms (2007) [math/0703237]
52. Brown, F., Levin, A.: Multiple elliptic polylogarithms (2011)
53. Broedel, J., Matthes, N., Schlotterer, O.: Relations between elliptic multiple zeta values and a special derivation algebra. J. Phys. A **49**(15), 155203 (2016) arXiv:1507.02254 [hep-th]
54. Green, M.B., Schwarz, J.H.: Supersymmetrical dual string theory. 3. Loops and renormalization. Nucl. Phys. B **198**, 441 (1982)
55. Mafra, C.R., Schlotterer, O.: The structure of n-point one-loop open superstring amplitudes. JHEP **1408**, 099 (2014). arXiv:1203.6215 [hep-th]
56. Mafra, C.R., Schlotterer, O.: Cohomology foundations of one-loop amplitudes in pure spinor superspace. arXiv:1408.3605 [hep-th]
57. Mafra, C.R., Schlotterer, O.: Towards one-loop SYM amplitudes from the pure spinor BRST cohomology. Fortsch. Phys. **63**(2), 105 (2015) arXiv:1410.0668 [hep-th]
58. Matthes, N.: Elliptic double zeta values. J. Number Theory **171**, 227 (2017). arXiv:1509.08760 [math-NT]
59. Manin, Y.I.: Iterated integrals of modular forms and noncommutative modular symbols. Algebraic Geom. Number Theory, 565 (2006)
60. Brown, F.: Multiple modular values for $SL_2(\mathbb{Z})$ (2014)
61. Tsunogai, H.: On some derivations of Lie algebras related to Galois representations. Publ. Res. Inst. Math. Sci. **31**, 113 (1995)
62. Calaque, D., Enriquez, B., Etingof, P.: Universal KZB equations: the elliptic case. Progr. Math. **269**, 165 (2009)
63. Hain, R.: Notes on the universal elliptic KZB equation. arXiv:1309.0580 [math-NT]
64. Pollack, A.: Relations between derivations arising from modular forms
65. Broedel, J., Matthes, N., Schlotterer, O. https://tools.aei.mpg.de/emzv
66. Green, M.B., Gutperle, M.: Effects of D instantons. Nucl. Phys. B **498**, 195 (1997). [hep-th/9701093]
67. Green, M.B., Kwon, H.H., Vanhove, P.: Two loops in eleven-dimensions. Phys. Rev. D **61**, 104010 (2000) [hep-th/9910055]

68. Green, M.B., Vanhove, P.: Duality and higher derivative terms in M theory. JHEP **0601**, 093 (2006). [hep-th/0510027]
69. Green, M.B., Russo, J.G., Vanhove, P.: Low energy expansion of the four-particle genus-one amplitude in type II superstring theory. JHEP **0802**, 020 (2008). arXiv:0801.0322 [hep-th]
70. D'Hoker, E., Green, M.B., Vanhove, P.: On the modular structure of the genus-one Type II superstring low energy expansion. JHEP **1508**, 041 (2015). arXiv:1502.06698 [hep-th]
71. Green, M.B., Vanhove, P.: The Low-energy expansion of the one loop type II superstring amplitude. Phys. Rev. D **61**, 104011 (2000). [hep-th/9910056]
72. Richards, D.M.: The One-loop five-graviton amplitude and the effective action. JHEP **0810**, 042 (2008). arXiv:0807.2421 [hep-th]
73. Green, M.B., Mafra, C.R., Schlotterer, O.: Multiparticle one-loop amplitudes and S-duality in closed superstring theory. JHEP **1310**, 188 (2013). arXiv:1307.3534 [hep-th]
74. D'Hoker, E., Green, M.B., Vanhove, P.: Proof of a modular relation between 1-, 2- and 3-loop Feynman diagrams on a torus. J. Number Theory (2018) arXiv:1509.00363 [hep-th]
75. D'Hoker, E., Green, M.B.: Zhang-Kawazumi invariants and superstring amplitudes. J. Number Theory **144**, 111 (2014). arXiv:1308.4597 [hep-th]
76. D'Hoker, E., Green, M.B., Pioline, B., Russo, R.: Matching the $D^6 R^4$ interaction at two-loops. JHEP **1501**, 031 (2015). arXiv:1405.6226 [hep-th]
77. Zhang, S.W.: Gross—Schoen cycles and dualising sheaves. Inventiones Mathematicae **179**, 1 (2010). arXiv:0812.0371
78. Kawazumi, N.: Johnson's homomorphisms and the Arakelov Green function. arXiv:0801.4218 [math.GT]

Overview on Elliptic Multiple Zeta Values

Nils Matthes

Abstract We give an overview of some work on elliptic multiple zeta values. First defined by Enriquez as the coefficients of the elliptic KZB associator, elliptic multiple zeta values are also special values of multiple elliptic polylogarithms in the sense of Brown and Levin. Common to both approaches to elliptic multiple zeta values is their representation as iterated integrals on a once-punctured elliptic curve. Having compared the two approaches, we survey various recent results about the algebraic structure of elliptic multiple zeta values, as well as indicating their relation to iterated integrals of Eisenstein series, and to a special algebra of derivations.

Keywords Elliptic multiple zeta values · Elliptic KZB equation · Multiple elliptic polylogarithms

1 Introduction

The purpose of this article is twofold. The first goal is to unify two different approaches to elliptic multiple zeta values in the literature: the approach of Enriquez, using elliptic associators [21, 22] on one hand, and the approach of Brown and Levin, using multiple elliptic polylogarithms [15] on the other. Comparing the two approaches also highlights the analogy between elliptic multiple zeta values and multiple zeta values. The second goal is to collect various results on the structure of the algebra of A-elliptic multiple zeta values, which appeared in work of the author [36], as well as in joint work [5, 6].[1]

[1] See also Sect. 1.4.

N. Matthes (✉)
Fachbereich Mathematik (AZ), Universität Hamburg, Bundesstrasse 55, 20146 Hamburg, Germany
e-mail: nils.matthes@uni-hamburg.de; nils.matthes@maths.ox.ac.uk

Mathematical Institute, University of Oxford, Andrew Wiles Building, Radcliffe Observatory Quarter Woodstock Road, Oxford OX2 6GG, UK

J. I. Burgos Gil et al. (eds.), *Periods in Quantum Field Theory and Arithmetic*, Springer Proceedings in Mathematics & Statistics 314,
https://doi.org/10.1007/978-3-030-37031-2_5

1.1 Multiple Zeta Values

Multiple zeta values are defined by nested sums

$$\zeta(n_1,\ldots,n_r) = \sum_{0<k_1<\ldots<k_r} \frac{1}{k_1^{n_1}\ldots k_r^{n_r}} \in \mathbb{R}, \tag{1}$$

for integers $n_1,\ldots,n_r \geq 1$, with $n_r \geq 2$. The *weight* of $\zeta(n_1,\ldots,n_r)$ is $n_1+\cdots+n_r$, and its *depth* is r. For $r=1$, they are the values of the Riemann zeta function at positive integers. Multiple zeta values appear in mathematics and physics in many different contexts, for example in the context of periods and motives [9, 19, 26]. Also, multiple zeta values arise naturally in the study of the Knizhnik-Zamolodchikov (KZ) equation from conformal field theory [20, 29], which in turn is important in both knot theory [30], as well as quantum groups [20] and associators [23]. On the side of mathematical physics, multiple zeta values occur in the computation of Feynman integrals in renormalizable quantum field theories [4], and in the computation of superstring amplitudes at tree-level [7, 8, 40].

In all of these contexts, a central problem is to find a complete description of all $\mathbb{Q}$-linear relations between multiple zeta values. The $\mathbb{Q}$-vector subspace $\mathscr{Z} \subset \mathbb{R}$, spanned by the multiple zeta values is a $\mathbb{Q}$-subalgebra of $\mathbb{R}$. Conjecturally, the weight is a grading on $\mathscr{Z}$, while the depth defines an ascending filtration on $\mathscr{Z}$. Set

$$\mathscr{D}_d\mathscr{Z}_N = \langle \zeta(n_1,\ldots,n_r) \in \mathscr{Z} \mid n_1+\cdots+n_r = N,\ r \leq d\rangle_{\mathbb{Q}}. \tag{2}$$

The Broadhurst–Kreimer conjecture [4] gives a formula for the dimensions of the associated graded $D_{N,d} = \dim_{\mathbb{Q}} \mathrm{gr}_d^{\mathscr{D}}(\mathscr{Z}_N) := \mathscr{D}_d(\mathscr{Z}_N)/\mathscr{D}_{d-1}(\mathscr{Z}_N)$, and suggests a relation between multiple zeta values and modular forms for $\mathrm{SL}_2(\mathbb{Z})$ [11, 12, 24, 25]. Work of Zagier [45] and Goncharov [25] implies that the Broadhurst–Kreimer formula gives an upper bound in depths $d \leq 3$, for arbitrary N.

1.2 The Algebra of A-elliptic Multiple Zeta Values

In [22], Enriquez introduced an elliptic analogue of multiple zeta values, which has found recent applications in superstring theory [5]. Elliptic multiple zeta values are defined by integrating the elliptic Knizhnik–Zamolodchikov–Bernard (KZB) connection [16] along paths on a once-punctured elliptic curve $E_\tau^\times = \mathbb{C}/(\mathbb{Z}+\mathbb{Z}\tau) \setminus \{0\}$ with $\tau \in \mathbb{H}$. There are essentially two choices of paths on $E_\tau^\times$, corresponding to the two natural homology cycles α and β on the elliptic curve, giving rise to two algebras of elliptic multiple zeta values, namely A- and B-elliptic multiple zeta values. The two types of elliptic multiple zeta values are related to each other by a modular transformation formula [22] and, for simplicity, we will only consider A-elliptic multiple zeta values.

The $\mathbb{Q}$-vector space spanned by A-elliptic multiple zeta values is a $\mathbb{Q}$-subalgebra $\mathcal{E}\mathcal{Z}^{\mathrm{A}}$ of the $\mathbb{C}$-algebra $\mathcal{O}(\mathbb{H})$ of holomorphic functions on the upper half-plane $\mathbb{H}$. The analogous notions of weight and depth for multiple zeta values are the *weight* and the *length* for A-elliptic multiple zeta values. Denote by $\mathcal{E}\mathcal{Z}^{\mathrm{A}}_N \subset \mathcal{E}\mathcal{Z}^{\mathrm{A}}$ the subspace spanned by A-elliptic multiple zeta values of weight N. The analogue of the weight grading conjecture for multiple zeta values is:

Conjecture 1 *The weight defines a grading on $\mathcal{E}\mathcal{Z}^{\mathrm{A}}$, i.e.*

$$\mathcal{E}\mathcal{Z}^{\mathrm{A}} = \bigoplus_{N\geq 0} \mathcal{E}\mathcal{Z}^{\mathrm{A}}_N. \tag{3}$$

Similar to the case of multiple zeta values, the length defines an ascending filtration $\mathcal{L}_\bullet(\mathcal{E}\mathcal{Z}^{\mathrm{A}}_N)$, and we denote by $\mathrm{gr}^{\mathcal{L}}_\bullet(\mathcal{E}\mathcal{Z}^{\mathrm{A}}_N)$ the associated graded. In light of the Broadhurst–Kreimer conjecture for multiple zeta values, it is natural to pose:

Problem 1 Compute the dimension

$$D^{\mathrm{ell}}_{N,l} = \dim_{\mathbb{Q}} \mathrm{gr}^{\mathcal{L}}_l(\mathcal{E}\mathcal{Z}^{\mathrm{A}}_N) \tag{4}$$

of the space of A-elliptic multiple zeta values of weight N and length l, for all $N, l \geq 0$.

A first step towards a solution to this problem is taken in [36].

Theorem 1 *We have*

$$D^{\mathrm{ell}}_{N,0} = \delta_{N,0}, \quad D^{\mathrm{ell}}_{N,1} = \begin{cases} 1 & N \geq 2 \text{ even} \\ 0 & \text{else,} \end{cases} \tag{5}$$

and

$$D^{\mathrm{ell}}_{N,2} = \begin{cases} 0 & N \text{ even} \\ \lfloor \frac{N}{3} \rfloor + 1 & N \text{odd.} \end{cases} \tag{6}$$

Also, we have the weight grading in lengths 0, 1 *and* 2*, i.e., for* $l = 0, 1, 2$, *we have*

$$\mathcal{L}_l(\mathcal{E}\mathcal{Z}^{\mathrm{A}}) = \bigoplus_{N\geq 0} \mathcal{L}_l(\mathcal{E}\mathcal{Z}^{\mathrm{A}}_N). \tag{7}$$

The proof uses a linear independence result for a certain family of elliptic multiple zeta values, as well some basic invariant theory. For more details, see Sect. 4.3, and also [6], in particular Sect. 4.5 for related work in higher lengths.

1.3 Structure of the Article

In Sect. 2, we give a very brief overview on multiple zeta values, multiple polylogarithms and the Drinfeld associator. This section is primarily intended to make the analogy with elliptic multiple zeta values transparent. In Sect. 3, we introduce elliptic multiple zeta values, comparing the works of Enriquez and of Brown and Levin. The main ingredient in both approaches, namely the universal elliptic KZB equation [16, 32], is introduced in Sect. 3.1.

In Sect. 4, we begin the study of A-elliptic multiple zeta values. All of the results of this section can be found in [5, 6, 36], and are primarily concerned with explicit algebraic relations between A-elliptic multiple zeta values. Finally, in Sect. 5, we draw the connection between A-elliptic multiple zeta values, iterated integrals of Eisenstein series, and a special algebra of derivations $\mathfrak{u}^{\mathrm{geom}}$ on a free Lie algebra on two generators [13, 16, 32, 37].

1.4 Note Added in Print

This article was written in 2015 and therefore does not reflect more recent progress on the subject of elliptic multiple zeta values. See in particular [33] for a study of double shuffle relations in the context of elliptic multiple zeta values and the author's PhD thesis [35] for a more detailed discussion of elliptic multiple zeta values.

2 Multiple Polylogarithms and the Drinfeld Associator

Multiple zeta values can be defined either as special values of multiple polylogarithms, or equivalently as the coefficients of Drinfeld's associator. A common theme of the two definitions is the theory of iterated integrals on $\mathbb{P}^1 \setminus \{0, 1, \infty\}$. Of particular importance is the concept of homotopy invariance of iterated integrals, because of the connection to the unipotent fundamental group of $\mathbb{P}^1 \setminus \{0, 1, \infty\}$.

2.1 Iterated Integrals

We begin with a brief reminder on iterated integrals. More background can be found for example in [17, 27].

Let M be a smooth manifold over k, where k denotes either the field of real or complex numbers, $\gamma : [0, 1] \to M$ a piecewise smooth path, and $(\omega_1, \ldots, \omega_r)$ an r-tuple of smooth differential one-forms on M. Write $f_i(t)dt$ for the pullback $\gamma^*(\omega_i)$ of ω_i along γ. Then we define

$$\int_\gamma \omega_1 \dots \omega_r = \int_0^1 f_1(t_1) \int_0^{t_1} f_2(t_2) \dots \int_0^{t_{r-1}} f_r(t_r) dt_r \dots dt_2 dt_1, \tag{8}$$

and call this integral an *iterated integral*. If $r = 0$, then we define $\int_\gamma \equiv 1$. Many properties of iterated integrals are known, for example the *shuffle product formula* [39]

$$\int_\gamma \omega_1 \dots \omega_r \int_\gamma \omega_{r+1} \dots \omega_{r+s} = \sum_{\sigma \in \Sigma_{r,s}} \int_\gamma \omega_{\sigma^{-1}(1)} \dots \omega_{\sigma^{-1}(r+s)}, \tag{9}$$

where $\Sigma_{r,s}$ denotes the set of all (r, s)-shuffles, i.e. $\Sigma_{r,s}$ consists of all permutations on the set $\{1, \dots, r+s\}$, which preserve the order of the first r elements and the order of the last s elements.

Now let PM denote the set of all piecewise smooth paths on M. Fixing an r-tuple $(\omega_1, \dots, \omega_r)$ of differential forms as above, we obtain a function

$$\begin{aligned} \int \omega_1 \dots \omega_r : PM &\to k \\ \gamma &\mapsto \int_\gamma \omega_1 \dots \omega_r. \end{aligned} \tag{10}$$

We call $\int \omega_1 \dots \omega_r$ a *homotopy invariant iterated integral*, if (10) depends only on the homotopy class of γ, i.e., if for every pair $\gamma_0, \gamma_1 \in PM$ of paths, which are homotopic relative to their extremal points, we have $\int_{\gamma_0} \omega_1 \dots \omega_r = \int_{\gamma_1} \omega_1 \dots \omega_r$. The importance of homotopy invariant iterated integrals can be seen from Chen's π_1 de Rham theorem: the affine ring of functions on the unipotent completion of the fundamental group of a smooth manifold M is given by the ring of all homotopy invariant iterated integrals on M [17, 27].

Finally, the following shorthand will come in useful later on. Denote by $\mathbb{C}\langle\langle x, y\rangle\rangle$ the $\mathbb{C}$-algebra of formal power series in non-commuting variables x, y, equipped with the concatenation product and let $I \subset \mathbb{C}\langle\langle x, y\rangle\rangle$ be the augmentation ideal, i.e. the two-sided ideal generated by x and y. For a differential one-form ω with values in I, we set

$$\exp\left[\int_\gamma \omega\right] = 1 + \sum_{k \geq 1} \int_\gamma \omega^k \in \mathbb{C}\langle\langle x, y\rangle\rangle, \tag{11}$$

where $\omega^k := \underbrace{\omega \dots \omega}_{k-times}$.

2.2 Multiple Polylogarithms

From now on, we work over the field $\mathbb{C}$ of complex numbers.

2.2.1 Sum Representation

For integers $n_1, \ldots, n_r \geq 1$, define the *multiple polylogarithm* to be

$$\mathrm{Li}_{n_1,\ldots,n_r}(z) = \sum_{0<k_1<\ldots<k_r} \frac{z^{k_r}}{k_1^{n_1}\ldots k_r^{n_r}}. \tag{12}$$

The sum converges absolutely and locally uniformly for all $z \in \mathbb{C}$ with $|z| < 1$, thus defines an analytic function on the open unit disk. Moreover, if $n_r \geq 2$, then (12) also converges for $z = 1$, and in that case, $\mathrm{Li}_{n_1,\ldots,n_r}(1)$ is equal to the *multiple zeta value* $\zeta(n_1, \ldots, n_r)$ (1).

2.2.2 Integral Representation

It follows directly from (12) that multiple polylogarithms satisfy the differential equation

$$\frac{\partial}{\partial z}\mathrm{Li}_{n_1,\ldots,n_r}(z) = \begin{cases} \frac{1}{z}\mathrm{Li}_{n_1,\ldots,n_r-1}(z) & n_r \geq 2 \\ \frac{1}{1-z}\mathrm{Li}_{n_1,\ldots,n_{r-1}}(z) & n_r = 1. \end{cases} \tag{13}$$

As a consequence multiple polylogarithms can be represented by iterated integrals. Consider the holomorphic differential one-forms $\omega_0 = \frac{dz}{z}$ and $\omega_1 = \frac{dz}{1-z}$ on $\mathbb{P}^1 \setminus \{0, 1, \infty\}$. Then for $|z| < 1$, one has

$$\mathrm{Li}_{n_1,\ldots,n_r}(z) = \int_0^z \omega_0^{n_r-1}\omega_1 \ldots \omega_0^{n_1-1}\omega_1, \tag{14}$$

where the path of integration is the straight line path from 0 to z. Note that although ω_0 has a pole at 0, the iterated integral is still well-defined, since the integration starts with ω_1, which is analytic at $z = 0$. Moreover, if $n_r \geq 2$, then (14) gives a representation of multiple zeta values as iterated integrals, namely

$$\zeta(n_1, \ldots, n_r) = \int_0^1 \omega_0^{n_r-1}\omega_1 \ldots \omega_0^{n_1-1}\omega_1. \tag{15}$$

The integral representation shows also that multiple polylogarithms can be extended to multi-valued functions on $\mathbb{P}^1 \setminus \{0, 1, \infty\}$. For many more properties of multiple polylogarithms, see for example the lecture notes [10].

2.3 *The Drinfeld Associator*

Now we let $\mathbb{C}'$ be the set of complex numbers with the two half-lines $(-\infty, 0]$ and $[1, \infty)$ removed. There exists a unique solution $G_0 : \mathbb{C}' \to \mathbb{C}\langle\langle x, y\rangle\rangle$ to the *Knizhnik-Zamolodchikov equation* (*KZ-equation* for short)

$$\frac{\partial}{\partial z} g(z) = \left(\frac{x}{z} + \frac{y}{z-1}\right) \cdot g(z), \tag{16}$$

which satisfies $G_0(z) \sim z^x$ as $z \to 0$ on $\mathbb{C}'$ [20]. Similarly, there exists a unique solution $G_1 : \mathbb{C}' \to \mathbb{C}\langle\langle x, y\rangle\rangle$ to the KZ-equation such that $G_1(z) \sim (1-z)^y$ as $z \to 1$ on $\mathbb{C}'$. The quotient

$$\Phi(x, y) = G_1^{-1}(z) G_0(z) \in \mathbb{C}\langle\langle x, y\rangle\rangle, \tag{17}$$

which does not depend on z, is called the *Drinfeld associator*. An explicit formula in terms of iterated integrals can be given as follows (cf. [41], Sect. 2). Let Ω_{KZ} denote the $\mathbb{C}\langle\langle x, y\rangle\rangle$-valued one-form given by

$$\Omega_{KZ} = \omega_0 \cdot x - \omega_1 \cdot y. \tag{18}$$

Then

$$\Phi(x, y) = \lim_{t \to 0} e^{-\log(t) y} \exp\left[\int_t^{1-t} \Omega_{KZ}\right] e^{\log(t) x}, \tag{19}$$

where the iterated integration is performed along the straight-line path from t to $1-t$, and exp was defined in (11). The coefficient of the word $x^{n_r - 1} y \dots x^{n_1 - 1} y$ in (19) is given by $(-1)^r \zeta(n_1, \dots, n_r)$, as can be seen from the integral representation of multiple zeta values (15).

3 Multiple Elliptic Polylogarithms and the Elliptic KZB Associator

Both multiple polylogarithms and the Drinfeld associator possess elliptic analogues. In the case of polylogarithms, one obtains *(multiple) elliptic polylogarithms* as functions on the once-punctured elliptic curve $E_\tau^\times \cong (\mathbb{C}^*/q^{\mathbb{Z}}) \setminus \{1\}$ by averaging the ordinary polylogarithms along the spiral $q^{\mathbb{Z}}$, where $q = e^{2\pi i \tau}$ with τ in the upper half-plane. This definition of elliptic polylogarithms was pioneered by Bloch [3] in the case of the single-valued dilogarithm, and later extended by Zagier [43] to all single-valued polylogarithms. After that, Levin, [31], following earlier joint work with Beilinson, introduced (multivalued) elliptic polylogarithms. Finally, Brown and Levin treated the case of multi-valued multiple polylogarithms [15]. Furthermore, they also established a representation of multiple elliptic polylogarithms as homotopy invariant iterated integrals, thus paralleling the dichotomy between the sum and the integral representation of multiple polylogarithms (cf. Sect. 2.2). While the integrands for multiple zeta values were given by the Knizhnik-Zamolodchikov equation, the integrands of the multiple elliptic polylogarithms are constructed from the *elliptic Knizhnik–Zamolodchikov–Bernard* (KZB) differential equation [16, 32].

On the other hand, Enriquez introduced the notion of an *elliptic associator* in the context of an elliptic version of Drinfeld's version of Grothendieck-Teichmüller theory [21]. Furthermore, he constructs an explicit elliptic associator $(\Phi, A(\tau), B(\tau))$ from the regularized monodromy of the KZB equation. Here, Φ denotes the Drinfeld associator, and $A(\tau)$, $B(\tau)$ are certain group-like elements of $\exp(\widehat{\mathfrak{t}_{1,2}})$ (where $\widehat{\mathfrak{t}}$ denotes the completion of the elliptic braid Lie algebra on two strands), depending holomorphically on the coordinate τ in the upper half-plane. The group-likeness implies that the coefficients of the series $A(\tau)$ and $B(\tau)$ give rise to $\mathbb{Q}$-algebras $\mathcal{E}\mathcal{Z}^{\mathrm{A}}$ and $\mathcal{E}\mathcal{Z}^{\mathrm{B}}$, which are called the algebra of *A-elliptic multiple zeta values* and *B-elliptic multiple zeta values* respectively.

3.1 An Elliptic Analogue of the KZ-Equation

The starting point for the construction of both multiple elliptic polylogarithms and the elliptic KZB associator is an elliptic analogue of the KZ-equation (16), namely the *universal elliptic KZB equation* [16, 32]. It is defined by a connection on a certain vector bundle over the universal elliptic curve. We will consider only its restriction to a fiber of the universal elliptic curve, and moreover work with a certain real analytic trivialization of this restricted bundle, which was introduced in [15].

3.1.1 A Meromorphic Jacobi Form

Let $\mathbb{H} = \{z \in \mathbb{C} \mid \Im(z) > 0\}$ denote the upper-half plane, and fix a point $\tau \in \mathbb{H}$. We consider the following Jacobi theta function

$$\theta_\tau(\xi) = \sum_{n\in\mathbb{Z}} (-1)^n q^{\frac{1}{2}(n+\frac{1}{2})^2} e^{(n+\frac{1}{2})2\pi i\xi}, \quad q = e^{2\pi i\tau}. \tag{20}$$

Definition 1 We define a meromorphic function on $\mathbb{C} \times \mathbb{C}$ by the formula

$$F_\tau(\xi, \alpha) = \frac{\theta'_\tau(0)\theta_\tau(\xi + \alpha)}{\theta_\tau(\xi)\theta_\tau(\alpha)}. \tag{21}$$

Following [15], we call $F_\tau(\xi, \alpha)$ the *Kronecker function*, [42]. The terminology for (21) varies in the literature. In [44], it is called "a meromorphic Jacobi form", while in [1], it is referred to as "the Kronecker theta function".

Since F_τ is meromorphic, it has a Laurent expansion in α. In what follows, we will consider $F_\tau(\xi, \alpha)$ as a formal series in α, whose coefficients are functions in ξ (with τ being fixed). Note that the Kronecker function has simple poles for $\xi \in \mathbb{Z} + \mathbb{Z}\tau$, and is holomorphic outside of that lattice.

3.1.2 Differential Forms on a Punctured Elliptic Curve

Now consider the complex elliptic curve $E_\tau = \mathbb{C}/(\mathbb{Z} + \mathbb{Z}\tau)$ with canonical coordinate $\xi = s + r\tau$, where $r, s \in \mathbb{R}$. Write $E_\tau^\times$ for E_τ with the origin 0 removed. Since the Kronecker function is quasi-periodic

$$F_\tau(\xi + 1, \alpha) = F_\tau(\xi, \alpha), \quad F_\tau(\xi + \tau, w) = e^{-2\pi i \frac{\Im(\xi)}{\Im(\tau)}\alpha} F_\tau(\xi, \alpha), \tag{22}$$

(cf. [15], Proposition 5), the smooth differential one-form (where α is viewed as a formal variable)

$$\Omega_\tau(\xi, \alpha) = e^{2\pi i r \alpha} F_\tau(\xi, \alpha) d\xi \tag{23}$$

descends to $E_\tau^\times$.

Definition 2 (*Brown-Levin*) Define a family $\{\omega^{(k)}\}_{k \geq 0}$ of real analytic, differential one-forms on $E_\tau^\times$ by

$$\Omega_\tau(\xi, \alpha) = \sum_{k \geq 0} \omega^{(k)} \alpha^{k-1}. \tag{24}$$

3.1.3 The Elliptic KZB Equation

With the differential form $\Omega_\tau(\xi, \alpha)$ in hand, we can now write down the elliptic KZB equation, following [15], Proposition 23. It is defined by a differential form J_{KZB}, which takes values in the lower central series completion $\widehat{\mathbb{L}}_{\mathbb{C}}(x, y)$ of the free Lie algebra $\mathbb{L}_{\mathbb{C}}(x, y)$ over $\mathbb{C}$, which we will now describe. Let $U \subset \mathbb{C}$ be an open subset.

Definition 3 For a function $g : U \to \mathbb{C}$, the *elliptic KZB equation* is the differential equation

$$dg(\xi) = J_{\mathrm{KZB}} \cdot g(\xi), \tag{25}$$

with

$$J_{\mathrm{KZB}} = -2\pi i dr \cdot x + \mathrm{ad}(x)\Omega_\tau(\xi, \mathrm{ad}(x))(y). \tag{26}$$

An important property of the form ω_{KZB} is that it satisfies the flatness condition

$$dJ_{\mathrm{KZB}} + J_{\mathrm{KZB}} \wedge J_{\mathrm{KZB}} = 0, \tag{27}$$

which is straightforward to verify.

3.2 The Elliptic KZB Associator

Having described an elliptic analogue of multiple polylogarithms, we now turn to an elliptic analogue of the Drinfeld associator, the *elliptic KZB associator* [21]. Again, fix $\tau \in \mathbb{H}$. The starting point for the definition of the elliptic KZB associator is also the elliptic KZB equation

$$dg(\xi) = -J_{\mathrm{KZB}} \cdot g(\xi), \tag{28}$$

here with an additional minus sign, where $g : S \to \mathbb{C}\langle\langle x, y\rangle\rangle$ is defined on the simply connected domain $S = \{u + vi \in \mathbb{C} \mid u, v \in (0, 1)\} \subset \mathbb{C}$. This equation has a unique solution G defined on S, which satisfies $G(\xi) \sim (-2\pi i \xi)^{-\mathrm{ad}(x)(y)}$, where the branch of the logarithm is chosen such that $\log(\pm i) = \pm\frac{\pi i}{2}$. The following definition can be found in [21], Sect. 5.

Definition 4 (*Enriquez*) The *elliptic KZB associator* is the triple $(\Phi, A(\tau), B(\tau))$, where Φ denotes the Drinfeld associator and $A(\tau), B(\tau) \in \mathbb{C}\langle\langle x, y\rangle\rangle$ are formal series, defined by the formulae

$$A(\tau) = G(\xi)^{-1}G(\xi + 1), \quad B(\tau) = G(\xi)^{-1}G(\xi + \tau). \tag{29}$$

By the same reasoning as for the Drinfeld associator, neither $A(\tau)$ nor $B(\tau)$ depend on the variable ξ. Since $G(\xi)$ can be expressed as an iterated integral

$$G(\xi) = \lim_{\varepsilon \to 0} \exp\left[\int_{\xi}^{\varepsilon} J_{\mathrm{KZB}}\right](-2\pi i \varepsilon)^{-\mathrm{ad}(x)(y)}, \tag{30}$$

one obtains explicit formulae for $A(\tau)$ and $B(\tau)$

$$A(\tau) = e^{\pi i \mathrm{ad}(x)(y)} \lim_{\varepsilon \to 0} (-2\pi i \varepsilon)^{\mathrm{ad}(x)(y)} \exp\left[\int_{\varepsilon}^{1-\varepsilon} -J_{\mathrm{KZB}}^{op}\right](-2\pi i \varepsilon)^{-\mathrm{ad}(x)(y)}, \tag{31}$$

$$B(\tau) = e^{-\pi i \mathrm{ad}(x)(y)} \lim_{\varepsilon \to 0} (-2\pi i \varepsilon)^{\mathrm{ad}(x)(y)} \exp\left[\int_{\varepsilon\tau}^{(1-\varepsilon)\tau} -J_{\mathrm{KZB}}^{op}\right](-2\pi i \varepsilon)^{-\mathrm{ad}(x)(y)}. \tag{32}$$

where J_{KZB} is now considered as a $\mathbb{C}\langle\langle x, y\rangle\rangle^{op}$-valued formal differential form, i.e. multiplication of words in x, y has been reversed.

The series $A(\tau)$ and $B(\tau)$ have a number of interesting properties, first found by Enriquez. They satisfy a number of functional equations, which relate them to elliptic braid Lie algebras, and as $\tau \to i\infty$, they degenerate to the Drinfeld associator. For all these properties and much more on elliptic associators in general, see [21].

3.3 Multiple Elliptic Polylogarithms

Multiple elliptic polylogarithms give a second perspective on elliptic multiple zeta values.

3.3.1 Series Representation

Roughly speaking, multiple elliptic polylogarithms are obtained from multiple polylogarithms by averaging along the spiral $q^{\mathbb{Z}}$, where $q = e^{2\pi i\tau}$. This uses the Jacobi uniformization $E_\tau \cong \mathbb{C}^*/q^{\mathbb{Z}}$ given by $\xi \mapsto z = e^{2\pi i\xi}$, and yields functions of the schematic form

$$\sum_{n\in\mathbb{Z}} \mathrm{Li}_{n_1,\ldots,n_r}(q^n z). \tag{33}$$

In order to make this approach rigorous, one has to employ a delicate regularization process, which in particular requires extensive knowledge about the singularities of (multi-variable) multiple polylogarithms. We cannot give the technical details here, and refer instead to Sects. 6 and 7 of [15].

3.3.2 Integral Representation

In [15], one also finds an approach to multiple elliptic polylogarithms via iterated integrals. Consider the formal series of iterated integrals on $E_\tau^\times$

$$T = \exp\left[\int J_{\mathrm{KZB}}\right] \in \mathrm{Hom}(PE_\tau^\times, \widehat{\mathbb{L}}_{\mathbb{C}}(x, y)), \tag{34}$$

where $\mathrm{Hom}(PE_\tau^\times, \mathbb{C})$ denotes the set of complex-valued functions on the set $PE_\tau^\times$ of piecewise smooth paths on $E_\tau^\times$, and the tensor product is completed. The flatness of (27) implies that in fact every coefficient of T, viewed as a power series in x, y, is a homotopy invariant iterated integral. One of the main results of [15] is:

Theorem 2 (Brown-Levin) *Every homotopy invariant iterated integral on $E_\tau^\times$ arises as a $\mathbb{C}$-linear combination of coefficients of T. Moreover, the $\mathbb{Q}$-vector space spanned by the coefficients of T is equal to the $\mathbb{Q}$-vector space spanned by the multiple elliptic polylogarithms, as subspaces of the $\mathbb{Q}$-vector space of multi-valued functions on $E_\tau^\times$.*

In this way, one obtains a complete description of multiple elliptic polylogarithms in terms of homotopy invariant iterated integrals on a once-punctured complex elliptic curve, which is the elliptic analogue of the iterated integral representation of the classical multiple polylogarithms (14).

3.3.3 Elliptic Associators via Elliptic Polylogarithms

The relation between [15] and the work of Enriquez is established by the fact that the series $A(\tau)$ and $B(\tau)$ are restrictions of (34) to certain natural paths on $E_\tau^\times$. This follows from the explicit description of the KZB-associator (31), (32). The construction of the elliptic KZB-associator using the function T has the advantage that elliptic multiple zeta values are exhibited as restrictions of homotopy invariant iterated integrals, which is a necessary prerequisite for any sort of relation to unipotent fundamental groups [18]. It also tightens the analogy with multiple zeta values, which are given by homotopy invariant iterated integrals on $\mathbb{P}^1 \setminus \{0, 1, \infty\}$.

4 The Algebra of A-elliptic Multiple Zeta Values

In the last section, we have given a very short introduction to multiple elliptic polylogarithms and the elliptic KZB associator, indicating how the two objects are related. On the other hand, in Sect. 2, we saw that multiple zeta values can be defined as special values of multiple polylogarithms, or equivalently as coefficients of the Drinfeld associator. In light of this fact we are lead to a definition of elliptic analogues of multiple zeta values as special values of multiple elliptic polylogarithms or as coefficients of the series $A(\tau)$ and $B(\tau)$. However, we proceed slightly differently, first giving a direct definition of A-elliptic multiple zeta values as iterated integrals $I^{\mathrm{A}}(n_1, \ldots, n_r; \tau)$ over the differential forms $\omega^{(k)}$ (24). That this $\mathbb{Q}$-vector space of iterated integrals is equal to the $\mathbb{Q}$-vector space spanned by the coefficients of (a slight variation of) the series $A(\tau)$ was shown in [36].

4.1 Definition of A-elliptic Multiple Zeta Values

Definition 5 We define $I^{\mathrm{A}}(n_1, \ldots, n_r; \tau)$ to be the shuffle-regularized[2] iterated integral

$$I^{\mathrm{A}}(n_1, \ldots, n_r; \tau) = \int_\alpha \omega^{(n_r)} \ldots \omega^{(n_1)}. \tag{35}$$

We call $I^{\mathrm{A}}(n_1, \ldots, n_r; \tau)$ an *A-elliptic multiple zeta value*. It is said to have *weight* $n_1 + \cdots + n_r$ and *length* r. The $\mathbb{Q}$-vector space $\mathscr{E}\mathscr{Z}^{\mathrm{A}}$ spanned by the $I^{\mathrm{A}}(n_1, \ldots, n_r; \tau)$ will be called the space of *A-elliptic multiple zeta values*.

By the shuffle product formula for iterated integrals, it follows that $\mathscr{E}\mathscr{Z}^{\mathrm{A}}$ is a $\mathbb{Q}$-subalgebra of the $\mathbb{C}$-algebra of holomorphic functions $\mathscr{O}(\mathbb{H})$ on $\mathbb{H}$. In particular, the

[2] Iterated integrals starting or ending with $\omega^{(1)}$ diverge, and need to be regularized, such that the shuffle product formula remains valid. See [22, 36] for details on this regularization procedure.

rational number $1 \in \mathbb{Q}$ is an A-elliptic multiple zeta value of weight and length equal to zero, in accordance with our conventions for the empty iterated integral.

It is known that the algebra of multiple zeta values is also the $\mathbb{Q}$-vector space spanned by the coefficients of the Drinfeld associator. There is an analogous result for A-elliptic multiple zeta values. Write

$$\widetilde{A}(\tau) = e^{-\pi i \mathrm{ad}(x)(y)} A(\tau) = \sum_{w \in \langle x, y \rangle} \widetilde{A}(\tau)_w \cdot w. \tag{36}$$

Note that the constant pre-factor $e^{-\pi i \mathrm{ad}(x)(y)}$, which is multiplied to $A(\tau)$, is an artifact of the difference between the regularization procedure for the KZB-associator, and the regularization of the iterated integrals (34).

Proposition 1 *We have an equality of vector spaces*

$$\mathcal{E}\mathcal{Z}^{\mathrm{A}} = \langle \widetilde{A}(\tau)_w \mid w \in \langle x, y \rangle \rangle_{\mathbb{Q}}. \tag{37}$$

Proof See [36], Proposition 2.3.

4.2 Relations Between A-elliptic Multiple Zeta Values

Recall that A-elliptic multiple zeta values are defined as iterated integrals over the differential forms $\omega^{(k)}$, which in turn are the coefficients of the Kronecker differential form

$$\sum_{k \geq 0} \omega^{(k)} \alpha^{k-1} = \Omega_\tau(\xi, \alpha) = e^{2\pi i r \alpha} \frac{\theta'_\tau(0)\theta_\tau(\xi + \alpha)}{\theta_\tau(\xi)\theta_\tau(\alpha)} d\xi, \tag{38}$$

where $\xi = s + r\tau$.

4.2.1 Shuffle Relations

From (9), one deduces a shuffle product formula for A-elliptic multiple zeta values, which is conveniently expressed using the generating series

$$\mathcal{I}^{\mathrm{A}}(X_1, X_2, \ldots, X_r; \tau) = \sum_{n_1, \ldots, n_r = 0}^{\infty} I^{\mathrm{A}}(n_1, \ldots, n_r; \tau) X_1^{n_1 - 1} X_2^{n_2 - 1} \ldots X_r^{n_r - 1}, \tag{39}$$

of A-elliptic multiple zeta values of length r, for $r \geq 0$. Explicitly, we have

$$\mathscr{I}^{\mathrm{A}}(X_1,\ldots,X_l;\tau)\mathscr{I}^{\mathrm{A}}(X_{l+1},\ldots,X_n;\tau)=\sum_{\sigma\in\Sigma_{l,n-l}}\mathscr{I}^{\mathrm{A}}(X_{\sigma^{-1}(1)},\ldots,X_{\sigma^{-1}(n)};\tau),\tag{40}$$

where $\Sigma_{l,n-l}$ denotes the set of $(l,n-l)$-shuffles. For example, in small lengths we get

$$\mathscr{I}^{\mathrm{A}}(X_1;\tau)\mathscr{I}^{\mathrm{A}}(X_2;\tau)=\mathscr{I}^{\mathrm{A}}(X_1,X_2;\tau)+\mathscr{I}^{\mathrm{A}}(X_2,X_1;\tau),\tag{41}$$

$$\begin{aligned}\mathscr{I}^{\mathrm{A}}(X_1;\tau)\mathscr{I}^{\mathrm{A}}(X_2,X_3;\tau)=\mathscr{I}^{\mathrm{A}}(X_1,X_2,X_3;\tau)&+\mathscr{I}^{\mathrm{A}}(X_2,X_1,X_3;\tau)\\&+\mathscr{I}^{\mathrm{A}}(X_2,X_3,X_1;\tau).\end{aligned}\tag{42}$$

4.2.2 Reflection Relations

Further relations between A-elliptic multiple zeta values can be inferred from functional equations for Ω_τ, for example, the symmetry equation (cf. [44], Theorem.(i) in Sect. 3)

$$\Omega_\tau(-\xi,-\alpha)=\Omega_\tau(\xi,\alpha).\tag{43}$$

Writing $\omega^{(k)}=f^{(k)}d\xi$, for some real-analytic function $f^{(k)}:E_\tau^\times\to\mathbb{C}$, (43) implies

$$f^{(k)}(-\xi)=(-1)^k f^{(k)}(\xi).\tag{44}$$

Using (44) and the reversal of paths formula for iterated integrals, one obtains the *reflection relation* for A-elliptic multiple zeta values (cf. [6], Eq. (2.13))

$$I^{\mathrm{A}}(n_1,\ldots,n_r;\tau)=(-1)^{n_1+\cdots+n_r}I^{\mathrm{A}}(n_r,\ldots,n_1;\tau),\tag{45}$$

which, on the level of generating series, simply becomes the functional equation

$$\mathscr{I}^{\mathrm{A}}(X_1,\ldots,X_r;\tau)=\mathscr{I}^{\mathrm{A}}(-X_r,\ldots,-X_1;\tau).\tag{46}$$

4.2.3 Fay Relations

The second, more interesting, functional equation satisfied by Ω_τ is the *Fay identity* (cf. [15], Proposition 4)

$$\begin{aligned}\Omega_\tau(\xi_1,\alpha_1)\wedge\Omega_\tau(\xi_2,\alpha_2)=\;&\Omega_\tau(\xi_1-\xi_2,\alpha_1)\wedge\Omega_\tau(\xi_2,\alpha_1+\alpha_2)\\&+\Omega_\tau(\xi_2-\xi_1,\alpha_2)\wedge\Omega_\tau(\xi_1,\alpha_1+\alpha_2).\end{aligned}\tag{47}$$

The Fay identity for $\Omega_\tau(\xi,\alpha)$ implies quadratic relations for the functions $f^{(k)}$, namely

$$\begin{aligned} f^{(m)}(\xi_1)f^{(n)}(\xi_2) = & -(-1)^n f^{(m+n)}(\xi_1-\xi_2) \\ & + \sum_{r=0}^{n} \binom{m+r-1}{m-1} f^{(n-r)}(\xi_2-\xi_1) f^{(m+r)}(\xi_1) \\ & + \sum_{r=0}^{m} \binom{n+r-1}{n-1} f^{(m-r)}(\xi_1-\xi_2) f^{(n+r)}(\xi_2). \end{aligned} \tag{48}$$

The Fay identity also yields $\mathbb{Q}$-linear relations between A-elliptic multiple zeta values, which are again most conveniently expressed as functional equations for the generating series $\mathcal{I}^{\mathrm{A}}$. In length two, we have (cf. [36], Eq. (2.40))

$$\mathcal{I}^{\mathrm{A}}(X, Y; \tau) + \mathcal{I}^{\mathrm{A}}(X+Y, -Y; \tau) + \mathcal{I}^{\mathrm{A}}(-X-Y, X; \tau) = 3\zeta(2). \tag{49}$$

Observe that the right hand side of (49) corresponds to a subset of divergent A-elliptic multiple zeta values, such as $I^{\mathrm{A}}(1, 1; \tau)$, which require a regularization procedure. While A-elliptic multiple zeta values have been regularized in a way compatible with the shuffle product, one can ask whether there is a "Fay-regularization" for a priori divergent A-elliptic multiple zeta values. In a similar vein, note that the stuffle regularization for multiple zeta values differs from the shuffle regularization, and that there is an explicit formula for the difference between the two regularizations (cf. [38], Proposition 2.4.14). For A-elliptic multiple zeta values, for example, the first equation in (49) suggests the convention

$$\mathcal{I}_*^{\mathrm{A}}(X, Y; \tau) := \mathcal{I}^{\mathrm{A}}(X, Y; \tau) - \zeta(2) = \mathcal{I}^{\mathrm{A}}(X, Y; \tau) + \frac{1}{2} I^{\mathrm{A}}(2; \tau), \tag{50}$$

such that $\mathcal{I}_*^{\mathrm{A}}(X, Y; \tau) + \mathcal{I}_*^{\mathrm{A}}(X+Y, -Y; \tau) + \mathcal{I}_*^{\mathrm{A}}(-X-Y, X; \tau) = 0$. See also [6], Sect. 2.2, for Fay relations in higher lengths.

4.3 *The Dimension of the Space of A-elliptic Multiple Zeta Values*

Now that we have seen that there are many relations between A-elliptic multiple zeta values, it is a natural next step to try and count these relations, in order to get upper bounds on the dimensions of the space of A-elliptic multiple zeta values. To arrive at a more precise formulation, consider for non-negative integers N, l the $\mathbb{Q}$-vector subspace

$$\mathcal{L}_l(\mathcal{E}\mathcal{Z}_N^{\mathrm{A}}) = \langle I^{\mathrm{A}}(n_1, \ldots, n_r; \tau) \in \mathcal{E}\mathcal{Z}^{\mathrm{A}} \mid n_1 + \cdots + n_r = N,\ r \leq l \rangle_{\mathbb{Q}} \subset \mathcal{E}\mathcal{Z}^{\mathrm{A}}, \tag{51}$$

spanned by A-elliptic multiple zeta values of weight N and length at most l. From (40), it is clear that the length filtration is compatible with the algebra structure of

$\mathscr{E}\mathscr{Z}^{\mathrm{A}}$, and that therefore $\mathscr{E}\mathscr{Z}^{\mathrm{A}}$ is a filtered $\mathbb{Q}$-algebra. Denote by $\mathrm{gr}_l^{\mathscr{L}}(\mathscr{E}\mathscr{Z}_N^{\mathrm{A}}) = \mathscr{L}_l(\mathscr{E}\mathscr{Z}_N^{\mathrm{A}})/\mathscr{L}_{l-1}(\mathscr{E}\mathscr{Z}_N^{\mathrm{A}})$ the associated graded. Since there are only finitely many A-elliptic multiple zeta values of a given weight and length, $\mathrm{gr}_l^{\mathscr{L}}(\mathscr{E}\mathscr{Z}_N^{\mathrm{A}})$ is a finite-dimensional $\mathbb{Q}$-vector space, and as outlined in Problem 1, we would like to compute the dimension

$$D_{N,l}^{\mathrm{ell}} = \dim_{\mathbb{Q}} \mathrm{gr}_l^{\mathscr{L}}(\mathscr{E}\mathscr{Z}_N^{\mathrm{A}}), \tag{52}$$

for all $N, l \geq 0$.

The case $l = 0$ is of course trivial: we have $D_{N,0}^{\mathrm{ell}} = \delta_{N,0}$, since the empty iterated integral $\int_\gamma = 1$ has weight zero, by definition. In the cases $l = 1, 2$, a complete solution to Problem 1 has been obtained in [36], on which we will report briefly in the rest of this section.

4.3.1 Elliptic Zeta Values

We begin our investigation of Problem 1 in length $l = 1$. It is clear from Definition 5 that for a given weight N, there is only one A-elliptic multiple zeta value of length one, namely $I^{\mathrm{A}}(N;\tau)$. Thus $D_{N,1}^{\mathrm{ell}} \leq 1$.

Now by (45), we have

$$I^{\mathrm{A}}(N;\tau) = (-1)^N I^{\mathrm{A}}(N;\tau), \tag{53}$$

which immediately yields $I^{\mathrm{A}}(N;\tau) = 0$, if N is odd, hence $D_{N,1}^{\mathrm{ell}} = 0$ in that case. In Sect. 5.4, we will see that for even $N \geq 0$

$$I^{\mathrm{A}}(N;\tau) = -2\zeta(N), \tag{54}$$

and we obtain a complete solution of Problem 1 in the length one case

$$D_{N,1}^{\mathrm{ell}} = \begin{cases} 0 & N = 0 \text{ or } N \text{ odd} \\ 1 & \text{else.} \end{cases} \tag{55}$$

Also, note that since the even zeta values $\zeta(2n)$ are linearly independent over $\mathbb{Q}$, it follows that the space of elliptic zeta values $\mathscr{L}_1(\mathscr{E}\mathscr{Z}^{\mathrm{A}}) = \sum_{N\geq 0} \mathscr{L}_1(\mathscr{E}\mathscr{Z}_N^{\mathrm{A}})$ is graded for the weight, i.e.

$$\mathscr{L}_1(\mathscr{E}\mathscr{Z}^{\mathrm{A}}) = \bigoplus_{N\geq 0} \mathscr{L}_1(\mathscr{E}\mathscr{Z}_N^{\mathrm{A}}). \tag{56}$$

4.3.2 Elliptic Double Zeta Values of Even Weight

We now investigate the case $l = 2$ and N even, i.e. A-elliptic multiple zeta values of length two and even weight. From (41) together with (54) and the fact that $\zeta(N) \in \langle(2\pi i)^N\rangle_{\mathbb{Q}}$, if N is even, we know that

$$I^{\mathrm{A}}(n_1, n_2; \tau) + I^{\mathrm{A}}(n_2, n_1; \tau) = I^{\mathrm{A}}(n_1; \tau) I^{\mathrm{A}}(n_2; \tau) \in \langle(2\pi i)^{n_1+n_2}\rangle_{\mathbb{Q}} \subset \mathscr{L}_1(\mathscr{E}\mathscr{Z}^{\mathrm{A}}_{n_1+n_2}). \quad (57)$$

On the other hand, if the weight $N = n_1 + n_2$ is even, then it follows from (45) that

$$I^{\mathrm{A}}(n_1, n_2; \tau) + I^{\mathrm{A}}(n_2, n_1; \tau) = 2I^{\mathrm{A}}(n_1, n_2; \tau), \quad (58)$$

and therefore in that case, by (54),

$$I^{\mathrm{A}}(n_1, n_2; \tau) = \begin{cases} 0 & n_1, n_2 \text{ odd} \\ 2\zeta(n_1)\zeta(n_2) & n_1, n_2 \text{ even.} \end{cases} \quad (59)$$

In particular, $D^{\mathrm{ell}}_{N,2} = 0$, if N is even.

4.3.3 Elliptic Double Zeta Values of Odd Weight

So far, we have seen that both the elliptic zeta values $I^{\mathrm{A}}(n; \tau)$, as well as the elliptic double zeta values $I^{\mathrm{A}}(n_1, n_2; \tau)$ of even weight are constant as functions in τ, and that they are equal to rational multiples of powers of $(2\pi i)^N$, where N is the weight. If the weight is odd, this is no longer true, since in that case (45) gives no new information about elliptic double zeta values of odd weight. Indeed, since the product of two elliptic zeta values of weights of different parities necessarily vanishes, the shuffle product formula becomes

$$I^{\mathrm{A}}(n_1, n_2; \tau) + I^{\mathrm{A}}(n_2, n_1; \tau) = 0, \quad (60)$$

which is also precisely the reflection relation in the case of odd weight.

In order to attack Problem 1 in the odd weight case, we first count the number of linearly independent Fay and shuffle relations. For this, we denote by $\widehat{V}_N \subset \mathbb{Q}(X, Y)$ the $\mathbb{Q}$-vector space of rational functions $P(X, Y)$ in the variables X and Y, such that the product $XY \cdot P$ is a homogeneous polynomial of degree $N + 2$.

Definition 6 We define the length two *Fay-shuffle space* $\mathrm{FSh}_2(N)$ of weight N to be the subspace $\mathrm{FSh}_2(N) \subset \widehat{V}_N$ of elements, satisfying the Fay and shuffle equations

$$P(X, Y) + P(X + Y, -Y) + P(-X - Y, X) = 0, \quad P(X, Y) + P(Y, X) = 0. \quad (61)$$

In particular, since in the length two case the right hand side of (49) and the left hand side of (41) vanish in $\mathrm{gr}^{\mathscr{L}}_2(\mathscr{E}\mathscr{Z}^{\mathrm{A}})$, as both are contained in $\mathscr{L}_1(\mathscr{E}\mathscr{Z}^{\mathrm{A}})$ by (54)), we see that the elliptic double zeta values satisfy the defining equations of $\mathrm{FSh}_2(N)$, modulo elliptic zeta values. As a consequence, we have:

Proposition 2 ([36], Proposition 3.10) *The dimension of the Fay-shuffle space gives an upper bound for* $D^{\mathrm{ell}}_{N,2}$, *i.e.*

$$D^{\mathrm{ell}}_{N,2} \leq \dim_{\mathbb{Q}} \mathrm{FSh}_2(N-2), \tag{62}$$

for all $N \geq 0$.

The dimensions on the right hand side can in fact be computed using invariant theory.

Theorem 3 ([36], Theorem 3.11) *We have*

$$\dim_{\mathbb{Q}} \mathrm{FSh}_2(N) = \begin{cases} 0 & \text{if } N \text{ is even} \\ \lfloor \frac{N}{3} \rfloor + 1 & \text{if } N \text{ is odd.} \end{cases} \tag{63}$$

It remains to show that $D^{\mathrm{ell}}_{N,2} \geq \lfloor \frac{N}{3} \rfloor + 1$ for odd N, and this follows from:

Theorem 4 ([36], Theorem 3.15) *Let* $N \geq 1$ *be odd. The family of elliptic double zeta values*

$$\{I^{\mathrm{A}}(r, N-r; \tau) \mid 0 \leq r \leq \lfloor N/3 \rfloor\} \tag{64}$$

is linearly independent over $\mathbb{Q}$.

The proof uses an explicit representation of elliptic double zeta values as indefinite integrals of Eisenstein series (87). Using in addition the fact that there are no non-trivial relations between elliptic double zeta values of different weights (cf. [36], Theorem 3.6), the two preceding theorems also imply that every relation between elliptic double zeta values is a consequence of Fay and shuffle relations.

5 Elliptic Multiple Zeta Values and Iterated Eisenstein Integrals

In this section, we will discuss the relation between A-elliptic multiple zeta values and indefinite iterated integrals of Eisenstein series [12, 34]. This relation has first been established by Enriquez [21, 22], who showed that the derivative of an A-elliptic multiple zeta value of length r can be expressed using A-elliptic multiple zeta values of length $r-1$ and Eisenstein series. Moreover, the boundary condition for Enriquez's differential equation is given explicitly in terms of multiple zeta values. Using Enriquez's ideas as a starting point, one finds in [6] completely explicit formulae for A-elliptic multiple zeta values as linear combinations of iterated integrals of Eisenstein series and multiple zeta values. It turns out that the precise linear combinations of iterated Eisenstein integrals, which appear as elliptic multiple zeta values are controlled by a certain well-studied Lie algebra $\mathfrak{u}^{\text{geom}}$ of derivations of a free Lie algebra on two generators [13, 16, 32, 37].

5.1 Reminder on Iterated Eisenstein Integrals

For $k \geq 1$, define the *Eisenstein series* $G_k(\tau)$ to be

$$G_{2k}(\tau) = \sum_{(m,n)\in\mathbb{Z}\setminus\{(0,0)\}} \frac{1}{(m+n\tau)^k}, \tag{65}$$

where for $k = 1, 2$, we use Eisenstein summation for double series, i.e.

$$\sum_{(m,n)\in\mathbb{Z}\setminus\{(0,0)\}} a_{m,n} = \lim_{N\to\infty} \lim_{M\to\infty} \sum_{n=-N}^{N} \sum_{m=-M}^{M} a_{m,n}. \tag{66}$$

We also set $G_0(\tau) \equiv -1$. It is well-known that the Eisenstein series $G_{2k}(\tau)$, $k \geq 1$, can be expanded as a Fourier series in $q = e^{2\pi i\tau}$

$$G_{2k}(\tau) = 2\zeta(2k) + 2\frac{(2\pi i)^{2k}}{(2k-1)!} \sum_{n\geq 1} \sigma_{2k-1}(n) q^n, \tag{67}$$

where $\sigma_k(n) = \sum_{d|n} d^k$ denotes the sum of the k-th powers of the divisors of n.

In [12], one finds a definition of iterated integrals of Eisenstein series (and even of iterated integrals of more general modular forms, the so-called *iterated Eichler* or *iterated Shimura integrals*)

$$\mathcal{G}(k_1, \ldots, k_r; \tau) = \int_\tau^{i\infty} G_{k_1}(\tau_1)d\tau_1 \ldots G_{k_r}(\tau_r)d\tau_r, \tag{68}$$

where the integral has to be regularized suitably at the boundary $i\infty$. Since the Eisenstein series are holomorphic functions on a one-dimensional complex manifold, the iterated integral (81) is independent of the choice of path from τ to $i\infty$. We refer to [12], Sect. 4, for the precise regularization scheme, which involves tangential base points, and confine ourselves with giving explicit formulae in lengths one and two:

$$\mathcal{G}(k; \tau) = -\frac{1}{2\pi i}\left(2\zeta(k)\log q + \frac{2(2\pi i)^k}{(k-1)!} \sum_{m,n=1}^{\infty} \frac{m^{k-2}}{n} q^{mn}\right), \tag{69}$$

and

$$\begin{aligned}\mathscr{G}(k_1,k_2;\tau) = \frac{1}{(2\pi i)^2}\Bigg(& 2\zeta(k_1)\zeta(k_2)(\log q)^2 \\ &+ 2\zeta(k_2)\frac{2(2\pi i)^{k_1}}{(k_1-1)!}\log q \sum_{m,n=1}^{\infty}\frac{m^{k_1-2}}{n}q^{mn} \\ &+ 2\zeta(k_1)\frac{2(2\pi i)^{k_2}}{(k_2-1)!}\sum_{m,n=1}^{\infty}\frac{m^{k_2-3}}{n^2}q^{mn} \\ &- 2\zeta(k_2)\frac{2(2\pi i)^{k_1}}{(k_1-1)!}\sum_{m,n=1}^{\infty}\frac{m^{k_1-3}}{n^2}q^{mn} \\ &+ \frac{4(2\pi i)^{k_1+k_2}}{(k_1-1)!(k_2-1)!}\sum_{m_i,n_i=1}^{\infty}\frac{m_1^{k_1-1}m_2^{k_2-2}}{(m_1n_1+m_2n_2)n_2}q^{m_1n_1+m_2n_2}\Bigg),\end{aligned} \tag{70}$$

which both follow from [12] Example 4.10, and from (67). Note that

$$\frac{\partial\mathscr{G}(k;\tau)}{\partial\tau} = -G_k(\tau), \qquad \frac{\partial\mathscr{G}(k_1,k_2;\tau)}{\partial\tau} = -G_{k_1}(\tau)\mathscr{G}(k_2;\tau), \tag{71}$$

which can be verified directly, and generalizes to

$$\frac{\partial\mathscr{G}(k_1,\ldots,k_r;\tau)}{\partial\tau} = -G_{k_1}(\tau)\mathscr{G}(k_2,\ldots,k_r;\tau), \tag{72}$$

using a general property of iterated integrals (cf. [12], Proposition 4.7, for the case of iterated Eichler integrals).

5.2 The Differential Equation for A-elliptic Multiple Zeta Values

As proved by Enriquez (cf. [22], Théorème 3.10), A-elliptic multiple zeta values satisfy a differential equation involving Eisenstein series. This differential equation is most conveniently written down in terms of the generating series $\mathscr{I}^{\mathrm{A}}(X_1,\ldots,X_r;\tau)$ of A-elliptic multiple zeta values of length r (39).

Theorem 5 (Enriquez) *For all $r \geq 0$, we have*

$$\begin{aligned}2\pi i\frac{\partial}{\partial\tau}\mathscr{I}^{\mathrm{A}}(X_1,\ldots,X_r;\tau) &= \wp_\tau^*(X_1)\mathscr{I}^{\mathrm{A}}(X_2,\ldots,X_r;\tau) - \wp_\tau^*(X_r)\mathscr{I}^{\mathrm{A}}(X_1,\ldots,X_{r-1};\tau) \\ &\quad + \sum_{i=1}^{r-1}(\wp_\tau^*(X_{i+1}) - \wp_\tau^*(X_i))\mathscr{I}^{\mathrm{A}}(X_1,\ldots,X_{i,i+1},\ldots,X_r;\tau),\end{aligned} \tag{73}$$

where $X_{i,j} := X_i + X_j$ and $\wp_\tau^*(\alpha) = \sum_{k=0}^{\infty}(2k-1)G_{2k}(\tau)\alpha^{2k-2}$. □

Note that $\wp_\tau^*$ is related to the Weierstrass $\wp_\tau$-function by the formula $\wp_\tau^*(\alpha) = \wp_\tau(\alpha) + G_2(\tau)$. The proof of Theorem 5 uses special properties of the Kronecker function, in particular that it satisfies the mixed heat equation (cf. [15], Proposition 4). The upshot is that the derivative of an A-elliptic multiple zeta value of length r can be expressed using A-elliptic multiple zeta values of lengths $r-1$ and Eisenstein series. This in turn identifies A-elliptic multiple zeta values as iterated integrals of Eisenstein series of length $r-1$, up to a constant term, the reconstruction of which we turn to next.

5.3 *Restoring the Constant Terms of A-elliptic Multiple Zeta Values*

By definition, the generating series of A-elliptic multiple zeta values is given by a version of Enriquez's A-associator [22, 36]

$$\widetilde{A}(\tau) = \sum_{r\geq 0}(-1)^r \sum_{n_1,\ldots,n_r\geq 0} I^{\mathrm{A}}(n_1,\ldots,n_r;\tau)\,\mathrm{ad}^{n_r}(x)(y)\ldots\mathrm{ad}^{n_1}(x)(y). \quad (74)$$

From [21], Proposition 5.4, it follows that $\widetilde{A}(\tau)$ satisfies $\widetilde{A}(\tau+1) = \widetilde{A}(\tau)$ and that it is holomorphic at infinity. Thus, it possesses a Fourier expansion in $q = e^{2\pi i\tau}$, and the same is true for every A-elliptic multiple zeta value. Moreover, Enriquez has shown that the coefficients of the Fourier expansion of A-elliptic multiple zeta values are essentially given by multiple zeta values. We state this result as a proposition (cf. [22], Proposition 5.2).

Proposition 3 (Enriquez) *Every A-elliptic multiple zeta values can be written as a Fourier series*

$$\sum_{n\geq 0} a_n q^n, \quad (75)$$

where $a_n \in \mathscr{Z}[(2\pi i)^{-1}]$, and $\mathscr{Z} \subset \mathbb{R}$ denotes the $\mathbb{Q}$-algebra spanned by the multiple zeta values.

An immediate consequence of the last proposition is that the constant term I_0^{A} of an A-elliptic multiple zeta value can be retrieved as the limit $\lim_{\tau\to i\infty} I^{\mathrm{A}}(n_1,\ldots,n_r;\tau)$. In order to compute this limit, we use the following result of Enriquez (cf. [22], Eq. (7)).

Theorem 6 (Enriquez) *The generating series of A-elliptic multiple zeta values satisfies*

$$\lim_{\tau\to i\infty} \widetilde{A}(\tau) = e^{\pi i t}\Phi(\tilde{y},t)e^{2\pi i\tilde{y}}\Phi(\tilde{y},t)^{-1}, \quad (76)$$

where $t = -\mathrm{ad}(x)(y)$, $\tilde{y} = -\frac{\mathrm{ad}(x)}{e^{2\pi i\,\mathrm{ad}(x)}-1}(y)$ and Φ denotes the Drinfeld associator.

We end this section by giving a few examples of constant terms of A-elliptic multiple zeta values (cf. [6], Sect. 2.3). If all of the indices $n_1, \dots, n_r \neq 1$, then a closed formula for the constant term $I_0^{\mathrm{A}}(n_1, \dots, n_r)$ of $I^{\mathrm{A}}(n_1, \dots, n_r; \tau)$ is easy to give, since in that case, only the factor $e^{2\pi i \tilde{y}}$ in (76) yields a non-trivial contribution, and therefore

$$I_0^{\mathrm{A}}(n_1, n_2, \dots, n_r)\Big|_{n_i \neq 1} = \begin{cases} 0 & \text{at least one } n_i \text{ is odd and all } n_i \neq 1 \\ \dfrac{1}{r!}\displaystyle\prod_{i=1}^{r}(-2\zeta(n_i)) & \text{all } n_i \text{ even.} \end{cases} \tag{77}$$

On the other hand, if several of the n_i are equal to one, then the formulae become more cumbersome to write down, since in that case the extraction of the relevant terms from (76) combined with the translation from the $\{\tilde{y}, t\}$ alphabet (in which (76) is expressed) to the x, y alphabet (which is used for A-elliptic multiple zeta values) requires many more steps. An implementation of this procedure, using Mathematica, yields for example

$$I_0^{\mathrm{A}}(1,0) = -\frac{i\pi}{2}, \quad I_0^{\mathrm{A}}(1,0,0) = -\frac{i\pi}{4}, \quad I_0^{\mathrm{A}}(1,0,0,0) = -\frac{i\pi}{12} + \frac{\zeta(3)}{(2\pi i)^2}, \tag{78}$$

which generalizes to

$$I_0^{\mathrm{A}}(\underbrace{1,0,\dots,0}_{r}) = -\frac{2\pi i}{4(r-1)!} + \sum_{k=1}^{\lfloor r/2 \rfloor - 1} \frac{1}{(r-(2k+1))!}\frac{\zeta(2k+1)}{(2\pi i)^{2k}}, \tag{79}$$

and shows that every odd Riemann zeta value arises as the constant term of some linear combination of A-elliptic multiple zeta value. Also, multiple zeta values, which cannot be written as polynomials in Riemann zeta values are found to appear as constant terms of higher length A-elliptic multiple zeta values. Conjecturally, the first such multiple zeta value appears in weight 8, for example $\zeta(3,5)$, and we have

$$\begin{aligned} I_0^{\mathrm{A}}(1,0,0,1,0,0,0,0,0) = \frac{1}{(2\pi i)^6}\Big(& -\zeta(3,5) - 2\zeta(2)\zeta(3)^2 - \frac{1}{15}i\pi^5\zeta(3) \\ & + \zeta(5)\zeta(3) + 2i\pi^3\zeta(5) - \frac{21}{2}i\pi\zeta(7) + \frac{\pi^8}{945}\Big), \end{aligned} \tag{80}$$

which casts $\zeta(3,5)$ as the constant term of some linear combination of A-elliptic multiple zeta values.

5.4 *Explicit Formulae for A-elliptic Multiple Zeta Values*

We now combine the results of the last two sections to write down explicit formulae for A-elliptic multiple zeta values in terms of iterated Eisenstein integrals and multiple zeta values. Denote by (cf. [6], Eq. (4.2))

$$\gamma(k_1, k_2, \ldots, k_r; \tau) = (2\pi i)^{-r}\mathscr{G}(k_r, \ldots, k_2, k_1; \tau) \tag{81}$$

the shuffle-regularized iterated integral of the Eisenstein series $G_{k_1}, \ldots, G_{k_r}$ as in Sect. 5.1 (the scaling factor $(-2\pi i)^{-r}$ is adapted to the differential equation satisfied by A-elliptic multiple zeta values (73)).

5.4.1 Length One

Let us begin by completely giving all A-elliptic multiple zeta values of length one. Comparing coefficients on both sides of (73), one sees that

$$2\pi i \frac{\partial}{\partial \tau} I^{\mathrm{A}}(n; \tau) = 0. \tag{82}$$

Thus $I^{\mathrm{A}}(n; \tau)$ is constant, and the precise value can be obtained from (77) as

$$I^{\mathrm{A}}(n; \tau) = \begin{cases} -2\zeta(n) & n \text{ even} \\ 0 & \text{else,} \end{cases} \tag{83}$$

in accordance with the results of Sect. 4.3.

5.4.2 Length Two

We now move on to the length two case. In even weight, we already know the answer by the results of Sect. 4.3.2: it is

$$I^{\mathrm{A}}(n_1, n_2; \tau) = \begin{cases} 0 & n_1, n_2 \text{ odd} \\ 2\zeta(n_1)\zeta(n_2) & n_1, n_2 \text{ even.} \end{cases} \tag{84}$$

Alternatively, this result could have also been obtained by a similar method as in the length one case, using the differential equation (73) and the constant term procedure (76). In the odd weight case, we encounter non-trivial iterated Eisenstein integrals for the first time: By (73), we see that for odd n

$$2\pi i \frac{\partial}{\partial \tau} I^{\mathrm{A}}(0, n; \tau) = -2n\zeta(n+1)G_0(\tau) - nG_{n+1}(\tau). \tag{85}$$

Together with (76), this gives, all in all

$$I^{\mathrm{A}}(0,n;\tau)=\delta_{1,n}\frac{\pi i}{2}+n\big(\gamma(n+1;\tau)+2\zeta(n+1)\gamma(0;\tau)\big). \tag{86}$$

The general case of an A-elliptic multiple zeta value of length two and odd weight can be reduced to this case, using the formula

$$\begin{aligned}
I^{\mathrm{A}}(n_1,n_2;\tau)&=(-1)^{n_1}I^{\mathrm{A}}(0,n_1+n_2;\tau)\\
&\quad+2\delta_{n_1,1}\zeta(n_2)I^{\mathrm{A}}(0,1;\tau)-2\delta_{n_2,1}\zeta(n_1)I^{\mathrm{A}}(0,1;\tau)\\
&\quad+2\sum_{p=1}^{\lceil\frac{1}{2}(n_2-3)\rceil}\binom{n_1+n_2-2p-2}{n_1-1}\zeta(n_1+n_2-2p-1)I^{\mathrm{A}}(0,2p+1;\tau)\\
&\quad-2\sum_{p=1}^{\lceil\frac{1}{2}(n_1-3)\rceil}\binom{n_1+n_2-2p-2}{n_2-1}\zeta(n_1+n_2-2p-1)I^{\mathrm{A}}(0,2p+1;\tau),
\end{aligned} \tag{87}$$

(cf. [6], Eq. (2.33)). This identifies A-elliptic multiple zeta values of length two as certain linear combinations of products of Eisenstein integrals and powers of $2\pi i$.

5.4.3 Length Three

We end this section by giving explicit formulae for A-elliptic multiple zeta values of length three. Given the simplicity of (86), it is natural to first compute $I^{\mathrm{A}}(0,0,n;\tau)$ in terms of iterated integrals of Eisenstein series. Using again the differential equation together with the constant term procedure, one finds

$$I^{\mathrm{A}}(0,0,n;\tau)=\begin{cases}\delta_{1,n}\frac{\pi i}{4}+n\big(\frac{1}{2}\gamma(n+1;\tau)+\zeta(n+1)\gamma(0;\tau)\big) & n\text{ odd}\\[2ex] -\frac{1}{3}\zeta(n)-n(n+1)(\gamma(n+2,0;\tau)+2\zeta(n+2)\gamma(0,0;\tau)) & n\text{ even.}\end{cases} \tag{88}$$

Using reflection and shuffle product formulae, the A-elliptic multiple zeta values $I^{\mathrm{A}}(0,n,0;\tau)$ and $I^{\mathrm{A}}(n,0,0;\tau)$ are readily expressed using $I^{\mathrm{A}}(0,0,n;\tau)$ and products of A-elliptic multiple zeta values of lower length:

$$I^{\mathrm{A}}(n,0,0;\tau)=(-1)^nI^{\mathrm{A}}(0,0,n;\tau), \tag{89}$$

$$I^{\mathrm{A}}(0,n,0;\tau)=I^{\mathrm{A}}(0,n;\tau)I^{\mathrm{A}}(0;\tau)-I^{\mathrm{A}}(0,0,n;\tau)-I^{\mathrm{A}}(n,0,0;\tau). \tag{90}$$

Moreover, up to and including weight 7, every A-elliptic multiple zeta value of weight n can be expressed as linear combinations of $I^{\mathrm{A}}(0,0,n;\tau)$ and products, homogeneous of weight n, of powers of $2\pi i$ and A-elliptic multiple zeta values of lower length (cf. [6], Sect. 2.2).

However, beginning with weight 8, the situation changes. For example,

$$I^{\mathrm{A}}(0,3,5;\tau) = -405\gamma(10,0;\tau) - 75\gamma(6,4;\tau) - \zeta(6)(150\gamma(0,4;\tau) + 90\gamma(4,0;\tau)) \\ - 1008\zeta(10)\gamma(0,0;\tau), \tag{91}$$

and since $\gamma(6,4;\tau)$ appears in (91), but neither in (88), nor in any $\mathbb{Q}$-linear combination of A-elliptic multiple zeta values of lengths one or two, this shows that $I^{\mathrm{A}}(0,3,5;\tau)$ cannot be expressed using $I^{\mathrm{A}}(0,0,8;\tau)$ and lower length A-elliptic multiple zeta values alone.

5.5 A Special Algebra of Derivations

It turns out that the precise linear combinations of iterated Eisenstein integrals, which appear in any linear combination of A-elliptic multiple zeta values are controlled by a special algebra of derivations $\mathfrak{u}^{\mathrm{geom}}$ of the free Lie algebra $\mathbb{L}(x,y)$ over $\mathbb{Q}$ on two generators, which we describe next.

Consider the Lie algebra $\mathrm{Der}^{\Theta}\,\mathbb{L}(x,y)$ of derivations of $\mathbb{L}(x,y)$, which map the commutator $[x,y]$ to zero. There exist distinguished elements $\varepsilon_{2k} \in \mathrm{Der}^{\Theta}(x,y)$, for $k \geq 0$, which satisfy

$$\varepsilon_{2k}(x) = \mathrm{ad}^{2k}(x)(y), \tag{92}$$

and their value on y is uniquely determined by demanding that they be homogeneous for the bigrading $(*,*)$ on $\mathbb{L}(x,y)$, under which x has bidegree $(1,0)$ and y has bidegree $(0,1)$. Explicitly (cf. [37], Eq. (2))

$$\varepsilon_0(y) = 0 \tag{93}$$

$$\varepsilon_{2k}(y) = \sum_{0\leq j<k} (-1)^j [\mathrm{ad}^j(x)(y), \mathrm{ad}^{2k-1-j}(x)(y)]. \tag{94}$$

Definition 7 Define $\mathfrak{u}^{\mathrm{geom}} \subset \mathrm{Der}^{\Theta}\,\mathbb{L}(x,y)$ to be the Lie subalgebra spanned by the ε_{2k}.

The derivations ε_{2k} occur for example in [2, 16, 32, 37]. They are related to universal mixed elliptic motives [28], as well as to polar solutions of the linearized double shuffle equations [13, 14].

By work of Pollack [37], the derivations ε_{2k} satisfy many relations, which are linked to the existence of cusp forms for $\mathrm{SL}_2(\mathbb{Z})$. In [6], Sect. 4.3, it is described how relations between ε_{2k} constrain the linear combinations of iterated Eisenstein integrals which can possibly appear as A-elliptic multiple zeta values. The starting point is the generating series of A-elliptic multiple zeta values

$$e^{-\pi i[x,y]}A(\tau) = \widetilde{A}(\tau) = \sum_{r\geq 0}(-1)^r \sum_{n_1,\ldots,n_r\geq 0} I^{\mathrm{A}}(n_1,\ldots,n_r;\tau)\,\mathrm{ad}^{n_r}(x)(y)\ldots\mathrm{ad}^{n_1}(x)(y). \tag{95}$$

By [22], Eq. (7), and since every ε_{2k} annihilates $[x, y]$, this series satisfies the differential equation

$$2\pi i \frac{\partial}{\partial \tau} \widetilde{A}(\tau) = -\sum_{k\geq 0} (2k-1) G_{2k}(\tau) \varepsilon_{2k}\big(\widetilde{A}(\tau)\big), \tag{96}$$

which is in fact equivalent to Theorem 5, as shown in [22], Sect. 4. Note that in (96), the derivation ε_{2k} is coupled to the Eisenstein series $G_{2k}(\tau)$. Iteratively integrating this differential equation leads to a coupling of iterated integrals of Eisenstein series of length r and commutators of derivations ε_{2k} of depth r.

As a simple example, consider the "Ihara-Takao relation" in $\mathfrak{u}^{\mathrm{geom}}$

$$[\varepsilon_{10}, \varepsilon_4] - 3[\varepsilon_8, \varepsilon_6] = 0, \tag{97}$$

whose existence can be traced back to the unique, up to a scalar, cusp form for $\mathrm{SL}_2(\mathbb{Z})$ of weight 12 [37]. It implies that the iterated Eisenstein integrals $\gamma(10, 4; \tau)$ and $\gamma(8, 6; \tau)$ only appear in a special ratio. More precisely, it follows from the differential equation (96) and from (97) that the only linear combination of the two double Eisenstein integrals, which appears as an A-elliptic multiple zeta value is, up to a scalar, given by

$$81\gamma(10, 4; \tau) + 35\gamma(8, 6; \tau). \tag{98}$$

If one introduces a different normalization of iterated Eisenstein integrals, namely

$$\underline{\gamma}(k_1, \ldots, k_r; \tau) = \prod_{i=1}^{r} (k_i - 1)\gamma(k_1, \ldots, k_r; \tau), \tag{99}$$

then (98) becomes

$$3\underline{\gamma}(10, 4; \tau) + \underline{\gamma}(8, 6; \tau). \tag{100}$$

Viewing $\underline{\gamma}(k_1, k_2; \tau)$ as dual to $[\varepsilon_{k_1}, \varepsilon_{k_2}]$, we see that (100) is orthogonal to the relation (97). In this way, the linear combinations of iterated Eisenstein integrals which appear in elliptic multiple zeta values are orthogonal to relations between commutators of the ε_{2k} (see also [12], Sect. 12). For much more detailed treatments of the relation between $\mathfrak{u}^{\mathrm{geom}}$ and elliptic multiple zeta values, see [6, 33, 35].

Acknowledgements Many thanks to the organizers of the *Research Trimester on Multiple Zeta Values*, held September-December 2014 at ICMAT, Madrid, where part of this research was carried out. This paper contains results obtained in joint work with Johannes Broedel, Carlos Mafra and Oliver Schlotterer, and I would like to thank them very much. Incidentally, that collaboration started after the author gave a talk at the ICMAT in September 2014, as part of this research trimester. Also, many thanks to Henrik Bachmann, Johannes Broedel, Ulf Kühn and Oliver Schlotterer for helpful comments, as well as the Albert-Einstein-Institute in Potsdam, the Department of Applied Mathematics and Theoretical Physics in Cambridge and the Mainz Institute for Theoretical Physics for hospitality. This work is part of the author's PhD thesis at Universität Hamburg, and I would like to thank my advisor Ulf Kühn for his constant support of my work and for his encouragement.

References

1. Bannai, K., Kobayashi, S., Tsuji, T.: On the de Rham and p-adic realizations of the elliptic polylogarithm for CM elliptic curves. Ann. Sci. École. Norm. Sup. (4) **43**(2), 185–234 (2010)
2. Baumard, S., Schneps, L.: On the derivation representation of the fundamental Lie algebra of mixed elliptic motives. Ann. Math. Qué. **41**(1), 43–62 (2017)
3. Bloch, S.J.: Higher regulators, algebraic K-theory, and zeta functions of elliptic curves. CRM Monograph Series, vol. 11. American Mathematical Society, Providence, RI (2000)
4. Broadhurst, D.J., Kreimer, D.: Association of multiple zeta values with positive knots via Feynman diagrams up to 9 loops. Phys. Lett. B **393**(3–4), 403–412 (1997)
5. Broedel, J., Mafra, C.R., Matthes, N., Schlotterer, O.: Elliptic multiple zeta values and one-loop superstring amplitudes. J. High Energy Phys. **7**, 112, front matter+41 pp 2015
6. Broedel, J., Matthes, N., Schlotterer, O.: Relations between elliptic multiple zeta values and a special derivation algebra. J. Phys. A **49**(15), 155203, 49 pp (2016)
7. Broedel, J., Schlotterer, O., Stieberger, S.: Polylogarithms, multiple zeta values and superstring amplitudes. Fortschr. Phys. **61**(9), 812–870 (2013)
8. Broedel, J., Schlotterer, O., Stieberger, S., Terasoma, T.: All order α'-expansion of superstring trees from the Drinfeld associator. Phys. Rev. D **89**(6), 066014 (2014)
9. Brown, F.: Mixed Tate motives over $\mathbb{Z}$. Ann. of Math. (2) **175**(2), 949–976 (2012)
10. Brown, F.: Iterated Integrals in Quantum Field Theory. Geometric and Topological Methods for Quantum Field Theory, pp. 188–240. Cambridge University Press, Cambridge (2013)
11. Brown, F.: Depth-graded motivic multiple zeta values. arXiv:1301.3053
12. Brown, F.: Multiple modular values and the relative completion of the fundamental group of $\mathcal{M}_{1,1}$. arXiv:1407.5167v3
13. Brown, F.: Zeta elements in depth 3 and the fundamental Lie algebra of the infinitesimal Tate curve. Forum Math. Sigma, **5:e1**(56) (2017)
14. Brown, F.: Anatomy of an associator. arXiv:1709.02765
15. Brown, F., Levin, A.: Multiple elliptic polylogarithms. arXiv:1110.6917
16. Calaque, D., Enriquez, B., Etingof, P.: Universal KZB equations: the elliptic case. In: Yu. I. (ed.) Manin Algebra, arithmetic, and geometry: in honor of Vol. I, volume 269 of Progr. Math., pages 165–266. Birkhäuser Boston, Inc., Boston, MA (2009)
17. Chen, K.T.: Iterated path integrals. Bull. Amer. Math. Soc. **83**(5), 831–879 (1977)
18. Deligne, P.: Le groupe fondamental de la droite projective moins trois points. In Galois groups over $\mathbb{Q}$ (Berkeley, CA, 1987), volume 16 of Math. Sci. Res. Inst. Publ., pages 79–297. Springer, New York (1989)
19. Deligne, P., Goncharov, A. B.: Groupes fondamentaux motiviques de Tate mixte. Ann. Sci. École Norm. Sup. (4) **38**(1), 1–56 (2005)
20. Drinfel'd, V.G.: On quasitriangular quasi-Hopf algebras and on a group that is closely connected with Gal($\overline{\mathbb{Q}}/\mathbb{Q}$). Leningrad Math. J. **2**(4), 829–860 (1991)
21. Enriquez, B.: Elliptic associators. Selecta Math. (N.S.) 20 (2014), no. 2, 491–584
22. Enriquez, B.: Analogues elliptiques des nombres multizétas. Bull. Soc. Math. France **144**(3), 395–427 (2016)
23. Furusho, H.: Double shuffle relation for associators. Ann. Math. (2) **174**(1), 341–360 (2011)
24. Gangl, H., Kaneko, M., Zagier, D.: Double zeta values and modular forms. In: Automorphic Forms and Zeta Functions, pp. 71–106. World Scientific Publishing, Hackensack, NJ (2006)
25. Goncharov, A.B.: Multiple polylogarithms, cyclotomy and modular complexes. Math. Res. Lett. **5**(4), 497–516 (1998)
26. Goncharov, A.B., Manin, Y.I.: Multiple ζ-motives and moduli spaces $\mathcal{M}_{0,n}$. Compos. Math. **140**(1), 1–14 (2004)
27. Hain, R.M.: The geometry of the mixed Hodge structure on the fundamental group. In: Algebraic geometry, Bowdoin, 1985 Brunswick, Maine, 1985, volume 46 of Proc. Sympos. Pure Math., pp. 247–282. Amer. Math. Soc., Providence, RI (1987)
28. Hain, R., Matsumoto M.: Universal mixed elliptic motives. J. Inst. Math. Jussieu 1–104 (2018). https://doi.org/10.1017/S1474748018000130

29. Knizhnik, V.G., Zamolodchikov, A.B.: Current algebra and Wess-Zumino model in two dimensions. Nuclear Phys. B **247**(1), 83–103 (1984)
30. Le, T.T.Q., Murakami, J.: Kontsevich's integral for the Kauffman polynomial. Nagoya Math. J. **142**, 39–65 (1996)
31. Levin, A.: Elliptic polylogarithms: an analytic theory. Compositio Math. **106**(3), 267–282 (1997)
32. Levin, A., Racinet, G.: Towards multiple elliptic polylogarithms. arXiv:math/0703237
33. Lochak, P., Matthes, N., Schneps, L.: Elliptic multizetas and the elliptic double shuffle relations, arXiv:1703.09410
34. Manin, Y. I.: Iterated integrals of modular forms and noncommutative modular symbols. In: Algebraic geometry and number theory, vol. 253 of Progr. Math., pages 565–597. Birkhäuser Boston, Boston, MA (2006)
35. Matthes, N.: Elliptic multiple zeta values. Ph.D. thesis, Universität Hamburg (2016)
36. Matthes, N.: Elliptic double zeta values. J. Number Theory **171**, 227–251 (2017)
37. Pollack, A.: Relations between derivations arising from modular forms. Master's thesis, Duke University (2009)
38. Racinet, G.: Doubles mélanges des polylogarithmes multiples aux racines de l'unité. Publ. Math. Inst. Hautes Études Sci. **95**, 185–231 (2002)
39. Ree, R.: Lie elements and an algebra associated with shuffles. Ann. Math. **2**(68), 210–2220 (1958)
40. Schlotterer, O., Stieberger, S.: Motivic multiple zeta values and superstring amplitudes. J. Phys. A **46**(47), 475401, 37 (2013)
41. Terasoma, T.: Geometry of multiple zeta values. In: International Congress of Mathematicians. Vol. II, pages 627–635. Eur. Math. Soc., Zürich (2006)
42. Weil, A.: Elliptic functions according to Eisenstein and Kronecker. Springer, Berlin-New York. Ergebnisse der Mathematik und ihrer Grenzgebiete, Band 88 (1976)
43. Zagier, D.: The Bloch-Wigner-Ramakrishnan polylogarithm function. Math. Ann. **286**(1–3), 613–624 (1990)
44. Zagier, D.: Periods of modular forms and Jacobi theta functions. Invent. Math. **104**(3), 449–465 (1991)
45. Zagier, D.: Values of zeta functions and their applications. In: First European Congress of Mathematics, Vol. II (Paris, 1992), volume 120 of Progr. Math., pages 497–512. Birkhäuser, Basel (1994)

The Elliptic Sunrise

Luise Adams, Christian Bogner and Stefan Weinzierl

Abstract In this talk, we discuss our recent computation of the two-loop sunrise integral with arbitrary non-zero particle masses in the vicinity of the equal mass point. In two space-time dimensions, we arrive at a result in terms of elliptic dilogarithms. Near four space-time dimensions, we obtain a result which furthermore involves elliptic generalizations of Clausen and Glaisher functions.

Keywords Feynman integrals · Special functions · Elliptic polylogarithms

1 Introduction

In the computation of many Feynman integrals the use of multiple polylogarithms[1] [1]

$$\mathrm{Li}_{(s_1,\,\ldots,\,s_k)}(z_1,\,\ldots,\,z_k) = \sum_{n_1>n_2\ldots>n_k\geq 1} \frac{z_1^{n_1}\ldots z_k^{n_k}}{n_1^{s_1}\ldots n_k^{s_k}},\ s_i \geq 1,\ |z_i| < 1$$

is very advantageous. In particular, these functions, shown as nested sums here, also have representations as iterated integrals, given by the classes of hyperlogarithms [2, 3] or by iterated integrals on moduli spaces of curves of genus zero (see [4]).

[1]Our summation convention is widely used in the physics literature, including our previous work. Notice that it differs from the convention in [1].

L. Adams · S. Weinzierl
PRISMA Cluster of Excellence, Institut für Physik, Johannes Gutenberg-Universität Mainz, 55099 Mainz, Germany
e-mail: ladams@students.uni-mainz.de

S. Weinzierl
e-mail: weinzierl@uni-mainz.de

C. Bogner (✉)
Institut für Physik, Humboldt-Universität zu Berlin, 10099 Berlin, Germany
e-mail: bogner@math.hu-berlin.de

J. I. Burgos Gil et al. (eds.), *Periods in Quantum Field Theory and Arithmetic*, Springer Proceedings in Mathematics & Statistics 314,
https://doi.org/10.1007/978-3-030-37031-2_6

Apparently, it is not possible to express every Feynman integral in terms of this framework of functions. This problem is expected to affect an entire class of massive integrals (see e.g. [5]) and was furthermore pointed out for certain massless integrals, arising in $\mathcal{N} = 4$ supersymmetric Yang-Mills theory [6, 7].

One of the simplest Feynman integrals where multiple polylogarithms are not sufficient to express the result is the massive two-loop sunrise integral

$$S(D, t) = \int \frac{d^D k_1 d^D k_2}{\left(i\pi^{D/2}\right)^2} \frac{1}{\left(-k_1^2 + m_1^2\right)\left(-k_2^2 + m_2^2\right)\left(-(p - k_1 - k_2)^2 + m_3^2\right)}.$$

In this talk, we consider this integral as a function of the three particle masses satisfying $0 < m_1 \leq m_2 \leq m_3 < m_1 + m_2$ and of the squared momentum $t = p^2$. The condition $m_3 < m_1 + m_2$ ensures that all pseudo-thresholds are positive. We omit an explicit mass-scale parameter μ in our equations. We discuss the computation of this Feynman integral at $D = 2$ and $D = 4$ dimensions in terms of the Laurent expansions

$$\begin{aligned} S(2 - 2\varepsilon, t) &= S^{(0)}(2, t) + S^{(1)}(2, t)\varepsilon + \mathcal{O}\left(\varepsilon^2\right), \\ S(4 - 2\varepsilon, t) &= S^{(-2)}(4, t)\varepsilon^{-2} + S^{(-1)}(4, t)\varepsilon^{-1} + S^{(0)}(4, t) + \mathcal{O}(\varepsilon). \end{aligned}$$

In the case of $D = 2$, the integral is finite and our result is the coefficient $S^{(0)}(2, t)$. In the case of $D = 4$, we compute the coefficient $S^{(0)}(4, t)$. The pole terms were already known and read

$$\begin{aligned} S^{(-2)}(4, t) &= -\frac{1}{2}\left(m_1^2 + m_2^2 + m_3^2\right), \\ S^{(-1)}(4, t) &= \frac{1}{4}t - \frac{3}{2}\left(m_1^2 + m_2^2 + m_3^2\right) + \sum_{i=1}^{3} m_i^2 \ln\left(m_i^2\right). \end{aligned}$$

In order to obtain $S^{(0)}(4, t)$, we compute the ε-coefficient $S^{(1)}(2, t)$ of the two-dimensional case and relate $S(2 - 2\varepsilon, t)$ with $S(4 - 2\varepsilon, t)$ by Tarasov's dimension shift relations [8, 9]. Our work on these integrals is motivated by the search for classes of functions beyond multiple polylogarithms, which are appropriate for the computation of Feynman integrals.

In Sect. 2 we briefly comment on three computational approaches which fail to provide a result in terms of multiple polylogarithms for the massive sunrise integral. We begin our computation with the integral in two dimensions and discuss our first solution of the differential equation for $S^{(0)}(2, t)$ in Sect. 3. In Sect. 4 we express this result in terms of an elliptic dilogarithm. Section 5 introduces further elliptic generalizations of polylogarithms, understood as elliptic generalizations of Clausen and Glaisher functions, which arise in our results for $S^{(1)}(2, t)$ and $S^{(0)}(4, t)$. Section 6 contains the conclusions of this talk.

2 Basic Properties of the Massive Sunrise Integral

The massive sunrise integral was extensively studied in the past [5, 10–28]. Let us recall some important aspects.

Firstly, in [15] the integral $S(D, t)$ is expressed as a linear combination of generalized hypergeometric functions of Lauricella type C, which are functions of t, of the squared particle masses and of the dimension D. While a wide range of generalized hypergeometric functions can be expanded in terms of multiple polylogarithms with today's methods, this has not been achieved for the mentioned result so far.

Secondly, one may attempt to compute the integral by integration over Feynman parameters. In terms of Feynman parameters, the integral in $D = 2$ dimensions reads

$$S(2, t) = \int_\sigma \frac{\omega}{\mathcal{F}},$$

with $\omega = x_1 dx_2 \wedge dx_3 + x_2 dx_3 \wedge dx_1 + x_3 dx_1 \wedge dx_2$ and $\sigma = \{[x_1 : x_2 : x_3] \in \mathbb{P}^2 | x_i \geq 0,\ i = 1, 2, 3\}$ while the second Symanzik polynomial is given as

$$\mathcal{F} = -x_1 x_2 x_3 t + \left(x_1 m_1^2 + x_2 m_2^2 + x_3 m_3^2\right)(x_1 x_2 + x_2 x_3 + x_1 x_3).$$

For an attempt to iteratively build up the result in terms of the mentioned iterated integrals which represent the multiple polylogarithms, the polynomial $\mathcal{F}$ would have to satisfy the criterion of linear reducibility [29]. The latter is a sufficient but not necessary criterion to obtain multiple polylogarithms in the result. However, the polynomial fails this criterion and a change of variables to restore linear reducibility for a new set of integration variables is unknown for this case.

Thirdly, the integral $S(D, t)$ for generic space-time dimension satisfies an inhomogeneous fourth-order differential equation in t:

$$\left(P_4 \frac{d^4}{dt^4} + P_3 \frac{d^3}{dt^3} + P_2 \frac{d^2}{dt^2} + P_1 \frac{d^1}{dt^1} + P_0\right) S(D, t) = c_{12} T_{12} + c_{13} T_{13} + c_{23} T_{23} \tag{1}$$

where the $T_{ij} = T\left(m_i^2, D\right) T\left(m_j^2, D\right)$ are products of tadpole integrals

$$T(m^2, D) = \int \frac{d^D k}{i\pi^{\frac{D}{2}}} \frac{1}{\left(-k^2 + m^2\right)} = \Gamma\left(1 - \frac{D}{2}\right)\left(m^2\right)^{\frac{D}{2}-1}.$$

All coefficients P_k and c_{ij} are polynomials in $m_1^2, m_2^2, m_3^2, t, D$. Their explicit expressions are provided in appendix A of [30]. Each of the functions $S^{(0)}(2, t)$, $S^{(1)}(2, t)$, $S^{(0)}(4, t)$ satisfies an inhomogeneous differential equation of second or higher order. If any of these operators would factorize into differential operators of first order the corresponding coefficient could be obtained as an iterated integral in a straightforward way (see e.g. Sect. 2 of [31]). However, this is not the case for any of these operators.

All of these points give rise to the expectation, that we need functions beyond multiple polylogarithms to express the integrals $S^{(0)}(2, t)$, $S^{(1)}(2, t)$, $S^{(0)}(4, t)$. This expectation is confirmed by our results for these functions.

3 The Differential Equation in Two Dimensions

We follow the approach of differential equations and begin with the Feynman integral in $D = 2$ dimensions. For the case of equal masses $m_1 = m_2 = m_3$, a differential equation of second order was already given in [14]. A full solution in terms of integrals over elliptic integrals was obtained in [19].

For the case of arbitrary masses, a differential equation of second order was found later in [23]:

$$\begin{aligned} L_2\, S(2, t) &= p_3(t), \\ L_2 &= p_2(t)\frac{d^2}{dt^2} + p_1(t)\frac{d}{dt} + p_0(t), \end{aligned} \tag{2}$$

where $p_0(t)$, $p_1(t)$, $p_2(t)$ are polynomials in t and in the m_i^2 and where $p_3(t)$ furthermore involves $\ln(m_i^2)$, $i = 1, 2, 3$. We take this equation as the starting point of our computation and make the classical ansatz

$$S(2, t) = C_1\psi_1(t) + C_2\psi_2(t) + \int_0^t dt_1 \frac{p_3(t_1)}{p_0(t_1)W(t_1)} \left(-\psi_1(t)\psi_2(t_1) + \psi_2(t)\psi_1(t_1)\right) \tag{3}$$

where ψ_1, ψ_2 are solutions of the homogeneous equation, C_1, C_2 are constants and

$$W(t) = \psi_1(t)\frac{d}{dt}\psi_2(t) - \psi_2(t)\frac{d}{dt}\psi_1(t)$$

is the Wronski determinant.

At this point, it is useful to consider the zero-set of the second Symanzik polynomial $\mathcal{F}$. This cubical curve intersects the integration domain σ of the Feynman integral at the three points

$$P_1 = [1 : 0 : 0],\ P_2 = [0 : 1 : 0],\ P_3 = [0 : 0 : 1].$$

We choose one of these points P_i as the origin and transform the curve to Weierstrass normal form

$$y^2 z - x^3 - g_2(t)xz^2 - g_3(t)z^3 = 0. \tag{4}$$

By this transformation, the chosen origin is mapped to the point $[x : y : z] = [0 : 1 : 0]$. In this way, we obtain three elliptic curves $E_{\mathcal{F},i}$ according to the three points P_i, $i = 1, 2, 3$.

In the chart $z = 1$ we write Eq. 4 as

$$y^2 = 4(x - e_1)(x - e_2)(x - e_3),$$

which defines the three roots e_1, e_2, e_3 with $e_1 + e_2 + e_3 = 0$. These provide the boundaries of the period integrals

$$\psi_1 = 2\int_{e_2}^{e_3} \frac{dx}{y} = \frac{4}{\tilde{D}^{\frac{1}{4}}} K(k), \quad \psi_2 = 2\int_{e_1}^{e_3} \frac{dx}{y} = \frac{4i}{\tilde{D}^{\frac{1}{4}}} K(k')$$

of the elliptic curve. Here the polynomial $\tilde{D}$ is given as

$$\tilde{D} = (t - (m_1 + m_2 - m_3)^2)(t - (m_1 - m_2 + m_3)^2)(t - (-m_1 + m_2 + m_3)^2)(t - (m_1 + m_2 + m_3)^2)$$

and we have obtained the complete elliptic integral of the first kind

$$K(x) = \int_0^1 dt \frac{1}{\sqrt{(1 - t^2)(1 - x^2 t^2)}}$$

with moduli $k = \sqrt{\frac{e_3 - e_2}{e_1 - e_2}}$, $k' = \sqrt{1 - k^2} = \sqrt{\frac{e_1 - e_3}{e_1 - e_3}}$. These period integrals ψ_1, ψ_2 are solutions of the homogeneous equation associated to Eq. 3.

We still have to fix the constants. It can be shown that C_2 has to vanish while the other constant C_1 is derived from a known result [32–34] for the zero-mass limit $S(2, 0)$. Now all pieces of our ansatz in Eq. 3 are determined. In order to simplify the integrand of the particular solution, we furthermore make use of the remaining two associated period integrals of $E_{\mathcal{F},i}$. In conclusion, we obtain a result [31] of the form

$$S(2, t) = S(2, 0) + \frac{\psi_1(t)}{\pi^2} \int_0^t dt_1 \rho(t_1) \tag{5}$$

where the integrand ρ involves elliptic integrals of the first and second kind.

4 The Massive Sunrise Integral in Two Dimensions

The general shape of our result of Eq. 5 has a disadvantage. While the involved elliptic integrals are well-studied functions, nicely related to the underlying elliptic curve of the problem, the integral over these functions in not a known function. This integral might remind us vaguely of an iterated integral, but in this form, it can not be recognized as a generalization of a polylogarithm. However, for the equal-mass case, it was shown more recently in [24], that the integral can be expressed in terms of an elliptic dilogarithm. Various notions of elliptic polylogarithms were previosly introduced in the mathematical literature [35–40].

Before we apply an elliptic generalization of a polylogarithm to the sunrise integral with arbitrary masses, let us briefly recall the basic concept of an elliptic function. With respect to a lattice $L = \mathbb{Z} + \tau\mathbb{Z}$ with $\tau \in \mathbb{C}$ and $\mathrm{Im}(\tau) > 0$, a function f is said to be elliptic, if it satisfies $f(x) = f(x + \lambda)$ for $\lambda \in L$. Accordingly, the corresponding function $\tilde{f}(z)$ of $z \in \mathbb{C}^\star$ defined by $\tilde{f}(e^{2\pi i x}) = f(x)$ is elliptic, if

$$\tilde{f}(z) = \tilde{f}(z \cdot q_\lambda), \quad q_\lambda \in e^{2\pi i \lambda} \text{ for } \lambda \in L. \tag{6}$$

Recall that a cell of the lattice with $\tau = \frac{\psi_2}{\psi_1}$ is isomorphic to an elliptic curve with the periods ψ_1, ψ_2.

A crucial idea for the construction of such elliptic functions is to consider sums of the form $\sum_{n\in\mathbb{Z}} g(z \cdot q^n)$ over some function g. If a sum of this type is well-defined, it clearly satisfies the condition of Eq. 6 by construction. This concept can serve for definitions of elliptic generalizations of polylogarithms. For example in [39] it is used to define the class of multiple elliptic polylogarithms. The elliptic dilogarithm in this framework reads

$$\tilde{E}_2(z;\ u;\ q) = \sum_{m\in\mathbb{Z}} u^m \mathrm{Li}_2(q^m z)$$

where u is a sufficiently small damping parameter to guarantee the convergence of the function.

Based on the same basic idea, we define the class of functions [41]

$$\mathrm{ELi}_{n;m}(x;y;q) = \sum_{j=1}^{\infty}\sum_{k=1}^{\infty} \frac{x^j}{j^n}\frac{y^k}{k^m} q^{jk} = \sum_{k=1}^{\infty} \frac{y^k}{k^m}\mathrm{Li}_n(q^k x),$$

$$\mathrm{E}_{n;\,m}(x;\ y;\ q) = \begin{cases} \frac{1}{i}\left(\frac{1}{2}\mathrm{Li}_n(x) - \frac{1}{2}\mathrm{Li}_n(x^{-1}) + \mathrm{ELi}_{n;\,m}(x;\ y;\ q) - \mathrm{ELi}_{n;\,m}(x^{-1};\ y^{-1};\ q)\right), \\ \frac{1}{2}\mathrm{Li}_n(x) + \frac{1}{2}\mathrm{Li}_n(x^{-1}) + \mathrm{ELi}_{n;\,m}(x;\ y;\ q) + \mathrm{ELi}_{n;\,m}(x^{-1};\ y^{-1};\ q) \end{cases} \tag{7}$$

with the first line for $n + m$ even and the second line for $n + m$ odd. Note that our elliptic dilogarithm

$$\mathrm{E}_{2;\,0}(x;\ y;\ q) = \frac{1}{i}\left(\frac{1}{2}\mathrm{Li}_2(x) - \frac{1}{2}\mathrm{Li}_2\left(x^{-1}\right) + \sum_{i=1}^{\infty} y^i \mathrm{Li}_2\left(q^i x\right) - \sum_{j=1}^{\infty} y^{-j}\mathrm{Li}_2\left(q^j x^{-1}\right)\right)$$

is closely related to the above function $\tilde{E}_2$. We obtain

$$\mathrm{E}_{2;\,0}(x;\; y;\; q) = \frac{1}{i}\left(\tilde{E}_2(x;\; y;\; q) - \frac{1}{2}\frac{1+y}{1-y}\zeta(2) - \frac{1}{4}\frac{1+y}{1-y}\ln^2(-x)\right.$$
$$\left. - \frac{y}{(1-y)^2}\ln(-x)\ln(q) - \frac{1}{2}\frac{y\,(1+y)}{(1-y)^3}\ln^2(q)\right)$$

in the region of parameters given by $x \in \mathbb{C}\backslash[0,\; \infty[,\; |y| > 1$ and real-valued q in the range $0 \leq q < \min\left(|x|,\; \frac{1}{|x|},\; |y|,\; \frac{1}{|y|}\right)$.

Using the function $\mathrm{E}_{2;\,0}$, we express our result for the massive sunrise integral in two space-time dimensions in a very compact way as[2]

$$S\,(2,\; t) = \frac{\psi_1(q)}{\pi}\sum_{i=1}^{3}\mathrm{E}_{2;\,0}(w_i(q);\; -1;\; -q) \text{ where } q = e^{\pi i \frac{\psi_2(t)}{\psi_1(t)}}. \tag{8}$$

Note that the dependence on t is now implicitly expressed in terms of q, which is defined by the periods of the elliptic curve. The arguments $w_1,\; w_2,\; w_3$ are functions of q and of the squared particle masses. They are directly obtained from the three intersection points $P_1,\; P_2,\; P_3$ by the consecutive transformations on the elliptic curves $E_{\mathcal{F},i}, i = 1,\; 2,\; 3$, indicated above. In this sense, every piece of the compact result Eq. 8 is nicely related to the underlying elliptic curves $E_{\mathcal{F},i}$.

In the case of equal masses, the result simplifies to

$$S\,(2,\; t) = 3\frac{\psi_1(q)}{\pi}\mathrm{E}_{2;\,0}\left(\exp\left(2\pi i/3\right);\; -1;\; -q\right).$$

5 The Massive Sunrise Integral Around Four Dimensions

By use of dimension shift relations [8, 9], we express the coefficient $S^{(0)}(4,\; t)$ of the sunrise integral near $D = 4$ dimensions in terms of coefficients of the $D = 2$ case [30]. We obtain $S^{(0)}(4,\; t)$ as a linear combination of terms $S^{(0)}(2,\; t),\; \frac{\partial}{\partial m_i^2}S^{(0)}(2,\; t),$ $S^{(1)}(2,\; t),\; \frac{\partial}{\partial m_i^2}S^{(1)}(2,\; t),\; i = 1,\; 2,\; 3$. Therefore, our remaining task is the computation of $S^{(1)}(2,\; t)$.

From Eq. 1 we obtain the differential equation

$$L_{1,a}\,L_{1,b}\,L_2\,S^{(1)}(2,\; t) = I_1(t). \tag{9}$$

Here $L_{1,a}$ and $L_{1,b}$ are differential operators of first order,

$$L_{1,a} = p_{1,a}\frac{d}{dt} + p_{0,a} \text{ and } L_{1,b} = p_{1,b}\frac{d}{dt} + p_{0,b},$$

[2]By a slight abuse of notation, we denote with ψ_1 the above function of t and the corresponding function of q.

where $p_{0,a},\ p_{1,a}$ are rational functions of t and the squared particle masses and $p_{0,b},\ p_{1,b}$ are polynomials in these variables. The homogeneous solutions $\psi_a,\ \psi_b$ of these operators, defined by

$$L_{1,a}\,\psi_a(t) = 0 \text{ and } L_{1,b}\,\psi_b(t) = 0$$

are easily obtained.

The operator L_2 in Eq. 9 is the one of Eq. 2 which already appeared in the differential equation of the two-dimensional case. The inhomogeneous term I_1 of Eq. 9 is a combination of certain differentiations of our result $S^{(0)}(2,\ t)$, of logarithms in the squared particle masses and of a polynomial in the squared masses and in t.

Solving Eq. 9 for the combination $L_2\, S^{(1)}(2,\ t)$, we obtain

$$L_2\, S^{(1)}(2,\ t) = I_2(t) \tag{10}$$

with

$$I_2(t) = \tilde{C}_1\psi_b(t) + \tilde{C}_2\psi_b(t)\int_0^t \frac{\psi_a(t_1)dt_1}{p_{1,b}(t_1)\psi_b(t_1)} + \psi_b(t)\int_0^t \frac{\psi_a(t_1)dt_1}{p_{1,b}(t_1)\psi_b(t_1)}\int_0^{t_1}\frac{I_1(t_2)dt_2}{p_{1,a}(t_2)\psi_a(t_2)}$$

where $\tilde{C}_1,\ \tilde{C}_2$ are integration constants.

Now with Eq. 10 we have to solve a similar differential equation as in the two-dimensional case, with the only difference that the inhomogeneous part is more complicated. However, we can make a similar ansatz and we have the same period integrals $\psi_1,\ \psi_2$ of $E_{\mathscr{F},i}$ as solutions of the homogeneous equation. Therefore, it is useful to introduce the variable q again in the same way as in Eq. 8. In terms of integrals over q, we obtain

$$S^{(1)}(2,\ t) = C_3\psi_1 + C_4\psi_2 - \frac{\psi_1}{\pi}\int_0^q \frac{dq_1}{q_1}\int_0^{q_1}\frac{dq_2}{q_2}\frac{I_2\,(q_2)\,\psi_1\,(q_2)^3}{\pi p_2\,(q_2)\,W\,(q_2)^2}.$$

The integration constants $C_3,\ C_4$ are determined from boundary conditions. Expanding the integrand, we can perform the integrations order by order and obtain a $q-$expansion of $S^{(1)}(2,\ t)$ to high orders. This step finally allows us to find a result for $S^{(1)}(2,\ t)$ in closed form, which can be confirmed to satisfy the differential Eq. 10.

Let us refer to [30] for the explicit result and just highlight some of its properties here. Apart from classical (multiple) polylogarithms, the result involves the functions $\mathrm{E}_{1;\,0}(x;\ y;\ q)$, $\mathrm{E}_{2;\,0}(x;\ y;\ q)$, $\mathrm{E}_{3;\,1}(x;\ y;\ q)$ as defined in Eq. 7 and furthermore a quadruple sum of the form

$$\Lambda\,(x_1, x_2; y_1, y_2; -q) = \sum_{j_1=1}^{\infty}\sum_{k_1=1}^{\infty}\sum_{j_2=1}^{\infty}\sum_{k_2=1}^{\infty}\frac{k_1^2\,(-q)^{j_1k_1+j_2k_2}}{j_2\,(j_1k_1+j_2k_2)^2}\left(x_1^{j_1}y_1^{k_1} - x_1^{-j_1}y_1^{-k_1}\right)\left(x_2^{j_2}y_2^{k_2} + x_2^{-j_2}y_2^{-k_2}\right).$$

For the arguments of these functions, we have $y,\ y_1,\ y_2 \in \{-1,\ 1\}$ and $x,\ x_1,\ x_2 \in \{w_1,\ w_2,\ w_3\}$, where the w_i again are the arguments obtained from the intersection points mentioned above.

The appearance of the functions $\mathrm{E}_{1;\,0}(x;\ y;\ q)$, $\mathrm{E}_{2;\,0}(x;\ y;\ q)$, $\mathrm{E}_{3;\,1}(x;\ y;\ q)$ shows that the framework of Eq. 7, set up for the coefficient $S^{(0)}(2,\ t)$, is also useful for $S^{(1)}(2,\ t)$ and hence also for the four-dimensional case. Furthermore, these functions can be viewed as elliptic generalizations of Clausen and Glaisher functions. Recall that the Clausen functions are defined by

$$\mathrm{Cl}_n(\varphi) = \begin{cases} \frac{1}{2i}\left(\mathrm{Li}_n\left(e^{i\varphi}\right) - \mathrm{Li}_n\left(e^{-i\varphi}\right)\right) & \text{for even } n, \\ \frac{1}{2}\left(\mathrm{Li}_n\left(e^{i\varphi}\right) + \mathrm{Li}_n\left(e^{-i\varphi}\right)\right) & \text{for odd } n, \end{cases}$$

and the Glaisher functions are given as

$$\mathrm{Gl}_n(\varphi) = \begin{cases} \frac{1}{2}\left(\mathrm{Li}_n\left(e^{i\varphi}\right) + \mathrm{Li}_n\left(e^{-i\varphi}\right)\right) & \text{for even } n, \\ \frac{1}{2i}\left(\mathrm{Li}_n\left(e^{i\varphi}\right) - \mathrm{Li}_n\left(e^{-i\varphi}\right)\right) & \text{for odd } n. \end{cases}$$

We therefore obtain as ‘non-elliptic limits’ of our functions:

$$\begin{aligned} \lim_{q\to 0}\mathrm{E}_{1;\,0}\left(e^{i\varphi};\ y;\ q\right) &= \mathrm{Cl}_1(\varphi), \\ \lim_{q\to 0}\mathrm{E}_{2;\,0}\left(e^{i\varphi};\ y;\ q\right) &= \mathrm{Cl}_2(\varphi), \\ \lim_{q\to 0}\mathrm{E}_{3;\,1}\left(e^{i\varphi};\ y;\ q\right) &= \mathrm{Gl}_3(\varphi). \end{aligned}$$

As a final remark, let us mention that $S^{(1)}(2,\ t)$ is a function of mixed weight. It shares this property with the function $\mathrm{E}_{3;\,1}(x;\ y;\ q)$ which has parts of weight three and of weight four.

6 Conclusions

We discussed the computation of the massive sunrise integral in two and around four space-time dimensions. We started with the computation of the $\mathcal{O}\left(\varepsilon^0\right)$-part of the integral in two dimensions and expressed our result in terms of an elliptic dilogarithm. In this form, the result is very compact and every part of it is nicely related to the underlying elliptic curve, given by the second Symanzik polynomial of the Feynman graph.

We continued with the computation of the $\mathcal{O}\left(\varepsilon^1\right)$-part in two dimensions. Apart from the elliptic dilogarithm, this result involves further elliptic generalizations of (multiple) polylogarithms, which can be understood as elliptic generalizations of Clausen and Glaisher functions. Due to well-known dimension shift relations, these results provide the $\mathcal{O}\left(\varepsilon^0\right)$-part of the Feynman integral in four dimensions.

Together with the results of [24, 42], our results give rise to the hope, that elliptic (multiple) polylogarithms may serve as an appropriate class of functions to compute further Feynman integrals beyond multiple polylogarithms. Some of our functions can be related to the functions of [39], where also a framework of iterated integrals, already applied in a different physics context [43], is provided.

References

1. Goncharov, A.B.: Math. Res. Lett. **5**, 497–516 (1998). arXiv:1105.2076 [math.AG]
2. Lappo-Danilevsky, J.A.: Rec. Math. Mosc. **34**(6), 113–146 (1927)
3. Lappo-Danilevsky, J.A.: Mémoires sur la théorie des systémes deséquations différen- tielles linéaires, Chelsea, vol. I–III (1953)
4. Brown, F.: Ann. Sci. Ec. Norm. Sup'er. (4) **42**, 371–489 (2009). arXiv:math.AG/0606419
5. Bauberger, S., Böhm, M., Weiglein, G., Berends, F.A., Buza, M.: Nucl. Phys. Proc. Suppl. **37B**, 95 (1994). arXiv:hep-ph/9406404
6. Caron-Huot, S., Larsen, K.J.: JHEP **1210**, 026 (2012). arXiv:1205.0801
7. Nandan, D., Paulos, M.F., Spradlin, M., Volovich, A.: JHEP **1305**, 105 (2013). arXiv:1301.2500
8. Tarasov, O.V.: Phys. Rev. D **54**, 6479 (1996). arXiv:hep-th/9606018
9. Tarasov, O.V.: Nucl. Phys. B **502**, 455 (1997). arXiv:hep-ph/9703319
10. Kalmykov, Y.M., Kniehl, B.A.: Nucl. Phys. B **809**, 365 (2009). arXiv:0807.0567
11. Broadhurst, D.: (2008). arXiv:0801.4813
12. Davydychev, A.I., Delbourgo, B.: J. Math. Phys. **39**, 4299 (1998). arXiv:hep-th/9709216
13. Smirnov, V.A.: Springer Tracts Mod. Phys. **211**, 1 (2004)
14. Broadhurst, D.J., Fleischer, J., Tarasov, O.: Z. Phys. C **60**, 287 (1993). arXiv:hep-ph/9304303
15. Berends, F.A., Buza, M., Böhm, M., Scharf, R.: Z. Phys. C **63**, 227 (1994)
16. Bauberger, S., Berends, F.A., Böhm, M., Buza, M.: Nucl. Phys. B **434**, 383 (1995). arXiv:hep-ph/9409388
17. Bauberger, S., Böhm, M.: Nucl. Phys. B **445**, 25 (1995). arXiv:hep-ph/9501201
18. Caffo, M., Czyz, H., Laporta, S., Remiddi, E.: Nuovo Cim. A **111**, 365 (1998). arXiv:hep-th/9805118
19. Laporta, S., Remiddi, E.: Nucl. Phys. B **704**, 349 (2005). arXiv:hep-ph/0406160
20. Groote, S., Körner, J.G., Pivovarov, A.A.: Ann. Phys. **322**, 2374 (2007). arXiv:hep-ph/0506286
21. Groote, S., Körner, J., Pivovarov, A.: Eur. Phys. J. C **72**, 2085 (2012). arXiv:1204.0694
22. Bailey, D.H., Borwein, J.M., Broadhurst, D., Glasser, M.L.: J. Phys. A **41**, 205203 (2008). arXiv:0801.0891
23. Müller-Stach, S., Weinzierl, S., Zayadeh, R.: Commun. Num. Theor. Phys. **6**, 203 (2012). arXiv:1112.4360
24. Bloch, S., Vanhove, P.: J. Numb. Theor. **148**, 328 (2015). arXiv:1309.5865
25. Remiddi, E., Tancredi, L.: Nucl. Phys. B **880**, 343 (2014). arXiv:1311.3342
26. Caffo, M., Czyz, H., Remiddi, E.: Nucl. Phys. B **634**, 309 (2002). arXiv:hep-ph/0203256
27. Pozzorini, S., Remiddi, E.: Comput. Phys. Commun. **175**, 381 (2006). arXiv:hep-ph/0505041
28. Caffo, M., Czyz, H., Gunia, M., Remiddi, E.: Comput. Phys. Commun. **180**, 427 (2009). arXiv:0807.1959
29. Brown, F.: Commun. Math. Phys. **287**(3), 925–958 (2009)
30. Adams, L., Bogner, C., Weinzierl, S.: J. Math. Phys. **56**, 072303 (2015). arXiv:1504.03255
31. Adams, L., Bogner, C., Weinzierl, S.: J. Math. Phys. **54**, 052303 (2013). arXiv:1302.7004
32. Ussyukina, N.I., Davydychev, A.I.: Phys. Lett. B **298**, 363 (1993)
33. Lu, H.J., Perez, C.A.: SLAC-PUB-5809
34. Bern, Z., Dixon, L., Kosower, D.A., Weinzierl, S.: Nucl. Phys. B **489**, 3 (1997). arXiv:hep-ph/9610370

35. Bloch, S.: Higher Regulators, Algebraic K-theory and Zeta-Functions of Elliptic Curves. University of Chicago—AMS, CRM (2000)
36. Beilinson, A., Levin, A.: In: U., Jannsen, S., Kleiman, J.-P. Serre Motives, Proceedings of Symposia in Pure Mathematics, vol. 55, Part 2, pp. 97–121. AMS (1994)
37. Levin, A.: Comput. Math. **106**, 267 (1997)
38. Levin, A., Racinet, G.: (2007). arXiv:math/0703237
39. Brown, F., Levin, A.: (2011). arXiv:1110.6917
40. Wildeshaus, J.: Lecture Notes in Mathematics, vol. 1650. Springer (1997)
41. Adams, L., Bogner, C., Weinzierl, S.: J. Math. Phys. **55**, 102301 (2014). arXiv:1405.5640
42. Bloch, S., Kerr, M., Vanhove, P.: (2014). arXiv:1406.2664
43. Broedel, J., Mafra, C.R., Matthes, N., Schlotterer, O.: JHEP **1507**, 112 (2015). arXiv:1412.5535

Polylogarithm Identities, Cluster Algebras and the $\mathcal{N} = 4$ Supersymmetric Theory

Cristian Vergu

Abstract Scattering amplitudes in $\mathcal{N} = 4$ super-Yang Mills theory can be computed to higher perturbative orders than in any other four-dimensional quantum field theory. The results are interesting transcendental functions. By a hidden symmetry (dual conformal symmetry) the arguments of these functions have a geometric interpretation in terms of configurations of points in $\mathbb{CP}^3$ and they turn out to be cluster coordinates. We briefly introduce cluster algebras and discuss their Poisson structure and the Sklyanin bracket. Finally, we present a 40-term trilogarithm identity which was discovered by accident while studying the physical results.

Keywords Scattering amplitudes · Polylogarithms · Cluster algebras · Twistors

1 Introduction

There is no doubt that quantum field theory and mathematics are deeply connected. There are many examples where field theory intuition helped formulate mathematical conjectures or even theorems (Seiberg-Witten theory in topology [62], Wilson loops in Chern-Simons theory for knot theory [61]). Similarly, progress in mathematics has stimulated progress in field theory (as a prime example we have ADHM construction [8] of instantons, but also work in index theory [7] which helped in the understanding of field theory anomalies). And these are just a few of many examples.

In this review we will focus on one of the many connecting bridges between quantum field theory and number theory: polylogarithms. In quantum field theory polylogarithms and the closely related multiple zeta values are ubiquitous. They arise in the perturbative computations of various quantities.

Expanded version of a talk given at the Opening Workshop of the Research Trimester on Multiple Zeta Values, Multiple Polylogarithms, and Quantum Field Theory, organized by José I. Burgos Gil, Kurusch Ebrahimi-Fard, D. Ellwood, Ulf Kühn, Dominique Manchon and P. Tempesta.

C. Vergu (✉)
Department of Mathematics, King's College London The Strand, London WC2R 2LS, UK
e-mail: c.vergu@gmail.com; c.vergu@nbi.ku.dk

J. I. Burgos Gil et al. (eds.), *Periods in Quantum Field Theory and Arithmetic*, Springer Proceedings in Mathematics & Statistics 314,
https://doi.org/10.1007/978-3-030-37031-2_7

There are many quantities one may attempt to compute and, moreover, there are many different quantum field theories. Many results are already available but frequently the complexity of the final answers (not to mention the complexity of the computation) is forbidding. We are then naturally led to ask which field theories and what quantities are most likely to be understood in simple terms.

These questions, while very natural, are not at all obvious, but in recent years an answer has began to emerge. As we will explain, the answer is somewhat surprising. The textbook example for the simplest interacting field theory is called the ϕ^4 theory. This is a theory of a single scalar field with a four-point interaction. The Feynman diagrams in this theory have internal vertices of degree four. Many results are known in this theory see, for example, Ref. [20, 59]. However, it has recently emerged that there is a better candidate for study, which we will discuss below.

Relativistic field theories are symmetric under the Poincaré group. The Poincaré group has the Lorentz group $O(1, 3)$ as a subgroup and particles are in correspondence with irreducible representations of these symmetry groups. The scalar particles transform in the trivial representation of $O(1, 3)$ so they realize the relativistic symmetry in the simplest possible way. As mentioned above, the ϕ^4 theory is a theory of scalar (or spin zero) fields.

Other representations of the Lorentz symmetry may appear: fermions which transform as a representation of the covering group $\mathrm{Spin}(1, 3)$, gauge fields which are vectors of $O(1, 3)$, the graviton which is rank two tensor representation, etc. In the case of the gauge fields and of the graviton the formulation of the quantum theory is complicated by the fact that states are defined modulo gauge transformations. This also complicates the computations since one has to make a choice of gauge (or a choice of representative in the equivalence class).

Despite these technical complications, in many cases the final results, when expressed in terms of appropriate variables, turn out to be strikingly simple (the computation of Parke and Taylor in Ref. [57] being a prime example). Then, we are led to suspect that there should be more efficient ways to find these answers.

We have briefly discussed the theories but we still haven't specified the types of quantities we are going to compute. We turn to this question next. The quantities which will be most relevant in the following discussion are scattering amplitudes. Let us give a rough definition of scattering amplitudes. A field theory of the kind we will consider is defined by a functional $S[\phi]$ called action, depending of functions $\phi(\mathbf{x}, t)$ called fields (here t is time, $\mathbf{x}$ is a three-dimensional vector and ϕ is a generic name for a field; in general the theory can contain several fields with different $O(1, 3)$ transformations). From this functional we can obtain by variational methods partial differential equations (called equations of motion) for the fields of the theory. Now, given some boundary conditions $\phi_\pm$ at $t = \pm\infty$ for the fields, from the solution ϕ_0 to the equation of motion satisfying these boundary conditions one can build a complex number $\exp(i S[\phi_0])$ which is called the tree level amplitude of transition between ϕ_- and ϕ_+ (if there is no solution for the prescribed boundary conditions, then the amplitude is defined to be zero). The name 'tree' is due to the fact that this quantity can be computed as a sum of tree-shaped Feynman diagrams.

The computation using the definition can be tedious in general, especially for gauge theories where one has to make an arbitrary choice of gauge (in the final result the dependence on this arbitrary choice must cancel; when this happens we call the answer 'gauge invariant'). The tree level amplitudes have two important properties: analyticity (in a certain domain) and factorization.[1] Factorization here means that the amplitude has certain poles whose residues are products of simpler amplitudes. The requirement of factorization is a very powerful constraint; using it, the BCFW [19] recursion relations allow the computation of all tree-level amplitudes of the $\mathcal{N} = 4$ theory we will describe in the next section.

In the quantum theory graphs with loops appear as well. Graphs with loops correspond to non-trivial integrals, which yield mathematically interesting results. It is an empirical observation that the transcendentality of an ℓ-loop result is bounded from above by 2ℓ; for a one-loop quantity the most complicated part can be expressed in terms of dilogarithms.

For theories relevant experimentally, like Quantum Chromodynamics (QCD), a one-loop answer will contain not only dilogarithms, but also logarithms and even rational terms. The transcendentality of the answer is not uniform. However, for the special case of $\mathcal{N} = 4$, the answers are of uniform transcendentality. In some cases, see Ref. [52], the $\mathcal{N} = 4$ answer can be obtained from the uniform transcendentality of the more complicated QCD result.

2 The Maximally Supersymmetric Theory

We mentioned previously that the theories with spin are in some sense simpler than theories of scalar (spinless) particles. Even so, there are many possible theories of particles with spin. Supersymmetry is a remarkable symmetry which can transform between particles of different spins. The maximal supersymmetry of a non-gravitational theory in three space and one time dimensions is called $\mathcal{N} = 4$ supersymmetry. The reason for the name is that $\mathcal{N} = 1$ supersymmetry is the minimal supersymmetry and the maximal supersymmetry has four times as many supersymmetries as the minimal one.

In Ref. [24], Coleman and Mandula proved a theorem about the possible symmetries of a relativistic theory. Under certain assumptions they showed that the symmetry group has the structure of a product between the Lorentz and some other 'internal' symmetry group. Later, Haag et al. [47] showed that a non-trivial symmetry structure is possible, but it has to be a supergroup symmetry, not a Lie group symmetry. A supergroup is obtained by exponentiating Lie superalgebra elements, where a Lie superalgebra is a $\mathbb{Z}_2$-graded algebra with a bracket satisfying graded commutativity and a graded version of Jacobi identity. The supergroup has a usual Lie group as

[1] Analyticity survives after adding quantum corrections, but factorization becomes more subtle in case there are infrared divergences (see Ref. [13]). Since scattering amplitudes in gauge theories are infrared divergent, exploiting factorization at loop level seems to be much harder.

a subgroup and, somewhat surprisingly, this is also enlarged with respect to a typical relativistic theory. In a relativistic theory the symmetry group is the Poincaré group, which now gets enhanced to a SO(2, 4) group, also known as the conformal group. The new symmetries are the dilatation D and four conformal transformations $K_0, \ldots, K_3$.

The theory with maximal supersymmetry was constructed shortly after in Ref. [18] by Brink, Schwarz and Scherk. This theory is uniquely defined by its symmetry. It is a theory of a connection A on an $\mathrm{SU}(N)$ principal bundle over Minkowski space $\mathbb{M}$, together with fermionic field Ψ and scalar fields Φ. The action functional is given by the Yang-Mills term together with other terms dictated by supersymmetry, which we do not write explicitly since they will not be important in the following

$$S[A, \Psi, \Phi] = \frac{1}{2g^2} \int_{\mathbb{M}} \mathrm{tr}(F \wedge *F + \cdots). \tag{1}$$

Here the trace is taken in the fundamental representation of $\mathrm{SU}(N)$ and g^2 is a real number, called coupling constant. $F = dA + A \wedge A$ is the curvature of the connection A and $*F$ is the Hodge dual. The scattering amplitudes, can be expanded as a power series in g.

Terms in the perturbative expansion are computed by summing Feynman graphs. The contribution of a Feynman graph can be factored in two different types of terms: the kinematic part, depending on the positions (or on the momenta after Fourier transform) and the 'color' part which depends on the Lie algebra $\mathfrak{su}(N)$ of the gauge group $\mathrm{SU}(N)$. The observables can then be decomposed on a basis of $\mathfrak{su}(N)$ invariants whose coefficients depend on N and g. If we select invariants which can be written as a single trace and, for these terms, we select the dominant behavior when $N \to \infty$, then the topology of the contributing graphs simplifies. We find that only planar graphs contribute. The way to select the planar graph contributions is to reorganize the perturbation theory as an expansion in $\lambda = g^2 N$ around $\lambda = 0$, with $N \to \infty$ and $g^2 \to 0$. This is the well-known 't Hooft limit [49].

From his study of the large N limit, 't Hooft conjectured that the result in the 't Hooft limit is the genus zero term in an expansion of a theory which sums over surfaces. A theory which sums over surfaces is a string theory (in a theory of particles, one sums[2] over particle paths, as instructed by the Feynman path integral). The conjecture also stated that subleading terms in N correspond to sums over surfaces of higher genera.

This conjecture of 't Hooft is very general, and was initially proposed for QCD, where the gauge group SU(3) was to be replaced by $\mathrm{SU}(N)$. It was hoped that understanding $N \to \infty$ case could shed some light on the $N = 3$ case. If instead of QCD we consider the $\mathcal{N} = 4$ supersymmetric theory, the conjecture was sharpened by the AdS/CFT correspondence of Maldacena (see Ref. [53]). The AdS/CFT

[2]The sum over particle histories is not well-defined mathematically. Nevertheless, we can use it formally to compute the perturbative expansion. A similar statement holds for a string theory, where we sum over string histories also called worldsheets.

correspondence identifies the precise measure on the space of surfaces. In fact, we should use super-strings, but if we set the fermions to zero we obtain a theory of a string moving in an $\mathrm{AdS}_5 \times \mathbb{S}^5$ geometry. Here CFT means Conformal Field Theory, which in this case is a theory with a symmetry group containing $\mathrm{SO}(2,4)$. The AdS_5 space is the five-dimensional hyperbolic space with a non-definite metric, which can be obtained by analytically continuing some coordinates to imaginary values (a procedure called Wick rotation in the Quantum Field Theory literature). This is similar to the relation between Euclidean space $\mathbb{R}^4$ and Minkowski space $\mathbb{M}$. The isometry group of AdS_5 is again $\mathrm{SO}(2,4)$. In fact, the full $\mathrm{PSU}(2,2|4)$ symmetry groups match on both sides of the correspondence.

The AdS/CFT duality describes a physical system in two different ways. When the 't Hooft coupling λ is small, the field theory perturbative expansion in powers of λ is reliable. When the 't Hooft coupling is large, instead, one should use string theory on the $\mathrm{AdS}_5 \times \mathbb{S}^5$ background. In this case, the expansion variable is $\lambda^{-1/2}$. Therefore, the duality is of strong-weak type; the strong coupling ($\lambda \to \infty$) in the CFT can be mapped to a weakly coupled description in the dual string theory.

The computation of the scattering amplitudes can also be done in the dual string theory, as described in Ref. [1]. In the dual string theory scattering amplitudes are given by the exponential of a minimal surface in AdS_5 which ends on the boundary of AdS_5 on a polygon whose sides are the momenta of the scattered particles (the polygon closes by momentum conservation).

3 Kinematics

In this section we describe the kinematics of a scattering process in terms of configurations of points in $\mathbb{CP}^3$. This was initiated in Ref. [48] for tree-level amplitudes, later extended to superspace in Ref. [54] and further studied in Ref. [6]. The usefulness of these variables for loop amplitudes was emphasized in Ref. [4] and also in Ref. [46] for an explicit two-loop result.

Consider an n-particle scattering process. The particle labeled by i is described by the on-shell momentum p_i (with $p_i^2 = 0$, where the norm is computed using the Minkowski metric), its helicity s_i and a gauge algebra generator $t_i \in \mathfrak{su}(N)$. The helicity labels the representation under the compact subgroup $\mathrm{U}(1)$ of the Lorentz group $\mathrm{O}(1,3)$ which preserves the momentum p_i. In fact, if our theory contains fermions we need to pass to the covering group $\mathrm{Spin}(1,3)$ of part of the Lorentz group connected to the identity. In the end, the representations turn out to be labeled by $s \in \mathbb{Z}/2$.

As we discussed above, in the 't Hooft limit $N \to \infty$, $g^2 N = \lambda$ fixed, only single-trace terms survive in the scattering amplitudes. If we look at one of these single-trace terms, we see that the scattered particles are cyclically ordered. We can therefore introduce a *dual* space with coordinates x such that the momenta p_i are expressed as $p_i = x_{i-1} - x_i$. The x_i coordinates are only defined up to a translation $x_i \sim x_i + a$. We denote by $\widetilde{\mathbb{M}}$ the space parametrized by dual coordinates x.

The $\mathcal{N} = 4$ super-Yang-Mills theory is superconformal invariant. Besides this superconformal symmetry, the $\mathcal{N} = 4$ super-Yang-Mills theory also has a surprising *dual* superconformal symmetry, whose bosonic subgroup acts on the dual coordinates x. In the following we will mostly be interested in the conformal subgroup of this dual superconformal group. The dual superconformal symmetry is a hidden symmetry, which only arises in the 't Hooft limit. In particular, it can not be verified on the Lagrangian of the theory.

Historically, this symmetry arose as follows. First, the authors of Ref. [34] noticed that integrals appearing in the perturbative computations of Refs. [2, 15] have a curious inversion property in the dual space. Together with the obvious Lorentz symmetry, this generates the conformal group. This symmetry was then confirmed, and in fact used to guide the computations, at higher loop orders and for larger numbers of external particles in Refs. [10, 11, 14]. In a parallel development [1], Alday and Maldacena showed how to compute scattering amplitudes in the dual string theory. This turned out to be closely related to the computation of a Wilson loop (in a language more familiar to mathematicians, a Wilson loop is the trace of the holonomy of the connection A around a curve). The strong coupling computation leads us to believe that there is a connection between scattering amplitudes and a Wilson loop around a polygonal contour with vertices x_i. This was confirmed also at weak coupling in several papers [12, 17, 31, 32, 35]. Under the duality the scattering amplitudes map to Wilson loops and the dual conformal symmetry of scattering amplitudes maps to the conformal symmetry of the Wilson loops. Reference [33] showed that in fact the scattering amplitudes enjoy a dual *super*-conformal symmetry. This corresponds in the dual side to the superconformal symmetry of a Wilson super-loop, which is the trace of the holonomy of a superconnection in superspace along a polygonal contour. The corresponding super-loops were first defined in Refs. [23, 55].

The dual space $\tilde{\mathbb{M}}$ is noncompact and it does not have an action of the conformal group since some points are sent to infinity under conformal transformations. This problem can be solved by compactifying $\tilde{\mathbb{M}}$ is a way compatible with the action of the conformal group. Moreover, $\tilde{\mathbb{M}}$ comes with a Minkowski signature. It is more convenient to use complex coordinates instead and to impose reality conditions when needed. Doing this, we can treat both the cases of Lorentz signature and of split signature. The complexified and compactified dual space can be represented as the $\mathbb{G}(2, 4)$ Grassmannian of two-planes in $\mathbb{C}^4$ containing the origin. Therefore, to each point in dual space $\tilde{\mathbb{M}}$ we can associate a two-plane in $\mathbb{C}^4$. Two points in dual space are light-like separated if their corresponding planes intersect in a line (it is easy to check that this imposes one constraint). If we projectivize this construction, to a line through the origin in $\mathbb{C}^4$ corresponds a point in $\mathbb{CP}^3$ and to a two-plane through the origin in $\mathbb{C}^4$ corresponds a projective line in $\mathbb{CP}^3$. We can do this for all pairs of points (x_{i-1}, x_i) and associate to each of them a point $Z_i \in \mathbb{CP}^3$. So instead of describing the kinematics by giving the momenta p_i subject to on-shell conditions $p_i^2 = 0$ and momentum conservation $\sum_{i=1}^n p_i = 0$, we can describe it by giving n

points $Z_i \in \mathbb{CP}^3$. The variables Z_i are known as momentum twistors[3] and were introduced in Ref. [48]. Unlike for the variables p_i or x_i, the momentum twistors are unconstrained.

The complexified dual conformal group acts as $\mathrm{SL}(4, \mathbb{C})$ on the momentum twistors $[Z] \to [MZ]$, where M is an $\mathrm{SL}(4, \mathbb{C})$ matrix and we have denoted by $[Z]$ the homogeneous coordinates of the point Z. The $\mathrm{SL}(4, \mathbb{C})$ is the double cover of the complexified orthogonal group $\mathrm{SO}(6, \mathbb{C})$. There is a small subtlety here. We defined the Lorentz group to be $\mathrm{O}(1, 3)$ and its complexification is $\mathrm{O}(4, \mathbb{C})$. However, the parity transformation in $\mathrm{O}(4, \mathbb{C})$ does not embed in $\mathrm{SO}(6, \mathbb{C})$, nor in its double cover $\mathrm{SL}(4, \mathbb{C})$. Then, the question is how does this discrete parity transformation act on the momentum twistor space. The answer is as follows. There is another space which, for lack of a better name, we call conjugate momentum twistor space whose points we label by W_i. There is a pairing between points in these two spaces, defined up to rescaling which we denote by $W \cdot Z$. Then we impose the rescaling invariant constraints $W_i \cdot Z_i = 0$ and $W_{i-1} \cdot Z_i = W_{i+1} \cdot Z_i = 0$ (here $i \pm 1$ are considered modulo n, the number of particles in the scattering process). Given the Z_i, the W_i are determined up to a rescaling. Then, parity acts as the discrete transformation $Z_i \leftrightarrow W_i$.

The translation of the kinematics to momentum twistor language makes it easy to build conformal invariants. In order to make $\mathrm{SL}(4, \mathbb{C})$ invariants, we can form four-brackets $\langle ijkl \rangle = \mathrm{Vol}(v_i, v_j, v_k, v_l)$, where v_i is a vector in $\mathbb{C}^4$ corresponding to Z_i and Vol is a volume form which is preserved by the action of $\mathrm{SL}(4, \mathbb{C})$.

So we have established that we can describe the kinematics of a scattering process by giving a configuration of n ordered points Z_i in $\mathbb{CP}^3$. The homogeneous coordinates of these points fit in a $4 \times n$ matrix. The conformal invariants are built from the 4×4 minors of this $4 \times n$ matrix.

The description above is very similar to the description of coordinates on a Grassmannian. For $k \leq n$, the Grassmannian $\mathbb{G}(k, n)$ of k-planes in an n-dimensional space can be described as the space of $k \times n$ matrices of full rank modulo the left action by $\mathrm{GL}(k)$. Given such a $k \times n$ matrix, we can form $\binom{n}{k}$ minors of type $k \times k$. They can be labeled by k integers $i_1, \ldots, i_k \in \{1, \ldots, n\}$, corresponding to the columns of the initial $k \times n$ matrix. We will denote the determinants of these minors by $\langle i_1, \ldots, i_k \rangle$. These determinants are also known as Plücker coordinates, and satisfy Plücker relations

$$\langle i, k, I \rangle \langle j, l, I \rangle = \langle i, j, I \rangle \langle k, l, I \rangle + \langle j, k, I \rangle \langle i, l, I \rangle, \tag{2}$$

where I is a multi-index with $k - 2$ entries. The Plücker relations define an embedding, called Plücker embedding, of the Grassmannian into a projective space of dimension $\binom{n}{k}$.

In the next section we will show that the Plücker relations in Eq. (2) are the same as the exchange relations in a cluster algebra (see Eq. (5, for example). This will also provide a way to build more complicated coordinates starting from simple

[3] A similar construction can be done for Minkowski space $\mathbb{M}$ instead, in which case we obtain the Penrose's twistor space (see Ref. [58]).

minors. Such combinations naturally appear in expressions for scattering amplitudes in $\mathcal{N} = 4$.

Grassmannians have the important property of duality which identifies $\mathbb{G}(k, n)$ with $\mathbb{G}(n-k, n)$. This is useful since it allows to simplify the geometric picture (as has been done in Refs. [41, 46]). Consider first the case $n = 6$. The kinematics is described by a configuration of six ordered points in $\mathbb{CP}^3$ or by the Grassmannian $\mathbb{G}(4, 6)$. By Grassmannian duality this is the same as $\mathbb{G}(2, 6)$ which then can be translated to a configuration of six ordered points in $\mathbb{CP}^1$, a much simpler-looking (though equivalent) geometric configuration.

A similar simplification can be performed for the case of $n = 7$, where a configuration of seven points in $\mathbb{CP}^3$ can be mapped to a configuration of seven points in $\mathbb{CP}^2$. In general, this means that the configurations of n ordered points in $\mathbb{CP}^{k-1}$ are the same as configurations of n ordered points in $\mathbb{CP}^{n-k-1}$. Therefore we can restrict to $2 \le k \le \lfloor \frac{n-1}{2} \rfloor$ without loss of generality.

4 Introduction to Cluster Algebras

In this section we present some useful facts about cluster algebras. In the next section we will make the connection with Grassmannians and Plücker coordinates. Cluster algebras have been introduced in a series of papers [9, 37–39] by Fomin and Zelevinsky.

Since the formal definition is a bit complicated, we will content ourselves with an informal description. Cluster algebras are characterized as follows: they are commutative algebras constructed from distinguished generators (called *cluster variables*) which are grouped into non-disjoint sets of constant cardinality (called *clusters*). The clusters are constructed recursively by an operation called *mutation* from an initial cluster. The number of variables in a cluster is called the rank of the cluster algebra.

Let us consider an example. The A_2 cluster algebra is defined by the following data:

- cluster variables: $x_m, \quad m \in \mathbb{Z}$
- clusters: $\{x_m, x_{m+1}\}$
- initial cluster: $\{x_1, x_2\}$
- rank: 2
- exchange relations: $x_{m-1}x_{m+1} = 1 + x_m$
- mutation: $\{x_{m-1}, x_m\} \to \{x_m, x_{m+1}\}$.

Using the exchange relations we find that

$$x_3 = \frac{1+x_2}{x_1}, \quad x_4 = \frac{1+x_1+x_2}{x_1 x_2}, \quad x_5 = \frac{1+x_1}{x_2}, \quad x_6 = x_1, \quad x_7 = x_2, \quad \ldots . \tag{3}$$

Therefore, the sequence x_m is periodic with period five and the number of cluster variables is finite.

When expressing the cluster variables x_m in terms of the variables (x_1, x_2), we encounter two unexpected features (which hold in general for arbitrary cluster algebras). First, the denominators of the cluster variables are always monomials. In general, we expect the cluster variables to be rational fractions of the initial cluster variables, but in fact the denominator is always a monomial. This is known under the name of "Laurent phenomenon" (see [37]). The second observation is that the numerator is a polynomial with positive coefficients.

As we alluded to before, this construction has a connection with Plücker relations. If we set $x_1 = \frac{\langle 23\rangle\langle 14\rangle}{\langle 12\rangle\langle 34\rangle}$ and $x_2 = \frac{\langle 13\rangle\langle 45\rangle}{\langle 34\rangle\langle 15\rangle}$, where $\langle ij\rangle$ are coordinates of the Grassmannian $\mathbb{G}(2,5)$, we can compute the rest of cluster variables by using the Plücker identities $\langle ik\rangle\langle jl\rangle = \langle ij\rangle\langle kl\rangle + \langle il\rangle\langle jk\rangle$, to obtain

$$x_1 = \frac{\langle 23\rangle\langle 14\rangle}{\langle 12\rangle\langle 34\rangle}, \quad x_2 = \frac{\langle 13\rangle\langle 45\rangle}{\langle 34\rangle\langle 15\rangle}, \quad x_3 = \frac{\langle 12\rangle\langle 35\rangle}{\langle 15\rangle\langle 23\rangle}, \quad x_4 = \frac{\langle 25\rangle\langle 34\rangle}{\langle 23\rangle\langle 45\rangle}, \quad x_5 = \frac{\langle 15\rangle\langle 24\rangle}{\langle 12\rangle\langle 45\rangle}.$$

In the following we will use a description of cluster algebras starting with quiver. We now describe how to obtain a cluster algebra from a quiver. A quiver is an oriented graph which we will require to be connected, finite, without loops (arrows with the same origin and target) and two-cycles (pairs of arrows going in opposite directions between two vertices).

Starting with a quiver with a given vertex k we define a new quiver obtained by mutating at vertex k. The new quiver is obtained by applying the following operations on the initial quiver:

- for each path $i \to k \to j$ we add an arrow $i \to j$,
- reverse all the arrows on the edges incident with k,
- remove all the two-cycles that may have formed.

The mutation at k is an involution; when applied twice in succession we obtain the initial cluster.

Each quiver of the restricted type defined above is in one-to-one correspondence with skew-symmetric matrices, once we fix an ordering of the vertices. The skew-symmetric matrix b is such that b_{ij} is the difference between the number of arrows $i \to j$ and the number of arrows $j \to i$. Since only one of the terms above is non-vanishing, $b_{ij} = -b_{ji}$. Under a mutation at vertex k the matrix b transforms to b' given by

$$b'_{ij} = \begin{cases} -b_{ij}, & \text{if } k \in \{i, j\}, \\ b_{ij}, & \text{if } b_{ik}b_{kj} \leq 0, \\ b_{ij} + b_{ik}b_{kj}, & \text{if } b_{ik}, b_{kj} > 0, \\ b_{ij} - b_{ik}b_{kj}, & \text{if } b_{ik}, b_{kj} < 0 \end{cases}. \tag{4}$$

If we start with a quiver with n vertices and associate to each vertex i a variable x_i, we can use the skew-symmetric matrix b to define a mutation relation at the vertex k by

$$x_k x_k' = \prod_{i|b_{ik}>0} x_i^{b_{ik}} + \prod_{i|b_{ik}<0} x_i^{-b_{ik}}, \tag{5}$$

with the understanding that an empty product is set to one. The mutation at k changes x_k to x_k' defined by Eq. (5) and leaves the other cluster variables unchanged.

The A_2 cluster algebra can be expressed by a quiver $x_1 \to x_2$. Then, a mutation at x_1 replaces it by $x_1' = \frac{1+x_2}{x_1} \equiv x_3$ and reverses the arrow. A mutation at x_2 replaces it by $x_2' = \frac{1+x_1}{x_2} \equiv x_5$. In the diagram (6) below we represent the quivers and the mutations for the A_2 cluster algebra (the arrows between quivers are labeled by the mutated variable).

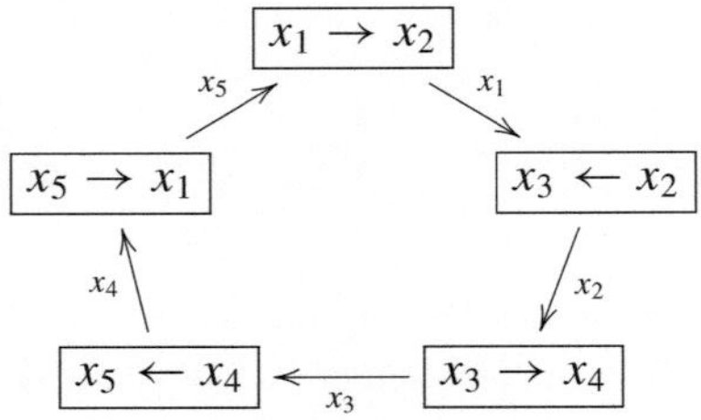

(6)

5 The Cluster Algebra for $\mathbb{G}(k, n)$

The Grassmannian $\mathbb{G}(k, n)$ has a cluster algebra structure which was described in Ref. [40] (this construction is also reviewed in Ref. [51]).

For $k < n$ we consider the description of the Grassmannian $\mathbb{G}(k, n)$ as the equivalence classes of $k \times n$ matrices of full rank, where two matrices are equivalent if they differ by the left action of a $\mathrm{GL}(k)$ matrix. If the leftmost $k \times k$ minor is non-singular, i.e. $\langle 1, \ldots, k\rangle \neq 0$ then, by left multiplication with an appropriate $\mathrm{GL}(k)$ matrix, we can transform it to the identity matrix. After this operation the representative $k \times n$ matrix has the form $(\mathbf{1}_k, Y)$, where $\mathbf{1}_k$ is the $k \times k$ identity matrix and Y is a $k \times l$ matrix with $l = n - k$. The entries y_{ij}, $1 \leq i \leq k$, $1 \leq j \leq l$ of the matrix Y are coordinates on the cell of the Grassmannian where $\langle 1, \ldots, k\rangle \neq 0$.

Now we define a matrix F_{ij} for $1 \leq i \leq k$, $1 \leq j \leq l$, which is the biggest square matrix which fits inside Y and whose lower-left corner is at position (i, j) inside Y. Then we define $l(i, j) = \min(i - 1, n - j - k)$ and

$$f_{ij} = (-1)^{(k-i)(l(i,j)-1)} \det F_{ij}. \tag{7}$$

According to Ref. [40], the initial quiver for the $\mathbb{G}(k, n)$ cluster algebra is given by[4]

[4]Here we are presented a flipped version of the quiver and with the arrows reversed with respect to the quivers of Refs. [40, 51].

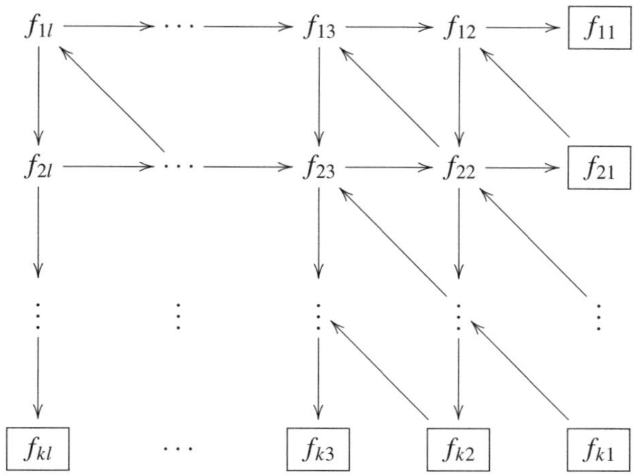

(8)

The quiver above has two types of vertices, boxed and unboxed. The boxed vertices are special and called *frozen vertices*. We do not allow mutations in the frozen vertices. The associated variables to the frozen vertices are called *coefficients* instead of *cluster variables*. We define the *principal part* of such a quiver to be the quiver obtained by erasing the frozen vertices and the edges incident with them.

For the case $n = 5$ and $k = 2$, we can compute $f_{11} = \langle 23\rangle$, $f_{12} = \langle 24\rangle$, $f_{13} = \langle 25\rangle$, $f_{21} = \langle 34\rangle$, $f_{22} = \langle 45\rangle$, $f_{23} = \langle 15\rangle$. Then, the the initial quiver diagram looks like below

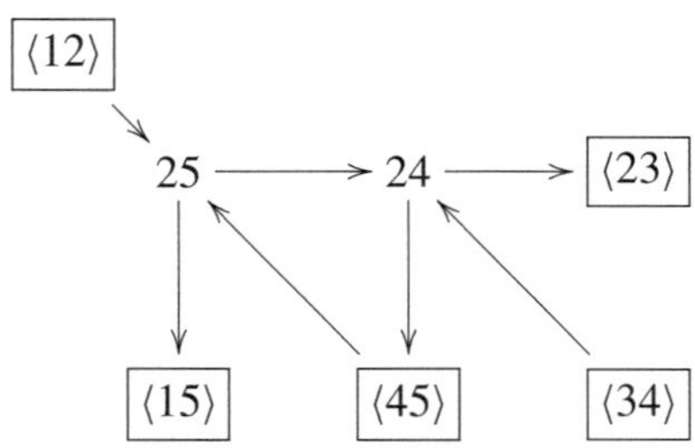

(9)

where we have also included explicitly a frozen variable $\langle 12\rangle$ which is equal to unity in the special parametrization we chose (on the part of the Grassmannian where $\langle 12\rangle \neq 0$).

After doing a mutation on the node $\langle 14\rangle$, we obtain a similar quiver diagram where the frozen vertex $\langle 15\rangle$ is special instead of $\langle 34\rangle$. Just like in the four-point case the arrows containing the mutated node get reversed and the link between $\langle 13\rangle$ and $\langle 34\rangle$ gets deleted and replaced with a link $\langle 13\rangle \to \langle 15\rangle$. It is easy to see that by mutating one gets the five similar quivers and nothing more.

The principal part of the quiver for configurations of five points in $\mathbb{CP}^1$ is the same as the Dynkin diagram of A_2 Lie algebra. Indeed, this *is* the A_2 cluster algebra we discussed in Sect. 4. The appearance of the A_2 Dynkin diagram provides the motivation for the name. We can define scaling invariant cross-ratios associated to any unfrozen node by taking the ratio of the product of coordinates in the quiver which can be reached by going against the arrows going in by the product of coordinates in the quiver which can be reached by following the arrows going out. For example, the cross-ratio corresponding to $\langle 13\rangle$ in the quiver (9) is given by $\frac{\langle 12\rangle\langle 34\rangle}{\langle 14\rangle\langle 23\rangle}$. A mutation reverses the arrows and therefore transforms these ratios to their inverse. These cross-ratios are the cluster variables of the A_2 algebra, and the exchange relations following from the quiver description can be shown to be the same as the exchange relations of the A_2 algebra.

More complicated cases appear for six points in $\mathbb{CP}^2$, where we obtain a D_4 Dynkin diagram. We can start with an initial quiver at the left below and mutate at vertex $\langle 236\rangle$ to obtain the principal part of the quiver shown at right, which is the same as the Dynkin diagram of D_4.

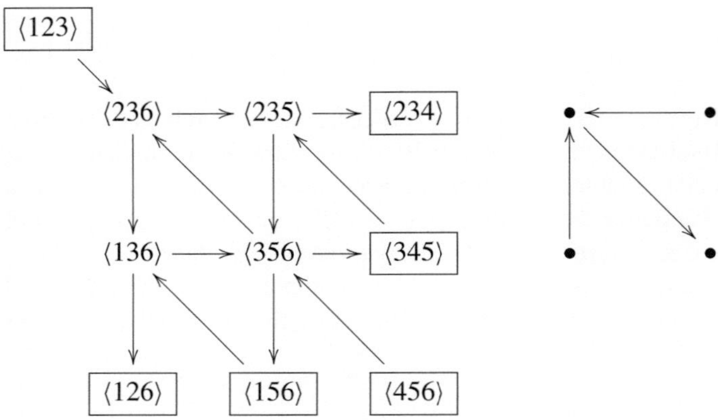

(10)

We should note that for the quiver in (10), the cross-ratio corresponding to the entry $\langle 356\rangle$ is given by $\frac{\langle 136\rangle\langle 235\rangle\langle 456\rangle}{\langle 156\rangle\langle 236\rangle\langle 345\rangle}$. This is more complicated than the cross-ratios which were obtained previously and it has some interesting properties. It appeared already in [45] (before the cluster algebras were discovered), in connection with functional equations for the trilogarithm. For a geometrical interpretation of this quantity see Sect. 7 and Figs. 4, 5 and 6.

In Ref. [38], Fomin and Zelevinsky showed that a cluster is of finite type (i.e. it has a finite number of cluster variables), if the principal part of its quiver can be transformed to a Dynkin diagram by a sequence of mutations. Furthermore, if the principal part of the quiver contains a subgraph which is an affine Dynkin diagram, then the cluster algebra is of infinite type. Using this characterization, one can show that the cluster algebras arising from $\mathbb{G}(2, n)$ and $\mathbb{G}(3, 6)$, $\mathbb{G}(3, 7)$ and $\mathbb{G}(3, 8)$ are of finite type. In Ref. [60], Scott has shown that all the other $\mathbb{G}(k, n)$ with $2 \le k \le \frac{n}{2}$ are of infinite type.

This has striking implications for scattering amplitudes in $\mathcal{N} = 4$ super-Yang-Mills theory which, as we have reviewed, are based on Grassmannians $\mathbb{G}(4, n)$, for $n \ge 6$. If $n = 6$ we obtain $\mathbb{G}(4, 6) = \mathbb{G}(2, 6)$ which is of finite type. If $n = 7$ we obtain $\mathbb{G}(4, 7) = \mathbb{G}(3, 7)$ which is again of finite type. However, starting at eight-point the cluster algebras are not of finite type anymore.

Notice that the seeds we have been using break the cyclic symmetry of the configuration of points. In order to see that the cyclic symmetry is preserved we need to show that two quivers whose labels are permuted by one unit are linked by a sequence of mutations. This can be shown in full generality (see Ref. [41] for details).

So far the most studied cases were $\mathbb{G}(4, n)$ for $n = 6, 7$. The case $n = 8$ is more complicated also because the cluster algebra is infinite. In the remainder of this section we will list a few of the cluster coordinates appearing for $\mathbb{G}(4, 8)$ and discuss their properties. By using mutations, one encounters

$$\langle 12(345) \cap (678)\rangle \equiv \langle 1345\rangle\langle 2678\rangle - \langle 2345\rangle\langle 1678\rangle. \tag{11}$$

Here, the $\cap$ notation emphasizes the following geometrical fact: the composite bracket $\langle 12(345) \cap (678)\rangle$ vanishes whenever the projective line $(345) \cap (678)$ obtained by intersecting two projective planes (345) and (678) and the points 1 and 2 lie in the same projective plane. This notation has been introduced in Ref. [4].

Already for $n = 7$ we encounter $\langle 12(345) \cap (567)\rangle$, when expressed in $\mathbb{CP}^3$ language. In previous work (see Ref. [45]) a different notation has been used for this quantity. First, a transformation to $\mathbb{CP}^2$ language was performed. Points in $\mathbb{CP}^2$ can be represented as vectors in $\mathbb{C}^3$, modulo rescalings. For two three-vectors v_1, v_2 we have a notion of vector product $v_1 \times v_2$ which is the vector orthogonal to the plane spanned by v_1 and v_2. Then, the composite brackets containing $\cap$ can be translated to

$$\langle v_1 \times w_1, v_2 \times w_2, v_3 \times w_3\rangle = \langle v_1 v_2 w_2\rangle\langle w_1 v_3 w_3\rangle - \langle w_1 v_2 w_2\rangle\langle v_1 v_3 w_3\rangle. \tag{12}$$

Above, the right-hand side does not have the same manifest symmetry as the left-hand side so more equivalent expressions can be found by applying permutations to the vector labels. Notice that the left-hand side vanishes when $v_1 \times w_1$ and $v_2 \times w_2$ differ by a rescaling. This is equivalent to the statement that the planes spanned by (v_1, w_1) and (v_2, w_2) are identical. Hence, $\langle v_1 v_2 w_2\rangle = 0$ and $\langle v_2 w_1 w_2\rangle = 0$ so the right-hand side vanishes as well.

Since the $\mathbb{G}(4, 8)$ cluster algebra is infinite, we are bound to find more and more complicated expressions. One remarkable feature of the mutations is that the denominator can always be canceled by the numerator, after using Plücker identities. Therefore, these coordinates always seem to be *polynomials* in the Plücker coordinates. This is an analog of the Laurent phenomenon, but this time we obtain polynomials.[5] As an example in $\mathbb{G}(4, 8)$, we have the following identity

$$\frac{\langle 1237\rangle\langle 1245\rangle\langle 1678\rangle + \langle 1278\rangle\langle 45(671) \cap (123)\rangle}{\langle 1267\rangle} = \langle 45(781) \cap (123)\rangle. \tag{13}$$

Here the left-hand side is the expression obtained following a mutation, while the right-hand side is the expression where the denominator has been canceled.

[5] This holds in many explicit examples, but I have not found a proof in the literature.

Even more complicated coordinates can be generated. As an example, we also find

$$\langle (123) \cap (345), (567) \cap (781) \rangle. \tag{14}$$

This vanishes when the lines $(123) \cap (345)$ and $(567) \cap (781)$ intersect. Equivalently, we can say that the lines $(345) \cap (567)$ and $(781) \cap (123)$ intersect.

6 Poisson Brackets

One can define a Poisson bracket on the cluster coordinates. It is enough to define the Poisson bracket between the coordinates in a given cluster. If X_i, X_j belong to the same cluster, i.e. they are vertices in the same quiver, then their Poisson bracket is defined as

$$\{X_i, X_j\} = b_{ij} X_i X_j, \tag{15}$$

where $b_{ij} = -b_{ji}$ is the b matrix of the cluster. The Poisson bracket is compatible with mutations. That is,

$$\{X'_i, X'_j\} = b'_{ij} X'_i X'_j, \tag{16}$$

where X'_i and b'_{ij} are obtained by a mutation from X_i and b_{ij}, respectively.

The Poisson structure is easiest to understand for $\mathbb{G}(2, n)$ cluster algebras (see Ref. [36] for a discussion). To a configuration of n points in $\mathbb{CP}^1$ with a cyclic ordering we associate a convex polygon. Each of the vertices of this polygon corresponds to one of the n points.

Then consider a complete triangulation of the polygon. Each of the $n-3$ diagonals in this triangulation determines a quadrilateral and therefore four points in $\mathbb{CP}^1$. Suppose a diagonal E determines a quadrilateral with vertices i, j, k, l where the ordering is the same as the ordering of the initial polygon. Using these four points we can form a cross-ratio $r(i, j, k, l) = \frac{z_{ij} z_{kl}}{z_{jk} z_{il}}$. We have $r(i, j, k, l) = r(k, l, i, j)$ which implies that the cross-ratio is uniquely determined by the diagonal E and we don't have to chose an orientation.

If we flip the diagonal E then the initial cross-ratio goes to its inverse, but the cross-ratios corresponding to neighboring quadrilaterals change in a more complicated way. In fact, they transform in the same way as the cluster coordinates, if the matrix b_{ij} is defined as follows. Two diagonals E and F in a given triangulation are called adjacent if they are the sides of one of the triangles of the triangulation. If the diagonals are adjacent we set $b_{EF} = 1$ if the diagonal E comes before F when listing the diagonals at the common vertex in clockwise order. Otherwise we set $b_{EF} = -1$. If two diagonals E and F are not adjacent we set $\epsilon_{EF} = 0$.

In general, it is hard to compute the Poisson bracket between two coordinates in different clusters. One approach is to express the second coordinate in terms of the coordinates of a cluster containing the first one. Then, we can use the definition. In general this is hard. Another approach is to use the Sklyanin bracket (see Ref. [40]). To explain this, we restrict again to the part of the Grassmannian $\mathbb{G}(k,n)$ where $\langle 1,\ldots,k\rangle \neq 0$ and we use a representative under the left GL(k) action which is $(\mathbf{1}_k, Y)$, where Y is a $k \times l$, $l = n-k$ matrix. We denote the entries of the matrix Y by y_{ij}, $i = 1,\ldots k$, $j = 1,\ldots,l$. On these coordinates we introduce a bracket called Sklyanin bracket given by

$$\{y_{ij}, y_{\alpha\beta}\}_S = (\text{sgn}(\alpha - i) - \text{sgn}(\beta - j))y_{i\beta}y_{\alpha j}. \tag{17}$$

In general, Sklyanin bracket is defined using an R-matrix, which is a solution of a modified classical Yang-Baxter equation (see Ref. [40] for details).

Now, we can extend the Sklyanin bracket to arbitrary functions of the variables y, in the usual way

$$\{f, g\}_S = \sum_{i,j,\alpha,\beta} \frac{\partial f}{\partial y_{ij}} \{y_{ij}, y_{\alpha\beta}\}_S \frac{\partial g}{\partial y_{\alpha\beta}}. \tag{18}$$

This bracket satisfies the Jacobi identity, as can be shown by direct computation, using the identity $\text{sgn}(x)\,\text{sgn}(y) + \text{sgn}(y)\,\text{sgn}(z) + \text{sgn}(z)\,\text{sgn}(x) = -1$ for $x+y+z = 0$ and $xyz \neq 0$.

The cluster coordinates can be expressed in terms of variables y and their bracket can be computed using the formula above. As an example, consider the case of the A_2 algebra again. There we have the cluster coordinates

$$X_1 = \frac{(12)(45)}{(15)(24)} = -\frac{y_{12}y_{23} - y_{13}y_{22}}{y_{12}y_{23}}, \qquad X_2 = \frac{(25)(34)}{(23)(45)} = \frac{y_{13}(y_{11}y_{22} - y_{12}y_{21})}{y_{11}(y_{12}y_{23} - y_{13}y_{22})}. \tag{19}$$

The computation of the bracket $\{X_1, X_2\}_S$ is a bit tedious, but straightforward. We find

$$\{X_1, X_2\}_S = 2X_1X_2. \tag{20}$$

Up to a factor of 2, we obtain the answer expected from the definition in terms of the b matrix of the quiver. Now, we can compute Poisson brackets between any cluster coordinates, even if they don't belong to the same cluster. Most of the Poisson brackets between coordinates which don't belong to the same cluster will be very complicated, but sometimes one obtains zero. This information combined with other physical requirements, can uniquely determine some parts of the amplitudes, as done for example in Ref. [44].

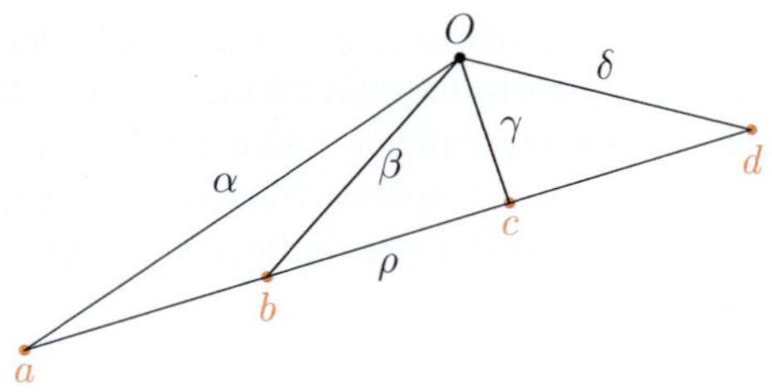

Fig. 1 The cross-ratio of four lines in $\mathbb{CP}^2$

7 Elements of Projective Geometry

It is very useful to understand the cross-ratios geometrically. For example, the A_2 cluster algebra described above involves the geometry of five points on $\mathbb{CP}^1$.

The simplest type of cross-ratio is the cross-ratio of four points (a, b, c, d) in $\mathbb{CP}^1$. If the points have have coordinates (z_a, z_b, z_c, z_d), then their cross-ratio is

$$r(a, b, c, d) = \frac{z_{ab} z_{cd}}{z_{bc} z_{da}}, \tag{21}$$

with $z_{ab} = z_a - z_b$. In the following we will try to reduce more complicated situations to configurations of four points on a projective line.

By duality, a point in $\mathbb{CP}^2$ is in correspondence with a line in $\mathbb{CP}^2$. A configuration of four points on a projective line in $\mathbb{CP}^2$ dualizes to a configuration of four lines intersecting in a point. Therefore, we can talk about the cross-ratio of four lines in $\mathbb{CP}^2$ (see Fig. 1).

The cross-ratios of four lines $(\alpha, \beta, \gamma, \delta)$ containing a point O can be related to the cross-ratio of four points by taking an arbitrary line ρ (not containing the point O) and computing the intersection points $a = \rho \cap \alpha, b = \rho \cap \beta, c = \rho \cap \gamma, d = \rho \cap \delta$. Then, the cross-ratio of the points (a, b, c, d) on ρ is independent on ρ and is equal to the cross-ratio of the lines $(\alpha, \beta, \gamma, \delta)$

$$r(\alpha, \beta, \gamma, \delta) = r(a, b, c, d). \tag{22}$$

If the lines are defined by pairs of points $\alpha = (OA)$, $\beta = (OB)$, $\gamma = (OC)$, $\delta = (OD)$, as in Fig. 2, then the cross-ratio of the four lines is

$$r(\alpha, \beta, \gamma, \delta) = r(a, b, c, d) = (O|A, B, C, D) \equiv \frac{\langle OAB \rangle \langle OCD \rangle}{\langle OBC \rangle \langle ODA \rangle}, \tag{23}$$

where $\langle XYZ \rangle$ is proportional to the oriented area of the triangle $\Delta(X, Y, Z)$.

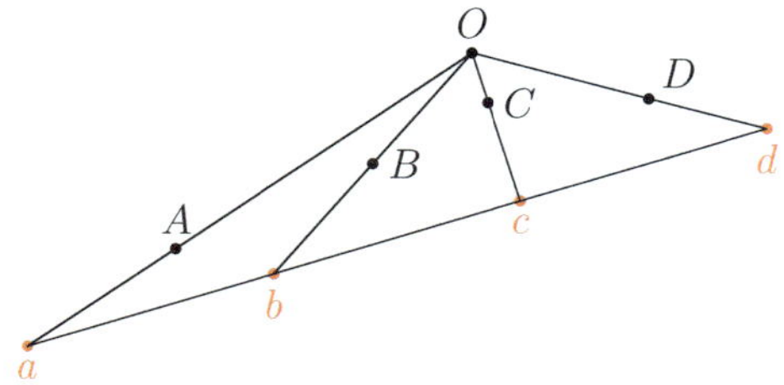

Fig. 2 The cross-ratio of four lines determined by their common intersection point O and another point on each on of them

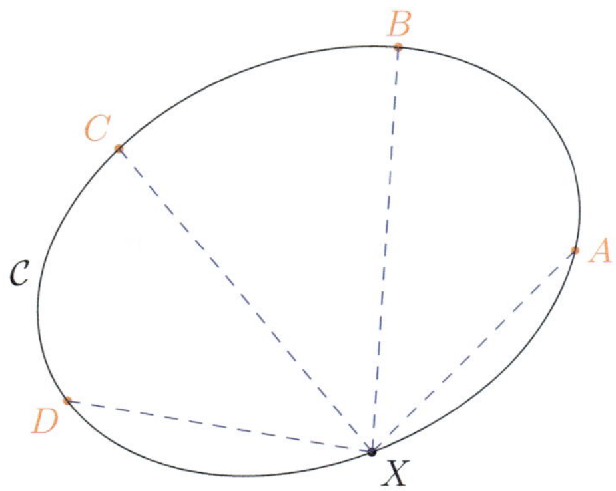

Fig. 3 The cross-ratio of points A, B, C, D with respect to the conic $\mathcal{C}$

If the four points A, B, C, D do not belong to a line we can't generically define their cross-ratio. However, given a conic $\mathcal{C}$ such that A, B, C, D belong[6] to $\mathcal{C}$, then we can define their cross-ratio as follows: pick a point X on the conic $\mathcal{C}$. Then, by Chasles' theorem the cross-ratio of the lines (XA), (XB), (XC) and (XD) is independent on the point X and is defined to be the cross-ratio of the points A, B, C, D (with respect to the conic $\mathcal{C}$). See Fig. 3.

Let us now discuss the triple ratio of six points in $\mathbb{CP}^2$ which was introduced by Goncharov. We take the six points to be A, B, C, X, Y, Z. Numerically, this triple ratio is given by

$$r_3(A, B, C; X, Y, Z) = \frac{\langle ABX\rangle\langle BCY\rangle\langle CAZ\rangle}{\langle ABY\rangle\langle BCZ\rangle\langle CAX\rangle}. \tag{24}$$

It turns out that this ratio has several geometrical interpretations. Consider first the situation in Fig. 4. There, we have four lines which are dashed and blue: $\alpha = (CB)$, $\beta = (Cb)$, $\gamma = (Cc)$, $\delta = (Cd)$, where $b = (AX) \cap (BY)$, $c = A$ and $d = (CZ) \cap (AX)$. Their cross-ratio, obtained by intersecting with the line (AX), is given by

[6]Any conic is determined by five points. Given four points there is an infinity of conics which contain them.

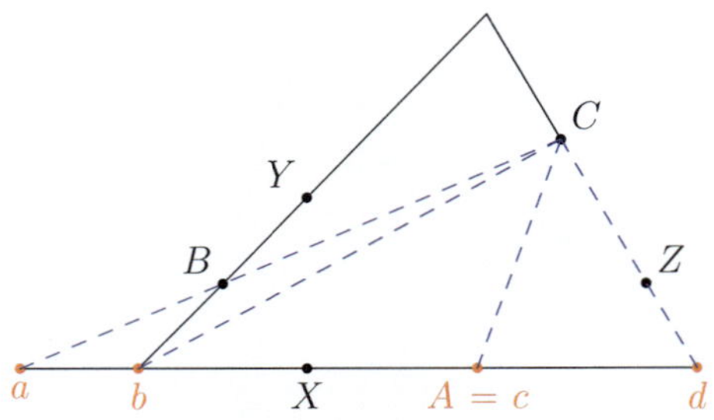

Fig. 4 Triple ratio, expressed as a cross-ratio of points on the line (AX)

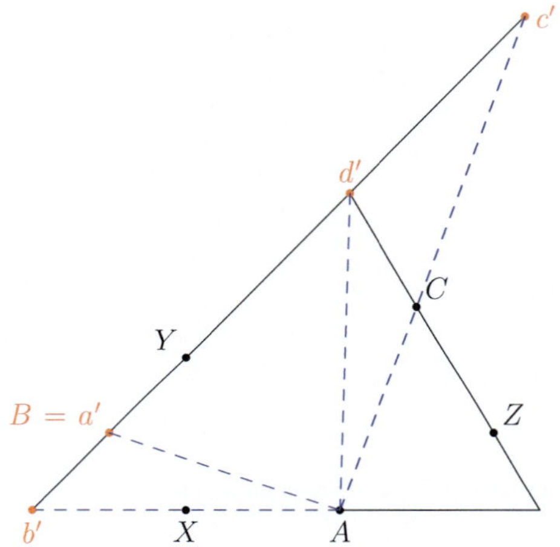

Fig. 5 Triple ratio, expressed as a cross-ratio of points on the line (BY)

$$r(\alpha, \beta, \gamma, \delta) = r(a, b, c, d) = (C|B, (AX) \cap (BY), A, Z). \tag{25}$$

But, instead of considering the intersections of the lines $(\alpha, \beta, \gamma, \delta)$ with the line (AX) as above, we can consider the intersection with the line (BY). The intersection points are

$$a' = \alpha \cap (BY) = B, \tag{26}$$
$$b' = \beta \cap (BY) = b = (AX) \cap (BY), \tag{27}$$
$$c' = \gamma \cap (BY) = (CA) \cap (BY), \tag{28}$$
$$d' = \delta \cap (BY) = (CZ) \cap (BY). \tag{29}$$

The corresponding figure is Fig. 5. If we denote by $\alpha' = (AB)$, $\beta' = (AX)$, $\gamma' = (AC)$, $\delta' = (Ad')$, we have

$$\begin{aligned} r(a, b, c, d) =& r(\alpha, \beta, \gamma, \delta) = r(a', b', c', d') = \\ &= r(\alpha', \beta', \gamma', \delta') = (A|B, X, C, (BY) \cap (CZ)). \end{aligned} \tag{30}$$

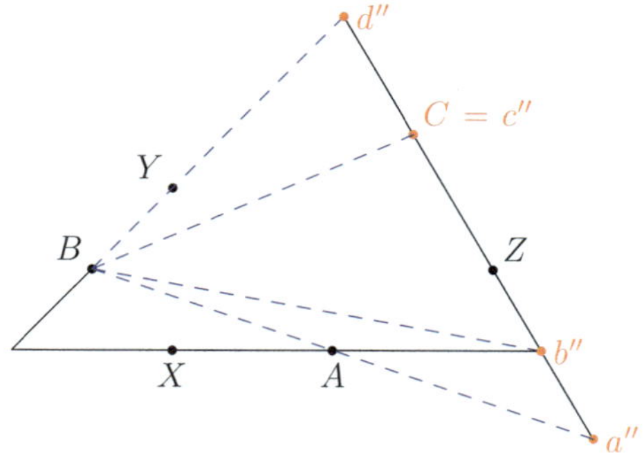

Fig. 6 Triple ratio, expressed as a cross-ratio of points on the line (CZ)

Now we can repeat the previous procedure. We compute the cross-ratio $r(\alpha', \beta', \gamma', \delta')$ by considering the intersection with (CZ). The intersection points are

$$a'' = \alpha' \cap (CZ) = (AB) \cap (CZ), \tag{31}$$
$$b'' = \beta' \cap (CZ) = (AX) \cap (CZ), \tag{32}$$
$$c'' = \gamma' \cap (CZ) = C, \tag{33}$$
$$d'' = \delta' \cap (CZ) = (BY) \cap (CZ). \tag{34}$$

See Fig. 6 for a geometrical representation. If we define the lines $\alpha'' = (BA)$, $\beta'' = (Bb'')$, $\gamma'' = (BC)$, $\delta'' = (Bd'')$, we have

$$(B|A, (CZ) \cap (AX), C, Y) = r(\alpha'', \beta'', \gamma'', \delta'') = r(a'', b'', c'', d'') = r(\alpha', \beta', \gamma', \delta'). \tag{35}$$

We have therefore shown that

$$(A|B, X, C, (BY) \cap (CZ)) = (B|A, (CZ) \cap (AX), C, Y) = (C|B, (AX) \cap (BY), A, Z). \tag{36}$$

Notice that this is also implied by the symmetry $r_3(A, B, C; X, Y, Z) = r_3(B, C, A; Y, Z, X)$.

Let us now show that the invariant $(A|B, X, C, (BY) \cap (CZ))$ has the same zeros and poles as $r_3(A, B, C; X, Y, Z)$. Form the definition, we know that $(A|B, X, C, (BY) \cap (CZ))$ vanishes when $\langle ABX \rangle = 0$ or $\langle AC(BY) \cap (CZ) \rangle = 0$. The second three-bracket vanishes if $\langle BCY \rangle = 0$ or $\langle CAZ \rangle = 0$. In the first case B, C, Y are collinear and therefore $(BY) \cap (CZ) = C$ so we have $\langle AC(BY) \cap (CZ) \rangle = \langle ACC \rangle = 0$. In the second case, when $\langle CAZ \rangle = 0$ we have that $A \in (CZ)$, $C \in (CZ)$ and $P \equiv (BY) \cap (CZ) \in (CZ)$. Since all the entries of the three-bracket are collinear, we find that $\langle AC(BY) \cap (CZ) \rangle = 0$. We have shown that $(A|B, X, C, (BY) \cap (CZ))$ vanishes if $\langle ABX \rangle = 0$ or $\langle BCY \rangle = 0$ or $\langle CAZ \rangle = 0$ which is the same as the numerator of $r_3(A, B, C; X, Y, Z)$. In order to find the poles we reason in the same way.

8 Polylogarithm Identities

In this section we provide some more mathematical details on transcendental functions and explain how to partially integrate them. We denote by $\mathcal{L}_n$ the Abelian group (under addition) of transcendental functions of transcendentality weight n. An important character in this story is the Bloch group B_n, also called the classical polylogarithm group: it is the subgroup of $\mathcal{L}_n$ generated by the classical polylogarithm functions Li_n and their products.

Consider first the simplest kind of transcendental function, the logarithm. If we are working modulo $2\pi i$, then we have that $\ln z + \ln w = \ln(zw)$, for any $z, w \in \mathbb{C}^*$. In order to express this simple functional relation formally, define $\mathbb{Z}[\mathbb{C}^*]$ to be the free Abelian group generated by $\{z\}$, with integer coefficients and z non-zero complex numbers. Concretely, elements of this group are quantities like $\{z\} + \{w\}$ and the group operation is defined in the obvious way. Then, we can quotient this group by the relations satisfied by the logarithm to obtain the logarithm group B_1,

$$B_1 = \mathbb{Z}[\mathbb{C}^*]/(\{z\} + \{w\} - \{zw\}). \tag{37}$$

This group is isomorphic to the multiplicative group of complex numbers, $\mathbb{C}^\times$.

The next simplest transcendental functions are the dilogarithms, Li_2. The dilogarithms satisfy a simple five-term functional relation. One way to express this functional relation is to consider five points on $\mathbb{CP}^1$ with coordinates $z_1, \ldots, z_5$. From any four such points we can form a cross-ratio $r(z_1, \ldots, \hat{z}_i, \ldots z_5)$, where the hatted argument is missing. We use the definition $r(i, j, k, l) = \frac{z_{ij} z_{kl}}{z_{jk} z_{li}}$ with $z_{ij} = z_i - z_j$. Then the five-term identity can be written as

$$\sum_{i=1}^{5} (-1)^i \mathrm{Li}_2(-r(z_1, \ldots, \hat{z}_i, \ldots, z_5)) = \text{logs}, \tag{38}$$

where we have denoted by logs the terms which can be written uniquely in terms of logarithms. There is a theorem (see Ref. [16]) that all the relations between dilogarithms are consequences of the five-term relations. We can now define the Bloch group B_2 by analogy to the logarithm case. We first define $\mathbb{Z}[\mathbb{C}]$ to be the free Abelian group generated by $\{z\}_2$, where z is a complex number. Then, we quotient be the five-term relations and the quotient is denoted by B_2

$$B_2 = \mathbb{Z}[\mathbb{C}]/(\text{five-term relations}). \tag{39}$$

In this case we have a group morphism δ, $B_2 \xrightarrow{\delta} \Lambda^2\mathbb{C}^*$ which is defined by $\delta(\{z\}_2) = (1 - z) \wedge z$. To check that this is a group morphism we need to show that $\delta(\text{five-term relation}) = 0$ or

$$\sum_{i=1}^{5}(-1)^i(1+r(z_1,\ldots,\hat{z}_i,\ldots,z_5))\wedge r(z_1,\ldots,\hat{z}_i,\ldots,z_5)=0, \tag{40}$$

which can be done by a short computation.

Let us now discuss Li_3 functions. There is a theorem stating that all transcendentality three functions can be written as a linear combination of Li_3 and products of lower transcendentality functions (see Ref. [45]).

Just like in the previous cases, we first need to find the functional relations satisfied by Li_3 functions. The identity satisfied by Li_3 is very similar to the one satisfied by Li_2 and can be described in terms of configurations of seven points on $\mathbb{CP}^2$. It is convenient to describe each of these points in terms of their homogeneous $v_i \in \mathbb{C}^3$ coordinates, with $i=1,\ldots,7$. For three such vectors v_i, v_j, v_k we can define a three-bracket $\langle\cdot,\cdot,\cdot\rangle : \mathbb{C}^3\times\mathbb{C}^3\times\mathbb{C}^3\to\mathbb{C}$ by the volume of the parallelepiped generated by them $\langle i,j,k\rangle = \mathrm{Vol}(v_i,v_j,v_k)$.

Given six points in $\mathbb{CP}^3$, we can form a cross-ratio

$$r_3(1,2,3,4,5,6)=\frac{\langle 124\rangle\langle 235\rangle\langle 316\rangle}{\langle 125\rangle\langle 236\rangle\langle 314\rangle}. \tag{41}$$

Such cross-ratios have been introduced and extensively used in Ref. [45] and we also discuss their geometric interpretation in Sect. 5. The Li_3 functional relations can be expressed in terms of this cross-ratio as

$$\sum_{i=1}^{7}(-1)^i\,\mathrm{Alt}_6\,\mathrm{Li}_3(-r_3(1,\ldots,\hat{\imath},\ldots,7))\approx 0, \tag{42}$$

where Alt_6 mean antisymmetrization in the six points on which r_3 depends and $\approx$ means that we have omitted the terms which are products of lower transcendentality functions.

Now we define

$$B_3=\mathbb{Z}[\mathbb{C}]/(\text{seven-term relations}). \tag{43}$$

There is a morphism $\delta : B_3 \to B_2\otimes\mathbb{C}^*$, $\delta(\{x\}_3)=\{x\}_2\otimes x$. In order to show that this morphism is well-defined, we need to show that that δ annihilates the seven-term relations.

It may seem that we can continue in the same way to higher transcendentality. However, this is not the case. At transcendentality four there are new functions which can not be expressed in terms of Li_4 and products of lower transcendentality functions. We can define B_n for $n\geq 4$ in the same way as before, but there is a bigger group $\mathcal{L}_n$ which is the Abelian group related to weight n polylogs, some of which are not classical polylogs.

We defined B_n to be the Abelian groups generated by classical polylogs and $\mathcal{L}_n$ to be the Abelian groups of all polylogs of weight n. Now we want to characterize them. The most mathematically concise way to describe their (conjectural!) connection is by an exact sequence, which for $n = 4$ reads

$$0 \to B_4 \to \mathcal{L}_4 \to \Lambda^2 B_2 \to 0. \tag{44}$$

An exact sequence is a sequence of maps between spaces such that the image of a map falls in the kernel of the next one. In the example above, the first arrow says that B_4 maps to $\mathcal{L}_4$ injectively, which is obvious since B_4 is contained in $\mathcal{L}_4$. The last arrow says that the map $\mathcal{L}_4 \to \Lambda^2 B_2$ is surjective. This is less obvious, but it means that for any element of $\Lambda^2 B_2$ one can find a weight four polylog with that $\Lambda^2 B_2$ projection.

Finally, the rest of the sequence means that $\ker(\mathcal{L}_4 \to \Lambda^2 B_2) = B_4$. This means that if a weight four polylog has zero $\Lambda^2 B_2$ projection, which is to say it belongs to $\ker(\mathcal{L}_4 \to \Lambda^2 B_2)$, then it is a classical polylog, and vice versa.

Notice that in Fig. 4, we have five points (a, b, X, c, d) on the line (AX). From five points $(z_1, \ldots, z_5)$ in $\mathbb{CP}^1$ we can produce a dilogarithm identity

$$\sum_{i=1}^{5} (-1)^i \{-r(z_1, \ldots, \widehat{z_i}, \ldots, z_5)\}_2 = 0. \tag{45}$$

This motivates us to find the expressions in terms of three-brackets for the other cross-ratios that can be constructed from these five points on (AX) (see Fig. 4):

$$r(b, X, A, d) = \frac{\langle BXY\rangle\langle ACZ\rangle}{\langle A \times X, B \times Y, C \times Z\rangle}, \tag{46}$$

$$r(a, X, A, d) = (C|B, X, A, Z), \tag{47}$$

$$r(a, b, A, d) = r_3(A, B, C; X, Y, Z), \tag{48}$$

$$r(a, b, X, d) = r_3(X, B, C; A, Y, Z), \tag{49}$$

$$r(a, b, X, A) = (B|C, Y, X, A). \tag{50}$$

This provides a geometric proof for the following dilogarithm identity

$$\begin{aligned} -\left\{\frac{\langle BXY\rangle\langle ACZ\rangle}{\langle A \times X, B \times Y, C \times Z\rangle}\right\}_2 + \left\{\frac{\langle CBX\rangle\langle CAZ\rangle}{\langle CXA\rangle\langle CZB\rangle}\right\}_2 - \left\{\frac{\langle ABX\rangle\langle BCY\rangle\langle CAZ\rangle}{\langle ABY\rangle\langle BCZ\rangle\langle CAX\rangle}\right\}_2 \\ + \left\{\frac{\langle XBA\rangle\langle BCY\rangle\langle CXZ\rangle}{\langle XBY\rangle\langle BCZ\rangle\langle CXA\rangle}\right\}_2 - \left\{\frac{\langle BCY\rangle\langle BXA\rangle}{\langle BYX\rangle\langle BAC\rangle}\right\}_2 = 0. \end{aligned} \tag{51}$$

Here is a 40-term trilogarithm identity which was discovered when analyzing results of two-loop computations in $\mathcal{N} = 4$ theory.

$$\left\{-\frac{\langle 125\rangle\langle 134\rangle}{\langle 123\rangle\langle 145\rangle}\right\}_3+\left\{-\frac{\langle 126\rangle\langle 145\rangle}{\langle 124\rangle\langle 156\rangle}\right\}_3+\left\{-\frac{\langle 126\rangle\langle 145\rangle\langle 234\rangle}{\langle 123\rangle\langle 146\rangle\langle 245\rangle}\right\}_3+$$
$$\frac{1}{3}\left\{-\frac{\langle 136\rangle\langle 145\rangle\langle 235}{\langle 123\rangle\langle 156\rangle\langle 345\rangle}\right\}_3+\text{(cyclic permutations)}-$$
$$\text{(anti-cyclic permutations)}=0. \qquad (52)$$

In order to check that the $B_2 \wedge \mathbb{C}^*$ projection of the 40-term trilogarithm identity is zero we need some dilogarithm identities. For example, one of the dilogarithm identities which is useful is

$$-\left\{-\frac{\langle 123\rangle\langle 456\rangle}{\langle 1\times 2,3\times 4,5\times 6\rangle}\right\}_2-\left\{-\frac{\langle 125\rangle\langle 134\rangle}{\langle 123\rangle\langle 145\rangle}\right\}_2-\left\{-\frac{\langle 123\rangle\langle 156\rangle\langle 345\rangle}{\langle 125\rangle\langle 134\rangle\langle 356\rangle}\right\}_2+$$
$$\left\{-\frac{\langle 124\rangle\langle 156\rangle\langle 345\rangle}{\langle 125\rangle\langle 134\rangle\langle 456\rangle}\right\}_2-\left\{-\frac{\langle 156\rangle\langle 345\rangle}{\langle 135\rangle\langle 456\rangle}\right\}_2=0. \qquad (53)$$

It can be interpreted geometrically as five points $(3, 4, (15)\cap(34), (12)\cap(34), (34)\cap(56))$ on the line (34).

The second useful dilogarithm identity is

$$\left\{-\frac{\langle 156\rangle\langle 234\rangle}{\langle 1\times 2,3\times 4,5\times 6\rangle}\right\}_2-\left\{-\frac{\langle 136\rangle\langle 234\rangle}{\langle 123\rangle\langle 346\rangle}\right\}_2-\left\{-\frac{\langle 156\rangle\langle 236\rangle}{\langle 126\rangle\langle 356\rangle}\right\}_2$$
$$+\left\{-\frac{\langle 123\rangle\langle 156\rangle\langle 346\rangle}{\langle 126\rangle\langle 134\rangle\langle 356\rangle}\right\}_2-\left\{-\frac{\langle 123\rangle\langle 256\rangle\langle 346\rangle}{\langle 126\rangle\langle 234\rangle\langle 356\rangle}\right\}_2=0. \qquad (54)$$

It can be interpreted geometrically as five points $(1, 2, (12)\cap(34), (12)\cap(36), (12)\cap(56))$ on the line (12).

The third useful dilogarithm identity is

$$-\left\{-\frac{\langle 156\rangle\langle 234\rangle}{\langle 1\times 2,3\times 4,5\times 6\rangle}\right\}_2+\left\{-\frac{\langle 145\rangle\langle 234\rangle}{\langle 124\rangle\langle 345\rangle}\right\}_2+\left\{-\frac{\langle 156\rangle\langle 245\rangle}{\langle 125\rangle\langle 456\rangle}\right\}_2-$$
$$\left\{-\frac{\langle 124\rangle\langle 156\rangle\langle 345\rangle}{\langle 125\rangle\langle 134\rangle\langle 456\rangle}\right\}_2+\left\{-\frac{\langle 124\rangle\langle 256\rangle\langle 345\rangle}{\langle 125\rangle\langle 234\rangle\langle 456\rangle}\right\}_2=0. \qquad (55)$$

It can be interpreted geometrically as five points $(1, 2, (12)\cap(34), (12)\cap(45), (12)\cap(56))$ on the line (12).

The fourth useful dilogarithm identity is

$$\left\{-\frac{\langle 123\rangle\langle 456\rangle}{\langle 1\times 2, 3\times 4, 5\times 6\rangle}\right\}_2 + \left\{-\frac{\langle 125\rangle\langle 234\rangle}{\langle 123\rangle\langle 245\rangle}\right\}_2 + \left\{-\frac{\langle 123\rangle\langle 256\rangle\langle 345\rangle}{\langle 125\rangle\langle 234\rangle\langle 356\rangle}\right\}_2 - \left\{-\frac{\langle 124\rangle\langle 256\rangle\langle 345\rangle}{\langle 125\rangle\langle 234\rangle\langle 456\rangle}\right\}_2 + \left\{-\frac{\langle 256\rangle\langle 345\rangle}{\langle 235\rangle\langle 456\rangle}\right\}_2 = 0. \tag{56}$$

It can be interpreted geometrically as five points $(3, 4, (12) \cap (34), (25) \cap (34), (34) \cap (56))$ on the line (34).

The identities above are the identities needed to show the vanishing of terms of type $* \otimes \langle 123\rangle$ in the projection to $B_2 \otimes \mathbb{C}^*$ of the 40-term trilogarithm identity. For the terms of type $* \otimes \langle 124\rangle$ the same identities are sufficient, but there is another, simpler identity too, written below

$$-\left\{-\frac{\langle 126\rangle\langle 145\rangle}{\langle 124\rangle\langle 156\rangle}\right\}_2 + \left\{-\frac{\langle 126\rangle\langle 245\rangle}{\langle 124\rangle\langle 256\rangle}\right\}_2 - \left\{-\frac{\langle 146\rangle\langle 245\rangle}{\langle 124\rangle\langle 456\rangle}\right\}_2 + \left\{-\frac{\langle 156\rangle\langle 245\rangle}{\langle 125\rangle\langle 456\rangle}\right\}_2 - \left\{-\frac{\langle 156\rangle\langle 246\rangle}{\langle 126\rangle\langle 456\rangle}\right\}_2 = 0. \tag{57}$$

This identity is special because it does not depend on point 3 at all. It can be more geometrically written as

$$\{(1|2654)\}_2 + \{(2|1456)\}_2 + \{(4|1652)\}_2 + \{(5|1246)\}_2 + \{(6|1542)\}_2 = 0. \tag{58}$$

Curiously, this simple-looking identity has a slightly more obscure geometrical interpretation. Through the five points 1, 2, 4, 5, 6 passes a unique conic $\mathcal{C}$. The cross-ratio $(1|2654)$ is the cross-ratio of the points $(2, 6, 5, 4)$ with respect to the conic $\mathcal{C}$. But we can pick another point $X \in \mathcal{C}$ and we have, by Chasles' theorem, that $(X|2654) = (1|2654)$. Then the previous identity becomes

$$\{(X|2456)\}_2 - \{(X|1456)\}_2 + \{(X|1256)\}_2 - \{(X|1246)\}_2 + \{(X|1245)\}_2 = 0, \tag{59}$$

which is the usual form of the dilogarithm identity, where the cross-ratios are cross-ratios of the lines $(X1)$, $(X2)$, $(X4)$, $(X5)$, $(X6)$.

9 Open Questions

The scattering amplitudes in $\mathcal{N} = 4$ theory split into sub-sectors which are not related by supersymmetry transformations. Scattering amplitudes in the simplest sectors are called MHV (maximally helicity violating) amplitudes, for historical reasons. More complicated sectors are called NMHV (next to MHV), etc. The six-point MHV amplitude has transcendentality four but, surprisingly, can be expressed in terms of classical polylogarithms only, as found in Ref. [46]. The next simplest amplitudes are the six-point NMHV, or the seven point MHV, which can not be written in terms of classical polylogarithms, since their $B_2 \wedge B_2$ projection does not vanish.

Consider the $\Lambda^2 B_2$ projection of the seven-point MHV amplitude computed in Ref. [21]. In $\mathbb{CP}^2$ language it is given by

$$
\begin{aligned}
-\left\{-\frac{\langle 2\times 3, 4\times 6, 7\times 1\rangle}{\langle 167\rangle\langle 234\rangle}\right\}_2 &\wedge \left\{-\frac{\langle 7\times 1, 2\times 3, 4\times 5\rangle}{\langle 127\rangle\langle 345\rangle}\right\}_2 \\
-\left\{-\frac{\langle 2\times 3, 4\times 6, 7\times 1\rangle}{\langle 167\rangle\langle 234\rangle}\right\}_2 &\wedge \left\{-\frac{\langle 234\rangle\langle 456\rangle}{\langle 246\rangle\langle 345\rangle}\right\}_2 \\
-\left\{-\frac{\langle 2\times 3, 4\times 6, 7\times 1\rangle}{\langle 167\rangle\langle 234\rangle}\right\}_2 &\wedge \left\{-\frac{\langle 146\rangle\langle 567\rangle}{\langle 167\rangle\langle 456\rangle}\right\}_2 \\
-\left\{-\frac{\langle 2\times 3, 4\times 6, 7\times 1\rangle}{\langle 167\rangle\langle 234\rangle}\right\}_2 &\wedge \left\{-\frac{\langle 5\times 6, 7\times 1, 2\times 3\rangle}{\langle 123\rangle\langle 567\rangle}\right\}_2 \\
+\left\{-\frac{\langle 137\rangle\langle 467\rangle}{\langle 167\rangle\langle 347\rangle}\right\}_2 \wedge \left\{-\frac{\langle 123\rangle\langle 347\rangle}{\langle 137\rangle\langle 234\rangle}\right\}_2 &- \left\{-\frac{\langle 137\rangle\langle 467\rangle}{\langle 167\rangle\langle 347\rangle}\right\}_2 \wedge \left\{-\frac{\langle 347\rangle\langle 456\rangle}{\langle 345\rangle\langle 467\rangle}\right\}_2 \\
&+ \text{cyclic permutations of } 1, 2, \ldots, 7. \qquad (60)
\end{aligned}
$$

Goncharov suggested to look at the Poisson bracket x, y for any $\{-x\}_2 \wedge \{-y\}_2 \in \Lambda^2 B_2$. This is well-defined since $\{-x\}_2 \wedge \{-y\}_2 = -\{-y\}_2 \wedge \{-x\}_2$ and a similar sign change appears from the Poisson bracket.

It is not understood why, but we find that these Poisson brackets are zero. We can show that for every term $\{-x\}_2 \wedge \{-y\}_2 \in \Lambda^2 B_2$ listed above there is at least one cluster containing x and y. In order to prove this, for every pair (x, y) we need to exhibit a quiver graph which contains them and which is such that there are no arrows between x and y. Alternatively, one can compute the Sklyanin bracket as in Sect. 6.

As mentioned in the introduction, scattering amplitudes have the property of factorization (see Ref. [3]). Formulating this precisely and studying its implications for the cluster algebra structure would be very interesting. A complete discussion would take us too far, but we want to mention only one important aspect: factorization only works if the transcendental functions satisfy some identities.

In mathematics one prefers to work with some *real* analytic functions, like

$$L_2(z) = \Im\left(\mathrm{Li}_2(z) + \ln|z| \ln(1-z)\right), \qquad (61)$$

$$L_3(z) = \Re\left(\mathrm{Li}_3(z) - \ln|z|\, \mathrm{Li}_2(z) - \frac{1}{3}\ln^2|z| \ln(1-z)\right), \qquad (62)$$

which have simple functional relations (modulo some additive constants, one can simply replace $\{z\}_2 \to L_2(z)$ and $\{z\}_3 \to L_3(z)$) to obtain an identity for functions. However, for physics we need to have *complex* analytic functions instead. Therefore, it is not yet clear what are the best building blocks for the scattering amplitudes.

The reader might be puzzled by the following fact: we have a big symmetry group $PSU(2, 2|4)$ but in terms of Grassmannians only the conformal group $SU(2, 2)$ or the complexified $SL(4)$ is visible. How to make the rest of the symmetry visible? This is not known at present. Maybe recent developments like the definition of cluster superalgebras in Ref. [56] hold the key to further progress.

Are there other polylogarithm identities of cluster type? As we have reviewed, the dilogarithm identity contains arguments which form an A_2 (or $\mathbb{G}(2,5)$ cluster algebra, while the trilogarithm identity contains arguments which form a D_4 (or $\mathbb{G}(3,6)$ cluster algebra. A computer search for a Li_4 identity with arguments in finite cluster algebra did not find anything. It is possible that there are such identities for infinite cluster algebras.

Before ending this brief review, let us point out some references which discuss complementary details. Cluster algebras appeared in Ref. [5] in connection with scattering amplitudes, but in a different way than we reviewed here. Reference [42] also reviews the connection between scattering amplitudes and cluster algebras, with an emphasis on the combinatorics of Stasheff polytopes. Reference [50] reviews the case of a three-dimensional analog of the $\mathcal{N}=4$ theory which we described here.

Many results were obtained by applying the bootstrap method (see Refs. [22, 25–30, 43, 44]).

Acknowledgements First, I would like to thank the organizers of the Opening Workshop of the Research Trimester on Multiple Zeta Values, Multiple Polylogarithms, and Quantum Field Theory: José I. Burgos Gil, Kurusch Ebrahimi-Fard, D. Ellwood, Ulf Kühn, Dominique Manchon and P. Tempesta.
I would also like to thank the participants and particularly Frédéric Chapoton and Herbert Gangl for discussions during the opening workshop Numbers and Physics (NAP2014). Finally, I am grateful to my coauthors in Refs. [41, 46] for collaboration.

References

1. Alday, L.F., Maldacena, J.M.: Gluon scattering amplitudes at strong coupling. *JHEP* **0706**, 064 (2007)
2. Anastasiou, C., Dixon, L., Bern, Z., Kosower, D.A.: Planar amplitudes in maximally supersymmetric yang-mills theory. Phys. Rev. Lett. **91**(25) (2003)
3. Anastasiou, C., Brandhuber, A., Heslop, P., Khoze, V.V., Spence, B., et al.: Two-loop polygon wilson loops in n = 4 SYM. JHEP **0905**, 115 (2009)
4. Arkani-Hamed, N., Bourjaily, J.L., Cachazo, F., Caron-Huot, S., Trnka, J.: The all-loop integrand for scattering amplitudes in planar n = 4 SYM. JHEP **1101**, 041 (2011)
5. Arkani-Hamed, N., Bourjaily, J.L., Cachazo, F., Goncharov, A.B., Postnikov, A., et al.: Scattering Amplitudes and the Positive Grassmannian (2012)
6. Arkani-Hamed, N., Cachazo, F., Cheung, C., Kaplan, J.: A duality for the s matrix. JHEP **1003**, 020 (2010)
7. Atiyah, M.F., Singer, I.M.: The index of elliptic operators on compact manifolds. Bull. Am. Math. Soc. **69**(3), 422–434 (1963)
8. Atiyah, M.F., Hitchin, N.J., Drinfeld, V.G., Manin, Y.I.: Construction of instantons. Phys. Lett. A **65**(3), 185–187 (1978)
9. Berenstein, A., Fomin, S., Zelevinsky, A.: Cluster algebras. iii: Upper bounds and double bruhat cells. Duke Math. J. **126**(1), 1–52 (2005)
10. Bern, Z., Carrasco, J.J.M., Johansson, H., Kosower, D.A.: Maximally supersymmetric planar yang-mills amplitudes at five loops. Phys. Rev. D **76**(12) (2007)

11. Bern, Z., Dixon, L.J., Kosower, D.A., Roiban, R., Spradlin, M., Vergu, C., Volovich, A.: Two-loop six-gluon maximally helicity violating amplitude in maximally supersymmetric yang-mills theory. Phys. Rev. D **78**(4) (2008)
12. Bern, Z., Dixon, L.J., Kosower, D.A., Roiban, R., Spradlin, M., et al.: The two-loop six-gluon mhv amplitude in maximally supersymmetric yang-mills theory. Phys. Rev. D **78**, 045007 (2008)
13. Bern, Z., Chalmers, G.: Factorization in one-loop gauge theory. Nucl. Phys. B **447**(2–3), 465–518 (1995)
14. Bern, Z., Czakon, M., Dixon, L.J., Kosower, D.A., Smirnov, V.A.: Four-loop planar amplitude and cusp anomalous dimension in maximally supersymmetric yang-mills theory. Phys. Rev. D **75**(8) (2007)
15. Bern, Z., Dixon, L.J., Smirnov, V.A.: Iteration of planar amplitudes in maximally supersymmetric yang-mills theory at three loops and beyond. Phys. Rev. D **72**, 085001 (2005)
16. Bloch, S.J.: Higher regulators, algebraic K-theory, and zeta functions of elliptic curves. CRM Monograph Series, vol. 11. American Mathematical Society, Providence, RI (2000)
17. Brandhuber, A., Heslop, P., Travaglini, G.: Mhv amplitudes in super-yang-mills and wilson loops. Nucl. Phys. B **794**(1–2), 231–243 (2008)
18. Brink, L., Schwarz, J.H., Scherk, J.: Supersymmetric yang-mills theories. Nucl. Phys. B **121**(1), 77–92 (1977)
19. Britto, R., Cachazo, F., Feng, B., Witten, E.: Direct proof of the tree-level scattering amplitude recursion relation in yang-mills theory. Phys. Rev. Lett. **94**(18) (2005)
20. Broadhurst, D.J., Kreimer, D.: Knots and numbers in ϕ^4 theory to 7 loops and beyond. Int. J. Mod. Phys. C **06**(04), 519–524 (1995)
21. Caron-Huot, S.: Superconformal symmetry and two-loop amplitudes in planar $n = 4$ super yang-mills. JHEP **1112**, 066 (2011)
22. Caron-Huot, S., He, S.: Jumpstarting the all-loop s-matrix of planar $\mathcal{N} = 4$ super yang-mills. J. High Energy Phys. **2012**(7) (2012)
23. Caron-Huot, S.: Notes on the scattering amplitude/wilson loop duality. JHEP **1107**, 058 (2011)
24. Coleman, S.R., Mandula, J.: All possible symmetries of the S matrix. Phys. Rev. **159**(5), 1251–1256 (1967)
25. Dixon, L.J., Drummond, J.M., Duhr, C., Pennington, J.: The four-loop remainder function and multi-regge behavior at nnlla in planar $\mathcal{N} = 4$ super-yang-mills theory. J. High Energy Phys. **2014**(6) (2014)
26. Dixon, L.J., Drummond, J.M., Henn, J.M.: Bootstrapping the three-loop hexagon. JHEP **1111**, 023 (2011)
27. Dixon, L.J., Drummond, J.M., Henn, J.M.: Analytic result for the two-loop six-point NMHV amplitude in $\mathcal{N} = 4$ super yang-mills theory. J. High Energy Phys. **2012**(1) (2012)
28. Dixon, L.J., Drummond, J.M., von Hippel, M., Pennington, J.: Hexagon functions and the three-loop remainder function. J. High Energy Phys. **2013**(12) (2013)
29. Dixon, L.J., von Hippel, M.: Bootstrapping an NMHV amplitude through three loops. J. High Energy Phys. **2014**(10) (2014)
30. Drummond, J.M., Papathanasiou, G., Spradlin, M.: A symbol of uniqueness: the cluster bootstrap for the 3-loop mhv heptagon. J. High Energy Phys. **2015**(3) (2015)
31. Drummond, J.M., Henn, J., Korchemsky, G.P., Sokatchev, E.: The hexagon wilson loop and the bds ansatz for the six-gluon amplitude. Phys. Lett. B **662**, 456–460 (2008)
32. Drummond, J.M., Henn, J., Korchemsky, G.P., Sokatchev, E.: Hexagon wilson loop = six-gluon mhv amplitude. Nucl. Phys. B **815**, 142–173 (2009)
33. Drummond, J.M., Henn, J., Korchemsky, G.P., Sokatchev, E.: Dual superconformal symmetry of scattering amplitudes in $n = 4$ super-yang-mills theory. Nucl. Phys. B **828**, 317–374 (2010)
34. Drummond, J.M., Henn, J., Smirnov, V.A., Sokatchev, E.: Magic identities for conformal four-point integrals. JHEP **0701**, 064 (2007)
35. Drummond, J.M., Korchemsky, G.P., Sokatchev, E.: Conformal properties of four-gluon planar amplitudes and wilson loops. Nucl. Phys. B **795**(1–2), 385–408 (2008)

36. Fock, V.V., Goncharov, A.B.: Cluster ensembles, quantization and the dilogarithm. Ann. Sci. Éc. Norm. Supér. (4) **42**(6), 865–930 (2009)
37. Fomin, S., Zelevinsky, A.: Cluster algebras. i: Foundations. J. Am. Math. Soc. **15**(2), 497–529 (2002)
38. Fomin, S., Zelevinsky, A.: Cluster algebras. ii: Finite type classification. Invent. Math. **154**(1), 63–121 (2003)
39. Fomin, S., Zelevinsky, A.: Cluster algebras. iv: coefficients. Compos. Math. **143**(1), 112–164 (2007)
40. Gekhtman, M., Shapiro, M., Vainshtein, A.: Cluster algebras and Poisson geometry. Mosc. Math. J. **3**(3), 899–934, 1199 (2003). Dedicated to Vladimir Igorevich Arnold on the occasion of his 65th birthday
41. Golden, J., Goncharov, A.B., Spradlin, M., Vergu, C., Volovich, A.: Motivic amplitudes and cluster coordinates. JHEP **1401**, 091 (2014)
42. Golden, J., Paulos, M.F., Spradlin, M., Volovich, A.: Cluster polylogarithms for scattering amplitudes. J. Phys. A: Math. Theor. **47**(47), 474005 (2014)
43. Golden, J., Spradlin, M.: An analytic result for the two-loop seven-point mhv amplitude in n = 4 SYM. J. High Energy Phys. **8**, 2014 (2014)
44. Golden, J., Spradlin, M.: A cluster bootstrap for two-loop MHV amplitudes. J. High Energy Phys. **2015**(2) (2015)
45. Goncharov, A.B.: Geometry of configurations, polylogarithms, and motivic cohomology. Adv. Math. **114**(2), 197–318 (1995)
46. Goncharov, A.B., Spradlin, M., Vergu, C., Volovich, A.: Classical polylogarithms for amplitudes and wilson loops. Phys. Rev. Lett. **105**(15), 11 (2010)
47. Haag, R., Lopuszanski, J.T., Sohnius, M.: All possible generators of supersymmetries of the s matrix. Nucl. Phys. B **88**(2), 257 (1975)
48. Hodges, A.: Eliminating spurious poles from gauge-theoretic amplitudes (2009)
49. 't Hooft, G.: A planar diagram theory for strong interactions. Nucl. Phys. B **72**(3), 461—473 (1974)
50. Huang, Y.-T., Wen, C., Xie, D.: The positive orthogonal grassmannian and loop amplitudes of abjm. J. Phys. A: Math. Theor. **47**(47), 474008 (2014)
51. Keller, B.: Cluster algebras, quiver representations and triangulated categories. In: Triangulated Categories. Cambridge University Press, Cambridge (2010)
52. Kotikov, A.V., Lipatov, L.N., Onishchenko, A.I., Velizhanin, V.N.: Three-loop universal anomalous dimension of the wilson operators in n = 4 susy yang-mills model. Phys. Lett. B **595**(1–4), 521–529 (2004)
53. Maldacena, J.: The large-n limit of superconformal field theories and supergravity. Int. J. Theor. Phys. **38**(4), 1113–1133 (1999)
54. Mason, L., Skinner, D.: Dual superconformal invariance, momentum twistors and grassmannians. J. High Energy Phys. **2009**(11), 045 (2009)
55. Mason, L.J., Skinner, D.: The complete planar s-matrix of n = 4 SYM as a wilson loop in twistor space. JHEP **1012**, 018 (2010)
56. Ovsienko, V.: Cluster superalgebras (2015)
57. Parke, S.J., Taylor, T.R.: Amplitude for n-gluon scattering. Phys. Rev. Lett. **56**, 2459–2460 (1986)
58. Penrose, R.: Twistor algebra. J. Math. Phys. **8**, 345 (1967)
59. Schnetz, O.: Quantum periods: a census of ϕ^4-transcendentals. Commun. Number Theory Phys. **4**(1), 1–47 (2010)
60. Scott, J.S.: Grassmannians and cluster algebras. Proc. Lond. Math. Soc. III. Ser. **92**(2), 345–380 (2006)
61. Witten, E.: Quantum field theory and the jones polynomial. Commun. Math. Phys. **121**(3), 351–399 (1989)
62. Witten, E.: Monopoles and four-manifolds. Math. Res. Lett. **1**(6), 769–796 (1994)

Multiple Eisenstein Series and q-Analogues of Multiple Zeta Values

Henrik Bachmann

Abstract This work is an example driven overview article of recent works on the connection of multiple zeta values, modular forms and q-analogues of multiple zeta values given by multiple Eisenstein series.

Keywords Multiple zeta values · q-analogues of multiple zeta values · Multiple Eisenstein series · Modular forms

1 Introduction

We study a specific connection of multiple zeta values and modular forms given by multiple Eisenstein series. This work is an example driven overview article and summary of the results obtained in the works [3, 6, 7, 9].

Multiple zeta values are real numbers that are natural generalizations of the Riemann zeta values. These are defined for integers $s_1 \geq 2$ and $s_2, \dots, s_l \geq 1$ by

$$\zeta(s_1, \dots, s_l) := \sum_{n_1 > n_2 > \cdots > n_l > 0} \frac{1}{n_1^{s_1} \dots n_l^{s_l}} .$$

Such real numbers were already studied by Euler in the $l = 2$ case in the 18th century. Because of its occurrence in various fields of mathematics and theoretical physics these real numbers had a comeback in the mathematical and physical research community in the late 1990s due to works by several people such as D. Broadhurst, F. Brown, P. Deligne, H. Furusho, A. Goncharov, M. Hoffman, M. Kaneko, D. Zagier et al.. Denote the $\mathbb{Q}$-vector space of all multiple zeta values of weight k by

$$\mathscr{Z}_k := \left\langle \zeta(s_1, \dots, s_l) \,\middle|\, s_1 + \cdots + s_l = k \text{ and } l > 0 \right\rangle_{\mathbb{Q}}$$

H. Bachmann (✉)
Graduate School of Mathematics, Nagoya University, Chikusa-ku, Nagoya 464-8602, Japan
e-mail: henrik.bachmann@math.nagoya-u.ac.jp

J. I. Burgos Gil et al. (eds.), *Periods in Quantum Field Theory and Arithmetic*, Springer Proceedings in Mathematics & Statistics 314,
https://doi.org/10.1007/978-3-030-37031-2_8

and write $\mathscr{Z}$ for the space of all multiple zeta values. One of the main interests is to understand the $\mathbb{Q}$-linear relations between these numbers. The first one is given by $\zeta(2,1) = \zeta(3)$ and there are several different ways to prove this relation [11]. Using the representation of multiple zeta values as an ordered sum, their product can be written as a linear combination of multiple zeta values of the same weight, i.e. the space $\mathscr{Z}$ has the structure of a $\mathbb{Q}$-algebra. For example it is

$$\zeta(2) \cdot \zeta(3) = \zeta(2,3) + \zeta(3,2) + \zeta(5)\,, \tag{1}$$

$$\zeta(3) \cdot \zeta(2,1) = \zeta(3,2,1) + \zeta(2,3,1) + \zeta(2,1,3) + \zeta(5,1) + \zeta(2,4)\,. \tag{2}$$

This way to express the product, which will be studied in Section 1 in more detail, is called the stuffle product (also named harmonic product). Besides this, a representation of multiple zeta values as iterated integrals yields another way to express the product of two multiple zeta values, which is called the shuffle product. For the above examples, this is given by

$$\zeta(2) \cdot \zeta(3) = \zeta(2,3) + 3\zeta(3,2) + 6\zeta(4,1)\,, \tag{3}$$

$$\begin{aligned}\zeta(3) \cdot \zeta(2,1) =& \zeta(2,1,3) + \zeta(2,2,2) + 2\zeta(2,3,1) + 2\zeta(3,1,2) \\ &+ 5\zeta(3,2,1) + 9\zeta(4,1,1)\,.\end{aligned} \tag{4}$$

Since (1) and (3) are two different expressions for the product $\zeta(2) \cdot \zeta(3)$ we obtain the linear relation $\zeta(5) = 2\zeta(3,2) + 6\zeta(4,1)$. These relations are called the double shuffle relations. Conjecturally all $\mathbb{Q}$-linear relations between multiple zeta values can be proven by using an extended version of these types of relations [24]. Often relations between multiple zeta values are not proven by using double shuffle relations, since there are easier ways to prove them in some cases. The relation $\zeta(4) = \zeta(2,1,1)$ for example, has an easy proof using the iterated integral expressions for multiple zeta values. A famous result by Euler is, that every even zeta value $\zeta(2k)$ is a rational multiple of π^{2k} and in particular we have, for example,

$$\zeta(2)^2 = \frac{5}{2}\zeta(4)\,, \quad \zeta(4)^2 = \frac{7}{6}\zeta(8)\,, \quad \zeta(6)^2 = \frac{715}{691}\zeta(12)\,. \tag{5}$$

The relations (5) can also be proven with the double shuffle relations, but for general k there is no explicit proof of Eulers relations using only double shuffle relations so far.

Since the double shuffle relations just give relations in a fixed weight it is conjectured that the space $\mathscr{Z}$ is a direct sum of the $\mathscr{Z}_k$, i.e. there are no relations between multiple zeta values with different weight.

Surprisingly there are several connections of these numbers to modular forms for the full modular group. Recall, modular forms are holomorphic functions in the complex upper half-plane fulfilling certain functional equations. One of the most famous connection is the Broadhurst-Kreimer conjecture.

Conjecture 1 (Broadhurst-Kreimer conjecture [15]) *The generating series of for the dimension* $\dim_{\mathbb{Q}}\left(\mathscr{Z}_{k,l}\right)$ *of weight k multiple zeta values of length l modulo lower lengths can be written as*

$$\sum_{\substack{k\geq 0\\ l\geq 0}} \dim_{\mathbb{Q}}\left(\mathscr{Z}_{k,l}\right) X^k Y^l = \frac{1+\mathbb{E}(X)Y}{1-\mathbb{O}(X)Y+\mathbb{S}(X)Y^2-\mathbb{S}(X)Y^4},$$

where

$$\mathbb{E}(X) = \frac{X^2}{1-X^2}, \quad \mathbb{O}(X) = \frac{X^3}{1-X^2}, \quad \mathbb{S}(X) = \frac{X^{12}}{(1-X^4)(1-X^6)}.$$

The connection to modular forms arises here, since

$$\mathbb{S}(X) = \sum_{k\geq 0} \dim S_k(\mathrm{SL}_2(\mathbb{Z}))X^k$$

is the generating function of the dimensions of cusp forms for the full modular group. In the formula of the Broadhurst-Kreimer conjecture one can see, that cusp forms give rise to relations between double zeta values, i.e. multiple zeta values in the length $l = 2$ case. For example in weight 12, the first weight where non-trivial cusp forms exist, there is the following famous relation

$$\frac{5197}{691}\zeta(12) = 168\zeta(5,7) + 150\zeta(7,5) + 28\zeta(9,3). \tag{6}$$

Even though we are not focused on this conjecture, the concept of obtaining relations of multiple zeta values by cusp forms also appears in our context of multiple Eisenstein series and q-analogues of multiple zeta values. It is known that every modular form for the full modular group can be written as a polynomial in classical Eisenstein series. These are for even $k > 0$ given by

$$G_k(\tau) = \frac{1}{2}\sum_{\substack{(m,n)\in\mathbb{Z}^2\\(m,n)\neq(0,0)}} \frac{1}{(m\tau+n)^k} = \zeta(k) + \frac{(-2\pi i)^k}{(k-1)!}\sum_{n=1}^{\infty}\sigma_{k-1}(n)q^n,$$

where $\tau \in \mathbb{H}$ is an element in the upper half-plane, $q = \exp(2\pi i\tau)$ and $\sigma_k(n) = \sum_{d|n} d^k$ denotes the classical divisor-sum. In [19] the authors introduced a direct connection of modular forms to double zeta values following ideas of Don Zagier introduced in [37]. They defined double Eisenstein series $G_{s_1,s_2} \in \mathbb{C}[[q]]$ which are a length two generalization of classical Eisenstein series and which are given by a double sum over ordered lattice points. These functions have a Fourier expansion given by sums of products of multiple zeta values and certain q-series with the double zeta value $\zeta(s_1, s_2)$ as their constant term. In [2] the author treated the multiple cases and calculated the Fourier expansion of multiple Eisenstein series $G_{s_1,\dots,s_l} \in \mathbb{C}[[q]]$. The result of [2] was that the Fourier expansion of multiple Eisenstein series is again a $\mathscr{Z}$-linear combination of multiple zeta values and the q-series $g_{t_1,\dots,t_m} \in \mathbb{C}[[q]]$ defined by $g_{t_1,\dots,t_m}(\tau) := (-2\pi i)^{t_1+\dots+t_m}[t_1,\dots,t_m]$ with $q = e^{2\pi i\tau}$ and

$$[t_1,\dots,t_m] := \sum_{\substack{u_1>\dots>u_m>0\\ v_1,\dots,v_m>0}} \frac{v_1^{t_1-1}\dots v_m^{t_m-1}}{(t_1-1)!\dots(t_m-1)!}\cdot q^{u_1v_1+\dots+u_mv_m}\,.$$

Theorem 2 ([2]) *For $s_1,\dots,s_l \geq 2$ the $G_{s_1,\dots,s_l}$ can be written as a $\mathscr{Z}$-linear combination of the above functions $g_{t_1,\dots,t_m}$.*

For example:

$$\begin{aligned} G_{3,2,2}(\tau) =&\,\zeta(3,2,2) + \left(\frac{54}{5}\zeta(2,3) + \frac{51}{5}\zeta(3,2)\right)g_2(\tau) + \frac{16}{3}\zeta(2,2)g_3(\tau)\\ &+ 3\zeta(3)g_{2,2}(\tau) + 4\zeta(2)g_{3,2}(\tau) + g_{3,2,2}(\tau)\,. \end{aligned}$$

The starting point of the thesis [4] was the fact that there are more multiple zeta values than multiple Eisenstein series, since $\zeta(s_1,\dots,s_l)$ exists for all $s_1 \geq 2, s_2,\dots,s_l \geq 1$ and the $G_{s_1,\dots,s_l}$ just exists when all $s_j \geq 2$. The main objective was to answer the following question.

Question 1 What is a "good" definition of a "regularized" multiple Eisenstein series, such that for each multiple zeta value $\zeta(s_1,\dots,s_l)$ with $s_1 > 1, s_2,\dots,s_l \geq 1$ there is a q-series

$$G^{reg}_{s_1,\dots,s_l} = \zeta(s_1,\dots,s_l) + \sum_{n>0} a_n q^n \in \mathbb{C}[[q]]$$

with this multiple zeta value as the constant term in its Fourier expansion and which equals the multiple Eisenstein series in the cases $s_1,\dots,s_l \geq 2$?

By "good" we mean that these regularized multiple Eisenstein series should have the same, or at least as close as possible, algebraic structure similar to multiple zeta values. Our answer to this question was approached in several steps which will be described in the following (i)–(iii). First (i) the algebraic structure of the functions g was studied. During this investigation it turned out, that these objects, or more precisely the q-series $[s_1,\dots,s_l]$ are very interesting objects in their own rights. It turned out that in order to understand their algebraic structure it was necessary to study a more general class of q-series, called bi-brackets in (ii). The results on bi-brackets and brackets then were used, together with a beautiful connection of the multiple Eisenstein series to the coproduct structure of formal iterated integrals, to answer the above question in (iii).

(i) To answer Question 1 the algebraic structure of the functions g or more precisely the algebraic structure of the q-series $[s_1,\dots,s_l]$ was studied in [6]. It turned out that these q-series, whose coefficients are given by weighted sums over partitions of n, are, independently to their appearance in the Fourier expansion of multiple Eisenstein series, very interesting objects. We will denote the $\mathbb{Q}$-vector space spanned by all these brackets and 1 by $\mathcal{MD}$. Since we also include the rational numbers, the normalized Eisenstein series $\widetilde{G}_k(\tau) := (-2\pi i)^{-k} G_k(\tau)$ are contained in $\mathcal{MD}$. For example we have

$$\widetilde{G}_2 = -\frac{1}{24} + [2]\,, \quad \widetilde{G}_4 = \frac{1}{1440} + [4]\,, \quad \widetilde{G}_6 = -\frac{1}{60480} + [6]\,.$$

The algebraic structure of the space $\mathcal{MD}$ was studied in [6] and one of the main result was the following

Theorem 3 ([6]) *The $\mathbb{Q}$-vector space spanned by all brackets equipped with the usual multiplication of formal q-series is a $\mathbb{Q}$-algebra, with the algebra of modular forms with rational coefficients as a subalgebra.*

In fact, the product fulfills a quasi-shuffle product and the notion of quasi-shuffle products will be made precise in Sect. 4.1. Roughly speaking, this means that the product of two brackets can be expressed as a linear combination of brackets similar to the stuffle product (1), (2) of multiple zeta values. For example we will see that

$$\begin{aligned} [2]\cdot[3] &= [3,2]+[2,3]+[5]-\frac{1}{12}[3]\,,\\ [3]\cdot[2,1] &= [3,2,1]+[2,3,1]+[2,1,3]+[5,1]+[2,4]\\ &\quad+\frac{1}{12}[2,2]-\frac{1}{2}[2,3]-\frac{1}{12}[3,1]\,, \end{aligned}$$

i.e. up to the lower weight term $-\frac{1}{12}[3]$ and $\frac{1}{12}[2,2]-\frac{1}{2}[2,3]-\frac{1}{12}[3,1]$ this looks exactly like (1) and (2). One might ask if there is also something which corresponds to the shuffle product (3) of multiple zeta values. It turned out that for the lowest length case, this has to do with the differential operator $\mathrm{d} = q\frac{d}{dq}$. In [6] it was shown that

$$[2]\cdot[3] = [2,3]+3[3,2]+6[4,1]-3[4]+\mathrm{d}[3]\,, \tag{7}$$

which, again up to the term $-3[4]+\mathrm{d}[3]$, looks exactly like the shuffle product (3) of multiple zeta values. In particular it follows that $\mathrm{d}[3]$ is again in the space $\mathcal{MD}$ and in general it was shown that

Theorem 4 ([6]) *The operator* $\mathrm{d} = q\frac{d}{dq}$ *is a derivation on* $\mathcal{MD}$.

(ii) Equation (7) above was the motivation to study a larger class of q-series, which will be called bi-brackets. While the quasi-shuffle product of brackets also exists in higher length, the second expression for the product, corresponding to the shuffle product, does not appear in higher length if one just allows derivatives as "error terms". The bi-brackets can be seen as a generalization of the derivative of brackets. For $s_1,\dots,s_l \geq 1, r_1,\dots,r_l \geq 0$ we define these bi-brackets by

$$\begin{bmatrix} s_1, \dots, s_l \\ r_1, \dots, r_l \end{bmatrix} := \sum_{\substack{u_1 > \dots > u_l > 0 \\ v_1, \dots, v_l > 0}} \frac{u_1^{r_1}}{r_1!} \cdots \frac{u_l^{r_l}}{r_l!} \cdot \frac{v_1^{s_1 - 1} \dots v_l^{s_l - 1}}{(s_1 - 1)! \dots (s_l - 1)!} \cdot q^{u_1 v_1 + \dots + u_l v_l} .$$

In the case $r_1 = \dots = r_l = 0$ these are just ordinary brackets. The products of these seemingly larger class of q-series have two representations similar to the stuffle and shuffle product of multiple zeta values in arbitrary length. For our example, the analogue of the shuffle product (4) for brackets can now be expressed as

$$\begin{aligned} [3] \cdot [2, 1] &= [2, 1, 3] + [2, 2, 2] + 2[2, 3, 1] + 2[3, 1, 2] + 5[3, 2, 1] + 9[4, 1, 1] \\ &+ \begin{bmatrix} 2, 3 \\ 0, 1 \end{bmatrix} + 2\begin{bmatrix} 3, 2 \\ 0, 1 \end{bmatrix} + 3\begin{bmatrix} 4, 1 \\ 1, 0 \end{bmatrix} - [2, 3] - 2[3, 2] - 6[4, 1] . \end{aligned}$$

We will see in Sect. 5.2 that these double shuffle structure can be described, using the so called partition relation, in a nice combinatorial way. This gives a large family of linear relations between bi-brackets. In fact numerical calculations show, that there are so many relations, that we have the following surprising conjecture.

Conjecture 5 ([3]) *Every bi-bracket can be written in terms of brackets, i.e.*

$$\mathcal{MD} = \mathcal{BD} .$$

Using the algebraic structure of the space of bi-brackets we now review the definition of shuffle brackets $[s_1, \dots, s_l]^{\sqcup\!\sqcup}$ and stuffle $[s_1, \dots, s_l]^*$ version of the ordinary brackets as certain linear combination of bi-brackets as introduced in [3]. These objects fulfill the same shuffle and stuffle products as multiple zeta values respectively. Both constructions use the theory of quasi-shuffle algebras first developed by Hoffman in [21] and later generalized in [22]. We summarize the results in the following Theorem.

Theorem 6 ([3])

(i) *The space $\mathcal{BD}$ spanned by all bi-brackets $\left[\begin{smallmatrix} s_1, \dots, s_l \\ r_1, \dots, r_l \end{smallmatrix}\right]$ forms a $\mathbb{Q}$-algebra with the space of (quasi-)modular forms and the space $\mathcal{MD}$ of brackets as subalgebras. There are two ways to express the product of two bi-brackets which correspond to the stuffle and shuffle product of multiple zeta values.*

(ii) *There are two subalgebras $\mathcal{MD}^{\sqcup\!\sqcup} \subset \mathcal{BD}$ and $\mathcal{MD}^* \subset \mathcal{MD}$ spanned by elements $[s_1, \dots, s_l]^{\sqcup\!\sqcup}$ and $[s_1, \dots, s_l]^*$ which fulfill the shuffle and stuffle products, respectively, and which are in the length one case given by the bracket $[s_1]$.*

For example, similarly to the relation between multiple zeta values above we have

$$[2, 3]^* + [3, 2]^* + [5] = [2] \cdot [3] = [2, 3]^{\sqcup\!\sqcup} + 3[3, 2]^{\sqcup\!\sqcup} + 6[4, 1]^{\sqcup\!\sqcup} .$$

(iii) A particular reason for studying the $[s_1, \ldots, s_l]^{ш}$ is due to their use in the regularization of multiple Eisenstein series, i.e. they are needed in the answer of the original Question 1. This was implicitly done in [9] by proving an explicit connection of the Fourier expansion of multiple Eisenstein series to the coproduct on formal iterated integrals introduced by Goncharov in [20]. This connection was already known to the authors of [19] in the length two case. Without knowing this connection it was then rediscovered independently by the authors of [9] during a research stay of the second author at the DFG Research training Group 1670 at the University of Hamburg in 2014. The result of this research stay was the work [9], in which the authors used this connection to give a definition of the shuffle regularized multiple Eisenstein series. Later, the present author combined the result of [9] and the algebraic structure of bi-brackets to give a more explicit definition of shuffle regularized multiple Eisenstein series using bi-brackets in [3].

Formal iterated integrals are symbols $I(a_0; a_1, \ldots, a_n; a_{n+1})$ with $a_j \in \{0, 1\}$ that fulfill identities like real iterated integrals. We will write $I(3, 2)$ for $I(1; 00101; 0)$ and we will see that the elements of the form $I(s_1, \ldots, s_l)$, obtained in the same way as $I(3, 2)$, form a basis of the space of formal iterated integrals in which we are interested. The space of these integrals has a Hopf algebra structure with the multiplication given by the shuffle product and the coproduct Δ given by an explicit formula which we will review in Sect. 6.1. For example it is

$$\Delta(I(3,2)) = 1 \otimes I(3,2) + 3I(2) \otimes I(3) + 2I(3) \otimes I(2) + I(3,2) \otimes 1\,.$$

Compare this with the Fourier expansion of the double Eisenstein series $G_{3,2}$

$$G_{3,2}(\tau) = \zeta(3,2) + 3g_2(\tau)\zeta(3) + 2g_3(\tau)\zeta(2) + g_{3,2}(\tau)\,.$$

Since $\Delta(I(s_1, \ldots, s_l))$ exists for all $s_1, \ldots, s_l \geq 1$ this comparison suggested a definition of shuffle regularized multiple Eisenstein series $G^{ш}_{s_1,\ldots,s_l}$ by sending the first component of the coproduct of $I(s_1, \ldots, s_l)$ to a $(-2\pi i)$-multiple of the shuffle bracket and the second component to shuffle regularized multiple zeta values. In [9] it was proven that this construction gives back the original multiple Eisenstein series in the cases $s_1, \ldots, s_l \geq 2$. Together with the results on the shuffle brackets in [3] we obtain the following

Theorem 7 ([3, 9]) *For all $s_1, \ldots, s_l \geq 1$ there exist shuffle regularized multiple Eisenstein series $G^{ш}_{s_1,\ldots,s_l} \in \mathbb{C}[[q]]$ with the following properties:*

(i) They are holomorphic functions on the upper half-plane (by setting $q = \exp(2\pi i \tau)$) having a Fourier expansion with the shuffle regularized multiple zeta values as the constant term.
(ii) They fulfill the shuffle product.
(iii) They can be written as a linear combination of multiple zeta values, powers of $(-2\pi i)$ and shuffle brackets $[\ldots]^{ш} \in \mathcal{BD}$.

(iv) *For integers* $s_1, \ldots, s_l \geq 2$ *they equal the multiple Eisenstein series*

$$G^{\sqcup\!\sqcup}_{s_1,\ldots,s_l}(\tau) = G_{s_1,\ldots,s_l}(\tau)$$

and therefore they fulfill the stuffle product in these cases.

We now study the $\mathbb{Q}$-algebra spanned by the $G^{\sqcup\!\sqcup}$ and its relation to multiple zeta values. Theorem 7 (iv) gives a subset of the double shuffle relations between the $G^{\sqcup\!\sqcup}$, since the stuffle product is just fulfilled for the case $s_1, \ldots, s_l \geq 2$. A natural question is, if they also fulfill the stuffle product when some indices s_j are equal to 1. For some cases this was proven in [3]. For example it was shown, that

$$G^{\sqcup\!\sqcup}_2 \cdot G^{\sqcup\!\sqcup}_{2,1} = G^{\sqcup\!\sqcup}_{2,1,2} + 2G^{\sqcup\!\sqcup}_{2,2,1} + G^{\sqcup\!\sqcup}_{2,3} + G^{\sqcup\!\sqcup}_{4,1}\,. \tag{8}$$

The method to prove this was to introduce stuffle regularized multiple Eisenstein series $G^*_{s_1,\ldots,s_l}$, which fulfill by construction the stuffle product and which equal the classical multiple Eisenstein series in the $s_1, \ldots, s_l \geq 2$ cases. Since both G^* and $G^{\sqcup\!\sqcup}$ can be written in terms of multiple zeta values and bi-brackets it was possible to compare these two regularization. It was shown that all $G^{\sqcup\!\sqcup}$ appearing in (31) equal the G^* ones, from which this equation followed. In contrast to the shuffle regularized multiple Eisenstein series the stuffle regularized ones could not be defined for all $s_1, \ldots, s_l \geq 1$, but we have the following results:

Theorem 8 ([3]) *For all* $s_1, \ldots, s_l \geq 1$ *and* $M \geq 1$ *there exists* $G^{*,M}_{s_1,\ldots,s_l} \in \mathbb{C}[[q]]$ *with the following properties*

(i) *They are holomorphic functions on the upper half-plane (by setting* $q = \exp(2\pi i\tau)$*) having a Fourier expansion with the stuffle regularized multiple zeta values as the constant term.*
(ii) *They fulfill the stuffle product.*
(iii) *In the case where the limit* $G^*_{s_1,\ldots,s_l} := \lim_{M\to\infty} G^{*,M}_{s_1,\ldots,s_l}$ *exists, the functions* $G^*_{s_1,\ldots,s_l}$ *are a linear combination of multiple zeta values, powers of* $(-2\pi i)$ *and bi-brackets.*
(iv) *For* $s_1, \ldots, s_l \geq 2$ *the* $G^*_{s_1,\ldots,s_l}$ *exist and equal the classical multiple Eisenstein series*

$$G_{s_1,\ldots,s_l}(\tau) = G^*_{s_1,\ldots,s_l}(\tau)\,.$$

It is still an open question which extended double shuffle relations of multiple zeta values are also fulfilled for the $G^{\sqcup\!\sqcup}$. Or equivalently, under what circumstances the product of two $G^{\sqcup\!\sqcup}$ can be expressed using the stuffle product formula. Clearly there are some double shuffle relations which can't be fulfilled by multiple Eisenstein series. For example not all of the Euler relations (5) are fulfilled since G_2^2 is not a multiple of G_4 as G_2 is not modular and G_6^2 is not a multiple of G_{12} as there are cusp forms in weight 12. In Sect. 6.3 we will explain this failure in terms of the double shuffle relations which are fulfilled by multiple Eisenstein series.

After the discussion above, we believe that Question 1 got a satisfying answer given by the regularized multiple Eisenstein series $G^{\sqcup\!\sqcup}$ and G^*. To go back from multiple Eisenstein series to multiple zeta values one can consider the projection to the constant term. But there is another direct connection of brackets, and therefore also of the subalgebra of modular forms, to multiple zeta values. The brackets can be seen as a q-analogue of multiple zeta values. A q-analogue of multiple zeta values is said to be a q-series which gives back multiple zeta values in the case $q \to 1$. Define for $k \in \mathbb{N}$ the map $Z_k : \mathbb{Q}[\![q]\!] \to \mathbb{R} \cup \{\infty\}$ by

$$Z_k(f) = \lim_{q \to 1} (1-q)^k f(q) .$$

Proposition 9 ([6, Proposition 6.4]) *For $s_1 \geq 2$ and $s_2, \ldots, s_l \geq 1$ the map Z_k sends a bracket to the corresponding multiple zeta value, i.e.*

$$Z_k([s_1, \ldots, s_l]) = \begin{cases} \zeta(s_1, \ldots, s_l) , & s_1 + \cdots + s_l = k, \\ 0 , & s_1 + \cdots + s_l < k . \end{cases}$$

Since every relation of multiple zeta values in a given weight k is, by Proposition 9, in the kernel of the map Z_k, this kernel was studied in [6] with the following result.

Theorem 10 ([6, Theorem 1.13])

(i) *For any $f \in \mathcal{MD}$ which can be written as a linear combination of brackets with weight $\leq k-2$ we have $\mathrm{d}f \in \ker Z_k$.*

(ii) *Any cusp form for $\mathrm{SL}_2(\mathbb{Z})$ of weight k is in the kernel of Z_k.*

We give an example for Theorem 10 (ii): Using the theory of brackets (Corollary 4.13) we can prove for the cusp form $\Delta = q\prod_{n>0}(1-q^n)^{24} \in S_{12}(\mathrm{SL}_2(\mathbb{Z}))$ the representation

$$-\frac{1}{2^6 \cdot 5 \cdot 691}\Delta = 168[5,7] + 150[7,5] + 28[9,3] \\ + \frac{1}{1408}[2] - \frac{83}{14400}[4] + \frac{187}{6048}[6] - \frac{7}{120}[8] - \frac{5197}{691}[12] . \quad (9)$$

Letting Z_{12} act on both sides of (9) one obtains a new proof for the relation (6), i.e.,

$$\frac{5197}{691}\zeta(12) = 168\zeta(5,7) + 150\zeta(7,5) + 28\zeta(9,3) .$$

Another reason for studying the enlargement of the brackets given by the bi-brackets is the following: In weight 4 one has the following relation of multiple zeta values

$\zeta(4) = \zeta(2,1,1)$, i.e. it is $[4] - [2,1,1] \in \ker Z_4$. But this element can't be written as a linear combination of cusp forms, lower weight brackets or derivatives. But one can show, by using the double shuffle relations of bi-brackets, that

$$[4] - [2,1,1] = \frac{1}{2}(\mathrm{d}[1] + \mathrm{d}[2]) - \frac{1}{3}[2] - [3] + \begin{bmatrix} 2,1 \\ 1,0 \end{bmatrix} \tag{10}$$

and $\left[\begin{smallmatrix} 2,1 \\ 1,0 \end{smallmatrix}\right] \in \ker Z_4$. To describe the kernel of the map Z_k was in fact our first motivation to study the bi-brackets. Equation (10) is also an example for the above mentioned Conjecture 5, since it shows that the bi-bracket $\left[\begin{smallmatrix} 2,1 \\ 1,0 \end{smallmatrix}\right]$ can be written as brackets and therefore is an element in $\mathcal{MD}$.

2 Outlook and Related Work

In the following paragraphs (a)–(g) we want to mention some related works and give an outlook to open questions.

(**a**) There are still a lot of open questions concerning multiple Eisenstein series as well as the space of (bi-)brackets. After the above mentioned works [3, 6, 9] we now have a good definition of regularized multiple Eisenstein series given by the $G^{ш}$. For the structure of the space spanned by these series there are still several open questions.

- (i) What exactly is the failure of the stuffle product for the $G^{ш}$ and when does it hold?
- (ii) For which indices $s_1, \dots, s_l \in \mathbb{N}$ do we have $G^{ш}_{s_1,\dots,s_r}(\tau) = G^{*}_{s_1,\dots,s_r}(\tau)$? Is there an explicit connection between these two regularizations similar to the regularized multiple zeta values given by the map ρ in [24]?
- (iii) What is the dimension of the space of (shuffle) regularised multiple Eisenstein series? Is there an explicit basis similar to the Hoffman basis of multiple zeta values (Which is given by all multiple zeta values $\zeta(s_1, \dots, s_l)$ with $s_j \in \{2,3\}$)?
- (iv) Which linear combinations of multiple Eisenstein series are modular forms for $\mathrm{SL}_2(\mathbb{Z})$? Is there an explicit way to describe the modular defect?
- (v) Is the space of multiple Eisenstein series closed under the derivative $\mathrm{d} = q\frac{d}{dq}$? Meanwhile this question was also already addressed in [5].
- (vi) What is the kernel of the projection to the constant term? Does it consist of more than derivatives and cusp forms?
- (vii) Is there a general theory behind the connection of the Fourier expansion of multiple Eisenstein series and the Goncharov coproduct? Can we equip the space of multiple Eisenstein series with a coproduct structure in an useful way?

Especially the last questions seems to be interesting since the connection to the coproduct of formal iterated integrals is quite mysterious and it seems that there might be a geometric interpretation for this connection.

(**b**) Several q-analogues of multiple zeta values were studied in recent years. The first works on this area are [14, 28, 31, 38]. Possible double shuffle structures are discussed for example in [18, 32, 33, 39], where the last one gives also a nice overview of various different q-analogue models. Often these q-analogues have a product structure similar to the stuffle product of multiple zeta values. To obtain something which corresponds to the shuffle product one usually needs to modify the space and add extra elements (like derivatives) or consider index sets $(s_1, \dots, s_r)$ with $s_j \in \mathbb{Z}$ or $s_j \geq 0$. The picture is similar for bi-brackets, where we consider double indices $\left[\begin{smallmatrix} s_1,\dots,s_l \\ r_1,\dots,r_l \end{smallmatrix}\right]$ to obtain an analogue for both products in a very natural way. This gives a lot of linear relations similar to the double shuffle relations. Numerical experiments suggest, that every bi-bracket can be written as a linear combination of brackets and therefore (conjecturally) every relation of bi-brackets gives rise to relations between multiple zeta values by applying the map Z_k.

(**c**) In the case of multiple zeta values one way to give upper bounds for the dimension is to study the double shuffle space [24, 25]. Similarly, one can study the partition shuffle space

$$\mathbb{PS}(k-l,l) = \left\{ f \in \mathbb{Q}[X_1,..,X_l,Y_1,..,Y_l] \,\middle|\, \deg f = k-l,\ f\big|_P - f = f\big|_{\mathrm{Sh}_j} = 0\ \forall j \right\},$$

for bi-brackets, where $|_P$ is the involution given by the partition relation (see Sect. 5.1, (23)) and $|_{\mathrm{Sh}_j}$ is given by the sum of all shuffles of type j similar to the one in [25]. Counting the number of these polynomials it is possible to give upper bounds for the dimensions of the space of bi-brackets. This approach therefore enabled us to prove the conjecture $\mathcal{MD} = \mathcal{BD}$ up to weight 7 in a current work in progress [8]. Therefore, considering the space $\mathbb{PS}(k-l,l)$ in more detail might be crucial to understand the structure of bi-brackets.

(**d**) In this work we were interested in modular forms for the full modular group and therefore studied the level 1 case. In [26] the authors studied double Eisenstein series and double zeta values of level 2. They also derive the Fourier expansion of these series which involves similar calculation as in the level 1 case. One result is, that they derive the dimension of the space of double Eisenstein series and give also an upper bound for the dimension of double zeta values of level 2, which involves the dimension of the spaces of cusp forms of level 2. Beside the work on Level 2 double Eisenstein series there are also work for level N double Eisenstein series of Yuan and Zhao in [34]. Later these authors also considered a level N version of the brackets in [35].

(**e**) At the end of [26] the authors give a proof for an upper bound of the dimension of double zeta values in even weight. We want to recall this result, since the presented results in the present work might be able to use these ideas for higher lengths. Consider the space spanned by all normalized double Eisenstein series $(-2\pi i)^{-r-s} G_{r,s}(\tau)$ in even weight $k = r + s$. Denote by π_i the projection of this space to the imaginary part. Using the Fourier expansion of double Eisenstein series the authors can write down the matrix representation of π_i explicitly. Together with well known results on period polynomials they obtain

$$\dim_{\mathbb{Q}}\langle \zeta(r,k-r) \mid 2 \le r \le k-1\rangle_{\mathbb{Q}} \le \frac{k}{2} - 1 - \dim S_k \,.$$

Due to the Broadhurst-Kreimer Conjecture 1 it is conjectured that this is actually an equality. The key fact here is, that it is possible to write down an explicit basis of the imaginary part and the matrix representation of π_i. To also obtain upper bounds for the dimensions of multiple zeta values in higher lengths, one might try to use the exact same method as in the length two case. The imaginary part of the (again normalized with the factor $(-2\pi i)^{-k}$) multiple Eisenstein series is more complicated since it involves the functions g in different length, where it is known that they are not linearly independent anymore. But the algebraic structure of the g or more precisely of the brackets [..] are subject of the current work. It is quite possible that the results on the brackets enable one to study the projection of the imaginary part of multiple Eisenstein series to obtain upper bounds for the Broadhurst-Kreimer conjecture.

(f) The multiple Eisenstein series and the bi-brackets itself also have connections to counting problems in enumerative geometry:

(i) In [1, 30] the author studies q-series $A_k(a) \in \mathbb{Q}[\![q]\!]$ which arises in counting certain types of hyperelliptic curves. One of the results is, that the $A_k(q)$ are contained in the ring of quasi-modular forms. The connection to the brackets is given by the fact that $A_k(q) = [\underbrace{2,\dots,2}_{k}]$. The results of [1] can also be obtained by using an explicit calculation of the Fourier expansion of $G_{2,\dots,2}$.
(ii) In [27, 29] the authors connect certain q-analogues of multiple zeta values to Hilbert schemes of points on surfaces. These q-analogues are just particular linear combinations of brackets as explained in [7] and Sect. 7.2.
(iii) The coefficients of bi-brackets also occur naturally when counting flat surfaces [40], i.e. certain covers of the torus.

(g) There also exists different "multiple"-versions of classical Eisenstein series. One of them is treated in [10], where the authors discuss the series defined by

$$\mathfrak{G}_{2p_1,\dots,2p_r}(\tau) = \sum_{m\in\mathbb{Z}} \sum_{\substack{n_1\in\mathbb{Z}\\(m,n_1)\neq(0,0)}} \cdots \sum_{\substack{n_r\in\mathbb{Z}\\(m,n_r)\neq(0,0)}} \prod_{j=1}^{r} \frac{1}{(m+n_j\tau)^{2p_j}}$$

for $r \in \mathbb{N}_{\ge 2}$ and $p_1,\dots,p_r \in \mathbb{N}$ and prove (Theorem 2) that for $r \in \mathbb{N}_{\ge 2}$ and $p_1,\dots,p_r \in \mathbb{N}$,

$$\tau^{2(p_1+\dots+p_r)}\mathfrak{G}_{2p_1,\dots,2p_r}(\tau) \in \mathbb{Q}\left[\tau^2,\, \pi^2,\, G_2(\tau),\, G_4(\tau),\, G_6(\tau)\right].$$

The methods used to prove these statements are similar to the methods used in the calculation of the Fourier expansion of multiple Eisenstein series. But besides this there does not seem to be a direct connection to the multiple Eisenstein series presented here.

3 Multiple Eisenstein Series

In this section we are going to introduce multiple zeta values and present the multiple Eisenstein series and their Fourier expansion. Especially the construction of the Fourier expansion of multiple Eisenstein series in Sect. 3.2 was rewritten for this survey. It will be a shortened version of the construction given in [2] using results by Bouillot obtained in [12]. This section is not part of the works [6, 7, 9]. Before we discuss multiple Eisenstein series, we give a short review of multiple zeta values and their algebraic structure given by the stuffle and shuffle product. In order to describe these two products we will use quasi-shuffle algebras, introduced by Hofmann in [21], which will also be needed later when we deal with the generating series of multiple divisor-sums (brackets) and their generalizations given by the bi-brackets.

3.1 Multiple Zeta Values and Quasi-shuffle Algebras

Multiple zeta values are natural generalizations of the Riemann zeta values that are defined[1] for integers $s_1 > 1$ and $s_i \geq 1$ for $i > 1$ by

$$\zeta(s_1, \ldots, s_l) := \sum_{n_1 > n_2 > \cdots > n_l > 0} \frac{1}{n_1^{s_1} \ldots n_l^{s_l}} .$$

We denote the $\mathbb{Q}$-vector space of all multiple zeta values of weight k by

$$\mathscr{Z}_k := \big\langle \zeta(s_1, \ldots, s_l) \,\big|\, s_1 + \cdots + s_l = k \text{ and } l > 0 \big\rangle_{\mathbb{Q}}.$$

It is well known that the product of two multiple zeta values can be written as a linear combination of multiple zeta values of the same weight by using the stuffle or shuffle relations (See for example [24, 42]). Thus they generate a $\mathbb{Q}$-algebra $\mathscr{Z}$. There are several connections of these numbers to modular forms for the full modular group. In the smallest length the stuffle product reads

$$\begin{aligned}
\zeta(s_1) \cdot \zeta(s_2) &= \sum_{n_1 > 0} \frac{1}{n_1^{s_1}} \sum_{n_2 > 0} \frac{1}{n_2^{s_2}} \\
&= \sum_{n_1 > n_2 > 0} \frac{1}{n_1^{s_1} n_2^{s_2}} + \sum_{n_2 > n_1 > 0} \frac{1}{n_1^{s_1} n_2^{s_2}} + \sum_{n_1 = n_2 > 0} \frac{1}{n_1^{s_1 + s_2}} \\
&= \zeta(s_1, s_2) + \zeta(s_2, s_1) + \zeta(s_1 + s_2) .
\end{aligned}$$

[1]Some authors use the opposite convention $0 < n_1 < \cdots < n_l$ in the definition of multiple zeta values. This is in particular the case for the work [9], where this opposite convention is used for multiple zeta values and multiple Eisenstein series.

For length 1 times length 2 the same argument gives

$$\zeta(s_1)\cdot\zeta(s_2,s_3) = \zeta(s_1,s_2,s_3)+\zeta(s_2,s_1,s_3)+\zeta(s_2,s_3,s_1) \\ +\zeta(s_1+s_2,s_3)+\zeta(s_2,s_1+s_3)\,.$$

The second expression for the product, the shuffle product, comes from the iterated integral expression of multiple zeta values. For example it is

$$\zeta(2,3) = \int_{1>t_1>\cdots>t_5>0} \underbrace{\frac{dt_1}{t_1}\cdot\frac{dt_2}{1-t_2}}_{2}\cdot\underbrace{\frac{dt_3}{t_3}\cdot\frac{dt_4}{t_4}\cdot\frac{dt_5}{1-t_5}}_{3}\,.$$

Multiplying two of these integrals one obtains again a linear combination of multiple zeta values as for example

$$\zeta(2)\cdot\zeta(3) = \zeta(2,3)+3\zeta(3,2)+6\zeta(4,1)\,.$$

More generally the smallest length case is given by

$$\zeta(s_1)\cdot\zeta(s_2) = \sum_{\substack{a+b=s_1+s_2\\ a>1}} \left(\binom{a-1}{s_1-1}+\binom{a-1}{s_2-1}\right)\zeta(a,b)\,. \tag{11}$$

To describe these two product structures precisely we will use the language of quasi-shuffle algebras as introduced in [21, 22].

Definition 3.1 Let A (the alphabet) be a countable set of letters, $\mathbb{Q}A$ the $\mathbb{Q}$-vector space generated by these letters and $\mathbb{Q}\langle A\rangle$ the noncommutative polynomial algebra over $\mathbb{Q}$ generated by words with letters in A. For a commutative and associative product $\diamond$ on $\mathbb{Q}A$, $a, b \in A$ and $w, v \in \mathbb{Q}\langle A\rangle$ we define on $\mathbb{Q}\langle A\rangle$ recursively a product by $1 \odot w = w \odot 1 = w$ and

$$aw \odot bv := a(w\odot bv) + b(aw\odot v) + (a\diamond b)(w\odot v)\,. \tag{12}$$

By a result of Hoffman [22, Theroem 2.1] $(\mathbb{Q}\langle A\rangle, \odot)$ is a commutative $\mathbb{Q}$-algebra which is called a *quasi-shuffle algebra*.

To describe the stuffle and the shuffle product for multiple zeta values we need to deal with two different alphabets A_{xy} and A_z. The first alphabet is given by $A_{xy} := \{x, y\}$ and we set $\mathfrak{H} = \mathbb{Q}\langle A_{xy}\rangle$ and $\mathfrak{H}^1 = 1\cdot\mathbb{Q} + \mathfrak{H}y$, with 1 being the empty word. It is easy to see that $\mathfrak{H}^1$ is generated by the elements $z_j = x^{j-1}y$ with $j \in \mathbb{N}$, i.e. $\mathfrak{H}^1 = \mathbb{Q}\langle A_z\rangle$ with the second alphabet $A_z := \{z_1, z_2, \dots\}$. Additionally, we define $\mathfrak{H}^0 = 1\mathbb{Q} + x\mathfrak{H}y$.

(i) On $\mathfrak{H}^1$ we have the following quasi-shuffle product with respect to the alphabet A_z, called the *stuffle product*. We denote it by $*$ and define it as the quasi-shuffle product with $z_j \diamond z_i = z_{j+i}$. For $a, b \in \mathbb{N}$ and $w, v \in \mathfrak{H}^1$ we therefore have:

$$z_a w * z_b v = z_a(w * z_b v) + z_b(z_a w * v) + z_{a+b}(w * v) .$$

By $(\mathfrak{H}^1, *)$ we denote the corresponding $\mathbb{Q}$-algebra.

(ii) On the alphabet A_{xy} we define the *shuffle product* as the quasi-shuffle product with $\diamond \equiv 0$, and by $(\mathfrak{H}^1, ш)$ we denote the corresponding $\mathbb{Q}$-algebra.

It is easy to check that $\mathfrak{H}^0$ is closed under both products $*$ and ш and therefore we have also the two algebras $(\mathfrak{H}^0, *)$ and $(\mathfrak{H}^0, ш)$.

By the definition of multiple zeta values as an ordered sum and by the iterated integral expression one obtains algebra homomorphisms $Z : (\mathfrak{H}^0, *) \to \mathcal{Z}$ and $Z : (\mathfrak{H}^0, ш) \to \mathcal{Z}$ by sending $w = z_{s_1} \dots z_{s_l}$ to $\zeta(w) = \zeta(s_1, \dots, s_l)$, since the words in $\mathfrak{H}^0$ correspond exactly to the indices for which the multiple zeta values are defined. It is a well known fact, that these algebra homomorphisms can be extended to $\mathfrak{H}^1$:

Proposition 3.2 ([24, Proposition 1]) *There exist algebra homomorphisms*

$$Z^* : (\mathfrak{H}^1, *) \longrightarrow \mathcal{Z} \quad \text{and} \quad Z^{ш} : (\mathfrak{H}^1, ш) \longrightarrow \mathcal{Z} ,$$

which are uniquely determined by $Z^*(w) = Z^{ш}(w) = \zeta(w)$ for $w \in \mathfrak{H}^0$ and by their images on the word z_1, which we set 0 here, i.e. $Z^*(z_1) = Z^{ш}(z_1) = 0$.

3.2 Multiple Eisenstein Series and the Calculation of Their Fourier Expansion

The Riemann zeta values appear as the constant term in the Fourier expansion of classical Eisenstein series. These series are defined for $\tau \in \mathbb{H}$ by

$$G_k(\tau) = \frac{1}{2} \sum_{\substack{(m,n)\in\mathbb{Z}^2 \\ (m,n)\neq(0,0)}} \frac{1}{(m\tau + n)^k}. \tag{13}$$

where $k > 2$ is the called the weight. Splitting the summation into the parts $m = 0$ and $m \in \mathbb{Z}\backslash 0$ we obtain for even k

$$G_k(\tau) = \frac{1}{2} \sum_{n\neq 0} \frac{1}{n^k} + \sum_{m=1}^{\infty} \left(\sum_{n\in\mathbb{Z}} \frac{1}{(m\tau + n)^k} \right) .$$

To calculate the Fourier expansion of the sum on the right, one uses the well known Lipschitz summation formula ($q = e^{2\pi i \tau}$)

$$\sum_{d\in\mathbb{Z}} \frac{1}{(\tau+d)^k} = \frac{(-2\pi i)^k}{(k-1)!}\sum_{m=1}^{\infty} m^{k-1} q^m\,, \tag{14}$$

which is valid for $k > 1$. With (14) we obtain

$$G_k(\tau) = \zeta(k) + \frac{(-2\pi i)^k}{(k-1)!}\sum_{m=1}^{\infty}\sum_{n=1}^{\infty} n^{k-1} q^{mn} = \zeta(k) + \frac{(-2\pi i)^k}{(k-1)!}\sum_{n=1}^{\infty} \sigma_{k-1}(n) q^n\,, \tag{15}$$

where $\sigma_k(n) = \sum_{d|n} d^k$ denote the divisor-sum. Formula (15) also makes sense for odd k but does not give a modular form, since there are no non trivial modular forms of odd weight. The sum in (13) vanishes for odd k, therefore instead of summing over the whole lattice, we restrict the summation to the positive lattice points, with positivity coming from an order on the lattice $\mathbb{Z}\tau + \mathbb{Z}$. This in turn will also enable us to give an multiple version of the Eisenstein series in an obvious way.

Let $\Lambda_\tau = \mathbb{Z}\tau + \mathbb{Z}$ be a lattice with $\tau \in \mathbb{H} := \{x + iy \in \mathbb{C} \mid y > 0\}$. An order $\succ$ on Λ_τ is defined by setting (see [19])

$$\lambda_1 \succ \lambda_2 :\Leftrightarrow \lambda_1 - \lambda_2 \in P$$

for $\lambda_1, \lambda_2 \in \Lambda_\tau$ and the following set P, which we call the set of positive lattice points

$$P := \{l\tau + m \in \Lambda_\tau \mid l > 0 \vee (l = 0 \wedge m > 0)\} = U \cup R$$

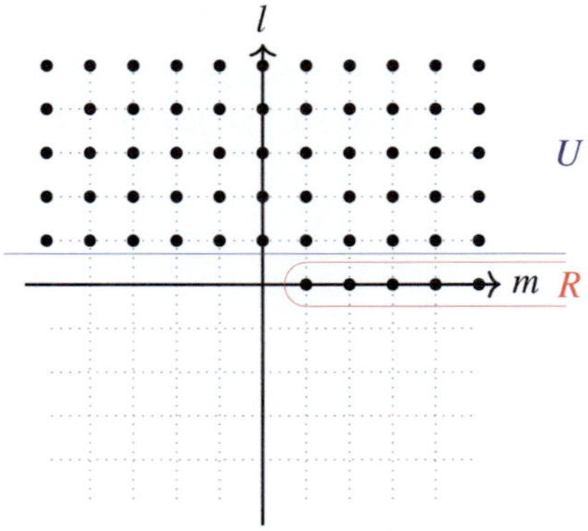

The set P for the case $\tau = i$.

Definition 3.3 For $s_1 \geq 3, s_2, \ldots, s_l \geq 2$ the *multiple Eisenstein series* is defined by

$$G_{s_1,\dots,s_l}(\tau) := \sum_{\substack{\lambda_1 \succ \cdots \succ \lambda_l \succ 0 \\ \lambda_i \in \Lambda_\tau}} \frac{1}{\lambda_1^{s_1} \dots \lambda_l^{s_l}} .$$

With $k = s_1 + \cdots + s_l$ we denote the weight and with l its length.

It is easy to see that these are holomorphic functions in the upper half-plane and that they fulfill the stuffle product, i.e. for example

$$G_3(\tau) \cdot G_4(\tau) = G_{4,3}(\tau) + G_{3,4}(\tau) + G_7(\tau) .$$

By definition it is $G_{s_1,\dots,s_l}(\tau + 1) = G_{s_1,\dots,s_l}(\tau)$, i.e. there exists a Fourier expansion of $G_{s_1,\dots,s_l}$ in $q = e^{2\pi i \tau}$. To write down the Fourier expansion of multiple Eisenstein series we need to introduce the following q-series which will be studied in detail in Sect. 4.1. For $s_1, \dots, s_l \geq 1$ we define

$$[s_1, \dots, s_l] := \sum_{\substack{u_1 > \cdots > u_l > 0 \\ v_1, \dots, v_l > 0}} \frac{v_1^{s_1 - 1} \dots v_l^{s_l - 1}}{(s_1 - 1)! \dots (s_l - 1)!} \cdot q^{u_1 v_1 + \cdots + u_l v_l} \in \mathbb{Q}[\![q]\!] .$$

and write $g_{s_1,\dots,s_l}(\tau) := (-2\pi i)^{s_1 + \cdots + s_l} [s_1, \dots, s_l]$, which is an holomorphic function in the upper half-plane by setting $q = e^{2\pi i \tau}$.

Theorem 3.4 ([2], Fourier expansion) *For $s_1 \geq 3, s_2, \dots, s_l \geq 2$ the $G_{s_1,\dots,s_l}(\tau)$ can be written as a $\mathscr{Z}$-linear combination of the functions g. More precisely there are rational numbers $\lambda_{r,j} \in \mathbb{Q}$, for $r = (r_1, \dots, r_l)$ and $1 \leq j \leq l - 1$, such that (with $k = s_1 + \cdots + s_l$)*

$$G_{s_1,\dots,s_l}(\tau) = \zeta(s_1, \dots, s_l) + \sum_{\substack{1 \leq j \leq l-1 \\ r_1 + \cdots + r_l = k}} \lambda_{r,j} \cdot \zeta(r_1, \dots, r_j) \cdot g_{r_{j+1},\dots,r_l}(\tau) + g_{s_1,\dots,s_l}(\tau) .$$

Even though the proof of this statement is the main result of [2] we will give a shortened version of it in the following.

The condition $s_1 \geq 3$ is necessary for the absolute convergence of the sum. Nevertheless we can also allow the case $s_1 = 2$ by using the Eisenstein summation as it was done in [9] Definition 2.1. This corresponds to the usual way of defining the quasi-modular form G_2 in length one. Since the construction of the Fourier expansion described below uses exactly this Eisenstein summation the Theorem 3.4 is also valid for $s_1 \geq 2$.

For example the triple Eisenstein series $G_{3,2,2}$ can be written as

$$G_{3,2,2}(\tau) = \zeta(3, 2, 2) + \left(\frac{54}{5} \zeta(2, 3) + \frac{51}{5} \zeta(3, 2) \right) g_2(\tau) + \frac{16}{3} \zeta(2, 2) g_3(\tau)$$
$$+ 3\zeta(3) g_{2,2}(\tau) + 4\zeta(2) g_{3,2}(\tau) + g_{3,2,2}(\tau) .$$

To derive the Fourier expansion we introduce the following functions, that can be seen as a multiple version of the term $\sum_{n\in\mathbb{Z}} \frac{1}{(x+n)^k}$ appearing in the calculation of the Fourier expansion of classical Eisenstein series.

Definition 3.5 For $s_1, \dots, s_l \geq 2$ we define the multitangent function of length l by

$$\Psi_{s_1,\dots,s_l}(x) = \sum_{\substack{n_1 > \dots > n_l \\ n_i \in \mathbb{Z}}} \frac{1}{(x+n_1)^{s_1} \dots (x+n_l)^{s_l}} .$$

In the case $l = 1$ we also refer to these as monotangent function.

These functions were introduced and studied in detail in [12]. One of the main results there, which is crucial for the calculation of the Fourier expansion presented here, is the following theorem which reduces the multitangent functions into monotangent functions.

Theorem 3.6 ([12, Theroem 3], Reduction of multitangent into monotangent functions) *For $s_1, \dots, s_l \geq 2$ and $k = s_1 + \dots + s_l$ the multitangent function can be written as a $\mathcal{Z}$-linear combination of monotangent functions, more precisely there are $c_{k,h} \in \mathcal{Z}_{k-h}$ such that*

$$\Psi_{s_1,\dots,s_l}(x) = \sum_{h=2}^{k} c_{k-h} \Psi_h(x) .$$

Proof An explicit formula for the coefficients c_k is given in Theorem 3 in [12]. The proof uses partial fraction and a non trivial relation between multiple zeta values to argue that the sum starts at $h = 2$. For example in length two it is

$$\begin{aligned}
\Psi_{3,2}(x) &= \sum_{m_1 > m_2} \frac{1}{(x+m_1)^3 (x+m_2)^2} \\
&= \sum_{m_1 > m_2} \left(\frac{1}{(m_1-m_2)^2 (x+m_1)^3} + \frac{2}{(m_1-m_2)^3 (x+m_1)^2} + \frac{3}{(m_1-m_2)^4 (x+m_1)} \right) \\
&+ \sum_{m_1 > m_2} \left(\frac{1}{(m_1-m_2)^3 (x+m_2)^2} - \frac{3}{(m_1-m_2)^4 (x+m_2)} \right) \\
&= 3\zeta(3)\Psi_2(x) + \zeta(2)\Psi_3(x) .
\end{aligned} \tag{16}$$

The connection between the functions g and the monotangent functions is given by the following

Proposition 3.7 *For $s_1, \dots, s_r \geq 2$ the functions g can be written as an ordered sum of monotangent functions*

$$g_{s_1,\dots,s_l}(\tau) = \sum_{m_1 > \dots > m_l > 0} \Psi_{s_1}(m_1\tau) \dots \Psi_{s_r}(m_r\tau) .$$

Proof This follows directly from the Lipschitz formula (14) and the definition of the functions g.

Preparation for the Proof of Theorem 3.4: We will now recall the construction of the Fourier expansion of multiple Eisenstein series introduced in [2], in order to prove Theorem 3.4. To calculate the Fourier expansion we rewrite the multiple Eisenstein series as

$$G_{s_1,\dots,s_l}(\tau) = \sum_{\lambda_1 \succ \dots \succ \lambda_l \succ 0} \frac{1}{\lambda_1^{s_1} \dots \lambda_l^{s_l}}$$
$$= \sum_{(\lambda_1,\dots,\lambda_l)\in P^l} \frac{1}{(\lambda_1 + \dots + \lambda_l)^{s_1}(\lambda_2 + \dots + \lambda_l)^{s_2} \dots \lambda_l^{s_l}}\,.$$

We decompose the set of tuples of positive lattice points P^l into the 2^l distinct subsets $A_1 \times \dots \times A_l \subset P^l$ with $A_i \in \{R, U\}$ and write

$$G^{A_1\dots A_l}_{s_1,\dots,s_l}(\tau) := \sum_{(\lambda_1,\dots,\lambda_l)\in A_1\times\dots\times A_l} \frac{1}{(\lambda_1 + \dots + \lambda_l)^{s_1}(\lambda_2 + \dots + \lambda_l)^{s_2} \dots \lambda_l^{s_l}}$$

this gives the decomposition

$$G_{s_1,\dots,s_l} = \sum_{A_1,\dots,A_l\in\{R,U\}} G^{A_1\dots A_l}_{s_1,\dots,s_l}\,.$$

In the following we identify the $A_1 \dots A_l$ with words in the alphabet $\{R, U\}$. In length $l = 1$ we have $G_k(\tau) = G^R_k(\tau) + G^U_k(\tau)$ and

$$G^R_k(\tau) = \sum_{\substack{m_1=0\\ n_1>0}} \frac{1}{(0\tau + n_1)^k} = \zeta(k)\,,$$
$$G^U_k(\tau) = \sum_{\substack{m_1>0\\ n_1\in\mathbb{Z}}} \frac{1}{(m_1\tau + n_1)^k} = \sum_{m_1>0} \Psi_k(m_1\tau)\,,$$

where Ψ_k is the monotangent function given by

$$\Psi_k(x) = \sum_{n\in\mathbb{Z}} \frac{1}{(x+n)^k}\,.$$

To calculate the Fourier expansion of G^U_k one uses the Lipschitz formula (14). In general the $G^{U^l}_{s_1,\dots,s_l}$ can be written as

$$\begin{aligned}
G^{U^l}_{s_1,\dots,s_l}(\tau) &= \sum_{\substack{m_1>\dots>m_l>0\\ n_1,\dots,n_l\in\mathbb{Z}}} \frac{1}{(m_1\tau+n_1)^{s_1}\dots(m_l\tau+n_l)^{s_l}}\\
&= \sum_{m_1>\dots>m_l>0} \Psi_{s_1}(m_1\tau)\dots\Psi_{s_l}(m_l\tau)\\
&= \frac{(-2\pi i)^{s_1+\dots+s_l}}{(s_1-1)!\dots(s_l-1)!} \sum_{\substack{m_1>\dots>m_l>0\\ d_1,\dots,d_l>0}} d_1^{s_1-1}\dots d_l^{s_l-1} q^{m_1d_1+\dots+m_ld_l}\\
&= g_{s_1,\dots,s_l}(\tau)\,.
\end{aligned}$$

The other special case $G^{R^l}_{s_1,\dots,s_l}$ can also be written down explicitly:

$$G^{R^l}_{s_1,\dots,s_l}(\tau) = \sum_{\substack{m_1=\dots=m_l=0\\ n_1>\dots>n_l>0}} \frac{1}{(0\tau+n_1)^{s_1}\dots(0\tau+n_l)^{s_l}} = \zeta(s_1,\dots,s_l)\,.$$

In length 2 we have $G_{s_1,s_2} = G^{RR}_{s_1,s_2} + G^{UR}_{s_1,s_2} + G^{RU}_{s_1,s_2} + G^{UU}_{s_1,s_2}$ and

$$\begin{aligned}
G^{UR}_{s_1,s_2} &= \sum_{\substack{m_1>0,m_2=0\\ n_1\in\mathbb{Z},n_2>0}} \frac{1}{(m_1\tau+n_1)^{s_1}(0\tau+n_2)^{s_1}}\\
&= \sum_{m_1>0}\Psi_{s_1}(m_1\tau)\sum_{n_2>0}\frac{1}{n_2^{s_2}} = g_{s_1}(\tau)\zeta(s_2)\,,
\end{aligned}$$

$$G^{RU}_{s_1,s_2}(\tau) = \sum_{\substack{m_1=0,m_2>0\\ n_1>n_2\\ n_i\in\mathbb{Z}}} \frac{1}{(m_1\tau+n_1)^{s_1}(m_1\tau+n_2)^{s_2}} = \sum_{m>0}\Psi_{s_1,s_2}(m\tau).$$

In the case G^{UR} we saw that we could write it as G^U multiplied with a zeta value. In general, having a word w of length l ending in the letter R, i.e. there is a word w' ending in U with $w = w'R^r$ and $1 \le r \le l$ we can write

$$G^w_{s_1,\dots,s_l}(\tau) = G^{w'}_{s_1,\dots,s_{l-r}}(\tau)\cdot\zeta(s_{l-r+1},\dots,s_l)\,.$$

Example: $G^{RUURR}_{3,4,5,6,7} = G^{RUU}_{3,4,5}\cdot\zeta(6,7)$
Hence one can concentrate on the words ending in U when calculating the Fourier expansion of a multiple Eisenstein series. Let w be a word ending in U then there are integers $r_1,\dots,r_j \ge 0$ with $w = R^{r_1}UR^{r_2}U\dots R^{r_j}U$. With this one can write

$$G^w_{s_1,\dots,s_l}(\tau) = \sum_{m_1>\dots>m_j>0} \Psi_{s_1,\dots,s_{r_1+1}}(m_1\tau)\cdot\Psi_{s_{r_1+2},\dots}(m_2\tau)\dots\Psi_{s_{l-r_j},\dots,s_l}(m_j\tau)\,.$$

Example: $w = RURRU$

$$G^{RURRU}_{s_1,\dots,s_l} = \sum_{m_1>m_2>0} \Psi_{s_1,s_2}(m_1\tau)\Psi_{s_3,s_4,s_5}(m_2\tau)$$

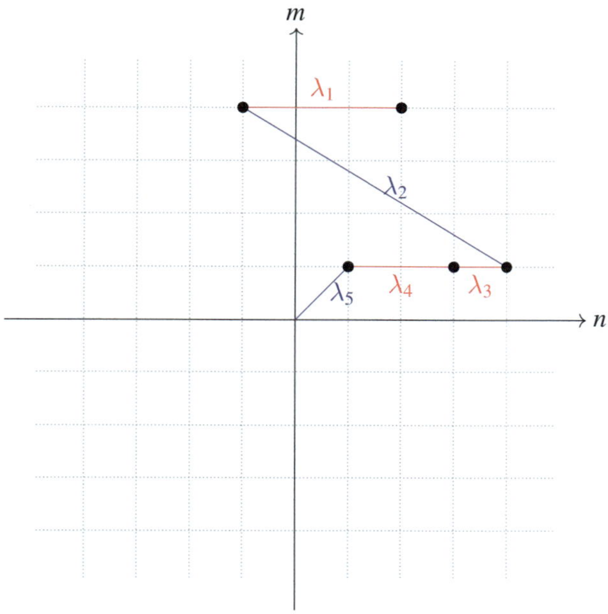

A summand of $G^{RURRU}_{s_1,\dots,s_l}$.

Proof of Theorem 3.4: For $s_1, \dots, s_l \geq 2$ the Fourier expansion of the multiple Eisenstein series $G_{s_1,\dots,s_l}$ can be computed in the following way

(i) Split up the summation into 2^l distinct parts $G^w_{s_1,\dots,s_l}$ where w are a words in $\{R, U\}$.

(ii) For w being a word ending in R one can write $G^w_{s_1,\dots,s_l}$ as $G^{w'}_{s_1,\dots} \cdot \zeta(\dots, s_l)$ with a word w' ending in U.

(iii) For w being a word ending in U one can write $G^w_{s_1,\dots,s_l}$ as

$$G^w_{s_1,\dots,s_l}(\tau) = \sum_{m_1>\dots>m_l>0} \Psi_{s_1,\dots}(m_1\tau)\dots\Psi_{\dots,s_l}(m_l\tau)\,.$$

(iv) Using the Theorem 3.6 we can write the multitangent functions in (iii) as a $\mathscr{Z}$-linear combination of monotangents. We therefore just have $\mathscr{Z}$-linear combinations with sums of the form

$$\sum_{m_1>\dots>m_j>0} \Psi_{k_1}(m_1\tau)\dots\Psi_{k_j}(m_j\tau) = g_{k_1,\dots,k_j}(\tau) = (-2\pi i)^{k_1+\dots+k_j}[k_1,\dots,k_j]\,.$$

□

An explicit formula for the Fourier expansion of the multiple Eisenstein series for arbitrary length can be found in [9] Proposition 2.4. (with a reversed order of indices). Here we just give the Fourier expansion for the length 2 and 3. For this we define for $n_1, n_2, k > 0$ the numbers $C^k_{n_1,n_2}$ by

$$C^k_{n_1,n_2} = (-1)^{n_2}\binom{k-1}{n_2-1} + (-1)^{k-n_1}\binom{k-1}{n_1-1}.$$

Proposition 3.8 *(i) ([2, 9, 19, Formula (52)]) For $s_1, s_2 \geq 2$ the Fourier expansion of the double Eisenstein series is given by*

$$G_{s_1,s_2}(\tau) = \zeta(s_1, s_2) + \zeta(s_2)g_{s_1}(\tau) + \sum_{\substack{k_1+k_2=s_1+s_2\\ k_1,k_2\geq 2}} C^{k_2}_{s_1,s_2}\zeta(k_2)g_{k_1}(\tau) + g_{s_1,s_2}(\tau).$$

(ii) ([2, 9]) For $s_1, s_2, s_3 \geq 2$ and $k = s_1 + s_2 + s_3$ the Fourier expansion of the triple Eisenstein series can be written as

$$\begin{aligned} G_{s_1,s_2,s_3}(\tau) &= \zeta(s_1, s_2, s_3) + \zeta(s_2, s_3)g_{s_1}(\tau) + \zeta(s_3)g_{s_1,s_2}(\tau) + g_{s_1,s_2,s_3}(\tau)\\ &+ \zeta(s_3) \sum_{k_1+k_2=s_1+s_2} C^{k_1}_{s_1,s_2}\zeta(k_1)g_{k_2}(\tau)\\ &+ \sum_{k_1+k_2=s_1+s_2} C^{k_2}_{s_1,s_2}\zeta(k_2)g_{k_1,s_3}(\tau) + \sum_{k_1+k_2=s_2+s_3} C^{k_2}_{s_2,s_3}\zeta(k_2)g_{s_1,k_1}(\tau)\\ &+ \sum_{k_1+k_2+k_3=k} (-1)^{s_2+s_3}\binom{k_2-1}{s_2-1}\binom{k_3-1}{s_3-1}\zeta(k_3, k_2)g_{k_1}(\tau)\\ &+ \sum_{k_1+k_2+k_3=k} (-1)^{s_1+s_2+k_2+k_3}\binom{k_2-1}{k_3-1}\binom{k_3-1}{s_2-1}\zeta(k_3, k_2)g_{k_1}(\tau)\\ &+ (-1)^{s_1+s_3} \sum_{k_1+k_2+k_3=k} (-1)^{k_2}\binom{k_2-1}{s_1-1}\binom{k_3-1}{s_3-1}\zeta(k_3)\zeta(k_2)g_{k_1}(\tau), \end{aligned}$$

where in the sums we sum over all $k_i \geq 2$.

We finish this section with a closer look at the stuffle product of two Eisenstein series. Since the product of multiple Eisenstein series can be written in terms of the stuffle product it is $G_2 \cdot G_3 = G_{2,3} + G_{3,2} + G_5$. On the other hand we have

$$G_2 \cdot G_3 = (\zeta(2) + g_2)(\zeta(3) + g_3) = \zeta(2)\zeta(3) + \zeta(3)g_2 + \zeta(2)g_3 + g_2 \cdot g_3.$$

and by Proposition 3.8 it is

$$G_{2,3} = \zeta(2,3) - 2\zeta(3)g_2 + \zeta(2)g_3 + g_{2,3}\,,$$
$$G_{3,2} = \zeta(3,2) + 3\zeta(3)g_2 + \zeta(2)g_3 + g_{3,2}\,.$$

In conclusion, we obtain a relation for the product of the g's namely $g_2 \cdot g_3 = g_{3,2} + g_{2,3} + g_5 + 2\zeta(2)g_3$ and dividing out $(-2\pi i)^5$ we get

$$[2] \cdot [3] = [3,2] + [2,3] + [5] - \frac{1}{12}[3]\,.$$

We conclude that a product of the q-series $[s_1, \dots, s_l] \in \mathbb{Q}[[q]]$ has an expression similar to the stuffle product and that conversely, a product structure on these q-series could be used, together with the Fourier expansion, to explain the stuffle product for multiple Eisenstein series.

One might now ask, if the multiple Eisenstein series also "fulfill" the shuffle product. As we saw above the shuffle product of $\zeta(2)$ and $\zeta(3)$ reads

$$\zeta(2) \cdot \zeta(3) = \zeta(2,3) + 3\zeta(3,2) + 6\zeta(4,1) \tag{17}$$

and since there is no definition of $G_{4,1}$ this question does not make sense when replacing ζ by G in (17). We will see that the understanding of the product structure of the brackets, explained in the next two sections, together with the Fourier expansion of multiple Eisenstein series will help to answer this question. This will be done by introducing shuffle regularized multiple Eisenstein series $G^{⧢}$ in Sect. 6.2. There we will see that we can replace the ζ in (17) by $G^{⧢}$ and that the $G^{⧢}$ are given by the original G, for the cases in which they are defined.

4 Multiple Divisor-Sums and Their Generating Functions

The classical divisor-sums $\sigma_r(n) = \sum_{d|n} d^r$ have a long history in number theory. They are well-known examples for multiplicative functions and appear in the Fourier expansion of Eisenstein series. This section is devoted to a larger class of functions, that can be seen as a multiple version of the divisor-sums and are therefore called multiple divisor-sums. For natural numbers $r_1, \dots, r_l \geq 0$ they are defined by

$$\sigma_{r_1,\dots,r_l}(n) = \sum_{\substack{u_1v_1+\dots+u_lv_l=n \\ u_1>\dots>u_l>0}} v_1^{r_1} \dots v_l^{r_l}\,. \tag{18}$$

Even though the definition of these arithmetic functions is not complicated and somehow canonical, the author could not find any results on these functions before he started studying them in his master thesis [2]. As mentioned in the introduction, the motivation to study them was due to their appearance in the Fourier expansion of multiple Eisenstein series, but as it turned out later in [6], they are very nice and interesting objects in their own rights. Similar to multiple zeta values they fulfill a lot of relations. For example it is

$$\frac{1}{2}\sigma_2(n) = \sigma_{1,0}(n) - \frac{1}{2}\sigma_1(n) + n\sigma_0(n)\,. \tag{19}$$

Having objects of this type it is natural to consider their generating functions, which we denote by

$$[s_1, \ldots, s_l] := \frac{1}{(s_1-1)!\ldots(s_l-1)!}\sum_{n>0}\sigma_{s_1-1,\ldots,s_l-1}(n)q^n$$

and which are, just for the sake of short notations, called brackets. The factorial factors and the "shift" of -1 are natural if one thinks about the Fourier expansion of Eisenstein series. With this notation the relation (19) reads as

$$[3] = [2, 1] - \frac{1}{2}[2] + q\frac{d}{dq}[1]\,, \tag{20}$$

which can be seen as a counterpart of the relation $\zeta(3) = \zeta(2, 1)$ between multiple zeta values.[2]

In this section, we want to focus on the algebraic structure of the space spanned by all brackets, which we will denote by $\mathscr{MD}$. This algebraic structure was studied in [6]. We will see that the space $\mathscr{MD}$ has the structure of a $\mathbb{Q}$-algebra and that the product of two brackets can be expressed in terms of brackets in a way that looks similar to the stuffle product of multiple zeta values. The operator $\mathrm{d} = q\frac{d}{dq}$ which appears in (20) plays an important role in the theory of (quasi-)modular forms. We will see that the space $\mathscr{MD}$ is closed under this operator and that this gives a second way of expressing the product of two brackets in length one similarly to the shuffle product of multiple zeta values. This second product expression in higher length will be discussed in Sect. 5.

4.1 Brackets

Definition 4.1 For any integers $s_1, \ldots, s_l > 0$ we define the generating function for the multiple divisor sum $\sigma_{s_1-1,\ldots,s_l-1}$ by the formal power series

[2]Further, one can prove the relation $\zeta(3) = \zeta(2, 1)$ between multiple zeta values by multiplying both sides in (20) with $(1-q)^3$ and then take the limit $q \to 1$. We will discuss this in Sect. 7.

$$
\begin{aligned}
[s_1, \dots, s_l] &:= \frac{1}{(s_1-1)!\dots(s_l-1)!} \sum_{n>0} \sigma_{s_1-1,\dots,s_l-1}(n) q^n \\
&= \sum_{\substack{u_1>\dots>u_l>0 \\ v_1,\dots,v_l>0}} \frac{v_1^{s_1-1}\dots v_l^{s_l-1}}{(s_1-1)!\dots(s_l-1)!} \cdot q^{u_1v_1+\dots+u_lv_l} \in \mathbb{Q}[\![q]\!] \,.
\end{aligned}
$$

In the first section, we saw that these series, by setting $q = \exp(2\pi i \tau)$, appear in the Fourier expansion of the multiple Eisenstein series but in this section we just view them as formal power series. We refer to these generating functions of multiple divisor sums as *brackets* and define the vector space $\mathcal{MD}$ to be the $\mathbb{Q}$ vector space generated by $1 \in \mathbb{Q}[\![q]\!]$ and all brackets $[s_1, \dots, s_l]$. It is important to notice that we also include the constants in the space $\mathcal{MD}$.

Example 4.2 We give a few examples:

$$
\begin{aligned}
[2] &= q + 3q^2 + 4q^3 + 7q^4 + 6q^5 + 12q^6 + 8q^7 + 15q^8 + \cdots \,, \\
[4, 2] &= \frac{1}{6} \left(q^3 + 3q^4 + 15q^5 + 27q^6 + 78q^7 + 135q^8 + \cdots\right) , \\
[4, 4, 4] &= \frac{1}{216} \left(q^6 + 9q^7 + 45q^8 + 190q^9 + 642q^{10} + 1899q^{11} + \cdots\right) , \\
[3, 1, 3, 1] &= \frac{1}{4} \left(q^{10} + 2q^{11} + 8q^{12} + 16q^{13} + 43q^{14} + 70q^{15} + \cdots\right) , \\
[1, 2, 3, 4, 5] &= \frac{1}{288} \left(q^{15} + 17q^{16} + 107q^{17} + 512q^{18} + 1985q^{19} + \cdots\right) .
\end{aligned}
$$

Notice that the first non vanishing coefficient of q^n in $[s_1, \dots, s_l]$ appears at $n = \frac{l(l+1)}{2}$, because it belongs to the "smallest" possible partition

$$
l \cdot 1 + (l-1) \cdot 1 + \dots + 1 \cdot 1 = n \,,
$$

i.e. $u_j = j$ and $v_j = 1$ for $1 \le j \le l$. The number $k = s_1 + \dots + s_l$ is called the *weight* of $[s_1, \dots, s_l]$ and l denotes the *length*.

We want to show that the brackets are closed under multiplication by proving that their product structure is an example for a quasi-shuffle product. To do this we first introduce some notations and quote some results which are needed for this.

Recall that for $s, z \in \mathbb{C}$, $|z| < 1$ the polylogarithm $\mathrm{Li}_s(z)$ of weight s is given by $\mathrm{Li}_s(z) = \sum_{n>0} \frac{z^n}{n^s}$. For $s \in \mathbb{N}$ the $\mathrm{Li}_{-s}(z)$ are rational functions in z with a pole in $z = 1$. More precisely for $|z| < 1$ they can be written as

$$
\mathrm{Li}_{-s}(z) = \sum_{n>0} n^s z^n = \frac{z P_s(z)}{(1-z)^{s+1}}
$$

where $P_s(z)$ is the s-th Eulerian polynomial. Such a polynomial is given by

$$P_s(X) = \sum_{n=0}^{s-1} A_{s,n} X^n \,,$$

where the Eulerian numbers $A_{s,n}$ are defined by

$$A_{s,n} = \sum_{i=0}^{n} (-1)^i \binom{s+1}{i} (n+1-i)^s \,.$$

For our purpose we write

$$\widetilde{\mathrm{Li}}_{1-s}(z) := \frac{\mathrm{Li}_{1-s}(z)}{(s-1)!}.$$

Lemma 4.3 ([6, Lemma 2.5]) *For $s_1, \dots, s_l \in \mathbb{N}$ we have*

$$\begin{aligned}
[s_1, \dots, s_l] &= \sum_{n_1 > \dots > n_l > 0} \widetilde{\mathrm{Li}}_{1-s_1}\left(q^{n_1}\right) \dots \widetilde{\mathrm{Li}}_{1-s_l}\left(q^{n_l}\right) \\
&= \frac{1}{(s_1-1)! \dots (s_l-1)!} \sum_{n_1 > \dots > n_l > 0} \prod_{j=1}^{l} \frac{q^{n_j} P_{s_j-1}\left(q^{n_j}\right)}{(1-q^{n_j})^{s_j}} \,.
\end{aligned}$$

Remark 4.4
(i) The second expression in terms of Eulerian Polynomials will be important for the interpretation of these series as q-analogues of multiple zeta values in Sect. 7.
(ii) This representation is also used for a fast implementation of these q-series in Pari GP. By doing so, the authors in [6] were able to give various results on the dimensions of the (weight and length filtered) spaces of $\mathcal{MD}$. These results can be found in Sect. 5 of [6].

The product of $[s_1]$ and $[s_2]$ can thus be written as

$$\begin{aligned}
[s_1] \cdot [s_2] &= \left(\sum_{n_1 > n_2 > 0} + \sum_{n_2 > n_1 > 0} \right) \widetilde{\mathrm{Li}}_{1-s_1}\left(q^{n_1}\right) \widetilde{\mathrm{Li}}_{1-s_2}\left(q^{n}\right) + \sum_{n_1 = n_2 > 0} \widetilde{\mathrm{Li}}_{1-s_1}\left(q^{n_1}\right) \widetilde{\mathrm{Li}}_{1-s_2}\left(q^{n_1}\right) \\
&= [s_1, s_2] + [s_2, s_1] + \sum_{n>0} \widetilde{\mathrm{Li}}_{1-s_1}\left(q^{n}\right) \widetilde{\mathrm{Li}}_{1-s_2}\left(q^{n}\right) \,.
\end{aligned}$$

In order to prove that this product is an element of $\mathcal{MD}$ the product $\widetilde{\mathrm{Li}}_{1-s_1}\left(q^n\right)$ $\widetilde{\mathrm{Li}}_{1-s_2}\left(q^n\right)$ must be a rational linear combination of $\widetilde{\mathrm{Li}}_{1-j}\left(q^n\right)$ with $1 \leq j \leq s_1 + s_2$. We therefore need the following

Lemma 4.5 *For $a, b \in \mathbb{N}$ we have*

$$\widetilde{\mathrm{Li}}_{1-a}(z) \cdot \widetilde{\mathrm{Li}}_{1-b}(z) = \sum_{j=1}^{a} \lambda_{a,b}^{j} \widetilde{\mathrm{Li}}_{1-j}(z) + \sum_{j=1}^{b} \lambda_{b,a}^{j} \widetilde{\mathrm{Li}}_{1-j}(z) + \widetilde{\mathrm{Li}}_{1-(a+b)}(z) \,,$$

where the coefficient $\lambda^j_{a,b} \in \mathbb{Q}$ *for* $1 \leq j \leq a$ *is given by*

$$\lambda^j_{a,b} = (-1)^{b-1}\binom{a+b-j-1}{a-j}\frac{B_{a+b-j}}{(a+b-j)!},$$

with B_k *being the* k*-th Bernoulli number.*[3]

Proof We prove this by using the generating function

$$L(X) := \sum_{k>0} \widetilde{\mathrm{Li}}_{1-k}(z) X^{k-1} = \sum_{k>0}\sum_{n>0} \frac{n^{k-1} z^n}{(k-1)!} X^{k-1} = \sum_{n>0} e^{nX} z^n = \frac{e^X z}{1-e^X z}.$$

With this one can see by direct calculation that

$$L(X)\cdot L(Y) = \frac{1}{e^{X-Y}-1}L(X) + \frac{1}{e^{Y-X}-1}L(Y).$$

By the definition of the Bernoulli numbers

$$\frac{X}{e^X-1} = \sum_{n\geq 0} \frac{B_n}{n!} X^n$$

this can be written as

$$L(X)\cdot L(Y) = \sum_{n>0} \frac{B_n}{n!}(X-Y)^{n-1}L(X) + \sum_{n>0}\frac{B_n}{n!}(Y-X)^{n-1}L(Y) + \frac{L(X)-L(Y)}{X-Y}.$$

The statement then follows by calculating the coefficient of $X^{a-1}Y^{b-1}$ in this equation.

Now we are able to interpret the product structure of brackets as an example for a quasi-shuffle product. We equip $\mathfrak{H}^1$ with a third product, beside the stuffle product $*$ and the shuffle product $ш$. This product will be denoted $\circledast$, since it can be seen as a "bracket version" of the stuffle product $*$. For $a, b \in \mathbb{N}$ and $w, v \in \mathfrak{H}^1$ we define recursively the product

$$z_a w \circledast z_b v = z_a(w \circledast z_b v) + z_b(z_a w \circledast v) + z_{a+b}(w \circledast v) + \sum_{j=1}^{a} \lambda^j_{a,b} z_j (w \circledast v)$$
$$+ \sum_{j=1}^{b} \lambda^j_{b,a} z_j (w \circledast v),$$

[3]For convenience we recall that the Bernoulli numbers B_k are defined by $\frac{X}{e^X-1} =: \sum_{k\geq 0} \frac{B_k}{k!} X^k$.

where the coefficients $\lambda^j_{a,b} \in \mathbb{Q}$ are the same as in Lemma 4.5. We equip $\mathcal{MD}$ with the usual multiplication of formal q-series and obtain the following:

Theorem 4.6 ([6, Prop 2.10]) *For the linear map* $[\,.\,] : (\mathfrak{H}^1, \circledast) \longrightarrow (\mathcal{MD}, \cdot)$ *defined on the generators* $w = z_{s_1} \dots z_{s_l}$ *by* $[w] := [s_1, \dots, s_l]$ *we have*

$$[w \circledast v] = [w] \cdot [v]$$

and therefore $\mathcal{MD}$ *is a* $\mathbb{Q}$*-algebra and* $[\,.\,]$ *an algebra homomorphism.*

Example 4.7 The first products of brackets are given by

$$\begin{aligned}
[1] \cdot [1] &= 2[1,1] + [2] - [1]\,,\\
[1] \cdot [2] &= [1,2] + [2,1] + [3] - \frac{1}{2}[2]\,,\\
[1] \cdot [2,1] &= [1,2,1] + 2[2,1,1] - \frac{3}{2}[2,1] + [2,2] + [3,1]\,,\\
[2] \cdot [3] &= [3,2] + [2,3] + [5] - \frac{1}{12}[3]\,,\\
[3] \cdot [2,1] &= [3,2,1] + [2,3,1] + [2,1,3] + [5,1] + [2,4] + \frac{1}{12}[2,2] - \frac{1}{2}[2,3] - \frac{1}{12}[3,1]\,.
\end{aligned}$$

We end this section by some notations which are needed for the rest of this paper.

Definition 4.8 On $\mathcal{MD}$ we have the increasing filtration $\mathrm{Fil}^{\mathrm{W}}_{\bullet}$ given by the weight and the increasing filtration $\mathrm{Fil}^{\mathrm{L}}_{\bullet}$ given by the length. For a subset $A \subset \mathcal{MD}$ we write[4]

$$\begin{aligned}
\mathrm{Fil}^{\mathrm{W}}_k(A) &:= \big\langle [s_1, \dots, s_l] \in A \,\big|\, l \geq 0\,, s_1 + \dots + s_l \leq k \big\rangle_{\mathbb{Q}}\,,\\
\mathrm{Fil}^{\mathrm{L}}_l(A) &:= \big\langle [s_1, \dots, s_r] \in A \,\big|\, 0 \leq r \leq l \big\rangle_{\mathbb{Q}}\,.
\end{aligned}$$

If we consider the length and weight filtration at the same time, we use the short notation $\mathrm{Fil}^{\mathrm{W,L}}_{k,l} := \mathrm{Fil}^{\mathrm{W}}_k \mathrm{Fil}^{\mathrm{L}}_l$.

Remark 4.9 As it can be seen by Theorem 4.6, the multiplication of two brackets respects these filtrations, i.e.

$$\mathrm{Fil}^{\mathrm{W,L}}_{k_1,l_1}(\mathcal{MD}) \cdot \mathrm{Fil}^{\mathrm{W,L}}_{k_2,l_2}(\mathcal{MD}) \subset \mathrm{Fil}^{\mathrm{W,L}}_{k_1+k_2,l_1+l_2}(\mathcal{MD}).$$

4.2 Derivatives and Subalgebras

In this section we want to give an overview of interesting subalgebras of the space $\mathcal{MD}$ and discuss the differential structure with respect to the differential $\mathrm{d} = q\frac{d}{dq}$. One of the main results in [6] is the following

[4] We set $[s_1, \dots, s_l] = 1$ for $l = 0$.

Theorem 4.10 ([6, Theroem 1.7]) *The operator* $\mathrm{d} = q\frac{d}{dq}$ *is a derivation on* $\mathcal{MD}$, *it maps* $\mathrm{Fil}^{\mathrm{W,L}}_{k,l}(\mathcal{MD})$ *to* $\mathrm{Fil}^{\mathrm{W,L}}_{k+2,l+1}(\mathcal{MD})$.

The proof of Theorem 4.10 uses generating functions of the brackets. It gives explicit formulas for the derivatives $\mathrm{d}[s_1, \ldots, s_l]$ for all l which we omit here, since they are complicated. For example we have

$$\begin{aligned} \mathrm{d}[2,1,1] = &-\frac{1}{6}[2,1,1] + \frac{1}{2}[2,1,2] - [2,1,2,1] + [2,1,3] + \frac{3}{2}[2,2,1] \\ &- 2\,[2,2,1,1] + [2,3,1] + 6[3,1,1] - 8[3,1,1,1] + [4,1,1]. \end{aligned}$$

In the following we give a list of subalgebras and review the results on whether they are also closed under d or not.

(i) (quasi-)modular forms: Next to the connection to modular forms due to their appearance in the Fourier expansion of multiple Eisenstein series, the brackets have a direct connection to quasi-modular forms for $\mathrm{SL}_2(\mathbb{Z})$ with rational coefficients. In the case $l = 1$ we get the divisor sums $\sigma_{k-1}(n) = \sum_{d|n} d^{k-1}$ and

$$[k] = \frac{1}{(k-1)!}\sum_{n>0} \sigma_{k-1}(n) q^n .$$

These simple brackets appear in the Fourier expansion of classical Eisenstein series with rational coefficients $\widetilde{G}_k(\tau) := (-2\pi i)^{-k} G_k(\tau)$ since we also included the rational numbers in $\mathcal{MD}$. For example we have

$$\widetilde{G}_2 = -\frac{1}{24} + [2], \quad \widetilde{G}_4 = \frac{1}{1440} + [4], \quad \widetilde{G}_6 = -\frac{1}{60480} + [6].$$

Denote by $M_{\mathbb{Q}}(\mathrm{SL}_2(\mathbb{Z})) = \mathbb{Q}[G_4, G_6]$ and $\widetilde{M}_{\mathbb{Q}}(\mathrm{SL}_2(\mathbb{Z})) = \mathbb{Q}[G_2, G_4, G_6]$ the algebras of modular forms and quasi-modular forms with rational coefficients.

It is a well-known fact that the space $\widetilde{M}_{\mathbb{Q}}(\mathrm{SL}_2(\mathbb{Z}))$ is closed under the operator $\mathrm{d} = q\frac{d}{dq}$.

(ii) Admissible brackets: We define the set of all admissible brackets $\mathrm{q}\mathcal{MZ}$ as the span of all brackets $[s_1, \ldots, s_l]$ with $s_1 > 1$ and 1. This space is a subalgebra of $\mathcal{MD}$ [6, Theorem 2.13] and every bracket can be written as a polynomial in the bracket [1] with coefficients in $\mathrm{q}\mathcal{MZ}$:

Theorem 4.11 ([6, Theorem 2.14, Proposition 3.14])

(i) We have $\mathcal{MD} = \mathrm{q}\mathcal{MZ}[[1]]$.
(ii) The algebra $\mathcal{MD}$ *is a polynomial ring over* $\mathrm{q}\mathcal{MZ}$ *with indeterminate* [1], *i.e.* $\mathcal{MD}$ *is isomorphic to* $\mathrm{q}\mathcal{MZ}[T]$ *by sending* [1] *to* T.
(iii) The space $\mathrm{q}\mathcal{MZ}$ *is closed under* d.

The elements in $\mathrm{q}\mathcal{MZ}$ are the ones, where the corresponding multiple zeta values exist. It will be reviewed in more detail in Sect. 7, when we consider the brackets as q-analogues of multiple zeta values.

(iii) Even brackets and brackets with no 1's: Denote by $\mathcal{MD}^{\text{even}}$ the space spanned by 1 and all $[s_1, \ldots, s_l]$ with s_j even for all $0 \leq j \leq l$ and by $\mathcal{MD}^\sharp$ the space spanned by 1 and all $[s_1, \ldots, s_l]$ with $s_j > 1$. Both spaces $\mathcal{MD}^{\text{even}}$ and $\mathcal{MD}^\sharp$ are subalgebras of $\mathcal{MD}$ [6, Proposition 2.15]. It is expected, that the space $\mathcal{MD}^{\text{even}}$ is not closed under d, since numerical calculation suggest, that for example $\mathrm{d}[4, 2] \notin \mathcal{MD}^{\text{even}}$. Whether the space $\mathcal{MD}^\sharp$ is closed under this operator is an open and interesting question. In [7] it is shown, that this is actually equivalent to one part of Conjecture 1 in [27] given by Okounkov.

To summarize, we have the following inclusion of $\mathbb{Q}$-algebras

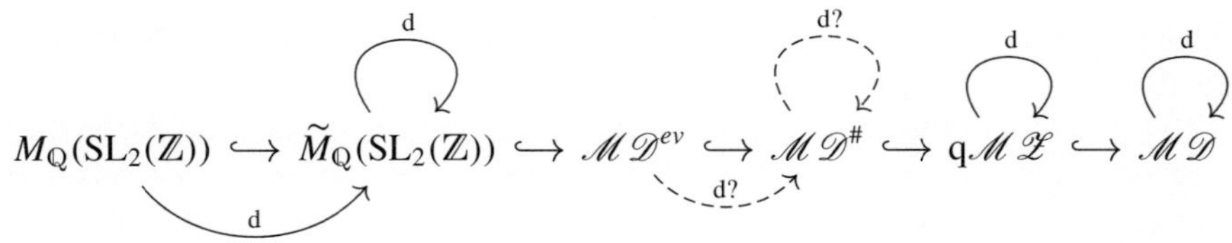

The dashed arrows indicate the conjectured behavior of the map d, whereas the other arrows are all known to be correct.

Though in length $l = 1$ we derive not just one but several expressions for $\mathrm{d}[s]$ given by the following Proposition.

Proposition 4.12 ([6, Proposition 3.3]) *For s_1, s_2 with $s_1 + s_2 > 2$ and $s = s_1 + s_2 - 2$ we have the following expression for* $\mathrm{d}[s]$*:*

$$\binom{s}{s_1 - 1}\frac{\mathrm{d}[s]}{s} = [s_1] \cdot [s_2] + \binom{s}{s_1 - 1}[s + 1] - \sum_{a+b=s+2}\left(\binom{a-1}{s_1 - 1} + \binom{a-1}{s_2 - 1}\right)[a, b]\,.$$

If you compare this formula with the shuffle product of multiple zeta values (11) in the length one times length one case you notice that Proposition 4.12 basically states that the brackets fulfill the shuffle product up to the term $\binom{s}{s_1-1}\frac{\mathrm{d}[s]}{s} - \binom{s}{s_1-1}[s + 1]$.

We end this section by using these formulas to prove the following identity

Proposition 4.13 *The unique normalized cusp form Δ in weight 12 can be written as*

$$-\frac{1}{2^6 \cdot 5 \cdot 691}\Delta = 168[5, 7] + 150[7, 5] + 28[9, 3]$$
$$+ \frac{1}{1408}[2] - \frac{83}{14400}[4] + \frac{187}{6048}[6] - \frac{7}{120}[8] - \frac{5197}{691}[12]\,.$$

Proof With the Eisenstein series $\widetilde{G}_6$ and $\widetilde{G}_{12}$ given by

$$\widetilde{G}_6 = (-2\pi i)^{-6}\zeta(6) + [6] = -\frac{1}{60480} + [6]\,,$$
$$\widetilde{G}_{12} = (-2\pi i)^{-12}\zeta(12) + [12] = \frac{691}{2615348736000} + [12]\,,$$

the cusp form Δ can be written as $\Delta = -3316800G_6^2 + 3432000G_{12}$. Using quasi-shuffle product of brackets one can derive

$$\Delta = \frac{3455}{198}[2] - \frac{691}{6}[4] + \frac{6910}{21}[6] + 115200[12] - 6633600[6,6]\,.$$

and therefore

$$-\frac{1}{2^6 \cdot 5 \cdot 691}\Delta = 30[6,6] - \frac{1}{12672}[2] + \frac{1}{1920}[4] - \frac{1}{672}[6] - \frac{360}{691}[12]\,. \tag{21}$$

Using Proposition 4.12 for $(s_1, s_2) = (4, 8), (5, 7), (6, 6)$ we get the following three expressions for d[10]

$$\begin{aligned}
\mathrm{d}[10] =& -\frac{1}{3}[5,7] - \frac{5}{6}[6,6] - \frac{5}{3}[7,5] - \frac{35}{12}[8,4] - \frac{16}{3}[9,3] - 10[10,2] - 20[11,1] \\
& - \frac{1}{4790016}[2] + \frac{1}{403200}[4] - \frac{1}{36288}[6] + \frac{1}{8640}[8] + 10[11] + \frac{1}{12}[12]\,, \\
\mathrm{d}[10] =& -\frac{5}{21}[6,6] - \frac{5}{7}[7,5] - 2[8,4] - \frac{14}{3}[9,3] - 10[10,2] - 20[11,1] \\
& + \frac{1}{4790016}[2] - \frac{1}{604800}[4] + \frac{1}{127008}[6] + 10[11] + \frac{1}{21}[12]\,, \\
\mathrm{d}[10] =& -\frac{10}{21}[7,5] - \frac{5}{3}[8,4] - \frac{40}{9}[9,3] - 10[10,2] - 20[11,1] \\
& - \frac{1}{4790016}[2] + \frac{1}{725760}[4] - \frac{1}{381024}[6] + 10[11] + \frac{5}{126}[12]\,.
\end{aligned}$$

Summing them up as $0 = -504\,\mathrm{d}[10] + 1890\,\mathrm{d}[10] - 1386\,\mathrm{d}[10]$ we get

$$\begin{aligned}
0 =& 168[5,7] - 30[6,6] + 150[7,5] + 28[9,3] \\
& + \frac{5}{6336}[2] - \frac{181}{28800}[4] + \frac{7}{216}[6] - \frac{7}{120}[8] - 7[12]
\end{aligned} \tag{22}$$

Combining (22) and (21), in order to eliminate the occurrence of [6, 6], we obtain the desired identity.

5 Bi-Brackets and a Second Product Expression for Brackets

In the previous section we have seen that the space $\mathcal{MD}$ of brackets has the structure of a $\mathbb{Q}$-algebra and that there is an explicit formula to express the product of two brackets as a linear combination of brackets similarly to the stuffle product of multiple zeta values. In this section we want to present a larger class of q-series, called bi-brackets. The quasi-shuffle product of brackets extend to this larger class and therefore the space of bi-brackets is also a $\mathbb{Q}$-algebra. The beautiful feature of bi-brackets is, that there is a relation, which we call partition relation, which enables one to express the product of two bi-brackets in a second different way. These two product expressions then give a large class of linear relations, similar to the double shuffle relations of multiple zeta values. A variation of the bi-brackets were also studied in [41]. Later, the bi-brackets will be used to define regularized multiple Eisenstein series in Sect. 6. All results in this section were studied and introduced in [3].

5.1 Bi-Brackets and Their Generating Series

As motivated in the introduction of this section we want to study the following q-series:

Definition 5.1 For $r_1, \dots, r_l \geq 0$, $s_1, \dots, s_l > 0$ and we define the following q-series

$$\begin{bmatrix} s_1, \dots, s_l \\ r_1, \dots, r_l \end{bmatrix} := \sum_{\substack{u_1 > \dots > u_l > 0 \\ v_1, \dots, v_l > 0}} \frac{u_1^{r_1}}{r_1!} \cdots \frac{u_l^{r_l}}{r_l!} \cdot \frac{v_1^{s_1-1} \dots v_l^{s_l-1}}{(s_1-1)! \dots (s_l-1)!} \cdot q^{u_1 v_1 + \dots + u_l v_l} \in \mathbb{Q}[[q]]$$

which we call *bi-brackets* of weight $r_1 + \dots + r_k + s_1 + \dots + s_l$, upper weight $s_1 + \dots + s_l$, lower weight $r_1 + \dots + r_l$ and length l. By $\mathcal{BD}$ we denote the $\mathbb{Q}$-vector space spanned by all bi-brackets and 1.

The factorial factors in the definition of bi-brackets will become natural when considering generating functions of bi-brackets and the connection to multiple zeta values.

For $r_1 = \dots = r_l = 0$ the bi-brackets are just the brackets

$$\begin{bmatrix} s_1, \dots, s_l \\ 0, \dots, 0 \end{bmatrix} = [s_1, \dots, s_l]$$

as defined in Sect. 4. Similarly to the Definition 4.8 of the filtration for the space $\mathcal{BD}$ we write for a subset $A \in \mathcal{BD}$

$$\begin{aligned}
\mathrm{Fil}_k^{\mathrm{W}}(A) &:= \left\langle \begin{bmatrix} s_1, \dots, s_l \\ r_1, \dots, r_l \end{bmatrix} \in A \,\middle|\, 0 \leq l \leq k\,,\ s_1 + \dots + s_l \leq k \right\rangle_{\mathbb{Q}} \\
\mathrm{Fil}_k^{\mathrm{D}}(A) &:= \left\langle \begin{bmatrix} s_1, \dots, s_l \\ r_1, \dots, r_l \end{bmatrix} \in A \,\middle|\, 0 \leq l \leq k\,,\ r_1 + \dots + r_l \leq k \right\rangle_{\mathbb{Q}} \\
\mathrm{Fil}_l^{\mathrm{L}}(A) &:= \left\langle \begin{bmatrix} s_1, \dots, s_t \\ r_1, \dots, r_t \end{bmatrix} \in A \,\middle|\, t \leq l \right\rangle_{\mathbb{Q}}.
\end{aligned}$$

and again if we consider the length and weight filtration at the same time we use the short notation $\mathrm{Fil}_{k,l}^{\mathrm{W,L}} := \mathrm{Fil}_k^{\mathrm{W}} \mathrm{Fil}_l^{\mathrm{L}}$ and similar for the other filtrations.

Proposition 5.2 ([3, Proposition 4.2]) *Let* $\mathrm{d} := q\frac{d}{dq}$, *then we have*

$$\mathrm{d}\begin{bmatrix} s_1, \dots, s_l \\ r_1, \dots, r_l \end{bmatrix} = \sum_{j=1}^{l} \left(s_j (r_j + 1) \begin{bmatrix} s_1, \dots, s_{j-1}, s_j + 1, s_{j+1}, \dots, s_l \\ r_1, \dots, r_{j-1}, r_j + 1, r_{j+1}, \dots, r_l \end{bmatrix} \right)$$

and therefore $\mathrm{d}\left(\mathrm{Fil}_{k,d,l}^{\mathrm{W,D,L}}(\mathcal{BD})\right) \subset \mathrm{Fil}_{k+1,d+1,l}^{\mathrm{W,D,L}}(\mathcal{BD})$.

Proof This is an easy consequence of the definition of bi-brackets and the fact that $\mathrm{d}\sum_{n>0} a_n q^n = \sum_{n>0} n a_n q^n$.

Proposition 5.2 suggests that the bi-brackets can be somehow viewed as partial derivatives of the brackets with total differential d.

In the following we now want to discuss the algebra structure of the space $\mathcal{BD}$. For this we extend the quasi-shuffle product $\circledast$ of $\mathfrak{H}^1$ to a larger space of words. Since we have double indices we replace the alphabet $A_z = \{z_1, z_2, \dots\}$ by $A_z^{\mathrm{bi}} := \{z_{s,r} \mid s \geq 1, r \geq 0\}$.

We consider on $\mathbb{Q}A_z^{\mathrm{bi}}$ the commutative and associative product

$$z_{s_1,r_1} \boxdot z_{s_2,r_2} = \binom{r_1 + r_2}{r_1} \sum_{j=1}^{s_1} \lambda_{s_1,s_2}^j z_{j,r_1+r_2} + \binom{r_1 + r_2}{r_1} \sum_{j=1}^{s_2} \lambda_{s_2,s_1}^j z_{j,r_1+r_2} + \binom{r_1 + r_2}{r_1} z_{s_1+s_2,r_1+r_2}$$

and on $\mathbb{Q}\langle A_z^{\mathrm{bi}}\rangle$ the commutative and associative quasi-shuffle product

$$z_{s_1,r_1} w \circledast z_{s_2,r_2} v = z_{s_1,r_1}(w \circledast z_{s_2,r_2} v) + z_{s_2,r_2}(z_{s_1,r_1} w \circledast v) + (z_{s_1,r_1} \boxdot z_{s_2,r_2})(w \circledast v),$$

where the the numbers $\lambda_{a,b}^j \in \mathbb{Q}$ for $1 \leq j \leq a$ are the same as before, i.e.

$$\lambda_{a,b}^j = (-1)^{b-1} \binom{a + b - j - 1}{a - j} \frac{B_{a+b-j}}{(a + b - j)!}.$$

Theorem 5.3 ([3, Theroem 3.6]) *The map* $[\cdot] : (\mathbb{Q}\langle A_z^{bi}\rangle, \circledast) \to (\mathcal{BD}, \cdot)$ *given by*

$$w = z_{s_1,r_1} \dots z_{s_l,r_l} \longmapsto [w] = \begin{bmatrix} s_1, \dots, s_l \\ r_1, \dots, r_l \end{bmatrix}$$

fulfills $[w \circledast v] = [w] \cdot [v]$ *and therefore* $\mathcal{BD}$ *is a* $\mathbb{Q}$*-algebra.*

Definition 5.4 For the generating function of the bi-brackets we write

$$\begin{vmatrix} X_1, \dots, X_l \\ Y_1, \dots, Y_l \end{vmatrix} := \sum_{\substack{s_1,\dots,s_l>0 \\ r_1,\dots,r_l>0}} \begin{bmatrix} s_1, \dots, s_l \\ r_1-1, \dots, r_l-1 \end{bmatrix} X_1^{s_1-1} \dots X_l^{s_l-1} \cdot Y_1^{r_1-1} \dots Y_l^{r_l-1} .$$

These are elements in the ring $\mathcal{BD}_{\text{gen}} = \varinjlim_j \mathcal{BD}[[X_1, \dots, X_j, Y_1, \dots, Y_j]]$ of all generating series of bi-brackets.

To derive relations between bi-brackets we will prove functional equations for their generating functions. The key fact for this is that there are two different ways of expressing these given by the following Theorem.

Theorem 5.5 ([3, Theroem 2.3]) *For $n \in \mathbb{N}$ set*

$$E_n(X) := e^{nX} \quad \text{and} \quad L_n(X) := \frac{e^X q^n}{1 - e^X q^n} \in \mathbb{Q}[[q, X]] .$$

Then for all $l \geq 1$ we have the following two different expressions for the generating functions:

$$\begin{aligned} \begin{vmatrix} X_1, \dots, X_l \\ Y_1, \dots, Y_l \end{vmatrix} &= \sum_{u_1 > \dots > u_l > 0} \prod_{j=1}^{l} E_{u_j}(Y_j) L_{u_j}(X_j) \\ &= \sum_{u_1 > \dots > u_l > 0} \prod_{j=1}^{l} E_{u_j}(X_{l+1-j} - X_{l+2-j}) L_{u_j}(Y_1 + \dots + Y_{l-j+1}) \end{aligned}$$

(with $X_{l+1} := 0$). In particular the partition relations[5] *holds:*

$$\begin{vmatrix} X_1, \dots, X_l \\ Y_1, \dots, Y_l \end{vmatrix} \overset{P}{=} \begin{vmatrix} Y_1 + \dots + Y_l, \dots, Y_1 + Y_2, Y_1 \\ X_l, X_{l-1} - X_l, \dots, X_1 - X_2 \end{vmatrix} . \qquad (23)$$

Remark 5.6 A nice combinatorial explanation for the partition relation (23) is the following: By a partition of a natural number n with l parts we denote a representation of n as a sum of l distinct natural numbers, i.e. $15 = 4 + 4 + 3 + 2 + 1 + 1$ is a partition of 15 with the 4 parts given by 4, 3, 2, 1. We identify such a partition with a tuple $(u, v) \in \mathbb{N}^l \times \mathbb{N}^l$ where the u_j's are the l distinct numbers in the partition and the v_j's count their appearance in the sum. The above partition of 15 is therefore given by the tuple $(u, v) = ((4, 3, 2, 1), (2, 1, 1, 2))$. By $P_l(n)$ we denote all partitions of n with l parts and hence we set

$$P_l(n) := \left\{ (u, v) \in \mathbb{N}^l \times \mathbb{N}^l \mid n = u_1 v_1 + \dots + u_l v_l \text{ and } u_1 > \dots > u_l > 0 \right\}$$

[5] The bi-brackets and their generating series also give examples of what is called a *bimould* by Ecalle in [16]. In his language the partition relation (23) states that the bimould of generating series of bi-brackets is swap invariant.

$((4,3,2,1),(2,1,1,2)) =$ 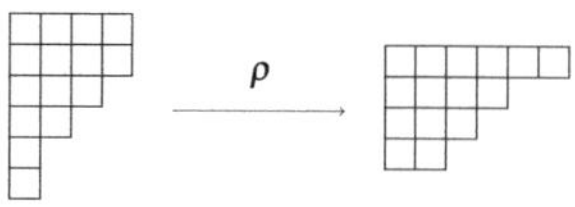 $= ((6,4,3,2),(1,1,1,1))$

Fig. 1 The conjugation of the partition $15 = 4+4+3+2+1+1$ is given by $\rho(((4,3,2,1),(2,1,1,2))) = ((6,4,3,2),(1,1,1,1))$ which can be seen by reflection the corresponding Young diagram at the main diagonal

On the set $P_l(n)$ one has an involution given by the conjugation ρ of partitions which can be obtained by reflecting the corresponding Young diagram across the main diagonal (Fig. 1).

On the set $P_l(n)$ the conjugation ρ is explicitly given by $\rho((u,v)) = (u',v')$ where $u'_j = v_1 + \dots + v_{l-j+1}$ and $v'_j = u_{l-j+1} - u_{l-j+2}$ with $u_{l+1} := 0$, i.e.

$$\rho : \begin{pmatrix} u_1, \dots, u_l \\ v_1, \dots, v_l \end{pmatrix} \longmapsto \begin{pmatrix} v_1 + \dots + v_l, \dots, v_1 + v_2, v_1 \\ u_l, u_{l-1} - u_l, \dots, u_1 - u_2 \end{pmatrix}. \tag{24}$$

By the definition of the bi-brackets its clear that with the above notation they can be written as

$$\begin{bmatrix} s_1, \dots, s_l \\ r_1, \dots, r_l \end{bmatrix} := \frac{1}{r_1!(s_1-1)! \dots r_l!(s_l-1)!} \sum_{n>0} \left(\sum_{(u,v) \in P_l(n)} u_1^{r_1} v_1^{s_1-1} \dots u_l^{r_l} v_l^{s_l-1} \right) q^n .$$

The coefficients are given by a sum over all elements in $P_l(n)$ and therefore it is invariant under the action of ρ. As an example, consider $[2,2]$ and apply ρ to the sum. Then we obtain

$$\begin{aligned} [2,2] &= \sum_{n>0} \left(\sum_{(u,v) \in P_2(n)} v_1 \cdot v_2 \right) q^n = \sum_{n>0} \left(\sum_{\rho((u,v))=(u',v') \in P_2(n)} u'_2 \cdot (u'_1 - u'_2) \right) q^n \\ &= \sum_{n>0} \left(\sum_{(u',v') \in P_2(n)} u'_2 \cdot u'_1 \right) q^n - \sum_{n>0} \left(\sum_{(u',v') \in P_2(n)} u'^2_2 \right) q^n = \begin{bmatrix} 1, 1 \\ 1, 1 \end{bmatrix} - 2 \begin{bmatrix} 1, 1 \\ 0, 2 \end{bmatrix}. \end{aligned} \tag{25}$$

This is exactly the relation one obtains by using the partition relation.

Corollary 5.7 ([3, Corollary 2.5]) *(Partition relation in length one and two) For $r, r_1, r_2 \geq 0$ and $s, s_1, s_2 > 0$ we have the following relations in length one and two*

$$\begin{bmatrix} s \\ r \end{bmatrix} = \begin{bmatrix} r+1 \\ s-1 \end{bmatrix},$$

$$\begin{bmatrix} s_1, s_2 \\ r_1, r_2 \end{bmatrix} = \sum_{\substack{0 \le j \le r_1 \\ 0 \le k \le s_2 - 1}} (-1)^k \binom{s_1 - 1 + k}{k} \binom{r_2 + j}{j} \begin{bmatrix} r_2 + j + 1\,, r_1 - j + 1 \\ s_2 - 1 - k\,, s_1 - 1 + k \end{bmatrix}.$$

Remark 5.8 (i) If we replace in the generating series in Definition 5.4 the bi-brackets by the corresponding bi-words in and enforce the partition relation (23) for this power series, we obtain an involution

$$P : \mathbb{Q}\langle A_z^{\mathrm{bi}} \rangle \to \mathbb{Q}\langle A_z^{\mathrm{bi}} \rangle\,.$$

By Corollary 5.7 it is for example $P(z_{s,r}) = z_{r+1,s-1}$. This will be needed to describe the second product structure in the next section.

(ii) In [41] the author introduces multiple q-zeta brackets $\mathfrak{Z}\left[\begin{smallmatrix} s_1,\dots,s_r \\ r_1,\dots,r_l \end{smallmatrix}\right]$, which can be written in terms of bi-brackets and vice versa. For these objects the partition relation has the nice form

$$\mathfrak{Z}\begin{bmatrix} s_1, \dots, s_r \\ r_1, \dots, r_l \end{bmatrix} = \mathfrak{Z}\begin{bmatrix} r_l, \dots, r_1 \\ s_l, \dots, s_1 \end{bmatrix},$$

which can be interpreted in terms of duality. This is also used in [41] to describe the second product structure for the $\mathfrak{Z}$. Similarly in [17] the authors use a duality by Zhao [39] to describe a second product structure for another model of q-analogues.

5.2 Double Shuffle Relations for Bi-Brackets

The partition relation together with the quasi-shuffle product can be used to obtain a second expression for the product of two bi-brackets. Before giving the general explanation this second product expression we illustrate it in two examples.

Example 5.9 (i) We want to given a second product expression for the product $[2] \cdot [3]$. By the partition relation we know that $[2] = \left[\begin{smallmatrix} 1 \\ 1 \end{smallmatrix}\right]$, $[3] = \left[\begin{smallmatrix} 1 \\ 2 \end{smallmatrix}\right]$ and using the quasi-shuffle product we have

$$\begin{bmatrix} 1 \\ 1 \end{bmatrix} \cdot \begin{bmatrix} 1 \\ 2 \end{bmatrix} = \begin{bmatrix} 1, 1 \\ 1, 2 \end{bmatrix} + \begin{bmatrix} 1, 1 \\ 2, 1 \end{bmatrix} - 3\begin{bmatrix} 1 \\ 3 \end{bmatrix} + 3\begin{bmatrix} 2 \\ 3 \end{bmatrix}.$$

The partition relations for the length two bi-brackets on the right is given by

$$\begin{bmatrix}1,1\\1,2\end{bmatrix} = \begin{bmatrix}3,2\\0,0\end{bmatrix} + 3\begin{bmatrix}4,1\\0,0\end{bmatrix} = [3,2] + 3[4,1]\,,$$
$$\begin{bmatrix}1,1\\2,1\end{bmatrix} = \begin{bmatrix}2,3\\0,0\end{bmatrix} + 2\begin{bmatrix}3,2\\0,0\end{bmatrix} + 3\begin{bmatrix}4,1\\0,0\end{bmatrix} = [2,3] + 2[3,2] + 3[4,1]\,.$$

Combining all of this we obtain

$$\begin{aligned}\begin{bmatrix}2\\0\end{bmatrix}\cdot\begin{bmatrix}3\\0\end{bmatrix} &= \begin{bmatrix}1\\1\end{bmatrix}\cdot\begin{bmatrix}1\\2\end{bmatrix}\\ &= \begin{bmatrix}1,1\\1,2\end{bmatrix} + \begin{bmatrix}1,1\\2,1\end{bmatrix} - 3\begin{bmatrix}1\\3\end{bmatrix} + 3\begin{bmatrix}2\\3\end{bmatrix}\\ &= [2,3] + 3[3,2] + 6[4,1] + 3\begin{bmatrix}4\\1\end{bmatrix} - 3[4]\,.\end{aligned}$$

Compare this to the shuffle product of multiple zeta values

$$\zeta(2)\zeta(3) = \zeta(2,3) + 3\zeta(3,2) + 6\zeta(4,1)\,.$$

Since $\mathrm{d}[3] = 3\left[\begin{smallmatrix}4\\1\end{smallmatrix}\right]$ this example exactly coincides with the formula in Proposition 4.12 for the derivative $\mathrm{d}[k]$.

(ii) In higher length, expressing the product of two bi-brackets in a similar way as in i) becomes interesting, since then the extra terms can't be expressed with the operator d anymore. Doing the same calculation for the product $[3]\cdot[2,1]$, i.e. using the partition relation, the quasi-shuffle product and again the partition relation we obtain

$$\begin{aligned}[3]\cdot[2,1] &= \begin{bmatrix}1\\2\end{bmatrix}\cdot\begin{bmatrix}1,1\\0,1\end{bmatrix}\\ &= \begin{bmatrix}1,1,1\\2,0,1\end{bmatrix} + \begin{bmatrix}1,1,1\\0,2,1\end{bmatrix} + \begin{bmatrix}1,1,1\\0,1,2\end{bmatrix} + 3\begin{bmatrix}1,2\\0,3\end{bmatrix} + \begin{bmatrix}2,1\\2,1\end{bmatrix} - 3\begin{bmatrix}1,1\\0,3\end{bmatrix} - \begin{bmatrix}1,1\\2,1\end{bmatrix}\\ &= [2,1,3] + [2,2,2] + 2[2,3,1] + 2[3,1,2] + 5[3,2,1] + 9[4,1,1]\\ &+ \begin{bmatrix}2,3\\0,1\end{bmatrix} + 2\begin{bmatrix}3,2\\0,1\end{bmatrix} + 3\begin{bmatrix}4,1\\1,0\end{bmatrix} - [2,3] - 2[3,2] - 6[4,1]\,.\end{aligned}$$

This product can be seen as the analogue of the shuffle product

$$\zeta(3)\cdot\zeta(2,1) = \zeta(2,1,3) + \zeta(2,2,2) + 2\zeta(2,3,1) + 2\zeta(3,1,2) + 5\zeta(3,2,1) + 9\zeta(4,1,1)\,.$$

Here the bi-brackets, which are not given as brackets, can not be written in terms of the operator d in an obvious way.

This works for arbitrary lengths and yields a natural way to obtain the second product expression for bi-brackets. To be more precise, denote by $P : \mathbb{Q}\langle A_z^{\mathrm{bi}}\rangle \to \mathbb{Q}\langle A_z^{\mathrm{bi}}\rangle$ the involution defined in Remark i. Using this convention the second product expression for bi-brackets can be written in $\mathbb{Q}\langle A_z^{\text{bi}}\rangle$ for two words $u, v \in \mathbb{Q}\langle A_z^{\text{bi}}\rangle$ as $P\left(P(u) \boxast P(v)\right)$, i.e. the two product expressions of bi-brackets which correspond to the stuffle and shuffle product of multiple zeta values are given by

$$[u] \cdot [v] = [u \boxast v], \qquad [u] \cdot [v] = [P\left(P(u) \boxast P(v)\right)]. \tag{26}$$

In contrast to multiple zeta values these two product expression are the same for some cases, as one can check for the example $[1] \cdot [1, 1]$. In the smallest length case, we have the following explicit formulas for the two products expressions.

Proposition 5.10 ([2, Proposition 3.3]) *For $s_1, s_2 > 0$ and $r_1, r_2 \geq 0$ we have the following two expressions for the product of two bi-brackets of length one:*

(i) ("Stuffle product analogue for bi-brackets")

$$\begin{aligned}
\begin{bmatrix} s_1 \\ r_1 \end{bmatrix} \cdot \begin{bmatrix} s_2 \\ r_2 \end{bmatrix} &= \begin{bmatrix} s_1, s_2 \\ r_1, r_2 \end{bmatrix} + \begin{bmatrix} s_2, s_1 \\ r_2, r_1 \end{bmatrix} + \binom{r_1 + r_2}{r_1} \begin{bmatrix} s_1 + s_2 \\ r_1 + r_2 \end{bmatrix} \\
&+ \binom{r_1 + r_2}{r_1} \sum_{j=1}^{s_1} \frac{(-1)^{s_2 - 1} B_{s_1 + s_2 - j}}{(s_1 + s_2 - j)!} \binom{s_1 + s_2 - j - 1}{s_1 - j} \begin{bmatrix} j \\ r_1 + r_2 \end{bmatrix} \\
&+ \binom{r_1 + r_2}{r_1} \sum_{j=1}^{s_2} \frac{(-1)^{s_1 - 1} B_{s_1 + s_2 - j}}{(s_1 + s_2 - j)!} \binom{s_1 + s_2 - j - 1}{s_2 - j} \begin{bmatrix} j \\ r_1 + r_2 \end{bmatrix}
\end{aligned}$$

(ii) ("Shuffle product analogue for bi-brackets")

$$\begin{aligned}
\begin{bmatrix} s_1 \\ r_1 \end{bmatrix} \cdot \begin{bmatrix} s_2 \\ r_2 \end{bmatrix} &= \sum_{\substack{1 \leq j \leq s_1 \\ 0 \leq k \leq r_2}} \binom{s_1 + s_2 - j - 1}{s_1 - j} \binom{r_1 + r_2 - k}{r_1} (-1)^{r_2 - k} \begin{bmatrix} s_1 + s_2 - j, j \\ k, r_1 + r_2 - k \end{bmatrix} \\
&+ \sum_{\substack{1 \leq j \leq s_2 \\ 0 \leq k \leq r_1}} \binom{s_1 + s_2 - j - 1}{s_1 - 1} \binom{r_1 + r_2 - k}{r_1 - k} (-1)^{r_1 - k} \begin{bmatrix} s_1 + s_2 - j, j \\ k, r_1 + r_2 - k \end{bmatrix} \\
&+ \binom{s_1 + s_2 - 2}{s_1 - 1} \begin{bmatrix} s_1 + s_2 - 1 \\ r_1 + r_2 + 1 \end{bmatrix} \\
&+ \binom{s_1 + s_2 - 2}{s_1 - 1} \sum_{j=0}^{r_1} \frac{(-1)^{r_2} B_{r_1 + r_2 - j + 1}}{(r_1 + r_2 - j + 1)!} \binom{r_1 + r_2 - j}{r_1 - j} \begin{bmatrix} s_1 + s_2 - 1 \\ j \end{bmatrix} \\
&+ \binom{s_1 + s_2 - 2}{s_1 - 1} \sum_{j=0}^{r_2} \frac{(-1)^{r_1} B_{r_1 + r_2 - j + 1}}{(r_1 + r_2 - j + 1)!} \binom{r_1 + r_2 - j}{r_2 - j} \begin{bmatrix} s_1 + s_2 - 1 \\ j \end{bmatrix}
\end{aligned}$$

Having these two expressions for the product of bi-brackets we obtain a large family of linear relations between them. Computer experiments suggest that actually every bi-bracket can be written in terms of brackets and that motivates the following surprising conjecture.

Conjecture 5.11 *The algebra* $\mathcal{BD}$ *of bi-brackets is a subalgebra of* $\mathcal{MD}$ *and in particular we have*

$$\mathrm{Fil}^{\mathrm{W,D,L}}_{k,d,l}(\mathcal{BD}) \subset \mathrm{Fil}^{\mathrm{W,L}}_{k+d,l+d}(\mathcal{MD})\,.$$

The results towards this conjecture, beside the computer experiments which have been done up to weight 8, are the following

Proposition 5.12 ([3, Proposition 4.4]) *For* $l = 1$ *the Conjecture 5.11 is true.*

In [8] it will be shown, that Conjecture 5.11 is also true for all length up to weight 7. For higher weights and lengths there are no general statements. The only general statement for the length two case is given by the following Proposition.

Proposition 5.13 ([3, Proposition 5.9]) *For all* $s_1, s_2 \geq 1$ *it is*

$$\begin{bmatrix} s_1, s_2 \\ 1, 0 \end{bmatrix}, \begin{bmatrix} s_1, s_2 \\ 0, 1 \end{bmatrix} \in \mathrm{Fil}^{\mathrm{W,L}}_{s_1+s_2+1,3}(\mathcal{MD})$$

5.3 *The Shuffle Brackets*

We now want to define a q-series which is an element in $\mathcal{BD}$ and whose products can be written in terms of the "real" shuffle product of multiple zeta values. For $e_1, \ldots, e_l \geq 1$ we generalize the generating function of bi-brackets to the following

$$\begin{vmatrix} X_1, & \ldots, & X_l \\ Y_1, & \ldots, & Y_l \\ e_1, & \ldots, & e_l \end{vmatrix} = \sum_{u_1 > \cdots > u_l > 0} \prod_{j=1}^{l} E_{u_j}(Y_j) L_{u_j}(X_j)^{e_j}\,. \tag{27}$$

In particular for $e_1 = \cdots = e_l = 1$ these are the generating functions of the bi-brackets. To show that the coefficients of these series are in $\mathcal{BD}$ for arbitrary e_j we need to define the differential operator $\mathcal{D}^{Y}_{e_1,\ldots,e_l} := D_{Y_1,e_1} D_{Y_2,e_2} \ldots D_{Y_l,e_l}$ with

$$D_{Y_j,e} = \prod_{k=1}^{e-1} \left(\frac{1}{k} \left(\frac{\partial}{\partial Y_{l-j+1}} - \frac{\partial}{\partial Y_{l-j+2}} \right) - 1 \right).$$

where we set $\frac{\partial}{\partial Y_{l+1}} = 0$.

Lemma 5.14 *Let* $\mathcal{A}$ *be an algebra spanned by elements* $a_{s_1,\ldots,s_l}$ *with* $s_1, \ldots, s_l \in \mathbb{N}$, *let* $H(X_1, \ldots, X_l) = \sum_{s_j} a_{s_1,\ldots,s_l} X_1^{s_1-1} \ldots X_1^{s_l-1}$ *be the generating functions of these elements and define for* $f \in \mathbb{Q}[[X_1, \ldots, X_l]]$

$$f^{\sharp}(X_1, \ldots, X_l) = f(X_1 + \cdots + X_l, X_2 + \cdots + X_l, \ldots, X_l)\,.$$

Then the following two statements are equivalent.

(i) *The map* $(\mathfrak{H}^1, \sqcup\!\sqcup) \to \mathscr{A}$ *given by* $z_{s_1} \dots z_{s_j} \mapsto a_{s_1,\dots,s_l}$ *is an algebra homomorphism.*

(ii) *For all* $r, s \in \mathbb{N}$ *it is*

$$H^\sharp(X_1,\dots,X_r)\cdot H^\sharp(X_{r+1},\dots,X_{r+s}) = H^\sharp(X_1,\dots,X_{r+s})_{|sh_r^{(r+s)}},$$

where $sh_r^{(r+s)} = \sum_{\sigma\in\Sigma(r,s)}\sigma$ *in the group ring* $\mathbb{Z}[\mathfrak{S}_{r+s}]$ *and the symmetric group* $\mathfrak{S}_r$ *acts on* $\mathbb{Q}[[X_1,\dots,X_r]]$ *by* $(f\big|\sigma)(X_1,\dots,X_r) = f(X_{\sigma^{-1}(1)},\dots, X_{\sigma^{-1}(r)})$.

Proof This can be proven by induction over l together with Proposition 8 in [23].

Theorem 5.15 ([3, Theroem 5.7]) *For* $s_1,\dots,s_l \in \mathbb{N}$ *define* $[s_1,\dots,s_l]^{\sqcup\!\sqcup} \in \mathscr{BD}$ *as the coefficients of the following generating function*

$$\begin{aligned} H_{\sqcup\!\sqcup}(X_1,\dots,X_l) &= \sum_{s_1,\dots,s_l\geq 1}[s_1,\dots,s_l]^{\sqcup\!\sqcup}X_1^{s_1-1}\dots X_l^{s_l-1} \\ &:= \sum_{\substack{1\leq m\leq l\\ i_1+\dots+i_m=l}} \frac{1}{i_1!\dots i_m!}\mathscr{D}^Y_{i_1,\dots,i_m}\left|\begin{matrix} X_1, X_{i_m+1}, X_{i_{m-1}+i_m+1},\dots,X_{i_2+\dots+i_m+1} \\ Y_1,\dots,Y_l\end{matrix}\right|_{|Y=0}. \end{aligned}$$

Then we have the following two statements

i) *The* $[s_1,\dots,s_l]^{\sqcup\!\sqcup}$ *fulfill the shuffle product, i.e.*

$$H^\sharp_{\sqcup\!\sqcup}(X_1,\dots,X_r)\cdot H^\sharp_{\sqcup\!\sqcup}(X_{r+1},\dots,X_{r+s}) = H^\sharp_{\sqcup\!\sqcup}(X_1,\dots,X_{r+s})_{|sh_r^{(r+s)}}.$$

ii) *For* $s_1 \geq 1,\ s_2,\dots,s_l\geq 2$ *we have* $[s_1,\dots,s_l]^{\sqcup\!\sqcup} = [s_1,\dots,s_l]$.

For low lengths we obtain the following examples:

Corollary 5.16 *It is* $[s_1]^{\sqcup\!\sqcup} = [s_1]$ *and for* $l = 2, 3, 4$ *the* $[s_1,\dots,s_l]^{\sqcup\!\sqcup}$*are given by*[6]

(i) $[s_1,s_2]^{\sqcup\!\sqcup} = [s_1,s_2] + \delta_{s_2,1}\cdot\frac{1}{2}\left(\begin{bmatrix}s_1\\1\end{bmatrix} - [s_1]\right)$,

(ii)
$$\begin{aligned} [s_1,s_2,s_3]^{\sqcup\!\sqcup} &= [s_1,s_2,s_3] + \delta_{s_3,1}\cdot\frac{1}{2}\left(\begin{bmatrix}s_1,s_2\\0,1\end{bmatrix} - [s_1,s_2]\right) \\ &+ \delta_{s_2,1}\cdot\frac{1}{2}\left(\begin{bmatrix}s_1,s_3\\1,0\end{bmatrix} - \begin{bmatrix}s_1,s_3\\0,1\end{bmatrix} - [s_1,s_3]\right) \\ &+ \delta_{s_2\cdot s_3,1}\cdot\frac{1}{6}\left(\begin{bmatrix}s_1\\2\end{bmatrix} - \frac{3}{2}\begin{bmatrix}s_1\\1\end{bmatrix} + [s_1]\right), \end{aligned}$$

[6]Here $\delta_{a,b}$ denotes the Kronecker delta, i.e $\delta_{a,b}$ is 1 for $a = b$ and 0 otherwise.

(iii)
$$[s_1,s_2,s_3,s_4]^{\sqcup\!\sqcup} = [s_1,s_2,s_3,s_4] + \delta_{s_4,1}\cdot\frac{1}{2}\left(\begin{bmatrix} s_1,s_2,s_3 \\ 0,0,1\end{bmatrix} - [s_1,s_2,s_3]\right)$$
$$+\delta_{s_3,1}\cdot\frac{1}{2}\left(\begin{bmatrix} s_1,s_2,s_4 \\ 0,1,0\end{bmatrix} - \begin{bmatrix} s_1,s_2,s_4 \\ 0,0,1\end{bmatrix} + [s_1,s_2,s_4]\right)$$
$$+\delta_{s_2,1}\cdot\frac{1}{2}\left(\begin{bmatrix} s_1,s_3,s_4 \\ 1,0,0\end{bmatrix} - \begin{bmatrix} s_1,s_3,s_4 \\ 0,1,0\end{bmatrix} + [s_1,s_3,s_4]\right)$$
$$+\delta_{s_2\cdot s_4,1}\cdot\frac{1}{4}\left(\begin{bmatrix} s_1,s_3 \\ 1,1\end{bmatrix} - 2\begin{bmatrix} s_1,s_3 \\ 0,2\end{bmatrix} - \begin{bmatrix} s_1,s_3 \\ 1,0\end{bmatrix} + [s_1,s_3]\right)$$
$$+\delta_{s_3\cdot s_4,1}\cdot\frac{1}{6}\left(\begin{bmatrix} s_1,s_2 \\ 0,2\end{bmatrix} - \frac{3}{2}\begin{bmatrix} s_1,s_2 \\ 0,1\end{bmatrix} + [s_1,s_2]\right)$$
$$+\delta_{s_2\cdot s_3,1}\cdot\frac{1}{6}\left(\begin{bmatrix} s_1,s_4 \\ 0,2\end{bmatrix} - \begin{bmatrix} s_1,s_4 \\ 1,1\end{bmatrix} + \frac{3}{2}\begin{bmatrix} s_1,s_4 \\ 0,1\end{bmatrix} + \begin{bmatrix} s_1,s_4 \\ 2,0\end{bmatrix} - \frac{3}{2}\begin{bmatrix} s_1,s_4 \\ 1,0\end{bmatrix} + [s_1,s_4]\right)$$
$$+\delta_{s_2\cdot s_3\cdot s_4,1}\cdot\frac{1}{24}\left(\begin{bmatrix} s_1 \\ 3\end{bmatrix} - 2\begin{bmatrix} s_1 \\ 2\end{bmatrix} + \frac{11}{6}\begin{bmatrix} s_1 \\ 1\end{bmatrix} - [s_1]\right).$$

Proof This follows by calculating the coefficients of the series $G_{\sqcup\!\sqcup}$ in Theorem 5.15.

The shuffle brackets will be used to define shuffle regularized multiple Eisenstein series in the next section.

6 Regularizations of Multiple Eisenstein Series

This section is devoted to Question 1 in the introduction, which was to find a regularization of the multiple Eisenstein series. We want to present two type of regularization: The shuffle regularized multiple Eisenstein series [3, 9] and stuffle regularized multiple Eisenstein series [3].

The definition of shuffle regularized multiple Eisenstein series uses a beautiful connection of the Fourier expansion of multiple Eisenstein series and the coproduct of formal iterated integrals. The other regularization, the stuffle regularized multiple Eisenstein series uses the construction of the Fourier expansion of multiple Eisenstein series together with a result on regularization of multitangent functions by Bouillot [12].

We start by reviewing the definition of formal iterated integrals and the coproduct defined by Goncharov. An explicit example in length two will make the above mentioned connection of multiple Eisenstein series and this coproduct clear. After doing this, we give the definition of shuffle and stuffle regularized multiple Eisenstein series as presented in [3, 9]. At the end of this section we compare these two regularizations with a help of a few examples.

6.1 Formal Iterated Integrals

Following Goncharov (Sect. 2 in [20]) we consider the algebra $\mathcal{I}$ generated by the elements

$$\mathbb{I}(a_0; a_1, \ldots, a_N; a_{N+1}), \quad a_i \in \{0, 1\}, N \geq 0.$$

together with the following relations

(i) For any $a, b \in \{0, 1\}$ the unit is given by $\mathbb{I}(a; b) := \mathbb{I}(a; \emptyset; b) = 1$.

(ii) The product is given by the shuffle product $⧢$

$$\begin{aligned} &\mathbb{I}(a_0; a_1, \ldots, a_M; a_{M+N+1})\mathbb{I}(a_0; a_{M+1}, \ldots, a_{M+N}; a_{M+N+1}) \\ &= \sum_{\sigma \in sh_{M,N}} \mathbb{I}(a_0; a_{\sigma^{-1}(1)}, \ldots, a_{\sigma^{-1}(M+N)}; a_{M+N+1}), \end{aligned}$$

where $sh_{M,N}$ is the set of $\sigma \in \mathfrak{S}_{M+N}$ such that $\sigma(1) < \cdots < \sigma(M)$ and $\sigma(M+1) < \cdots < \sigma(M+N)$.

(iii) The path composition formula holds: for any $N \geq 0$ and $a_i, x \in \{0, 1\}$, one has

$$\mathbb{I}(a_0; a_1, \ldots, a_N; a_{N+1}) = \sum_{k=0}^{N} \mathbb{I}(a_0; a_1, \ldots, a_k; x)\mathbb{I}(x; a_{k+1}, \ldots, a_N; a_{N+1}).$$

(iv) For $N \geq 1$ and $a_i, a \in \{0, 1\}$ it is $\mathbb{I}(a; a_1, \ldots, a_N; a) = 0$.

(v) The path inversion is satisfied:

$$\mathbb{I}(a_0; a_1, \ldots, a_N; a_{N+1}) = (-1)^N \mathbb{I}(a_{N+1}; a_N, \ldots, a_1; a_0).$$

Definition 6.1 (*Coproduct*) Define the coproduct Δ on $\mathcal{I}$ by

$$\begin{aligned} &\Delta\left(\mathbb{I}(a_0; a_1, \ldots, a_N; a_{N+1})\right) := \\ &\sum \left(\mathbb{I}(a_0; a_{i_1}, \ldots, a_{i_k}; a_{N+1}) \otimes \prod_{p=0}^{k} \mathbb{I}(a_{i_p}; a_{i_p+1}, \ldots, a_{i_{p+1}-1}; a_{i_{p+1}}) \right), \end{aligned}$$

where the sum on the right runs over all $i_0 = 0 < i_1 < \cdots < i_k < i_{k+1} = N+1$ with $0 \leq k \leq N$.

Proposition 6.2 ([20, Proposition 2.2]) *The triple $(\mathcal{I}, ⧢, \Delta)$ is a commutative graded Hopf algebra over $\mathbb{Q}$.*

To calculate $\Delta\left(\mathbb{I}(a_0; a_1, \ldots, a_8; a_9)\right)$ one sums over all possible diagrams of the following form (Fig. 2).

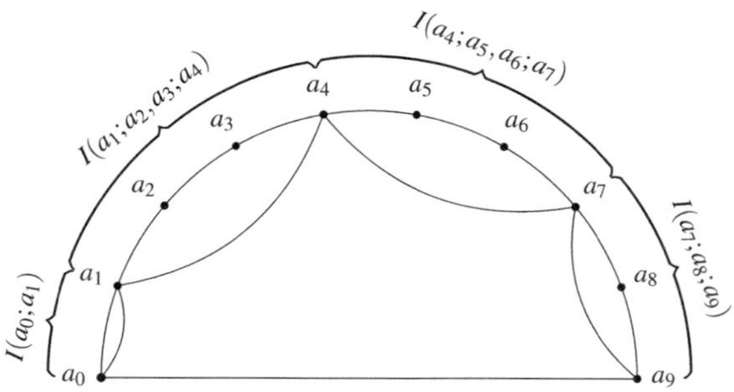

Fig. 2 One diagram for the calculation of $\Delta(\mathbb{I}(a_0; a_1, \ldots, a_8; a_9))$. It gives the term $I(a_0; a_1, a_4, a_7; a_9) \otimes I(a_0; a_1) I(a_1; a_2, a_3; a_4) I(a_4; a_5, a_6; a_7) I(a_7; a_8; a_9)$

For our purpose it will be important to consider the quotient space[7]

$$\mathscr{I}^1 = \mathscr{I} / \mathbb{I}(1; 0; 0)\mathscr{I} .$$

Let us denote by

$$I(a_0; a_1, \ldots, a_N; a_{N+1})$$

an image of $\mathbb{I}(a_0; a_1, \ldots, a_N; a_{N+1})$ in $\mathscr{I}^1$. The quotient map $\mathscr{I} \to \mathscr{I}^1$ induces a Hopf algebra structure on $\mathscr{I}^1$, but for our application we just need that for any $w_1, w_2 \in \mathscr{I}^1$, one has $\Delta(w_1 \sqcup\!\sqcup\, w_2) = \Delta(w_1) \sqcup\!\sqcup\, \Delta(w_2)$. The coproduct on $\mathscr{I}^1$ is given by the same formula as before by replacing $\mathbb{I}$ with I. For integers $n \geq 0, s_1, \ldots, s_r \geq 1$, we set

$$I_n(s_1, \ldots, s_r) := I(1; \underbrace{0, 0, \ldots, 1}_{s_1}, \ldots, \underbrace{0, 0, \ldots, 1}_{s_r}, \underbrace{0, \ldots, 0}_{n}; 0).$$

In particular, we write[8] $I(s_1, \ldots, s_r)$ to denote $I_0(s_1, \ldots, s_r)$.

Proposition 6.3 ([9, Eqs. (3.5), (3.6) and Proposition 3.5])

(i) We have $I_n(\emptyset) = 0$ if $n \geq 1$ or 1 if $n = 0$.
(ii) For integers $n \geq 0, s_1, \ldots, s_r \geq 1$,

$$I_n(s_1, \ldots, s_r) = (-1)^n \sum^{*} \left(\prod_{j=1}^{r} \binom{k_j - 1}{s_j - 1} \right) I(k_1, \ldots, k_r),$$

[7]If one likes to interpret the integrals as real integrals, then the passage from $\mathscr{I}$ to $\mathscr{I}^1$ regularizes these integrals such that "$-\log(0) = \int_{1>t>0} \frac{dt}{t} := 0$".

[8]This notion fits well with the iterated integral expression of multiple zeta values. Recall that

$$\zeta(2, 3) = \int_{1>t_1>\cdots>t_5>0} \underbrace{\frac{dt_1}{t_1} \cdot \frac{dt_2}{1-t_2}}_{2} \cdot \underbrace{\frac{dt_3}{t_3} \cdot \frac{dt_4}{t_4} \cdot \frac{dt_5}{1-t_5}}_{3} .$$

This corresponds to $I(2, 3)$ (but is of course not the same since the I are formal symbols).

where the sum runs over all $k_1 + \cdots + k_r = s_1 + \cdots + s_r + n$ with $k_1, \ldots, k_r \geq 1$.

(iii) *The set $\{I(s_1, \ldots, s_r) \mid r \geq 0, s_i \geq 1\}$ forms a basis of the space $\mathscr{I}^1$.*

We give an example for ii): In $\mathscr{I}^1$ it is $I(1; 0; 0) = 0$ and therefore

$$\begin{aligned} 0 &= I(1;0;0)I(1;0,1;0) \\ &= I(1;0,0,1;0) + I(1;0,0,1;0) + I(1;0,1,0;0) \\ &= 2I(3) + I_1(2) \end{aligned}$$

which gives $I_1(2) = -2I(3) = (-1)^1\binom{2}{1}I(3)$.

Remark 6.4 Statement (iii) in Proposition 6.3 basically states that we can identify $\mathscr{I}^1$ with $\mathfrak{H}^1$ by sending $I(s_1, \ldots, s_l)$ to $z_{s_1} \ldots z_{s_l}$. In other words we can equip $\mathfrak{H}^1$ with the coproduct Δ. Instead of working with I we will use this identification in the next section, when defining the shuffle regularized multiple Eisenstein series.

Example 6.5 In the following we are going to calculate $\Delta(I(3, 2)) = \Delta(I(1; 0, 0, 1, 0, 1; 0))$. Therefore we have to determine all possible markings of the diagram

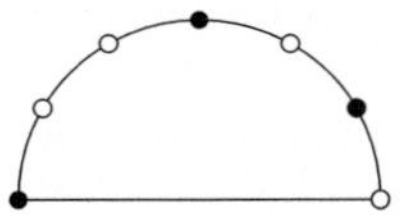

where the corresponding summand in the coproduct does not vanish. For simplicity we draw $\circ$ to denote a 0 and $\bullet$ to denote a 1. We will consider the $4 = 2^2$ ways of marking the two $\bullet$ in the top part of the circle separately. As mentioned in the introduction, we want to compare the coproduct to the Fourier expansion of multiple Eisenstein series. Therefore, in this case we also calculate the expansion of $G_{3,2}(\tau)$ using the construction described in Sect. 3.2. Recall that we also had the 4 different parts $G^{RR}_{3,2}$, $G^{UR}_{3,2}$, $G^{RU}_{3,2}$ and $G^{UU}_{3,2}$. We will see that the number and positions of the marked $\bullet$ correspond to the number and positions of the letter U in the word w of G^w.

(i) Diagrams with no marked $\bullet$:

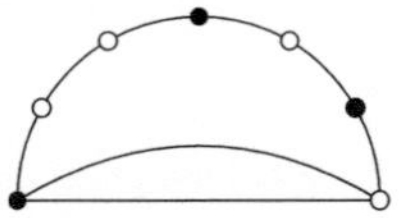

Corresponding sum in the coproduct:

$$I(0; \emptyset; 1) \otimes I(1; 0, 0, 1, 0, 1; 0) = 1 \otimes I(3, 2)\,.$$

The part of the Fourier expansion of $G_{3,2}$ which is associated to this, is the one with no U "occurring", i.e. $G^{RR}_{3,2}(\tau) = \zeta(3, 2)$.

(ii) Diagrams with the first • marked:

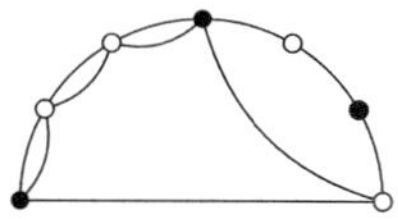

Corresponding sum in the coproduct:

$$I(1; 0, 0, 1; 0) \otimes \big(I(1; 0) \cdot I(0; 0) \cdot I(0; 1) \cdot I(1; 0, 1; 0)\big) = I(3) \otimes I(2)\,.$$

The associated part of the Fourier expansion of $G_{3,2}$ is $G_{3,2}^{UR}(\tau) = g_3(\tau) \cdot \zeta(2)$.

(iii) Diagrams with the second • marked:

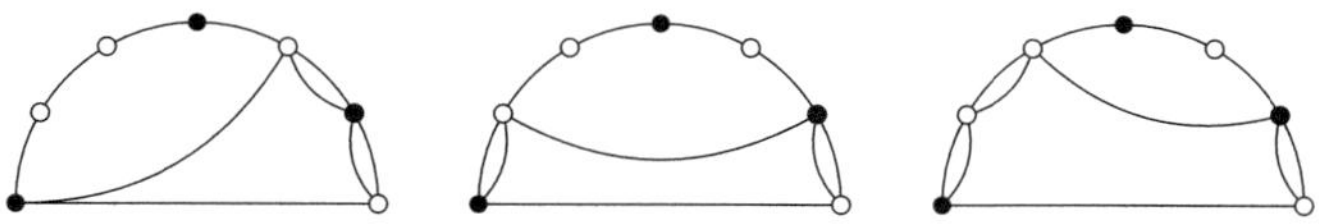

Corresponding sum in the coproduct:

$$\begin{aligned} &I(1; 0, 1; 0) \otimes \big(I(1; 0, 0, 1; 0) \cdot I(0; 1) \cdot I(1; 0)\big) \\ &+I(1; 0, 1; 0) \otimes \big(I(1; 0) \cdot I(0; 0, 1, 0; 1) \cdot I(1; 0)\big) \\ &+I(1; 0, 0, 1; 0) \otimes \big(I(1; 0) \cdot I(0; 0) \cdot I(0; 1, 0; 1) \cdot I(1; 0)\big) \\ &= I(2) \otimes I(3) - I(2) \otimes I_1(2) + I(3) \otimes I(2)\,, \end{aligned}$$

where we used $I(0, 0, 1, 0; 1) = -I_1(2)$ and $I(0; 1, 0; 1) = (-1)^2 I(1; 0, 1; 0) = I(2)$. Together with $I_1(2) = -2I(3)$ this gives

$$3I(2) \otimes I(3) + I(3) \otimes I(2)\,.$$

Also the associated part of the Fourier expansion is the most complicated one. We had $G_{3,2}^{RU}(\tau) = \sum_{m>0} \Psi_{3,2}(m\tau)$ and with (16) we derived $\Psi_{3,2}(x) = 3\Psi_2(x) \cdot \zeta(3) + \Psi_3(x) \cdot \zeta(2)$, i.e.

$$G_{3,2}^{RU}(\tau) = 3g_2(\tau) \cdot \zeta(3) + g_3(\tau) \cdot \zeta(2)\,.$$

(iv) Diagrams with both • marked:

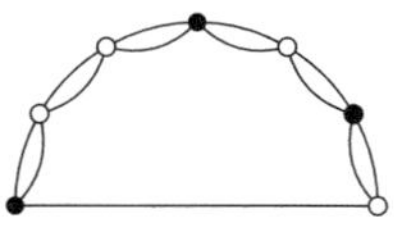

Corresponding sum in the coproduct: $I(3, 2) \otimes 1$. The associated part of the Fourier expansion of $G_{3,2}$ is $G_{3,2}^{UU}(\tau) = g_{3,2}(\tau)$.

Summing all 4 parts together we obtain for the coproduct

$$\Delta(I(3,2)) = 1 \otimes I(3,2) + 3I(2) \otimes I(3) + 2I(3) \otimes I(2) + I(3,2) \otimes 1$$

and for the Fourier expansion of $G_{2,3}(\tau)$:

$$G_{3,2}(\tau) = \zeta(3,2) + 3g_2(\tau)\zeta(3) + 2g_3(\tau)\zeta(2) + g_{3,2}(\tau)\,.$$

This shows that the left factors of the terms in the coproduct corresponds to the functions g and the right factors side to the multiple zeta values. We will use this in the next section to define shuffle regularized multiple Eisenstein series.

6.2 *Shuffle Regularized Multiple Eisenstein Series*

In this section we present the definition of shuffle regularized multiple Eisenstein series as it was done in [9] together with the simplification developed in [3]. We use the observation of the section before and use the coproduct Δ of formal iterated integrals to define these series. As mentioned in Remark 6.4 we can equip the space $\mathfrak{H}^1$ with the coproduct Δ instead of working with the space $\mathscr{I}^1$. Denote by $\mathscr{MZB} \subset \mathbb{C}[\![q]\!]$ the space of all formal power series in q which can be written as a $\mathbb{Q}$-linear combination of products of multiple zeta values, powers of $(-2\pi i)$ and bi-brackets. In the following, we set $q = \exp(2\pi i\tau)$ with τ being an element in the upper half-plane. Since the coefficient of bi-brackets just have polynomials growth, the elements in $\mathscr{MZB}$ and $\mathscr{BD}$ can be viewed as holomorphic functions in the upper half-plane with this identification.

In analogy to the map $Z^{ш} : (\mathfrak{H}^1, ш) \to \mathscr{Z}$ of shuffle regularized multiple zeta values (Proposition 3.2), the map $\mathfrak{g}^{ш} : (\mathfrak{H}^1, ш) \to \mathbb{Q}[2\pi i][\![q]\!]$ defined on the generators $z_{t_1} \dots z_{t_l}$ by

$$\mathfrak{g}^{ш}(z_{t_1} \dots z_{t_m}) = g^{ш}_{t_1,\dots,t_m}(\tau) := (-2\pi i)^{t_1+\dots+t_m}[t_1, \dots, t_m]^{ш}\,,$$

is also an algebra homomorphism by Theorem 5.15.

With this notation we can recall the definition of $G^{ш}$ from [3] (which is a variant of the definition in [9], where the authors did not use bi-brackets and the shuffle bracket).

Definition 6.6 For integers $s_1, \dots, s_l \geq 1$, define the functions $G^{ш}_{s_1,\dots,s_l}(\tau) \in \mathscr{MZB}$, called *shuffle regularized multiple Eisenstein series*, as

$$G^{ш}_{s_1,\dots,s_l}(\tau) := m\left((\mathfrak{g}^{ш} \otimes Z^{ш}) \circ \Delta\big(z_{s_1} \dots z_{s_l}\big)\right)\,,$$

where m denotes the multiplication given by $m : a \otimes b \mapsto a \cdot b$ and $Z^{ш}$ denotes the map for shuffle regularized multiple zeta values given in Proposition 3.2.

We can view $G^{ш}$ as an algebra homomorphism $G^{ш} : (\mathfrak{H}^1, ш) \to \mathcal{MZB}$ such that the following diagram commutes

$$\begin{array}{ccc} (\mathfrak{H}^1, ш) & \xrightarrow{\Delta} & (\mathfrak{H}^1, ш) \otimes (\mathfrak{H}^1, ш) \\ {\scriptstyle G^{ш}} \big\downarrow & & \big\downarrow {\scriptstyle \mathfrak{g}^{ш} \otimes Z^{ш}} \\ \mathcal{MZB} & \xleftarrow[m]{} & \mathbb{Q}[2\pi i][[q]] \otimes \mathcal{Z} \end{array}$$

Theorem 6.7 ([3, Theroem 6.5], [9, Theroem 1.1, 1.2]) *For all $s_1, \dots, s_l \geq 1$ the shuffle regularized multiple Eisenstein series $G^{ш}_{s_1,\dots,s_l}$ have the following properties:*

(i) They are holomorphic functions on the upper half-plane having a Fourier expansion with the shuffle regularized multiple zeta values as the constant term.
(ii) They fulfill the shuffle product.
(iii) For integers $s_1, \dots, s_l \geq 2$ they equal the multiple Eisenstein series

$$G^{ш}_{s_1,\dots,s_l}(\tau) = G_{s_1,\dots,s_l}(\tau)$$

and therefore they fulfill the stuffle product in these cases.

Parts (i) and (ii) in this theorem follow directly by definition. The important part here is (iii), which states that the connection of the Fourier expansion and the coproduct, as illustrated in Example 6.5, holds in general. It also proves that the shuffle regularized multiple Eisenstein series fulfill the stuffle product in many cases. Though the exact failure of the stuffle product of these series is unknown so far.

6.3 Stuffle Regularized Multiple Eisenstein Series

Motivated by the calculation of the Fourier expansion of multiple Eisenstein series described in Sect. 3.2 we consider the following construction.

Construction 6.8 *Given a $\mathbb{Q}$-algebra $(A, \cdot)$ and a family of homomorphism*

$$\{w \mapsto f_w(m)\}_{m \in \mathbb{N}}$$

from $(\mathfrak{H}^1, *)$ to $(A, \cdot)$, we define for $w \in \mathfrak{H}^1$ and $M \in \mathbb{N}$

$$F_w(M) := \sum_{\substack{1 \leq k \leq l(w) \\ w_1 \dots w_k = w \\ M > m_1 > \dots > m_k > 0}} f_{w_1}(m_1) \dots f_{w_k}(m_k) \in A\,,$$

where $l(w)$ denotes the length of the word w and $w_1 \dots w_k = w$ is a decomposition of w into k words in $\mathfrak{H}^1$.

Proposition 6.9 ([3, Proposition 6.8]) *For all $M \in \mathbb{N}$ the assignment $w \mapsto F_w(M)$, described above, determines an algebra homomorphism from $(\mathfrak{H}^1, *)$ to $(A, \cdot)$. In particular $\{w \mapsto F_w(m)\}_{m\in\mathbb{N}}$ is again a family of homomorphism as used in Construction 6.8.*

For a word $w = z_{s_1} \dots z_{s_l} \in \mathfrak{H}^1$ we also write in the following $f_{s_1,\dots,s_l}(m) := f_w(m)$ and similarly $F_{s_1,\dots,s_l}(M) := F_w(M)$.

Example 6.10 Let $f_w(m)$ be as in Construction 6.8. In small lengths the F_w are given by

$$F_{s_1}(M) = \sum_{M>m_1>0} f_{s_1}(m_1)\,, \quad F_{s_1,s_2}(M) = \sum_{M>m_1>0} f_{s_1,s_2}(m_1) + \sum_{M>m_1>m_2>0} f_{s_1}(m_1) f_{s_2}(m_2)$$

and one can check directly by the use of the stuffle product for the f_w that

$$\begin{aligned}
F_{s_1}(M) \cdot F_{s_2}(M) &= \sum_{M>m_1>0} f_{s_1}(m_1) \cdot \sum_{M>m_2>0} f_{s_2}(m_2) \\
&= \sum_{M>m_1>m_2>0} f_{s_1}(m_1) f_{s_2}(m_2) + \sum_{M>m_2>m_1>0} f_{s_2}(m_2) f_{s_1}(m_1) + \sum_{M>m_1>0} f_{s_1}(m_1) f_{s_2}(m_1) \\
&= \sum_{M>m_1>m_2>0} f_{s_1}(m_1) f_{s_2}(m_2) + \sum_{M>m_2>m_1>0} f_{s_2}(m_2) f_{s_1}(m_1) \\
&\quad + \sum_{M>m_1>0} \left(f_{s_1,s_2}(m_1) + f_{s_2,s_1}(m_1) + f_{s_1+s_2}(m_1) \right) \\
&= F_{s_1,s_2}(M) + F_{s_2,s_1}(M) + F_{s_1+s_2}(M)\,.
\end{aligned}$$

Let us now give an explicit example for maps f_w in which we are interested. Recall (Definition 3.5) that for integers $s_1, \dots, s_l \geq 2$ we defined the multitangent function by

$$\Psi_{s_1,\dots,s_l}(z) = \sum_{\substack{n_1>\dots>n_l \\ n_j\in\mathbb{Z}}} \frac{1}{(z+n_1)^{s_1} \cdots (z+n_l)^{s_l}}\,.$$

In [12], where these functions were introduced, the author uses the notation $\mathcal{T}e^{s_1,\dots,s_l}(z)$ which corresponds to our notation $\Psi_{s_1,\dots,s_l}(z)$. It was shown there that the series $\Psi_{s_1,\dots,s_l}(z)$ converges absolutely when $s_1, \dots, s_l \geq 2$. These functions fulfill (for the cases they are defined) the stuffle product. As explained in Sect. 3.2 the multitangent functions appear in the calculation of the Fourier expansion of the multiple Eisenstein series $G_{s_1,\dots,s_l}$, for example in length two it is

$$\begin{aligned}
G_{s_1,s_2}(\tau) =& \zeta(s_1,s_2) + \zeta(s_1) \sum_{m_1>0} \Psi_{s_2}(m_1\tau) + \sum_{m_1>0} \Psi_{s_1,s_2}(m_1\tau) \\
&+ \sum_{m_1>m_2>0} \Psi_{s_1}(m_1\tau) \Psi_{s_2}(m_2\tau)\,.
\end{aligned}$$

One nice result of [12] is a regularization of the multitangent function to get a definition of $\Psi_{s_1,\dots,s_l}(z)$ for all $s_1,\dots,s_l \in \mathbb{N}$. We will use this result together with the above construction to recover the Fourier expansion of the multiple Eisenstein series.

Theorem 6.11 ([12]) *For all $s_1,\dots,s_l \in \mathbb{N}$ there exist holomorphic functions $\Psi_{s_1,\dots,s_l}$ on $\mathbb{H}$ with the following properties*

(i) Setting $q = e^{2\pi i\tau}$ for $\tau \in \mathbb{H}$ the map $w \mapsto \Psi_w(\tau)$ defines an algebra homomorphism from $(\mathfrak{H}^1, *)$ to $(\mathbb{C}[\![q]\!], \cdot)$.

(ii) In the case $s_1,\dots,s_l \geq 2$ the $\Psi_{s_1,\dots,s_l}$ are given by the multitangent functions in Definition 3.5.

(iii) The monotangents functions have the q-expansion given by

$$\Psi_1(\tau) = \frac{\pi}{\tan(\pi\tau)} = (-2\pi i)\left(\frac{1}{2} + \sum_{n>0} q^n\right)$$

and for $k \geq 2$ by

$$\Psi_k(\tau) = \frac{(-2\pi i)^k}{(k-1)!}\sum_{n>0} n^{k-1} q^n \,.$$

(iv) (Reduction into monotangent function) Every $\Psi_{s_1,\dots,s_l}(\tau)$ can be written as a $\mathcal{Z}$-linear combination of monotangent functions. There are explicit $\epsilon_{i,k}^{s_1,\dots,s_l} \in \mathcal{Z}$ s.th.

$$\Psi_{s_1,\dots,s_l}(\tau) = \delta^{s_1,\dots,s_l} + \sum_{i=1}^{l}\sum_{k=1}^{s_i} \epsilon_{i,k}^{s_1,\dots,s_l}\Psi_k(\tau)\,,$$

where $\delta^{s_1,\dots,s_l} = \frac{(\pi i)^l}{l!}$ if $s_1 = \dots = s_l = 1$ and l even and $\delta^{s_1,\dots,s_l} = 0$ otherwise. For $s_1 > 1$ and $s_l > 1$ the sum on the right starts at $k = 2$, i.e. there are no $\Psi_1(\tau)$ appearing and therefore there is no constant term in the q-expansion.

Proof This is just a summary of the results in Section 6 and 7 of [12]. The last statement (iv) is given by Theorem 6 in [12].

Due to iv) in the Theorem the calculation of the Fourier expansion of multiple Eisenstein series, where ordered sums of multitangent functions appear, reduces to ordered sums of monotangent functions. The connection of these sums to the brackets, i.e. to the functions g, is given by the following fact which can be seen by using iii) of the above Theorem. For $n_1,\dots,n_r \geq 2$ it is

$$g_{s_1,\dots,s_r}(\tau) = \sum_{m_1>\dots>m_l>0} \Psi_{s_1}(m_1\tau)\dots\Psi_{s_l}(m_l\tau)\,.$$

For $w \in \mathfrak{H}^1$ we now use the Construction 6.8 with $A = \mathbb{C}[\![q]\!]$ and the family of homomorphism $\{w \mapsto \Psi_w(n\tau)\}_{n\in\mathbb{N}}$ (see Theorem 6.11 (i)) to define

$$\mathfrak{g}^{*,M}(w) := (-2\pi i)^{|w|} \sum_{\substack{1\le k\le l(w)\\ w_1\dots w_k = w}} \sum_{M>m_1>\dots>m_k>0} \Psi_{w_1}(m_1\tau)\dots\Psi_{w_k}(m_k\tau)\,.$$

From Proposition 6.9 it follows that for all $M \in \mathbb{N}$ the map $\mathfrak{g}^{*,M}$ is an algebra homomorphism from $(\mathfrak{H}^1, *)$ to $\mathbb{C}[\![q]\!]$.

To define stuffle regularized multiple Eisenstein series we need the following: For an arbitrary quasi-shuffle algebra $\mathbb{Q}\langle A\rangle$ define the following coproduct for a word w

$$\Delta_H(w) = \sum_{uv=w} u \otimes v\,.$$

Then it is known due to Hoffman ([21]) that the space $(\mathbb{Q}\langle A\rangle, \odot, \Delta_H)$ has the structure of a bialgebra. With this we try to mimic the definition of the $G^{\sqcup\!\sqcup}$ and use the coproduct structure on the space $(\mathfrak{H}^1, *, \Delta_H)$ to define for $M \ge 0$ the function $G^{*,M}$ and then take the limit $M \to \infty$ to obtain the stuffle regularized multiple Eisenstein series. For this we consider the following diagram

$$\begin{array}{ccc} (\mathfrak{H}^1, *) & \xrightarrow{\Delta_H} & (\mathfrak{H}^1, *) \otimes (\mathfrak{H}^1, *) \\ {\scriptstyle G^{*,M}}\big\downarrow & & \big\downarrow{\scriptstyle \mathfrak{g}^{*,M}\otimes Z^*} \\ \mathbb{C}[\![q]\!] & \xleftarrow[m]{} & \mathbb{C}[\![q]\!] \otimes \mathcal{Z} \end{array}$$

with the above algebra homomorphism $\mathfrak{g}^{*,M} : (\mathfrak{H}^1, *) \to \mathbb{C}[\![q]\!]$ and the map Z^* for stuffle regularized multiple zeta values given in Proposition 3.2.

Definition 6.12 For integers $s_1, \dots, s_l \ge 1$ and $M \ge 1$, we define the q-series $G^{*,M}_{s_1,\dots,s_r} \in \mathbb{C}[\![q]\!]$ as the image of the word $w = z_{s_1} \dots z_{s_l} \in \mathfrak{H}^1$ under the algebra homomorphism $(\mathfrak{g}^{*,M} \otimes Z^*) \circ \Delta_H$:

$$G^{*,M}_{s_1,\dots,s_l}(\tau) := m\left((\mathfrak{g}^{*,M} \otimes Z^*) \circ \Delta_H\big(w\big)\right) \in \mathbb{C}[\![q]\!]\,.$$

For $s_1, \dots, s_l \ge 2$ the limit

$$G^{*}_{s_1,\dots,s_l}(\tau) := \lim_{M\to\infty} G^{*,M}_{s_1,\dots,s_l}(\tau) \tag{28}$$

exists and we have $G_{s_1,\dots,s_l} = G^{*}_{s_1,\dots,s_l} = G^{\sqcup\!\sqcup}_{s_1,\dots,s_l}$ ([3, Proposition 6.13]).

Remark 6.13 The open question is for what general $s_1, \dots, s_l$ the limit in (28) exists. It is believed that this is exactly the case for $s_1 \ge 2$ and $s_2, \dots, s_l \ge 1$ as explained in Remark 6.14 in [3]. This would be the case if $\Psi_{1,\dots,1}$ are the only multitangent functions with a constant term in the decomposition of Theorem 6.11 (iv). That this is the case is remarked, without a proof, in [13] in the last sentence of page 3.

Theorem 11 ([3]) *For all $s_1, \dots, s_l \in \mathbb{N}$ and $M \in \mathbb{N}$ the $G^{*,M}_{s_1,\dots,s_l} \in \mathbb{C}[\![q]\!]$ have the following properties:*

(i) *Their product can be expressed in terms of the stuffle product.*
(ii) *In the case where the limit* $G^*_{s_1,\dots,s_l} := \lim_{M\to\infty} G^{*,M}_{s_1,\dots,s_l}$ *exists, the functions* $G^*_{s_1,\dots,s_l}$ *are elements in* $\mathcal{MZB}$.
(iii) *For* $s_1,\dots,s_l \geq 2$ *the* $G^*_{s_1,\dots,s_l}$ *exist and equal the classical multiple Eisenstein series*

$$G_{s_1,\dots,s_l}(\tau) = G^*_{s_1,\dots,s_l}(\tau)\,.$$

6.4 Double Shuffle Relations for Regularized Multiple Eisenstein Series

By Theorem 6.7 we know that the product of two shuffle regularized multiple Eisenstein series $G^{⧢}_{s_1,\dots,s_l}$ with $s_1,\dots,s_l \geq 1$ can be expressed by using the shuffle product formula. This means we can for example replace every ζ by $G^{⧢}$ in the shuffle product (4) of multiple zeta values and obtain

$$G^{⧢}_2 \cdot G^{⧢}_3 = G^{⧢}_{2,3} + 3G^{⧢}_{3,2} + 6G^{⧢}_{4,1}\,. \tag{29}$$

Due to Theorem 6.7 (iii) we know that $G^{⧢}_{s_1,\dots,s_l} = G_{s_1,\dots,s_l}$ whenever $s_1,\dots,s_l \geq 2$. Since the product of two multiple Eisenstein series $G_{s_1,\dots,s_l}$ can be expressed using the stuffle product formula we also have

$$\begin{aligned} G^{⧢}_2 \cdot G^{⧢}_3 &= G_2 \cdot G_3 = G_{2,3} + G_{3,2} + G_5 \\ &= G^{⧢}_{2,3} + G^{⧢}_{3,2} + G^{⧢}_5\,. \end{aligned} \tag{30}$$

Combining (29) and (30) we obtain the relation $G^{⧢}_5 = 2G^{⧢}_{3,2} + 6G^{⧢}_{4,1}$. In the following we will call these relations, i.e. the relations obtained by writing the product of two $G^{⧢}_{s_1,\dots,s_l}$ with $s_1,\dots,s_l \geq 2$ as the stuffle and shuffle product, *restricted double shuffle relations*.

We know that multiple zeta values fulfill even more linear relations, in particular we can express the product of two multiple zeta values $\zeta(s_1,\dots,s_l)$ in two different ways whenever $s_1 \geq 2$ and $s_2,\dots,s_l \geq 1$. A natural question therefore is, in which cases the $G^{⧢}$ also fulfill these additional relations. The answer to this question is that some are satisfied and some are not, as the following will show.

In [3, Example 6.15] it is shown that $G^{⧢}_{2,1,2} = G^*_{2,1,2}$, $G^{⧢}_{2,1} = G^*_{2,1}$, $G^{⧢}_{2,2,1} = G^*_{2,2,1}$ and $G^{⧢}_{4,1} = G^*_{4,1}$. Since the product of two G^* can be expressed using the stuffle product we obtain

$$\begin{aligned} G^{⧢}_2 \cdot G^{⧢}_{2,1} &= G^*_2 \cdot G^*_{2,1} \\ &= G^*_{2,1,2} + 2G^*_{2,2,1} + G^*_{4,1} + G^*_{2,3} \\ &= G^{⧢}_{2,1,2} + 2G^{⧢}_{2,2,1} + G^{⧢}_{4,1} + G^{⧢}_{2,3}\,. \end{aligned} \tag{31}$$

Using also the shuffle product to express $G^{⧢}_2 \cdot G^{⧢}_{2,1}$ we obtain a linear relation in weight 5 which is not covered by the restricted double shuffle relations. This linear relation was numerically observed in [9] but could not be proven there. So far it is not known exactly which products of the $G^{\sqcup\!\sqcup}$ can be written in terms of stuffle products.

We end this section by comparing different versions of the double shuffle relations and explain, why multiple Eisenstein series can't fulfill every double shuffle relation of multiple zeta values. For this we write for words $u, v \in \mathfrak{H}^1$

$$\mathrm{ds}(u, v) := u \sqcup\!\sqcup v - u * v \in \mathfrak{H}^1 .$$

Recall that by $\mathfrak{H}^0$ we denote the algebra of all admissible words, i.e. $\mathfrak{H}^0 = 1 \cdot \mathbb{Q} + x\mathfrak{H}y$. Additionally we set $\mathfrak{H}^2 = \mathbb{Q}\langle\{z_2, z_3, \dots\}\rangle$ to be the span of all words in $\mathfrak{H}^1$ with no z_1 occurring, i.e. the words for which the multiple Eisenstein series G exists. These are also the words for which the product of two multiple Eisenstein series can be expressed as the shuffle and stuffle product by Theorem 6.7. Denote by $|w| \in \mathfrak{H}^1$ the length of the word w with respect to the alphabet $\{x, y\}$ and define

$$\begin{aligned}
\mathsf{eds}_k &:= \left\{ \mathrm{ds}(u, v) \in \mathfrak{H}^0 \mid |u| + |v| = k,\ u \in \mathfrak{H}^0, v \in \mathfrak{H}^0 \cup \{z_1\} \right\},\\
\mathsf{fds}_k &:= \left\{ \mathrm{ds}(u, v) \in \mathfrak{H}^0 \mid |u| + |v| = k,\ u, v \in \mathfrak{H}^0 \right\},\\
\mathsf{rds}_k &:= \left\{ \mathrm{ds}(u, v) \in \mathfrak{H}^0 \mid |u| + |v| = k,\ u, v \in \mathfrak{H}^2 \right\}.
\end{aligned}$$

Also set $\mathsf{eds} = \bigcup_{k>0} \mathsf{eds}_k$ and similarly fds and rds. These spaces can be seen as the words in $\mathfrak{H}^0$ corresponding to the extended[9]-, finite- and the restricted double shuffle relations. We have the inclusions

$$\mathsf{rds}_k \subset \mathsf{fds}_k \subset \mathsf{eds}_k \ .$$

View ζ as a map $\mathfrak{H}^0 \to \mathscr{Z}$ by sending the word $z_{s_1} \dots z_{s_l}$ to $\zeta(s_1, \dots, s_l)$. It is known ([24, Theroem 2]), that eds_k is in the kernel of the map ζ and it is expected (Statement (3) after Conjecture 1 in [24]) that actually $\mathsf{eds}_k = \ker(\zeta)$. Viewing $G^{\sqcup\!\sqcup}$ in a similar way as a map $\mathfrak{H}^0 \to \mathscr{MZB}$, we know that rds_k is contained in the kernel of this map (Theorem 6.7 (iv)). But due to (31) we also have $\mathrm{ds}(z_2, z_2 z_1) \in \ker(G^{\sqcup\!\sqcup})$ which is not an element of rds_5. In [2] Example 6.15 ii) it is shown that there are also elements in $\mathsf{fds}_k \subset \mathsf{eds}_k$, that are not in the kernel of $G^{\sqcup\!\sqcup}$. We therefore expect

$$\mathsf{rds} \subsetneq \ker G^{\sqcup\!\sqcup} \subsetneq \mathsf{eds}$$

and the above examples show, that it seems to be crucial to understand for which indices we have $G^{\sqcup\!\sqcup} = G^*$ to answer these questions.

We now discuss applications of the extended double shuffle relations to the classical theory of (quasi-)modular forms. As we have seen in the introduction it is known due to Euler that

[9] In [24] the authors introduced the notion of extended double shuffle relations. We use this notion here for smaller subset of these relations given there as the relations described in statement (3) on page 315.

$$\zeta(2)^2 = \frac{5}{2}\zeta(4)\,, \quad \zeta(4)^2 = \frac{7}{6}\zeta(8)\,, \quad \zeta(6)^2 = \frac{715}{691}\zeta(12)\,. \tag{32}$$

In the following, we want to show how to prove these relations using extended double shuffle relations and argue why for multiple Eisenstein series the second is fulfilled but the first and the last equation of (32) are not.

(i) The relation $\zeta(2)^2 = \frac{5}{2}\zeta(4)$ can be proven in the following way by using double shuffle relations. It is $z_2 * z_2 = 2\,\mathrm{ds}(z_3, z_1) - \frac{1}{2}\,\mathrm{ds}(z_2, z_2) + \frac{5}{2}z_4$, since

$$\begin{aligned} \mathrm{ds}(z_3, z_1) &= z_3 z_1 + z_2 z_2 - z_4\,, \\ \mathrm{ds}(z_2, z_2) &= 4 z_3 z_1 - z_4\,, \\ z_2 * z_2 &= 2 z_2 z_2 + z_4\,. \end{aligned}$$

Applying the map ζ we therefore deduce

$$\zeta(2)^2 = \zeta(z_2 * z_2) = \zeta\left(2\,\mathrm{ds}(z_3, z_1) - \frac{1}{2}\,\mathrm{ds}(z_2, z_2) + \frac{5}{2}z_4\right) = \frac{5}{2}\zeta(4)\,.$$

This relation is not true for Eisenstein series. Though $\mathrm{ds}(z_2, z_2)$ is in the kernel of $G^{\sqcup\!\sqcup}$ the element $\mathrm{ds}(z_3, z_1)$ is not. In fact, using the explicit formula for the Fourier expansion of $G^{\sqcup\!\sqcup}_{3,1}$ and $G^{\sqcup\!\sqcup}_{2,2}$ together with Proposition 4.12 for d[2] we obtain $G^{\sqcup\!\sqcup}(\mathrm{ds}(z_3, z_1)) = 6\zeta(2)\,\mathrm{d}\,G_2$, where as before $\mathrm{d} = q\frac{d}{dq}$. Using this we get

$$G_2^2 = G^{\sqcup\!\sqcup}(z_2 * z_2) = G^{\sqcup\!\sqcup}\left(2\,\mathrm{ds}(z_3, z_1) - \frac{1}{2}\,\mathrm{ds}(z_2, z_2) + \frac{5}{2}z_4\right) = 12\zeta(2)\,\mathrm{d}\,G_2 + \frac{5}{2}G_4\,.$$

This is a well-known fact in the theory of quasi-modular forms ([36]).

(ii) Similarly to the above example one can prove the relation $\zeta(4)^2 = \frac{7}{6}\zeta(8)$ by checking that

$$z_4 * z_4 = \frac{2}{3}\,\mathrm{ds}(z_4, z_4) - \frac{1}{2}\,\mathrm{ds}(z_3, z_5) + \frac{7}{6}z_8$$

and since $\mathrm{ds}(z_4, z_4), \mathrm{ds}(z_3, z_5) \in \mathsf{rds}_8 \subset \ker G^{\sqcup\!\sqcup}$ we also derive ${G_4}^2 = \frac{7}{6}G_8$ by applying the map $G^{\sqcup\!\sqcup}$ to this equation.

(iii) To prove the relation $\zeta(6)^2 = \frac{715}{691}\zeta(12)$ in addition to the double shuffles of the form $\mathrm{ds}(z_a, z_b)$ double shuffles of the form $\mathrm{ds}(z_a z_b, z_c)$ are needed as well. This follows indirectly from the results obtained in [19]. Using the computer one can check that

$$z_6 * z_6 = 2 z_6 z_6 + z_{12} = \frac{715}{691}z_{12} + \frac{1}{2^2 \cdot 19 \cdot 113 \cdot 691} \cdot (R + E)$$

with $R \in \mathsf{rds}_{12}$ and $E \in \mathsf{eds}_{12} \setminus \mathsf{rds}_{12}$ being the quite complicated elements

$$\begin{aligned}
R = {} & 2005598\,\mathrm{ds}(z_6, z_6) - 8733254\,\mathrm{ds}(z_7, z_5) + 8128450\,\mathrm{ds}(z_8, z_4) + 5121589\,\mathrm{ds}(z_9, z_3) \\
& + 16364863\,\mathrm{ds}(z_{10}, z_2) + 2657760\,\mathrm{ds}(z_2 z_8, z_2) + 5220600\,\mathrm{ds}(z_3 z_7, z_2) \\
& + 12711531\,\mathrm{ds}(z_4 z_6, z_2) + 10460184\,\mathrm{ds}(z_5 z_5, z_2) + 18601119\,\mathrm{ds}(z_6 z_4, z_2) \\
& + 33877826\,\mathrm{ds}(z_7 z_3, z_2) + 39496002\,\mathrm{ds}(z_8 z_2, z_2) - 13288800\,\mathrm{ds}(z_2 z_2, z_8) \\
& - 5220600\,\mathrm{ds}(z_2 z_7, z_3) - 5734750\,\mathrm{ds}(z_3 z_6, z_3) - 84659\,\mathrm{ds}(z_4 z_5, z_3) \\
& + 2820467\,\mathrm{ds}(z_5 z_4, z_3) - 5486485\,\mathrm{ds}(z_6 z_3, z_3) + 8462489\,\mathrm{ds}(z_7 z_2, z_3) \\
& - 6067131\,\mathrm{ds}(z_2 z_6, z_4) - 7532671\,\mathrm{ds}(z_3 z_5, z_4) - 10879336\,\mathrm{ds}(z_4 z_3, z_5) \\
& - 5151234\,\mathrm{ds}(z_4 z_4, z_4) + 3440519\,\mathrm{ds}(z_5 z_3, z_4) - 1458819\,\mathrm{ds}(z_6 z_2, z_4) \\
& + 2259096\,\mathrm{ds}(z_5 z_2, z_5) - 4319105\,\mathrm{ds}(z_3 z_4, z_5) - 778598\,\mathrm{ds}(z_5 z_2, z_5) \\
& + 7609581\,\mathrm{ds}(z_2 z_4, z_6) + 13064898\,\mathrm{ds}(z_3 z_3, z_6) - 1281420\,\mathrm{ds}(z_3 z_2, z_7)\,,
\end{aligned}$$

$$\begin{aligned}
E = {} & -22681134\,\mathrm{ds}(z_{11}, z_1) + 10631040\,\mathrm{ds}(z_3 z_8, z_1) + 4241200\,\mathrm{ds}(z_7 z_1, z_4) \\
& + 31893120\,\mathrm{ds}(z_4 z_7, z_1) + 58185960\,\mathrm{ds}(z_5 z_6, z_1) + 78309000\,\mathrm{ds}(z_6 z_5, z_1) \\
& + 77976780\,\mathrm{ds}(z_7 z_4, z_1) + 44849700\,\mathrm{ds}(z_8 z_3, z_1) - 13288800\,\mathrm{ds}(z_9 z_2, z_1) \\
& - 15946560\,\mathrm{ds}(z_{10} z_1, z_1) + 75052824\,\mathrm{ds}(z_9 z_1, z_2) + 19477164\,\mathrm{ds}(z_8 z_1, z_3) \\
& - 12951740\,\mathrm{ds}(z_6 z_1, z_5) - 10631040\,\mathrm{ds}(z_2 z_1, z_9)
\end{aligned}$$

Here the elements E and R are in the kernel of ζ but E, in contrast to R, is not in the kernel of $G^{ш}$. The defect here is given by the cusp form Δ in weight 12 as one can derive

$$G^{ш}(E) = -\frac{2147}{1200}(-2\pi i)^{12}\Delta\,.$$

It is still an open problem how to derive these Euler relations in general by using double shuffle relations. The last example shows that this also seems to be very complicated. But as the examples above show, this might be of great interest to understand the connection of modular forms and multiple zeta values. This together with the question which double shuffle relations are fulfilled by multiple Eisenstein series will be considered in upcoming works by the author.

7 *q*-Analogues of Multiple Zeta Values

In general, a q-analogue of an mathematical object is a generalization involving a new parameter q that returns the original object in the limit as $q \to 1$. The easiest example of such an generalization is the q-analogue of a natural number $n \in \mathbb{N}$ given by

$$[n]_q := \frac{1-q^n}{1-q} = 1 + q + \cdots + q^{n-1}\,.$$

Clearly this gives back the original number n as $\lim_{q\to 1}[n]_q = n$.

Several different models for q-analogues of multiple zeta values have been studied in recent years. A good overview of them can be found in [39]. There are different motivations to study q-analogues of multiple zeta values.

That our brackets can be seen as q-analogue of multiple zeta values somehow occurred by accident since their original motivation was their appearance in the Fourier expansion of multiple Eisenstein series. But as turned out, seeing them as q-analogues gives a direct connection to multiple zeta values. In this section we first show how the brackets can be seen as a q-analogue of multiple zeta values and then discuss how one can obtain relations between multiple zeta values using the results obtained in [6]. The second section will be devoted to connecting the brackets to other q-analogues.

7.1 Brackets as q-Analogues of MZV and the Map Z_k

Define for $k \in \mathbb{N}$ the map $Z_k : \mathbb{Q}[[q]] \to \mathbb{R} \cup \{\infty\}$ by

$$Z_k(f) = \lim_{q \to 1} (1-q)^k f(q) \,.$$

Since we have seen that the brackets can be written as

$$[s_1, \dots, s_l] = \frac{1}{(s_1-1)! \dots (s_l-1)!} \sum_{n_1 > \dots > n_l > 0} \prod_{j=1}^{l} \frac{q^{n_j} P_{s_j-1}(q^{n_j})}{(1-q^{n_j})^{s_j}}$$

and using $P_{k-1}(1) = (k-1)!$ and interchanging the summation and the limit we derive ([6, Proposition 6.4]), that for $s_1 > 1$, i.e. $[s_1, \dots, s_l] \in \mathrm{q}\mathcal{MZ}$

$$Z_k([s_1, \dots, s_l]) = \begin{cases} \zeta(s_1, \dots, s_l)\,, & k = s_1 + \dots + s_l\,, \\ 0\,, & k > s_1 + \dots + s_l\,. \end{cases}$$

Due to $\mathcal{MD} = \mathrm{q}\mathcal{MZ}[\,[1]\,]$ (Theorem 4.11) we can define a well-defined map[10] on the whole space $\mathcal{MD}$ by

$$\begin{aligned} Z_k^{alg} &: \mathrm{Fil}_k^{\mathrm{W}}(\mathcal{MD}) \to \mathbb{R}[T] \\ Z_k^{alg}\left(\sum_{j=0}^{k} g_j [1]^{k-j}\right) &= \sum_{j=0}^{k} Z_j(g_j) T^{k-j} \in \mathbb{R}[T] \end{aligned}$$

where $g_j \in \mathrm{Fil}_j^{\mathrm{W}}(\mathrm{q}\mathcal{MZ})$.

[10] This map is similar to the evaluation map $Z^* : \mathfrak{H}^1 \to \mathbb{R}[T]$, of stuffle regularized multiple zeta values, given in Proposition 1 in [24]. We used this map in the previous sections (Proposition 3.2) with $T = 0$.

Every relation between multiple zeta values of weight k is contained in the kernel of the map Z_k. Therefore the kernel of Z_k was studied in [6].

Theorem 7.1 ([6, Theroem 1.13]) *For the kernel of* $Z_k^{alg} \in \mathrm{Fil}_k^{\mathrm{W}}(\mathcal{MD})$ *we have*

(i) If for $[s_1, \dots, s_l]$ *it holds* $s_1 + \cdots + s_l < k$, *then* $Z_k^{alg}[s_1, \dots, s_l] = 0$.
(ii) For any $f \in \mathrm{Fil}_{k-2}^{\mathrm{W}}(\mathcal{MD})$ *we have* $Z_k^{alg}\, \mathrm{d}(f) = 0$, *i.e.,* $\mathrm{d}\,\mathrm{Fil}_{k-2}^{\mathrm{W}}(\mathcal{MD}) \subseteq \ker Z_k$.
(iii) If $f \in \mathrm{Fil}_k^{\mathrm{W}}(\mathcal{MD})$ *is a cusp form for* $\mathrm{SL}_2(\mathbb{Z})$, *then* $Z_k^{alg}(f) = 0$.

Example 7.2 We illustrate some applications for Theorem 7.1. For this we recall identities for the derivatives and relations of brackets as they were given in [6]. All of them can be obtained by using the results explained in Sect. 4.

$$\mathrm{d}[1] = [3] + \frac{1}{2}[2] - [2,1]\,, \tag{33}$$

$$\mathrm{d}[2] = [4] + 2[3] - \frac{1}{6}[2] - 4[3,1]\,, \tag{34}$$

$$\mathrm{d}[2] = 2[4] + [3] + \frac{1}{6}[2] - 2[2,2] - 2[3,1]\,, \tag{35}$$

$$\mathrm{d}[1,1] = [3,1] + \frac{3}{2}[2,1] + \frac{1}{2}[1,2] + [1,3] - 2[2,1,1] - [1,2,1]\,, \tag{36}$$

$$[8] = \frac{1}{40}[4] - \frac{1}{252}[2] + 12[4,4]\,. \tag{37}$$

Using Theorem 7.1 as immediate consequences and without any difficulties we recover the following well-known identities for multiple zeta values.

(i) If we apply Z_3 to (33) we deduce $\zeta(3) = \zeta(2,1)$.
(ii) If we apply Z_4 to (34) and (35) we deduce $\zeta(4) = 4\zeta(3,1) = \frac{4}{3}\zeta(2,2)$.
(iii) The identity (36) reads in $\mathrm{q}\mathcal{MZ}[[1]]$ as

$$\mathrm{d}[1,1] = \left([3] - [2,1] + \frac{1}{2}[2]\right) \cdot [1] + 2[3,1] - \frac{1}{2}[4] - \frac{1}{2}[2,1] - \frac{1}{2}[3] + \frac{1}{3}[2]\,.$$

Applying Z_4^{alg} we deduce again the two relations $\zeta(3) = \zeta(2,1)$ and $4\zeta(3,1) = \zeta(4)$, since by Theorem 7.1 we have

$$Z_4^{alg}(\mathrm{d}[1,1]) = (\zeta(3) - \zeta(2,1))\,T - \frac{1}{2}\zeta(4) + 2\zeta(3,1) = 0\,.$$

(iv) If we apply Z_8 to (37) we deduce $\zeta(8) = 12\zeta(4,4)$.
(v) As we have seen in Proposition 4.13 the cusp form Δ can be written as

$$-\frac{1}{2^6 \cdot 5 \cdot 691}\Delta = 168[5,7] + 150[7,5] + 28[9,3] + \frac{1}{1408}[2] - \frac{83}{14400}[4] + \frac{187}{6048}[6] - \frac{7}{120}[8] - \frac{5197}{691}[12]\,. \tag{38}$$

Letting Z_{12} act on both sides of (38) one obtains the relation (6)

$$\frac{5197}{691}\zeta(12) = 168\zeta(5,7) + 150\zeta(7,5) + 28\zeta(9,3)\,.$$

But as mentioned in the introduction there are also elements in the kernel of Z_k that are not covered by Theorem 7.1. In weight 4 one has the following relation of multiple zeta values $\zeta(4) = \zeta(2,1,1)$, i.e. it is $[4] - [2,1,1] \in \ker Z_4$. But this element can't be written as a linear combination of cusp forms, lower weight brackets or derivatives. But using the double shuffle relations for bi-brackets described in Sect. 5.2 one can prove[11] that

$$[4] - [2,1,1] = \frac{1}{2}\left(\mathrm{d}[1] + \mathrm{d}[2]\right) - \frac{1}{3}[2] - [3] + \begin{bmatrix} 2,1 \\ 1,0 \end{bmatrix}. \tag{39}$$

Another way to see that many of the bi-brackets of weight k are in the kernel of the map Z_k is the following. Assume that $s_1 > r_1 + 1$ and $s_j \geq r_j + 1$ for $j = 2, \ldots, l$, then using again the representation with the Eulerian polynomials (See also Proposition 1 [41]) we get

$$Z_{s_1+\cdots+s_l}\left(\begin{bmatrix} s_1, \ldots, s_l \\ r_1, \ldots, r_l \end{bmatrix}\right) = \frac{1}{r_1! \ldots r_l!}\zeta(s_1 - r_1, \ldots, s_l - r_l)$$

and in particular with this assumption it is $\left[\begin{smallmatrix} s_1,\ldots,s_l \\ r_1,\ldots,r_l \end{smallmatrix}\right] \in \ker Z_{s_1+\cdots+s_l+1}$.

The study of the kernel Z_k is of great interest since it contains every relation of weight k. We expect that every element in the kernel of Z_k can be described using bi-brackets of a "certain kind" and it seems to be a really interesting question to specify this "certain kind" explicitly. To determine which bi-brackets are exactly in the kernel of the map Z_k and also which bi-brackets can be written in terms of brackets in q$\mathcal{MZ}$ is an open problem. The naive guess, that exactly the bi-brackets

---

[11] That the last term $\left[\begin{smallmatrix} 2,1 \\ 1,0 \end{smallmatrix}\right]$ in (39) is in the kernel of Z_4 can be proven in the following way: In Proposition 7.2 [6] it is shown, that an element $f = \sum_{n>0} a_n q^n$ with $a_n = O(n^m)$ and $m < k-1$ is in the kernel of Z_k. Here we have

$$\begin{bmatrix} 2,1 \\ 1,0 \end{bmatrix} = \sum_{\substack{u_1 > u_2 > 0 \\ v_1, v_2 > 0}} v_1 u_1 q^{v_1 u_1 + v_2 u_2} < \sum_{\substack{u_1, u_1 0 \\ v_1, v_2 > 0}} v_1 u_1 q^{v_1 u_1 + v_2 u_2} = \mathrm{d}[1] \cdot [1]\,,$$

where the $<$ is meant to be coefficient wise. Since the coefficients of $\mathrm{d}[1] \cdot [1]$ grow like $n^2 \log(n)^2$ we conclude $\left[\begin{smallmatrix} 2,1 \\ 1,0 \end{smallmatrix}\right] \in \ker Z_4$.

$\left[\begin{smallmatrix} s_1,\dots,s_l \\ r_1,\dots,r_l \end{smallmatrix}\right]$ where at least one $r_j > 0$ are elements in the kernel of $Z_{s_1+\dots+s_l+r_1+\dots+r_l}$ is wrong, since for example

$$\lim_{q\to 1}(1-q)^3\begin{bmatrix}1,1\\1,0\end{bmatrix} = \infty\,.$$

7.2 Connection to Other q-Analogues

In [39] the author gives an overview over several different q-analogues of multiple zeta values. Here we complement his work and focus on aspects related to our brackets. To compare the brackets to other q-analogues we first generalize the notion of a q-analogue of multiple zeta values as it was done in [7]. This notion of a q-analogue does cover many but not all q-analogues described in [39].

In the following we fix a subset $S \subset \mathbb{N}$, which we consider as the support for index entries, i.e. we assume $s_1, \dots, s_l \in S$. For each $s \in S$ we let $Q_s(t) \in \mathbb{Q}[t]$ be a polynomial with $Q_s(0) = 0$ and $Q_s(1) \neq 0$. We set $Q = \{Q_s(t)\}_{s\in S}$. A sum of the form

$$Z_Q(s_1, \dots, s_l) := \sum_{n_1>\dots>n_l>0}\prod_{j=1}^{l}\frac{Q_{s_j}(q^{n_j})}{(1-q^{n_j})^{s_j}} \tag{40}$$

with polynomials Q_s as before, defines a q-analogue of a multiple zeta-value of weight $k = s_1 + \dots + s_l$ and length l. Observe only because of $Q_{s_1}(0) = 0$ this defines an element of $\mathbb{Q}[[q]]$. That these objects are in fact a q-analogue of a multiple zeta-value is justified by the following calculation.

$$\begin{aligned}\lim_{q\to 1}(1-q)^k Z_Q(s_1,\dots,s_l) &= \sum_{n_1>\dots>n_l>0}\prod_{j=1}^{l}\lim_{q\to 1}\left(Q_{s_j}(q^{n_j})\frac{(1-q)^{s_j}}{(1-q^{n_j})^{s_j}}\right)\\ &= Q_{s_1}(1)\dots Q_{s_l}(1)\cdot\zeta(s_1,\dots,s_l)\,.\end{aligned}$$

Here we used that $\lim_{q\to 1}(1-q)^s/(1-q^n)^s = 1/n^s$ and with the same arguments as in [6] Proposition 6.4, the above interchange of the limit with the sum can be justified for all $(s_1, \dots, s_l)$ with $s_1 > 1$. Related definitions for q-analogues of multiple zeta values are given in [14, 28, 33, 42]. It is convenient to define $Z_Q(\emptyset) = 1$ and then we denote the vector space spanned by all these elements by

$$Z(Q,S) := \big\langle Z_Q(s_1,\dots,s_l)\,\big|\, l \geq 0 \text{ and } s_1,\dots,s_l \in S\big\rangle_{\mathbb{Q}}\,. \tag{41}$$

Note by the above convention we have, that $\mathbb{Q}$ is contained in this space.

Lemma 7.3 ([7, Lemma 2.1]) *If for each $r, s \in S$ there exists numbers $\lambda_j(r,s) \in \mathbb{Q}$ such that*

$$Q_r(t) \cdot Q_s(t) = \sum_{\substack{j \in S \\ 1 \le j \le r+s}} \lambda_j(r,s)(1-t)^{r+s-j} Q_j(t), \tag{42}$$

then the vector space $Z(Q, S)$ *is a* $\mathbb{Q}$*-algebra.*

Theorem 7.4 ([7, Theroem 2.4]) *Let* $Z(Q, \mathbb{N}_{>1})$ *be any family of q-analogues of multiple zeta values as in* (41)*, where each* $Q_s(t) \in Q$ *is a polynomial with degree at most* $s-1$*, then*

$$Z(Q, \mathbb{N}_{>1}) = \mathcal{MD}^\sharp,$$

where $\mathcal{MD}^\sharp$ *was the in Sect. 4.2 defined subalgebra of* $\mathcal{MD}$ *spanned by all brackets* $[s_1, \ldots, s_l]$ *with* $s_j \ge 2$*. Therefore, all such families of q-analogues of multiple zeta values are* $\mathbb{Q}$*-subalgebras of* $\mathcal{MD}$*.*

The following proposition allows one to write an arbitrary element in $Z(Q, \mathbb{N}_{>1})$ as an linear combination of $[s_1, \ldots, s_l] \in \mathcal{MD}^\sharp$.

Proposition 7.5 ([7, Proposition 2.5]) *Assume* $k \ge 2$*. For* $1 \le i, j \le k-1$ *define the numbers* $b^k_{i,j} \in \mathbb{Q}$ *by*

$$\sum_{j=1}^{k-1} \frac{b^k_{i,j}}{j!} t^j := \binom{t+k-1-i}{k-1}.$$

With this it is for $1 \le i \le k-1$ *and* $Q^E_j(t) = \frac{1}{(j-1)!} t P_j(t)$

$$t^i = \sum_{j=2}^{k} b^k_{i,j-1} (1-t)^{k-j} Q^E_j(t).$$

We give some examples of q-analogues of multiple zeta values, with some being of the above type.

(**i**) To write the brackets in the above way we choose $Q^E_s(t) = \frac{1}{(s-1)!} t P_{s-1}(t)$, where the $P_s(t)$ are the Eulerian polynomials defined earlier by

$$\frac{t P_{s-1}(t)}{(1-t)^s} = \sum_{d=1}^{\infty} d^{s-1} t^d$$

for $s \ge 0$. With this we have for all $s_1, \ldots, s_l \in \mathbb{N}$

$$[s_1, \ldots, s_l] := \sum_{n_1 > \ldots > n_l > 0} \prod_{j=1}^{l} \frac{Q^E_{s_j}(q^{n_j})}{(1-q^{n_j})^{s_j}}.$$

and $\mathcal{MD} = Z(\{Q^E_s(t))\}_s, \mathbb{N})$.

(**ii**) The polynomials $Q_s^T(t) = t^{s-1}$ are considered in [33, 42] and sums of the form (40) with $s_1 > 1$ and $s_2, \dots, s_l \geq 1$ are studied there. Using Proposition 7.5 every q-analogue of this type can be written explicitly in terms of brackets.

(**iii**) Okounkov chooses the following polynomials in [27]

$$Q_s^O(t) = \begin{cases} t^{\frac{s}{2}} & s = 2, 4, 6, \dots \\ t^{\frac{s-1}{2}}(1+t) & s = 3, 5, 7, \dots . \end{cases}$$

and defines for $s_1, \dots, s_l \in S = \mathbb{N}_{>1}$

$$\mathsf{Z}(s) = \sum_{n_1 > \dots > n_l > 0} \prod_{j=0}^{l} \frac{Q_{s_j}^O(q^{n_j})}{(1-q^{n_j})^{s_j}} .$$

We write for the space of the Okounkov q-multiple zetas

$$\mathsf{qMZV} = Z(\{Q_s^O(t)\}_s, \mathbb{N}_{>1}) .$$

Due to Theorem 7.4 we have $\mathsf{qMZV} = \mathcal{MD}^\sharp$. In [27] Okounkov conjectures, that the space qMZV is closed under the operator d. In length 1 this is proven in Proposition 2.9 [7].

(**iv**) There are also q-analogues which are not of the type as in (40). For example, the model introduced in [28] and further studied in [18]. For $s_1, \dots, s_l \geq 1$ they are define by

$$\mathfrak{z}_q(s_1, \dots, s_l) = \sum_{n_1 > \dots > n_l > 0} \frac{q^{n_1}}{(1-q^{n_1})^{s_1} \dots (1-q^{n_l})^{s_l}} .$$

It is easy to see, that every $\mathfrak{z}_q(s_1, \dots, s_l)$ can be written in terms of bi-brackets. For example

$$\begin{aligned} \mathfrak{z}_q(2,1) &= \sum_{n_1 > n_2 > 0} \frac{q^{n_1}}{(1-q^{n_1})^2(1-q^{n_2})} = \sum_{n_1 > n_2 > 0} \frac{q^{n_1}(q^{n_2} + 1 - q^{n_2})}{(1-q^{n_1})^2(1-q^{n_2})} \\ &= \sum_{n_1 > n_2 > 0} \frac{q^{n_1} q^{n_2}}{(1-q^{n_1})^2(1-q^{n_2})} + \sum_{n_1 > n_2 > 0} \frac{q^{n_1}}{(1-q^{n_1})^2} \\ &= [2,1] + \sum_{n_1 > 0} \frac{(n_1 - 1) q^{n_1}}{(1-q^{n_1})^2} = [2,1] + \begin{bmatrix} 2 \\ 1 \end{bmatrix} - [2] . \end{aligned}$$

Similarly one can prove $\mathfrak{z}_q(2,1,1) = [2,1,1] - 2[2,1] + \begin{bmatrix} 2,1 \\ 1,0 \end{bmatrix} + \begin{bmatrix} 2 \\ 2 \end{bmatrix} - \frac{3}{2}\begin{bmatrix} 2 \\ 1 \end{bmatrix} + [2]$. For higher weights this also works as illustrated in the following

$$\mathfrak{z}_q(2,2) = \sum_{n_1>n_2>0} \frac{q^{n_1}}{(1-q^{n_1})^2(1-q^{n_2})^2} = \sum_{n_1>n_2>0} \frac{q^{n_1}(q^{n_2}+1-q^{n_2})}{(1-q^{n_1})^2(1-q^{n_2})^2}$$
$$= [2,2] + \mathfrak{z}_q(2,1) = [2,2] + [2,1] + \begin{bmatrix}2\\1\end{bmatrix} - [2].$$

Using again Proposition 7.5 it becomes clear for arbitrary weights $s_1, \dots, s_l \geq 2$ we can write $\mathfrak{z}_q(s_1, \dots, s_l)$ in terms of bi-brackets.

Writing any q-analogue in terms of bi-brackets enables us to use the double shuffle structure explained in Sect. 5 to obtain linear relations for all of these q-analogues. Though it might be difficult to compare our double shuffle relations to the double shuffle relations of other models. For example in the case of $\mathfrak{z}_q(s_1, \dots, s_l)$ the authors in [18] consider $s_1, \dots, s_l \in \mathbb{Z}$ to describe their double shuffle relations. See [8] for further details on the comparison between different models of q-analogues and bi-brackets.

Acknowledgements This paper has served as the introductory part of my cumulative thesis written at the University of Hamburg. First of all I would like to thank my supervisor Ulf Kühn for his continuous, encouraging and patient support during the last years. Besides this I also want to thank several people for supporting me during my PhD project by whether giving me suggestion and ideas, letting me give talks on conferences and seminars, proof reading papers or having general discussions on this topic with me. A big "thank you" goes therefore to Olivier Bouillot, Kathrin Bringmann, David Broadhurst, Kurusch Ebrahimi-Fard, Herbert Gangl, José I. Burgos Gil, Masanobu Kaneko, Dominique Manchon, Nils Matthes, Martin Möller, Koji Tasaka, Don Zagier, Jianqiang Zhao and Wadim Zudilin. Finally I would like to thank the referee for various helpful comments and remarks.

References

1. Andrews, G., Rose, S.: MacMahon's sum-of-divisors functions, Chebyshev polynomials, and Quasi-modular forms. J. Reine Angew. Math. **676**, 97–103 (2013)
2. Bachmann, H.: Multiple Zeta-Werte und die Verbindung zu Modulformen durch multiple Eisensteinreihen. Master thesis, Hamburg University (2012). http://www.henrikbachmann.com
3. Bachmann, H.: The algebra of bi-brackets and regularized multiple Eisenstein series. J. Number Theory **200**, 260–294 (2019)
4. Bachmann, H.: Multiple Eisenstein series and q-analogues of multiple zeta values. Thesis, Hamburg University (2015). http://www.henrikbachmann.com
5. Bachmann, H.: Double shuffle relations for q-analogues of multiple zeta values, their derivatives and the connection to multiple Eisenstein series. RIMS Kôyûroku **2017**, 22–43 (2015)
6. Bachmann, H., Kühn, U.: The algebra of generating functions for multiple divisor sums and applications to multiple zeta values. Ramanujan J. **40**(3), 605–648 (2016)
7. Bachmann, H., Kühn, U.: A short note on a conjecture of Okounkov about a q-analogue of multiple zeta values. arXiv:1309.3920 [math.NT]
8. Bachmann, H., Kühn, U.: A dimension conjecture for q-analogues of multiple zeta values. In This Volume
9. Bachmann, H., Tasaka, K.: The double shuffle relations for multiple Eisenstein series. Nagoya Math. J. **230**, 1–33 (2017)
10. Bachmann, H., Tsumura, H.: Multiple series of Eisenstein type. Ramanujan J. **42**(2), 479–489 (2017)

11. Borwein, J., Bradley, D.: Thirty-two Goldbach variations. Int. J. Number Theory **02**, 65–103 (2006)
12. Bouillot, O.: The algebra of multitangent functions. J. Algebra **410**, 148–238 (2014)
13. Bouillot, O.: Table of reduction of multitangent functions of weight up to 10 (2012). http://www-igm.univ-mlv.fr/~bouillot/Tables_de_multitangentes.pdf
14. Bradley, D.M.: Multiple q-zeta values. J. Algebra **283**, 752–798 (2005)
15. Broadhurst, D., Kreimer, D.: Association of multiple zeta values with positive knots via Feynman diagrams up to 9 loops. Phys. Lett. B **393**, 403–412 (1997)
16. Ecalle, J.: The flexion structure and dimorphy: flexion units, singulators, generators, and the enumeration of multizeta irreducibles. In: Asymptotics in Dynamics, Geometry and PDEs, Generalized Borel Summation, vol. II, pp. 27–211 (2011)
17. Ebrahimi-Fard, K., Manchon, D., Singer, J.: Duality and (q-)multiple zeta values. Adv. Math. **298**, 254–285 (2016)
18. Ebrahimi-Fard, K., Manchon, D., Medina, J.C.: Unfolding the double shuffle structure of q-multiple zeta values. Bull. Austral. Math. Soc. **91**(3), 368–388 (2015)
19. Gangl, H., Kaneko, M., Zagier, D.: Double zeta values and modular forms. Automorphic Forms and Zeta Functions, pp. 71–106. World Science Publisher, Hackensack, NJ (2006)
20. Goncharov, A.B.: Galois symmetries of fundamental groupoids and noncommutative geometry. Duke Math. J. **128**(2), 209–284 (2005)
21. Hoffman, M.E.: Quasi-shuffle products. J. Algebraic Combin. **11**(1), 49–68 (2000)
22. Hoffman, M.E., Ihara, K.: Quasi-shuffle products revisited. J. Algebra **481**, 293–326 (2017)
23. Ihara, K.: Derivation and double shuffle relations for multiple zeta values, joint work with M. Kaneko, D. Zagier. RIMS Kôyûroku **1549**, 47–63
24. Ihara, K., Kaneko, M., Zagier, D.: Derivation and double shuffle relations for multiple zeta values. Compos. Math. **142**, 307–338 (2006)
25. Ihara, K., Ochiai, H.: Symmetry on linear relations for multiple zeta values. Nagoya Math. J. **189**, 49–62 (2008)
26. Kaneko, M., Tasaka, K.: Double zeta values, double Eisenstein series, and modular forms of level 2. Math. Ann. **357**(3), 1091–1118 (2013)
27. Okounkov, A.: Hilbert schemes and multiple q-zeta values. Funct. Anal. Appl. **48**, 138–144 (2014)
28. Ohno, Y., Okuda, J., Zudilin, W.: Cyclic q-MZSV sum. J. Number Theory **132**(1), 144–155 (2012)
29. Qin, Z., Yu, F.: On Okounkov's conjecture connecting Hilbert schemes of points and multiple q-zeta values. Int. Math. Res. Not. **2**, 321–361 (2018)
30. Rose, S.: Quasi-modularity of generalized sum-of-divisors functions. Res. Number Theory **1**, Art. 18, 11 pp (2015)
31. Schlesinger, K.-G.: Some remarks on q-deformed multiple polylogarithms. arXiv:math/0111022 [math.QA]
32. Singer, J.: On q-analogues of multiple zeta values. Funct. Approx. Comment. Math. **53**(1), 135–165 (2015)
33. Takeyama, Y.: The algebra of a q-analogue of multiple harmonic series. SIGMA Symmetry Integrability Geom. Methods Appl. **9**, Paper 061 (2013)
34. Yuan, H., Zhao, J.: Double shuffle relations of double zeta values and double Eisenstein series of level N. J. Lond. Math. Soc. (2) **92**(3), 520–546 (2015)
35. Yuan, H., Zhao, J.: Multiple Divisor Functions and Multiple Zeta Values at Level N. arXiv:1408.4983 [math.NT]
36. Zagier, D.: Elliptic modular forms and their applications. The 1-2-3 of Modular Forms, pp. 1–103. Universitext Springer, Berlin (2008)
37. Zagier, D.: Periods of modular forms, traces of Hecke operators, and multiple zeta values. RIMS Kôyûroku **843**, 162–170 (1993)
38. Zhao, J.: Multiple q-zeta functions and multiple q-polylogarithms. Ramanujan J. **14**(2), 189–221 (2007)

39. Zhao, J.: Uniform approach to double shuffle and duality relations of various q-analogs of multiple zeta values via Rota-Baxter algebras. arXiv:1412.8044 [math.NT]
40. Zorich, A.: Flat surfaces. Frontiers in Number Theory, Physics, and Geometry, vol. I, Springer (2006)
41. Zudilin, W.: Multiple q-zeta brackets. Math. 3:1, Spec. Issue Math. Phys. 119–130 (2015)
42. Zudilin, W.: Algebraic relations for multiple zeta values, (Russian. Russian summary) Uspekhi Mat. Nauk **58** (2003)

A Dimension Conjecture for q-Analogues of Multiple Zeta Values

Henrik Bachmann and Ulf Kühn

Abstract We study a class of q-analogues of multiple zeta values given by certain formal q-series with rational coefficients. After introducing a notion of weight and depth for these q-analogues of multiple zeta values we present dimension conjectures for the spaces of their weight- and depth-graded parts, which have a similar shape as the conjectures of Zagier and Broadhurst-Kreimer for multiple zeta values.

Keywords Multiple zeta values · q-Analogues of multiple zeta values · Modular forms · Dimension conjecture

1 Introduction

Multiple zeta values are real numbers appearing in various areas of mathematics and theoretical physics. By a q-analogue of these numbers one usually understands q-series, which degenerate to multiple zeta values as $q \to 1$. The algebraic structure of several models of q-analogues has been the subject of recent research (see [28] for an overview). Besides a conjecture of Okounkov in [17] for the dimension of the weight-graded spaces for a specific such model, no conjectures for the dimensions of the spaces of any of these q-analogues in a given weight and depth have occurred in the literature. The purpose of this work is to introduce a space of q-series which contains a lot of these models and to present conjectures on the dimensions of their weight- and depth-graded parts. For natural numbers $s_1 \geq 2, s_2, \dots, s_l \geq 1$ define the multiple zeta value (MZV)

H. Bachmann (✉)
Graduate School of Mathematics, Nagoya University, Chikusa-ku, Nagoya 464-8602, Japan
e-mail: henrik.bachmann@math.nagoya-u.ac.jp

U. Kühn
Universität Hamburg Fachbereich Mathematik, Bundesstrasse 55, 20146 Hamburg, Germany
e-mail: kuehn@math.uni-hamburg.de

J. I. Burgos Gil et al. (eds.), *Periods in Quantum Field Theory and Arithmetic*, Springer Proceedings in Mathematics & Statistics 314,
https://doi.org/10.1007/978-3-030-37031-2_9

$$\zeta(s_1,\ldots,s_l)=\sum_{n_1>\cdots>n_l>0}\frac{1}{n_1^{s_1}\ldots n_l^{s_l}}.$$

By $s_1+\cdots+s_l$ we denote its weight, by l its depth and we write $\mathcal{Z}$ for the $\mathbb{Q}$-vector space spanned by all MZVs. It is a well-known fact that the space $\mathcal{Z}$ is a $\mathbb{Q}$-algebra and that there are two different ways, known as the stuffle and shuffle product formulas respectively, to express the product of two MZVs as a $\mathbb{Q}$-linear combination of MZVs. These two ways of writing the product give a large family of $\mathbb{Q}$-linear relations between MZVs in a fixed weight, known as the double shuffle relations. Conjecturally all relations between MZVs follow from this type of relations. In particular it is conjectured that the algebra $\mathcal{Z}$ is graded by the weight. Let $\mathcal{Z}_k$ denote the $\mathbb{Q}$-vector space spanned by the MZVs of weight k, then there is the following famous dimension conjecture due to Zagier:

Conjecture 1 (Zagier [26]) *We have the following generating series*

$$\sum_{k\geq 0}\dim_{\mathbb{Q}}(\mathcal{Z}_k)\,x^k=\frac{1}{1-x^2-x^3}.$$

A stronger version of this conjecture was later proposed by Hoffman [13], which states that the $\zeta(s_1,\ldots,s_l)$ with $s_j\in\{2,3\}$ form a basis of $\mathcal{Z}$. So far it is only known, due to a result of Brown [7], that these MZVs span the space $\mathcal{Z}$. Conjecture 1 has a refinement by Broadhurst and Kreimer who proposed the following conjecture on the dimension of the weight- and depth-graded parts.

Conjecture 2 (Broadhurst-Kreimer [6]) *The generating series of the dimensions of the weight- and depth-graded parts of multiple zeta values is given by*

$$\sum_{k,l\geq 0}\dim_{\mathbb{Q}}\left(\mathrm{gr}_l^{\mathrm{D}}\,\mathcal{Z}_k\right)x^k y^l=\frac{1+\mathsf{E}_2(x)y}{1-\mathsf{O}_3(x)y+\mathsf{S}(x)y^2-\mathsf{S}(x)y^4},$$

where

$$\mathsf{E}_2(x)=\frac{x^2}{1-x^2},\quad \mathsf{O}_3(x)=\frac{x^3}{1-x^2},\quad \mathsf{S}(x)=\frac{x^{12}}{(1-x^4)(1-x^6)}.$$

Observe that $\mathsf{E}_2(x)$ (resp. $\mathsf{O}_3(x)$) is the generating series of the number of even (resp. odd) zeta values and $\mathsf{S}(x)$ is the generating series for the dimensions of cusp forms for $\mathrm{SL}_2(\mathbb{Z})$. Furthermore, by setting $y=1$ on the right-hand side of the Broadhurst-Kreimer conjecture one obtains precisely the right-hand side in the Zagier conjecture. We are interested in conjectures similar to the above in the context of q-analogues of multiple zeta values. There are various different models of q-analogues for multiple zeta values. For most of these models the algebraic setup, i.e. analogues of the stuffle and the shuffle product, is well understood (See for example [3, 9, 20, 21, 28]). The problem of understanding the dimension of the weight-graded spaces has been considered in [3, 4, 9, 17, 20, 21, 28]. On the other hand possible analogues of the Broadhurst-Kreimer conjecture for these q-analogues have not been proposed yet.

Now we will define the q-analogues of multiple zeta values we consider in this paper. For $s_1, \dots, s_l \geq 1$ and polynomials $Q_1(t) \in t\mathbb{Q}[t]$ and $Q_2(t) \dots, Q_l(t) \in \mathbb{Q}[t]$ we define

$$\zeta_q(s_1, \dots, s_l; Q_1, \dots, Q_l) = \sum_{n_1 > \dots > n_l > 0} \frac{Q_1(q^{n_1}) \dots Q_l(q^{n_l})}{(1-q^{n_1})^{s_1} \cdots (1-q^{n_l})^{s_l}} . \tag{1}$$

This series can be seen as a q-analogue[1] of $\zeta(s_1, \dots, s_l)$, since we have for $s_1 > 1$

$$\lim_{q \to 1} (1-q)^{s_1 + \dots + s_l} \zeta_q(s_1, \dots, s_l; Q_1, \dots, Q_l) = Q_1(1) \dots Q_l(1) \cdot \zeta(s_1, \dots, s_l) .$$

We only consider the case where $\deg(Q_j) \leq s_j$ and consider the following $\mathbb{Q}$-algebra:

$$\mathcal{Z}_q := \left\langle \zeta_q(s_1, \dots, s_l; Q_1, \dots, Q_l) \;\middle|\; l \geq 0,\, s_1, \dots, s_l \geq 1,\, \deg(Q_j) \leq s_j \right\rangle_{\mathbb{Q}} .$$

Contrary to the case of MZVs, the number $s_1 + \dots + s_l$ does not give a good notion of weight for the ζ_q, since for example $\zeta_q(s; Q) = \zeta_q(s+1, Q \cdot (1-t))$. Also the number l will not be used to define the depth. Instead we will consider a class of q-series which also span the space $\mathcal{Z}_q$ and use these series to define a weight and a depth filtration on $\mathcal{Z}_q$. For $s_1, \dots, s_l \geq 1, r_1, \dots, r_l \geq 0$ these q-series are given by

$$\begin{bmatrix} s_1, \dots, s_l \\ r_1, \dots, r_l \end{bmatrix} := \sum_{\substack{u_1 > \dots > u_l > 0 \\ v_1, \dots, v_l > 0}} \frac{u_1^{r_1}}{r_1!} \cdots \frac{u_l^{r_l}}{r_l!} \cdot \frac{v_1^{s_1 - 1} \dots v_l^{s_l - 1}}{(s_1-1)! \dots (s_l-1)!} \cdot q^{u_1 v_1 + \dots + u_l v_l} \in \mathbb{Q}[[q]] . \tag{2}$$

We refer to these q-series as *bi-brackets* of depth l and weight $s_1 + \dots + s_l + r_1 + \dots + r_l$. They were introduced by the first author in [1] and their algebraic structure is well-understood and described in the papers [1–3, 31]. The bi-brackets have a natural connection to quasi-modular forms (for $\mathrm{SL}_2(\mathbb{Z})$), since for even k the Fourier expansion of the classical Eisenstein series G_k of weight k is given by $\begin{bmatrix} k \\ 0 \end{bmatrix}$ plus an appropriate constant term. In particular the space of quasi-modular forms with rational coefficients, which is given by $\mathbb{Q}[G_2, G_4, G_6]$, is a sub-algebra of the space $\mathcal{Z}_q$.

As we will see in Theorem 1 the bi-brackets span the space $\mathcal{Z}_q$ and therefore we can define a weight and a depth filtration by using the notion of weight and depth of bi-brackets. We point out the fact that $\mathcal{Z}_q$ is not graded by the weight, i.e. the weight graded spaces $\mathrm{gr}_k^W \mathcal{Z}_q$ are in general not isomorphic to the $\mathbb{Q}$-vector spaces

[1]These type of series are often called *modified* q-analogues of multiple zeta values, since one needs to multiply by $(1-q)^{s_1 + \dots + s_l}$ before taking the limit $q \to 1$.

spanned by bi-brackets of weight k. In analogy to the Zagier and Broadhurst-Kreimer conjecture we conjecture the following.

Conjecture 3 *(i) The dimension of the weight graded parts of* $\mathcal{Z}_q$ *is given by*

$$\sum_{k\geq 0} \dim_{\mathbb{Q}} \left(\mathrm{gr}_k^W \mathcal{Z}_q\right) x^k = \frac{1}{1-x-x^2-x^3+x^6+x^7+x^8+x^9}.$$

(ii) The dimension of the weight and depth graded parts of $\mathcal{Z}_q$ *is given by*

$$\sum_{k,l\geq 0} \dim_{\mathbb{Q}} \left(\mathrm{gr}_{k,l}^{\mathrm{W,D}} \mathcal{Z}_q\right) x^k y^l = \frac{1+\mathsf{D}(x)\,\mathsf{E}_2(x)y+\mathsf{D}(x)\,\mathsf{S}(x)y^2}{1-\mathsf{a}_1(x)\,y+\mathsf{a}_2(x)\,y^2-\mathsf{a}_3(x)\,y^3-\mathsf{a}_4(x)\,y^4+\mathsf{a}_5(x)\,y^5},$$

where $\mathsf{D}(x)=1/(1-x^2)$, $\mathsf{O}_1(x)=x/(1-x^2)$ *and* $\mathsf{E}_2(x), \mathsf{S}(x)$ *are as in Conjecture 2 and*

$$\mathsf{a}_1(x)=\mathsf{D}(x)\,\mathsf{O}_1(x)\,, \qquad \mathsf{a}_2(x)=\mathsf{D}(x)\sum_{k\geq 1}\dim_{\mathbb{Q}}(M_k(\mathrm{SL}_2(\mathbb{Z}))^2\,x^k\,,$$

$$\mathsf{a}_3(x)=\mathsf{a}_5(x)=\mathsf{O}_1(x)\,\mathsf{S}(x)\,, \qquad \mathsf{a}_4(x)=\mathsf{D}(x)\sum_{k\geq 1}\dim_{\mathbb{Q}}(S_k(\mathrm{SL}_2(\mathbb{Z}))^2\,x^k\,.$$

Here $M_k(\mathrm{SL}_2(\mathbb{Z}))$ *and* $S_k(\mathrm{SL}_2(\mathbb{Z}))$ *denote the spaces of modular forms and cusp forms for* $\mathrm{SL}_2(\mathbb{Z})$ *of weight* k.

Note that setting $y=1$ in (ii) implies (i). This holds because of the formula

$$\sum_{k\geq 0}\dim_{\mathbb{Q}}(M_k(\mathrm{SL}_2(\mathbb{Z}))^2\,x^k = \frac{1+x^{12}}{(1-x^4)(1-x^6)(1-x^{12})},$$

which is straightforward to prove.[2]

In the Broadhurst-Kreimer conjecture the numerator $1+\mathsf{E}_2(x)y$ can be interpreted as the generating series of $\dim_{\mathbb{Q}} \mathrm{gr}_{k,l}^{W,D}\, \mathbb{Q}[\zeta(2)]$, i.e.

$$\sum_{k,l\geq 0}\dim_{\mathbb{Q}} \mathrm{gr}_{k,l}^{\mathrm{W,D}}\,(\mathbb{Q}[\zeta(2)])\,x^k y^l = 1+\mathsf{E}_2(x)y\,.$$

As we will see in Proposition 1 the numerator in Conjecture 3 (ii) is essentially the generating series for the weight- and depth-graded dimensions of the quasi-modular forms, since $\mathsf{D}(x)\,\mathsf{E}_2(x)$ counts the number of Eisenstein series and their derivatives and $\mathsf{D}(x)\,\mathsf{S}(x)$ corresponds to the number of cusp forms and their derivatives.

---

[2]Recall the series expansion $\frac{1}{(1-x)^2}=1+2x+3x^2+4x^3+\dots$ and $\frac{1}{(1-x)^2}\frac{1+x}{1-x}=1+4x+9x^2+16x^3+\dots$. Now, since we have $\sum_{k\geq 0}\dim_{\mathbb{Q}}(M_k(\mathrm{SL}_2(\mathbb{Z}))\,x^k=\frac{1}{(1-x^4)(1-x^6)}=(1+x^4+x^6+x^8+x^{10}+x^{14})\frac{1}{(1-x^{12})^2}$, the claim follows by replacing x by x^{12} in the second series expansion.

Therefore it is reasonable to expect that

$$\sum_{k,l\geq 0} \dim_{\mathbb{Q}} \mathrm{gr}^{\mathrm{W,D}}_{k,l} \left(\mathbb{Q}[G_2, G_4, G_6]\right) x^k y^l \stackrel{?}{=} 1 + \mathsf{D}(x)\,\mathsf{E}_2(x) y + \mathsf{D}(x)\,\mathsf{S}(x) y^2 .$$

In some sense the quasi-modular forms in the context of q-analogues of multiple zeta values play the role of the even single zeta values (see also [11, 29]).

For $k \leq 15$ we determined, by calculating a large number of coefficients, lower bounds for $\dim_{\mathbb{Q}} \left(\mathrm{gr}^W_k \mathscr{Z}_q\right)$, which equal the expected dimensions in Conjecture 3 (i). Furthermore, Conjecture 3 (i) actually holds for $k \leq 7$ by Theorem 2 below. For the refined Conjecture 3 (ii) our computer experiments provide us with lower bounds, which again equal the expected dimensions, in the range given by Table 4 on page 18.

2 q-analogues of MZVs and Bi-Brackets

Usually a function $f(q)$ is called a q-analogue of multiple zeta value, if $\lim\limits_{q\to 1} f(q)$ is a multiple zeta value. There are various different models of q-analogues in the literature (See [28] for a nice overview). One of the first models was studied by Bradley [5] and Zhao [27] independently. This model is given for $s_1 \geq 2, s_2, \dots s_l \geq 1$ by the q-series

$$\sum_{n_1>\dots>n_l>0} \frac{q^{(s_1-1)n_1} \dots q^{(s_l-1)n_l}}{\{n_1\}_q^{s_1} \cdots \{n_l\}_q^{s_l}} , \tag{3}$$

with $\{n\}_q = \frac{1-q^n}{1-q}$ being the usual q-integer. Taking the limit $q \to 1$ in above sum one obtains $\zeta(s_1, \dots, s_l)$. For a cleaner description of the algebraic structure and (in our case) a connection to modular forms it is convenient to consider a modified version of (3) by removing the factor $(1-q)^{s_1+\dots+s_l}$, i.e. to consider the series

$$\zeta_q^{\mathrm{BZ}}(s_1, \dots, s_l) = \sum_{n_1>\dots>n_l>0} \frac{q^{(s_1-1)n_1} \dots q^{(s_l-1)n_l}}{(1-q^{n_1})^{s_1} \cdots (1-q^{n_l})^{s_l}} , \tag{4}$$

which then satisfies $\lim\limits_{q\to 1} (1-q)^{s_1+\dots+s_l} \zeta_q(s_1, \dots, s_l) = \zeta(s_1, \dots, s_l)$.

In a greater generality we will consider, for $s_1, \dots, s_l \geq 1$ and polynomials $Q_1(t) \in t\mathbb{Q}[t]$ and $Q_2(t) \dots, Q_l(t) \in \mathbb{Q}[t]$, sums of the form

$$\zeta_q(s_1, \dots, s_l; Q_1, \dots, Q_l) = \sum_{n_1>\dots>n_l>0} \frac{Q_1(q^{n_1}) \dots Q_l(q^{n_l})}{(1-q^{n_1})^{s_1} \cdots (1-q^{n_l})^{s_l}} . \tag{5}$$

The condition $Q_1(t) \in t\mathbb{Q}[t]$ ensures that this is an element in $\mathbb{Q}[[q]]$. In contrast to (4) we also allow $s_1 = 1$ in our setup, i.e., we also include q-analogues of the non-convergent multiple zeta values. In the case $s_1 > 1$ we can (by the same arguments as in [4] Proposition 6.4) again take the limit $q \to 1$ after multiplying by $(1-q)^{s_1+\cdots+s_l}$, which gives

$$\lim_{q\to 1}(1-q)^{s_1+\cdots+s_l}\zeta_q(s_1,\ldots,s_l;Q_1,\ldots,Q_l) = Q_1(1)\ldots Q_l(1)\cdot\zeta(s_1,\ldots,s_l)\,.$$

Almost all models of q-analogues in the literature are given by sums of the form (5). In the following we always set $\zeta_q(s_1,\ldots,s_l;Q_1,\ldots,Q_l) = 1$ for the case $l = 0$. We will consider the following spaces spanned by the series (5) of a particular kind

$$\mathscr{Z}_q = \left\langle \zeta_q(s_1,\ldots,s_l;Q_1,\ldots,Q_l) \;\middle|\; l \geq 0,\, s_1,\ldots,s_l \geq 1,\, \deg(Q_j) \leq s_j \right\rangle_{\mathbb{Q}},$$

where as before we always assume $Q_1(t) \in t\mathbb{Q}[t]$ and $Q_2(t)\ldots,Q_l(t) \in \mathbb{Q}[t]$. As we will see below $\mathscr{Z}_q$ is the space in which we are interested the most. For $d \geq 0$ we define the subspace $\mathscr{Z}_{q,d} = \left\langle \zeta_q(s_1,\ldots,s_l;Q_1,\ldots,Q_l) \in \mathscr{Z}_q \;\middle|\; \deg(Q_j) \leq s_j - d \right\rangle_{\mathbb{Q}}$. So in particular we have $\mathscr{Z}_q = \mathscr{Z}_{q,0}$ and $\mathscr{Z}_{q,d+1} \subset \mathscr{Z}_{q,d}$. We also restrict to the case in which all polynomials Q_j (not just Q_1) have no constant terms and therefore are elements in $t\mathbb{Q}[t]$. The resulting space is denoted by

$$\mathscr{Z}_q^\circ = \left\langle \zeta_q(s_1,\ldots,s_l;Q_1,\ldots,Q_l) \in \mathscr{Z}_q \;\middle|\; Q_1,\ldots,Q_l \in t\mathbb{Q}[t] \right\rangle_{\mathbb{Q}}.$$

For the spaces $\mathscr{Z}_{q,d}^\circ$ given by $\mathscr{Z}_q^\circ \cap \mathscr{Z}_{q,d}$ it holds $\mathscr{Z}_q^\circ = \mathscr{Z}_{q,0}^\circ$ and $\mathscr{Z}_{q,d+1}^\circ \subset \mathscr{Z}_{q,d}^\circ$. Notice that all of these spaces are closed under multiplication. In depth one for example we have

$$\zeta_q(s_1;Q_1)\cdot\zeta_q(s_2;Q_2) = \zeta_q(s_1,s_2;Q_1,Q_2) + \zeta_q(s_2,s_1;Q_2,Q_1) + \zeta_q(s_1+s_2;Q_1\cdot Q_2)\,,$$

and clearly $\deg Q_1\cdot Q_2 \leq s_1+s_2-d$ if $\deg Q_j \leq s_j - d$ for $j = 1, 2$.

In [28] Zhao considers for $s_1,\ldots,s_l,d_1,\ldots,d_l \in \mathbb{Z}$ the series

$$\mathfrak{z}_q^{d_1,\ldots,d_l}(s_1,\ldots,s_l) = \sum_{n_1>\cdots>n_l>0} \frac{q^{n_1d_1}\ldots q^{n_ld_l}}{(1-q^{n_1})^{s_1}\cdots(1-q^{n_l})^{s_l}}\,, \tag{6}$$

which gives an even more general setup than our ζ_q. Especially these series can be seen as natural generators of the spaces $\mathscr{Z}_{q,d}$ and $\mathscr{Z}_{q,d}^\circ$ by choosing the appropriate conditions on the d_j. We will now give a short overview of different q-analogues of multiple zeta values, which can be written in terms of the ζ_q and relate them to the spaces $\mathscr{Z}_{q,d}$ and $\mathscr{Z}_{q,d}^\circ$.

(i) The space spanned by the Bradley-Zhao model $\zeta_q^{\mathrm{BZ}} = \zeta_q(s_1, \dots, s_l; t^{s_1-1}, \dots, t^{s_l-1})$, defined in (4), is given by[3]

$$\mathscr{Z}_{q,1} = \left\langle \zeta_q^{\mathrm{BZ}}(s_1, \dots, s_l) \mid l \geq 0\,, s_1 \geq 2, s_2, \dots, s_l \geq 1 \right\rangle_Q .$$

(ii) Another interesting case is the Schlesinger-Zudilin model. These q-analogues are for $s_1 \geq 1, s_2, \dots, s_l \geq 0$ defined by

$$\begin{aligned} \zeta_q^{\mathrm{SZ}}(s_1, \dots, s_l) &= \sum_{n_1 > \dots > n_l > 0} \frac{q^{n_1 s_1} \dots q^{n_l s_l}}{(1-q^{n_1})^{s_1} \cdots (1-q^{n_l})^{s_l}} \\ &= \zeta_q(s_1, \dots, s_l; t^{s_1}, \dots, t^{s_j}) . \end{aligned} \tag{7}$$

The space spanned by these series is, using the same argument as in (i), given by

$$\mathscr{Z}_q = \left\langle \zeta_q^{\mathrm{SZ}}(s_1, \dots, s_l) \,\middle|\, l \geq 0,\ s_1 \geq 1, s_2, \dots, s_l \geq 0 \right\rangle_{\mathbb{Q}} .$$

Originally defined by Schlesinger [18] and Zudilin [30] for the cases $s_1 \geq 2, s_2, \dots, s_l \geq 1$, it was observed in [20] and further discussed in [9] that the algebraic setup for this model, especially the shuffle product analogue, can be described nicely by allowing $s_1 \geq 1, s_2, \dots, s_l \geq 0$. Restricting to $s_1, \dots, s_l \geq 1$ we get the subspace

$$\mathscr{Z}_q^{\circ} = \left\langle \zeta_q^{\mathrm{SZ}}(s_1, \dots, s_l) \,\middle|\, l \geq 0,\ s_1, \dots, s_l \geq 1 \right\rangle_{\mathbb{Q}} .$$

(iii) In [22] Ohno, Okuda and Zudilin define for $s_1, \dots, s_l \in \mathbb{Z}$ the series

$$\zeta_q^{\mathrm{OOZ}}(s_1, \dots, s_l) = \sum_{n_1 > \dots > n_l > 0} \frac{q^{n_1}}{(1-q^{n_1})^{s_1} \cdots (1-q^{n_l})^{s_l}} . \tag{8}$$

In the case $s_1, \dots, s_l \geq 1$ these can be written as $\zeta_q(s_1, \dots, s_l; t, 1, \dots, 1) \in \mathscr{Z}_q$, but the space spanned by (8) for $s_1, \dots, s_j \geq 1$ is a priori not given by one of the $\mathscr{Z}_{q,d}$ or $\mathscr{Z}_{q,d}^{\circ}$.

(iv) For $s_1, \dots, s_l \geq 2$ Okounkov chooses the following polynomials in [17]

$$Q_j^O(t) = \begin{cases} t^{\frac{s_j}{2}} & s_j = 2, 4, 6, \dots \\ t^{\frac{s_j - 1}{2}}(1+t) & s_j = 3, 5, 7, \dots . \end{cases}$$

[3]This follows easily from the fact that $t^{j-1}(1-t)^{s-j}$ with $j = 1, \dots, s$ (resp. $j = 2, \dots, s$) forms a basis of $\{Q \in \mathbb{Q}[t] \mid \deg Q \leq s-1\}$ (resp. $\{Q \in t\mathbb{Q}[t] \mid \deg Q \leq s-1\}$).

and defines $\mathsf{Z}(s_1,\dots,s_l) = \zeta_q(s_1,\dots,s_l; Q_1^O,\dots,Q_l^O)$. With the same arguments as before (see also the proof of Theorem 1 (iii)) the span of these series is given by

$$\mathscr{Z}_{q,1}^\circ = \left\langle \mathsf{Z}(s_1,\dots,s_l) \,\middle|\, l \geq 0,\ s_1,\dots,s_l \geq 2 \right\rangle_{\mathbb{Q}}.$$

Although the space $\mathscr{Z}_q$ seems to be much larger than the space $\mathscr{Z}_q^\circ$, we expect that they both coincide (Conjecture 5 (B2) below) and therefore every ζ_q^{SZ} should be expressible as a linear combination of $\zeta_q^{\mathrm{SZ}}(s_1,\dots,s_l)$ with $s_1,\dots,s_l \geq 1$. In [9] (Theorem 5.5) such an expression for ζ_q^{OOZ} in terms of ζ_q^{SZ} is given, which in turn can be seen as a special case of that conjecture.

Remark 1 As seen in the example above, the polynomials Q_j often depend just on s_j. For these types of q-analogues one can also define subspaces of $\mathscr{Z}_q$ in the following way: Suppose that $\{Q_s\}_{s\geq 1}$ is a family of polynomials, where for all $s_1, s_2 \geq 1$ there exists numbers $\lambda_j^{s_1,s_2} \in \mathbb{Q}$ with $j \geq 1$ and $\lambda_j^{s_1,s_2} = 0$ for almost all j, such that

$$Q_{s_1}(t) \cdot Q_{s_2}(t) = \sum_{j=1}^{\infty} \lambda_j^{s_1,s_2} Q_j(t)(1-t)^{s_1+s_j-j}.$$

Then the space spanned by all $\zeta_q(s_1,\dots,s_l; Q_{s_1},\dots,Q_{s_l})$ is a sub-algebra of $\mathscr{Z}_q$. This also gives an example of a so called quasi-shuffle algebra as described in [14]. For this one can define for $a, b \geq 1$ the product $z_a \diamond z_b = \sum_{j=1}^{\infty} \lambda_j^{a,b} z_j$ with the same notation as used in the first section of [14]. This was for example done in [4] for the space $\mathscr{Z}_q^\circ$.

2.1 Bi-Brackets as q-Analogues of MZVs

In this section we will consider the q-series from the introduction in more detail and explain their connection to q-analogues of multiple zeta values in the section before.

Definition 1 (i) For $s_1,\dots,s_l \geq 1$, $r_1,\dots,r_l \geq 0$ we define the following q-series

$$\begin{bmatrix} s_1,\dots,s_l \\ r_1,\dots,r_l \end{bmatrix} := \sum_{\substack{u_1>\dots>u_l>0 \\ v_1,\dots,v_l>0}} \frac{u_1^{r_1}}{r_1!}\cdots\frac{u_l^{r_l}}{r_l!} \cdot \frac{v_1^{s_1-1}\dots v_l^{s_l-1}}{(s_1-1)!\dots(s_l-1)!} \cdot q^{u_1v_1+\dots+u_lv_l} \in \mathbb{Q}[[q]].$$

We refer to these q-series as *bi-brackets* of depth l and of weight $s_1+\dots+s_l+r_1+\dots+r_l$.

(ii) For $r_1 = \dots = r_l = 0$ we write

$$[s_1,\dots,s_l] := \begin{bmatrix} s_1,\dots,s_l \\ 0,\dots,0 \end{bmatrix}.$$

The series $[s_1, \dots, s_l]$, which we call *brackets*, were introduced and studied in [4].

The bi-brackets also have an alternative form, which we will use now. For this recall that the Eulerian polynomials (cf. [10, (3.2)]) are defined by

$$\frac{t P_{s-1}(t)}{(1-t)^s} = \sum_{d=1}^{\infty} d^{s-1} t^d .$$

For $s > 1$ the polynomials $t P_{s-1}(t)$ have degree $s-1$ and in the case $s = 1$ we have $t P_0(t) = t$. By definition of the bi-brackets it is then clear that

$$\begin{bmatrix} s_1, \dots, s_l \\ r_1, \dots, r_l \end{bmatrix} = \sum_{n_1 > \dots > n_l > 0} \prod_{j=1}^{l} \left(\frac{n_j^{r_j}}{r_j!} \cdot \frac{q^{n_j} P_{s_j-1}(q^{n_j})}{(s_j-1)! \cdot (1-q^{n_j})^{s_j}} \right) . \tag{9}$$

We will now see that the spaces spanned[4] by the bi-brackets and brackets are exactly given by the spaces $\mathcal{Z}_q$ and $\mathcal{Z}_q^\circ$ respectively.

Theorem 1 *The following equalities hold*

(i) $$\mathcal{Z}_q = \left\langle \begin{bmatrix} s_1, \dots, s_l \\ r_1, \dots, r_l \end{bmatrix} \mid l \geq 0, s_1, \dots, s_l \geq 1,\ r_1, \dots, r_l \geq 0 \right\rangle_{\mathbb{Q}} .$$

(ii) $$\mathcal{Z}_q^\circ = \left\langle [s_1, \dots, s_l] \mid l \geq 0 \,, s_1, \dots, s_l \geq 1 \right\rangle_Q .$$

(iii) $$\mathcal{Z}_{q,1}^\circ = \left\langle [s_1, \dots, s_l] \mid l \geq 0 \,, s_1, \dots, s_l \geq 2 \right\rangle_Q .$$

Proof Since for all $s \geq 1$ we have $P_{s-1}(1) \neq 0$ the polynomials $t P_{j-1}(t)(1-t)^{s-j}$ with $j = 1, \dots, s$ form a basis of the space $\{Q \in t\mathbb{Q}[t] \mid \deg Q \leq s\}$. In particular for every polynomial Q in this space there exist coefficients $\alpha_j \in \mathbb{Q}$ with

$$\frac{Q(t)}{(1-t)^s} = \sum_{j=1}^{s} \alpha_j \frac{t P_{j-1}(t)}{(1-t)^j} , \tag{10}$$

from which the statement (ii) follows. Also (iii) follows, since for $d = 1$ the condition $\mathbb{Q}_j(t) \in t\mathbb{Q}[t]$ and $\deg Q_j \leq s_j - 1$ implies $s_j \geq 2$ for all $j = 1, \dots, l$. One can also see that

$$\left\langle [s_1, \dots, s_l] \mid l \geq 0 \,, s_1, \dots, s_l \geq 2 \right\rangle_Q = \left\langle \zeta_q^{\mathrm{BZ}}(s_1, \dots, s_l) \mid l \geq 0 \,, s_1, \dots, s_l \geq 2 \right\rangle_Q .$$

To prove (i) we will first show the inclusion '$\subseteq$' , i.e. that every $\zeta_q(s_1, \dots, s_l; Q_1, \dots, Q_l)$ can be written in terms of bi-brackets. For this we need to see what happens if one of the $Q_2, \dots, Q_l$ has a constant term. Without loss of generality we can, by the proof of (ii), focus on the cases $Q_i(t) = 1$ for a $2 \leq i \leq l$. Since for all $s \geq 1$ we have

[4]In the articles [1, 2, 4] these spaces were denoted $\mathcal{BD}$ and $\mathcal{MD}$.

$$\frac{1}{(1-t)^s} = 1 + \sum_{m=1}^{s} \frac{t}{(1-t)^m}\,,$$

we can write

$$\sum_{n_1>\cdots>n_l>0} \prod_{j=1}^{l} \frac{Q_j(q^{n_j})}{(1-q^{n_j})^{s_j}} = \sum_{n_1>\cdots>n_l>0} \prod_{\substack{j=1\\ j\neq i}}^{l} \frac{Q_j(q^{n_j})}{(1-q^{n_j})^{s_j}} + \sum_{\substack{n_1>\cdots>n_l>0\\ 1\le m\le s_i}} \frac{q^{n_i}}{(1-q^{n_i})^m} \prod_{\substack{j=1\\ j\neq i}}^{l} \frac{Q_j(q^{n_j})}{(1-q^{n_j})^{s_j}}\,.$$

For the the second sum on the right-hand side we can again use (10). For the first sum we obtain (by setting $n_{l+1}=0$)

$$\sum_{n_1>\cdots>n_l>0} \prod_{\substack{j=1\\ j\neq i}}^{l} \frac{Q_j(q^{n_j})}{(1-q^{n_j})^{s_j}} = \sum_{n_1>\cdots>n_{i-1}>n_{i+1}>\cdots>n_l>0} (n_{i-1}-n_{i+1}-1) \prod_{\substack{j=1\\ j\neq i}}^{l} \frac{Q_j(q^{n_j})}{(1-q^{n_j})^{s_j}}\,.$$

Repeating this for all $2 \le i \le l$ with $Q_i(t)=1$ we obtain sums of the form (9) from which we deduce '$\subseteq$'.

Now to prove '$\supseteq$' we first define for $m \ge 0$ the polynomials $p_m(n)$ by $p_0(n)=1$ and

$$p_m(n) = \sum_{n>N_1>\cdots>N_m>0} 1 = \binom{n-1}{m}. \tag{11}$$

The $p_m(n)$ is a polynomial in n of degree m and therefore we can always find $c_m(r) \in \mathbb{Q}$ with $n^r = \sum_{m=0}^{r} c_m(r)\, p_m(n)$. The idea is now to replace $n_j^{r_j}$ in the definition of the bi-brackets by $\sum_{m_j=0}^{r_j} c_{m_j}(r_j)\, p_{m_j}(n_j)$ and then use (11) to get sums which can be written in terms of the ζ_q. We illustrate this in the depth two case from which the general case becomes clear. We have with $\kappa = (s_1-1)!(s_2-1)!r_1!r_2!$

$$\begin{aligned}
\kappa \cdot \begin{bmatrix} s_1, s_2 \\ r_1, r_2 \end{bmatrix} &= \sum_{n_1>n_2>0} \frac{n_1^{r_1} q^{n_1} P_{s_1-1}(q^{n_1})}{(1-q^{n_1})^{s_1}} \frac{n_2^{r_2} q^{n_2} P_{s_2-1}(q^{n_2})}{(1-q^{n_2})^{s_2}} \\
&= \sum_{0\le m_2\le r_2} c_{m_2}(r_2) \sum_{n_1>n_2>N_1>\cdots>N_{m_2}>0} \frac{n_1^{r_1} q^{n_1} P_{s_1-1}(q^{n_1})}{(1-q^{n_1})^{s_1}} \frac{q^{n_2} P_{s_2-1}(q^{n_2})}{(1-q^{n_2})^{s_2}} \\
&= \sum_{\substack{0\le m_1\le r_1\\ 0\le m_2\le r_2}} c_{m_1}(r_1) c_{m_2}(r_2) \sum_{\substack{n_1>n_2>N_1>\cdots>N_{m_2}>0\\ n_1>N'_1>\cdots>N'_{m_1}>0}} \frac{q^{n_1} P_{s_1-1}(q^{n_1})}{(1-q^{n_1})^{s_1}} \frac{q^{n_2} P_{s_2-1}(q^{n_2})}{(1-q^{n_2})^{s_2}}\,.
\end{aligned}$$

Now considering all the possible shuffles, and possible equalities of the N and the N' it is clear that this sum can be written as a linear combination of ζ_q by interpreting appearing 1 as $(1-q^N)(1-q^N)^{-1}$. For general depth l the idea is the same and therefore we obtain '$\supseteq$' from which (i) follows.

As an example of how to write a bi-bracket in terms of ζ_q, we give the following.

$$\begin{aligned}
\begin{bmatrix}1,1\\0,1\end{bmatrix} &= \sum_{n_1>n_2>0} \frac{q^{n_1}}{(1-q^{n_1})}\frac{n_2 q^{n_2}}{(1-q^{n_2})} \\
&= \sum_{n_1>n_2>0} \frac{q^{n_1}}{(1-q^{n_1})}\frac{q^{n_2}}{(1-q^{n_2})} + \sum_{n_1>n_2>n_3>0} \frac{q^{n_1}}{(1-q^{n_1})}\frac{q^{n_2}}{(1-q^{n_2})}\frac{1-q^{n_3}}{(1-q^{n_3})} \\
&= \zeta_q(1,1;t,t) + \zeta_q(1,1,1;t,t,1-t)\,.
\end{aligned} \tag{12}$$

2.2 Bi-Brackets and Quasi-modular Forms

We now define the weight and the depth filtration for the space $\mathcal{Z}_q$ by writing for a subset $A \subseteq \mathcal{Z}_q$

$$\begin{aligned}
\mathrm{Fil}_k^{\mathrm{W}}(A) &:= \Big\langle \begin{bmatrix}s_1,\dots,s_l\\r_1,\dots,r_l\end{bmatrix} \in A \,\Big|\, 0 \le l \le k\,,\ s_1+\dots+s_l+r_1+\dots+r_l \le k \Big\rangle_{\mathbb{Q}} \\
\mathrm{Fil}_l^{\mathrm{D}}(A) &:= \Big\langle \begin{bmatrix}s_1,\dots,s_t\\r_1,\dots,r_t\end{bmatrix} \in A \,\Big|\, t \le l \Big\rangle_{\mathbb{Q}}\,.
\end{aligned}$$

If we consider the depth and weight filtration at the same time we use the short notation $\mathrm{Fil}_{k,l}^{\mathrm{W,D}} := \mathrm{Fil}_k^{\mathrm{W}}\,\mathrm{Fil}_l^{\mathrm{D}}$ and similar for the other filtrations. The associated graded spaces will be denoted by $\mathrm{gr}_k^{\mathrm{W}}$ and $\mathrm{gr}_{k,l}^{\mathrm{W,D}}$

Remark 2 (i) We point to the fact that the filtration by depth coming from bi-brackets is different from the naive notion of depth for the $\zeta_q(s_1,\dots,s_l)$, given as the number of variables s_i. For example, as indicated by (12), the $\zeta_q(1,1,1;t,t,1-t)$ is an element in $\mathrm{Fil}_2^{\mathrm{D}}(\mathcal{Z}_q)$.

(ii) As seen before the Schlesinger-Zudilin model $\zeta_q^{\mathrm{SZ}}(s_1,\dots,s_l)$, defined in (7) for $s_1 \ge 1, s_2,\dots,s_l \ge 0$, span the space $\mathcal{Z}_q$ and therefore we also obtain a depth and weight filtration for these series. By the proof of Theorem 1 we see that $\zeta_q^{\mathrm{SZ}}(s_1,\dots,s_l) \in \mathrm{Fil}_{K,L}^{\mathrm{W,D}}(\mathcal{Z}_q)$ with $K = s_1+\dots+s_l+z$ and $L = l+z$, where $z = \#\{j \mid s_j = 0\}$ is the number of s_j which are zero.

For several reasons one should consider these filtrations to be the natural ones. First of all the multiplication in $\mathcal{Z}_q$ respects the depth as well as the weight grading. Secondly, on $\mathcal{Z}_q$ we have the derivation given by $q\frac{d}{dq}$, which increases the weight by 2 and keeps the depth, since we obtain directly from the definition that

$$q\frac{d}{dq}\begin{bmatrix}s_1,\dots,s_l\\r_1,\dots,r_l\end{bmatrix} = \sum_{j=1}^{l}\left(s_j(r_j+1)\begin{bmatrix}s_1\,,\dots,s_{j-1}\,,s_j+1\,,s_{j+1},\dots,s_l\\r_1\,,\dots,r_{j-1}\,,r_j+1\,,r_{j+1}\,,\dots,r_l\end{bmatrix}\right).$$

Thirdly, the classical Eisenstein series are contained in $\mathcal{Z}_q^\circ \subset \mathcal{Z}_q$. For example we have

$$G_2 = -\frac{1}{24} + [2], \quad G_4 = \frac{1}{1440} + [4], \quad G_6 = -\frac{1}{60480} + [6],$$

since in depth one we have for $k > 0$

$$[k] = \sum_{\substack{u>0\\v>0}} \frac{v^{k-1}}{(k-1)!} q^{uv} = \frac{1}{(k-1)!} \sum_{n>0} \sum_{d|n} d^{k-1} q^n = \frac{1}{(k-1)!} \sum_{n>0} \sigma_{k-1}(n) q^n .$$

The space of quasi-modular forms for $\mathrm{SL}_2(\mathbb{Z})$ with rational coefficients is given by $\widetilde{M}(\mathrm{SL}_2(\mathbb{Z}))_\mathbb{Q} = \mathbb{Q}[G_2, G_4, G_6]$ (see [16]) and therefore it is a sub-algebra of $\mathcal{Z}_q^\circ$ and $\mathcal{Z}_q$. It is graded by the weight, in the classical sense, and obviously $\widetilde{M}_k(\mathrm{SL}_2(\mathbb{Z}))_\mathbb{Q} \subset \mathrm{Fil}_k^{\mathrm{W}}(\mathcal{Z}_q)$. The derivation $q\frac{d}{dq}$ increases the weight by 2, i.e.

$$q\frac{d}{dq} : \widetilde{M}_k(\mathrm{SL}_2(\mathbb{Z}))_\mathbb{Q} \to \widetilde{M}_{k+2}(\mathrm{SL}_2(\mathbb{Z}))_\mathbb{Q}.$$

The space of quasi-modular forms has the decomposition

$$\widetilde{M}_k(\mathrm{SL}_2(\mathbb{Z}))_\mathbb{Q} = \left\langle G_k, q\frac{d}{dq}G_{k-2}, \dots, \left(q\frac{d}{dq}\right)^{k/2-1} G_2 \right\rangle_\mathbb{Q} \oplus \bigoplus_{i=0}^{k/2} \left(q\frac{d}{dq}\right)^i S_{k-2i}(\mathrm{SL}_2(\mathbb{Z}))_\mathbb{Q}, \tag{13}$$

which follows from [16] Proposition 1 together with the fact that $q\frac{d}{dq}$ respects the decomposition $M_k(\mathrm{SL}_2(\mathbb{Z}))_\mathbb{Q} = G_k\mathbb{Q} \oplus S_k(\mathrm{SL}_2(\mathbb{Z}))_\mathbb{Q}$.

Proposition 1 *Set* $\widetilde{\mathsf{M}}(x,t) = 1 + \mathsf{D}(x)\,\mathsf{E}_2(x)\,t + \mathsf{D}(x)\,\mathsf{S}(x)\,t^2$, *then the generating series for the weight- and depth-graded dimensions of* $\widetilde{M}(\mathrm{SL}_2(\mathbb{Z}))_\mathbb{Q} \subset \mathcal{Z}_q$ *satisfies the coefficient-wise inequality*

$$\sum_{k,l\geq 0} \dim_\mathbb{Q} \mathrm{gr}_{k,l}^{\mathrm{W,D}}(\widetilde{M}(\mathrm{SL}_2(\mathbb{Z}))_\mathbb{Q}) x^k t^l \leq \widetilde{\mathsf{M}}(x,t). \tag{14}$$

Proof The Eisenstein series and their derivatives are in the depth one subspaces. For the space of cusp forms of weight k we have

$$S_k(\mathrm{SL}_2(\mathbb{Z}))_\mathbb{Q} \subset \left\langle G_{k-a}G_a \,|\, a = 0, \dots, k/2 \right\rangle_\mathbb{Q} \subset \mathrm{Fil}_{k,2}^{\mathrm{W,D}}(\mathcal{Z}_q).$$

This consequence of a result of Rankin was observed by Zagier in [25, p. 146]. Finally since $q\frac{d}{dq}$ does not alter the depth we get the claim by the decomposition (13).

The expected equality in Proposition 1 would hold if the brackets [2, 4, 6] and the odd brackets [1, 3],.. together with all of their derivatives were algebraically independent, but by now only partial results for linear independence are available [24, 29].

Conjecture 4 *We have a decomposition of $\mathbb{Q}$-algebras*

$$\mathcal{Z}_q \cong \widetilde{M}(\mathrm{SL}_2(\mathbb{Z}))_{\mathbb{Q}} \otimes \mathscr{A}\,.$$

This decomposition is respected by the operator $q\frac{d}{dq}$. Moreover $\mathscr{A}$ is a free polynomial algebra that is bi-graded with respect to weight and depth compatible with those of $\mathcal{Z}_q$. In particular it equals the graded dual of the universal enveloping algebra of a bi-graded Lie algebra.[5]

This decomposition of algebras should be seen as an analogue of [11, Conjecture 1.1. b)] in our context. The conjecture above implies the weaker claim, that the algebra $\mathcal{Z}_q$ is isomorphic to a free polynomial algebra graded by the weight. It also implies that in Proposition 1 the equality holds.

Remark 3 In [17] Okounkov gives the following conjecture for the dimension of the weight-graded parts of $\mathcal{Z}^{\circ}_{q,1}$.

$$\sum_{k\geq 0} \dim_{\mathbb{Q}} \left(\mathrm{gr}^{\mathrm{W}}_k \mathcal{Z}^{\circ}_{q,1}\right) x^k \overset{?}{=} \frac{1}{1-x^2-x^3-x^4-x^5+x^8+x^9+x^{10}+x^{11}+x^{12}}. \tag{15}$$

We expect that the decomposition of Conjecture 4 induces also a decomposition for $\mathcal{Z}^{\circ}_{q,1}$. Indeed, keeping the previous notation, this is compatible with the factorization

$$\frac{1}{1-x^2-\ldots-x^5+x^8+\ldots+x^{12}} = \widetilde{\mathsf{M}}(x,1)\frac{1}{1-\mathsf{D}(x)\,\mathsf{O}_3(x)+2\,\mathsf{D}(x)\,\mathsf{S}(x)}\,.$$

Our Conjecture 3 (i) for $\mathcal{Z}_q$ yields with $\mathsf{E}_4(x) = x^4/(1-x^2)$

$$\frac{1}{1-x-x^2-x^3+x^6+\ldots+x^9} = \widetilde{\mathsf{M}}(x,1)\frac{1}{1-\mathsf{D}(x)\,\mathsf{O}_1(x)+\mathsf{D}(x)\big(\mathsf{E}_4(x)+2\,\mathsf{S}(x)\big)}\,.$$

Thus we may think of the Lie algebra behind $\mathcal{Z}_q$ compared to that behind $\mathcal{Z}^{\circ}_{q,1}$ as having additional generators induced by the derivatives of a generator in weight 1 and having additional relations being counted by the number of Eisenstein series for $\mathrm{SL}_2(\mathbb{Z})$ and their derivatives.

[5] Some authors prefer to denote this as the symmetric algebra of a Lie algebra.

3 Computational Evidences for the Conjectures

In this section we want to describe how to implement the bi-brackets to obtain the numerical results, which were used to obtain Conjecture 3 in the introduction and further conjectures stated below. A similar method to perform such calculations has been communicated to us by Don Zagier.

Using (9) we define for a fixed $N \in \mathbb{N}$ an approximate version of bi-brackets by

$$\begin{bmatrix} s_1, \dots, s_l \\ r_1, \dots, r_l \end{bmatrix}_N := \sum_{N \geq n_1 > \dots > n_l > 0} \prod_{j=1}^{l} \left(\frac{n_j^{r_j}}{r_j!} \cdot \frac{q^{n_j} P_{s_j - 1}(q^{n_j})}{(s_j - 1)! \cdot (1 - q^{n_j})^{s_j}} \right) \in \mathbb{Q}[[q]]. \quad (16)$$

Observe that $\begin{bmatrix} s_1, \dots, s_l \\ r_1, \dots, r_l \end{bmatrix}_N = 0$ for $N < l$. It is clear that at least the first N coefficients of these approximate versions are identical to the bi-brackets, i.e.

$$\begin{bmatrix} s_1, \dots, s_l \\ r_1, \dots, r_l \end{bmatrix}_N \equiv \begin{bmatrix} s_1, \dots, s_l \\ r_1, \dots, r_l \end{bmatrix} \mod q^{N+1}.$$

To calculate the first N coefficients of the bi-brackets we use the following recursive formula for these approximate versions.

Lemma 1 *For all $s_1, \dots, s_l, r_1, \dots, r_l$ and $N \geq l$ we have*

$$\begin{bmatrix} s_1, \dots, s_l \\ r_1, \dots, r_l \end{bmatrix}_N = \begin{bmatrix} s_1, \dots, s_l \\ r_1, \dots, r_l \end{bmatrix}_{N-1} + \frac{N^{r_1}}{r_1!} \frac{q^N P_{s_1 - 1}(q^N)}{(s_1 - 1)! \cdot (1 - q^N)^{s_1}} \begin{bmatrix} s_2, \dots, s_l \\ r_2, \dots, r_l \end{bmatrix}_{N-1},$$

where we set $\begin{bmatrix} s_2, \dots, s_l \\ r_2, \dots, r_l \end{bmatrix}_{N-1} = 1$ for $l = 1$.

Proof This follows by splitting up the summation $N \geq n_1 > \dots > n_l > 0$ into the parts where $N > n_1$ and $N = n_1$ to get the first and the second term respectively.

We implemented an algorithm based on Lemma 1 in parallel PARI/GP [23]. On a computer with 32 cores it takes several hours to obtain each of the following Tables 1 and 2:

In fact for these tables we calculated approximated bi-brackets with coefficients modulo some large prime and determined the dimension they span at least. Experimentally the choice of a sufficiently large prime does not alter these dimensions.[6] We have similar tables for various subspaces like the *positive bi-brackets*

$$\mathcal{Z}_q^+ = \left\langle \begin{bmatrix} s_1, \dots, s_l \\ r_1, \dots, r_l \end{bmatrix} \in \mathcal{Z}_q \;\middle|\; l \geq 0,\ s_1 > r_1, \dots, s_l > r_l \right\rangle_{\mathbb{Q}}$$

[6]More precisely, we checked this for a few primes between k and 10007.

Table 1 lower bounds $\mathrm{fil}^{\mathrm{num}}_{k,l}(\mathcal{Z}_q)$ for $\dim_{\mathbb{Q}} \mathrm{Fil}^{\mathrm{W,D}}_{k,l}(\mathcal{Z}_q)$ with depth ≤ 14

$k\backslash l$	1	2	3	4	5	6	7	8	9	10	11	12	13	14
1	2	0	0	0	0	0	0	0	0	0	0	0	0	0
2	3	4	0	0	0	0	0	0	0	0	0	0	0	0
3	5	7	8	0	0	0	0	0	0	0	0	0	0	0
4	7	12	14	15	0	0	0	0	0	0	0	0	0	0
5	10	19	25	27	28	0	0	0	0	0	0	0	0	0
6	13	30	41	48	50	51	0	0	0	0	0	0	0	0
7	17	44	68	81	89	91	92	0	0	0	0	0	0	0
8	21	65	106	138	153	162	164	165	0	0	0	0	0	0
9	26	90	167	223	264	281	291	293	294	0	0	0	0	0
10	31	126	249	366	439	490	509	520	522	523	0	0	0	0
11	37	167	376	571	738	830	892	913	925	927	928	0	0	0
12	43	222	537	905	1190	1418	1531	1605	1628	1641	1643	1644	0	0
13	50	285	778	1364	1948	2344	2645	2781	2868	2893	2907	2909	2910	0
14	57	368	1075	2090	3051	3923	4453	4840	5001	5102	5129	5144	5146	5147

or the space of 123-*brackets* given by

$$\Big\langle [s_1, \ldots, s_l] \,\Big|\, l \geq 0,\ s_1, \ldots, s_l \in \{1, 2, 3\} \Big\rangle_{\mathbb{Q}} \subset \mathcal{Z}_q^{\circ}$$

and for sub-algebras like $\mathcal{Z}_q^{\circ}$ or $\mathcal{Z}_{q,1}^{\circ}$. This lead us to the following conjectures

Conjecture 5 *(B1) Every bi-bracket equals a linear combination of positive bi-brackets*

(B1) More precisely, the space of positive bi-brackets* $\mathcal{Z}_q^{+}$ *satisfies*

$$\mathrm{Fil}^{\mathrm{W,L}}_{w,l}(\mathcal{Z}_q^{+}) = \mathrm{Fil}^{\mathrm{W,L}}_{w,l}(\mathcal{Z}_q)\,.$$

(B2) Every bi-bracket equals a linear combination of brackets, i.e. $\mathcal{Z}_q^{\circ} = \mathcal{Z}_q$.
(B3) Every bracket equals a linear combination of 123*-brackets.*

Although our experiments support conjectures (B1) and (B3), we were not able to prove the weaker claims that the positive bi-brackets respectively the 123-brackets generate sub-algebras of $\mathcal{Z}_q$. In [1] the conjecture (B2) was stated the first time and therein examples which complement those in [9] (Theorem 5.5) were given.

Theorem 2 *For all weights $k \leq 7$ the coefficients on both sides of Conjecture 3 (i) coincide and the Conjectures 5 (B1), (B2) and (B3) hold for these weights.*

We will give a proof of this theorem at the end of this section.

Table 2 lower bounds $\mathrm{fil}^{\mathrm{num}}_{k,l}(\mathscr{Z}_q)$ for $\dim_{\mathbb{Q}} \mathrm{Fil}^{\mathrm{W,D}}_{k,l}(\mathscr{Z}_q)$ with depth ≤ 4

$l\backslash k$	1	2	3	4	5	6	7	8	9	10	11	12	13	14	15	16	17	18	19	20	21
1	2	3	5	7	10	13	17	21	26	31	37	43	50	57	65	73	82	91	101	111	122
2	0	4	7	12	19	30	44	65	90	126	167	222	285	368	460	577	706	866	1041	1254	1485
3	0	0	8	14	25	41	68	106	167	249	376	537	778	1075	1503	2017	2737	3584	4739	6077	7859
4	0	0	0	15	27	48	81	138	223	366	571	905	1364	2090	3053	4535	6440	9293	?	?	?

The Conjecture 3 is based on the assumption that the above lower bounds were the actual dimensions. In other words, for the quantities

$$\mathrm{gr}_{k,l}^{\mathrm{num}} = \mathrm{fil}_{k,l}^{\mathrm{num}}(\mathscr{Z}_q) - \mathrm{fil}_{k,l-1}^{\mathrm{num}}(\mathscr{Z}_q) - \mathrm{fil}_{k-1,l}^{\mathrm{num}}(\mathscr{Z}_q) + \mathrm{fil}_{k-1,l-1}^{\mathrm{num}}(\mathscr{Z}_q)$$

we expect the equalities $\mathrm{gr}_{k,l}^{\mathrm{num}} = \dim_{\mathbb{Q}} \mathrm{gr}_{k,l}^{\mathrm{W,D}}(\mathscr{Z}_q)$. Now we check if the generating series of the weight- and depth-graded parts of $\mathscr{Z}_q$ can be of the shape implied by the conjectures. For example, if we assume that there is a decomposition $\mathscr{Z}_q \cong \widetilde{M}(\mathrm{SL}_2(\mathbb{Z})) \otimes \mathscr{A}$, where the algebra $\mathscr{A}$ is a free polynomial algebra, then there must hold an equation of the form

$$\sum_{k,l\geq 0} \dim_{\mathbb{Q}}(\mathrm{gr}_{k,l}(\mathscr{Z}_q))\, x^k y^l = \widetilde{\mathsf{M}}(x,y) \cdot \prod_{k,l\geq 1} \frac{1}{(1-x^k y^l)^{g_{k,l}}},$$

where the $g_{k,l}$ equal the number of generators of $\mathscr{A}$ in weight k and depth l. Solving such an equation with with $\mathrm{gr}_{k,l}^{\mathrm{num}}$ on the left-hand side, give us numerical $g_{k,l}^{\mathrm{num}}$ and within the range of our experiments (See Table 3 on page 18) these are positive and satisfy a parity pattern.

If we assume that there is a decomposition $\mathscr{Z}_q \cong \widetilde{M}(\mathrm{SL}_2(\mathbb{Z})) \otimes \mathscr{A}$, where the algebra $\mathscr{A}$ is the graded dual to the universal enveloping algebra of a bi-graded Lie algebra, then there must hold an equation of the form

$$\sum_{k,l\geq 0} \dim_{\mathbb{Q}}(\mathrm{gr}_{k,l}^{\mathrm{W,D}}(\mathscr{Z}_q))\, x^k y^l = \widetilde{\mathsf{M}}(x,y) \cdot \frac{1}{1-\sum_{k,l\geq 1} b_{k,l}\, x^k y^l}$$

with $b_{k,l} \in \mathbb{Z}$. Solving such an equation with with $\mathrm{gr}_{k,l}^{\mathrm{num}}(\mathscr{Z}_q)$ on the left-hand side, give us numerical $b_{k,l}^{\mathrm{num}}$. Within the range of our experiments (See Table 4 on page 18) these are as expected in Conjecture 3 (ii).

Whereas it is known that the numbers from Zagier's conjecture give upper bound for the dimensions in question, the knowledge about the Broadhurst-Kreimer conjecture is very little. The only known results are the following:

Theorem 3 (Euler, Ihara-Kaneko-Zagier, Goncharov, Ihara-Ochiai) *For $1 \leq l \leq 3$ the numbers $g_{k,l}$ of generators for $\mathscr{Z}$ of weight k and depth l are not bigger than implied by the Broadhurst-Kreimer conjecture.*

The proof of this result for $l = 1$ is a trivial consequence of Euler's formula for even zeta values. For $l = 2, 3$ one can bound the number of generators by the dimension of the so called double shuffle spaces, see e.g. [11, 12, 15] .

We now want to use a similar technique to obtain upper bounds of the number of algebra generators for bi-brackets.

Table 3 Evidence for $\mathscr{A}$ being a free polynomial algebra

$\mathrm{gr}^{\mathrm{num}}_{k\backslash l}$	1	2	3	4	5	6	7	8	9	10	11	12	13	14
1	1	0	0	0	0	0	0	0	0	0	0	0	0	0
2	1	1	0	0	0	0	0	0	0	0	0	0	0	0
3	2	1	1	0	0	0	0	0	0	0	0	0	0	0
4	2	3	1	1	0	0	0	0	0	0	0	0	0	0
5	3	4	4	1	1	0	0	0	0	0	0	0	0	0
6	3	8	5	5	1	1	0	0	0	0	0	0	0	0
7	4	10	13	6	6	1	1	0	0	0	0	0	0	0
8	4	17	17	19	7	7	1	1	0	0	0	0	0	0
9	5	20	36	24	26	8	8	1	1	0	0	0	0	0
10	5	31	46	61	32	34	9	9	1	1	0	0	0	0
11	6	35	86	78	94	41	43	10	10	1	1	0	0	0
12	6	49	106	173	118	136	51	53	11	11	1	1	0	0
13	7	56	178	218	299	168	188	62	64	12	12	1	1	0
14	7	76	214	429	377	476	229	251	74	76	13	13	1	1
$g^{\mathrm{num}}_{k\backslash l}$	1	2	3	4	5	6	7	8	9	10	11	12	13	14
1	1	0	0	0	0	0	0	0	0	0	0	0	0	0
2	0	0	0	0	0	0	0	0	0	0	0	0	0	0
3	2	0	0	0	0	0	0	0	0	0	0	0	0	0
4	0	1	0	0	0	0	0	0	0	0	0	0	0	0
5	3	0	1	0	0	0	0	0	0	0	0	0	0	0
6	0	2	0	1	0	0	0	0	0	0	0	0	0	0
7	4	0	3	0	1	0	0	0	0	0	0	0	0	0
8	0	7	0	3	0	1	0	0	0	0	0	0	0	0
9	5	0	8	0	4	0	1	0	0	0	0	0	0	0
10	0	12	0	11	0	4	0	1	0	0	0	0	0	0
11	6	0	22	0	14	0	5	0	1	0	0	0	0	0
12	0	20	0	31	0	17	0	5	0	1	0	0	0	0
13	7	0	47	0	44	0	21	0	6	0	1	0	0	0
14	0	31	0	81	0	58	0	25	0	6	0	1	0	0

Table 4 Evidence for $\mathscr{A}$ being a symmetric algebra of a Lie algebra

l	1	2	3	4	5	6	7	8	9	10	11	12	13	14
$k \leq$	63	31	21	19	15	14	14	14	14	14	14	14	14	14

For the generating function of the bi-brackets we write

$$\begin{vmatrix} X_1, \dots, X_l \\ Y_1, \dots, Y_l \end{vmatrix} := \sum_{\substack{s_1,\dots,s_l>0 \\ r_1,\dots,r_l>0}} \begin{bmatrix} s_1, \dots, s_l \\ r_1-1, \dots, r_l-1 \end{bmatrix} X_1^{s_1-1} \dots X_l^{s_l-1} \cdot Y_1^{r_1-1} \dots Y_l^{r_l-1}.$$

As shown in [1] this satisfies the partition relation

$$\begin{vmatrix} X_1, \dots, X_l \\ Y_1, \dots, Y_l \end{vmatrix} = \begin{vmatrix} X_1, \dots, X_l \\ Y_1, \dots, Y_l \end{vmatrix}\Bigg|_P,$$

with $f(X_1, \dots, X_l, Y_1, \dots, Y_l)\big|_P = f(Y_1 + \dots + Y_l, \dots, Y_1 + Y_2, Y_1, X_l, X_l - X_{l-1}, \dots, X_2 - X_1)$. Up to terms of depth less than l their product is given by

$$\begin{vmatrix} X_1, \dots, X_j \\ Y_1, \dots, Y_j \end{vmatrix} \cdot \begin{vmatrix} X_{j+1}, \dots, X_l \\ Y_{j+1}, \dots, Y_l \end{vmatrix} = \begin{vmatrix} X_1, \dots, X_l \\ Y_1, \dots, Y_l \end{vmatrix}\Bigg|_{\mathrm{Sh}_{j,l}} + \dots,$$

where, if $\Sigma_{j,l} \subset \Sigma_n$ denotes the shuffles of ordered sets with j and $l - j$ elements, we have

$$f(X_1, \dots, X_l, Y_1, \dots, Y_l)|_{\mathrm{Sh}_{j,l}} = \sum_{\sigma \in \Sigma_{j,l}} f(X_{\sigma^{-1}(1)}, \dots, X_{\sigma^{-1}(l)}, Y_{\sigma^{-1}(1)}, \dots, Y_{\sigma^{-1}(l)}).$$

Hence we get modulo products and lower depth bi-brackets

$$\begin{vmatrix} X_1, \dots, X_l \\ Y_1, \dots, Y_l \end{vmatrix} \equiv \sum_{\alpha} \alpha F_\alpha(X_1, \dots, X_l, Y_1, \dots, Y_l),$$

where α runs through a vector space basis of the depth l algebra generators of $\mathscr{Z}_q$ and F_α is a polynomial in the partition shuffle space, which is defined as follows.

Definition 2 Define for $l, k \geq 0$ the partition shuffle space by

$$\mathbb{PS}(k-l, l) = \{f \in \mathbb{Q}[x_1, .., x_l, y_1, .., y_l] |\deg f = k - l,\ f\big|_P - f = f\big|_{\mathrm{Sh}_j} = 0\ \forall j\}.$$

Using the same argument as in [12] the above discussion leads to the following upper bounds.

Corollary 1 *The number $g_{k,l}$ of generators of weight k and depth l for the $\mathbb{Q}$-algebra $\mathscr{Z}_q$ is bounded by*

$$g_{k,l} \leq \dim_{\mathbb{Q}} \mathbb{PS}(k-l, l).$$

The bounds obtained via the partition shuffle spaces for the number of generators in depth 1 and even weights are not optimal, as it is well-known that the ring of quasi-modular forms is generated in weight 2, 4 and 6. We view this as the analogue

Table 5 $p_{k,l} = \dim_{\mathbb{Q}} \mathbb{PS}(k-l, l)$

$p_l \backslash k$	1	2	3	4	5	6	7	8	9	10	11	12	13	14	15	16	17
1	1	1	2	1	3	1	4	0	5	0	6	0	7	0	8	0	9
1	1	1	2	2	3	3	4	4	5	5	6	6	7	7	8	8	9
2	–	0	0	1	0	2	0	8	0	14	0	23	0	38	0	58	0
3	–	–	0	0	1	0	3	0	9	0	27	0	62	0	125	0	238
4	–	–	–	0	0	1	0	3	0	12	0	37	?	?	?	?	?
5	–	–	–	–	0	0	1	0	4	0	15	?	?	?	?	?	?
6	–	–	–	–	–	0	0	1	?	?	?	?	?	?	?	?	?

to the fact that Euler's relation for even zeta values is not seen by the depth 1 double shuffle spaces as defined in [12].

Proof of Theorem 2 Using the structure of the ring of quasi-modular forms and the data of Table 5 we get the coefficient-wise upper bounds

$$\sum_{k\geq 0} \dim_{\mathbb{Q}} \mathrm{Fil}_k^{\mathrm{W}}(\mathscr{Z}_q)x^k \leq \frac{1}{1-x}\frac{1}{(1-x^2)(1-x^4)(1-x^6)}\frac{x}{(1-x^2)^2}\cdot\prod_{k,l\geq 2}\frac{1}{(1-x^k)^{p_{k,l}}}$$
$$\leq 1+2\,x+4\,x^2+8\,x^3+15\,x^4+28\,x^5+51\,x^6+92\,x^7+166\,x^8+\dots$$

In addition, since "123-brackets" $\subseteq \mathscr{Z}_q^{\circ} \subseteq \mathscr{Z}_q$, we get by the data of our tables

$$1+2\,x+4\,x^2+8\,x^3+15\,x^4+28\,x^5+51\,x^6+92\,x^7+165\,x^8+\dots$$
$$\leq \sum_{k\geq 0}\dim_{\mathbb{Q}} \mathrm{Fil}_k^{\mathrm{W}}(\text{"123-brackets"})x^k \leq \sum_{k\geq 0}\dim_{\mathbb{Q}} \mathrm{Fil}_k^{\mathrm{W}}(\mathscr{Z}_q)x^k.$$

The claim of the theorem follows as the lower and upper bounds coincide for $k \leq 7$. □

Remark 4 In contrast to the multiple zeta values we expect that the upper bounds for the number of generators obtained by the partition shuffle spaces are not optimal for all $l \geq 2$, i.e. we don't expect equality in Corollary 1. We think that this reflects the existence of cusp forms as distinguished elements in depth 2, whereas even zeta values just live in depth 1. By work of Ecalle we know that there is a Lie algebra structure on the partition shuffle spaces, see, for e.g., [8] or [19]. In forthcoming work we will study a sub Lie algebra which conjecturally has the algebra $\mathscr{A}$ as its symmetric algebra, which might give another explanation of this effect. Another optimistic hope is that a coproduct structure on $\mathscr{Z}_q$, which allows to mimic Brown's proof in order to obtain conjecture (B3), exists.

Acknowledgements We would like to thank N. Matthes and the referees for their careful reading of our manuscript and their valuable comments. The first author would also like to thank the Max-Planck Institute for Mathematics in Bonn for hospitality and support.

References

1. Bachmann, H.: The algebra of bi-brackets and regularized multiple Eisenstein series. J. Number Theory **200**, 260–294 (2019)
2. Bachmann, H.: Multiple Eisenstein series and q-analogues of multiple zeta values, In this volume
3. Bachmann, H.: Double shuffle relations for q-analogues of multiple zeta values, their derivatives and the connection to multiple Eisenstein series. RIMS Kôyûroku No. **2017**, 22–43 (2015)
4. Bachmann, H., Kühn, U.: The algebra of generating functions for multiple divisor sums and applications to multiple zeta values. Ramanujan J. **40**, 605–648 (2016)
5. Bradley, D.M.: Multiple q-zeta values. J. Algebra **283**, 752–798 (2005)
6. Broadhurst, D., Kreimer, D.: Association of multiple zeta values with positive knots via Feynman diagrams up to 9 loops. Phys. Lett. B **393**, 403–412 (1997)
7. Brown, F.: Mixed Tate motives over $\mathbb{Z}$. Ann. Math. (2) **175**, 949–976 (2012)
8. Ecalle, J.: The flexion structure and dimorphy: flexion units, singulators, generators, and the enumeration of multizeta irreducibles. Asymptotics in dynamics, geometry and PDEs, generalized Borel summation **II**, 27–211 (2011)
9. Ebrahimi-Fard, K., Manchon, D., Singer, J.: Duality and (q-)multiple zeta values. Adv. Math. **298**, 254–285 (2016)
10. Foata, D.: Eulerian polynomials: from Euler's Time to the Present, The legacy of Alladi Ramakrishnan in the mathematical sciences, pp. 253–273. Springer, New York (2010)
11. Goncharov, A.B.: Multiple ζ-values, Galois groups and geometry of modular varieties. Progr. Math. **201**, 361–392 (2001)
12. Ihara, K., Kaneko, M., Zagier, D.: Derivation and double shuffle relations for multiple zeta values. Compositio Math. **142**, 307–338 (2006)
13. Hoffman, M.E.: The algebra of multiple harmonic series. J. Algebra **194**, 477–495 (1997)
14. Hoffman, M.E., Ihara, K.: Quasi-shuffle products revisited. J. Algebra **481**, 293–326 (2017)
15. Ihara, K., Ochiai, H.: Symmetry on linear relations for multiple zeta values. Nagoya Math. J. **189**, 49–62 (2008)
16. Kaneko, M., Zagier, D.: A generalized Jacobi theta function and quasimodular forms, The moduli space of curves. Progr. Math. **129**, 165–172 (1995)
17. Okounkov, A.: Hilbert schemes and multiple q-zeta values. Funct. Anal. Appl. **48**, 138–144 (2014)
18. Schlesinger, K.: Some remarks on q-deformed multiple polylogarithms. arXiv:math/0111022 [math.QA]
19. Schneps, L.: ARI, GARI, Zig and Zag: An introduction to Ecalle's theory of multiple zeta values. arXiv:1507.01534 [math.NT]
20. Singer, J.: On q-analogues of multiple zeta values. Funct. Approx. Comment. Math. **53**, 135–165 (2015)
21. Takeyama, Y.: The algebra of a q-analogue of multiple harmonic series. SIGMA 9 Paper 061, 1–15 (2013)
22. Ohno, Y., Okuda, J., Zudilin, W.: Cyclic q- MZSV sum. J. Number Theory **132**, 144–155 (2012)
23. The PARI Group, PARI/GP version `2.10.0`, Univ. Bordeaux (2017). http://pari.math.u-bordeaux.fr/
24. Pupyrev, Y.: On the linear and algebraic independence of q-zeta values, (Russian. Russian summary) Mat. Zametki **78**(4), 608–613 (2005); translation in Math. Notes **78**(3–4), 563–568 (2005)
25. Zagier, D.: Modular forms whose Fourier coefficients involve zeta-functions of quadratic fields. Modular functions of one variable VI, Lecture Notes in Math. 627, Springer, Berlin, 105–169 (1977)
26. Zagier, D.: Values of zeta functions and their applications. First European Congress of Mathematics, Volume II, Progress in Math. 120, Birkhäuser-Verlag, Basel, 497–512 (1994)

27. Zhao, J.: Multiple q-zeta functions and multiple q-polylogarithms. Ramanujan J. **14**(2), 189–221 (2007)
28. Zhao, J.: Uniform approach to double shuffle and duality relations of various q-analogs of multiple zeta values via Rota-Baxter algebras. arXiv:1412.8044 [math.NT]
29. Zudilin, W.: Diophantine problems for q-zeta values, (Russian) Mat. Zametki **72**(6), 936–940 (2002); translation in Math. Notes **72**, 858–862 (2002)
30. Zudilin, W.: Algebraic relations for multiple zeta values. Russian Math. Surveys **58**(1), 1–29 (2003)
31. Zudilin, W.: Multiple q-zeta brackets, Mathematics 3:1, special issue Mathematical physics, 119–130 (2015)

Uniform Approach to Double Shuffle and Duality Relations of Various q-Analogs of Multiple Zeta Values via Rota–Baxter Algebras

Jianqiang Zhao

Abstract The multiple zeta values (MZVs) have been studied extensively in recent years. Currently there exist a few different types of q-analogs of the MZVs (q-MZVs) defined and studied by mathematicians and physicists. In this paper, we give a uniform treatment of these q-MZVs by considering their double shuffle relations (DBSFs) and duality relations. The main idea is a modification and generalization of the one used by Castillo Medina et al. to a few other types of q-MZVs including the one defined by the author in 2003. With different approach, Takeyama already studied this type by "regularization" and observed that there exist a new family of $\mathbb{Q}$-linear relations which are not consequences of the DBSFs. We call these duality relations in this paper and generalize them to all other types of q-MZVs. Since there are still some missing relations we further define the most general type of q-MZVs together with a new kind of relations called **P**-**R** relations which are used to lower the deficiencies further. As an application, we will confirm a conjecture of Okounkov on the dimensions of certain q-MZV spaces, either theoretically or numerically, for the weight up to 12. Some relevant numerical data are provided at the end.

Keywords Multiple zeta values · q-analog of multiple zeta values · Double shuffle relations · Duality relations · Rota-Baxter algebras

1 Introduction

The multiple zeta values are iterated generalizations of the Riemann zeta values to the multiple variable setting. Euler [8] first studied the double zeta values in the 18th century. Hoffman [13] and Zagier [28] independently considered systematically the following more general form in the early 1990's. Let $\mathbb{N}$ be the set of positive integers. For any $d \in \mathbb{N}$ and $\mathbf{s} = (s_1, \ldots, s_d) \in \mathbb{N}^d$ with $s_1 \geq 2$ one defines the *multiple zeta values* (MZVs) as the d-fold sum

J. Zhao (✉)
Department of Mathematics, The Bishop's School, La Jolla, CA 92037, USA
e-mail: zhaoj@ihes.fr

J. I. Burgos Gil et al. (eds.), *Periods in Quantum Field Theory and Arithmetic*, Springer Proceedings in Mathematics & Statistics 314,
https://doi.org/10.1007/978-3-030-37031-2_10

$$\zeta(\mathbf{s}) = \sum_{k_1>\cdots>k_d>0} \frac{1}{k_1^{s_1}\cdots k_d^{s_d}}.$$

In 1980s, Ecalle studied some quite general mathematical objects called "moulds" (functions with variable number of variables) of which the MZVs are one of the examples [9, p. 429]. He even mentioned their iterated integral representation [9, p. 431, Remark 4] without explicitly producing it.

A lot of important and sometimes surprising applications of MZVs have been found in many areas in mathematics and theoretical physics in recent years; see [4, 5, 10, 18, 19]. One of the most powerful ideas is to consider the so-called double shuffle relations (DBSFs). The stuffle relations are obtained directly by using the above series definition when multiplying two MZVs. The other, the shuffle relations, can be produced by multiplying their integral representations and using Chen's theory of iterated integrals [6]. The interested reader is referred to the seminal paper [16] for more details.

Lagging behind the above development for about a decade, a few q-analogs were proposed and studied by different mathematicians and physicists. All of these q-analogs enjoy the property that when $q \to 1$ one can recover the ordinary MZVs defined in the above if no divergence occurs. In this paper, by modifying and generalizing an idea in [7] we give a uniform treatment of these q-analogs by using some suitable Rota–Baxter algebras which reflect the properties of Jackson's iterated integral representations of these q-analogs.

Recall that for any fixed q with $0 < q < 1$ one can define the q-analog of positive integers by setting $[k] = [k]_q := 1 + q + \cdots + q^{k-1} = (1-q^k)/(1-q)$ for all $k \in \mathbb{N}$. To summarize the various versions of q-analog of MZVs (q-MZVs for abbreviation), we first define a general type of q-MZV of $2d$ variables $s_1, \ldots, s_d, a_1, \ldots, a_d \in \mathbb{Z}$ by

$$\zeta_q^{\mathbf{a}}[\mathbf{s}] := \sum_{k_1>\cdots>k_d>0} \frac{q^{k_1a_1+\cdots+k_da_d}}{[k_1]^{s_1}\cdots[k_d]^{s_d}} = (1-q)^{|\mathbf{s}|} \sum_{k_1>\cdots>k_d>0} \frac{q^{k_1a_1+\cdots+k_da_d}}{(1-q^{k_1})^{s_1}\cdots(1-q^{k_d})^{s_d}}, \tag{1}$$

where $|\mathbf{s}| = s_1 + \cdots + s_d$ is called the *weight* and d the *depth*. The variables of $\mathbf{a}$ are called *auxiliary variables*. Also, it is often convenient to study its modified form by dropping the power of $1-q$, i.e.,

$$\mathfrak{z}_q^{\mathbf{a}}[\mathbf{s}] := \sum_{k_1>\cdots>k_d>0} \frac{q^{k_1a_1+\cdots+k_da_d}}{(1-q^{k_1})^{s_1}\cdots(1-q^{k_d})^{s_d}},$$

In the following table, we list a few different versions of q-MZVs that have been studied so far by different authors, except for one new type (type IV in the table). We only write down their modified form although sometimes the original authors considered only ζ_q. We note that in 2004, Bradley [3] apparently defined $\zeta_q^{(s_1-1,\ldots,s_d-1)}[s_1,\ldots,s_d]$ independently, and later, Okuda and Takeyama also studied some of the relations among this type of q-MZVs in [22]. Additionally, it is not hard

Table 1 A time line of different versions of q-MZVs. $\star$=this paper

Type	Year	Authors	q-MZV	DBSF
	2001	Schlesinger [23]	$\mathfrak{z}_q^{(0,\dots,0)}[s_1,\dots,s_d]$	See (2)
I	2002	Kaneko et al. [17]	$\mathfrak{z}_q^{(s-1)}[s]$ (depth=1)	N/A
I	2003	Zhao [29]	$\mathfrak{z}_q^{(s_1-1,\dots,s_d-1)}[s_1,\dots,s_d]$	[25], [26], $\star$
II	2003	Zudilin [30]	$\mathfrak{z}_q^{(s_1,\dots,s_d)}[s_1,\dots,s_d]$	[24], $\star$
III	2012	Ohno et al. [20]	$\mathfrak{z}_q^{(1,0,\dots,0)}[s_1,\dots,s_d]$	[7], $\star$
IV	2014	Zhao $\star$	$\mathfrak{z}_q^{(s_1-1,s_2,\dots,s_d)}[s_1,\dots,s_d]$	$\star$
BK	2013	Bachmann & Kühn [1]	$\mathfrak{z}_q^{\mathrm{BK}}[s_1,\dots,s_d]$	[31]
O	2014	Okounkov [21]	$\mathfrak{z}_q^{\mathrm{O}}[s_1,\dots,s_d]$, $s_j\geq 2$	$\star$
G	2003	Zhao [29]	$\mathfrak{z}_q^{(a_1,\dots,a_d)}[s_1,\dots,s_d]$	$\star$

to see that Schlesinger's version diverges when $|q|<1$ but can converge if $|q|>1$. In fact, for $\mathbf{s}\in\mathbb{Z}^d$

$$\mathfrak{z}_{q^{-1}}^{(0,\dots,0)}[s_1,\dots,s_d]=(-1)^{s_1+\dots+s_d}\mathfrak{z}_q^{(s_1,\dots,s_d)}[s_1,\dots,s_d]=(-1)^{s_1+\dots+s_d}\mathfrak{z}_q^{\mathrm{II}}[s_1,\dots,s_d]. \tag{2}$$

So it suffices to consider type II in order to understand Schlesinger's q-MZVs. The last column of Table 1 provides the references where DBSFs are considered systematically (not only the stuffle), some of which are apparently different from our approach in this paper.

In this paper, we will use suitable Rota–Baxter algebras to study types I-IV q-MZVs listed in Table 1 in details. We also briefly consider the general type G and Okounkov's type O q-MZVs. Note that the numerators inside the summands of ζ_q^{BK} and ζ_q^{O} are not exact powers of q, but some polynomials of q enjoying nice properties. Further, for ζ_q^{O} the polynomial numerator is at worst a sum of two q-powers so our method can still work. See Corollary 6.6. It may be difficult to use the approach here to study the Bachmann and Kühn type since the numerators are much more complicated.

In the classical setting, the so-called regularized DBSFs play extremely important roles in discovering and proving $\mathbb{Q}$-linear relations among the MZVs. The first serious attempt to discover the DBSFs among q-MZVs was carried out by the author in [29] by using Jackson's iterated q-integrals. However, the computation there was too complicated so only very few relations were found successfully. The real breakthrough came with Takeyama's successful application of Hoffman's algebras to study type I q-MZVs in [26]. However, his approach to the shuffle relations relies on some auxiliary multiple polylogarithm functions and consequently it is very hard to see why these relations should hold.

The situation looks much better with the appearance of the recent paper [7] by Castillo Medina, Ebrahimi-Fard and Manchon who generalized Chen's iterated integrals to Jackson's iterated q-integrals to study type III q-MZVs by using Rota–Baxter algebra techniques. Later, Singer [24, 25] adopted the algebraic setup of the DBSFs to study type I and type II q-MZVs. Motivated by this new idea, in this paper we will consider all the q-MZVs of type I, II, III and IV by finding/using their correct realizations in terms of Jackson's iterated q-integrals. Then by combining the Rota–Baxter algebra technique and Hoffman's algebra of words we are able to study the DBSFs of all of these q-MZVs.

When one considers the $\mathbb{Q}$-linear relations among the ordinary MZVs, the main difficulty lies in the insufficiency of DBSFs produced by only admissible arguments. In the q-analog setting, the situation is only partially similar and sometimes much more complicated.

For type I q-MZVs, our computation shows that the DBSFs CAN provide all the $\mathbb{Q}$-linear relations. However, in order to study these relations, as Takeyama noticed first, one has to enlarge the set of type I q-MZVs to something we call type $\tilde{\mathrm{I}}$ q-MZVs which are a kind of "regularized" q-MZVs in the sense that one needs to consider some convergent versions of q-MZVs when $s_1 = 1$ by modifying the auxiliary variables of $\mathbf{a}$. But for these type $\tilde{\mathrm{I}}$ q-MZVs themselves, DBSFs are insufficient to provide all the $\mathbb{Q}$-linear relations and a certain "Resummation Identity" defined by Takeyama is required. In this paper, we will adopt the term "duality" due to its similarity to the duality relations of the ordinary MZVs. Moreover, for type $\tilde{\mathrm{I}}$ q-MZVs of weight bounded by w there are often still missing relations even after we consider both DBSFs and duality relations within the same weight and depth range. These missing relations can be recovered only after we increase the weight and depth. This phenomenon is not unique to type $\tilde{\mathrm{I}}$ q-MZVs. We have recorded this fact by using the "deficiency" numbers listed in the tables in the last section of this paper.

Similar to type I, we find that type IV q-MZVs also need to be "regularized" when $s_1 = 1$. Again, we achieve this by introducing some convergent versions of the q-MZVs by modifying the auxiliary variables in $\mathbf{a}$.

It turns out that type II q-MZVs behave the most regularly and enjoy some properties closest to those of the classical MZVs. For example, their duality relations (see Theorem 8.4) have the cleanest form. Moreover, every other type of q-MZVs considered in this paper can be converted to type II. But still, there are relations that cannot be proved by DBSFs and dualities, at least when one is confined within the same weight and depth range. In fact, we find three independent $\mathbb{Q}$-linear relations in weight 4 that can only be proved when we consider weight 5 DBSFs and dualities.

All type III q-MZVs are convergent, even for negative arguments. For simplicity, in this paper we consider only those with nonnegative arguments $s_1, \dots, s_d$ with $s_1 \geq 1$. In this case, the DBSFs are still insufficient. In the last section, we will see that in weight 3 there is already a missing relation which can be recovered by the duality. Essentially because of the need to apply the duality relations, we have to modify the original Jackson's iterated integral representation given in [7]. See the remarks after Theorem 6.1. In contrast to the other types of q-MZVs, we cannot suppress the deficiency for type III even if we consider more DBSFs and duality relations by increasing the weight and depth. This might be caused by our restriction of only nonnegative arguments and thus further investigations are called for.

On the other hand, we can improve the above situation by considering the more general type G values. All the missing relations are thus proved up to and including weight 4 and at the same time both deficiencies are decreased in weight 5 and 6. The key idea here is to convert all type G values to type II values by using a new kind of relations called **P-R** relations.

We point out that our method can be easily adapted to study q-MZVs of the following general forms:

$$\mathfrak{z}_q^{(s_1-a_1,\dots,s_d-a_d)}[s_1,\dots,s_d], \qquad \mathfrak{z}_q^{(a_1,\dots,a_d)}[s_1,\dots,s_d],$$

where $a_1 \geq a_2 \geq \cdots \geq a_d \geq 0$ are all fixed integers. Furthermore, when the weight is not too large, our method can be programmed to compute all the relations among q-MZVs of the general form $\mathfrak{z}_q^{\mathbf{a}}[\mathbf{s}]$ when $\mathbf{a}$ is taken within a certain range. This will be carried out in Sect. 9.

As an application, for small weight cases it is possible to confirm Okounkov's conjecture [21] on the dimension of the q-MZVs $\mathfrak{z}_q^{\mathrm{O}}[\mathbf{s}]$ using Corollary 6.6. We do this numerically up to weight 12 and give rigorous proof up to weight 6 (both inclusive).

Throughout the paper we will use the modified form $\mathfrak{z}_q$ most of the time. All the results can be translated into the standard form ζ_q by inserting the correct powers of $(1-q)^w$, where w is the corresponding weight, into the formulas.

2 Convergence Domain for q-MZVs

We need the following result to find the convergence domain for different types of q-MZVs. It is Proposition 2.2 of [29] where the order of the indices in the definition of $\zeta_q^{(a_1,\dots,a_d)}[s_1,\dots,s_d]$ (denoted by $f_q(s_d,\dots,s_1;a_d,\dots,a_1)$ in loc. cit.) is opposite to this paper.

Proposition 2.1 *The function $\zeta_q^{(a_1,\dots,a_d)}[s_1,\dots,s_d]$ converges if $\mathrm{Re}(a_1+\cdots+a_j) > 0$ for all $j=1,\dots,d$. It can be analytically continued to a meromorphic function over $\mathbb{C}^{2d}$ via the series expansion*

$$\zeta_q^{(a_1,\dots,a_d)}[s_1,\dots,s_d] = (1-q)^{|\mathbf{s}|} \sum_{b_1,\dots,b_d=0}^{\infty} \prod_{j=1}^{d} \left[\binom{s_j+b_j-1}{b_j} \frac{q^{(d+1-j)(b_j+a_j)}}{1-q^{b_1+a_1+\cdots+b_j+a_j}} \right]. \tag{3}$$

It has the following (simple) poles: $a_1+\cdots+a_j \in \mathbb{Z}_{\leqslant 0} + \frac{2\pi i}{\log q}\mathbb{Z}$ for $j=1,\dots,d$.

Corollary 2.2 *Let $\mathbf{s}=(s_1,\dots,s_d) \in \mathbb{Z}^d$. Then*

(i) *$\zeta_q^{\mathrm{I}}[\mathbf{s}]$ converges if $s_1+\cdots+s_j > j$ for all $j=1,\dots,d$.*
(ii) *$\zeta_q^{\mathrm{II}}[\mathbf{s}]$ converges if $s_1+\cdots+s_j > 0$ for all $j=1,\dots,d$.*
(iii) *$\zeta_q^{\mathrm{III}}[\mathbf{s}]$ always converges.*
(iv) *$\zeta_q^{\mathrm{IV}}[\mathbf{s}]$ converges if $s_1+\cdots+s_j > 1$ for all $j=1,\dots,d$.*

Definition 2.3 For convenience, a composition $\mathbf{s} \in \mathbb{Z}_{\geq 0}$ is said to be *type τ-admissible* if $\mathbf{s}$ satisfies the condition for type τ q-MZVs in the corollary. Here and in what follows, $\tau = \mathrm{I}, \mathrm{II}, \mathrm{III}$, or IV.

3 Rota–Baxter Algebra

In this section we briefly review some fundamental facts of Rota–Baxter algebras which will be crucial in the study of the q-analog of *shuffle* relations for all of q-MZVs considered in this paper. For a good introduction to the Rota–Baxter algebras, see [11].

Definition 3.1 Fix an algebra A over a commutative ring R and an element $\lambda \in R$. We call A a Rota–Baxter R-algebra and $\mathscr{P}$ a Rota–Baxter operator of weight λ if the operator $\mathscr{P}$ satisfies the following Rota–Baxter relation of weight λ:

$$\mathscr{P}(x)\mathscr{P}(y) = \mathscr{P}(\mathscr{P}(x)y) + \mathscr{P}(x\mathscr{P}(y)) + \lambda\mathscr{P}(xy) \quad \forall x, y \in A. \tag{4}$$

Recall that for any continuous function $f(x)$ on $[\alpha, \beta]$, Jackson's q-integral is defined by

$$\int_\alpha^\beta f(x)\, d_q x := \sum_{k\geq 0} f\big(\alpha + q^k(\beta - \alpha)\big)(q^k - q^{k+1})(\beta - \alpha). \tag{5}$$

Taking $\alpha = 0$ and $\beta = t$ in (5), we set

$$\mathbf{J}[f](t) := (1-q)\sum_{k\geq 0} f(q^k t)q^k t = (1-q)\sum_{k\geq 0} \mathbf{E}^k\big[\mathbf{M}[f]\big](t) = (1-q)\mathbf{P}\big[\mathbf{M}[f]\big](t) \tag{6}$$

where the multiplication operator $\mathbf{M}[f](t) := tf(t)$,

$$\mathbf{E}[f](t) := \mathbf{E}_q[f](t) := f(qt), \text{ and } \mathbf{P}[f](t) := \mathbf{P}_q[f](t) := f(t) + f(qt) + f(q^2 t) + \cdots$$

are the *q-expanding* and the (*principle*) *q-summation operators*, respectively. We also need to define the (*remainder*) *q-summation operator*

$$\mathbf{R}[f](t) := \mathbf{R}_q[f](t) := f(qt) + f(q^2 t) + \cdots = (\mathbf{P}[f] - f)(t).$$

So, $\mathbf{P}$ is the principle part (i.e. the whole thing) while $\mathbf{R}$ is the remainder (i.e., without the first term). Clearly, $\mathbf{P} = \mathbf{R} + \mathbf{I}$ where, as an operator, $\mathbf{I}[f] = f$. This implies $\mathbf{PR} = \mathbf{RP}$.

Let $t\mathbb{Q}[\![t,q]\!]$ be the ring of formal series in two variables with $t>0$. Then **J**, **E**, **P** and **R** are all $\mathbb{Q}[\![q]\!]$-linear endomorphisms of $t\mathbb{Q}[\![t,q]\!]$. We can further define the inverse to **P** which is called the *q-difference operator*:

$$\mathbf{D} := \mathbf{I} - \mathbf{E}. \tag{7}$$

The following results extend those of [7, (21)–(23)]. In the final computation we will not need **D** since we will consider only nonnegative arguments in all the q-MZVs. But in the theoretical part of this paper we do need to use **D** for type Ш q-MZVs.

Proposition 3.2 *For any $f, g \in t\mathbb{Q}[\![t,q]\!]$ we have*

$$\mathbf{P}[f]\mathbf{P}[g] = \mathbf{P}\big[\mathbf{P}[f]g\big] + \mathbf{P}\big[f\mathbf{P}[g]\big] - \mathbf{P}[fg], \tag{8}$$

$$\mathbf{R}[f]\mathbf{R}[g] = \mathbf{R}\big[\mathbf{R}[f]g\big] + \mathbf{R}\big[f\mathbf{R}[g]\big] + \mathbf{R}[fg], \tag{9}$$

$$\mathbf{R}[f]\mathbf{P}[g] = \mathbf{R}\big[\mathbf{R}[f]g\big] + \mathbf{R}\big[f\mathbf{R}[g]\big] + \mathbf{R}[f]g + \mathbf{R}[fg], \tag{10}$$

$$\mathbf{J}[f]\mathbf{J}[g] = \mathbf{J}\big[\mathbf{J}[f]g\big] + \mathbf{J}\big[f\mathbf{J}[g]\big] - (1-q)\mathbf{J}\big[\mathbf{I}fg\big], \tag{11}$$

$$= \mathbf{J}\big[f\mathbf{J}[g]\big] + q\mathbf{J}\big[\mathbf{J}\big[E[f]\big]g\big], \tag{12}$$

$$\mathbf{D}[f]\mathbf{D}[g] = \mathbf{D}[f]g + f\mathbf{D}[g] - \mathbf{D}[fg], \tag{13}$$

$$\mathbf{D}[f]\mathbf{P}[g] = \mathbf{D}\big[f\mathbf{P}[g]\big] + \mathbf{D}[f]g - fg, \tag{14}$$

$$\mathbf{D}[f]\mathbf{R}[g] = \mathbf{D}\big[f\mathbf{R}[g]\big] + \mathbf{D}[fg] - fg, \tag{15}$$

$$\mathbf{DP} = \mathbf{PD} = \mathbf{I}, \qquad \mathbf{PR} = \mathbf{RP}. \tag{16}$$

Proof The identities (8), (13) and (14) are just (21), (23) and (26) of [7], respectively. All the others follow from $\mathbf{R} = \mathbf{P} - \mathbf{I}$ easily.

By Proposition 3.2 we see that **P** and **R** are both Rota–Baxter operators on $t\mathbb{Q}[\![t,q]\!]$ (of weight -1 and 1, respectively) but **D** is not. In fact, **D** satisfies the condition (13) of a differential Rota–Baxter operator [12]. Moreover, it is *invertible* in the sense that Rota–Baxter operator **P** and the differential **D** are mutually inverse by (16).

We end this section with an identity which will be used to interpret Takeyama's Resummation Identity in [26]. For any $n \in \mathbb{N}$, set

$$\mathbf{P}^n = \underbrace{\mathbf{P} \circ \cdots \circ \mathbf{P}}_{n \text{ times}} \quad \text{and} \quad \mathbf{R}^n = \underbrace{\mathbf{R} \circ \cdots \circ \mathbf{R}}_{n \text{ times}}.$$

Theorem 3.3 *Let $d \in \mathbb{N}$ and $\alpha_j, \beta_j \in \mathbb{N}$ for all $j = 1, \dots, d$. Let $\mathbf{y}(t) = \frac{t}{1-t}$. Then we have*

$$\mathbf{R}^{\alpha_1}\mathbf{y}^{\beta_1}\cdots\mathbf{R}^{\alpha_\ell}\mathbf{y}^{\beta_\ell}(t) = \sum_{\substack{j_1\ge\beta_1,\dots,j_\ell\ge\beta_\ell\\ k_1\ge\alpha_1,\dots,k_\ell\ge\alpha_\ell}} \prod_{r=1}^{\ell}\left[\binom{j_r-1}{\beta_r-1}\binom{k_r-1}{\alpha_r-1} q^{k_r\sum_{s=r}^{\ell} j_s} t^{j_r}\right]. \tag{17}$$

Proof First we show that

$$\mathbf{R}^{\alpha}(t^j) = \frac{q^{\alpha j} t^j}{(1-q^j)^{\alpha}}. \tag{18}$$

Indeed, if $\alpha = 1$ then

$$\mathbf{R}(t^j) = \sum_{k\geq 1} q^{kj} t^j = \frac{q^j t^j}{1-q^j}.$$

So (18) can be proved easily by induction.

Now we proceed to prove that for any integer $m \geq 0$

$$\mathbf{R}^{\alpha_1}\mathbf{y}^{\beta_1}\cdots\mathbf{R}^{\alpha_\ell}\Big(\mathbf{y}^{\beta_\ell}(t)\cdot t^m\Big) = \sum_{\substack{j_1\geq\beta_1,\dots,j_\ell\geq\beta_\ell\\ k_1\geq\alpha_1,\dots,k_\ell\geq\alpha_\ell}} t^m \prod_{r=1}^{\ell}\left[\binom{j_r-1}{\beta_r-1}\binom{k_r-1}{\alpha_r-1} q^{k_r(m+\sum_{s=r}^{\ell} j_s)} t^{j_r}\right]. \tag{19}$$

If $\ell = 1$ then we have

$$\begin{aligned}
\mathbf{R}^{\alpha}\Big(\mathbf{y}^{\beta}(t)\cdot t^m\Big) &= \mathbf{R}^{\alpha}\left(\left(\frac{t}{1-t}\right)^{\beta} t^m\right) = \mathbf{R}^{\alpha}\sum_{j\geq 0}\binom{\beta+j-1}{j} t^{m+\beta+j}\\
&= \mathbf{R}^{\alpha}\sum_{j\geq\beta}\binom{j-1}{\beta-1} t^{m+j}\\
&= \sum_{j\geq\beta}\binom{j-1}{\beta-1}\frac{q^{\alpha(m+j)}t^{m+j}}{(1-q^{m+j})^{\alpha}} \qquad \text{(by (18))}\\
&= \sum_{j\geq\beta}\binom{j-1}{\beta-1}\sum_{k\geq 0}\binom{\alpha+k-1}{k} q^{(\alpha+k)(m+j)} t^{m+j}\\
&= \sum_{j\geq\beta}\sum_{k\geq\alpha}\binom{j-1}{\beta-1}\binom{k-1}{\alpha-1} q^{k(m+j)} t^{m+j}.
\end{aligned}$$

This proves (19) when $\ell = 1$. In general

$$\begin{aligned}
&\mathbf{R}^{\alpha_1}\mathbf{y}^{\beta_1}\cdots\mathbf{R}^{\alpha_{\ell-1}}\Big(\mathbf{y}^{\beta_{\ell-1}}(t)\Big(\mathbf{R}^{\alpha_\ell}\Big(\mathbf{y}^{\beta_\ell}(t)\cdot t^m\Big)\Big)\Big)\\
&= \sum_{j_\ell\geq\beta_\ell}\sum_{k_\ell\geq\alpha_\ell}\binom{j_\ell-1}{\beta_\ell-1}\binom{k_\ell-1}{\alpha_\ell-1} q^{k_\ell(m+j_\ell)}\mathbf{R}^{\alpha_1}\mathbf{y}^{\beta_1}\cdots\mathbf{R}^{\alpha_{\ell-1}}\Big(\mathbf{y}^{\beta_{\ell-1}}(t)\cdot t^{m+j_\ell}\Big).
\end{aligned}$$

So (19) follows immediately by induction. We can now finish the proof of the theorem by taking $m = 0$.

Corollary 3.4 *Let $d \in \mathbb{N}$ and $\alpha_j, \beta_j \in \mathbb{N}$ for all $j = 1, \dots, d$. Then we have*

$$\mathbf{R}^{\alpha_1}\mathbf{y}^{\beta_1}\cdots\mathbf{R}^{\alpha_\ell}\mathbf{y}^{\beta_\ell}(1) = \mathbf{R}^{\beta_\ell}\mathbf{y}^{\alpha_\ell}\cdots\mathbf{R}^{\beta_1}\mathbf{y}^{\alpha_1}(1). \tag{20}$$

Proof In (17) we use the substitutions $j_r \leftrightarrow k_{\ell+1-r}$ for all $r = 1, \dots, \ell$. Then the power of q in the term of (17) indexed by $(j_1, ..., j_\ell, k_1, ..., k_\ell)$ is equal to

$$\sum_{r=1}^{\ell}\sum_{s=r}^{\ell} j_s k_r \longrightarrow \sum_{r=1}^{\ell}\sum_{s=r}^{\ell} j_{\ell+1-r}k_{\ell+1-s} = \sum_{s=1}^{\ell}\sum_{r=1}^{s} j_{\ell+1-r}k_{\ell+1-s}$$
$$= \sum_{s=1}^{\ell}\sum_{r=1}^{\ell+1-s} j_{\ell+1-r}k_s = \sum_{s=1}^{\ell}\sum_{r=s}^{\ell} j_r k_s = \sum_{r=1}^{\ell} k_r \sum_{s=r}^{\ell} j_s.$$

which follows from the substitution $s \leftrightarrow \ell + 1 - s$ followed by $r \leftrightarrow \ell + 1 - r$ and $r \leftrightarrow s$. This proves the corollary.

4 q-Analogs of Hoffman algebras

We know that (regularized) DBSFs lead to many (and conjecturally all) $\mathbb{Q}$-linear relations among the MZVs. The key idea here was first suggested by Hoffman [14] who used some suitable algebra of words to codify both the stuffle (also called harmonic shuffle [27] or quasi-shuffle [15]) relations coming from the series representation of MZVs and the shuffle relations coming from the iterated integral expressions of MZVs. The detailed regularization process can be found in [16]. To study similar relations of the q-MZVs we should modify the Hoffman algebras in the q-analog setting.

The following definition for type I q-MZVs was first proposed by Takeyama [26]. We adopt different notations here in hoping to give a uniform and more transparent presentation for all types of q-MZVs.

First we consider some algebras which will be used to define the stuffle relations later.

Definition 4.1 Let X_θ^* be the set of words on the alphabet $X_\theta = \{a, a^{-1}, b, \theta\}$. Denote by $\mathfrak{A}_\theta = \mathbb{Q}\langle a, a^{-1}, b, \theta\rangle$ the noncommutative polynomial $\mathbb{Q}$-algebra of words from X_θ^*. Set

$$\gamma := b - \theta, \qquad z_s := a^{s-1}b, \quad z'_s := a^{s-1}\theta, \quad s \in \mathbb{Z}.$$

Let $Y_{\tilde{\mathrm{I}}} := \{\theta\} \cup \{z_k\}_{k\ge 1}$, $Y_{\mathrm{II}} := \{z'_k\}_{k\ge 0}$, $Y_{\mathrm{III}} := \{z_k\}_{k\in\mathbb{Z}}$ and $Y_{\widetilde{\mathrm{IV}}} := Y_{\mathrm{II}}$. We point out that $z_0, z'_0 \neq \mathbf{1}$ where $\mathbf{1}$ is the empty word. We put a tilde on top of both I and IV since we need to consider some kind of regularization due to convergence issues involved in type I and IV q-MZVs. This is realized by the introduction of the letter θ. Again, we use Y_τ^* to denote the set of words generated on Y_τ for any type τ.

Let $\mathfrak{A}^1_{\tilde{\mathrm{I}}}$, $\mathfrak{A}^1_{\mathrm{II}}$, $\mathfrak{A}_{\mathrm{III}}$ and $\mathfrak{A}_{\widetilde{\mathrm{IV}}}$ be the subalgebras of $\mathfrak{A}_\theta$ freely generated by the sets $Y_{\tilde{\mathrm{I}}}$, Y_{II}, Y_{III} and $Y_{\widetilde{\mathrm{IV}}}$, respectively. Set

$$\mathfrak{A}^1_{\mathrm{III}} := \mathbb{Q}\mathbf{1} + \sum_{k\in\mathbb{Z}} z'_k \mathfrak{A}_{\mathrm{III}} \not\subset \mathfrak{A}_{\mathrm{III}}, \qquad \mathfrak{A}^1_{\widetilde{\mathrm{IV}}} := \mathbb{Q}\mathbf{1} + \theta\mathfrak{A}_{\widetilde{\mathrm{IV}}} + \sum_{k\geq 1} z_k \mathfrak{A}_{\widetilde{\mathrm{IV}}} \not\subset \mathfrak{A}_{\widetilde{\mathrm{IV}}}.$$

Here, all integer subscripts are allowed in Y_{III} because type III q-MZVs converge for all integer arguments. Further, we define the following subalgebras corresponding to the convergent values:

$$\begin{aligned}
\mathfrak{A}^0_{\mathrm{I}} &:= \mathbb{Q}\mathbf{1} + \sum_{k\geq 2} z_k \mathfrak{A}^1_{\mathrm{I}}, & \mathfrak{A}^0_{\tilde{\mathrm{I}}} &:= \mathbb{Q}\mathbf{1} + \theta\mathfrak{A}^1_{\tilde{\mathrm{I}}} + \sum_{k\geq 2} z_k \mathfrak{A}^1_{\tilde{\mathrm{I}}} \subsetneq \mathfrak{A}^1_{\tilde{\mathrm{I}}}, \\
\mathfrak{A}^0_{\mathrm{II}} &:= \mathbb{Q}\mathbf{1} + \sum_{k\geq 1} z'_k \mathfrak{A}^1_{\mathrm{II}} \subsetneq \mathfrak{A}^1_{\mathrm{II}}, & \mathfrak{A}^0_{\mathrm{III}} &:= \mathfrak{A}^1_{\mathrm{III}}, \\
\mathfrak{A}^0_{\mathrm{IV}} &:= \mathbb{Q}\mathbf{1} + \sum_{k\geq 2} z_k \mathfrak{A}_{\widetilde{\mathrm{IV}}}, & \mathfrak{A}^0_{\widetilde{\mathrm{IV}}} &:= \mathbb{Q}\mathbf{1} + \theta\mathfrak{A}_{\widetilde{\mathrm{IV}}} + \sum_{k\geq 2} z_k \mathfrak{A}_{\widetilde{\mathrm{IV}}} \subsetneq \mathfrak{A}^1_{\widetilde{\mathrm{IV}}}.
\end{aligned}$$

For each type τ the words in $\mathfrak{A}^0_\tau$ are called *type τ-admissible*. This is consistent with Definition 2.3 since we consider only non-negative compositions $\mathbf{s}$.

Definition 4.2 To define the stuffle product for type $\tau = \tilde{\mathrm{I}}$ and II, similar to the MZV case, we define a commutative product $[-,-]_\tau$ first:

$$[z_k, z_l]_{\tilde{\mathrm{I}}} = z_{k+l} + z_{k+l-1}, \quad [\theta, z_k]_{\tilde{\mathrm{I}}} = z_{k+1}, \quad [\theta,\theta]_{\tilde{\mathrm{I}}} = z_2 - \theta, \quad [z'_k, z'_l]_{\mathrm{II}} = z'_{k+l} \tag{21}$$

for all $k, l \geq 1$. Now we define the stuffle product $*_\tau$ on $\mathfrak{A}^1_\tau$ inductively as follows. For any words $\mathbf{u}, \mathbf{v} \in \mathfrak{A}^1_\tau$ and letters $\alpha, \beta \in Y_\tau$, we set $\mathbf{1} *_\tau \mathbf{u} = \mathbf{u} = \mathbf{u} *_\tau \mathbf{1}$ and

$$(\alpha\mathbf{u}) *_\tau (\beta\mathbf{v}) = \alpha(\mathbf{u} *_\tau \beta\mathbf{v}) + \beta(\alpha\mathbf{u} *_\tau \mathbf{v}) + [\alpha, \beta]_\tau (\mathbf{u} *_\tau \mathbf{v}). \tag{22}$$

Remark 4.3 (i). The definition for $*_{\tilde{\mathrm{I}}}$ is the same as in [26].

(ii). One can check that $*_\tau$ is well-defined for $\tau = \tilde{\mathrm{I}}$ and II. Namely, $\mathbf{u} *_\tau \mathbf{v} \in \mathfrak{A}^1_\tau$ if $\mathbf{u}, \mathbf{v} \in \mathfrak{A}^1_\tau$.

(iii). It is not hard to check that for $\tau = \tilde{\mathrm{I}}$ and II, $(\mathfrak{A}^0_\tau, *_\tau) \subset (\mathfrak{A}^1_\tau, *_\tau)$ as subalgebras.

(iv). In the following, we will need to define stuffle product $*_{\mathrm{ord}}$ on $\mathfrak{A}_{\mathrm{III}}$ by setting $\tau = \mathrm{ord}$ and $[z_r, z_s]_{\mathrm{ord}} = z_{r+s}$ for all $r, s \in \mathbb{Z}$ in (22).

In [7], the stuffle product $\uplus$ for type III q-MZVs is defined. We will modify this in the following way (see the remarks after Theorem 6.1). Our modified stuffle product for type III q-MZVs will be denoted by $*_{\mathrm{III}}$.

Definition 4.4 We now define a stuffle product $*_{\mathrm{III}}$ on $\mathfrak{A}^1_{\mathrm{III}}$. First, we define an injective shifting operator $\mathscr{S}_-$ on any word of $\mathfrak{A}^1_{\mathrm{III}}$ by acting on the first letter:

$$\mathscr{S}_-(z'_n\mathbf{w}) := z_n\mathbf{w} - z_{n-1}\mathbf{w} \in \mathfrak{A}_{\mathrm{III}} \quad \text{for all } n \in \mathbb{Z} \text{ and } \mathbf{w} \in Y^*_{\mathrm{III}}. \tag{23}$$

For any $k, l \in \mathbb{Z}$ and any $\mathbf{u}, \mathbf{v} \in Y^*_{\mathrm{III}}$, define the stuffle product $*_{\mathrm{III}}$ by

$$\mathbf{1} *_{\text{III}} \mathbf{1} = \mathbf{1}, \qquad \mathbf{1} *_{\text{III}} z'_k\mathbf{u} = z'_k\mathbf{u} *_{\text{III}} \mathbf{1} = z'_k\mathbf{u},$$
$$z'_k\mathbf{u} *_{\text{III}} z'_l\mathbf{v} = z'_k\big(\mathbf{u} *_{\text{ord}} \mathscr{S}_-(z'_l\mathbf{v})\big) + z'_l\big(\mathscr{S}_-(z'_k\mathbf{u}) *_{\text{ord}} \mathbf{v}\big) + (z'_{k+l} - z'_{k+l-1})(\mathbf{u} *_{\text{ord}} \mathbf{v}).$$

Here $*_{\text{ord}}$ is the ordinary stuffle defined in Remark 4.3 (iv).

For type $\widetilde{\text{IV}}$, we provide a definition similar to type III.

Definition 4.5 We now define a stuffle product $*_{\widetilde{\text{IV}}}$ on $\mathfrak{A}^1_{\widetilde{\text{IV}}}$. First, we define a shifting operator $\mathscr{S}_+$ similar to (23) by

$$\mathscr{S}_+(z_n\mathbf{w}) := z'_n\mathbf{w} + z'_{n-1}\mathbf{w} \in \mathfrak{A}_{\widetilde{\text{IV}}} \quad \text{for all } n \in \mathbb{N} \text{ and } \mathbf{w} \in Y^*_{\widetilde{\text{IV}}}.$$

Then, for any $k, l \geq 1$ and any $\mathbf{u}, \mathbf{v} \in Y^*_{\widetilde{\text{IV}}}$ we set

$$\mathbf{1} *_{\widetilde{\text{IV}}} \mathbf{1} = \mathbf{1}, \qquad \mathbf{1} *_{\widetilde{\text{IV}}} z_k\mathbf{u} = z_k\mathbf{u} *_{\widetilde{\text{IV}}} \mathbf{1} = z_k\mathbf{u},$$
$$z_k\mathbf{u} *_{\widetilde{\text{IV}}} z_l\mathbf{v} = z_k\big(\mathbf{u} *_{\text{II}} \mathscr{S}_+(z_l\mathbf{v})\big) + z_l\big(\mathscr{S}_+(z_k\mathbf{u}) *_{\text{II}} \mathbf{v}\big) + (z_{k+l} + z_{k+l-1})(\mathbf{u} *_{\text{II}} \mathbf{v}),$$
$$z_k\mathbf{u} *_{\widetilde{\text{IV}}} \theta\mathbf{v} = \theta\mathbf{v} *_{\widetilde{\text{IV}}} z_k\mathbf{u} = z_k\big(\mathbf{u} *_{\text{II}} \theta\mathbf{v}\big) + \theta\big(\mathscr{S}_+(z_k\mathbf{u}) *_{\text{II}} \mathbf{v}\big) + z_{k+1}(\mathbf{u} *_{\text{II}} \mathbf{v}),$$
$$\theta\mathbf{u} *_{\widetilde{\text{IV}}} \theta\mathbf{v} = \theta\big(\mathbf{u} *_{\text{II}} \theta\mathbf{v}\big) + \theta\big(\theta\mathbf{u} *_{\text{II}} \mathbf{v}\big) + (z_2 - \theta)(\mathbf{u} *_{\text{II}} \mathbf{v}),$$

where $*_{\text{II}}$ is the stuffle product on $\mathfrak{A}^1_{\text{II}} = \mathfrak{A}_{\widetilde{\text{IV}}}$ defined in Definition 4.2.

Lemma 4.6 *The stuffle products* $*_{\text{III}}$ *and* $*_{\widetilde{\text{IV}}}$ *are both well-defined. Namely, if* $\mathbf{u}, \mathbf{v} \in \mathfrak{A}^1_\tau$ *then* $\mathbf{u} *_\tau \mathbf{v} \in \mathfrak{A}^1_\tau$ *for* $\tau = \text{III}$ *or* $\widetilde{\text{IV}}$.

Proof We prove the lemma for type $\widetilde{\text{IV}}$ only. Type III is similar but simpler.

First we note that $k + l - 1 \geq 1$ if $k, l \geq 1$. So the first letter of each of the terms of $\mathbf{u} *_{\widetilde{\text{IV}}} \mathbf{v}$ has the right form, i.e., either θ or z_k for $k \geq 1$. We need to show that after truncating the first letter each term lies in $\mathfrak{A}_{\widetilde{\text{IV}}}$. Notice that $\mathscr{S}_+(z_l\mathbf{v}), \mathscr{S}_+(z_k\mathbf{u}) \in \mathfrak{A}_{\widetilde{\text{IV}}}$ and $*_{\text{II}}$ does not decrease the size the subscripts (which are all non-negative). The lemma is now proved.

Proposition 4.7 *Let* $\tau = \tilde{\text{I}}, \text{II}, \text{III}$ *or* $\widetilde{\text{IV}}$. *Then the stuffle algebras* $(\mathfrak{A}^1_\tau, *_\tau)$ *are all commutative and associative.*

Proof This follows from the fact that the product $[-, -]_\tau$ are all commutative and associative which can be verified easily.

We now turn to the shuffle algebra which is an analog of the corresponding algebra for MZVs reflecting the properties of their representations using iterated integrals.

Definition 4.8 Let $X_\pi = \{\pi, \delta, y\}$ be an alphabet and X^*_π be the set of words generated by X_π. Define $\mathfrak{A}_\pi = \mathbb{Q}\langle \pi, \delta, y\rangle$ to be the noncommutative polynomial $\mathbb{Q}$-algebra of words of X^*_π. We may embed $\mathfrak{A}_\theta$ defined by Definition 4.1 as a subalgebra of $\mathfrak{A}_\pi$ in two different ways: put $\rho = \pi - \mathbf{1}$ and let

$$
\begin{aligned}
&(A) \quad a := \pi, \quad a^{-1} := \delta, \quad b := \pi y, \quad \theta = \rho y \quad \Longrightarrow \quad \gamma := y,\\
&(B) \quad a := \rho, \quad a^{-1} := -, \quad b := \pi y, \quad \theta = \rho y \quad \Longrightarrow \quad \gamma := y.
\end{aligned}
$$

We denote the image of the embedding by $\mathfrak{A}_\theta^{(A)}$ and $\mathfrak{A}_\theta^{(B)}$, respectively. The dash − for the image of a^{-1} in (B) means it does not matter what image we choose since a^{-1} appears only when we consider type III q-MZVs using (A). We will use embedding (B) for the other three types for which a^{-1} will not be utilized essentially because of convergence issues.

5 q-Stuffle relations

First we define the $\mathbb{Q}$-linear realization maps $\mathfrak{z}_q : \mathfrak{A}_\tau^0 \to \mathbb{C}$ ($\tau = \tilde{\mathrm{I}}, \mathrm{II}$) by $\mathfrak{z}_q[\mathbf{1}] = 1$ and

$$
\mathfrak{z}_q[y_1^\tau \dots y_d^\tau] := \sum_{k_1 > \cdots > k_d > 0} M_{k_1}^\tau(y_1^\tau) \dots M_{k_d}^\tau(y_d^\tau),
$$

for every admissible word $y_1^\tau \dots y_d^\tau \in \mathfrak{A}_\tau^0$ where the $\mathbb{Q}$-linear maps

$$
M_m^{\tilde{\mathrm{I}}}(\theta) := \frac{q^m}{1-q^m}, \quad M_m^{\tilde{\mathrm{I}}}(z_s) := \frac{q^{(s-1)m}}{(1-q^m)^s}, \quad M_m^{\mathrm{II}}(z_s') := \frac{q^{sm}}{(1-q^m)^s},
$$

for all $m \in \mathbb{N}$. Note that $M_k^{\tilde{\mathrm{I}}}(\gamma) = M_k^{\tilde{\mathrm{I}}}(z_1 - \theta) = 1$. For example, we have

$$
\mathfrak{z}_q[z_2 z_5 \gamma^2 z_1] = \mathfrak{z}_q^{(1,4,0,0,0)}[2,5,0,0,1], \quad \mathfrak{z}_q[\theta z_7 \theta z_4] = \mathfrak{z}_q^{(1,6,1,3)}[1,7,1,4],
$$

which are not q-MZVs of type I.

For type $\tau = \mathrm{III}$ or $\widetilde{\mathrm{IV}}$, we similarly define the $\mathbb{Q}$-linear realization maps $\mathfrak{z}_q : \mathfrak{A}_\tau^0 \to \mathbb{C}$ by $\mathfrak{z}_q[\mathbf{1}] = 1$ and

$$
\mathfrak{z}_q[y_1^\tau \dots y_d^\tau] := \sum_{k_1 > \cdots > k_d > 0} M_{k_1}^{1,\tau}(y_1^\tau) M_{k_2}^\tau(y_2^\tau) \dots M_{k_d}^\tau(y_d^\tau),
$$

for every admissible word $y_1^\tau \dots y_d^\tau \in \mathfrak{A}_\tau^0$ where the $\mathbb{Q}$-linear maps

$$
\begin{aligned}
&M_m^{1,\mathrm{III}}(z_s') := \frac{q^m}{(1-q^m)^s}, \quad M_m^{\mathrm{III}}(z_s) := \frac{1}{(1-q^m)^s},\\
&M_m^{\widetilde{\mathrm{IV}}}(\theta) = M_m^{1,\widetilde{\mathrm{IV}}}(\theta) := \frac{q^m}{1-q^m}, \quad M_m^{1,\widetilde{\mathrm{IV}}}(z_s) := \frac{q^{m(s-1)}}{(1-q^m)^s}, \quad M_m^{\widetilde{\mathrm{IV}}}(z_s') := \frac{q^{sm}}{(1-q^m)^s},
\end{aligned}
$$

for all $m \in \mathbb{N}$.

The following theorem is parallel to [7, Proposition 9] and includes [26, Theorem 1].

Theorem 5.1 *Let* $\tau = \tilde{\mathrm{I}}, \mathrm{II}, \mathrm{III}$ *or* $\widetilde{\mathrm{IV}}$. *For any* $\mathbf{u}_\tau, \mathbf{v}_\tau \in \mathfrak{A}_\tau^0$ *we have*

$$\mathfrak{z}_q[\mathbf{u}_\tau *_\tau \mathbf{v}_\tau] = \mathfrak{z}_q[\mathbf{u}_\tau]\mathfrak{z}_q[\mathbf{v}_\tau]. \tag{24}$$

Proof Since type $\tilde{\mathrm{I}}$ case is just [26, Theorem 1], we only need to consider the other three types. The proof is basically the same as that of [26, Theorem 1]. In fact, it suffices to observe that

$$\begin{aligned}
&M_m^{\mathrm{II}}(z'_k)M_m^{\mathrm{II}}(z'_l) = M_m^{\mathrm{II}}(z'_{k+l}), && M_m^{1,\mathrm{III}}(z'_k)M_m^{\mathrm{III}}(z_l) = M_m^{\mathrm{III}}(z_{k+l} - z_{k+l-1}),\\
&M_m^{\mathrm{III}}(z_k)M_m^{\mathrm{III}}(z_l) = M_m^{\mathrm{III}}(z_{k+l}), && M_m^{1,\mathrm{III}}(z'_k)M_m^{1,\mathrm{III}}(z'_l) = M_m^{1,\mathrm{III}}(z'_{k+l} - z'_{k+l-1}),\\
&M_m^{\widetilde{\mathrm{IV}}}(z'_k)M_m^{\widetilde{\mathrm{IV}}}(z'_l) = M_m^{\widetilde{\mathrm{IV}}}(z'_{k+l}), && M_m^{1,\widetilde{\mathrm{IV}}}(z_k)M_m^{\widetilde{\mathrm{IV}}}(z'_l) = M_m^{\widetilde{\mathrm{IV}}}(z'_{k+l} + z'_{k+l-1}),\\
&M_m^{\widetilde{\mathrm{IV}}}(\theta)M_m^{\widetilde{\mathrm{IV}}}(z'_k) = M_m^{\widetilde{\mathrm{IV}}}(z'_{k+1}), && M_m^{1,\widetilde{\mathrm{IV}}}(z_k)M_m^{1,\widetilde{\mathrm{IV}}}(z_l) = M_m^{1,\widetilde{\mathrm{IV}}}(z_{k+l} + z_{k+l-1}),\\
&M_m^{1,\widetilde{\mathrm{IV}}}(\theta)M_m^{\widetilde{\mathrm{IV}}}(z'_k) = M_m^{\widetilde{\mathrm{IV}}}(z'_{k+1}), && M_m^{\widetilde{\mathrm{IV}}}(\theta)M_m^{\widetilde{\mathrm{IV}}}(\theta) = M_m^{1,\widetilde{\mathrm{IV}}}(\theta)M_m^{\widetilde{\mathrm{IV}}}(\theta) = M_m^{\widetilde{\mathrm{IV}}}(z'_2),\\
&M_m^{1,\widetilde{\mathrm{IV}}}(\theta)M_m^{1,\widetilde{\mathrm{IV}}}(z_k) = M_m^{1,\widetilde{\mathrm{IV}}}(z_{k+1}), && M_m^{1,\widetilde{\mathrm{IV}}}(\theta)M_m^{1,\widetilde{\mathrm{IV}}}(\theta) = M_m^{1,\widetilde{\mathrm{IV}}}(z_2 - \theta),
\end{aligned}$$

for all $k, l \geq 0$, $m \geq 1$. Of course, we need to assume $k, l \geq 2$ for $M_m^{1,\widetilde{\mathrm{IV}}}(z_k)$ and $M_m^{1,\widetilde{\mathrm{IV}}}(z_l)$.

6 Jackson's Iterated q-Integrals

Set

$$x_0 := x_0(t) = \frac{1}{t}, \quad x_1 := x_1(t) = \frac{1}{1-t}, \quad \mathbf{y} := \mathbf{y}(t) = \frac{t}{1-t}.$$

Recall that for $a = x_0(t)dt$ and $b = x_1(t)dt$, we can express MZVs by Chen's iterated integrals:

$$\zeta(s_1, \ldots, s_d) = \int_0^1 a^{s_1-1}b \cdots a^{s_d-1}b.$$

Replacing the Riemann integrals by Jackson's q-integrals (6) one gets

Theorem 6.1 ([7, (29)]) *For* $\mathbf{s} = (s_1, \ldots, s_d) \in \mathbb{N}^d$ *set* $w = |\mathbf{s}|$ *and*

$$\tilde{\zeta}_q^{\mathrm{III}}[\mathbf{s}; t] := \mathbf{J}\Big[c_1\mathbf{J}\big[c_2 \cdots \mathbf{J}[c_w] \cdots\big]\Big](t),$$

where $c_i = x_1$ *if* $i \in \{u_1, u_2, \ldots, u_d\}$, $u_j := s_1 + s_2 + \cdots + s_j$, *and* $c_i = x_0$ *otherwise. Or, equivalently, set* $\mathbf{w} = \pi^{s_1}y\pi^{s_2}y \ldots \pi^{s_d}y$ *and*

$$\tilde{\mathfrak{z}}_q^{\mathrm{III}}[\mathbf{w}; t] := \mathbf{P}^{s_1}\big[\mathbf{y} \cdots \mathbf{P}^{s_d}[\mathbf{y}] \cdots\big](t).$$

Then

$$\mathfrak{z}_q^{\text{Ш}}[\mathbf{s}] = \tilde{\mathfrak{z}}_q^{\text{Ш}}[\mathbf{w}; q]$$

However, the representation of $\zeta_q^{\text{Ш}}[\mathbf{s}]$ using $\tilde{\mathfrak{z}}_q^{\text{Ш}}$ in Theorem 6.1 is not ideal in the sense that one has to evaluate t at q. We would like to use Corollary 3.4 so we need to set $t = 1$. This leads to the idea of replacing the first factor $\mathbf{P}^{s_1}$ by $\mathbf{P}^{s_1-1}\mathbf{R}$ and, more generally, the following two generalizations. For any $\mathbf{a} = (a_1, \ldots, a_d) \in (\mathbb{Z}_{\geq 0})^d$, define

$$\mathfrak{z}_q[\rho^{a_1} y \ldots \rho^{a_d} y; t] := \mathbf{R}^{a_1}\big[\mathbf{y}\mathbf{R}^{a_2}[\mathbf{y} \cdots \mathbf{R}^{a_d}[\mathbf{y}] \cdots]\big](t).$$

Theorem 6.2 *Let* $\mathbf{s} = (s_1, \ldots, s_d) \in \mathbb{N}^d$ *and* $\mathbf{a} = (a_1, \ldots, a_d) \in (\mathbb{Z}_{\geq 0})^d$. *Put* $w = |\mathbf{s}|$ *and* $\mathbf{w} = \mathbf{w}^{\mathbf{a}}(\mathbf{s}) = \rho^{a_1}\pi^{s_1-a_1} y \ldots \rho^{a_d}\pi^{s_d-a_d} y$. *Then*

$$\mathfrak{z}_q[\mathbf{w}^{\mathbf{a}}(\mathbf{s}); t] := \mathbf{R}^{a_1}\big[\mathbf{P}^{s_1-a_1}[\mathbf{y}\mathbf{R}^{a_2}[\mathbf{P}^{s_2-a_2}[\mathbf{y} \cdots \mathbf{R}^{a_d}[\mathbf{P}^{s_d-a_d}[\mathbf{y}]] \cdots]]]\big](t). \tag{25}$$

Suppose $a_1 + \cdots + a_j > 0$ *for all* $j = 1, \ldots, d$. *Then we have*

$$\zeta_q^{\mathbf{a}}[\mathbf{s}] = (1-q)^w \mathfrak{z}_q[\mathbf{w}^{\mathbf{a}}(\mathbf{s});\ 1], \qquad \mathfrak{z}_q^{\mathbf{a}}[\mathbf{s}] = \mathfrak{z}_q[\mathbf{w}^{\mathbf{a}}(\mathbf{s})] := \mathfrak{z}_q[\mathbf{w}^{\mathbf{a}}(\mathbf{s});\ 1]. \tag{26}$$

Proof First we observe three important facts: for any $k \geq 1$ we have

$$\mathbf{P}(t^k) = \sum_{j\geq 0} q^{kj} t^k = \frac{t^k}{1-q^k}, \quad \mathbf{D}(t^k) = t^k(1-q^k), \quad \text{and} \quad \mathbf{R}(t^k) = \sum_{j\geq 1} q^{kj} t^k = \frac{q^k t^k}{1-q^k}$$

by the definition of the two summation operators and the difference operator. Repeatedly applying this we get

$$\mathbf{P}^m(t^k) = \mathbf{P}\left(\frac{t^k}{(1-q^k)^{m-1}}\right) = \frac{t^k}{(1-q^k)^m}, \qquad \forall m \in \mathbb{Z}, \tag{27}$$

$$\mathbf{R}^m(t^k) = \mathbf{R}\left(\left(\frac{q^k}{1-q^k}\right)^{m-1} t^k\right) = \frac{q^{mk} t^k}{(1-q^k)^m} \qquad \forall m \in \mathbb{Z}_{\geq 0}. \tag{28}$$

Thus

$$\mathbf{P}\big(\mathbf{y}(t) \cdot t^k\big) = \sum_{j\geq 0} \frac{q^{j(k+1)} t^{k+1}}{1-q^j t} = \sum_{j\geq 0}\sum_{\ell\geq 0} q^{j(k+\ell+1)} t^{k+\ell+1} = \sum_{\ell>k} \frac{t^\ell}{1-q^\ell}.$$

Similarly, we have

$$\mathbf{D}\big(\mathbf{y}(t) \cdot t^k\big) = \frac{t^{k+1}}{1-t} - \frac{q^{k+1} t^{k+1}}{1-qt} = \sum_{\ell\geq 0} (1-q^{k+\ell+1}) t^{k+\ell+1} = \sum_{\ell>k} (1-q^\ell) t^\ell,$$

and

$$\mathbf{R}\big(\mathbf{y}(t)\cdot t^k\big)=\sum_{j\geq 1}\frac{q^{j(k+1)}t^{k+1}}{1-q^jt}=\sum_{j\geq 1}\sum_{\ell\geq 0}q^{j(k+\ell+1)}t^{k+\ell+1}=\sum_{\ell>k}\frac{q^\ell t^\ell}{1-q^\ell}.$$

It follows from (27) and (28) that

$$\mathbf{P}^m\big(\mathbf{y}(t)\cdot t^k\big)=\sum_{\ell>k}\frac{t^\ell}{(1-q^\ell)^m} \qquad \forall m\in\mathbb{Z}, \tag{29}$$

$$\mathbf{R}^m\big(\mathbf{y}(t)\cdot t^k\big)=\sum_{\ell>k}\frac{q^{m\ell}t^\ell}{(1-q^\ell)^m} \qquad \forall m\in\mathbb{Z}_{\geq 0}. \tag{30}$$

We now prove by induction on the the depth d that for all $\mathbf{s}=(s_1,\dots,s_d)\in\mathbb{N}^d$,

$$\mathfrak{z}_q[\mathbf{w}^{\mathbf{a}}(\mathbf{s});t]=\sum_{k_1>\cdots>k_d>0}\frac{t^{k_1}q^{k_1a_1}\dots q^{k_da_d}}{(1-q^{k_1})^{s_1}\cdots(1-q^{k_d})^{s_d}}. \tag{31}$$

When $d=1$, i.e., $\mathbf{s}=s$, then by (29) followed by (28)

$$\mathfrak{z}_q[\mathbf{w}^a(s);t]=\mathbf{R}^a\mathbf{P}^{s-a}[\mathbf{y}](t)=\sum_{k>0}\frac{\mathbf{R}^a\big(t^k\big)}{(1-q^k)^{s-a}}=\sum_{k>0}\frac{q^{ak}t^k}{(1-q^k)^s}.$$

This proof works even when $s=a$ because of (30) (take $k=0$ and $m=a$ there).

Turning to the general case, we let $d\geq 2$ and assume (31) holds for smaller depths. Then by the inductive assumption

$$\begin{aligned}
\mathfrak{z}_q[\mathbf{w}^{\mathbf{a}}(\mathbf{s});t]&=\mathbf{R}^{a_1}\mathbf{P}^{s_1-a_1}\big[\mathbf{y}\mathbf{R}^{a_2}\mathbf{P}^{s_2-a_2}[\mathbf{y}\cdots\mathbf{R}^{a_d}\mathbf{P}^{s_d-a_d}[\mathbf{y}]\cdots]\big](t)\\
&=\sum_{k_2>\cdots>k_d>0}\frac{\mathbf{R}^{a_1}\mathbf{P}^{s_1-a_1}\big(\mathbf{y}(t)\cdot t^{k_2}\big)q^{k_2a_2}\dots q^{k_da_d}}{(1-q^{k_2})^{s_2}\cdots(1-q^{k_d})^{s_d}}\\
&=\sum_{k_1>\cdots>k_d>0}\frac{\mathbf{R}^{a_1}\big(t^{k_1}\big)q^{k_2a_2}\dots q^{k_da_d}}{(1-q^{k_1})^{s_1-a_1}(1-q^{k_2})^{s_2}\cdots(1-q^{k_d})^{s_d}} \qquad \text{(by (29))}\\
&=\sum_{k_1>\cdots>k_d>0}\frac{t^{k_1}q^{k_1a_1}\dots q^{k_da_d}}{(1-q^{k_1})^{s_1}\cdots(1-q^{k_d})^{s_d}}
\end{aligned}$$

by (28). Again, if $s_1=a_1$ the proof is still valid. This completes the proof of (31). Setting $t=1$ we arrive at (26).

By change of variables $a_j\to s_j-a_j$ for all $j=1,\dots,d$ we immediately obtain the next result. Observe that by (25), we have

$$\mathfrak{z}_q[\mathbf{w}^{\mathbf{s}-\mathbf{a}}(\mathbf{s});t]:=\mathbf{R}^{s_1-a_1}\big[\mathbf{P}^{a_1}[\mathbf{y}\mathbf{R}^{s_2-a_2}[\mathbf{P}^{a_2}[\mathbf{y}\cdots\mathbf{R}^{s_d-a_d}[\mathbf{P}^{a_d}[\mathbf{y}]]\cdots]]]\big](t),$$

where $\mathbf{s}-\mathbf{a}=(s_1-a_1,\dots,s_d-a_d)$.

Theorem 6.3 *Let* $\mathbf{s} = (s_1, \ldots, s_d) \in \mathbb{N}^d$, $\mathbf{a} = (a_1, \ldots, a_d) \in (\mathbb{Z}_{\geq 0})^d$, *and* $w = |\mathbf{s}|$. *Suppose* $s_1 + \cdots + s_j > a_1 + \cdots + a_j$ *for all* $j = 1, \ldots, d$. *Then we have*

$$\zeta_q^{\mathbf{s}-\mathbf{a}}[\mathbf{s}] = (1-q)^w \mathfrak{z}_q[\mathbf{w}^{\mathbf{s}-\mathbf{a}}(\mathbf{s});\ 1], \qquad \mathfrak{z}_q^{\mathbf{s}-\mathbf{a}}[\mathbf{s}] = \mathfrak{z}_q[\mathbf{w}^{\mathbf{s}-\mathbf{a}}(\mathbf{s})] := \mathfrak{z}_q[\mathbf{w}^{\mathbf{s}-\mathbf{a}}(\mathbf{s});\ 1]. \tag{32}$$

By specializing the preceding two theorems to the four types of q-MZVs in Table 1 we quickly find the following corollary. For future reference, we will say $\mathbf{w}_\tau$ has the *typical type* τ *form* for each type τ.

Corollary 6.4 *For* $\mathbf{s} = (s_1, \ldots, s_d) \in \mathbb{N}^d$, *we set*

$$\begin{aligned}
\mathbf{w}_{\mathrm{I}}(\mathbf{s}) &:= \rho^{s_1-1}\pi y \ldots \rho^{s_d-1}\pi y = z_{s_1} \ldots z_{s_d} \in \mathfrak{A}_\theta^{(B)} \subset \mathfrak{A}_\pi y \qquad (s_1 \geq 2),\\
\mathbf{w}_{\mathrm{II}}(\mathbf{s}) &:= \rho^{s_1} y \ldots \rho^{s_d} y = z'_{s_1} \ldots z'_{s_d} \in \mathfrak{A}_\theta^{(B)} \subset \mathfrak{A}_\pi y,\\
\mathbf{w}_{\mathrm{III}}(\mathbf{s}) &:= \pi^{s_1-1}\rho y \pi^{s_2} y \ldots \pi^{s_d} y = z'_{s_1} z_{s_2} \ldots z_{s_d} \in \mathfrak{A}_\theta^{(A)} \subset \mathfrak{A}_\pi y,\\
\mathbf{w}_{\mathrm{IV}}(\mathbf{s}) &:= \rho^{s_1-1}\pi y \rho^{s_2} y \ldots \rho^{s_d} y = z_{s_1} z'_{s_2} \ldots z'_{s_d} \in \mathfrak{A}_\theta^{(B)} \subset \mathfrak{A}_\pi y \qquad (s_1 \geq 2),
\end{aligned}$$

and

$$\begin{aligned}
\mathfrak{z}_q[\mathbf{w}_{\mathrm{I}}(\mathbf{s}); t] &:= \mathbf{R}^{s_1-1}\big[\mathbf{P}[\mathbf{y}\mathbf{R}^{s_2-1}[\mathbf{P}[\mathbf{y} \cdots \mathbf{R}^{s_d-1}[\mathbf{P}[\mathbf{y}]] \cdots]]]\big](t),\\
\mathfrak{z}_q[\mathbf{w}_{\mathrm{II}}(\mathbf{s}); t] &:= \mathbf{R}^{s_1}\big[\mathbf{y}\mathbf{R}^{s_2}[\mathbf{y} \cdots \mathbf{R}^{s_d}[\mathbf{y}] \cdots]\big](t),\\
\mathfrak{z}_q[\mathbf{w}_{\mathrm{III}}(\mathbf{s}); t] &:= \mathbf{P}^{s_1-1}\Big[\mathbf{R}\big[\mathbf{y}[\mathbf{P}^{s_2}[\mathbf{y}[\mathbf{P}^{s_3}[\mathbf{y} \cdots \mathbf{P}^{s_d}[\mathbf{y}] \cdots]]]]\big]\Big](t),\\
\mathfrak{z}_q[\mathbf{w}_{\mathrm{IV}}(\mathbf{s}); t] &:= \mathbf{R}^{s_1-1}\Big[\mathbf{P}\big[\mathbf{y}\mathbf{R}^{s_2}[\mathbf{y}\mathbf{R}^{s_3}[\mathbf{y} \cdots \mathbf{R}^{s_d}[\mathbf{y}] \cdots]]\big]\Big](t).
\end{aligned}$$

Then for all the types $\tau = \mathrm{I}, \mathrm{II}, \mathrm{III}$ *and* IV, *we have*

$$\zeta_q^\tau[\mathbf{s}] = (1-q)^w \mathfrak{z}_q[\mathbf{w}_\tau(\mathbf{s});\ 1], \qquad \mathfrak{z}_q^\tau[\mathbf{s}] = \mathfrak{z}_q[\mathbf{w}_\tau(\mathbf{s})] := \mathfrak{z}_q[\mathbf{w}_\tau(\mathbf{s});\ 1].$$

Moreover, similar results hold for type $\widetilde{\mathrm{I}}$ *and* $\widetilde{\mathrm{IV}}$ *q-MZVs. We may replace any of the consecutive strings* $\rho^{s_j-1}\pi$ *by a single* ρ *in* $\mathbf{w}_{\widetilde{\mathrm{I}}}(\mathbf{s})$ *and* $\mathbf{w}_{\widetilde{\mathrm{IV}}}(\mathbf{s})$, *and replace the corresponding operator string* $\mathbf{P}^{s_j-1}\mathbf{R}$ *by a single* $\mathbf{R}$.

We now apply the above to Okounkov's q-MZVs. For any $n \in \mathbb{N}$ we let n^- and n^+ be the two nonnegative integers such that

$$\frac{n-1}{2} \leq n^- \leq \frac{n}{2} \leq n^+ \leq \frac{n+1}{2}.$$

Clearly we have $n^+ + n^- = n$ always, $n^+ = n^-$ if n is even, and $n^+ = n^- + 1$ if n is odd. We can now define a variation of Okounkov's q-MZVs. Let $\mathbf{s} \in (\mathbb{Z}_{\geq 2})^d$. Then

$$\zeta_q^{\mathrm{O}}[\mathbf{s}] := \sum_{k_1 > \cdots > k_d > 0} \prod_{j=1}^d \frac{q^{k_j s_j^+} + q^{k_j s_j^-}}{[k_j]^{s_j}} = (1-q)^{|\mathbf{s}|} \sum_{k_1 > \cdots > k_d > 0} \prod_{j=1}^d \frac{q^{k_j s_j^+} + q^{k_j s_j^-}}{(1-q^{k_j})^{s_j}}.$$

Again, its modified form is:

$$\mathfrak{z}_q^{\mathrm{O}}[\mathbf{s}] := \sum_{k_1>\cdots>k_d>0} \prod_{j=1}^{d} \frac{q^{k_j s_j^+} + q^{k_j s_j^-}}{(1-q^{k_j})^{s_j}}.$$

Remark 6.5 The above variation is equal to Okounkov's original q-MZVs up to a suitable 2-power. More precisely, the power is given by the number of even arguments in $\mathbf{s}$.

Corollary 6.6 *For* $\mathbf{s} = (s_1, \ldots, s_d) \in \mathbb{N}^d$, *we set*

$$\mathbf{w}_{\mathrm{O}}(\mathbf{s}) = (\rho^{s_1^-}\pi^{s_1^+} + \rho^{s_1^+}\pi^{s_1^-})y \ldots (\rho^{s_d^-}\pi^{s_d^+} + \rho^{s_d^+}\pi^{s_d^-})y \in \mathfrak{A}_\theta^{(B)} \subset \mathfrak{A}_\pi y$$

and

$$\mathfrak{z}_q[\mathbf{w}_{\mathrm{O}}(\mathbf{s}); t] = (\mathbf{R}^{s_1^-}\mathbf{P}^{s_1^+} + \mathbf{R}^{s_1^+}\mathbf{P}^{s_1^-})\big[\mathbf{y} \cdots (\mathbf{R}^{s_d^-}\mathbf{P}^{s_d^+} + \mathbf{R}^{s_d^+}\mathbf{P}^{s_d^-})[\mathbf{y}] \cdots\big](t).$$

Then we have

$$\zeta_q^{\mathrm{O}}[\mathbf{s}] = (1-q)^w \mathfrak{z}_q[\mathbf{w}_{\mathrm{O}}(\mathbf{s});\ 1], \qquad \mathfrak{z}_q^{\mathrm{O}}[\mathbf{s}] = \mathfrak{z}_q[\mathbf{w}_{\mathrm{O}}(\mathbf{s})] := \mathfrak{z}_q[\mathbf{w}_{\mathrm{O}}(\mathbf{s});\ 1].$$

It is possible to obtain the shuffle relations among $\mathfrak{z}_q^{\mathrm{O}}[\mathbf{s}]$-values using Corollary 6.6. The stuffle relations among $\mathfrak{z}_q^{\mathrm{O}}[\mathbf{s}]$ is mentioned implicitly in Okounkov's original paper. For our modified version, they can be derived from the following fact (cf. Proposition 2.2 (ii) of [2]). Let $F_n^{\mathrm{O}}(t) = (t^{n^+} + t^{n^-})/(1-t)^n$ for all $n \geq 2$. Then for all $r, s \in \mathbb{Z}_{\geq 2}$, we have

$$F_r^{\mathrm{O}}(t) \cdot F_s^{\mathrm{O}}(t) = \begin{cases} 2F_{r+s}^{\mathrm{O}}(t), & \text{if } r \text{ or } s \text{ is even;} \\ 2F_{r+s}^{\mathrm{O}}(t) + \frac{1}{2}F_{r+s-2}^{\mathrm{O}}(t), & \text{if } r \text{ and } s \text{ are odd.} \end{cases}$$

For example,

$$\mathfrak{z}_q^{\mathrm{O}}[2,3]\mathfrak{z}_q^{\mathrm{O}}[2] = 2\mathfrak{z}_q^{\mathrm{O}}[2,2,3] + \mathfrak{z}_q^{\mathrm{O}}[2,3,2] + 2\mathfrak{z}_q^{\mathrm{O}}[4,3] + 2\mathfrak{z}_q^{\mathrm{O}}[2,5],$$
$$\mathfrak{z}_q^{\mathrm{O}}[2,3]\mathfrak{z}_q^{\mathrm{O}}[3] = 2\mathfrak{z}_q^{\mathrm{O}}[2,3,3] + \mathfrak{z}_q^{\mathrm{O}}[3,2,3] + 2\mathfrak{z}_q^{\mathrm{O}}[5,3] + 2\mathfrak{z}_q^{\mathrm{O}}[2,6] + \frac{1}{2}\mathfrak{z}_q^{\mathrm{O}}[2,4].$$

7 q-Shuffle Relations

In contrast to the MZV case, the q-shuffle product is much more difficult to define than the q-stuffle product. In this section we will use the Rota–Baxter algebra approach to define this for type $\widetilde{\mathrm{I}}$, II, III, and $\widetilde{\mathrm{IV}}$ q-MZVs. Note that this has been done for type III q-MZVs in [7] which we recall first.

The *q-shuffle product* on $\mathfrak{A}_\pi$ (see Definition 4.8) is defined recursively as follows: for any words $\mathbf{u}, \mathbf{v} \in X_\pi^*$ we define $\mathbf{1} \sqcup\!\sqcup \mathbf{u} = \mathbf{u} \sqcup\!\sqcup \mathbf{1} = \mathbf{u}$ and

$$(y\mathbf{u}) \sqcup\!\sqcup \mathbf{v} = \mathbf{u} \sqcup\!\sqcup (y\mathbf{v}) = y(\mathbf{u} \sqcup\!\sqcup \mathbf{v}), \tag{33}$$

$$\pi\mathbf{u} \sqcup\!\sqcup \pi\mathbf{v} = \pi(\mathbf{u} \sqcup\!\sqcup \pi\mathbf{v}) + \pi(\pi\mathbf{u} \sqcup\!\sqcup \mathbf{v}) - \pi(\mathbf{u} \sqcup\!\sqcup \mathbf{v}), \tag{34}$$

$$\delta\mathbf{u} \sqcup\!\sqcup \delta\mathbf{v} = \mathbf{u} \sqcup\!\sqcup \delta\mathbf{v} + \delta\mathbf{u} \sqcup\!\sqcup \mathbf{v} - \delta(\mathbf{u} \sqcup\!\sqcup \mathbf{v}), \tag{35}$$

$$\delta\mathbf{u} \sqcup\!\sqcup \pi\mathbf{v} = \pi\mathbf{v} \sqcup\!\sqcup \delta\mathbf{u} = \delta(\mathbf{u} \sqcup\!\sqcup \pi\mathbf{v}) + \delta\mathbf{u} \sqcup\!\sqcup \mathbf{v} - \mathbf{u} \sqcup\!\sqcup \mathbf{v} \tag{36}$$

for any words $\mathbf{u}, \mathbf{v} \in X_\pi^*$. Equation (33) reflects the fact that when $\mathbf{y}(t)$ is multiplied in front of either of the two factors in a product, it can be multiplied after taking the product. Equations (34)–(36) formalize (8), (13) and (14), respectively.

Corollary 7.1 *For any words* $\mathbf{u}, \mathbf{v} \in X_\pi^*$, *we have*

$$\rho\mathbf{u} \sqcup\!\sqcup \rho\mathbf{v} = \rho(\mathbf{u} \sqcup\!\sqcup \rho\mathbf{v}) + \rho(\rho\mathbf{u} \sqcup\!\sqcup \mathbf{v}) + \rho(\mathbf{u} \sqcup\!\sqcup \mathbf{v}), \tag{37}$$

$$\rho\mathbf{u} \sqcup\!\sqcup \pi\mathbf{v} = \pi\mathbf{v} \sqcup\!\sqcup \rho\mathbf{u} = \rho(\rho\mathbf{u} \sqcup\!\sqcup \mathbf{v}) + \rho(\mathbf{u} \sqcup\!\sqcup \rho\mathbf{v}) + \rho\mathbf{u} \sqcup\!\sqcup \mathbf{v} + \rho(\mathbf{u} \sqcup\!\sqcup \mathbf{v}), \tag{38}$$

$$\delta\mathbf{u} \sqcup\!\sqcup \rho\mathbf{v} = \rho\mathbf{v} \sqcup\!\sqcup \delta\mathbf{u} = \delta(\mathbf{u} \sqcup\!\sqcup \pi\mathbf{v}) - \mathbf{u} \sqcup\!\sqcup \mathbf{v} = \delta(\mathbf{u} \sqcup\!\sqcup \rho\mathbf{v}) + \delta(\mathbf{u} \sqcup\!\sqcup \mathbf{v}) - \mathbf{u} \sqcup\!\sqcup \mathbf{v}. \tag{39}$$

Proof These follows easily from (33)–(36) and the relation $\rho = \pi - \mathbf{1}$.

Corollary 7.2 *For* $j = 1, 2$ *let* $X_\theta^{(j)}$ *and* $X_\theta^{(j),*}$ *be the embedding of* X_θ *and* X_θ^* *into* X_π^*, *respectively, by Definition 4.8. For any* $\alpha, \beta \in X_\theta^{(j)}$ *and* $\mathbf{u}, \mathbf{v} \in X_\theta^{(j),*}$, *we have* $1 \sqcup\!\sqcup \mathbf{u} = \mathbf{u} \sqcup\!\sqcup 1 = \mathbf{u}$ *and*

$$\alpha\mathbf{u} \sqcup\!\sqcup \beta\mathbf{v} = \alpha(\mathbf{u} \sqcup\!\sqcup \beta\mathbf{v}) + \beta(\alpha\mathbf{u} \sqcup\!\sqcup \mathbf{v}) + [\alpha, \beta]_j(\mathbf{u} \sqcup\!\sqcup \mathbf{v}),$$

where $[\alpha, \beta]_j$ *is determined by* $[a, b]_1 = [b, a]_1 = -b$, $[a, b]_2 = [b, a]_2 = 0$ *and*

$$[a, a]_j = (-1)^j a, \quad [b, b]_j = -b\gamma, \quad [\alpha, \gamma]_j = [\gamma, \alpha]_j = -\alpha\gamma. \tag{40}$$

Proof All of these identities follow from straightforward computation using (33)–(39). For example,

$$\begin{aligned} b\mathbf{u} \sqcup\!\sqcup b\mathbf{v} = \pi y\mathbf{u} \sqcup\!\sqcup \pi y\mathbf{v} &= \pi(y\mathbf{u} \sqcup\!\sqcup \pi y\mathbf{v}) + \pi(\pi y\mathbf{u} \sqcup\!\sqcup y\mathbf{v}) - \pi(y\mathbf{u} \sqcup\!\sqcup y\mathbf{v}) \\ &= \pi y(\mathbf{u} \sqcup\!\sqcup \pi y\mathbf{v}) + \pi y(\pi y\mathbf{u} \sqcup\!\sqcup \mathbf{v}) - \pi yy(\mathbf{u} \sqcup\!\sqcup \mathbf{v}) \\ &= b(\mathbf{u} \sqcup\!\sqcup b\mathbf{v}) + b(b\mathbf{u} \sqcup\!\sqcup \mathbf{v}) - b\gamma(\mathbf{u} \sqcup\!\sqcup \mathbf{v}). \end{aligned} \tag{41}$$

Similarly,

$$\begin{aligned} \theta\mathbf{u} \sqcup\!\sqcup \theta\mathbf{v} = \rho y\mathbf{u} \sqcup\!\sqcup \rho y\mathbf{v} &= \rho(y\mathbf{u} \sqcup\!\sqcup \rho y\mathbf{v}) + \rho(\rho y\mathbf{u} \sqcup\!\sqcup y\mathbf{v}) + \rho(y\mathbf{u} \sqcup\!\sqcup y\mathbf{v}) \\ &= \rho y(\mathbf{u} \sqcup\!\sqcup \rho y\mathbf{v}) + \rho y(\rho y\mathbf{u} \sqcup\!\sqcup \mathbf{v}) + \rho yy(\mathbf{u} \sqcup\!\sqcup \mathbf{v}) \\ &= \theta(\mathbf{u} \sqcup\!\sqcup \theta\mathbf{v}) + \theta(\theta\mathbf{u} \sqcup\!\sqcup \mathbf{v}) + \theta\gamma(\mathbf{u} \sqcup\!\sqcup \mathbf{v}). \end{aligned} \tag{42}$$

The rest of the proof is left to the interested reader.

Proposition 7.3 *The algebra* $(\mathfrak{A}_\pi, ш)$ *is commutative and associative.*

Proof See [7, Theorem 7].

The following corollary generalizes [26, Proposition 1].

Corollary 7.4 *For* $j = 1$ *or* 2 *the algebras* $(\mathfrak{A}_\theta^{(j)}, ш)$ *are commutative and associative.*

Proof This follows immediately from Proposition 7.3 since $(\mathfrak{A}_\theta^{(j)}, ш)$ are sub-algebras of $(\mathfrak{A}_\pi, ш)$ if ш for $\mathfrak{A}_\theta^{(j)}$ is defined as in Corollary 7.2.

Our next theorem shows that we may use the shuffle algebra structure defined above to describe the q-shuffle relations among different types of q-MZVs. Before doing so, we need to show that the q-shuffle products really make sense for all the types.

Proposition 7.5 *Embed* $\mathfrak{A}^0_{\tilde{\mathrm{I}}}, \mathfrak{A}^0_{\mathrm{II}}, \mathfrak{A}^0_{\widetilde{\mathrm{IV}}} \subset \mathfrak{A}_\theta^{(B)}$ *and* $\mathfrak{A}^0_{\mathrm{III}} \subset \mathfrak{A}_\theta^{(A)}$. *Then for each type* τ, *if the two words* $\mathbf{u}, \mathbf{v} \in \mathfrak{A}^0_\tau$ *have the typical type* τ *form listed in Corollary 6.4 then there is an algorithm to express* $\mathbf{u}\, ш\, \mathbf{v}$ *using only those words in the same form.*

Proof The case for type $\tilde{\mathrm{I}}$ is proved by [26, Proposition 2].

Type II is in fact the easiest since we can restrict ourselves to use only (33) and (37) to compute the shuffle and therefore π never comes into the picture. Clearly all such words must start with ρ and end with y.

For type III let's assume $\mathbf{u} = \pi^{s_1-1}\rho y \pi^{s_2} y \dots \pi^{s_d} y$ and $\mathbf{v} = \pi^{a_1-1}\rho y \pi^{a_2} y \dots \pi^{a_d} y$. If we use the definition (34) repeatedly then in each word appearing in $\mathbf{u}\, ш\, \mathbf{v}$ the first ρ always appears before all the y's. Such a word can be written in the form $\pi^s \rho^r y \cdots$ for some $s \in \mathbb{Z}$ and $r \geq 1$ (notice that if ρ and π are commutative). Now we can rewrite this as $\pi^s(\pi - \mathbf{1})^{r-1}\rho y \cdots$ and replace all the ρ's after the first y by $\pi - \mathbf{1}$. This produces a word of typical type III form.

Type $\widetilde{\mathrm{IV}}$ is similar to type III except that we need to take θ into account. Notice that by definition if $\mathbf{w} \in \mathfrak{A}^0_{\widetilde{\mathrm{IV}}}$ then it can be written as $\theta \mathbf{w}'$, or $z_k \mathbf{w}'$ ($k \geq 2$, $\mathbf{w}' \in Y^*_{\widetilde{\mathrm{IV}}}$) or a finite linear combination of these. So we have three cases to check. First, we prove that for all $k, l \geq 2$ and $\mathbf{u}, \mathbf{v} \in Y^*_{\widetilde{\mathrm{IV}}}$

$$z_k \mathbf{u}\, ш\, z_l \mathbf{v} \in \mathfrak{A}^0_{\widetilde{\mathrm{IV}}}. \tag{43}$$

Indeed, putting $k = r + 1$ and $l = s + 1$ we have

$$\begin{aligned} &\rho^r \pi y \mathbf{u}\, ш\, \rho^s \pi y \mathbf{v} \\ &\quad = \rho(\rho^{r-1}\pi y \mathbf{u}\, ш\, \rho^s \pi y \mathbf{v}) + \rho(\rho^r \pi y \mathbf{u}\, ш\, \rho^{s-1}\pi y \mathbf{v}) + \rho(\rho^{r-1}\pi y\, ш\, \rho^{s-1}\pi y \mathbf{v}). \end{aligned}$$

Now inside each of the three parentheses we replace every π by $\rho + \mathbf{1}$ and use only (33) and (37) to expand (recall that $\theta = \rho y$). We see that every term in the expansion has the form $\rho^n y\mathbf{w}$ for some $n \geq 1$ and $\mathbf{w} \in Y^*_{\widetilde{\mathrm{IV}}}$. If $n = 1$ then we have $\rho^n y\mathbf{w} = \theta\mathbf{w} \in \mathfrak{A}^0_{\widetilde{\mathrm{IV}}}$. If $n \geq 2$ we can write it as

$$\rho^{n-1}(\pi - \mathbf{1})y\mathbf{w} = \sum_{j=1}^{n-1}(-1)^{j-1}\rho^{n-j}\pi y\mathbf{w} + (-1)^{n-1}\theta\mathbf{w} \in \mathfrak{A}^0_{\widetilde{\mathrm{IV}}} \tag{44}$$

with each word of typical type $\widetilde{\mathrm{IV}}$ form.

Now we assume $k = r + 1 \geq 2$ and $\mathbf{u}, \mathbf{v} \in Y^*_{\widetilde{\mathrm{IV}}}$. Then

$$\begin{aligned} z_k\mathbf{u} \sqcup\!\sqcup\, \theta\mathbf{v} =& \rho^r \pi y\mathbf{u} \sqcup\!\sqcup\, \rho y\mathbf{v} \\ &= \rho(\rho^{r-1}\pi y\mathbf{u} \sqcup\!\sqcup\, \rho y\mathbf{v}) + \rho y(\rho^r \pi y\mathbf{u} \sqcup\!\sqcup\, \mathbf{v}) + \rho y(\rho^{r-1}\pi y \sqcup\!\sqcup\, \mathbf{v}) \in \mathfrak{A}^0_{\widetilde{\mathrm{IV}}} \end{aligned}$$

since $\rho y = \theta$ and the first term can be dealt with as in the proof of (43).

Finally,

$$\theta\mathbf{u} \sqcup\!\sqcup\, \theta\mathbf{v} \in \mathfrak{A}^0_{\widetilde{\mathrm{IV}}}$$

follows from (42) immediately. This completes the proof of the proposition.

The following theorem generalizes [26, Theorem 2] but it does not contain [7, Theorem 7] since our word representation of type III q-MZVs is different from that given in [7].

Theorem 7.6 *Embed $\mathfrak{A}^0_{\widetilde{\mathrm{I}}}, \mathfrak{A}^0_{\mathrm{II}}, \mathfrak{A}^0_{\widetilde{\mathrm{IV}}} \subset \mathfrak{A}^{(B)}_\theta$ and $\mathfrak{A}^0_{\mathrm{III}} \subset \mathfrak{A}^{(A)}_\theta$. Then for each type τ and for any $\mathbf{u}_\tau, \mathbf{v}_\tau \in \mathfrak{A}^0_\tau$, we have*

$$\mathfrak{z}_q[\mathbf{u}_\tau]\mathfrak{z}_q[\mathbf{v}_\tau] = \mathfrak{z}_q[\mathbf{u}_\tau \sqcup\!\sqcup\, \mathbf{v}_\tau]. \tag{45}$$

Proof For each type τ we observe that $\mathfrak{z}_q[\mathbf{w}_\tau; t]$ satisfies (45) because of the identities in Proposition 3.2. Then the theorem follows from the fact that $\mathfrak{z}_q[\mathbf{w}_\tau] = \mathfrak{z}_q[\mathbf{w}_\tau;\ 1]$ for any word $\mathbf{w}_\tau \in \mathfrak{A}^0_\tau$ by Corollary 6.4.

8 Duality Relations

The DBSFs do not contain all linear relations among the various types of q-MZVs. In [26], Takeyama discovered the following relations which provides some of the missing relations for type $\widetilde{\mathrm{I}}$ q-MZVs, at least in the small weight cases. He called them *Resummation Identities*. We would rather call them "duality" relations because of their similarity to the duality relations for the classical MZVs.

Theorem 8.1 ([26, Theorem 4]) *For a positive integer k, set*

$$\varphi_k := (-1)^k \left(\sum_{j=2}^{k} (-1)^j z_j - \theta \right),$$

where $\varphi_1 = \theta = \rho y \in \mathfrak{A}_\theta^{(B)}$. *Let* $\ell \in \mathbb{N}$ *and* $\alpha_j, \beta_j \in \mathbb{Z}_{\geq 0}$ *for all* $j = 1, \dots, \ell$. *Then we have*

$$\mathfrak{z}_q^{\tilde{\mathrm{I}}}[\varphi_{\alpha_1+1}\gamma^{\beta_1} \cdots \varphi_{\alpha_\ell+1}\gamma^{\beta_\ell}] = \mathfrak{z}_q^{\tilde{\mathrm{I}}}[\varphi_{\beta_\ell+1}\gamma^{\alpha_\ell} \cdots \varphi_{\beta_1+1}\gamma^{\alpha_1}]. \tag{46}$$

We can use the Rota–Baxter algebra approach to give a new proof of this result.

Proof Notice that $\gamma = y, z_j = \rho^{j-1}\pi y$ and $\theta = \rho y$ with the embedding $\mathfrak{A}_{\tilde{\mathrm{I}}}^0 \subset \mathfrak{A}_\theta^{(B)}$. Since $\pi = \rho + \mathbf{1}$, for all $k \geq 1$, we have (cf. (44))

$$\begin{aligned} \varphi_k &= (-1)^k \left(\sum_{j=2}^{k} (-1)^j \rho^{j-1}(\rho + \mathbf{1})y - \rho y \right) \\ &= (-1)^k \left(\sum_{j=2}^{k} (-1)^j \rho^j y + \sum_{j=2}^{k} (-1)^j \rho^{j-1} y - \rho y \right) = \rho^k y. \end{aligned} \tag{47}$$

Thus the theorem follows from Corollaries 3.4 and 6.4 easily.

Remark 8.2 Although not mentioned explicitly in [26], there is a subtle point in applying Theorem 8.1. Notice that in the expression of φ_k the letter θ appears. However, q-MZVs of the form such as $\zeta_q^{\mathrm{I}}[\theta\gamma z_2\gamma] = \zeta_q^{\mathrm{I}}[\rho y^2 \rho^2 \pi y]$ is not really defined. In fact, it should be denoted by $\zeta_q^{\tilde{\mathrm{I}}}[\theta\gamma z_2\gamma] = \zeta_q^{(1,0,1,0)}[1, 0, 2, 0]$ (and such values always converge by Proposition 2.1 because of the leading 1 in the auxiliary variable **a**). But, suitable $\mathbb{Q}$-linear combinations of (46) may lead to identities in which only z_k's appear. Then all terms can be written as honest ζ_q^{I}-values. This explains the use of two admissible structures $\widehat{\mathfrak{H}}^0$ and $\mathfrak{H}^0$ in [26]. For an illuminating example, see the proof of Proposition 7 of op. cit. This remark also applies to Theorem 8.5 for the duality of type $\widetilde{\mathrm{IV}}$ q-MZVs.

Similar relations for type II q-MZVs have the most aesthetic appeal and is the primary reason why we prefer to call it by the name "duality".

Theorem 8.3 *Let* $\ell \in \mathbb{N}$ *and* $\alpha_j, \beta_j \in \mathbb{N}$ *for all* $j = 1, \dots, \ell$. *Then we have*

$$\mathfrak{z}_q^{\mathrm{II}}[\rho^{\alpha_1} y^{\beta_1} \cdots \rho^{\alpha_\ell} y^{\beta_\ell}] = \mathfrak{z}_q^{\mathrm{II}}[\rho^{\beta_\ell} y^{\alpha_\ell} \cdots \rho^{\beta_1} y^{\alpha_1}].$$

Proof This follows from Corollaries 3.4 and 6.4 immediately.

Of course we may apply the same idea to type III and $\widetilde{\mathrm{IV}}$ q-MZVs.

Theorem 8.4 *Let* $\ell \in \mathbb{N}$ *and* $\alpha_j, \beta_j \in \mathbb{N}$ *for all* $j = 1, \dots, \ell$. *Then we have*

$$\mathfrak{z}_q^{\mathrm{III}}[(\pi-\mathbf{1})^{\alpha_1-1}\rho y^{\beta_1}(\pi-\mathbf{1})^{\alpha_2}y^{\beta_2}\cdots(\pi-\mathbf{1})^{\alpha_\ell}y^{\beta_\ell}]$$
$$=\mathfrak{z}_q^{\mathrm{III}}[(\pi-\mathbf{1})^{\beta_\ell-1}\rho y^{\alpha_\ell}(\pi-\mathbf{1})^{\beta_{\ell-1}}y^{\alpha_{\ell-1}}\cdots(\pi-\mathbf{1})^{\beta_1}y^{\alpha_1}].$$

Proof Since $\rho=\pi-\mathbf{1}$ this follows from Corollaries 3.4 and 6.4 easily.

Theorem 8.5 *Let $\ell\in\mathbb{N}$ and $\alpha_j,\beta_j\in\mathbb{N}$ for all $j=1,\dots,\ell$. Then we have*

$$\mathfrak{z}_q^{\widetilde{\mathrm{IV}}}[\varphi_{\alpha_1}y^{\beta_1-1}\rho^{\alpha_2}y^{\beta_2}\cdots\rho^{\alpha_\ell}y^{\beta_\ell}]=\mathfrak{z}_q^{\widetilde{\mathrm{IV}}}[\varphi_{\beta_\ell}y^{\alpha_\ell-1}\rho^{\beta_{\ell-1}}y^{\alpha_{\ell-1}}\cdots\rho^{\beta_1}y^{\alpha_1}].$$

Here $\varphi_1=\theta=\rho y\in\mathfrak{A}_\theta^{(B)}$.

Proof This follows from (47), Corollaries 3.4 and 6.4.

9 The General Type G q-MZVs

All of the q-MZVs of type $\tilde{\mathrm{I}}$, II, III and $\widetilde{\mathrm{IV}}$ considered in the above are some special forms of the q-MZVs $\mathfrak{z}_q^{(a_1,\dots,a_d)}[s_1,\dots,s_d]$ where $1\le a_1\le s_1$, $0\le a_j\le s_j$ for all $j\ge 2$, all of which are convergent by Proposition 2.1. We call these *type G* q-MZVs. Similar to the first four types, we may use words to encode these values according to Theorem 6.2. Namely, we can define

$$\mathfrak{z}_q[\rho^{a_1}y\rho^{a_2}y\cdots\rho^{a_d}y;t]:=\mathbf{R}^{a_1}\mathbf{y}\big[\mathbf{R}^{a_2}\mathbf{y}[\cdots[\mathbf{R}^{a_d}\mathbf{y}]\cdots]\big](t).$$

By the relation $\pi=\rho+\mathbf{1}$ and $\mathbf{P}=\mathbf{R}+\mathbf{I}$, we get

$$\mathfrak{z}_q[\rho^{a_1}\pi^{b_1}y\rho^{a_2}\pi^{b_2}y\cdots\rho^{a_d}\pi^{b_d}y;t]=\mathbf{R}^{a_1}\big[\mathbf{P}^{b_1}[\mathbf{y}\mathbf{R}^{a_2}[\mathbf{P}^{b_2}[\mathbf{y}\cdots\mathbf{R}^{a_d}[\mathbf{P}^{b_d}[\mathbf{y}]]\cdots]]]\big](t).$$

Thus we have

$$\mathfrak{z}_q[\mathbf{w^a}(\mathbf{s})]:=\mathfrak{z}_q[\mathbf{w^a}(\mathbf{s});\ 1]=\mathfrak{z}_q^{(a_1,\dots,a_d)}[s_1,\dots,s_d],$$

where $\mathbf{w^a}(\mathbf{s})=\rho^{a_1}\pi^{s_1-a_1}y\cdots\rho^{a_d}\pi^{s_d-a_d}y\in X_\pi^*$. The shuffle product structure is reflected by $(X_\pi^*,ш)$ where the ш is defined by (33), (34), (37) and (38).

We observe that there is often more than one way to express a type G q-MZV using words because of the relation $\pi=\rho+\mathbf{1}$. For example, using the relations

$$\pi^2\rho y=\pi\rho^2y+\pi\rho y=\rho^3y+2\rho^2y+\rho y$$

we get immediately the relations

$$\mathfrak{z}_q^{(1)}[3]=\mathfrak{z}_q^{(2)}[3]+\mathfrak{z}_q^{(1)}[2]=\mathfrak{z}_q^{(3)}[3]+2\mathfrak{z}_q^{(2)}[2]+\mathfrak{z}_q^{(1)}[1].$$

We call all such relations **P**-**R** relations.

Proposition 9.1 *For all* $\mathbf{u}, \mathbf{v} \in \mathfrak{A}_{\mathrm{G}}^0$, *we have* $\mathbf{u} \sqcup\!\sqcup \mathbf{v} \in \mathfrak{A}_{\mathrm{G}}^0$.

Proof Notice that admissible words in $\mathfrak{A}_{\mathrm{G}}^0$ must end with y and have at least one ρ before the first y. Moreover, the converse is also true. This is rather straightforward if we use the **P**-**R** relations repeatedly to get rid of all the π's.

Now, by using the definition of $\sqcup\!\sqcup$ it is not hard to see that $\mathbf{u} \sqcup\!\sqcup \mathbf{v}$ ends with y and and has at least one ρ before the first y if both $\mathbf{u}$ and $\mathbf{v}$ are admissible. So $\mathbf{u} \sqcup\!\sqcup \mathbf{v} \in \mathfrak{A}_{\mathrm{G}}^0$ and the proposition is proved.

To define the stuffle product we let

$$Y_{\mathrm{G}} = \{z_{a,s} \mid a, s \in \mathbb{Z}_{\geq 0}, a \leq s\},$$

and let $\mathfrak{A}_{\mathrm{G}}$ be the the noncommutative polynomial $\mathbb{Q}$-algebra of words of Y_{G}^* built on the alphabet Y_{G}. Define the type G-admissible words as those in

$$\mathfrak{A}_{\mathrm{G}}^0 = \bigcup_{1 \leq a \leq s} z_{a,s}\mathfrak{A}_{\mathrm{G}}.$$

We can regard $\mathfrak{A}_{\mathrm{G}}$ as a subalgebra of X_π^* by setting $z_{ta,s} = \rho^a \pi^{s-a} y$. Then stuffle product $*_{\mathrm{G}}$ on $\mathfrak{A}_{\mathrm{G}}^0$ can be defined inductively as follows. For any words $\mathbf{u}, \mathbf{v} \in \mathfrak{A}_{\mathrm{G}}^0$ and letters $z_{a,s}, z_{a',s'} \in Y_{\mathrm{G}}$ with $1 \leq a \leq s$ and $1 \leq a' \leq s'$ we set $\mathbf{1} *_{\mathrm{G}} \mathbf{u} = \mathbf{u} = \mathbf{u} *_{\mathrm{G}} \mathbf{1}$ and

$$(z_{a,s}\mathbf{u}) *_{\mathrm{G}} (z_{a',s'}\mathbf{v}) = z_{a,s}(\mathbf{u} *_{\mathrm{G}} z_{a',s'}\mathbf{v}) + z_{a',s'}(z_{a,s}\mathbf{u} *_{\mathrm{G}} \mathbf{v}) + z_{a+a',s+s'}(\mathbf{u} *_{\mathrm{G}} \mathbf{v}).$$

It is easy to show that $(\mathfrak{A}_{\mathrm{G}}^0, *_{\mathrm{G}})$ is a commutative and associative algebra.

We leave the proof of the following theorems to the interested readers. The first result clearly provides the DBSFs of type G q-MZVs.

Theorem 9.2 *For any* $\mathbf{u}, \mathbf{v} \in \mathfrak{A}_{\mathrm{G}}^0 \subset X_\pi^*$ *we have*

$$\mathfrak{z}_q[\mathbf{u} *_{\mathrm{G}} \mathbf{v}] = \mathfrak{z}_q[\mathbf{u} \sqcup\!\sqcup \mathbf{v}] = \mathfrak{z}_q[\mathbf{u}]\mathfrak{z}_q[\mathbf{v}]. \tag{48}$$

The duality relations are given in the cleanest form by Theorem 8.4 which can be translated into the following.

Theorem 9.3 *Let* $\ell \in \mathbb{N}$ *and* $\alpha_j, \beta_j \in \mathbb{N}$ *for all* $j = 1, \dots, \ell$. *Set*

$$\begin{aligned}\mathbf{s} &= (\alpha_1, 0^{\beta_1 - 1}, \alpha_2, 0^{\beta_2 - 1}, \dots, \alpha_\ell, 0^{\beta_\ell - 1}),\\ \mathbf{s}^\vee &= (\beta_\ell, 0^{\alpha_\ell - 1}, \beta_{\ell-1}, 0^{\alpha_{\ell-1} - 1}, \dots, \beta_1, 0^{\alpha_1 - 1}).\end{aligned}$$

Then we have

$$\mathfrak{z}_q^{\mathbf{s}}[\mathbf{s}] = \mathfrak{z}_q^{\mathbf{s}^\vee}[\mathbf{s}^\vee].$$

Remark 9.4 We point out that the duality in Theorem 9.3 cannot be used to derive any relation among MZVs. For example, when $\ell = 1$ the MZV $\zeta(\alpha, 0^{\beta-1})$ converges if $\alpha > \beta$ while the MZV $\zeta(\beta, 0^{\alpha-1})$ converges if $\beta > \alpha$.

10 Numerical Data

In this last section, we compute the $\mathbb{Q}$-linear relations among various types of q-MZVs of small weight by using the DBSFs and the duality relations. Most of the computation is carried out with the computer algebra system MAPLE, version 16. My laptop has Intel Core i7 with CPU speed at 2.4 GHz and 16 GB RAM.

For each type τ we will define the set of type τ-admissible words $\mathsf{W}^{\tau}_{\leqslant w}$ with both weight and depth bounded by w. This is necessary since we allow 0 in some types of q-MZVs. We have to control the number of 0's occurring as arguments in q-MZVs since otherwise the dimensions to be considered becomes infinite. Another reason that the depth has to be bounded is because the duality essentially swaps the depth and the weight.

We denote by $\mathsf{Z}^{\tau}_{\leqslant w}$ the $\mathbb{Q}$-space generated by q-MZVs of type τ corresponding to the type τ-admissible words $\mathsf{W}^{\tau}_{\leqslant w}$, $\mathsf{DS}^{\tau}_{\leqslant w}$ the space generated by the DBSFs, and $\mathsf{DU}^{\tau}_{\leqslant w}$ the space generated by the duality relations. Hence $\mathsf{DU}^{\tau}_{\leqslant w} \setminus \mathsf{DS}^{\tau}_{\leqslant w}$ gives the duality relations that are not contained in $\mathsf{DS}^{\tau}_{\leqslant w}$.

Type I. We have seen that it is necessary to consider q-MZVs of the form $\mathfrak{z}_q^{\mathbf{a}}(\mathbf{s})$ with $(a_j, s_j) = (s_j - 1, s_j)$ or $(a_j, s_j) = (1, 1)$. The latter case corresponds to the words containing the letter θ. We have called all of these values *type* $\tilde{\mathrm{I}}$ *q-MZVs*.

Proposition 10.1 *Let $F_{-1} = 0$, $F_0 = 1$, $F_1 = 1$, ... be the Fibonacci sequence. Then for all $w \geq 1$ we have*

$$\sharp\mathsf{W}^{\mathrm{I}}_{\leqslant w} = 2^{w-1} - 1 \quad \textit{and} \quad \sharp\mathsf{W}^{\tilde{\mathrm{I}}}_{\leqslant w} = F_{2w} - 1.$$

Proof The first equation follows from the same argument as that for MZVs. It is given by the number of integer solutions to the inequality

$$s_1 + \cdots + s_d \leq w, \quad d \geq 1, s_1 \geq 2, s_2, \ldots, s_d \geq 1.$$

Or, more directly and perhaps much easier, we can count the corresponding admissible words. Clearly, there are 2^{w-1} ways to form a word consisting of $w - 1$ letters where the letters can be either ρ or π. Let S_w be the set of such words. We now show that there is a one-to-one correspondence between S_w and the set A_w of admissible word of weight w. First, from each word $\mathbf{w} \in S_w$ we can obtain a word in A_w by inserting a letter y after each π in $\mathbf{w}$ and attach πy at the end. On the other hand, for each word in A_w we may chop off the ending πy and removing all the y's to get a word in $\mathbf{w}$. This establishes the one-to-one correspondence.

Table 2 Dimension of q-MZVs of type $\tilde{\mathrm{I}}$

w	2	3	4	5	6	7
$\sharp \mathsf{W}^{\tilde{\mathrm{I}}}_{\leqslant w}$	4	12	33	88	232	609
Lower bound of dim $\mathsf{Z}^{\tilde{\mathrm{I}}}_{\leqslant w}$	3	7	14	27	50	91
dim $\mathsf{DS}^{\tilde{\mathrm{I}}}_{\leqslant w}$	1	4	17	56	171	497
dim $\mathsf{DU}^{\tilde{\mathrm{I}}}_{\leqslant w} \setminus \mathsf{DS}^{\tilde{\mathrm{I}}}_{\leqslant w}$	0	0	1	2	3	6
Deficiency	0	1,**0**	1,**0**	3	8	15

We now prove the second equation. Let a_n (resp. b_n) be the number of type $\tilde{\mathrm{I}}$ q-MZVs of weight n beginning with $(a_j, s_j) = (1, 1)$ (resp. $(a_j, s_j) = (s_j - 1, s_j)$). Let's call the two different beginnings 1-initial and 2-initial, respectively. Then $a_1 = 1$ and $b_1 = 0$. Now to produce weight $n+1$ 1-initials one can attach $(t, s) = (1, 1)$ to the beginning of any weight n type $\tilde{\mathrm{I}}$ q-MZVs. Moreover, one can change the beginning of any weight n 1-initial to $(t, s) = (0, 1)$ and then attach $(t, s) = (1, 1)$. Thus $a_{n+1} = 2a_n + b_n$. To obtain 2-initials of weight $n+1$ one either changes a 1-initial of weight n to begin with $(t, s) = (1, 2)$ or changes a 2-initial value of weight n to begin with $(s, s+1)$ from $(s-1, s)$ (i.e., increases the first argument by 1). Hence $b_{n+1} = a_n + b_n$. Thus it is easy to see that $a_n = F_{2n-2}$ and $b_n = F_{2n-3}$ for all $n \geq 1$. Therefore

$$\sharp \mathsf{W}^{\tilde{\mathrm{I}}}_{\leqslant w} = \sum_{n=0}^{2w-2} F_n = F_{2w} - 1$$

which can be proved easily by induction.

We find up to weight 3 the following identity (49) cannot be proved by DBSFs and dualities up to weight 3. Let 1_n denote the string where 1 is repeated n times. Then

$$\mathfrak{z}_q^{(1,1)}[2, 1] = \mathfrak{z}_q^{(1,1)}[1, 1] - \mathfrak{z}_q^{(1_3)}[1_3] + \mathfrak{z}_q^{(1,1,0)}[1_3]. \tag{49}$$

Interestingly, (49) can be proved using weight 4 DBSFs and dualities. This is why we put **0** as the final deficiency (Table 2).

Having proved (49), we find, up to weight 4, the only one missing relation is

$$\begin{aligned} \mathfrak{z}_q^{(2,1)}[3, 1] =& \mathfrak{z}_q^{(1,0,1)}[1_3] - 2\mathfrak{z}_q^{(1_3)}[1, 2, 1] + \mathfrak{z}_q^{(1_2,0)}[1, 2, 1] \\ &+ \mathfrak{z}_q^{(1_3)}[2, 1_2] - \mathfrak{z}_q^{(1,0,1)}[2, 1_2] - \mathfrak{z}_q^{(1_3,0)}[1_4] + \mathfrak{z}_q^{(1_2,0_2)}[1_4]. \end{aligned} \tag{50}$$

In weight 5, there are three missing relations:

$$\begin{aligned} \mathfrak{z}_q^{(1_4)}[1_3, 2] =& \mathfrak{z}_q^{\mathbf{t}_1}[1_4] - \mathfrak{z}_q^{\mathbf{t}_2}[1_4] - \mathfrak{z}_q^{(1_4)}[\mathbf{s}_1] - \mathfrak{z}_q^{\mathbf{t}_1}[1_3, 2] - 2\mathfrak{z}_q^{\mathbf{t}_3}[1_5] - 2\mathfrak{z}_q^{\mathbf{t}_4}[\mathbf{s}_2] + 2\mathfrak{z}_q^{\mathbf{t}_5}[1_5], \\ \mathfrak{z}_q^{\mathbf{t}_4}[1_4] =& \mathfrak{z}_q^{(1_3)}[2, 1_2] - \mathfrak{z}_q^{\mathbf{t}_2}[\mathbf{s}_1] - 2\mathfrak{z}_q^{(1_4)}[\mathbf{s}_1] - \mathfrak{z}_q^{\mathbf{t}_6}[\mathbf{s}_2] + \mathfrak{z}_q^{(1_4)}[\mathbf{s}_2] + \mathfrak{z}_q^{\mathbf{t}_7}[1_5] - \mathfrak{z}_q^{\mathbf{t}_1}[1_4] \\ &- \mathfrak{z}_q^{(1_3)}[2, 1, 2] - \mathfrak{z}_q^{\mathbf{t}_5}[1_5] + \mathfrak{z}_q^{\mathbf{t}_8}[1_4] + 2\mathfrak{z}_q^{\mathbf{t}_2}[1_4] - \mathfrak{z}_q^{\mathbf{t}_2}[2, 1_3] - \mathfrak{z}_q^{\mathbf{t}_8}[2, 1_3], \end{aligned}$$

Table 3 Dimension of q-MZVs of type I

w	2	3	4	5	6	7	8	9
$\sharp \mathsf{W}^{\mathrm{I}}_{\leqslant w}$	1	3	7	15	31	63	127	255
Lower bound of dim $\mathsf{Z}^{\mathrm{I}}_{\leqslant w}$	1	2	4	7	11	18	27	42
dim $\mathsf{DS}^{\mathrm{I}}_{\leqslant w}$	0	1	3	8	20	45		
dim $\mathsf{DU}^{\mathrm{I}}_{\leqslant w} \setminus \mathsf{DS}^{\mathrm{I}}_{\leqslant w}$	0	0	0	0	0	0		
Deficiency	0	0	0	0	0	0		

$$\begin{aligned}\mathfrak{z}_q^{(1_4)}[1_3, 2] =& 3\mathfrak{z}_q^{(1_3)}[2, 1_2] - 3\mathfrak{z}_q^{(1_3)}[2, 1, 2] - 3\mathfrak{z}_q^{\mathbf{t}_4}[1_4] + \mathfrak{z}_q^{\mathbf{t}_2}[1_4] - \mathfrak{z}_q^{\mathbf{t}_3}[1_5] \\ &-\mathfrak{z}_q^{(1_4)}[\mathbf{s}_1] - \mathfrak{z}_q^{\mathbf{t}_4}[\mathbf{s}_1] - 2\mathfrak{z}_q^{\mathbf{t}_2}[\mathbf{s}_1] - \mathfrak{z}_q^{(1_3,0_2)}[1_5] - 2\mathfrak{z}_q^{(1_4)}[\mathbf{s}_2] + \mathfrak{z}_q^{\mathbf{t}_4}[\mathbf{s}_2] \\ &+\mathfrak{z}_q^{\mathbf{t}_4}[2, 1_3] + 2\mathfrak{z}_q^{\mathbf{t}_1}[1_4] + \mathfrak{z}_q^{\mathbf{t}_1}[1_3, 2] + 2\mathfrak{z}_q^{(1_2,0_2,1)}[1_5],\end{aligned}$$

where $\mathbf{s}_1 = (1, 2, 1_2)$, $\mathbf{s}_2 = (1_2, 2, 1)$, $\mathbf{t}_1 = (1, 0, 1_2)$, $\mathbf{t}_2 = (1_2, 0, 1)$, $\mathbf{s}_3 = (\mathbf{t}_2, 0)$, $\mathbf{t}_4 = (1_3, 0)$, $\mathbf{t}_5 = (1, 0_2, 1_2)$, $\mathbf{t}_6 = (1, 0, 1, 0)$, $\mathbf{t}_7 = (1, 0_3, 1)$, and $\mathbf{t}_8 = (1, 0_2, 1)$.

Equation (50) was initially verified numerically. Even with all the DBSFs and dualities from weight 5 and 6 this still would not follow. Fortunately, we will see in a moment that this relation can be proved using type G q-MZVs. However, the three missing relations in weight 5 are only proved numerically, since, unfortunately, there are too many type G q-MZVs of weight 5 so the computer computation requires too much memory to provide a solution at the moment.

Using the relations obtained above for type $\tilde{\mathrm{I}}$ q-MZVs we can compute the following data for type I q-MZVs (Table 3).

It is consistent with Takeyama's computation at the end of [26]. However, our computation shows that the DBSFs from type $\tilde{\mathrm{I}}$ q-MZVs already imply all the relations among type I q-MVZs, at least when the weight is less than 8. We thus can think these type $\tilde{\mathrm{I}}$ DBSFs as "regularized" DBSFs for type I q-MVZs.

Conjecture 10.2 *All the $\mathbb{Q}$-linear relations of type I q-MZVs can be derived by the regularized DBSFs, i.e., by the DBSFs for type $\tilde{\mathrm{I}}$ q-MZVs.*

Type II. For each fixed weight $w \geq 1$ we collect all the type II-admissible words of the following form since we want to use the duality relations to its maximal utility. Such admissible words must consist of letters ρ and y only, begin with ρ, end with y, and the occurrence of ρ and y is at most w each. For example, we have the duality

$$\mathfrak{z}_q^{\mathrm{II}}(\rho^3 y^2 \rho y^4) = \mathfrak{z}_q^{\mathrm{II}}(\rho^4 y \rho^2 y^3) \Longrightarrow \mathfrak{z}_q^{\mathrm{II}}(3, 0, 1, 0_3) = \mathfrak{z}_q^{\mathrm{II}}(4, 2, 0_2)$$

when we consider weight 6.

Proposition 10.3 *For all $w \geq 1$, the number of type II-admissible words is*

Table 4 Dimension of q-MZVs of type II

w	1	2	3	4	5	6
$\sharp\mathsf{W}^{\mathrm{II}}_{\leqslant w}$	1	5	19	69	251	923
Lower bound of dim $\mathsf{Z}^{\mathrm{II}}_{\leqslant w}$	1	3	12	30	73	173
dim $\mathsf{DS}^{\mathrm{II}}_{\leqslant w}$	0	1	5	28	124	536
dim $\mathsf{DU}^{\mathrm{II}}_{\leqslant w} \setminus \mathsf{DS}^{\mathrm{II}}_{\leqslant w}$	0	1	2	8	35	127
Deficiency	0	0	0	3,**0**	19,**6**	87

$$\sharp\mathsf{W}^{\mathrm{II}}_{\leqslant w} = \sum_{i=0}^{w-1}\sum_{j=0}^{w-1}\binom{i+j}{j} = \binom{2w}{w} - 1.$$

Remark 10.4 This is the sequence A030662 according to the On-Line Encyclopedia of Integer Sequences http://oeis.org.

Proof For the first equality, note that if $i+1$ (resp. $j+1$) is the number of occurrence of ρ (resp. y) in an admissible word of $\mathsf{W}^{\mathrm{II}}_{\leqslant w}$ then we can put one ρ at the beginning and one y at the end, then put i of the other ρ's and j of the other y's in between in arbitrary order. Thus, by a well-known binomial identity

$$1+\sharp\mathsf{W}^{\mathrm{II}}_{\leqslant w} = 1+\sum_{i=0}^{w-1}\sum_{j=0}^{w-1}\binom{i+j}{j} = 1+\sum_{i=0}^{w-1}\binom{w+i}{w-1} = \sum_{i=0}^{w}\binom{w+i-1}{i} = \binom{2w}{w}.$$

This completes the proof of the proposition.

Up to weight 4, the following three independent relations cannot be proved using DBSFs and dualities up to weight 4 (Table 4).

$$\begin{aligned}
\mathfrak{z}_q^{\mathrm{II}}[1,0,3] =& \mathfrak{z}_q^{\mathrm{II}}[2,2]+3\mathfrak{z}_q^{\mathrm{II}}[1_2,2]+2\mathfrak{z}_q^{\mathrm{II}}[1,0,2,0]-2\mathfrak{z}_q^{\mathrm{II}}[1_2,0,1]\\
&+\mathfrak{z}_q^{\mathrm{II}}[1_2,0,2]+\mathfrak{z}_q^{\mathrm{II}}[1_2,1,0]-\mathfrak{z}_q^{\mathrm{II}}[1,2,0,1]+2\mathfrak{z}_q^{\mathrm{II}}[2,0,1_2],\\
\mathfrak{z}_q^{\mathrm{II}}[3,0] =& \mathfrak{z}_q^{\mathrm{II}}[2,2]-2\mathfrak{z}_q^{\mathrm{II}}[3,1]+\mathfrak{z}_q^{\mathrm{II}}[1,0,2,0]-2\mathfrak{z}_q^{\mathrm{II}}[1_2,0,1]+2\mathfrak{z}_q^{\mathrm{II}}[1_2,1,0]\\
&-\mathfrak{z}_q^{\mathrm{II}}[2,0,2,0]+\mathfrak{z}_q^{\mathrm{II}}[3,0_2,0]+2\mathfrak{z}_q^{\mathrm{II}}[3,0_2,1]-\mathfrak{z}_q^{\mathrm{II}}[3,0,1,0]+2\mathfrak{z}_q^{\mathrm{II}}[3,1,0_2],\\
\mathfrak{z}_q^{\mathrm{II}}[1,0,3] =& \mathfrak{z}_q^{\mathrm{II}}[2,2]+2\mathfrak{z}_q^{\mathrm{II}}[3,1]+\mathfrak{z}_q^{\mathrm{II}}[1_2,2]+4\mathfrak{z}_q^{\mathrm{II}}[1_2,0,1]+\mathfrak{z}_q^{\mathrm{II}}[1_2,0,2]\\
&+\mathfrak{z}_q^{\mathrm{II}}[1_2,1,0]+\mathfrak{z}_q^{\mathrm{II}}[1,2,0,1]+4\mathfrak{z}_q^{\mathrm{II}}[2,0,1,0]+2\mathfrak{z}_q^{\mathrm{II}}[2,1,0,1]+2\mathfrak{z}_q^{\mathrm{II}}[2,1_2,0].
\end{aligned}$$

But using DBSFs and dualities in weight 5, these can all be verified. In weight 5, we have to use the relations from weight 6 to push the deficiency from 19 down to 6. It is very likely that relations from weight 7 (or even higher) can reduce this further down to 0. But our computer runs out of memories so this is not proved.

Type III. The set of type III-admissible words $\mathsf{W}^{\mathrm{III}}_{\leqslant w}$ up to weight w consist of those of the form $\pi^{s_1-1}\rho y \pi^{s_2} y \cdots \pi^{s_d} y$ with $d \le w$, $|\mathbf{s}| \le w$, $s_1 \ge 1$ and $s_2, \dots, s_d \ge 0$. First we have

Table 5 Dimension of q-MZVs of type III

w	1	2	3	4	5	6
$\sharp \mathsf{W}^{\mathrm{III}}_{\leqslant w}$	1	5	19	69	251	923
Lower bound of dim $\mathsf{Z}^{\mathrm{III}}_{\leqslant w}$	1	4	12	30	73	173
dim $\mathsf{DS}^{\mathrm{III}}_{\leqslant w}$	0	1	5	28	124	536
dim $\mathsf{DU}^{\mathrm{III}}_{\leqslant w} \setminus \mathsf{DS}^{\mathrm{III}}_{\leqslant w}$	0	0	1	1	5	4
Deficiency	0	0	1,**0**	10,**0**	49,**6**	210,**87**

Proposition 10.5 *For all $w \geq 1$, we have*

$$\sharp \mathsf{W}^{\mathrm{III}}_{\leqslant w} = \binom{2w}{w} - 1.$$

Proof Notice there is an onto map from $\mathsf{W}^{\mathrm{III}}_{\leqslant w}$ to $\mathsf{W}^{\mathrm{II}}_{\leqslant w}$ by changing the all the π's to ρ. For the inverse map, we can change all the ρ's to π except for the one immediately before the first y. Thus this is a one-to-one correspondence and therefore the proposition follows from Proposition 10.3.

We find that the deficiency is not zero when the weight $w = 3, 4, 5, 6$. Moreover, none of these missing $\mathbb{Q}$-linear relations can be recovered even if we consider all the DBSFs and dualities of weight up to 6.

The only missing relation in weight 3 that cannot be proved is

$$\mathfrak{z}_q^{\mathrm{III}}[1,0,1] = 2\mathfrak{z}_q^{\mathrm{III}}[1,1,0] - \mathfrak{z}_q^{\mathrm{III}}[1,2,0] - \mathfrak{z}_q^{\mathrm{III}}[2,0,0] + \mathfrak{z}_q^{\mathrm{III}}[2,0,1]. \tag{51}$$

Up to weight 4 there are 10 missing, up to weight 5, 49, and up to weight 6, 210. Below, we will see that all of the 10 missing relations up to weight 4 including (51) can be proved using type G q-MZVs. Similarly, the deficiency up to weight 5 and 6 can be reduced to 6 and 87, respectively.

Type IV. To study type IV q-MZVs $\mathfrak{z}_q^{(s_1-1,s_2,\dots,s_d)}[s_1,\dots,s_d]$ we have used the special type II values $\mathfrak{z}_q^{\mathrm{II}}[1,s_2,\dots,s_d]$ to facilitate us (which can be thought as a kind of regularization). Type IV q-MZVs together with these values have been called type $\widetilde{\mathrm{IV}}$ q-MZVs.

Proposition 10.6 *For all $w \geq 1$, we have*

$$\sharp \mathsf{W}^{\mathrm{IV}}_{\leqslant w} = \binom{2w-1}{w} - 1, \qquad \sharp \mathsf{W}^{\widetilde{\mathrm{IV}}}_{\leqslant w} = \binom{2w}{w} - 1.$$

Remark 10.7 The first number gives the sequence A010763 according to the On-Line Encyclopedia of Integer Sequences http://oeis.org.

Proof Notice that type IV-admissible q-MZVs are in one-to-one correspondence to the set $\{(x_1,\dots,x_l) \in (\mathbb{Z}_{\geq 0})^l \mid x_1+\cdots+x_l = j, 0 \leq j \leq w-2, 1 \leq l \leq w\}$. For

Table 6 Dimension of q-MZVs of type $\widetilde{\mathrm{IV}}$

w	1	2	3	4	5	6
$\sharp \mathsf{W}^{\widetilde{\mathrm{IV}}}_{\leqslant w}$	1	5	19	69	251	923
Lower bound of dim $\mathsf{Z}^{\widetilde{\mathrm{IV}}}_{\leqslant w}$	1	4	12	30	73	173
dim $\mathsf{DS}^{\widetilde{\mathrm{IV}}}_{\leqslant w}$	0	1	5	28	124	536
dim $\mathsf{DU}^{\widetilde{\mathrm{IV}}}_{\leqslant w} \setminus \mathsf{DS}^{\widetilde{\mathrm{IV}}}_{\leqslant w}$	0	0	1	1	4	4
Deficiency	0	0	1,**0**	10,**0**	50,**6**	210,**87**

each fixed j we see that the number of nonnegative integer solutions of $x_1 + \cdots + x_l = j$ is given by $\binom{l+j-1}{l-1}$. But

$$\sum_{l=1}^{w} \binom{l+j-1}{l-1} = \binom{w+j}{w-1}$$

by a well-known binomial identity. By the proof similar to that of Proposition 10.3 we see that

$$\sharp \mathsf{W}^{\mathrm{IV}}_{\leqslant w} = \sum_{j=0}^{w-2} \binom{w+j}{w-1} = \binom{2w-1}{w} - 1.$$

For the second equation, we note that in the word form we have the additional contribution of the following words: ρy and $\rho y \rho^{s_1} y \ldots \rho^{s_d} y$, $|\mathbf{s}| < w$, $1 \le d < w$. The number of such words is given by (i = number of ρ's, j = number of y's)

$$1 + \sum_{j=0}^{w-2} \sum_{i=0}^{w-1} \binom{i+j}{i} = 1 + \sum_{j=0}^{w-2} \binom{w+j}{w-1} = 1 + \sharp \mathsf{W}^{\mathrm{IV}}_{\leqslant w}.$$

Therefore

$$\sharp \mathsf{W}^{\widetilde{\mathrm{IV}}}_{\leqslant w} = 1 + 2\sharp \mathsf{W}^{\mathrm{IV}}_{\leqslant w} = 2\binom{2w-1}{w} - 1 = \binom{2w}{w} - 1.$$

The proposition is now proved.

Type $\widetilde{\mathrm{IV}}$ q-MZVs are similar to type II and III in the sense that the deficiency is often nonzero, at least when the weight is less than 6. For example, in weight 3 we have the following identity which cannot be proved using the DBSFs and dualities if we only restrict to type $\widetilde{\mathrm{IV}}$ q-MZVs of weight and depth no greater than 3.

$$\mathfrak{z}_q^{\mathrm{IV}}[2,0,1] = \mathfrak{z}_q^{\mathrm{II}}[1,0,1] + \mathfrak{z}_q^{\mathrm{II}}[1,2,0]$$

However this identity follows from weight 4 DBSFs and dualities.

Table 7 Dimension of q-MZVs of type IV

w	2	3	4	5	6
$\sharp\mathsf{W}^{\mathrm{IV}}_{\leqslant w}$	2	9	34	125	461
Lower bound of dim $\mathsf{Z}^{\mathrm{IV}}_{\leqslant w}$	2	7	20	55	141
dim $\mathsf{DS}^{\mathrm{IV}}_{\leqslant w}$	0	7	9	51	205
dim $\mathsf{DU}^{\mathrm{IV}}_{\leqslant w} \setminus \mathsf{DS}^{\mathrm{IV}}_{\leqslant w}$	0	0	0	2	24
Deficiency	0	0	5,**0**	17,**0**	91,**56**

Comparing Tables 5 and 6 we observe that there should be some hidden relations between type III and $\widetilde{\mathrm{IV}}$ q-MZVs. Although the dimensions seem to be the same, at least for lower weight, the deficiencies are very different. But using the most general type G values to be considered in a moment, we can make all the deficiencies smaller.

We can now use all of the relations among type $\widetilde{\mathrm{IV}}$ q-MZVs to deduce those for type IV and collect the data in Table 7. Furthermore, by converting all the missing relations using type II values we can reduce all the deficiencies up to weight 5 to 0. For weight 6, using type II values we can only reduce the deficiency from 91 to 56. It is possible that this can be further reduced to 0 using weight 7 relations of type II values.

Type G. To study the general type G q-MZVs $\mathfrak{z}_q^{(a_1,\dots,a_d)}[s_1,\dots,s_d]$ we need all of the following relations we have defined so far: DBSFs, **P**-**R** and duality relations.

Proposition 10.8 *For all $w \geq 1$, we have*

$$\sharp\mathsf{W}^{\mathrm{G}}_{\leqslant w} = \sum_{1\leq d\leq k\leq w} \sum_{\substack{x_1+\cdots+x_d=d+k-1\\ x_1,\dots,x_d\geq 1}} x_1x_2\cdots x_d.$$

Proof For each fixed depth d and weight $k \leq w$, let $\mathfrak{z}_q^{(a_1,\dots,a_d)}[s_1,\dots,s_d]$ be a type G-admissible q-MZV satisfying $s_1+\cdots+s_d = k$, $1 \leq a_1 \leq s_1$, $0 \leq a_j \leq s_j$ for all $j \geq 2$. When $s_1,\dots,s_d$ are fixed and $a_1,\dots,a_d$ vary, the number of such values is given by

$$s_1(s_2+1)(s_3+1)\cdots(s_d+1).$$

Hence the proposition follows by setting $x_1 = s_1, x_2 = s_2+1, \dots, x_d = s_d+1$.

Let $\mathbf{P}\text{-}\mathbf{R}^{\mathrm{G}}_{\leqslant w}$ be the space generated by all the **P**-**R** relations of weight bounded by w. Then we see that DBSFs are far from enough and both **P**-**R** relations and duality relations contribute non-trivially. Table 8 provides our computational data for the lower weight cases. One can see that the number of admissible words increases very fast so that it is very difficult to prove relations of other type q-MZVs by first finding all the relations for type G q-MZVs. This is possible theoretically, but not feasible with our current computer powers.

Table 8 Dimension of q-MZVs of type G

w	1	2	3	4	5	6
$\sharp \mathsf{W}^{\mathrm{G}}_{\leqslant w}$	1	8	49	294	1791	11087
Lower bound of dim $\mathsf{Z}^{\mathrm{G}}_{\leqslant w}$	1	4	12	30	73	173
dim $\mathsf{DS}^{\mathrm{G}}_{\leqslant w}$	0	1	8	76	≤608	
dim $\mathbf{P}$-$\mathbf{R}^{\mathrm{G}}_{\leqslant w} \setminus (\mathsf{DS}^{\mathrm{G}}_{\leqslant w} \bigcup \mathsf{DU}^{\mathrm{G}}_{\leqslant w})$	0	3	27	177	≤1540	
dim $\mathsf{DU}^{\mathrm{G}}_{\leqslant w} \setminus (\mathbf{P}$-$\mathbf{R}^{\mathrm{G}}_{\leqslant w} \bigcup \mathsf{DS}^{\mathrm{G}}_{\leqslant w})$	0	0	2	8	≤219	
Deficiency	0	0	0	3, **0**		

Table 9 Dimension of type O q-MZVs, proved rigorously for $w \le 6$ and numerically for $w \le 12$

w	2	3	4	5	6	7	8	9	10	11	12
$\sharp \mathsf{W}^{\mathrm{O}}_{\leqslant w}$	1	2	4	7	12	20	33	54	88	143	232
Lower bound of dim $\mathsf{Z}^{\mathrm{O}}_{\leqslant w}$	1	2	4	7	11	18	27	42	63	95	142
dim $\mathsf{DS}^{\mathrm{O}}_{\leqslant w} \cup \mathsf{DU}^{\mathrm{O}}_{\leqslant w}$	0	0	0	0	1	2	6	12	25	48	90

Fortunately, by using **P**-**R** relations, all the type G q-MZVs can be converted to $\mathbb{Q}$-linear combinations of type II values. Therefore, the three missing relations in weight 4 must be provable using weight 5 DBSFs, **P**-**R** and duality relations.

Hence, as we expected, the missing relation (51) for type III q-MZVs of weight 3 and the 9 missing relations of weight 4 can now be proved. And furthermore, the only one missing relation (50) for type $\tilde{\mathrm{I}}$ q-MZVs of weight 4 can now be proved. We can also obtain the lower bound of dim $\mathsf{Z}^{\mathrm{G}}_{\leqslant w}$ from that of type II q-MZVs.

Type O. Using Corollary 6.6 we may regard Okounkov's q-MZVs as $\mathbb{Q}$-linear combinations of the q-MZVs $\mathfrak{z}_q^{\mathbf{a}}[\mathbf{s}]$ for suitable auxiliary variable **a**. Further by using the **P**-**R** relations we may further reduce this to type II q-MZVs where we don't need the letter π (Table 9).

Applying the same idea as above it is possible to verify the following Okounkov's dimension conjecture, at least when the weight is small.

Conjecture 10.9 *Let* $\mathbf{Z}^{\mathrm{O}}_{w}$ *be the* $\mathbb{Q}$*-vector space generated by* $\mathfrak{z}_q^{\mathrm{O}}[\mathbf{s}]$, $|\mathbf{s}| \le w$. *Then*

$$\sum_{w=0}^{\infty} t^w \dim \mathbf{Z}^{\mathrm{O}}_{\leqslant w} = \frac{1}{1-t-t^2+t^6+t^8-t^{13}} - \frac{1}{1-t}$$
$$= t^2 + 2t^3 + 4t^4 + 7t^5 + 11t^6 + 18t^7 + 27t^8 + 42t^9 + 63t^{10} + 95t^{11} + 142t^{12} + O(t^{13}).$$

For example, we have verified all of the following $\mathbb{Q}$-linearly independent relations in the lower weight cases up to q^{100}, and we can rigorously prove the first identity (52) involving only weight 4 and 6 values by using the relations we have found for type II q-MZVs: ($\mathfrak{z} = \mathfrak{z}_q^{\mathrm{O}}$)

$$\begin{aligned}
4\mathfrak{z}[6] &= \mathfrak{z}[2,2] + 12\mathfrak{z}[3,3] - 6\mathfrak{z}[4,2], \\
4\mathfrak{z}[7] &= \mathfrak{z}[2,3] + \mathfrak{z}[3,2] + 8\mathfrak{z}[3,4] + 6\mathfrak{z}[4,3] - 4\mathfrak{z}[5,2], \\
\mathfrak{z}[8] &= \mathfrak{z}[2,4] - \mathfrak{z}[6] + 2\mathfrak{z}[3,3] + 6\mathfrak{z}[4,4], \\
9\mathfrak{z}[8] &= \mathfrak{z}[6] - 6\mathfrak{z}[3,3] + 3\mathfrak{z}[4,2] + 20\mathfrak{z}[3,5] + 16\mathfrak{z}[5,3] - 10\mathfrak{z}[6,2], \\
\mathfrak{z}[8] &= 2\mathfrak{z}[2,6] - \mathfrak{z}[6] + 2\mathfrak{z}[3,3] + 4\mathfrak{z}[3,5] - 16\mathfrak{z}[5,3] \\
&\quad -6\mathfrak{z}[2,3,3] + 3\mathfrak{z}[2,4,2] - 6\mathfrak{z}[3,2,3] - 3\mathfrak{z}[4,2,2], \\
4\mathfrak{z}[3,6] &= \mathfrak{z}[2,5] + 4\mathfrak{z}[5,2] + 3\mathfrak{z}[3,4] + 6\mathfrak{z}[4,5] + 8\mathfrak{z}[5,4] + 2\mathfrak{z}[7,2], \\
8\mathfrak{z}[9] &= \mathfrak{z}[3,4] - 5\mathfrak{z}[2,5] - 8\mathfrak{z}[5,2] - 30\mathfrak{z}[4,5] - 2\mathfrak{z}[4,3] - 36\mathfrak{z}[5,4] - 10\mathfrak{z}[6,3], \\
6\mathfrak{z}[4,2] &= 10\mathfrak{z}[6] + 42\mathfrak{z}[8] - 60\mathfrak{z}[2,6] - 12\mathfrak{z}[3,3] - 120\mathfrak{z}[3,5] + 312\mathfrak{z}[5,3] \\
&\quad -15\mathfrak{z}[2,2,2] + 180\mathfrak{z}[2,3,3] - 90\mathfrak{z}[2,4,2] + 180\mathfrak{z}[3,2,3] + 60\mathfrak{z}[3,3,2], \\
72\mathfrak{z}[9] &= 62\mathfrak{z}[5,2] + 40\mathfrak{z}[2,5] - 4\mathfrak{z}[3,4] + 40\mathfrak{z}[3,6] - 2\mathfrak{z}[4,3] + 240\mathfrak{z}[4,5] \\
&\quad +264\mathfrak{z}[5,4] - 5\mathfrak{z}[2,2,3] - 60\mathfrak{z}[3,3,3] - 30\mathfrak{z}[4,2,3], \\
16\mathfrak{z}[9] &= 2\mathfrak{z}[3,4] - 10\mathfrak{z}[2,5] - 12\mathfrak{z}[2,7] - 8\mathfrak{z}[5,2] - 60\mathfrak{z}[4,5] - 24\mathfrak{z}[5,4] \\
&\quad +4\mathfrak{z}[2,3,2] + 4\mathfrak{z}[3,2,2] + 3\mathfrak{z}[2,2,3] + 24\mathfrak{z}[2,3,4] + 18\mathfrak{z}[2,4,3] \\
&\quad +12\mathfrak{z}[3,3,3] - 12\mathfrak{z}[2,5,2] + 24\mathfrak{z}[3,2,4] + 6\mathfrak{z}[4,3,2], \\
64\mathfrak{z}[9] &= 40\mathfrak{z}[2,5] + 20\mathfrak{z}[2,7] - 8\mathfrak{z}[3,4] + 44\mathfrak{z}[5,2] + 20\mathfrak{z}[3,6] - 4\mathfrak{z}[4,3] \\
&\quad +240\mathfrak{z}[4,5] + 168\mathfrak{z}[5,4] - 5\mathfrak{z}[2,3,2] - 5\mathfrak{z}[2,2,3] - 40\mathfrak{z}[2,3,4] - 30\mathfrak{z}[2,4,3] \\
&\quad +20\mathfrak{z}[2,5,2] - 5\mathfrak{z}[3,2,2] - 40\mathfrak{z}[3,2,4] - 100\mathfrak{z}[3,3,3] + 10\mathfrak{z}[3,4,2], \\
56\mathfrak{z}[9] &= 30\mathfrak{z}[2,5] + 20\mathfrak{z}[2,7] + 26\mathfrak{z}[5,2] - \mathfrak{z}[3,4] + 40\mathfrak{z}[3,6] - 6\mathfrak{z}[4,3] \\
&\quad +180\mathfrak{z}[4,5] + 112\mathfrak{z}[5,4] - 5\mathfrak{z}[2,2,3] - 5\mathfrak{z}[2,3,2] - 5\mathfrak{z}[3,2,2] - 40\mathfrak{z}[2,3,4] \\
&\quad +20\mathfrak{z}[5,2,2] - 40\mathfrak{z}[3,2,4] - 30\mathfrak{z}[2,4,3] + 20\mathfrak{z}[2,5,2] - 140\mathfrak{z}[3,3,3].
\end{aligned} \tag{52}$$

Therefore, Conjecture 10.9 is proved rigorously up to weight 6 (inclusive), and verified numerically up to weight 12 (inclusive). The list of relations for weight 10 to 12 is too long to be presented here.

11 Conclusions

We have studied various q-analogs of MZVs in this paper using the uniform method of Rota–Baxter algebras. Among these q-MZVs, there are many $\mathbb{Q}$-linear relations, most of which can be proved using DBSFs, **P**-**R** and duality relations.

From the data collected in Sect. 10, we have seen that for all of the type $\widetilde{\mathrm{I}}$, II, III and $\widetilde{\mathrm{IV}}$ q-MZVs duality relations are necessary to generate some $\mathbb{Q}$-linear relations among q-MZVs that are missed by the DBSFs, at least when the weight is large enough. However, the combination of all the DBSFs and dualities are often not exhaustive yet. Sometimes, this difficulty can be overcome by increasing the weight and depth. But this seems to fail in some other cases, for example, for type $\widetilde{\mathrm{I}}$ q-MZVs of weight 4.

We can improve the above situation by considering the more general type G values. The advantage is that we have the new **P-R** relations which provide a lot of new relations between type G q-MZVs, much more than the DBSFs and dualities combined. The disadvantage is that there are too many type G values so that even when the weight is 5 our computer power is too weak to produce all the necessary relations. However, by using **P-R** relations all type G values can be converted to $\mathbb{Q}$-linear combinations of type II values which can be handled by computer a lot easier.

As we mentioned in the introduction our method can be easily adapted to study the q-MZVs of the following forms:

$$\mathfrak{z}_q^{(a_1,\dots,a_d)}[s_1,\dots,s_d], \qquad \mathfrak{z}_q^{(s_1-a_1,\dots,s_d-a_d)}[s_1,\dots,s_d],$$

where $a_1 \ge a_2 \ge \cdots \ge a_d \ge 0$ are all integers. The monotonicity guarantees that a good stuffle structure can be defined. For $\mathfrak{z}_q^{(a_1,\dots,a_d)}[s_1,\dots,s_d]$, we need to use embedding (A) together with shifting operator $\mathscr{S}_-$ in defining the stuffle and, for $\mathfrak{z}_q^{(s_1-a_1,\dots,s_d-a_d)}[s_1,\dots,s_d]$, we need (B) together with $\mathscr{S}_+$.

As an application, we are able to prove Okounkov's Conjecture 10.9 rigorously up to weight 6 (inclusive), and verify it numerically up to weight 12 (inclusive). It would be more effective if one can define a shuffle structure for type O values themselves and find a relation to the differential operator $q\frac{d}{dq}$ which should play an important role in the study of these values.

Acknowledgements This work, supported by NSF grant DMS-1162116, was done while the author was visiting Max Planck Institute for Mathematics and ICMAT at Madrid, Spain. He is very grateful to both institutions for their hospitality and support. He also would like to thank Kurusch Ebrahimi-Fard for a few enlightening conversations and his detailed explanation of their joint paper [7]. The anonymous referees provided many valuable comments and suggestions which greatly improved the clarity of the paper.

References

1. Bachmann, H., Kühn, U.: The algebra of multiple divisor functions and applications to multiple zeta values. Ramanujan J. **40**(3), 605–648 (2016)
2. Bachmann, H., Kühn, U.: A short note on a conjecture of Okounkov about a q-analogue of multiple zeta values. arXiv:1407.6796
3. Bradley, D.M.: Multiple q-zeta values. J. Algebra **283**, 752–798 (2005)
4. Broadhurst, D.J.: Conjectured enumeration of irreducible multiple zeta values, from knots and Feynman diagrams. arXiv:hep-th/9612012
5. Brown, F.: Mixed Tate motives over Spec($\mathbb{Z}$). Ann. Math. **175**, 949–976 (2012)
6. Chen, K.-T.: Algebras of iterated path integrals and fundamental groups. Trans. Amer. Math. Soc. **156**, 359–379 (1971)
7. Castillo Medina, J., Ebrahimi-Fard, K., Manchon, D.: Unfolding the double shuffle structure of q-multiple zeta values. Bull. Aust. Math. Soc. **91**, 368–388 (2015)
8. Euler, L.: Meditationes circa singulare serierum genus. Novi Comm. Acad. Sci. Petropol. **20** , 140–186 (1776); reprinted in Opera Omnia, Ser. I, Vol. **15**, B. Teubner, Berlin, 217–267 (1927)

9. Ecalle, J.: Les fonctions résurgentes, Vol. 2: Les fonctions résurgentes appliquées à l'itération (in French), Publ. Math. Orsay **81.06** (1981), # 283 pp. Available online: www.math.u-psud.fr/~ecalle/publi.html
10. Goncharov, A.B., Manin, Y.I.: Multiple ζ-motives and moduli spaces $M_{0,n}$. Compos. Math. **140**, 1–14 (2004)
11. Guo, L.: An Introduction to Rota-Baxter Algebra, Surveys of Modern Mathematics series, vol. 4. Press of Boston Inc., Intl (2012)
12. Guo, L., Keigher, W.: On differential Rota-Baxter algebras. J. Pure Appl. Alg. **212**, 522–540 (2008)
13. Hoffman, M.E.: Multiple harmonic series. Pacific J. Math. **152**, 275–290 (1992)
14. Hoffman, M.E.: The algebra of multiple harmonic series. J. Alg. **194**, 477–495 (1997)
15. Hoffman, M.E.: Quasi-shuffle products. J. Alg. Combin. **11**, 49–68 (2000)
16. Ihara, K., Kaneko, M., Zagier, D.: Derivation and double shuffle relations for multiple zeta values. Compos. Math. **142**, 307–338 (2006)
17. Kaneko, M., Kurokawa, N., Wakayama, M.: A variation of Euler's approach to values of the Riemann zeta function. Kyushu J. Math. **57**, 175–192 (2003). arXiv:math/0206024
18. Kurokawa, N., Lalin, M., Ochiai, H.: Higher Mahler measures and zeta functions. Acta Arith. **135**(3), 269–297 (2008)
19. Le, T.Q.T., Murakami, J.: Kontsevich's integral for the Homfly polynomial and relations between values of multiple zeta functions. Topology Appl. **62**, 193–206 (1995)
20. Ohno, Y., Okuda, J., Zudilin, W.: Cyclic q-MZSV sum. J. Number Theory **132**, 144–155 (2012)
21. Okounkov, A.: Hilbert schemes and multiple q-zeta values. Func. Ana. Appl. **48**(2), 138–144 (2014)
22. Okuda, J., Takeyama, Y.: On relations for the q-multiple zeta values. Ramanujan J. **14**, 379–387 (2007)
23. Schlesinger, K.-G.: Some remarks on q-deformed multiple polylogarithms. arXiv:math/0111022
24. Singer, J.: On q-analogues of multiple zeta values. Funct. Approx. Comment. Math. **53**(1), 135–165 (2015)
25. Singer, J.: On Bradley's q-MZVs and a generalized Euler decomposition formula. J. Alg. **454**, 92–122 (2016)
26. Takeyama, Y.: The algebra of a q-analogue of multiple harmonic series. SIGMA **9**, Paper 0601, 15 pp (2013)
27. Terasoma, T.: Geometry of multiple zeta values. In: Proceedings International Congress of Mathematicians (Madrid, 2006), Vol. **II**, pp. 627–635, European Mathe. Soc., Zürich (2006)
28. Zagier, D.: Values of zeta functions and their applications. In: First European Congress of Mathematics (Paris, 1992), Vol. **II**, 497–512, A. Joseph et al. (eds.), Birkhäuser, Basel (1994)
29. Zhao, J.: Multiple q-zeta functions and multiple q-polylogarithms. Ramanujan J. **14**, 189–221 (2007). arXiv:math/0304448
30. Zudilin, W.: Algebraic relations for multiple zeta values. Russian Math. Surv. **58**(1), 1–29 (2003)
31. Zudilin, W.: Multiple q-zeta brackets. Mathematics **3**, 119–130 (2015)

q-Analogues of Multiple Zeta Values and Their Application in Renormalization

Johannes Singer

Abstract In this paper we report on recent developments on q-analogues of multiple zeta values (MZVs), which are power series in a formal parameter q that reduce to classical MZVs in the limit $q \to 1$. First of all, we systematically develop the double shuffle relations of three q-models, whose shuffle products rely on a description of iterated Rota–Baxter operators. In the second part we use two of these q-models to construct solutions to the renormalization problem of MZVs, i.e., a systematic extension of MZVs to negative integers. In one case the renormalized MZVs satisfy the quasi-shuffle relations and in the other case the shuffle relations are verified.

Keywords q-Analogues of multiple zeta values · Rota–Baxter operators · Renormalization

1 Introduction

The *multiple zeta function* is defined by the nested series

$$\zeta_n(s_1, \dots, s_n) := \sum_{m_1 > \cdots > m_n > 0} \frac{1}{m_1^{s_1} \cdots m_n^{s_n}} \tag{1}$$

for $s_1, \dots, s_n \in \mathbb{C}$ with $\sum_{i=1}^{j} \mathrm{Re}(s_i) > j$, $j = 1, \dots, n$ [26]. In the special case $n = 1$ we obtain the well-known *Riemann zeta function* given by the Dirichlet series

$$\zeta_1(s) := \sum_{m>0} \frac{1}{m^s} = \prod_{p \text{ prime}} \frac{1}{1 - p^{-s}} \tag{2}$$

J. Singer (✉)
Department Mathematik, Friedrich–Alexander–Universität Erlangen–Nürnberg, Cauerstraße 11, 91058 Erlangen, Germany
e-mail: singer@math.fau.de

J. I. Burgos Gil et al. (eds.), *Periods in Quantum Field Theory and Arithmetic*, Springer Proceedings in Mathematics & Statistics 314,
https://doi.org/10.1007/978-3-030-37031-2_11

for $s \in \mathbb{C}$ with $\mathrm{Re}(s) > 1$. The first study of this function can be traced back to Euler, e.g., he calculated the explicit formula for ζ_1 at even positive integers

$$\zeta_1(2k) = -\frac{(2\pi i)^{2k} B_{2k}}{2(2k)!} \qquad (k \in \mathbb{N}),$$

where $B_m \in \mathbb{Q}$ denotes the m-th Bernoulli number defined by the generating series

$$\frac{te^t}{e^t - 1} = \sum_{m \geq 0} B_m \frac{t^m}{m!}.$$

Furthermore the Riemann zeta function can be meromorphically continued to $\mathbb{C}$ with a single pole in $s = 1$. It is a well-known fact that for $s \in \mathbb{C} \setminus \{0, 1\}$ we have the following identity

$$\zeta_1(s) = 2^s \pi^{s-1} \sin\left(\frac{\pi s}{2}\right) \Gamma(1-s)\zeta_1(1-s),$$

which is known as the functional equation of the Riemann zeta function, where $\Gamma(s)$ denotes the meromorphic continuation of the Gamma function. As an immediate consequence we deduce for $k \in \mathbb{N}_0$ that $\zeta_1(-k) = -B_{k+1}/(k+1)$, which unveils the trivial zeros of the Riemann zeta function at $-2, -4, -6, \ldots$, since $B_{2k+1} = 0$ for $k \in \mathbb{N}$. In contrast to this only very few results on the arithmetic nature of the Riemann zeta function at odd positive integers are known. For example, Apéry proved that $\zeta_1(3)$ is irrational [3], Zudilin showed that at least one of the four values $\zeta_1(5), \zeta_1(7), \zeta_1(9), \zeta_1(11)$ is irrational [38] and Ball and Rivoal proved that for infinitely many $k \in \mathbb{N}$ the value $\zeta_1(2k+1)$ is irrational [4]. It is even conjectured that for any $k \in \mathbb{N}$ and any non-zero polynomial $p \in \mathbb{Q}[T_1, \ldots, T_k]$

$$p(\pi, \zeta_1(3), \zeta_1(5), \ldots, \zeta_1(2k-1)) \neq 0.$$

Essentially, this would imply that there are no interesting $\mathbb{Q}$-algebraic relations among the positive Riemann zeta values.

However, regarding *multiple zeta values* (MZVs)—which are the functional characteristics of the multiple zeta function at positive integer arguments—the situation is completely different from the one-dimensional case. Subsequently, we denote MZVs by

$$\zeta(k_1, \ldots, k_n) := \zeta_n(k_1, \ldots, k_n)$$

for $k_1, \ldots, k_n \in \mathbb{N}$ with $k_1 \geq 2$ and call n the *length* and $k_1 + \cdots + k_n$ the *weight*. Let $\mathscr{M}$ denote the $\mathbb{Q}$-vector space spanned by the MZVs. First of all one observes that $\mathscr{M}$ is an algebra. For instance, multiplying the sums

$$\sum_{m>0}\frac{1}{m^a}\sum_{n>0}\frac{1}{n^b}=\sum_{m>n>0}\frac{1}{m^an^b}+\sum_{n>m>0}\frac{1}{n^bm^a}+\sum_{m>0}\frac{1}{m^{a+b}} \tag{3}$$

leads to *Nielsen's reflexion formula* $\zeta(a)\zeta(b)=\zeta(a,b)+\zeta(b,a)+\zeta(a+b)$ which exemplifies the so called *quasi-shuffle product* (also called *harmonic* or *stuffle product*).

The MZVs are in the spotlight of several far-reaching conjectures which underline their significance. Let $w\in\mathbb{N}_{\geq 2}$. Then $\mathcal{M}_w$ denotes the $\mathbb{Q}$-subspace of $\mathcal{M}$ spanned by all MZVs of weight w. The dimension of $\mathcal{M}_w$ is denoted by d_w.

Conjecture 1 (Zagier)

(a) The weight defines a graduation of $\mathcal{M}$, i.e.,

$$\mathcal{M}=\bigoplus_{w\geq 2}\mathcal{M}_w.$$

(b) For $w\geq 3$ we have $d_w=d_{w-2}+d_{w-3}$, where $d_0:=1$, $d_1:=0$ and $d_2:=1$.

The previous conjecture has also very strong implications on the arithmetic nature of the Riemann zeta values, e.g., as an immediate consequence of (a) we can deduce that $\zeta(k)$ $(k\geq 2)$ is transcendental which implies the irrationality of all odd positive zeta values. Currently, a proof of these properties seems to be out of range [9]. Moreover Hoffman conjectured a basis for the space of MZVs.

Conjecture 2 *The numbers $\zeta(k_1,\dots,k_n)$ with $n\geq 1$ and $k_i\in\{2,3\}$ form a $\mathbb{Q}$-basis of $\mathcal{M}$.*

It is known by a theorem of Brown [7, 35] that $\mathcal{M}$ is spanned by $\zeta(k_1,\dots,k_n)$ with $n\geq 1$ and $k_i\in\{2,3\}$.

Furthermore there is also a geometric interpretation of MZVs. The following theorem of Kontsevich shows that MZVs are periods:

Theorem 1 *For $k_1\geq 2$ and $k_2,\dots,k_n\geq 1$ we have*

$$\zeta(k_1,\dots,k_n)=\int\limits_{1>t_1>\dots>t_k>0}\omega_1(t_1)\cdots\omega_k(t_k), \tag{4}$$

where $k:=k_1+\dots+k_n$ and $\omega_i(t):=dt/(1-t)$ if $i\in\{k_1,k_1+k_2,\dots,k_1+\dots+k_n\}$ and $\omega_i(t):=dt/t$ otherwise.

The motivic aspects of MZVs are based on this geometric interpretation which turned out to be very fruitful (see [7, 8]).

The integral representation (4) of MZVs induces a second product on the $\mathbb{Q}$-vector space $\mathcal{M}$, which is called *shuffle product*. For example we obtain by decomposing the domain

$$1 > t_1 > t_2 > 0, \quad 1 > \tilde{t}_1 > \tilde{t}_2 > 0$$

into six disjoint domains that

$$\zeta(2)^2 = \int_{\substack{1>t_1>t_2>0 \\ 1>\tilde{t}_1>\tilde{t}_2>0}} \frac{dt_1}{t_1} \frac{dt_2}{1-t_2} \frac{d\tilde{t}_1}{\tilde{t}_1} \frac{d\tilde{t}_2}{1-\tilde{t}_2} = 4\zeta(3,1) + 2\zeta(2,2). \tag{5}$$

Combining the quasi-shuffle relations induced by (3) and the shuffle relations induced by (4) we obtain the so-called *double shuffle relations*. As a concrete example we observe from (3) and (5) the relation $\zeta(4) = 4\zeta(3,1)$. It is conjectured that all $\mathbb{Q}$-linear relations among MZVs are obtained by (extended) double shuffle relations [23]. Moreover MZVs and their generalizations, i.e., multiple polylogarithms, appear to play an important role in quantum field theory [6, 30].

In this paper we study q-analogues of multiple zeta values (q-MZVs). The paramount aim is always to achieve a better understanding of a classical object—in our case the MZVs—by considering a functional analogue of it in the form of a formal power series. In a first step one has to identify suitable q-analogues of MZVs. The following general requirements are seen to be reasonable:

(A) In the limit $q \to 1$ the q-MZVs, which are power series in q, should always reduce to the MZVs, which are real numbers.
(B) The algebraic structure, which is one of the dominating properties of MZVs, should be reflected by an appropriate q-model.
(C) The q-MZVs should emphasis particular properties of MZVs, e.g., an intrinsic regularization or algebraic aspect, which lead to new insights into the theory of MZVs.

Item (C) supports the view that it is not the primary aim to find the "true" q-analogue of MZVs. In fact it is more beneficial to have several different q-models at hand and finally the specific choice of the model should then depend on the particular properties of the classical object one is interested in.

Following this philosophy we regard three different q-models. We always assume $0 < q < 1$. A q-analogue of the positive integer $m \in \mathbb{N}$ can be defined by

$$[m]_q := \frac{1-q^m}{1-q} = 1 + q + \cdots + q^{m-1}.$$

All the models we consider have the same basic structure. In the denominator of the defining series (1) of MZVs the summation variables m_i are replaced by the corresponding q-integer $[m_i]_q$. Then we define

$$\zeta_q^{\bullet}(k_1, \ldots, k_n) := \sum_{m_1 > \cdots > m_n > 0} \frac{f_q^{\bullet}(m_1, \ldots, m_n; k_1, \ldots, k_n)}{[m_1]_q^{k_1} \cdots [m_n]_q^{k_n}} \tag{6}$$

where $\bullet \in \{\text{BZ}, \text{SZ}, \text{OOZ}\}$. The only difference lies in the choice of the function $f_q^\bullet$ in the numerator of the defining nested sum, which is defined for the

- Bradley–Zhao (BZ) model [5, 36] by

$$f_q^{\text{BZ}}(m_1, \dots, m_n; k_1, \dots, k_n) := q^{m_1(k_1-1)+\cdots+m_n(k_n-1)};$$

- Schlesinger–Zudilin (SZ) model [31, 39] by

$$f_q^{\text{SZ}}(m_1, \dots, m_n; k_1, \dots, k_n) := q^{m_1k_1+\cdots+m_nk_n};$$

- Ohno–Okuda–Zudilin (OOZ) model [29] by

$$f_q^{\text{OOZ}}(m_1, \dots, m_n; k_1, \dots, k_n) := q^{m_1}.$$

Now we discuss the previously stated requirements on q-analogues. Item (A) is verified by all models. Since $\lim_{q\to 1}[m]_q = m$ and $\lim_{q\to 1} f_q^\bullet(m_1, \dots, m_n; k_1, \dots, k_n) = 1$ we easily observe for integers $k_1 \geq 2, k_2, \dots, k_n \geq 1$ that

$$\lim_{q\to 1} \zeta_q^\bullet(k_1, \dots, k_n) = \zeta(k_1, \dots, k_n).$$

Addressing the above stated item (B) we first remark that all three q-models presented here exhibit a natural quasi q-shuffle product which is directly induced by their defining series. In order to establish double q-shuffle relations an appropriate q-analogue of the shuffle product is necessary. We will borrow ideas from [11]—where the authors provided q-shuffle products for the OOZ-model using techniques from Rota–Baxter algebra—to derive q-shuffle relations for the BZ- and SZ-model. Therefore all q-models discussed here reflect the double shuffle structure of the MZVs as requested in (B).

We discuss the requirement (C) separately for each model. We will see that the different choices of the function $f_q^\bullet$ lead to very specific algebraic properties concerning the quasi q-shuffle and q-shuffle relations as well as their intrinsic regularization properties.

- The motivation for the introduction of the BZ-model is based on an analytical observation concerning the Riemann zeta function. In [25] Kaneko, Kurokawa and Wakayama considered the one-parameter function

$$f_q(s, t) := \sum_{n\geq 1} \frac{q^{nt}}{[n]_q^s}.$$

For $\text{Re}(s) > 1$ and $\text{Re}(t) > 0$ one has $\lim_{q\to 1} f_q(s, t) = \zeta_1(s)$. Regarding the meromorphic continuation of $f_q(s, t)$, they proved that $\lim_{q\to 1} f_q(s, s-1) = \zeta_1(s)$ for any $s \in \mathbb{C}$. Like the Riemann zeta function, $f_q(s, s-1)$ has only a pole at one on any compact set in the $\mathbb{C}$-plane if $q < 1$ is sufficiently large. This is only true for the case $t = s - 1$. Therefore this one-dimensional model is distinguished from others. Hence, the BZ-model is then considered as the natural extension to higher length.

- The characteristic property of the SZ-model lies in its quasi-shuffle structure. It is easily seen that ζ_q^{SZ} can be directly obtained from ζ by the substitution $m_i \mapsto [m_i]_q/q^{m_i}$ in Eq. (1) ($i = 1, \ldots, n$). Hence, the quasi-shuffle relations of these q-MZVs and the classical MZVs coincide. Moreover the q-series of this model is convergent even if we allow $k_1 = 1$. Therefore we observe better convergence properties than for classical multiple zeta values.
- Since f_q^{OOZ} is independent of $k_1, \ldots, k_n$ it is easily verified that the OOZ q-MZVs are convergent for arbitrary integer arguments. Additionally, Castillo-Medina, Ebrahimi-Fard and Manchon constructed in [11] a q-shuffle product that is also defined for negative and mixed sign integer arguments. Therefore these q-MZVs occupy an intrinsic regularization in contrast to the classical MZVs and exhibit a compatible q-shuffle structure.

The second part of this paper deals with the values of the multiple zeta function at negative integer arguments. As we have seen in the case of the Riemann zeta function the meromorphic continuation completely determines all Riemann zeta values at negative integers. In contrast to this the meromorphic continuation of the multiple zeta function (1) only partially describes the values for negative integer arguments because most of them are points of indeterminacy due to poles (see Sect. 4). This raises the question of how to systematically extend the MZVs to negative integer arguments. On the one hand the extension should verify the values obtained by the meromorphic continuation whenever it is defined. Therefore there is a certain degree of freedom concerning the extension of the MZVs. Since the algebraic aspects of MZVs play a decisive role it is reasonable to additionally expect that the extended MZVs should at least verify certain algebraic properties, e.g., the quasi-shuffle or shuffle relations. The extension problem is called the *renormalization problem of MZVs* (Problem 1). In order to construct solutions to this problem we exploit the aforementioned specific properties of the SZ- and OOZ-model. The key observation is that we can interpret q-analogues as a natural q-perturbation of MZVs to which we apply renormalization methods from perturbative quantum field theory. Here we realize that the specific algebraic structures of the corresponding q-models carry over to the extended MZVs: For the quasi-shuffle case we will use the SZ-model because its quasi-shuffle product coincides with that of classical MZVs. In the shuffle case we work with the OOZ-model since it exhibits a natural shuffle product also for negative integer arguments.

The paper is organized as follows. In Sect. 2 we review the main aspects of the algebraic theory of MZVs. Section 3 is devoted to the study of the double q-shuffle relations of q-MZVs. This section essentially relies on the works [11, 32, 33]. In Sect. 4 we regard two q-models as natural q-perturbations of MZVs and implement a renormalization procedure which was carried out in the papers [14, 15] as a joint project with K. Ebrahimi-Fard and D. Manchon.

2 Linear Relations Among Multiple Zeta Values

In this section we review the essential aspects of the theory of $\mathbb{Q}$-linear relations among MZVs. Following [21] it is convenient to introduce a polynomial algebra in two non-commutative variables x_0 and x_1 which is denoted by $\mathfrak{h} := \mathbb{Q}\langle x_0, x_1\rangle$. The empty word $\mathbf{1}$ is the unit of $\mathfrak{h}$. The subalgebra $\mathfrak{h}^1 := \mathbb{Q} \oplus \mathfrak{h}x_1$ consists of all words not ending in x_0. Furthermore we denote by $\mathfrak{h}^0 := \mathbb{Q} \oplus x_0\mathfrak{h}x_1$ the linear span of all words not beginning with x_1 and not ending in x_0. Note that the subalgebra $\mathfrak{h}^1$ is generated by words with letters $u_k := x_0^{k-1}x_1$ ($k \in \mathbb{N}$) and $\mathfrak{h}^0$ is generated by words $u_{k_1} \cdots u_{k_n}$ with $k_1 \geq 2, k_2, \ldots, k_n \geq 1$.

On $\mathfrak{h}^0$ we define an evaluation map $\zeta : \mathfrak{h}^0 \to \mathbb{R}$ by $\zeta(\mathbf{1}) := 1$ and

$$\zeta(u_{k_1} \cdots u_{k_n}) := \zeta(k_1, \ldots, k_n). \tag{7}$$

2.1 Double Shuffle Relations

As remarked in the introduction the double shuffle relations of MZVs rely on two non-compatible products: the quasi-shuffle product and the shuffle product. In this section we review their algebraic structure in terms of the polynomial algebra $\mathfrak{h}^1$ introduced above [21, 34].

Generalizing Nilsen's reflexion formula (3) to arbitrary depth we define the *quasi-shuffle product* $*\colon \mathfrak{h}^1 \otimes \mathfrak{h}^1 \to \mathfrak{h}^1$, iteratively by

(i) $\mathbf{1} * w := w * \mathbf{1} := w$,
(ii) $u_m v * u_n w := u_m(v * u_n w) + u_n(u_m v * w) + u_{m+n}(v * w)$

for words $v, w \in \mathfrak{h}^1$ and $m, n \in \mathbb{N}$. For example

$$u_a * u_b = u_a u_b + u_b u_a + u_{a+b}. \tag{8}$$

Furthermore the *deconcatenation coproduct* $\Delta\colon \mathfrak{h}^1 \to \mathfrak{h}^1 \otimes \mathfrak{h}^1$ is defined by $\Delta(\mathbf{1}) := \mathbf{1} \otimes \mathbf{1}$ and

$$\Delta(u_{k_1} \cdots u_{k_n}) := \mathbf{1} \otimes u_{k_1} \cdots u_{k_n} + u_{k_1} \cdots u_{k_n} \otimes \mathbf{1} + \sum_{l=1}^{n-1} u_{k_1} \cdots u_{k_l} \otimes u_{k_{l+1}} \cdots u_{k_n}.$$

Then the triple $(\mathfrak{h}^1, *, \Delta)$ forms a bialgebra. Since the bialgebra is connected and filtered it is automatically a Hopf algebra which is called the *quasi-shuffle Hopf algebra*. The *antipode* S is given by the following iterative formula:

$$S(w) = -w - \sum_{(w)} S(w') * w'',$$

where we use Sweedler's notation $\widetilde{\Delta}(w) := \Delta(w) - \mathbf{1} \otimes w - w \otimes \mathbf{1} = \sum_{(w)} w' \otimes w''$. The quasi-shuffle Hopf algebra will be of great importance in view of Theorem 11.

Now we address the shuffle product of MZVs. For this we define *multiple polylogarithms* (MPLs) in a single variable for $k_1, \ldots, k_n \in \mathbb{N}$ by

$$\mathrm{Li}_{k_1,\ldots,k_n}(z) := \sum_{m_1 > \cdots > m_n > 0} \frac{z^{m_1}}{m_1^{k_1} \cdots m_n^{k_n}}, \tag{9}$$

generalizing $z \mapsto -\log(1-z)$ for $k_1 = 1$ and $n = 1$, where z is a complex number. The function $\mathrm{Li}_{k_1,\ldots,k_n}(z)$ is of depth $\mathrm{dpt}(\mathbf{k}) := n \geq 1$ and weight $\mathrm{wt}(\mathbf{k}) := k_1 + \cdots + k_n$, for $\mathbf{k} := (k_1, \ldots, k_n)$. It is analytic in the open unit disk and, in the case $k_1 > 1$, continuous on the closed unit disk. In this case we observe the connection of MPLs and MZVs $\zeta(k_1, \ldots, k_n) = \mathrm{Li}_{k_1,\ldots,k_n}(1)$. Equivalently, we can define MPLs by induction on the weight $\mathrm{wt}(\mathbf{k})$ as follows:

$$z \frac{d}{dz} \mathrm{Li}_{k_1,\ldots,k_n}(z) = \mathrm{Li}_{k_1-1,k_2,\ldots,k_n}(z) \quad \text{if} \quad k_1 > 1, \tag{10}$$

$$(1-z) \frac{d}{dz} \mathrm{Li}_{1,k_2,\ldots,k_n}(z) = \mathrm{Li}_{k_2,\ldots,k_n}(z) \quad \text{if} \quad n > 1, \tag{11}$$

$$\mathrm{Li}_{k_1,\ldots,k_n}(0) = 0. \tag{12}$$

Therefore we observe an integral formula for MPLs using iterated Chen integrals. Indeed, setting

$$\omega_0(t) := \frac{dt}{t} \quad \text{and} \quad \omega_1(t) := \frac{dt}{1-t} \tag{13}$$

the differential equations (10), (11) and the initial conditions (12) lead to

$$\mathrm{Li}_{k_1,\ldots,k_n}(z) = \int_0^z \omega_0^{k_1-1} \omega_1 \cdots \omega_0^{k_n-1} \omega_1,$$

using the convention $\mathrm{Li}_{\emptyset}(z) = 1$. Here

$$\int_x^y \varphi_1 \cdots \varphi_p := \int_x^y \varphi_1(t) \int_x^t \varphi_2 \cdots \varphi_p$$

is defined for complex-valued differential 1-forms $\varphi_1, \ldots, \varphi_p$ on a compact interval and real numbers x and y. This representation proves Theorem 1 and gives rise to the well-known shuffle products of MPLs and MZVs. The shuffle product counterpart of Nielsen's reflexion formula (3) is *Euler's decomposition formula*

$$\zeta(a)\zeta(b) = \sum_{l=0}^{a-1} \binom{b+l-1}{b-1} \zeta(b+l, a-l) + \sum_{l=0}^{b-1} \binom{a+l-1}{a-1} \zeta(a+l, b-l) \tag{14}$$

for $a, b \geq 2$, which solely relies on the integral representation of MZVs. Based on the integral representation (4) we introduce, by imitating the integration by parts formula for iterated Chen integrals, the *shuffle product* $ш\colon \mathfrak{h}^1 \otimes \mathfrak{h}^1 \to \mathfrak{h}^1$ inductively by

(i) $\mathbf{1} ш w := w ш \mathbf{1} := w$,
(ii) $av ш bw := a(v ш bw) + b(av ш w)$

for words $v, w \in \mathfrak{h}^1$ and $a, b \in \{x_0, x_1\}$. For example, we observe

$$x_0x_1 ш x_0x_1 = 4x_0^2x_1^2 + 2x_0x_1x_0x_1. \tag{15}$$

Theorem 2 ([22, 23]) *We have:*

(a) The evaluation maps $\zeta\colon (\mathfrak{h}^0, *) \to \mathbb{R}$ *and* $\zeta\colon (\mathfrak{h}^0, ш) \to \mathbb{R}$ *are morphisms of algebras.*
(b) The regularized double shuffle relations *hold, i.e.,*

$$\zeta(v * w - v ш w) = 0 \quad \text{and} \quad \zeta(u_1 * w - x_1 ш w) = 0$$

for any words $v, w \in \mathfrak{h}^0$.

Using the previous theorem, Eq. (8) immediately implies Nielsen's reflexion formula stated in (3). Together with Eq. (15) we obtain the following double shuffle relation:

$$0 = \zeta(u_2 * u_2 - x_0x_1 ш x_0x_1) = \zeta(4) - 4\zeta(3, 1).$$

In addition, in [23] the authors conjecture that all linear relations among MZVs follow from the regularized double shuffle relations.

2.2 Duality and Derivation Relations

Another way to deduce linear relations among MZVs is given in terms of *duality*. These relations can be deduced from the integral representation (4) by a transformation of variables. We start with the algebraic description. Define an antiautomorphism $\tau\colon \mathfrak{h}^0 \to \mathfrak{h}^0$ by $\tau(\mathbf{1}) := \mathbf{1}$ and

$$\tau(x_0) := x_1 \quad \text{and} \quad \tau(x_1) := x_0.$$

For example we have $\tau(x_0^2x_1^3x_0x_1) = x_0x_1x_0^3x_1^2$. The transformation of variables given by $\Psi(t_1, \dots, t_n) := (1-t_n, \dots, 1-t_1)$ applied to the integral (4) implies the following theorem:

Theorem 3 ([21]) *For any word $w \in \mathfrak{h}^0$ we have $\zeta(w) = \zeta(\tau(w))$.*

For instance Euler's relation $\zeta(3) = \zeta(2,1)$ can be deduced from $\tau(x_0^2x_1) = x_0x_1^2$.

Finally, we consider the so-called derivation relations, *which also give rise to linear relations among MZVs. For $n \in \mathbb{N}$ we introduce the derivation $\partial_n : \mathfrak{h}^0 \to \mathfrak{h}^0$ with respect to the concatenation product on the generators of $\mathfrak{h}^0$ by*

$$\partial_n(x_0) := x_0(x_0+x_1)^{n-1}x_1 \qquad \text{and} \qquad \partial_n(x_1) := -x_0(x_0+x_1)^{n-1}x_1.$$

The following theorem shows that the image of words under the derivation lies in the kernel of ζ:

Theorem 4 ([23]) *For any $w \in \mathfrak{h}^0$ and $n \in \mathbb{N}$ we have $\zeta(\partial_n(w)) = 0$.*

In Sect. 3.3 we show a surprising connection between derivation and the duality relations in terms of a dual product.

3 q-Analogues of Multiple Zeta Values

Now we study the algebraic properties of the q-analogues introduced in Sect. 1. In the q-case we are confronted with some specific phenomenons.

As shown in Sect. 2.1 the quasi-shuffle and shuffle product of MZVs always preserve the weight. This is also in accordance with Conjecture 1. However, considering q-MZVs the situation is different. For example we observe for the BZ-model for $a, b \geq 2$

$$\begin{aligned}\sum_{m>0} \frac{q^{m(a-1)}}{(1-q^m)^a} \sum_{n>0} \frac{q^{n(b-1)}}{(1-q^n)^b} &= \sum_{m>n>0} \frac{q^{m(a-1)+n(b-1)}}{(1-q^m)^a(1-q^n)^b} + \sum_{n>m>0} \frac{q^{n(b-1)+m(a-1)}}{(1-q^n)^b(1-q^m)^a} \\ &\quad + \sum_{m>0} \frac{q^{m(a+b-1)}}{(1-q^m)^{a+b}} + \sum_{m>0} \frac{q^{m(a+b-2)}}{(1-q^m)^{a+b-1}}.\end{aligned}$$

Multiplying this equation by $(1-q)^{a+b}$ we obtain

$$\zeta_q^{\mathrm{BZ}}(a)\zeta_q^{\mathrm{BZ}}(b) = \zeta_q^{\mathrm{BZ}}(a,b) + \zeta_q^{\mathrm{BZ}}(b,a) + \zeta_q^{\mathrm{BZ}}(a+b) + (1-q)\zeta_q^{\mathrm{BZ}}(a+b-1).$$

Hence, the last summand shows that a decrease of weight arises. In the limit $q \to 1$ we rediscover the classical Nilsen reflexion formula (3) since the last terms vanishes because of the prefactor $(1-q)$. The loss of weight occurs also for the q-shuffle product. Similar phenomenons are visible for the OOZ- and SZ-model.

In order to establish the regularized double shuffle relations for MZVs (Theorem 2) it is necessary to identify the divergent zeta value "$\zeta(1)$" with a formal variable T (see [23]). But in contrast to classical MZVs the q-MZV $\zeta_q^\bullet(1)$ is convergent for $\bullet \in \{\text{SZ}, \text{OOZ}\}$. Therefore the double q-shuffle relations for those two models have intrinsically implemented the regularized double shuffle relations.

The last observation addresses the integral representation of MZVs (Theorem 1) which encodes the geometric information of MZVs. At the first glance there is no obvious substitute in the q-case. As we will show in the next paragraph one can define an appropriate analogue in terms of Rota–Baxter operators (RBO). Structurally, the integral operator, which is a RBO of weight 0, can be substituted by a RBO of weight unequal to 0. This naturally induces a q-shuffle product for the q-models which we describe in the next section.

3.1 From Jackson's Integral to Rota–Baxter Operators

In this paragraph we motivate the Rota–Baxter operator characterization of q-MZVs by imitating the integral representation of classical MZVs (4) for the OOZ q-model by the ordinary Riemann integral. See [11] for more details. The *Jackson integral*

$$I_q[f](t) := \int_0^t f(x)\, d_q x := (1-q) \sum_{n \geq 0} f(q^n t) q^n t,$$

is the q-analogue of the classical integral operator $I[f](t) := \int_0^t f(x)\, dx$ in the sense that under appropriate regularity conditions on f we have $I[f](t) = \lim_{q \to 1} I_q[f](t)$, [24].

The key observation of [10, 11] was that the OOZ-model is related to an iteration of Jackson integrals. Recall the differential one-forms introduced in (13). In analogy to this we define the functions

$$\widetilde{\omega}_0(t) := \frac{1}{t} \qquad \text{and} \qquad \widetilde{\omega}_1(t) := \frac{1}{1-t}.$$

Replacing the integral operator I by its q-analogue I_q in Eq. (4) we obtain for positve integers $k_1, \ldots, k_n \in \mathbb{N}$

$$\zeta_q^{\text{OOZ}}(k_1, \ldots, k_n) = I_q[\alpha_1 I_q[\alpha_2 \cdots I_q[\alpha_{k_1 + \cdots + k_n}] \cdots]](q), \tag{16}$$

where $\alpha_i := \widetilde{\omega}_1$ if $i \in \{k_1, k_1 + k_2, \ldots, k_1 + \cdots + k_n\}$ and $\alpha_i := \widetilde{\omega}_0$ otherwise. For classical MZVs the product rule of iterated Chen integrals naturally induces the shuffle product. In the q-case this role is transferred to a modified Jackson integral in terms of a Rota–Baxter operator. Therefore we shortly review the main aspects of Rota–Baxter algebra [18].

Let k be a ring, $\mathscr{A}$ a k-algebra and $\lambda \in k$. A *Rota–Baxter operator (RBO) of weight* λ *on* $\mathscr{A}$ *over* k is a k-vector space endomorphism L of $\mathscr{A}$ which satisfies

$$L(x)L(y) = L(xL(y)) + L(L(x)y) + \lambda L(xy) \tag{17}$$

for any $x, y \in \mathscr{A}$. A *Rota–Baxter k-algebra of weight* λ is a pair $(L, \mathscr{A})$ with a k-algebra $\mathscr{A}$ and a RBO of weight λ on $\mathscr{A}$ over k.

Next we provide several examples of Rota–Baxter algebras, which will be important in the context of this work.

(a) Let $C(\mathbb{R})$ denote the algebra of continuous functions from $\mathbb{R}$ to $\mathbb{R}$. The integration operator

$$I : C(\mathbb{R}) \to C(\mathbb{R}), \qquad I[f](t) := \int_0^t f(x)\, dx$$

is a RBO of weight 0, which is an obvious consequence of the integration by parts formula.

(b) We regard the $\mathbb{C}$-algebra

$$\mathscr{P}_{\geq 1} := \left\{ f(z) := \sum_{k \geq 1} a_k z^k : R_f \geq 1 \right\} \subseteq z\mathbb{C}[\![z]\!]$$

of power series with non-constant terms and radius of convergence of at least 1. On $\mathscr{P}_{\geq 1}$ we define the operator

$$J : \mathscr{P}_{\geq 1} \to \mathscr{P}_{\geq 1}, \qquad J[f](z) := \int_0^z f(t) \frac{dt}{t}.$$

Again, the integration by parts formula implies that $(J, \mathscr{P}_{\geq 1})$ is a Rota–Baxter algebra of weight 0.

(c) The vector space $t\mathbb{C}[\![t, q]\!]$ of power series in two variables t, q and strictly positive valuation in t can interpreted as the $\mathbb{C}[\![q]\!]$-algebra $t\mathbb{C}[\![t, q]\!]$, which we denote by $\mathscr{A}$. The operator

$$P_q : \mathscr{A} \to \mathscr{A}, \qquad P_q[f](t) := \sum_{n > 0} f(q^n t)$$

is a RBO of weight 1.

(d) Furthermore, on the same algebra $\mathscr{A}$ we define the operator

$$\overline{P}_q : \mathscr{A} \to \mathscr{A}, \qquad \overline{P}_q[f](t) := \sum_{n \geq 0} f(q^n t),$$

which is a RBO of weight -1.

(e) On the algebra $\mathbb{C}[z^{-1}, z]]$ of Laurent series the projector $\pi\colon \mathbb{C}[z^{-1}, z]] \to \mathbb{C}[z^{-1}, z]]$ given by

$$\pi\left(\sum_{n=-l}^{\infty} a_n z^n\right) := \sum_{n=-l}^{-1} a_n z^n$$

with the convention that the empty sum is zero is a RBO of weight -1.

(f) The operator

$$D_q\colon \mathscr{A} \to \mathscr{A}, \qquad D_q[f](t) := f(t) - f(qt)$$

is a weight -1 differential RBO [19], i. e., for any $f, g \in \mathscr{A}$

$$D_q[fg] = D_q[f]g + f D_q[g] - D_q[f]D_q[g].$$

Let us return to the iterated Jackson integral (16), which characterizes the OOZ q-MZVs. Now we show that the iterated Jackson integrals are replaced by an interleaving of Rota–Baxter operators. The Jackson integral operator and the previously defined RBO $\overline{P}_q$ are related by

$$\widetilde{I}_q[f](t) := \int_0^t \frac{f(x)}{x}\, d_q x = (1-q)\overline{P}_q[f](t).$$

Therefore (16) leads to

$$\zeta_q^{\mathrm{OOZ}}(k_1, \dots, k_n) = (1-q)^{k_1+\cdots+k_n} \overline{P}_q^{k_1}[y\overline{P}_q^{k_2}[y \cdots \overline{P}_q^{k_n}[y] \cdots]](q), \tag{18}$$

where $y(t) := t/(1-t)$. In the light of the prefactor $(1-q)^{k_1+\cdots+k_n}$ in Eq. (18) it is reasonable to introduce modified q-MZVs in the sense

$$\overline{\zeta}_q^{\bullet}(k_1, \dots, k_n) := (1-q)^{-(k_1+\cdots+k_n)} \zeta_q^{\bullet}(k_1, \dots, k_n)$$

with $\bullet \in \{\mathrm{BZ}, \mathrm{SZ}, \mathrm{OOZ}\}$.

Example 1 For instance, using the defining Rota–Baxter relation for $\overline{P}_q$, Eq. (18) implies

$$\begin{aligned}
\overline{\zeta}_q^{\mathrm{OOZ}}(1)\overline{\zeta}_q^{\mathrm{OOZ}}(1) &= \overline{P}_q[y]\overline{P}_q[y]\big|_{t=q} = 2\overline{P}_q[y\overline{P}_q[y]]\big|_{t=q} - \overline{P}_q[y^2]|_{t=q} \\
&= 2\overline{\zeta}_q^{\mathrm{OOZ}}(1,1) - \overline{\zeta}_q^{\mathrm{OOZ}}(1,0).
\end{aligned}$$

In contrast to this we observe using the defining series of the q-MZVs

$$
\begin{aligned}
&\overline{\zeta}_q^{\mathrm{OOZ}}(1)\overline{\zeta}_q^{\mathrm{OOZ}}(1)\\
&= 2\sum_{m>n>0}\frac{q^{m+n}}{(1-q^m)(1-q^n)} + \sum_{m>0}\frac{q^{2m}}{(1-q^m)^2}\\
&= 2\sum_{m>n>0}\frac{q^{m}}{(1-q^m)(1-q^n)} - 2\sum_{m>n>0}\frac{q^m}{1-q^m} + \sum_{m>0}\frac{q^m}{(1-q^m)^2} - \sum_{m>0}\frac{q^m}{(1-q^m)}\\
&= 2\overline{\zeta}_q^{\mathrm{OOZ}}(1,1) - 2\overline{\zeta}_q^{\mathrm{OOZ}}(1,0) + \overline{\zeta}_q^{\mathrm{OOZ}}(2) - \overline{\zeta}_q^{\mathrm{OOZ}}(1).
\end{aligned}
$$

Therefore we obtain the following linear relations among q-MZVs:

$$
\overline{\zeta}_q^{\mathrm{OOZ}}(1,0) = \overline{\zeta}_q^{\mathrm{OOZ}}(2) - \overline{\zeta}_q^{\mathrm{OOZ}}(1).
$$

Even though the Rota–Baxter operator description of the OOZ-model is directly related to the integral representation of MZVs by replacing the integral operator by its q-analogue, we extend this framework to MPLs as well as to the other q-models introduced above.

Proposition 1 ([11, 15, 32, 33]) *Let $n \in \mathbb{N}$ and $y(t) := \frac{t}{1-t} \in t\mathbb{Q}[\![t]\!]$. Then we have*

(a) for $\mathbf{k} := (k_1, \ldots, k_n) \in \mathbb{Z}^n$ and $|z| < 1$

$$
\mathrm{Li}_{\mathbf{k}}(z) = J^{k_1}[yJ^{k_2}[y\cdots J^{k_n}[y]\cdots]](z),
$$

where $J^{-1} = t\frac{d}{dt}$;

(b) for $\mathbf{k} := (k_1, \ldots, k_n) \in \mathbb{N}_0^n$ with $k_1 > 0$

$$
\overline{\zeta}_q^{\mathrm{SZ}}(\mathbf{k}) = P_q^{k_1}[yP_q^{k_2}[y\cdots P_q^{k_n}[y]\cdots]](1);
$$

(c) for $\mathbf{k} := (k_1, \ldots, k_n) \in \mathbb{Z}^n$

$$
\overline{\zeta}_q^{\mathrm{OOZ}}(\mathbf{k}) = \overline{P}_q^{k_1}[y\overline{P}_q^{k_2}[y\cdots \overline{P}_q^{k_n}[y]\cdots]](q),
$$

where $\overline{P}_q^{-1} = D_q$;

(d) for $\mathbf{k} := (k_1, \ldots, k_n) \in \mathbb{N}^n$ with $k_1 > 1$

$$
\overline{\zeta}_q^{\mathrm{BZ}}(\mathbf{k}) = P_q^{k_1-1}\overline{P}_q[yP_q^{k_2-1}\overline{P}_q[y\cdots P_q^{k_n-1}\overline{P}_q[y]\cdots]](1).
$$

Since for the OOZ-model the operators $\overline{P}_q, D_q$ and for the BZ-model the operators $P_q, \overline{P}_q$ interact, we need the following compatibility relations:

Lemma 1 ([33]) *For the mixed products we have*

$$
\begin{aligned}
P_q[f]\overline{P}_q[g] &= \overline{P}_q[P_q[f]g] + P_q[f\overline{P}_q[g]],\\
D_q[f]\overline{P}_q[g] &= D_q[f\overline{P}_q[g]] + D_q[f]g - fg
\end{aligned}
$$

for any $f, g \in \mathscr{A}$.

3.2 Algebraic Setting of Double Shuffle Structure

In this paragraph we translate the operator theoretic setting of Proposition 1 and Lemma 1 into an algebraic framework which allows us to describe the double q-shuffle relations for all these models.

3.2.1 Schlesinger–Zudilin Model

Let $\mathscr{R} := \mathbb{Q}\langle p, y\rangle$ be the polynomial algebra with two non-commutative variables p and y. The empty word is denoted by $\mathbf{1}$. The subalgebra of words not ending in p is denoted by $\mathscr{R}^1 := \mathbb{Q} \oplus \mathscr{R}y$. Moreover the subalgebra $\mathscr{R}^0 := \mathbb{Q} \oplus p\mathscr{R}y$ consists of all words not beginning with y and not ending with p. Note that $\mathscr{R}^1$ is generated by words with letters $u_k := p^k y$ with $k \in \mathbb{N}_0$.

We define an evaluation map $\overline{\zeta}_q^{\mathrm{SZ}} : \mathscr{R}^0 \to \mathbb{Q}[[q]]$ by $\overline{\zeta}_q^{\mathrm{SZ}}(\mathbf{1}) := 1$ and

$$\overline{\zeta}_q^{\mathrm{SZ}}(u_{k_1} \cdots u_{k_n}) := \overline{\zeta}_q^{\mathrm{SZ}}(k_1, \ldots, k_n).$$

The quasi q-shuffle product $*_{\mathrm{SZ}} : \mathscr{R}^1 \otimes \mathscr{R}^1 \to \mathscr{R}^1$ is defined recursively by

(i) $w *_{\mathrm{SZ}} \mathbf{1} := \mathbf{1} *_{\mathrm{SZ}} w := w$,
(ii) $u_s v *_{\mathrm{SZ}} u_t w := u_s(v *_{\mathrm{SZ}} u_t w) + u_t(u_s v *_{\mathrm{SZ}} w) + u_{s+t}(v *_{\mathrm{SZ}} w)$

for all words $v, w \in \mathscr{R}^1$ and $s, t \in \mathbb{N}_0$.

The q-shuffle product $\sqcup\!\sqcup_{\mathrm{SZ}} : \mathscr{R}^1 \otimes \mathscr{R}^1 \to \mathscr{R}^1$ is defined iteratively by

(i) $\mathbf{1} \sqcup\!\sqcup_{\mathrm{SZ}} w := w \sqcup\!\sqcup_{\mathrm{SZ}} \mathbf{1} := w$,
(ii) $yv \sqcup\!\sqcup_{\mathrm{SZ}} w := v \sqcup\!\sqcup_{\mathrm{SZ}} yw := y(v \sqcup\!\sqcup_{\mathrm{SZ}} w)$,
(iii) $pv \sqcup\!\sqcup_{\mathrm{SZ}} pw := p(v \sqcup\!\sqcup_{\mathrm{SZ}} pw) + p(pv \sqcup\!\sqcup_{\mathrm{SZ}} w) + p(v \sqcup\!\sqcup_{\mathrm{SZ}} w)$

for any words $v, w \in \mathscr{R}^1$. Note that item (iii) reflects the RBO-property of the operator P_q. We have the following result:

Theorem 5 ([32]) *We have:*

(a) $(\mathscr{R}^0, *_{\mathrm{SZ}})$ *and* $(\mathscr{R}^0, \sqcup\!\sqcup_{\mathrm{SZ}})$ *are commutative and associative algebras.*
(b) The maps $\overline{\zeta}_q^{\mathrm{SZ}} : (\mathscr{R}^0, *_{\mathrm{SZ}}) \to \mathbb{Q}[[q]]$ *and* $\overline{\zeta}_q^{\mathrm{SZ}} : (\mathscr{R}^0, \sqcup\!\sqcup_{\mathrm{SZ}}) \to \mathbb{Q}[[q]]$ *are morphisms of algebras. Especially we have double q-shuffle relations given by*

$$\overline{\zeta}_q^{\mathrm{SZ}}(v *_{\mathrm{SZ}} w - v \sqcup\!\sqcup_{\mathrm{SZ}} w) = 0$$

for any $v, w \in \mathscr{R}^0$.

For example we observe

$$u_2 *_{\mathrm{SZ}} u_1 = u_2u_1 + u_1u_2 + u_3,$$
$$p^2y \sqcup\!\sqcup_{\mathrm{SZ}} py = 2p^2ypy + pyp^2y + p^2y^2 + pypy$$

and Theorem 5 implies

$$\overline{\zeta}_q^{\mathrm{SZ}}(u_3 - u_2u_1 - u_2u_0 - u_1u_1) = 0.$$

Multiplying the previous equation with $(1-q)^3$ we obtain

$$\zeta_q^{\mathrm{SZ}}(3) = \zeta_q^{\mathrm{SZ}}(2,1) + (1-q)\zeta_q^{\mathrm{SZ}}(2,0) + (1-q)\zeta_q^{\mathrm{SZ}}(1,1),$$

which descends in the limit $q \to 1$ to the well-known Euler relation $\zeta(3) = \zeta(2,1)$.

3.2.2 Ohno–Okuda–Zudilin Model

The particular feature of this model is that it is defined for any integer arguments. This is also reflected by the shuffle product.

Let $\mathscr{R}$ denote the polynomial algebra with respect to the non-commutative variables p, d and y subject to $pd = dp = \mathbf{1}$. Further let $\mathscr{R}^1 := \mathbb{Q} \oplus \mathscr{R}y$ be the subalgebra of $\mathscr{R}$ consisting of word not ending in p or d. Note that $\mathscr{R}^1$ is generated by words consisting of letters $u_k := p^k y$ with $k \in \mathbb{Z}$ where $p^{-1} := d$.

The evaluation map $\overline{\zeta}_q^{\mathrm{OOZ}} : \mathscr{R}^1 \to \mathbb{Q}[\![q]\!]$ is defined by $\overline{\zeta}_q^{\mathrm{OOZ}}(\mathbf{1}) := 1$ and

$$\overline{\zeta}_q^{\mathrm{OOZ}}(u_{k_1} \cdots u_{k_n}) := \overline{\zeta}_q^{\mathrm{OOZ}}(k_1, \ldots, k_n). \tag{19}$$

The quasi q-shuffle product $*_{\mathrm{OOZ}} : \mathscr{R}^1 \otimes \mathscr{R}^1 \to \mathscr{R}^1$ is defined recursively

(i) $w *_{\mathrm{OOZ}} \mathbf{1} := \mathbf{1} *_{\mathrm{OOZ}} w := w,$
(ii) $u_s v *_{\mathrm{OOZ}} u_t w := u_s(v * u_t w) + u_t(u_s v * w) - u_s(v * u_{t-1} w) - u_t(u_{s-1} v * w) + (u_{s+t} - u_{s+t-1})(v * w)$

for all words $v, w \in \mathscr{R}^1$ and $s, t \in \mathbb{Z}$. Note that $*$ denotes the ordinary quasi-shuffle product of MZVs introduced in Sect. 2.1 with the natural extension to non-positive integers.

The q-shuffle product $\sqcup\!\sqcup_{\mathrm{OOZ}} : \mathscr{R}^1 \otimes \mathscr{R}^1 \to \mathscr{R}^1$ is defined iteratively by

(i) $\mathbf{1} \sqcup\!\sqcup_{\mathrm{OOZ}} w := w \sqcup\!\sqcup_{\mathrm{OOZ}} \mathbf{1} := w,$
(ii) $yv \sqcup\!\sqcup_{\mathrm{OOZ}} w := v \sqcup\!\sqcup_{\mathrm{OOZ}} yw := y(v \sqcup\!\sqcup_{\mathrm{OOZ}} w),$
(iii) $pv \sqcup\!\sqcup_{\mathrm{OOZ}} pw := p(v \sqcup\!\sqcup_{\mathrm{OOZ}} pw) + p(pv \sqcup\!\sqcup_{\mathrm{OOZ}} w) - p(v \sqcup\!\sqcup_{\mathrm{OOZ}} w),$
(iv) $dv \sqcup\!\sqcup_{\mathrm{OOZ}} dw := v \sqcup\!\sqcup_{\mathrm{OOZ}} dw + dv \sqcup\!\sqcup_{\mathrm{OOZ}} w - d(v \sqcup\!\sqcup_{\mathrm{OOZ}} w),$
(v) $dv \sqcup\!\sqcup_{\mathrm{OOZ}} pw := pw \sqcup\!\sqcup_{\mathrm{OOZ}} dv := d(v \sqcup\!\sqcup_{\mathrm{OOZ}} pw) + dv \sqcup\!\sqcup_{\mathrm{OOZ}} w - v \sqcup\!\sqcup_{\mathrm{OOZ}} w$

for any words $v, w \in \mathscr{R}^1$.

Remark 1 In the previous definition item (iii) reflects the RBO-property of the operator $\overline{P}_q$ and (iv) the differential RBO-property of D_q. The compatibility relation of $\overline{P}_q$ and D_q stated in the second equation of Lemma 1 corresponds to (v).

We have the following result:

Theorem 6 ([11]) *We have:*

(a) $(\mathscr{R}^1, *_{\mathrm{OOZ}})$ *and* $(\mathscr{R}^1, ш_{\mathrm{OOZ}})$ *are commutative and associative algebras.*
(b) The maps $\overline{\zeta}_q^{\mathrm{OOZ}} : (\mathscr{R}^1, *_{\mathrm{OOZ}}) \to \mathbb{Q}[[q]]$ *and* $\overline{\zeta}_q^{\mathrm{OOZ}} : (\mathscr{R}^1, ш_{\mathrm{OOZ}}) \to \mathbb{Q}[[q]]$ *are morphisms of algebras. Especially we have double q-shuffle relations given by*

$$\overline{\zeta}_q^{\mathrm{OOZ}}(v *_{\mathrm{OOZ}} w - v \, ш_{\mathrm{OOZ}} \, w) = 0$$

for any $v, w \in \mathscr{R}^1$.

As shown in [15] this algebraic framework is closely relate to the classical shuffle product by Proposition 1 (a) which will be used in Sect. 4.2.

3.2.3 Bradley–Zhao Model

The algebraic description of the BZ-model is more involved since we have to establish an appropriate algebra extension. We motivate the extension by the example of Euler's decomposition formula. As previously remarked, Euler's decomposition formula of MZVs (14) is solely related to the integral representation of MZVs and therefore of pure shuffle nature. For the BZ-model the corresponding Euler decomposition formula is

$$\begin{aligned}\overline{\zeta}_q^{\mathrm{BZ}}(a)\overline{\zeta}_q^{\mathrm{BZ}}(b) = &\sum_{s=0}^{a-1}\sum_{t=0}^{a-1-s}\binom{s+b-1}{b-1}\binom{b-1}{t}\overline{\zeta}_q^{\mathrm{BZ}}(b+s, a-s-t)\\ &+\sum_{s=0}^{b-1}\sum_{t=0}^{b-1-s}\binom{s+a-1}{a-1}\binom{a-1}{t}\overline{\zeta}_q^{\mathrm{BZ}}(a+s, b-s-t)\\ &-\sum_{s=1}^{\min(a,b)}\frac{(a+b-s-1)!}{(a-s)!(b-s)!(s-1)!}\varphi_q(a+b-s),\end{aligned}$$

for $a, b \geq 2$, where

$$\varphi_q(k) := \sum_{n>0}\frac{(n-1)q^{(k-1)n}}{(1-q^n)^k} = \sum_{n>0}\frac{nq^{(k-1)n}}{(1-q^n)^k} - \overline{\zeta}_q^{\mathrm{BZ}}(k).$$

Therefore, for a coherent algebraic framework, an extension of the model is necessary. Kronecker's delta is denoted by δ_0, i.e., $\delta_0(0) = 1$ and $\delta_0(k) = 0$ for $k \neq 0$. We regard

$$\overline{\zeta}_q^{\mathrm{BZ,ex}}(k_1,\dots,k_n) := \sum_{m_1>\dots>m_n>0} \frac{q^{(k_1-1)m_1+(k_2-1+\delta_0(k_2))m_2+\dots+(k_n-1+\delta_0(k_n))m_n}}{(1-q^{m_1})^{k_1}\dots(1-q^{m_n})^{k_n}} \tag{20}$$

for $k_1 \geq 2, k_2, \dots, k_n \geq 0$. In order to derive a q-shuffle product we need a corresponding analogue of Proposition 1 (d).

Lemma 2 ([33]) *Let $n \in \mathbb{N}$. Then we have*

$$\overline{\zeta}_q^{\mathrm{BZ,ex}}(\mathbf{k}) = P_q^{k_1-1}\overline{P}_q[yP_q^{k_2-1+\delta_0(k_2)}\overline{P}_q^{1-\delta_0(k_2)}[y\cdots P_q^{k_n-1+\delta_0(k_n)}\overline{P}_q^{1-\delta_0(k_n)}[y]\cdots]](1) \tag{21}$$

for $\mathbf{k} := (k_1,\dots,k_n) \in \mathbb{N}_0^n$ with $k_1 > 1$ and $y(t) := t/(1-t)$.

Obviously, we recognize that $\varphi_q(k) = \overline{\zeta}_q^{\mathrm{BZ,ex}}(k,0)$ for $k > 1$. Therefore in the light of the extended BZ-model all terms in the q-analogue of Euler's decomposition formula have a natural explanation. Again, we translate Lemma 2 in an algebraic framework.

The polynomial algebra with three non-commutative variables p and $\overline{p}$, y is defined by $\mathscr{R} := \mathbb{Q}\langle p, \overline{p}, y\rangle$ with empty word $\mathbf{1}$. The subalgebra generated by words with letters of the form $u_k := p^{k-1+\delta_0(k)}\overline{p}^{1-\delta_0(k)}y$ with $(k \in \mathbb{N}_0)$ is denoted by $\mathscr{R}^1$. Additionally we regard the following subalgebra of $\mathscr{R}^1$:

$$\mathscr{R}^0 := \mathbb{Q} \oplus \bigoplus_{k\geq 2} u_k\mathscr{R}^1.$$

Now we define an evaluation map $\overline{\zeta}_q^{\mathrm{BZ}} : \mathscr{R}^0 \to \mathbb{Q}[[q]]$ by $\overline{\zeta}_q^{\mathrm{BZ}}(\mathbf{1}) := 1$ and

$$\overline{\zeta}_q^{\mathrm{BZ}}(u_{k_1}\cdots u_{k_n}) := \overline{\zeta}_q^{\mathrm{BZ,ex}}(k_1,\dots,k_n).$$

Now we define the quasi q-shuffle product $*_{\mathrm{BZ}} : \mathscr{R}^1 \otimes \mathscr{R}^1 \to \mathscr{R}^1$ recursively by

(i) $w *_{\mathrm{BZ}} \mathbf{1} := \mathbf{1} *_{\mathrm{BZ}} w := w$,
(ii) $u_s v *_{\mathrm{BZ}} u_t w := u_s(v *_{\mathrm{BZ}} u_t w) + u_t(u_s v *_{\mathrm{BZ}} w) + \langle u_s, u_t\rangle(v *_{\mathrm{BZ}} w)$

for words $v, w \in \mathscr{R}^1$ and $s, t \in \mathbb{N}_0$, where $\langle u_s, u_0\rangle := u_s$ for all $s \in \mathbb{N}_0$ and $\langle u_s, u_t\rangle := u_{s+t} + u_{s+t-1}$ for $s, t > 0$.

The q-shuffle product $⧢_{\mathrm{BZ}} : \mathscr{R}^1 \otimes \mathscr{R}^1 \to \mathscr{R}^1$ is defined iteratively by

(i) $\mathbf{1} ⧢_{\mathrm{BZ}} w := w ⧢_{\mathrm{BZ}} \mathbf{1} := w$,
(ii) $yv ⧢_{\mathrm{BZ}} w := v ⧢_{\mathrm{BZ}} yw := y(v ⧢_{\mathrm{BZ}} w)$,
(iii) $av ⧢_{\mathrm{BZ}} bw := a(v ⧢_{\mathrm{BZ}} bw) + b(av ⧢_{\mathrm{BZ}} w) + [a,b]a(v ⧢_{\mathrm{BZ}} w)$

for words $v, w \in \mathscr{R}^1$ and $a, b \in \{p, \overline{p}\}$, where $[p,p] := -[\overline{p},\overline{p}] := 1$ and $[p,\overline{p}] := [\overline{p},p] := 0$.

Remark 2 Note that the definition of $[\cdot,\cdot]$ exactly reflects the RBO properties of P_q and $\overline{P}_q$ as well as their compatibility relation stated in the first equation of Lemma 1.

All in all we have the following result:

Theorem 7 ([33]) *We have:*

(a) $(\mathscr{R}^0, *_{\mathrm{BZ}})$ *and* $(\mathscr{R}^0, ⧢_{\mathrm{BZ}})$ *are commutative and associative algebras.*
(b) The maps $\overline{\zeta}_q^{\mathrm{BZ}} : (\mathscr{R}^0, *_{\mathrm{BZ}}) \to \mathbb{Q}[[q]]$ *and* $\overline{\zeta}_q^{\mathrm{BZ}} : (\mathscr{R}^0, ⧢_{\mathrm{BZ}}) \to \mathbb{Q}[[q]]$ *are morphisms of algebras. Especially we have double q-shuffle relations given by*

$$\overline{\zeta}_q^{\mathrm{BZ}}(v *_{\mathrm{BZ}} w - v ⧢_{\mathrm{BZ}} w) = 0$$

for any $v, w \in \mathscr{R}^0$.

3.3 Duality and q-Shuffle

The previous paragraph illustrated that the Rota–Baxter operator approach for the q-shuffle product is a very powerful tool and can be generalized to many different models. Following an idea of [40], in some cases the q-shuffle product can be obtained by a duality construction. We shortly present this approach for the SZ-model, where it turns out to be equivalent to the RBO approach.

Using the algebraic framework introduced in Paragraph 3.2.1 we define an antiautomorphism $\tau_q : \mathscr{R}^0 \to \mathscr{R}^0$ by $\tau_q(\mathbf{1}) := \mathbf{1}$, $\tau_q(p) := y$ and $\tau_q(y) := p$. Similar to the duality theorem for classical MZVs (Theorem 3) we have a corresponding q-version at hand:

Theorem 8 ([37]) *For any* $w \in \mathscr{R}^0$ *we have* $\overline{\zeta}_q^{\mathrm{SZ}}(w) = \overline{\zeta}_q^{\mathrm{SZ}}(\tau_q(w))$.

Now we define a dual product $\square_{\mathrm{SZ}} : \mathscr{R}^0 \otimes \mathscr{R}^0 \to \mathscr{R}^0$ in terms of the antiautomorphism τ and the quasi q-shuffle product $*_{\mathrm{SZ}}$ such that the following diagram commutes:

$$\begin{array}{ccc} \mathscr{R}^0 \otimes \mathscr{R}^0 & \xrightarrow{\ \square_{\mathrm{SZ}}\ } & \mathscr{R}^0 \\ \Big\downarrow{\scriptstyle \tau_q \otimes \tau_q} & & \Big\uparrow{\scriptstyle \tau_q} \\ \mathscr{R}^0 \otimes \mathscr{R}^0 & \xrightarrow{\ *_{\mathrm{SZ}}\ } & \mathscr{R}^0 \end{array}$$

Due to Theorems 5 and 8 the map $\overline{\zeta}_q^{\mathrm{SZ}} : (\mathscr{R}^0, \square_{\mathrm{SZ}}) \to \mathbb{Q}[[q]]$ is an algebra morphism. Finally, we obtain:

Theorem 9 ([13]) *The products* $\square_{\mathrm{SZ}} : \mathscr{R}^0 \otimes \mathscr{R}^0 \to \mathscr{R}^0$ *and* $⧢_{\mathrm{SZ}} : \mathscr{R}^0 \otimes \mathscr{R}^0 \to \mathscr{R}^0$ *coincide.*

Remark 3 A similar construction can also be applied to the OOZ-model. However, the duality approach is limited to non-negative arguments whereas the RBO approach applies to arbitrary integer arguments which will be exploited in the next section.

Moreover we explore the dual product in the classical case. As in the q-case above for MZVs a duality theorem (Theorem 3) and a quasi-shuffle product (Sect. 2.1) are available. Again, we define the dual product $\square\colon \mathfrak{h}^0 \otimes \mathfrak{h}^0 \to \mathfrak{h}^0$ such that the following diagram commutes:

$$\begin{array}{ccc} \mathfrak{h}^0 \otimes \mathfrak{h}^0 & \xrightarrow{\quad\square\quad} & \mathfrak{h}^0 \\ {\scriptstyle \tau\otimes\tau}\big\downarrow & & \big\uparrow{\scriptstyle \tau} \\ \mathfrak{h}^0 \otimes \mathfrak{h}^0 & \xrightarrow{\quad *\quad} & \mathfrak{h}^0 \end{array}$$

It has already been remarked by Zudilin [40] that the dual product $\square$ does not coincide with the shuffle product. Nevertheless it can be related to a derivation relation (recall Sect. 2.2) of MZVs:

Theorem 10 ([13]) *For $w \in \mathfrak{h}^0$ we have*

$$\partial_2(w) = w\square u_2 - w * u_2.$$

Remark 4 This result should be compared with an identify of Hoffman and Ohno [22]. They proved that

$$\partial_1(w) = w ⧢ u_1 - w * u_1$$

for $w \in \mathfrak{h}^0$. For more details the reader is referred to [13].

4 q-Renormalization of Multiple Zeta Values

In this section we discuss the renormalization problem of multiple zeta values. The defining series (1) of the MZVs $\zeta(k_1, \ldots, k_n)$ is convergent if $k_1 + \cdots + k_j > j$ for $j = 1, \ldots, n$. Hence, this raises the question of how to extend the MZVs to non-positive integer arguments. In length one this question can be easily answered by using the meromorphic continuation of the Riemann zeta function which leads to $\zeta_1(-k) = -B_{k+1}/(k+1)$ for $k \in \mathbb{N}_0$. Although the multiple zeta function ζ_n can be meromorphically continued to $\mathbb{C}^n$ as shown in [1] the situation is more involved than in the dimension one case. The subvariety $\mathcal{S}_n$ of singularities of ζ_n is given by the hyperplanes

$$\begin{aligned} s_1 &= 1, \\ s_1 + s_2 &= 2, 1, 0, -2, -4, \dots, \\ s_1 + \cdots + s_j &\in \mathbb{Z}_{\leq j}\ (3 \leq j \leq n). \end{aligned}$$

Hence, in contrast to the length one case, we observe for $k_1, k_2 \in \mathbb{N}_0$ with $k_1 + k_2$ odd that

$$\zeta_2(-k_1, -k_2) = \frac{1}{2}\big(1 + \delta_0(k_2)\big)\frac{B_{k_1+k_2+1}}{k_1 + k_2 + 1},$$

whereas for $k_1 + k_2$ even we do not have any information due to the existence of poles in the points $(-k_1, -k_2)$. In length greater than two there are solely points of indeterminacy for any non-positive integer arguments, i.e., $(\mathbb{Z}_{\leq 0})^n \subseteq \mathscr{S}_n, n \geq 3$. Hence, the meromorphic continuation does not prescribe all values of the MZVs at non-positive integers. However, the vector space $\mathscr{M}$ of MZVs is an algebra with two non-compatible products. Therefore it is reasonable to expect the extended MZVs to satisfy at least the quasi-shuffle or the shuffle relations.

These requirements give rise to the so-called *renormalization problem of MZVs.*

Problem 1 (Renormalization problem of MZVs) How to extend the MZVs such that

(A) the meromorphic continuation and
(B) the quasi-shuffle or the shuffle relations

are verified?

Remark 5 The requirement (B) of the previous problem needs some comments. On the one hand the quasi-shuffle product is induced by the series representation of MZVs. Therefore the combinatorics is essentially the same as for positive arguments, e.g., if we interpret (1) as a formal series we still observe

$$\zeta(-a)\zeta(-b) = \zeta(-a, -b) + \zeta(-b, -a) + \zeta(-a - b)$$

for $a, b \in \mathbb{N}_0$. Similarly, the definition of the quasi-shuffle product in Sect. 2.1 extends to negative arguments. On the other hand a characterization of the shuffle product at non-positive integers is a crucial point. The shuffle product for positive indices is induced by the integral representation (4) which encodes the geometric aspects of MZVs. The combinatorics behind this product comes from the shuffling of integration variables. It could be illustrated by the shuffling of two decks of cards, say a deck of red and blue cards, each consecutively numbered such that the internal numbering of red and blue cards is preserved. In this approach, however, it is unclear how to handle non-positive arguments, which correspond to a non-positive number of cards. To overcome this difficulty we make use of the RBO characterization of MZVs given in Proposition 1 (a). Since the corresponding RBO J is invertible we obtain a shuffle product for non-positive arguments which is naturally induced by the inverse operator of J.

There are several approaches aiming at obtaining explicit values at points of indeterminacy. The first one was presented by Akiyama et al. [2]. It describes a limiting process to define MZVs at non-positive integer arguments. However, it turns out that the values depend on the order of conducting the limits. Additionally, the quasi-shuffle relations are not verified.

Furusho et al. proposed in [17] an approach called *desingularisation of MZVs.* They showed that certain finite sums of multiple zeta functions display non-trivial cancellations of all singularities involved, which therefore provide entire functions. This yields finite values for any tuple of integer arguments. Albeit, questions concerning the algebraic properties have not been addressed.

Furthermore there are two approaches that provide solutions for Problem 1. In [20] Guo and Zhang introduced directional regularized MZVs and Manchon and Paycha [28] discussed a specific cut of procedure for MZVs to obtain renormalized MZVs. Both sets of values satisfy the quasi-shuffle relation of MZVs and verify the meromorphic continuation.

The primary aim of this section is to construct explicit solutions to Problem 1. In Sect. 4.1 we provide a one-parameter family of solutions satisfying the quasi-shuffle relations. Our ansatz essentially relies on the Schlesinger–Zudilin q-model. In contrast to the works [20] and [28] the parameter q allows a very clear and transparent setting. Moreover in Sect. 4.2 we derive a set of numbers satisfying the shuffle relations.

Subsequently we sketch our construction principle. Both approaches apply the following factorization theorem of Connes and Kreimer:

Theorem 11 ([12, 27]) *Let $(\mathscr{H}, m_{\mathscr{H}}, \Delta)$ be a graded, connected (or filtered) Hopf algebra and $(\mathscr{A}, m_{\mathscr{A}})$ a commutative unital algebra equipped with a renormalization scheme $\mathscr{A} = \mathscr{A}_- \oplus \mathscr{A}_+$ and the corresponding idempotent Rota–Baxter operator π, where $\mathscr{A}_- = \pi(\mathscr{A})$ and $\mathscr{A}_+ = (\mathrm{Id} - \pi)(\mathscr{A})$. Further let $\psi : \mathscr{H} \to \mathscr{A}$ be a Hopf algebra character, i.e., a multiplicative linear map from $\mathscr{H}$ to $\mathscr{A}$. Then the character ψ admits a unique decomposition*

$$\psi = \psi_-^{\star(-1)} \star \psi_+ \tag{22}$$

called algebraic Birkhoff decomposition, *in which $\psi_- : \mathscr{H} \to \mathbb{Q} \oplus \mathscr{A}_-$ and $\psi_+ : \mathscr{H} \to \mathscr{A}_+$ are characters. The product on the right hand side of* (22) *is the convolution product defined on the vector space $L(\mathscr{H}, \mathscr{A})$ of linear maps from the $\mathscr{H}$ to $\mathscr{A}$ by $\phi \star \psi := m_{\mathscr{A}} \circ (\phi \otimes \psi) \circ \Delta$.*

Recall that the vector space $L(\mathscr{H}, \mathscr{A})$ together with the convolution product $\phi \star \psi := m_{\mathscr{A}} \circ (\phi \otimes \psi) \circ \Delta : \mathscr{H} \to \mathscr{A}$, where $\phi, \psi \in L(\mathscr{H}, \mathscr{A})$, is an unital associative algebra. The set of characters is denoted by $G_{\mathscr{A}}$ and forms a (pro-unipotent) group for the convolution product with (pro-nilpotent) Lie algebra $g_{\mathscr{A}}$ of infinitesimal characters. The latter are linear maps $\xi \in L(\mathscr{H}, \mathscr{A})$ such that for elements $x, y \in \mathscr{H}$, both different from $\mathbf{1}$, $\xi(xy) = 0$. The exponential map $\exp^\star$ restricts to a bijection between $g_{\mathscr{A}}$ and $G_{\mathscr{A}}$. The inverse of a character $\psi \in G_{\mathscr{A}}$ is given by composition with the Hopf algebra antipode $S : \mathscr{H} \to \mathscr{H}$, e.g., $\psi_-^{\star(-1)} = \psi_- \circ S$.

The maps ψ_+ and ψ_- of Eq. (22) are recursively given by

$$\psi_-(x) = -\pi\left(\psi(x) + \sum_{(x)} \psi_-(x')\psi(x'')\right), \tag{23}$$

$$\psi_+(x) = (\mathrm{Id} - \pi)\left(\psi(x) + \sum_{(x)} \psi_-(x')\psi(x'')\right) \tag{24}$$

for $x \in \mathscr{H} \cap \ker(\varepsilon)$, where ε denotes the counit of the graded (or filtered) Hopf algebra $\mathscr{H}$ and $\psi_\pm \in G_{\mathscr{A}}$. Note that we used Sweedler's notation for the reduced coproduct $\Delta'(x) := \sum_{(x)} x' \otimes x'' := \Delta(x) - \mathbf{1} \otimes x - x \otimes \mathbf{1}$.

Using Theorem 11 we proceed as follows:

First we have to construct a Hopf algebra $(\mathscr{H}, m_{\mathscr{H}}, \Delta)$. In the quasi-shuffle case the deconcatenation coproduct leads to a Hopf algebra which can be naturally extended to negative integer arguments. For the renormalization with respect to the shuffle product one of the main tasks is the construction of a Hopf algebra which reflects the algebraic structure induced by Proposition 1 (a).

The next step addresses the construction of a regularized character ψ. For this we provide a deformation of the MZV-character (7), which is a priori defined on the algebra $\mathfrak{h}^0$. In our approach the regularization relies on the q-deformation of the Schlesinger–Zudilin model and the MPLs in the quasi-shuffle and shuffle case respectively. After substituting the respective regulator q or z by e^z this defines an algebra morphism for all words related to negative integers which takes values in the commutative unital Rota–Baxter algebra of Laurent series $\mathscr{A} := \mathbb{C}[z^{-1}, z]]$. The corresponding projector $\pi : \mathscr{A} \to \mathscr{A}_-$, which is a RBO of weight -1, is defined by *minimal subtraction*

$$\pi\left(\sum_{n=-l}^{\infty} a_n z^n\right) := \sum_{n=-l}^{-1} a_n z^n,$$

where $\mathscr{A}_- := z^{-1}\mathbb{C}[z^{-1}]$ and $\mathscr{A}_+ := \mathbb{C}[[z]]$.

Then we apply the algebraic Birkhoff decomposition of Theorem 11, which is a systematic subtraction method to eliminate the pole terms of the deformed MZV characters, in order to obtain a character ψ_+. Finally, we have to ensure that the values extracted from ψ_+ verify the meromorphic continuation.

4.1 *Quasi-shuffle Renormalization Problem*

In this section we construct a one-parameter family of extensions of MZVs to negative integer arguments satisfying the quasi-shuffle relations and verifying the meromorphic continuation. Therefore there exist infinitely many solutions to Problem 1. Our approach is based on a very concrete and transparent construction using a modification of the Schlesinger–Zudilin q-model. The main feature we exploit is the fact that the SZ-model satisfies the ordinary quasi-shuffle relations of MZVs.

We define the *regularised Schlesinger–Zudilin* model by

$$\zeta_q(k_1, \dots, k_n) := \sum_{m_1 > \dots > m_n > 0} \frac{q^{|k_1|m_1 + \dots + |k_n|m_n}}{[m_1]_q^{k_1} \cdots [m_n]_q^{k_n}} \in \mathbb{Q}[\![q]\!]$$

for $k_1, \dots, k_n \in \mathbb{Z}$ with $k_1 \neq 0$. This permits us to consider the q-parameter as a natural regulator in view of ordinary multiple zeta values which assures convergence of the corresponding sums.

Let $D := \{s \in \mathbb{C} \colon \operatorname{Re}(s) > 0\}$. For $t \in D$ we define a one-parameter family of *modified Schlesinger–Zudilin q-MZVs* by

$$\overline{\zeta}_q^{(t),*}(k_1, \dots, k_n) := \sum_{m_1 > \dots > m_n > 0} \frac{q^{(|k_1|m_1 + \dots + |k_n|m_n)t}}{(1 - q^{m_1})^{k_1} \cdots (1 - q^{m_n})^{k_n}}. \tag{25}$$

For $0 < q < 1$ and $t \in D$, convergence of the previous series is always ensured for any $k_1, \dots, k_n \in \mathbb{Z}$ with $k_1 \neq 0$.

The key observation is the fact that the quasi-shuffle product carries over to the one-parameter family of modified Schlesinger–Zudilin q-MZVs. Indeed, the series of nested sums in (25) satisfy the quasi-shuffle relations if all arguments are either strictly positive or strictly negative integers. For example the corresponding Nielsen reflexion formula for the one-parameter family of modified Schlesinger–Zudilin q-MZVs defined in (25) for negative integers is

$$\overline{\zeta}_q^{(t),*}(-a)\overline{\zeta}_q^{(t),*}(-b) = \overline{\zeta}_q^{(t),*}(-a, -b) + \overline{\zeta}_q^{(t),*}(-b, -a) + \overline{\zeta}_q^{(t),*}(-a - b)$$

for $a, b \in \mathbb{N}$. One should note that the quasi-shuffle product is not preserved if we allow integer arguments with mixed signs in (25).

It is easily seen that the quasi-shuffle product $*$ defined in Sect. 3.2.1 extends to words with letters $Y := \{u_n : n \in \mathbb{Z}\}$. As stated in Theorem 2 the map $\zeta : (\mathfrak{h}^0, *) \to (\mathbb{R}, \cdot)$ is an algebra morphism. Let $\mathfrak{h}^- := \mathbb{Q}\langle Y^- \rangle$ with $Y^- := \{u_n : n \in \mathbb{Z}_{<0}\}$. Then the map $\overline{\zeta}_q^{(t)} : (\mathfrak{h}^-, *) \to (\mathbb{C}[\![q]\!], \cdot)$ defined by $\overline{\zeta}_q^{(t)}(\mathbf{1}) := 1$ and

$$\overline{\zeta}_q^{(t)}(u_{k_1} \cdots u_{k_n}) := \overline{\zeta}_q^{(t),*}(k_1, \dots, k_n)$$

for any $k_1, \dots, k_n \in \mathbb{Z}_{<0}$ and $t \in D$ is a morphism of algebras. Furthermore the deconcatenation coproduct defined in Sect. 2.1 extends to arbitrary integer arguments, i.e., we have $\Delta \colon \mathbb{Q}\langle Y \rangle \to \mathbb{Q}\langle Y \rangle \otimes \mathbb{Q}\langle Y \rangle$

$$\Delta(w) = \sum_{uv=w} u \otimes v \tag{26}$$

for $w \in Y^*$. Together with the quasi-shuffle product $(\mathbb{Q}\langle Y\rangle, *, \Delta)$ becomes a Hopf algebra. This construction directly applies to Y^- and $\mathfrak{h}^-$ and therefore the triple $(\mathfrak{h}^-, *, \Delta)$ is a filtered, connected Hopf algebra, in which the filtration is given by the depth, i.e., the number of letters of a word. Furthermore we use the notation $|\mathbf{k}| := k_1 + \cdots + k_n$ for $\mathbf{k} := (k_1, \ldots, k_n) \in \mathbb{Z}^n$.

For the Hopf algebra $(\mathfrak{h}^-, *, \Delta)$ we construct a character $\psi^{(t)} : \mathfrak{h}^- \to \mathbb{C}[z^{-1}, z]]$ using the one-parameter family of modified Schlesinger–Zudilin q-MZVs defined in (25).

The map $\psi^{(t)} : (\mathfrak{h}^-, *) \to \mathbb{C}[z^{-1}, z]]$ is defined by the following composition of maps:

$$\begin{array}{ccccccc} (\mathfrak{h}^-, *) & \longrightarrow & (\mathbb{C}[[q]], \cdot) & \longrightarrow & (\mathbb{C}[z^{-1}, z]], \cdot) & \longrightarrow & (\mathbb{C}[z^{-1}, z]], \cdot) \\ u_{k_1} \cdots u_{k_n} & \longmapsto & \overline{\zeta}_q^{(t)}(u_{k_1} \cdots u_{k_n}) & \longmapsto & \overline{\zeta}_{e^z}^{(t)}(u_{k_1} \cdots u_{k_n}) & \longmapsto & (-z)^{|\mathbf{k}|}\overline{\zeta}_{e^z}^{(t)}(u_{k_1} \cdots u_{k_n}), \end{array}$$

where $k_1, \ldots, k_n \in \mathbb{Z}_{<0}$ or equivalently for $k_1, \ldots, k_n \in \mathbb{N}$

$$\psi^{(t)}(u_{-k_1} \cdots u_{-k_n})(z) := \frac{(-1)^{k_1+\cdots+k_n}}{z^{k_1+\cdots+k_n}} \overline{\zeta}_{e^z}^{(t)}(u_{-k_1} \cdots u_{-k_n}). \tag{27}$$

Since all the maps involved in the previous composition are algebra morphisms, $\psi^{(t)}$ is a character. Therefore we are in the position to apply Theorem 11. The algebraic Birkhoff decomposition (22) leads to an algebra morphism $\psi_+^{(t)} : (\mathfrak{h}^-, *) \to \mathbb{C}[[z]]$. Hence, $\zeta_+^{(t)}$ defined by

$$\zeta_+^{(t)}(-k_1, \ldots, -k_n) := \lim_{z \to 0} \psi_+^{(t)}(u_{-k_1} \cdots u_{-k_n})(z)$$

is well-defined for any $k_1, \ldots, k_n \in \mathbb{N}$. We call $\zeta_+^{(t)}(-k_1, \ldots, -k_n)$ the *renormalized MZVs*.

Now we can state the main result of this section:

Theorem 12 ([14]) *Let $t \in D$.*

(a) The renormalized MZVs $\zeta_+^{(t)}$ satisfy the quasi-shuffle product.
(b) The renormalized MZVs $\zeta_+^{(t)}$ verify the meromorphic continuation of MZVs, i.e., for any $\mathbf{k} \in (\mathbb{Z}_{<0})^n \setminus \mathscr{S}_n$ we have $\zeta_+^{(t)}(\mathbf{k}) = \zeta(\mathbf{k}) \in \mathbb{Q}$.
(c) The renormalized MZVs $\zeta_+^{(t)}$ are rational functions in t over $\mathbb{Q}$ without singularities in D.

Proof Theorem 11 ensures the algebraic properties of the renormalized MZVs. Therefore (a) is verified. For (b) we have to check that $\zeta_+^{(t)}$ coincides with the values given by the meromorphic continuation of MZVs in length two with odd weight. For this one has to perform an explicit calculation using the formulas (23) and (24).

Finally, one verifies that all coefficients of the power series $\psi^{(t)}$ are rational functions in t over $\mathbb{Q}$ without singularities in D. Since the algebraic Birkhoff decomposition is calculated by multiplying and adding these coefficients, (c) follows. □

As a consequence we obtain:

Corollary 1 *(a) Let $t \in D \cap \mathbb{Q}$. Then $\zeta_+^{(t)}(\mathbf{k}) \in \mathbb{Q}$ for any $\mathbf{k} \in (\mathbb{Z}_{<0})^n$, $n \in \mathbb{N}$.*
(b) Let $t \in D \cap \mathbb{R}$ be transcendental over $\mathbb{Q}$. Then there exists a $\mathbf{k} \in (\mathbb{Z}_{<0})^n$ such that $\zeta_+^{(t)}(\mathbf{k}) \in \mathbb{R} \setminus \mathbb{Q}$.

Proof The first claim follows from Theorem 12 c). Since not all renormalized MZVs $\zeta_+^{(t)}$ are constant as a rational function in t over $\mathbb{Q}$ (see Eqs. (28) and (29)) the choice of a transcendental $t \in D$ leads to irrational values for $\zeta_+^{(t)}$ which proves (b). □

In Table 1 we list the renormalized MZVs in the case $t = 1$ for depth two. For depth one the renormalized MZVs are always rational as well as for $k_1, k_2 \in \mathbb{N}$ with $k_1 + k_2$ odd, due to Theorem 12 (b). Because of the quasi-shuffle relation

$$\zeta_+^{(t)}(-k)\zeta_+^{(t)}(-k) = 2\zeta_+^{(t)}(-k, -k) + \zeta_+^{(t)}(-2k)$$

for $k \in \mathbb{N}$ the diagonal entries $\zeta_+^{(t)}(-k, -k)$ are also always rational and do not depend on the parameter t. The first case for which we obtain a non-constant rational function in t over $\mathbb{Q}$ is the case $\zeta_+^{(t)}(-1, -3)$. The quasi-shuffle product implies

$$\zeta_+^{(t)}(-1)\zeta_+^{(t)}(-3) = \zeta_+^{(t)}(-1, -3) + \zeta_+^{(t)}(-3, -1) + \zeta_+^{(t)}(-4).$$

Therefore a priori $\zeta_+^{(t)}(-1, -3)$ and $\zeta_+^{(t)}(-3, -1)$ are not explicitly given by that relation. They only have to satisfy $\zeta_+^{(t)}(-1, -3) + \zeta_+^{(t)}(-3, -1) = \zeta_+^{(t)}(-1)\zeta_+^{(t)}(-3)$. Using (23) and (24) we find

$$\zeta_+^{(t)}(-1, -3) = \frac{1}{8064} \frac{166t^2 + 166t + 31}{(4t + 3)(4t + 1)}, \tag{28}$$

$$\zeta_+^{(t)}(-3, -1) = -\frac{1}{40320} \frac{1278t^2 + 1278t + 239}{(4t + 3)(4t + 1)}. \tag{29}$$

4.2 *Shuffle Renormalization Problem*

Now we address Problem 1 where we demand for (B) the verification of the shuffle product relations. In the first step we need an appropriate deformation of the MZV character. For the classical case we choose the MPL in a single variable defined in (9) and for the q-case we take the modified OOZ-model defined in (19). This is motivated by the fact that—as shown in Proposition 1 (a) and (c)—both models admit

Table 1 The renormalized MZVs $\zeta_{+}^{(1)}(k_1, k_2)$

$k_1 \diagdown k_2$	-1	-2	-3	-4	-5
-1	$\frac{1}{288}$	$-\frac{1}{240}$	$\frac{121}{94080}$	$\frac{1}{504}$	$-\frac{31093}{17740800}$
-2	$-\frac{1}{240}$	0	$\frac{1}{504}$	$-\frac{48529}{66528000}$	$-\frac{1}{480}$
-3	$-\frac{559}{282240}$	$\frac{1}{504}$	$\frac{1}{28800}$	$-\frac{1}{480}$	$\frac{941347763}{1150753443840}$
-4	$\frac{1}{504}$	$\frac{48529}{66528000}$	$-\frac{1}{480}$	0	$\frac{1}{264}$
-5	$\frac{110879}{53222400}$	$-\frac{1}{480}$	$-\frac{979401779}{1150753443840}$	$\frac{1}{264}$	$\frac{1}{127008}$

a characterization via iterated RBOs where the corresponding RBOs are invertible. This is a natural approach to attack the shuffle problem described in Remark 5. The geometric properties of the given models, which are encoded in terms of iterated integrals, are translated to an algebraic property. Therefore we convert Proposition 1 (a) and (c) into an appropriate word algebraic setting. The superordinate aim is then to construct a graded, connected Hopf algebra to which we apply Theorem 11.

Let $L := \{d, y\}$ be the set of letters, L^* the set of words with empty word $\mathbf{1}$ and $\mathbb{Q}\langle L\rangle$ the free polynomial algebra with to non-commutative variables d, y. The subspace of $\mathbb{Q}\langle L\rangle$ spanned by *non-admissible words* is denoted by $\mathcal{T} := \langle\{wd : w \in L^*\}\rangle_{\mathbb{Q}}$ and the set of *admissible words* by $W := L^*y \cup \{\mathbf{1}\}$. Let $w \in W$. Then we define the *weight* $\mathrm{wt}(w)$ by the number of letters of w and the *depth* $\mathrm{dpt}(w)$ by the number of y in w.

We establish an algebraic framework for the shuffle product at non-positive integers. In order to provide a uniform presentation we introduce a parameter $\lambda \in \mathbb{Q}$. The case $\lambda = 0$ corresponds to classical MZVs and the case $\lambda = -1$ to the OOZ q-model.

Based on Proposition 1 (a) and (b) we define the product $ш_\lambda : \mathbb{Q}\langle L\rangle \otimes \mathbb{Q}\langle L\rangle \to \mathbb{Q}\langle L\rangle$ iteratively by

(P1) $\mathbf{1} ш_\lambda w := w ш_\lambda \mathbf{1} := w,$

(P2) $yu ш_\lambda v := u ш_\lambda yv := y(u ш_\lambda v),$

(P3) $$du ш_\lambda dv := \begin{cases} \frac{1}{\lambda}\left(d(u ш_\lambda v) - du ш_\lambda v - u ш_\lambda dv\right) & \lambda \neq 0, \\ d(u ш_0 dv) - u ш_0 d^2 v & \lambda = 0, \end{cases}$$

for words $u, v, w \in L^*$.

Remark 6

- By induction on the number of letters it is easily seen that $ш_\lambda$ defines a product for $\lambda \neq 0$. The case $\lambda = 0$ is more involved. We shift ds from the left to the right until we hit a y or $\mathbf{1}$. Using the rule (P1) and (P2) we can go on iteratively.
- Note that for $\lambda = -1$ the above shuffle product $ш_{-1}$ coincides for non-positive arguments with the product $ш_{\mathrm{OOZ}}$ introduced in Sect. 3.2.2.
- For $\lambda = 0$, (P3) reflects the Leibniz rule since the inverse of J is given by the Euler derivation. The map $ш_0$ is a magma and therefore a priori not necessarily associative and commutative.

In the next step we prove that the non-admissible words form an ideal with respect to $\sqcup\!\sqcup_\lambda$.

Lemma 3 ([15]) *For $\lambda \in \mathbb{Q}$ the subspace $\mathscr{T}$ is a two-sided ideal of* $(\mathbb{Q}\langle L\rangle, \sqcup\!\sqcup_\lambda)$.

As state in Remark 6 the map $\sqcup\!\sqcup_0$ is a magma. In order to obtain a product structure we establish an ideal-theoretic variant of the Leibniz rule. By $\mathscr{L}_0$ we denote the ideal of $(\mathbb{Q}\langle L\rangle, \sqcup\!\sqcup_0)$ generated by the elements

$$d(u \sqcup\!\sqcup_0 v) - du \sqcup\!\sqcup_0 v - u \sqcup\!\sqcup_0 dv), \qquad u, v \in L^*$$

and stable under left concatenation with d, i.e., $d\mathscr{L}_0 \subseteq \mathscr{L}_0$. For $\lambda \neq 0$ we define $\mathscr{L}_\lambda := \{0\}$.

On a corresponding quotient space we obtain the desired product structure for any $\lambda \in \mathbb{Q}$:

Lemma 4 ([15]) *For $\lambda \in \mathbb{Q}$ we define $\mathscr{H}_\lambda := \mathbb{Q}\langle L\rangle/(\mathscr{T} + \mathscr{L}_\lambda)$. Then $(\mathscr{H}_\lambda, \sqcup\!\sqcup_\lambda)$ is an associative and commutative algebra.*

After establishing the shuffle product structure for non-positive integers we provide a corresponding coalgebra structure on $\mathscr{H}_\lambda$. In contrast to the product structure we do not have to distinguish between $\lambda = 0$ and $\lambda \neq 0$.

We define the coproduct $\overline{\Delta}_\lambda : \mathbb{Q}\langle L\rangle \to \mathbb{Q}\langle L\rangle \otimes \mathbb{Q}\langle L\rangle$ by

(C1) $\overline{\Delta}_\lambda(\mathbf{1}) := \mathbf{1} \otimes \mathbf{1}$,
(C2) $\overline{\Delta}_\lambda(y) := \mathbf{1} \otimes y + y \otimes \mathbf{1}$,
(C3) $\overline{\Delta}_\lambda(d) := \mathbf{1} \otimes d + d \otimes \mathbf{1} + \lambda d \otimes d$,

which extends uniquely to an algebra morphism (with respect to concatenation) on the free algebra $\mathbb{Q}\langle L\rangle$. In the next lemma we prove that the ideals $\mathscr{T}$ and $\mathscr{L}_\lambda$ are also compatible with the coproduct $\overline{\Delta}_\lambda$:

Lemma 5 ([15]) *For $\lambda \in \mathbb{Q}$ the double $(\mathbb{Q}\langle L\rangle, \overline{\Delta}_\lambda)$ is a cocommutative coalgebra. The subspaces $\mathscr{T}$ and $\mathscr{L}_\lambda$ are coideals of $\mathbb{Q}\langle L\rangle$.*

The algebra and coalgebra on $\mathscr{H}_\lambda$ form a Hopf algebra which is the first main step in the construction of shuffle renormalized multiple zeta values:

Theorem 13 ([15]) *Let $\lambda \in \mathbb{Q}$. Then $(\mathscr{H}_\lambda, \sqcup\!\sqcup_\lambda, \Delta_\lambda)$ is a graded, connected Hopf algebra with*

$$\Delta_\lambda([w]) := \overline{\Delta}_\lambda(w) \mod ((\mathscr{T} + \mathscr{L}_\lambda) \otimes \mathbb{Q}\langle L\rangle + \mathbb{Q}\langle L\rangle \otimes (\mathscr{T} + \mathscr{L}_\lambda))$$

for any word $w \in W$.

After the construction of the shuffle Hopf algebra $\mathscr{H}_\lambda$ we need to specify characters Ψ^c and Ψ^q which deform the divergent (q-)MZVs to Laurent series in $\mathbb{Q}[z^{-1}, z]]$. For this we define the following maps:

$$\begin{array}{rcccc} \Psi^c\colon & (\mathscr{H}_0, ш_0) & \longrightarrow & (\mathbb{Q}[\![t]\!], \cdot) & \longrightarrow (\mathbb{Q}[z^{-1}, z]\!], \cdot) \\ & [d^{k_1}y\cdots d^{k_n}y] & \longmapsto & \mathrm{Li}_{-k_1,\ldots,-k_n}(t) & \longmapsto \mathrm{Li}_{-k_1,\ldots,-k_n}(e^z) \end{array}$$

and

$$\begin{array}{rcccc} \Psi^q\colon & (\mathscr{H}_{-1}, ш_{-1}) & \longrightarrow & (\mathbb{Q}[\![q]\!], \cdot) & \longrightarrow (\mathbb{Q}[z^{-1}, z]\!], \cdot) \\ & [d^{k_1}y\cdots d^{k_n}y] & \longmapsto & \overline{\zeta}_q^{\mathrm{OOZ}}(-k_1,\ldots,-k_n) & \longmapsto \overline{\zeta}_{e^z}^{\mathrm{OOZ}}(-k_1,\ldots,-k_n) \end{array}$$

Lemma 6 ([15]) *The maps* $\Psi^c\colon (\mathscr{H}_0, ш_0) \to (\mathbb{Q}[z^{-1}, z]\!], \cdot)$ *and* $\Psi^q\colon (\mathscr{H}_{-1}, ш_{-1}) \to (\mathbb{Q}[z^{-1}, z]\!], \cdot)$ *are well-defined and morphisms of algebras.*

Now we are in the position to apply the algebraic Birkhoff decomposition of Theorem 11 to Ψ^c and Ψ^q. Hence, we define for $k_1, \ldots, k_n \in \mathbb{N}_0$ *renormalized MZVs* by

$$\zeta_+(-k_1,\ldots,-k_n) := \lim_{z\to 0} \Psi_+^c([d^{k_1}y\cdots d^{k_n}y])(z)$$

and *renormalized q-MZVs* by

$$\zeta_+^{\mathrm{OOZ}}(-k_1,\ldots,-k_n) := \lim_{z\to 0} \frac{(-1)^{k_1+\cdots+k_n}}{z^{k_1+\cdots+k_n}} \Psi_+^q([d^{k_1}y\cdots d^{k_n}y])(z). \qquad (30)$$

Remark 7 A priori Ψ_+^q takes values in $\mathbb{Q}[\![z]\!]$. However, the subsequent Theorem 14 together with an explicit computation of Ψ_+^q in the length one case ensures that (30) is well-defined.

Further we obtain a factorization theorem which plays a key role on the conceptual level of the shuffle renormalization.

Theorem 14 (Shuffle factorization, [15]) *Let* $\lambda \in \mathbb{Q}$. *Then for all* $w \in W$ *we have*

$$ш_\lambda \circ \Delta_\lambda([w]) = 2^{\mathrm{dpt}([w])}[w].$$

This theorem shows that any (q-)MZV related to non-positive arguments, which is given by a certain equivalence class $[w]$, can be written as a linear combination of factorizations of (q-)ZVs with respect to the shuffle product $ш_\lambda$. The factorization is described by the coproduct Δ_λ.

All in all we have the following main result:

Theorem 15 ([15])

(a) *The renormalization process is compatible with the meromorphic continuation, i.e.,* ζ_+ *coincides with the meromorphic continuation* ζ_n *whenever it is defined.*
(b) *The map* ζ_+ *satisfies the shuffle product* $ш_0$.
(c) *The map* ζ_+^{OOZ} *is well-defined and for any* $\mathbf{k} \in (\mathbb{Z}_{\le 0})^n$ *we have* $\zeta_+(\mathbf{k}) = \zeta_+^{\mathrm{OOZ}}(\mathbf{k})$.
(d) *For any* $\mathbf{k} \in (\mathbb{Z}_{\le 0})^n$ *we have* $\zeta_+(\mathbf{k}) \in \mathbb{Q}$.

Table 2 The renormalized MZVs $\zeta_+(k_1, k_2)$

$k_1 \backslash k_2$	0	−1	−2	−3
0	$\frac{1}{4}$	$\frac{1}{24}$	0	$-\frac{1}{240}$
−1	$\frac{1}{12}$	$\frac{1}{144}$	$-\frac{1}{240}$	$-\frac{1}{1440}$
−2	$\frac{1}{72}$	$-\frac{1}{240}$	$-\frac{1}{720}$	$\frac{1}{504}$
−3	$-\frac{1}{120}$	$-\frac{1}{360}$	$\frac{1}{504}$	$\frac{107}{100800}$

(e) For any character $\psi\colon (\mathscr{H}_0, ш_0) \to (\mathbb{Q}[z^{-1}, z]], \cdot)$ *with*

$$\lim_{z\to 0} \psi_+([d^k y])(z) = \zeta_1(-k)$$

for $k \in \mathbb{N}_0$ *we have*

$$\zeta_+(-k_1, \ldots, -k_n) = \lim_{z\to 0} \psi_+([d^{k_1} y \cdots d^{k_n} y])(z)$$

for $k_1, \ldots, k_n \in \mathbb{N}_0$.

In Table 2 we provide some numerical examples for ζ_+ in depth 2.

4.3 Comparison of Different Solutions to the Renormalization Problem

In a nutshell the structure of the solutions to the renormalization problem is very different depending on whether we demand the verification of the shuffle or quasi-shuffle product. Theorem 15 (e) shows that for the shuffle product $ш_0$ a unique solution to Problem 1 exists. In contrast to this the one-parameter family of renormalized MZVs constructed in Sect. 4.1 exemplifies that there are infinitely many solutions to Problem 1 in the quasi-shuffle case. Additionally there are the solutions presented in [20] and [27].

Therefore naturally the question of the relations between different renormalized MZVs in the quasi-shuffle case arises. A comprehensive answer is given by Ebrahimi-Fard, Manchon, Zhao and the author in terms of a transitive and free group action [16]. Let $\mathscr{H} := \mathbb{Q}\langle u_k : k \in \mathbb{Z}\rangle$ be the non-commutative polynomial algebra generated by the letters u_k, $k \in \mathbb{Z}$. It is easily seen that $(\mathscr{H}, *, \Delta)$ is a Hopf algebra, where $*$ and Δ denote the natural extensions of the quasi-shuffle product and the deconcatenation coproduct, respectively (see Sect. 2.1). We call a word $w := u_{k_1} \cdots u_{k_n} \in \mathscr{H}$ *non-singular* if

$$k_1 \neq 1 \quad \text{and} \quad k_1 + k_2 \notin \{2, 1, 0, -2, -4, \ldots\} \quad \text{and} \quad k_1 + \cdots + k_j \notin \mathbb{Z}_{\leq j} \quad (j = 3, \ldots, n).$$

We denote with $N \subseteq \mathscr{H}$ the vector space spanned by non-singular words. Moreover we recall that $G_{\mathbb{C}}$ denotes the set of quasi-shuffle characters from $\mathscr{H}$ to $\mathbb{C}$. Then the set of all solutions to Problem 1 is given by

$$X_{\mathbb{C},\zeta} := \{\phi \in G_{\mathbb{C}} \colon \phi|_N = \zeta\},$$

where ζ denotes the meromorphic continuation of the multiple zeta function. Furthermore we define the *renormalization group*

$$T_{\mathbb{C}} := \{\phi \in G_{\mathbb{C}} \colon \phi|_N = 0\}.$$

Theorem 16 ([16]) *We have:*

(a) The set $T_{\mathbb{C}}$ is a subgroup of $(G_{\mathbb{C}}, \star)$.
(b) The set $X_{\mathbb{C},\zeta}$ is a $T_{\mathbb{C}}$-principal homogenous space. More precisely, the left group action

$$\begin{aligned} T_{\mathbb{C}} \times X_{\mathbb{C},\zeta} &\longrightarrow X_{\mathbb{C},\zeta}, \\ (\alpha, \phi) &\longmapsto \alpha \star \phi \end{aligned}$$

is free and transitive.
(c) The set $X_{\mathbb{C},\zeta}$ is of infinite cardinality.

Therefore geometrically speaking, all solutions to the renormalization problem lie on the same orbit of the previously defined group action. By restricting the Hopf algebra $(\mathscr{H}, *, \Delta)$ to words consisting only of non-positive or negative letters, we can compare all the values obtained in Sect. 4.1 with those of [20, 28] by restricting them to the corresponding Hopf subalgebra. The reader is referred to [16] for more details.

A further natural question concerns a possible double shuffle structure of renormalized MZVs. Currently this question is still open. Two observations indicate problems which have to be overcome. First of all, the quasi-shuffle relation

$$\zeta(0)^2 = 2\zeta(0,0) + \zeta(0)$$

implies $\zeta(0,0) = 3/8$ since the meromorphic continuation of the Riemann zeta function forces $\zeta_1(0) = -1/2$. As we observe from Table 2 in the shuffle case we have $\zeta(0,0) = 1/4$. Therefore these structures are superficially in conflict with each other. Maybe this point can be resolved by a transformation map between shuffle and quasi-shuffle renormalized MZVs similar to that used for multiple polylogarithms in several variables. However, this poses a second problem. Although Theorem 16 clarifies the relation between all different quasi-shuffle renormalized MZVs it is not clear which element of the set $X_{\mathbb{C},\zeta}$ should be chosen to compare with the shuffle renormalized MZVs.

Acknowledgements Some results presented in this work are based on a project which was carried out during the Research Trimester on Multiple Zeta Values, Multiple Polylogarithms, and Quantum Field Theory (September–December 2014) at ICMAT, Madrid. I have greatly enjoyed the hospitality and the nice and stimulating atmosphere at ICMAT. Furthermore I am very grateful to Andreas Knauf for his comments which significantly improved the paper.

References

1. Akiyama, S., Egami, S., Tanigawa, Y.: Analytic continuation of multiple zeta-functions and their values at non-positive integers. Acta Arith. **98**(2), 107–116 (2001)
2. Akiyama, S., Tanigawa, Y.: Multiple zeta values at non-positive integers. Ramanujan J. **5**(4), 327–351 (2001)
3. Apéry, R.: Irrationalité de $\zeta(2)$ et $\zeta(3)$. Astérisque **61**, 11–13 (1979)
4. Ball, K., Rivoal, T.: Irrationalité d'une infinité de valeurs de la fonction zêta aux entiers impairs. Invent. Math. **146**(1), 193–207 (2001)
5. Bradley, D.: Multiple q-zeta values. J. Algebra, 752–798 (2005)
6. Broadhurst, D.: Multiple zeta values and modular forms in quantum field theory. In: Computer Algebra in Quantum Field Theory, Texts & Monographs in Symbolic Computation, pp. 33–73. Springer (2013)
7. Brown, F.: Mixed Tate motives over $\mathbb{Z}$. Ann. Math. **175**, 949–976 (2012)
8. Brown, F.: On the decomposition of motivic multiple zeta values. In: Galois-Teichmüller theory and arithmetic geometry, vol. 63 of Adv. Stud. Pure Math., pp. 31–58. Math. Soc. Japan, Tokyo (2012)
9. Brown, F.: Irrationality proofs for zeta values, moduli spaces and dinner parties. arXiv:1412.6508 (2014)
10. Castillo Medina, J., Ebrahimi-Fard, K., Manchon, D.: On Euler's decomposition formula for qMZVs. Ramanujan J. **37**(2), 365–389 (2015)
11. Castillo Medina, J., Ebrahimi-Fard, K., Manchon, D.: Unfolding the double shuffle structure of q-multiple zeta values. Bull. Aust. Math. Soc. **91**(3), 368–388 (2015)
12. Connes, A., Kreimer, D.: Renormalization in quantum field theory and the Riemann–Hilbert problem. I. The Hopf algebra structure of graphs and the main theorem. Comm. Math. Phys. **210**(1), 249–273 (2000)
13. Ebrahimi-Fard, K., Manchon, D., Singer, J.: Duality and (q-)multiple zeta values. Adv. Math. **298**, 254–285 (2016)
14. Ebrahimi-Fard, K., Manchon, D., Singer, J.: Renormalisation of q-regularised multiple zeta values. Lett. Math. Phys. **106**(3), 365–380 (2016)
15. Ebrahimi-Fard, K, Manchon, D., Singer, J.: The Hopf algebra of (q-)Multiple Polylogarithms with non-positive arguments. Int. Math. Res. Notices 2016 rnw128 (2016)
16. Ebrahimi-Fard, K., Manchon, D., Singer, J., Zhao, J.: Renormalisation group for multiple zeta values. Commun. Number Theory **12**(1), 75–96 (2018)
17. Furusho, H., Komori, Y., Matsumoto, K., Tsumura, H.: Desingularization of complex multiple zeta-functions. To appear in Amer. J. Math. (2016)
18. Guo, L.: An introduction to Rota–Baxter Algebra, vol. IV of Surveys of Modern Mathematics. International Press (2010)
19. Guo, L., Keigher, W.: On differential Rota-Baxter algebras. J. Pure Appl. Algebra **212**(3), 522–540 (2008)
20. Guo, L., Zhang, B.: Renormalization of multiple zeta values. J. Algebra **319**(9), 3770–3809 (2008)
21. Hoffman, M.: The Algebra of multiple harmonic series. J. Algebra **194**(2), 477–495 (1997)
22. Hoffman, M., Ohno, Y.: Relations of multiple zeta values and their algebraic expression. J. Algebra **262**(2), 332–347 (2003)

23. Ihara, K., Kaneko, M., Zagier, D.: Derivation and double shuffle relations for multiple zeta values. Compos. Math. **142** (2006)
24. Kac, V., Cheung, P.: Quantum Calculus. Universitext, Springer(2002)
25. Kaneko, M., Kurokawa, N., Wakayama, M.: A variation of Euler's approach to values of the Riemann zeta function. Kyushu J. Math., 175–192 (2003)
26. Krattenthaler, C., Rivoal, T.: An identity of Andrews, multiple integrals, and very-well-poised hypergeometric series. Ramanujan J. **13**(1–3), 203–219 (2007)
27. Manchon, D.: Hopf algebras in renormalisation. In: Handbook of algebra. Vol. 5, volume 5 of Handb. Algebr., pp. 365–427. Elsevier/North-Holland, Amsterdam (2008)
28. Manchon, D., Paycha, S.: Nested sums of symbols and renormalized multiple zeta values. Int. Math. Res. Not. IMRN **24**, 4628–4697 (2010)
29. Ohno, Y., Okuda, J., Zudilin, W.: Cyclic q-MZSV sum. J. Number Theory **132**(1), 144–155 (2012)
30. Panzer, E.: The parity theorem for multiple polylogarithms. J. Number Theory **172**, 93–113 (2017)
31. Schlesinger, K.-G.: Some remarks on q-deformed multiple polylogarithms. arXiv:0111022 (2001)
32. Singer, J.: On q-analogues of multiple zeta values. Funct. Approx. Comment. Math. **53**(1), 135–165 (2015)
33. Singer, J.: On Bradley's q-MZVs and a generalized Euler decomposition formula. J. Algebra **454**, 92–122 (2016)
34. Zagier, D.: Values of zeta functions and their applications. First European Congress of Mathematics. volume 120, pp. 497–512. Birkhäuser-Verlag, Basel (1994)
35. Zagier, D.: Evaluation of the multiple zeta values $\zeta(2, \ldots, 2, 3, 2, \ldots, 2)$. Ann. Math. **175**, 977–1000 (2012)
36. Zhao, J.: Multiple q-zeta functions and multiple q-polylogarithms. Ramanujan J. **14**(2), 189–221 (2007)
37. Zhao, J.: Uniform approach to double shuffle and duality relations of various q-analogs of multiple zeta values via rota–baxter algebras. arXiv:1412.8044 (2014)
38. Zudilin, W.: One of the numbers $\zeta(5)$, $\zeta(7)$, $\zeta(9)$, $\zeta(11)$ is irrational. Uspekhi Mat. Nauk **56**(4(340)), 149–150 (2001)
39. Zudilin, W.: Algebraic relations for multiple zeta values. Uspekhi Mat. Nauk **58**, 3–32 (2003)
40. Zudilin, W.: Multiple q-zeta brackets. Mathematics **3**(1), 119 (2015)

Vertex Algebras and Renormalization

Nikolay M. Nikolov

Abstract The Operator Product Expansion (OPE) and Renormalization Group (RG) are two of the most advanced and sophisticated structures in Quantum Field Theory (QFT). With this work we aim to show that the complexity in those areas is contained in one and the same universal operad structure. In more detail, this is a symmetric operad (with derivations) and its universality means that it is model independent within a large class of QFT models. The latter operad we call expansion operad. In the context of renormalization theory we find an isomorphic operad, which we call renormalization operad. The applications of the latter are for the description of the so called renormalization group and its action on the space of physical coupling constants via formal diffeomorphisms.

Keywords Quantum Field Theory · Vertex Algebras · Operads · Renormalization Group

This paper is organized as follows. In the first two sections we outline the notion of a vertex algebra and its generalization to an OPE algebra. In the third section we start with a brief introduction of the notion of operad. The remaining part of the paper contains the new ideas, which in many cases are given in sketch and are published in a final form in [19].

1 The General Concept of OPE in the Axiomatic QFT

According to the Wightman axiomatic framework to QFT [8, 22] the quantum fields are operator valued distributions, i.e., they are distributions taking values that are operators acting on an invariant dense subspace in a Hilbert space. In particular, this implies that the product of two local quantum fields $\phi(\mathrm{x})\psi(\mathrm{y})$ in general do not

N. M. Nikolov (✉)
Institute for Nuclear Research and Nuclear Energy of Bulgarian Academy of Sciences, Tzarigradsko Chaussee 72, BG-1784 Sofia, Bulgaria
e-mail: mitov@inrne.bas.bg

J. I. Burgos Gil et al. (eds.), *Periods in Quantum Field Theory and Arithmetic*, Springer Proceedings in Mathematics & Statistics 314,
https://doi.org/10.1007/978-3-030-37031-2_12

allow a restriction for coinciding arguments $\mathrm{x} = \mathrm{y}$. Here, x and y are points in the Minkowski space, but later we will use the same notations also for D–dimensional vectors in the complexified Euclidian space $\mathbb{C}^D$, where also the real D–dimensional Minkowski space can be embedded as a real subspace (hence, D stands for the space–time dimension). The concept of OPE was introduced by Wilson in [23] for studying the short distance behavior of such products, which further has a relation to the behaviour of various quantities in QFT at high energies. In more detail, the existence of OPE means that the product $\phi(\mathrm{x})\psi(\mathrm{y})$ posses an (asymptotic)[1] expansion at short distances $\mathrm{x} - \mathrm{y} \to 0$ of the form

$$\phi(\mathrm{x})\,\psi(\mathrm{y}) \underset{\mathrm{x}\to\mathrm{y}}{\sim} \sum_{A} \theta_A(\mathrm{y})\, C_A(\mathrm{x}-\mathrm{y}), \tag{1}$$

for a suitable system of two-point numerical functions (distributions) $C_A(\mathrm{x}-\mathrm{y})$ that describe the local behavior of the product; the coefficients $\theta_A(\mathrm{y})$ are again local fields (the sign $\underset{\mathrm{x}\to\mathrm{y}}{\sim}$ stands for the asymptotic, or, other kind of expansions at short distances). For instance, in perturbative massless QFT one can choose

$$C_A(\mathrm{x}-\mathrm{y}) = \big((\mathrm{x}-\mathrm{y})^2\big)^{\nu}\,\big((\log(\mathrm{x}-\mathrm{y}))\big)^{\ell}\, h_{m,\sigma}(\mathrm{x}-\mathrm{y})\,, \quad A = (\nu,\ell,m,\sigma)\,,$$

where $\nu \in \mathbb{R}$, $\ell \in \{0, 1, \ldots\}$ and $\{h_{m,\sigma}(\mathrm{x})\}_{\sigma}$ is a basis of harmonic homogeneous polynomials (spherical functions) of degree $m = 0, 1, \ldots$. If we can choose a basic system of two–point functions $\{C_A(\mathrm{x}-\mathrm{y})\}_A$ in (1), which is independent of the local fields $\phi(\mathrm{x})$ and $\psi(\mathrm{y})$, as in the latter case, then for every index A we obtain a binary operation

$$\theta_A =: \phi \underset{A}{*} \psi \quad \Longrightarrow \quad \{\underset{A}{*}\}_A \tag{2}$$

in the vector space of all local quantum fields (this space is called "Borchers class"). Hence, under the above assumptions we obtain a new kind of algebraic structure in the space of all local quantum fields, which is defined by the obtained infinite system of binary products $\{\underset{A}{*}\}_A$ (2). This structure is generally called *an OPE algebra*. One of the main conditions on the system of operations $\{\underset{A}{*}\}_A$ comes from the associativity of the fields' products:

$$\phi_1(\mathrm{x}_1)\,\big(\phi_2(\mathrm{x}_2)\,\phi_3(\mathrm{x}_3)\big) = \big(\phi_1(\mathrm{x}_1)\,\phi_2(\mathrm{x}_2)\big)\,\phi_3(\mathrm{x}_3)\,.$$

However, in the general case of QFT it is rather nontrivial to derive in an mathematically rigorous frame the existence of the system of products $\{\underset{A}{*}\}_A$ and formulate the above associativity in terms of them.

[1]The analyticity properties of local quantum fields indicate that these expansions have more strong convergence than as asymptotic series. This is clearly seen in the so called Globally Conformal Invariant QFT (see [16, Sect. 8] and Sect. 2). It can be achieved even in the most general case of Wightman QFT [17].

According to the Wightman axiomatic QFT, any product of *Lorentz covariant, local quantum fields*, $\phi_1(\mathrm{x}_1) \cdots \phi_n(\mathrm{x}_n)$, acting on the *vacuum* $\hat{1}$ is a boundary value of an analytic function in the so called *forward tume domain* (cf. [8, Sect. IV.2]),

$$\phi_1(\mathrm{x}_1) \cdots \phi_n(\mathrm{x}_n)\, \hat{1} \;=\; \underset{\text{as } \mathrm{y}_1,\dots,\mathrm{y}_n \to 0}{\text{boundary value of}} \;\; \phi_1(\mathrm{x}_1 + i\mathrm{y}_1) \cdots \phi_n(\mathrm{x}_n + i\mathrm{y}_n)\, \hat{1}\,. \quad (3)$$

There is a slight abuse in the notations in the right hand side of (3) as $\phi_j(\mathrm{x}_j + i\mathrm{y}_j)$ does not exist as an operator valued function in any domain in $\mathbb{C}^D$ ($\ni \mathrm{x}_j + i\mathrm{y}_j$), but the product in (3) exists only when it is applied on the vacuum (or, to more general, suitable, state vectors). Due to this general feature of QFT one can try to derive, or impose an existence of an asymptotic expansion for the analytically continued product of fields acting on the vacuum, $\phi_1(\mathrm{z}_1) \cdots \phi_n(\mathrm{z}_n)\, \hat{1}$ ($\mathrm{z}_j = \mathrm{x}_j + i\mathrm{y}_j$):

$$\phi_1(\mathrm{z}_1) \cdots \phi_n(\mathrm{z}_n)\, \hat{1} \quad \underset{\substack{\mathrm{z}_1 - \mathrm{z}_n \to 0,\, \dots, \\ \mathrm{z}_{n-1} - \mathrm{z}_n \to 0}}{\sim} \quad \sum_{\alpha=0}^{\infty} G_\alpha(\mathrm{z}_1 - \mathrm{z}_n, \dots, \mathrm{z}_{n-1} - \mathrm{z}_n)\, \theta_\alpha(\mathrm{z}_n)\, \hat{1}\,. \quad (4)$$

To specify in what sense the asymptotic expansion in (4) is understood, one introduces a grading (or, et least a filtration) in the space of all Lorentz covariant, local fields (the Borhers class), which we shall call *scaling dimension* of a field. We shall consider this grading as a *negative integral grading* since it will correspond to the *scaling behaviour* of a product of fields at short distances in position space.[2] Then, the asymptotic expansion (4) means that there exists $K \in \mathbb{Z}$ such that if we take the difference of the left hand side and the partial sum in the right hand side up to $\alpha = N$ it should follow that the remainder has an asymptotic behaviour at $\mathrm{z}_1 = \cdots = \mathrm{z}_n$ bounded by

$$\text{const} \cdot \text{dist}(\mathrm{z}_1, \dots, \mathrm{z}_n)^{-K + d_1 + \cdots + d_n + N + 1}\,. \quad (5)$$

where $d_1, \dots, d_n$ are the scaling dimensions of the fields $\phi_1, \dots, \phi_n$, respectively, and

$$\text{dist}(\mathrm{z}_1, \dots, \mathrm{z}_n) := \max_{j<k} \|\mathrm{z}_j - \mathrm{z}_k\|\,, \quad (6)$$

for some subsidiary norm distance $\|\mathrm{z}_j - \mathrm{z}_k\|$. Furthermore, each of the resulting fields θ_α has a scaling dimension $-\alpha$ and the coefficient function G_α has an asymptotic behaviour bounded by

$$\text{const} \cdot \text{dist}(\mathrm{z}_1, \dots, \mathrm{z}_n)^{-K + d_1 + \cdots + d_n + \alpha}\,. \quad (7)$$

---

[2] In physics literature the minus scaling dimension corresponds to the so called *an energy dimension* of a field.

The corresponding number $-K + d_1 + \cdots + d_n + \alpha$ in the exponent in (7) we shall call a *scaling degree of* G_α and so, the expansion (4) becomes a *graded expansion*, i.e., we have

$$\sum_{j=1}^{n} \text{scaling dimension of } \phi_j$$
$$= \text{scaling degree of } G_\alpha \ + \ \text{scaling dimension of } \theta_\alpha \,. \tag{8}$$

Relying on some general arguments, the main of which being the positivity of the scalar product in the Hilbert space (reflected by the so called *Wightman positivity*), one can expect that the grading by the scaling degree in the space of quantum fields is *finitely degenerated* and the only fields of dimension zero are the *constant fields*. For the expansion (4) the latter means that the most singular (admissible) term in the expansion is proportional to the vacuum vector. All the above assumptions (4)–(8) look sufficiently general and are valid at least within the *perturbative* QFT. Let us assume now that for some $1 < m < n$ we have a similar expansion

$$\phi_{n-m+1}(\mathrm{z}_{n-m+1}) \cdots \phi_n(\mathrm{z}_n)\, \hat{1} \underset{\substack{\mathrm{z}_{n-m+1} - \mathrm{z}_n \,\to\, 0,\, \dots, \\ \mathrm{z}_{n-1} - \mathrm{z}_n \,\to\, 0}}{\sim} \tag{9}$$
$$\sim \sum_{\alpha=0}^{\infty} \sum_{a} G''_{\alpha,a}(\mathrm{z}_{n-m+1} - \mathrm{z}_n, \dots, \mathrm{z}_{n-1} - \mathrm{z}_n)\, \theta_{\alpha,a}(\mathrm{z}_n)\, \hat{1}\,,$$

where $\{\theta_{\alpha,a}\}_a$ is some finite basis of fields of scaling dimension $-\alpha$, and similarly

$$\phi_1(\mathrm{z}_1) \cdots \phi_{n-m}(\mathrm{z}_{n-m})\, \theta_{\beta,b}(\mathrm{z}_n)\, \hat{1} \tag{10}$$
$$\underset{\substack{\mathrm{z}_1 - \mathrm{z}_n \,\to\, 0,\, \dots, \\ \mathrm{z}_{n-m} - \mathrm{z}_n \,\to\, 0}}{\sim} \sum_{\alpha=0}^{\infty} \sum_{a} G'_{\alpha,a;\beta,b}(\mathrm{z}_1 - \mathrm{z}_n, \dots, \mathrm{z}_{n-m} - \mathrm{z}_n)\, \theta_{\alpha,a}(\mathrm{z}_n)\, \hat{1}\,.$$

Then, we obtain as a consequence of the associativity,

$$\phi_1(\mathrm{x}_1) \cdots \phi_n(\mathrm{x}_n) \tag{11}$$
$$= \phi_1(\mathrm{x}_1) \cdots \phi_{n-m}(\mathrm{x}_{n-m}) \big(\phi_{n-m+1}(\mathrm{x}_{n-m+1}) \cdots \phi_n(\mathrm{x}_n)\big)$$

that the following relation holds:

$$G_{\alpha,a}(\mathrm{z}_1 - \mathrm{z}_n, \dots, \mathrm{z}_{n-1} - \mathrm{z}_n) = \sum_{\beta} \sum_{b} G'_{\alpha,a;\beta,b}(\mathrm{z}_1 - \mathrm{z}_n, \dots, \mathrm{z}_{n-m} - \mathrm{z}_n)$$
$$\times\, G''_{\beta,b}(\mathrm{z}_{n-m+1} - \mathrm{z}_n, \dots, \mathrm{z}_{n-1} - \mathrm{z}_n)\,, \tag{12}$$

where $G_{\alpha,a}$ are the corresponding coefficient functions for (4) if we perform an expansion in the basis $\theta_{\alpha,a}$.

Now, the expansion (12) makes sense at the first place again as a kind of asymptotic expansion but we can consider it also as a *formal series expansion*, which is graded with respect to the scaling degree. **From now on, we shall replace everywhere the asymptotic expansion notation $\sim$ with the equality sign as in** (12).

Let us consider in more detail the nature of the expansion (12). To this end we shall use another property (Osterwalder–Schrader theorem) of the Wightman fields stating that the scalar product

$$\langle \hat{1} \mid \phi_1(\mathrm{z}_1) \cdots \phi_n(\mathrm{z}_n)\, \hat{1} \rangle \tag{13}$$

(the so called *nth Wightman function*) admits an analytic continuation to a domain in $(\mathbb{C}^D)^{\times n}$ which contains as a real subspace the *off–diagonal n–point Euclidean space*:

$$(\mathbb{R}^D)^{\times n} \backslash \widetilde{\Delta}_n \,, \tag{14}$$

$$\begin{aligned} \{(\mathrm{x}_1, \dots, \mathrm{x}_n) \in (\mathbb{R}^D)^{\times n} \mid \mathrm{x}_1 = \cdots = \mathrm{x}_n\} &=: \Delta_n \\ \subsetneqq \widetilde{\Delta}_n := \{(\mathrm{x}_1, \dots, \mathrm{x}_n) \in (\mathbb{R}^D)^{\times n} \mid \mathrm{x}_j = \mathrm{x}_k \text{ for some } j \neq k\} & \end{aligned} \tag{15}$$

(Δ_n and $\widetilde{\Delta}_n$ are the so called *small/thin/total diagonal* and *large diagonal*, respectively). As a consequence, one can expect that the functions G_α in (4) will contain in their analyticity domain also the *off–diagonal n–point Euclidean space* (14) and let us further assume that all these functions belong to some *algebra of functions*

$$\mathrm{Tr.I.Func}_n \subseteq C^\infty\Big(\frac{(\mathbb{R}^D)^{\times n} \backslash \widetilde{\Delta}_n}{\mathbb{R}^D}\Big) \tag{16}$$

($n = 2, 3, \dots$), where the quotient is by the translations (since G_α are translation invariant functions). Further arguments coming from perturbative QFT suggest that we can replace the scaling degree on $\mathrm{Tr.I.Func}_n$ with the more explicit degree of *associate homogeneity*,

$$\begin{gathered} G \in \mathrm{Tr.I.Func}_n \text{ has a degree of associate homogeneity } \alpha \text{ iff} \\ \Big(-\alpha + \sum_{j=1}^{n} \mathrm{z}_j \cdot \partial_{\mathrm{z}_j}\Big)^N G = 0 \text{ for a sufficiently large positive integer } N, \end{gathered} \tag{17}$$

where $\mathrm{z} \cdot \partial_{\mathrm{z}} \equiv \sum\limits_{\mu=1}^{D} z^\mu \frac{\partial}{\partial z^\mu}$ is the Euler operator on $\mathbb{R}^D$. We shall call an **expansion system** (or, **OPE functional system**) of algebras any sequence of subalgebras $(\mathrm{Tr.I.Func}_n)_{n=2}^\infty$ (16), which allow for every $G \in \mathrm{Tr.I.Func}_n$ a graded expansion

$$G(\mathrm{z}_1,\ldots,\mathrm{z}_n) \qquad (18)$$
$$= \sum_{\alpha=0}^{\infty}\sum_{j=1}^{N_\alpha} G'_{\alpha,j}(\mathrm{z}_1,\ldots,\mathrm{z}_{j-1},\mathrm{z}_{j+m-1},\ldots,\mathrm{z}_n)\, G''_{\alpha,j}(\mathrm{z}_j,\ldots,\mathrm{z}_{j+m-1})$$

for $G'_{\alpha,j} \in \mathrm{Tr.I.Func}_{n-m+1}$ and $G''_{\alpha,j} \in \mathrm{Tr.I.Func}_m$, so that (18) is absolutely convergent in the domain where $|\mathrm{z}_a - \mathrm{z}_{j+m-1}| \ll |\mathrm{z}_b - \mathrm{z}_{j+m-1}|$ for $a \in \{j,\ldots,j+m-1\} \not\ni b$ (note that in (18) we regard G, $G'_{\alpha,j}$ and $G''_{\alpha,j}$ as translation invariant functions without specifying basic differences between the points).

The OPE algebras are, to some extent, a generalization of associative algebras. Like in the latter case, an OPE algebra can be determined if we know know all the two–point OPE's. In particular, if we fix again a basis $\{\theta_{\alpha,a}\}_{,a}$ of local fields for every scaling dimension $-\alpha$ one can consider the OPE

$$\theta_{\alpha,a}(\mathrm{x})\theta_{\beta,b}(\mathrm{y}) = \sum_{\gamma}\sum_{d} C^{\gamma,d}_{\alpha,a;\beta,b}(\mathrm{x}-\mathrm{y})\,\theta_{\gamma,d}(\mathrm{y}) \qquad (19)$$

and regard the two–point functions $C^{\gamma,d}_{\alpha,a;\beta,b}(\mathrm{x}-\mathrm{y})$ as a kind of *OPE structure functions* (see [5–7]).

Up to this point we did not pay much attention to one of the most important features of the Wightman fields: the *locality*. It reflects the fundamental principle of special relativity that no cause can propagate faster than the speed of light. As a consequence of this principle it turns out that the analytically continued Wightman functions (13) are *symmetric* under the simultaneous exchange of the fields and their arguments, i.e.,

$$\langle \hat{1} \,|\, \phi_1(\mathrm{z}_1)\cdots\phi_n(\mathrm{z}_n)\,\hat{1}\rangle = \langle \hat{1} \,|\, \phi_{\sigma(1)}(\mathrm{z}_{\sigma(1)})\cdots\phi_{\sigma(n)}(\mathrm{z}_{\sigma(n)})\,\hat{1}\rangle \qquad (20)$$

for any permutations σ. If we combine the latter symmetry with the OPE (4) we should allow expansions not only around the last (the right) point x_n but also around any other point z_s $(s = 1,\ldots,n)$,

$$\phi_1(\mathrm{z}_1)\cdots\phi_n(\mathrm{z}_n)\,\hat{1} = \sum_{\alpha=0}^{\infty}\sum_{a} G^{(s)}_{\alpha,a}(\mathrm{z}_1-\mathrm{z}_s,\ldots,\mathrm{z}_{n-1}-\mathrm{z}_s)\,\theta_{\alpha,a}(\mathrm{z}_s)\,\hat{1}\,. \qquad (21)$$

Note that one can pass from one center of the expansion (21) to another even on a level of formal series:

$$\sum_{\alpha=0}^{\infty} \sum_{a} G_{\alpha,a}^{(s)}(\mathrm{z}_1 - \mathrm{z}_s, \ldots, \mathrm{z}_{n-1} - \mathrm{z}_s)\, \theta_{\alpha,a}(\mathrm{z}_s)\, \hat{1} \tag{22}$$
$$= \sum_{\alpha=0}^{\infty} \sum_{a} G_{\alpha,a}^{(s)}(\mathrm{z}_1 - \mathrm{z}_n, \ldots, \mathrm{z}_{n-1} - \mathrm{z}_n)\, \exp\Big((\mathrm{z}_s - \mathrm{z}_n) \cdot \partial_{\mathrm{z}_n}\Big)\, \theta_{\alpha,a}(\mathrm{z}_n)\, \hat{1}$$
$$= \sum_{\alpha=0}^{\infty} \sum_{a} G_{\alpha,a}^{(n)}(\mathrm{z}_1 - \mathrm{z}_n, \ldots, \mathrm{z}_{n-1} - \mathrm{z}_n)\, \theta_{\alpha,a}(\mathrm{z}_n)\, \hat{1}\,.$$

Indeed, one first applies the (formal) Taylor expansion $\theta_{\alpha,a}(\mathrm{z}_s)\,\hat{1} = \exp\big((\mathrm{z}_s - \mathrm{z}_n) \cdot \partial_{\mathrm{z}_n}\big)\,\theta_{\alpha,a}(\mathrm{z}_n)\,\hat{1}$. Then, we note that by taking partial derivatives of the fields $\theta_{\alpha,a}$ we *decrease* their scaling dimension,

$$\text{scaling dimension of } \partial_{x^{\mu_1}} \cdots \partial_{x^{\mu_k}} \phi(\mathrm{x}) \;=\; -k + \text{scaling dimension of } \phi\,. \tag{23}$$

Thus, in order to pass from the system of OPE functions $\{G_{\alpha,a}^{(s)}\}$ to the system $\{G_{\alpha,a}^{(n)}\}$ one needs just to reexpress $\partial_{x^{\mu_1}} \cdots \partial_{x^{\mu_k}} \phi(\mathrm{x})$ as a (finite) linear combination of $\theta_{\alpha+k,a}(\mathrm{x})$ (of scaling dimension $-k-\alpha$); then, we can express every $G_{\alpha,a}^{(s)}$ by a finite number of $G_{\beta,b}^{(n)}$ using a finite number of algebraic operations. In this way, we see that Eq. (22) indeed makes sense as an equality of formal series.

2 Huygens Locality and Vertex Algebras

In the previous subsection we outlined the general notion of an OPE algebra and we needed some additional technical assumptions to the Wightman axioms that cannot be justified from fundamental physical principles. The main justification for the additional conditions is that they are satisfied for the free field models and for the models of perturbative QFT. However, in some classes of QFTs with stronger spatial symmetry, as *conformal* QFT, one can obtain in a natural way restrictions on the Wightman functions, which allow then to achieve the OPE conditions. In particular, one of the strongest conditions of conformal symmetry, called *Global Conformal Invariance* implies rationality of all the Wightman functions [21]. As it was shown in [21], the Global Conformal Invariance leads, at first place, to strengthening the locality property of the quantum fields to the so called *Huygens locality*, which is presented by an algebraic condition:

$$\big((\mathrm{x}-\mathrm{y})^2\big)^N \big[\phi(\mathrm{x}), \phi(\mathrm{y})\big] \;=\; 0 \tag{24}$$

satisfied for a sufficiently large positive integer N (depending also on the field ϕ). In (24), $(\mathrm{x}-\mathrm{y})^2$ is the *light cone quadric*, which in the theory of special relativity is also call the *relativistic interval*.

The Huygens locality (24) is exactly this additional condition to the Wightman axioms, which implies that all the (analytically continued) Wightman functions in such a theory are rational functions with light–cone singularities, i.e., they are of the form

$$\frac{P(\mathrm{z}_1 - \mathrm{z}_n, \ldots, \mathrm{z}_{n-1} - \mathrm{z}_n)}{\prod\limits_{1 \leqslant j < k \leqslant n} \big((\mathrm{z}_j - \mathrm{z}_k)^2\big)^{\nu_{j,k}}} \tag{25}$$

for some integers $\mu_{j,k}$. We shall consider the functions (25) on $(\mathbb{C}^D)^{\times n}$, where we fix the light cone quadric $(\mathrm{z}_j - \mathrm{z}_k)^2$ as

$$(\mathrm{z}_j - \mathrm{z}_k)^2 = \sum_{\mu=1}^{D} (z_j^\mu - z_k^\mu)^2 \tag{26}$$

and we stress that this is an analytic function in z_j and z_k, which is in fact, a homogeneous quadratic polynomial, and it should not be mixed with the square of the (subsidiary) norm distance (that we introduced in Eq. (6)), i.e.,

$$(\mathrm{z}_j - \mathrm{z}_k)^2 \neq \|\mathrm{z}_j - \mathrm{z}_k\|^2 := \sum_{\mu=1}^{D} |z_j^\mu - z_k^\mu|^2 . \tag{27}$$

The rationality of Wightman functions is what in turn allows us to obtain a nice and purely algebraic reformulation of QFT with Global Conformal Invariance in terms of OPE algebras. In this special case the OPE algebras are usually called **vertex algebras** and have been considered in [16]. In particular, the Huygens locality and the rationality allow us to start with formal series expansions for the OPE and achieve the convergence as a consequence.

Two dimensional conformal QFTs split into tensor products of theories on two light rays, which are called chiral conformal QFTs. The latter correspond to the $D = 1$ case of the vertex algebras based on the Huygens locality (24). Then, the OPE (1) takes the form[3]

$$\phi(z)\,\psi(w) = \sum_{n=-N_{\phi,\psi}}^{\infty} \big(\phi_{(-n-1)}\psi\big)(w) \cdot (z-w)^n \tag{28}$$

($z, w \in \mathbb{C}$, $N_{\phi,\psi} \in \mathbb{N}$). In Eq. (28) the role of the basic system of two–point functions $\{C_A(\mathrm{x}-\mathrm{y})\}_A$ of Eq. (1) is played by $\{(z-w)^n\}_{n\in\mathbb{Z}}$ and according to Eq. (2) one should denote by $\phi \underset{n}{*} \psi$ the field coefficient to the term $(z-w)^n$. However, for some technical reasons in dimension $D = 1$ people usually denote $\phi \underset{n}{*} \psi$ by $\phi_{(-n-1)}\psi$. The associativity and further properties of the OPE (28) in this case was first axiomatized

[3] The series (28) is absolutely convergent on vector states with bounded conformal energy but in this case it is enough to axiomatize the OPE on a level of formal power series.

by Borcherds [1] and the resulting structure was called (*chiral*) *vertex algebra*[4] and it is widely studied nowadays in mathematics (see [3, 4, 9, 11]; in [9, Chap. 1] the reader can find a connection with the Wightman axiomatic approach). Before [16] generalizations of vertex algebras to higher space–time dimensions D (>2) have been considered in [2] (see also [10, 12]). However, as we pointed out, the multidimensional generalization of vertex algebras of [16] was introduced within the context of QFT. In more detail, it has been shown in the latter paper that these algebras are in one–to–one correspondence with models of Wightman axioms obeying the Global Conformal Invariance [21, Sect. 9].

Thus, a vertex algebra is a completely fixed mathematical structure, while the notion of an OPE algebra is still developing. In particular, one of the aims of this paper is to give an insight to such a notion of an OPE algebra that goes beyond the case of vertex algebras.

Taking into account the rationality of Wightman functions in Globally Conformal Invariant QFT one can specify further the OPE algebras (16) as:

$$\begin{aligned} \text{Tr.I.Func}_n \,:=\, \mathscr{O}_n \,:=\, & \text{ The algebra of rational } n\text{–point functions} \\ & \text{on } \mathbb{R}^D \ni \mathrm{x}_1, \ldots, x_n, \text{ which are of the form (25).} \end{aligned} \tag{29}$$

In this case the grading is provided by the degree of homogeneity.

There is an additional simplification in the theory of vertex algebras provided by the so called *state–field correspondence*. As we have pointed out already, any Wightman field acting on the vacuum $\phi(\mathrm{x})\hat{1}$ posseses an analytic extension in the forward tube. It turns out that the latter tube domain is a homogeneous space for the real conformal group and there exists a complex conformal transformation which maps the whole tube into a bounded open subset of $\mathbb{C}^D$. The resulting coordinates on $\mathbb{C}^D$ are known as a *compact picture* in conformal QFT. As the center of the coordinates is now in the domain of holomorphy of the vector–valued function $\phi(\mathrm{z})\hat{1}$ we can set $\mathrm{z} = 0$ and in this way we obtain a linear map

$$\left\{ \begin{array}{c} \text{the space of all} \\ \text{Wightman fields} \end{array} \right\} \ni \phi \mapsto \phi(\mathrm{z})\hat{1}\Big|_{\mathrm{z}=0} \in \left\{ \begin{array}{c} \text{the Hilbert space} \\ \text{of states} \end{array} \right\}, \tag{30}$$

which is called the state–field correspondence. As a consequence of the *Reeh–Schlieder theorem* from the axiomatic QFT it follows that (30) is an injection, whose image is dense.

Free massless quantum fields (for example, the free electromagnetic field) obey the above Huygens locality. Like for the general QFT also in the case of Huygens local fields there are yet no known models in more than three space–time dimensions, which cannot be realized by free fields (models with such a realization physicists call trivial). The main difficulty for finding nontrivial models of Huygens local fields is the lack of relation between the theory of vertex algebras in higher dimensions and

[4]Some authors call them also *vertex operator algebras*; the adjective "chiral" indicates that the $D = 1$ case is ment.

the theory of infinite dimensional Lie algebras and their representations. The latter relation is present in two space–time dimensions and it is a rich source for constructing models. One of the aims of the operadic point of view on vertex algebras is to provide a substitute for the infinite dimensional Lie algebras and their representations for creating QFT models in higher dimensions.

3 Operads and OPE Algebras

We aim to consider the vertex algebras, or more generally, the OPE algebras, as algebras over an *operad*. Besides one of the first references on this topic [15] we shall mention one recent book [14], from which we follow the definitions and conventions.

One can think of an operad as a generalized type of algebra. An algebra of a certain type is determined by introducing a set of n-ary operations for some positive integers n. In the case of *algebraic operads* these are multilinear maps over the underlying vector space of the described algebra. The defining operations can be further subject to certain identities, in which one uses compositions of the operations, eventually combined with permutations of the input arguments. Alternatively, one can consider the spaces of all possible multilinear operations obtained by taking compositions and actions of permutations, and finally, quotients by the relations. In this way, we obtain the operad corresponding to the considered type of algebra.

In more detail, an operad includes

- a sequence of vector spaces $\{\mathcal{M}(n)\}_{n=1}^{\infty}$ ($\mathcal{M}(2)$ being the space of binary operations, ...).
- The structure is provided by various structure maps called *operadic compositions*,

$$\begin{array}{rcl} \mathcal{M}(k) \otimes \mathcal{M}(j_1) \otimes \cdots \otimes \mathcal{M}(j_k) & \longrightarrow & \mathcal{M}(n) \\ \mu'' \otimes \mu'_1 \otimes \cdots \otimes \mu'_k & \longmapsto & \mu'' \circ (\mu'_1, \ldots, \mu'_k) \,, \end{array} \tag{31}$$

where $n = j_1 + \cdots + j_k$, and right permutation actions

$$\mathcal{M}(n) \times \mathscr{S}_n \ni \mu \times \sigma \mapsto \mu^{\sigma} \in \mathcal{M}(n) \,, \quad (\mu^{\sigma_1})^{\sigma_2} = \mu^{\sigma_1 \sigma_2} \,. \tag{32}$$

- A prescribed element $1_{\mathcal{M}} \in \mathcal{M}(1)$ called an *operadic unit*.

All the above data are subject to three sets of conditions:

- associativity for the operadic compositions;
- equivariance of the operadic compositions with respect to the action of permutations;
- unit property of the operadic unit with respect to the operadic compositions.

One of the main examples of an operad is the *endomorphism operad* $\mathcal{E}nd_V$:

$$\mathcal{E}nd_V(n) := Hom\big(V^{\otimes n}, V\big) \,,$$

where $\mu'' \circ (\mu'_1, \ldots, \mu'_k)$ is the actual composition of multilinear maps and

$$\mu^\sigma(v_1, \ldots, v_n) := \mu(v_{\sigma_1}, \ldots, v_{\sigma_n}) .$$

In this example, the reader who does not have experience in operad theory can try to recover the above mentioned three sets of axioms as a simple exercise.

Instead of the full set of operadic compositions (31) one can introduce the operad structure by a smaller piece of data given by the so called *partial compositions*:

$$\mu \circ_j \nu := \mu \circ (1_{\mathcal{M}}, \ldots, 1_{\mathcal{M}}, \underset{\stackrel{\uparrow}{j}}{\nu}, 1_{\mathcal{M}}, \ldots, 1_{\mathcal{M}}) . \tag{33}$$

It is easy to see in the example of the operad $\mathcal{E}nd_V$ that one can recover the full operadic compositions (31) by the partial compositions by the formula:

$$\mu'' \circ (\mu'_1, \ldots, \mu'_k) = \big(\cdots \big((\, \mu'' \circ_1 \mu'_1) \circ_{j_1+1} \mu'_2\big) \circ_{j_1+j_2+1} \cdots \big) \circ_{j_1+\cdots+j_{k-1}+1} \mu'_k$$

(for more detail see [14, 5.3.7]).

Morphisms of operads are defined by sequences of linear maps:

$$\{\mathcal{M}(n)\}_{n=1}^{\infty} \to \{\mathcal{N}(n)\}_{n=1}^{\infty} \equiv \big\{\mathcal{M}(n) \to \mathcal{N}(n)\big\}_{n=1}^{\infty}$$

plus compatibility with all structure maps (31) and (32). In particular, morphisms from an operad to the endomorphism operads have a meaning of "representations" but are called *algebras over the corresponding operad*:

$$\begin{aligned} \text{Representation} &\equiv \text{Algebra over an operad,} \\ \text{i.e., } \{\mathcal{M}(n)\}_n \to \{\mathcal{E}nd_V(n)\}_n &- \text{morphism of operads,} \\ \text{i.e., } \mathcal{M}(n) &\to Hom\big(V^{\otimes n}, V\big) . \end{aligned}$$

The reason for using this terminology is because in this way the abstract operations in $\mathcal{M}(n)$ become actual n–linear maps on V that is the underlined space of the algebra.

Example. The Lie operad $\mathcal{L}ie$ corresponds the class of Lie algebras and is defined as:

$$\begin{aligned} \mathcal{L}ie(1) &= Span_{\mathbb{K}}\{1\} \xrightarrow{\pi_1} Hom(V, V) , \\ \mathcal{L}ie(2) &= Span_{\mathbb{K}}\{\lambda\} \xrightarrow{\pi_2} Hom(V^{\otimes 2}, V) , \\ \pi_2(\lambda)(a, b) &= [a, b] , \\ \mathcal{L}ie(3) &= Span_{\mathbb{K}}\{\, \lambda \circ (1, \lambda) \ , \ \ \lambda \circ (\lambda, 1) \,\} \\ & \qquad\qquad\qquad\qquad \downarrow \qquad\qquad\quad \downarrow \\ \big(\lambda \circ (\lambda, 1)\big)^{(1,3,2)} &\to [[a, c], b] = [a, [b, c]] - [[a, b], c] , \\ \big(\lambda \circ (\lambda, 1)\big)^{(1,3,2)} &= \lambda \circ (\lambda, 1) - \lambda \circ (1, \lambda) \end{aligned}$$

(where μ^σ, for $\mu = \lambda \circ (\lambda, 1)$ and $\sigma = (1, 3, 2)$, is the (right) permutation action (32)).

If we drop the symmetry structure (32) from the definition of an operad we obtain the so called *nonsymmetric operad*. In this case, following the notations of [14] we shall denote the operadic spaces by a subscript (i.g., $\mathcal{M} = (\mathcal{M}_n)_{n=1}^{\infty}$).

The main construction in this work is based on a particular example of an operad, which we call the **expansion operad** $\mathcal{E} = \big\{\mathcal{E}(n)\big\}_n$. It can be defined whenever we have a sequence of algebras $\{\text{Tr.I.Func}_n\}_{n=2}^{\infty}$ (16) satisfying the conditions (17) and (18) and in particular, for the case (29), which corresponds to the vertex algebras. The expansion operad contains a nonsymmetric suboperad, $\mathcal{E}^{\text{n.s.}} = (\mathcal{E}_n^{\text{n.s.}})_{n=1}^{\infty}$, which is easier to define:

$$\mathcal{E}_n^{\text{n.s.}} = \text{Tr.I.Func}'_n \,, \qquad \mathcal{E}_1^{\text{n.s.}} := \mathbb{C} \cdot 1_{\mathcal{E}} \,, \tag{34}$$

where $\text{Tr.I.Func}'_n$ is the *graded dual* of Tr.I.Func_n. Then, if $\Gamma' \in \text{Tr.I.Func}'_{n-m+1}$, $\Gamma'' \in \text{Tr.I.Func}'_m$ and $G \in \text{Tr.I.Func}_n$ we set

$$\big(\Gamma' \circ_j \Gamma''\big)(G) := \sum_{\alpha=0}^{\infty} \Gamma'\big(G'_\alpha\big)\, \Gamma''\big(G''_\alpha\big) \,, \tag{35}$$

where the sum in the right hand side has only a finite number of nonzero terms according to the grading conditions (in fact, if G, Γ' and Γ'' have a fixed grading then only one term of this sum will contribute).

Now, if we have an OPE algebra with an underlying vector space V and V' is the graded dual of V (with respect to the scaling dimensions) then we obtain a graded linear map for every $n = 2, 3, \ldots$:

$$\begin{aligned} V' \otimes V^{\otimes n} &\longrightarrow \text{Tr.I.Func}_n \\ \lambda \otimes \phi_1 \otimes \cdots \otimes \phi_n &\longmapsto \sum_{\alpha=0}^{\infty} \sum_{a} G_{\alpha,a}(\mathrm{z}_1 - \mathrm{z}_n, \ldots, \mathrm{z}_{n-1} - \mathrm{z}_n)\, \lambda(\theta_{\alpha,a}) \end{aligned} \tag{36}$$

defined under the OPE (4), where the sum in the right hand side of the second row again has only a finite number of nonzero terms due to the grading condition (8) (namely, there will be only a finite number of nonzero $\lambda(\theta_{\alpha,a})$). Hence, dualizing (36) we get a graded linear map

$$\mathcal{E}_n^{\text{n.s.}} \equiv \text{Tr.I.Func}'_n \longmapsto (V')^{\otimes n} \otimes V \equiv Hom\big(V^{\otimes n}, V\big) \equiv \mathcal{E}nd_V(n) \,, \tag{37}$$

and the equalities

$$\big(V' \otimes V^{\otimes n}\big)' = (V')^{\otimes n} \otimes V = Hom\big(V^{\otimes n}, V\big)$$

are due to the fact that we are working with graded duals and V is supposed to be graded with finite dimensional subspaces (however, this finite dimensionality is not crucial to get the map (37)). In this way we obtain a natural candidate for an operadic morphism, the sequence $\mathcal{E}_n^{\text{n.s.}} \to \mathcal{E}nd_V(n)$ (37), which would present the OPE algebra V as an algebra over the (nonsymmetric) operad $\mathcal{E}^{\text{n.s.}}$ (34). The sequence of linear maps $\mathcal{E}_n^{\text{n.s.}} \to \mathcal{E}nd_V(n)$ (37) indeed is an operadic morphism due to the compatibility relations of type (12) on the one hand, and the definition of the operadic compositions (35), on the other hand.

Finally, if we want to incorporate the permutation symmetry in the above operadic construction we have to work with expansions at different base points (not only the right one), as in (21). To this end, we refine the definition of the OPE algebras Tr.I.Func_n (16) by explicitly specifying a basic set of point differences around a base point z_s:

$$\begin{aligned} \text{Tr.I.Func}_{n|s} := \big\{ G(\mathrm{z}_1 - \mathrm{z}_s, \dots, \mathrm{z}_n - \mathrm{z}_s) \in C^\infty\big((\mathbb{R}^D)^{\times(n-1)}\big) \\ \big| \, G \in \text{Tr.I.Func}_n \big\} \subseteq C^\infty\big((\mathbb{R}^D)^{\times(n-1)}\big), \\ \text{Tr.I.Func}_{n|n} \equiv \text{Tr.I.Func}_n\,, \end{aligned} \tag{38}$$

for every $s = 1, \dots, n$. Then, we set

$$\mathcal{E}(n) := \bigoplus_{s=1}^{\infty} \text{Tr.I.Func}'_{n|s}\,, \tag{39}$$

and the operadic compositions go as

$$\begin{aligned} - \circ_j - : \text{Tr.I.Func}'_{n-m+1|s'} \otimes \text{Tr.I.Func}'_{m|s''} \to \text{Tr.I.Func}'_{n|s} \\ \text{with} \quad s := s' + \delta_{j,s'}(s'' - 1) \end{aligned} \tag{40}$$

for $s' = 1, \dots, n - m + 1$ and $s'' = 1, \dots, m$ (note that the assignment

$$- \circ_j - : \big((n - m + 1, s'), (m, s'')\big) \longmapsto (n, s)$$

defines a set theoretic operad on the sets of *punctured finite sets* $(n, s) \equiv (\{1, \dots, n\}, s)$). Each of the graded linear maps (40) is provided similarly to (35) by expanding the elements $G \in \text{Tr.I.Func}_{n|s}$ around the base point $\mathrm{z}_{s''}$ in a formal series of products $G'_\alpha G''_\alpha$ for $G'_\alpha \in \text{Tr.I.Func}'_{n-m+1|s'}$ and $G''_\alpha \in \text{Tr.I.Func}''_{m|s''}$.

Remark 1 The sequence of spaces $\big(\text{Tr.I.Func}_n\big)_n \equiv \big(\text{Tr.I.Func}_{n|n}\big)_n$ has not only a structure of a nonsymmetric operad but also of a *shuffle operad* (for the definition of the shuffle operads the reader may look at [14, Sect. 8.2] and the references therein). Indeed, this is achieved if we introduce expansions of the form (18) where the role of the sequence $j, \dots, j + m - 1$ is played by an arbitrary sequence $1 \leqslant j_1 < \cdots < j_m \leqslant n$.

4 Connection with the Renormalization Theory

Passing to the renormalization let us mention first that the same rational functions belonging to $\mathcal{O}_n$ (29) appear as "Feynman amplitudes" (=integrands in the Feynman integrals) in massless field theories. Here is an example of such a Feynman amplitude in the ϕ^4–theory:

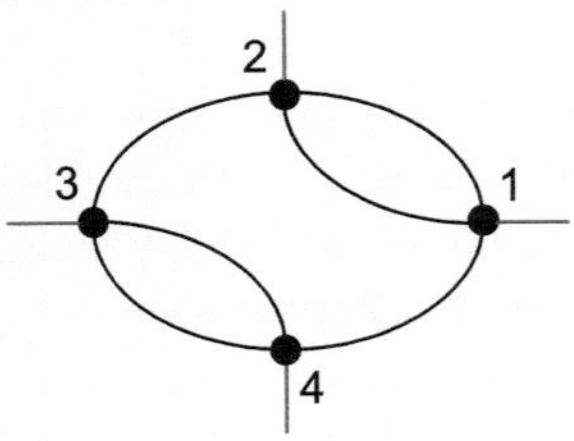

$$\longleftrightarrow \quad \frac{1}{\left((\mathrm{x}_1-\mathrm{x}_2)^2\right)^2}\frac{1}{(\mathrm{x}_2-\mathrm{x}_3)^2} \times \frac{1}{\left((\mathrm{x}_3-\mathrm{x}_4)^2\right)^2}\frac{1}{(\mathrm{x}_1-\mathrm{x}_4)^2} \in \mathcal{O}_4$$

The above assignment, which maps Feynman graphs to Feynman amplitudes is called a Feynman rule. The Feynman rules are explained in every QFT textbook that contains the perturbation theory but the reader who is not familiar with them can find a concise combinatorial definition in [13, Sect. 4.g].

It is important for the physical applications of the present construction that we consider the ultraviolet renormalization on *configuration space*. In terms of Feynman amplitudes the renormalization is given by a system of linear maps

$$\mathcal{O}_n \to \mathcal{D}'\left((\mathbb{R}^D)^{\times(n-1)}\right),$$

called *renormalization maps* and subject to (recursive) conditions (see [18, 20] and references therein). It is crucial that the latter conditions do not fix completely the above linear maps $\mathcal{O}_n \to \mathcal{D}'\left((\mathbb{R}^D)^{\times(n-1)}\right)$. There is a remaining ambiguity that is called *renormalization ambiguity*. It is shown in [18] that, the renormalization ambiguity at order n is described by a linear map: $\mathcal{O}_n \to \mathcal{D}'[0_n]$, where $\mathcal{D}'[0_n]$ stands for the space of distributions on $(\mathbb{R}^D)^{\times(n-1)}$ supported at the origin. More precisely, given a sequence[5] of renormalization maps $\left\{\mathcal{O}_n \to \mathcal{D}'\left((\mathbb{R}^D)^{\times(n-1)}\right)\right\}_{n=1}^{\infty}$ there is a map from the the set of all sequences of maps

$$\left\{Q_n \in \mathcal{R}(n)\right\}_{n=1}^{\infty}$$

to the set of all sequences of renormalization maps, where we introduced the vector spaces

$$\mathcal{R}(n) := \left\{Q : \mathcal{O}_n \to \mathcal{D}'[0_n] \,\middle|\, \text{commuting with multiplication by polynomials}\right\}, \tag{41}$$

[5] All the sequences below are trivial at $n = 1$, but nevertheless it is convenient to start from $n = 1$.

and the defining condition comes from the requirements on the renormalization maps (as explained in [18, 20]). In this way, one can think of the sequences $\big\{Q_n \in \mathcal{R}(n)\big\}_{n=1}^{\infty}$ as "acting"on the set of all sequences of renormalization maps, or, in other words, that every passage from one sequence $\big\{\mathscr{O}_n \to \mathcal{D}'\big((\mathbb{R}^D)^{\times(n-1)}\big)\big\}_{n=1}^{\infty}$ of renormalization maps to another is induced by a sequence $\big\{Q_n \in \mathcal{R}(n)\big\}_{n=1}^{\infty}$.

On the other hand, we would like to see the passage from one system of renormalization maps $\big\{\mathscr{O}_n \to \mathcal{D}'\big((\mathbb{R}^D)^{\times(n-1)}\big)\big\}_{n=1}^{\infty}$ to another as resulting from an action of a certain group. The latter group is called the *renormalization group* (*of Petermann–Stückelberg–Bogolubov*). Hence, there should be a group structure on the sequence of vector spaces (41) with a group multiplication coming from the diagram:

$$
\left.\begin{array}{c}
\big\{\mathscr{O}_n \to \mathcal{D}'\big((\mathbb{R}^D)^{\times n}\big)\big\}_{n=1}^{\infty} \\
\downarrow \{Q'_n\}_{n=1}^{\infty} \\
\big\{\mathscr{O}_n \to \mathcal{D}'\big((\mathbb{R}^D)^{\times n}\big)\big\}_{n=1}^{\infty} \\
\downarrow \{Q''_n\}_{n=1}^{\infty} \\
\big\{\mathscr{O}_n \to \mathcal{D}'\big((\mathbb{R}^D)^{\times n}\big)\big\}_{n=1}^{\infty}
\end{array}\right) \{Q'''_n \in \mathcal{R}(n)\}_{n=1}^{\infty},
$$

$$
\{Q'''_n\}_{n=1}^{\infty} =: \{Q''_n\}_{n=1}^{\infty} \bullet \{Q'_n\}_{n=1}^{\infty}
$$

Now comes the role of operad theory in renormalization. In [13] (see also [14, Sect. 5.3.14]) it was constructed a functor

$$
\big\{\text{Operads}\big\} \longrightarrow \big\{\text{Groups}\big\}.
$$

Furthermore, the formula for the group multiplication (42), which was derived in [18, Sect. 2.6], indicates that it is induced by an operad according to the above functor of [13]. In this way we encounter an operad whose operadic compositions are schematically given in the following figure:

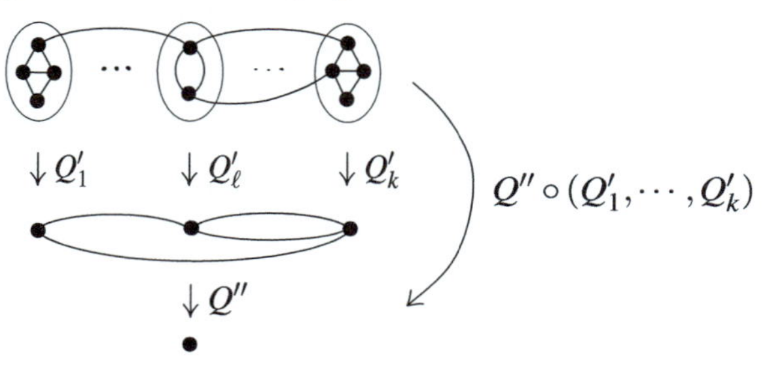

and its combinatorial version was described in [13]

Thus, the role of the operad $\mathcal{R}$ in renormalization theory is that it describes the Petermann–Stückelberg–Bogolubov renormalization group. Furthermore, in physics there is another action of the renormalization group that is on the space of so called physical coupling constants. The latter are the parameters that characterize a particular perturbative QFT model and they appear as constants in the Lagrangean of the model. According to the main theorem of renormalization theory any change of the renormalization scheme (i.e., the renormalization maps) should be equivalent to a formal diffeomorphism in the space of physical coupling constants. This is the renormalization group action. Another application of the functor of [13] is that it produces the group of formal diffeomorphisms around the zero of a vector space V when it is applied to $\mathcal{E}nd_V$ and the renormalization group action comes from an operadic morphism $\mathcal{R} \to \mathcal{E}nd_V$.

The bridge between the theory of the vertex algebras and renormalization is based on an existence of a natural isomorphism [18]

$$\mathcal{E}(n) \cong \mathcal{R}(n) \,.$$

Furthermore, this is an isomorphism of operads.

Our conclusion is summarized in the following scheme:

$$\begin{array}{ccc} \text{The } \mathcal{E} - \text{algebras are vertex algebras} & & \\ \uparrow & & \\ \text{Expansion operad } \mathcal{E} & \cong & \text{Renormalization operad } \mathcal{R} \\ & & \downarrow \\ & & \text{The group associated to } \mathcal{R} \text{ is the renormalization group} \end{array}$$

The constructed here operad can be studied with methods of various fields in mathematics as: Algebraic Geometry, Differential Algebra and Topology. In this way, we obtain also a relation to purely mathematical problems, among which is in particular, the structure of multizeta values and possible their extensions (within the ring of periods), which are needed for the (analytic) calculations in perturbative QFT.

Acknowledgements The author thanks for the useful discussions with Spencer Bloch, Francis Brown, Pierre Cartier and Maxim Kontsevich on various topics related to this work during his visits at Institut des Hautes Études Scientifiques (IHÉS, Bures-sur-Yvette, France). The author is grateful for the support and hospitality by Instituto de Ciencias Matemáticas (ICMAT, Madrid), where this work was presented during the Research Trimester on Multiple Zeta Values, Multiple Polylogarithms, and Quantum Field Theory (ICMAT, September 15–December 19, 2014). The author thanks the referees for their careful reading of the manuscript an for suggesting many corrections and improvements. This work was supported in part by the Bulgarian National Science Fund under research grant DN-18/3.

References

1. Borcherds, R.E.: Vertex algebras, Kac-Moody algebras, and the Monster. Proc. Natl. Acad. Sci. USA **83**, 3068–3071 (1986)
2. Borcherds, R.E.: Vertex algebras. In: Topological Field Theory, Primitive Forms and Related Topics, (Kyoto 1996). Progress in Mathematics, vol. 160, pp. 35–77. Birkhäuser Boston, Boston, MA (1998)
3. Frenkel, E., Ben-Zvi, D.: Vertex Algebras and Algebraic Curves. Mathematical Surveys and Monographs, vol. 88. American Mathematical Society, Providence, RI (2001)
4. Frenkel, I.B., Lepowsky, J., Meurman, A.: Vertex Operator Algebras and the Monster. Pure and Applied Mathematics, vol. 134. Academic, Boston, MA (1988)
5. Holland, J., Hollands, S.: Operator product expansion algebra. J. Math. Phys. **54**, 72302 (2013)
6. Hollands, S.: Quantum field theory in terms of consistency conditions. I. General framework, and perturbation theory via Hochschild cohomology, vol. 5, p. 90. SIGMA (2009)
7. Hollands, S., Olbermann, H.: Perturbative quantum field theory via vertex algebras. J. Math. Phys. **50**, 112304 (2009)
8. Jost, R.: The General Theory of Quantized Fields. American Mathematical Society, Providence, RI (1965)
9. Kac, V.G.: Vertex Algebras for Beginners, 2nd ed. University Lecture Series, vol. 10. American Mathematical Society, Providence, RI (1998)
10. Kapustin, A., Orlov, D.: Vertex algebras, mirror symmetry, and D-branes: the case of complex tori. Commun. Math. Phys. **233**, 79–136 (2003)
11. Lepowsky, J., Li, H.: Introduction to Vertex Operator Algebras and Their Representations. Progress in Mathematics, vol. 227. Birkhäuser, Boston, MA (2004)
12. Li, H.: A higher-dimensional generalization of the notion of vertex algebra. J. Algebra **262**, 1–41 (2003)
13. Loday, J.-L., Nikolov, N.M.: Operadic construction of the renormalization group, In: Proceedings of the IX International Workshop “Lie Theory and Its Applications in Physics”. Springer Proceedings in Mathematics, pp. 169–189 (2012)
14. Loday, J.-L., Vallette, B.: Algebraic Operads. Springer (2012)
15. May, J.-P.: The Geometry of Iterated Loop Spaces. Lecture Notes in Mathematics, vol. 271. Springer (1972)
16. Nikolov, N.M.: Vertex algebras in higher dimensions and globally conformal invariant quantum field theory. Commun. Math. Phys. **253**, 283–322 (2005)
17. Nikolov, N.M.: Germs of Distributions and Construction of Composite Fields in Axiomatic QFT, unpublished, presented at Seminare über Quantenfeldtheorie at the Institut für Theoretische Physik of Goettingen University (2006)
18. Nikolov, N.M.: Anomalies in quantum field theory and cohomologies in configuration spaces. arXiv:0903.0187 [math-ph]; Talk on anomaly in quantum field theory and cohomologies of configuration spaces. arXiv:0907.3735 [hep-th]
19. Nikolov, N.M.: Semi-differential operators and the algebra of operator product expansion of quantum fields. arXiv:1911.01412
20. Nikolov, M.N., Stora, R., Todorov, I.: Renormalization of massless Feynman amplitudes in configuration space. Preprint CERN-TH-PH/2013-107 (2013). arXiv:1307.6854 [hep-th]
21. Nikolov, N.M., Todorov, I.T.: Rationality of conformally invariant local correlation functions on compactified Minkowski space. Commun. Math. Phys. **218**, 417–436 (2001)
22. Streater, R.F., Wightman, A.S.: PCT, Spin and Statististics and All That. Second Printing. Benjamin/Cummings, Reading, MA (1978)
23. Wilson, K.G.: Non-lagrangian models of current algebra. Phys. Rev. **179**, 1499–1512 (1969)

Renormalization and Periods in Perturbative Algebraic Quantum Field Theory

Kasia Rejzner

Abstract In this paper I give an overview of mathematical structures appearing in perturbative algebraic quantum field theory (pAQFT) in the case of the massless scalar field on Minkowski spacetime. I also show how these relate to Kontsevich-Zagier periods. Next, I review the pAQFT version of the renormalization group flow and reformulate it in terms of Feynman graphs. This allows me to relate Kontsevich-Zagier periods to numbers appearing in computing the pAQFT β-function.

Keywords Quantum field theory · Periods · Epstein-Glaser renormalization

1 Introduction

Perturbative AQFT is a mathematically rigorous framework that allows to build models of physically relevant quantum field theories on a large class of Lorentzian manifolds. The basic objects in this framework are functionals on the space of field configurations and renormalization method used is the Epstein-Glaser (EG) renormalization [20]. The main idea in the EG approach is to reformulate the renormalization problem, using functional analytic tools, as a problem of extending almost homogeneously scaling distributions that are well defined outside some partial diagonals in $\mathbb{R}^n$. Such an extension is not unique, but it gives rise to a unique "residue", understood as an obstruction for the extended distribution to scale almost homogeneously. Physically, such scaling violations are interpreted as contributions to the β function.

The main result of this paper is Proposition 2, where we show how a large class of residues relevant for computing the β function in the pAQFT framework, is related to Kontsevich-Zagier periods. Following [35] we define:

K. Rejzner (✉)
Department of Mathematics, University of York, Heslington, York YO10 5DD, UK
e-mail: kasia.rejzner@york.ac.uk; kr763@york.ac.uk

J. I. Burgos Gil et al. (eds.), *Periods in Quantum Field Theory and Arithmetic*, Springer Proceedings in Mathematics & Statistics 314,
https://doi.org/10.1007/978-3-030-37031-2_13

Definition 1 A period is a complex number whose real and imaginary parts are values of absolutely convergent integrals of rational functions with rational coefficients, over domains in $\mathbb{R}^n$ given by polynomial inequalities with rational coefficients.

A very accessible introduction to periods and their relation to Feynman integrals can be found for example in [9, 12].

In Sect. 5 we review the main ideas behind the pAQFT renormalization group (following [4]) and propose a reformulation in terms of Feynman graphs. The latter allows then to relate the numbers appearing in the computation of the pAQFT β function to periods discussed in Sect. 4.

2 Functionals

Let $\mathbb{M}$ be the D-dimensional Minkowski spacetime, i.e. $\mathbb{R}^D$ with the metric

$$\eta = \mathrm{diag}(1, \underbrace{-1, \ldots, -1}_{D-1}) .$$

Define the *configuration space* $\mathscr{E}$ of the theory as the space of smooth sections of a vector bundle E over $\mathbb{M}$, i.e. $\mathscr{E} \doteq \Gamma(E \xrightarrow{\pi} \mathbb{M})$. Fixing E specifies the particle content of the model under consideration. In this paper we will consider only the scalar field, i.e. $\mathscr{E} = \mathscr{C}^\infty(\mathbb{M}, \mathbb{R})$. The field configurations are denoted by φ. For future reference, define $\mathscr{D} \doteq \mathscr{C}^\infty_c(\mathbb{M}, \mathbb{R})$ the space of smooth compactly supported functions on $\mathbb{M}$ and more generally, $\mathscr{D}(\mathscr{O}) \doteq \mathscr{C}^\infty_c(\mathscr{O}, \mathbb{R})$, where $\mathscr{O}$ is an open subset of $\mathbb{R}^n$.

Let $\mathscr{C}^\infty(\mathscr{E}, \mathbb{C})$ denote the space of smooth [2, 36] functionals on $\mathscr{E}$. An important class of functionals is provided by the local ones.

Definition 2 A functional $F \in \mathscr{C}^\infty(\mathscr{E}, \mathbb{C})$ is called local (an element of $\mathscr{F}_{\mathrm{loc}}$) if for each $\varphi \in \mathscr{E}$ there exists $k \in \mathbb{N}$ such that

$$F(\varphi) = \int_{\mathbb{M}} f(j^k_x(\varphi)) , \tag{1}$$

where $j^k_x(\varphi)$ is the k-th jet prolongation of φ and f is a density-valued function on the jet bundle.

The following definition introduces the notion of spacetime localization of a functional.

Definition 3 The spacetime support $\mathrm{supp}\, F$ of a functional $F \in \mathscr{C}^\infty(\mathscr{E}, \mathbb{C})$ is defined by

$$\mathrm{supp}\, F \doteq \{x \in \mathbb{M} | \forall \text{ neighborhoods } U \text{ of } x\ \exists \varphi, \psi \in \mathscr{E}, \mathrm{supp}\, \psi \subset U ,$$
$$\text{such that } F(\varphi + \psi) \neq F(\varphi)\} .$$

Derivatives of smooth compactly-supported functionals are distributions with compact support,[1] i.e.

$$F^{(n)}(\varphi) \in \mathscr{E}'(\mathbb{M}^n, \mathbb{C}) \equiv \mathscr{E}'^{\mathbb{C}}(\mathbb{M}^n), \quad \forall \varphi \in \mathscr{E}, n \in \mathbb{N}.$$

If F is local then each $F^{(n)}(\varphi)$ is a distribution supported on the thin diagonal

$$D_n \doteq \{(x_1, \ldots, x_n) \in \mathbb{M}^n, x_1 = \cdots = x_n\}. \tag{2}$$

Local functionals are important, since they are used to model interactions in perturbative QFT. In the Epstein-Glaser approach, interaction is first restricted to a compact region to avoid the IR problem and subsequently extended by taking the *adiabatic limit*. In this work we are interested only in the UV (i.e. short distance) behavior of the theory, so we leave this last step out.

One can define various important classes of functionals by formulating conditions on the singularity structure of their derivatives $F^{(n)}(\varphi) \in \mathscr{E}'^{\mathbb{C}}(\mathbb{M}^n)$. A notion used in this context is that of a *wavefront set*. For a given distribution $u \in \mathscr{D}'(\mathbb{R}^n)$, its wavefront set $\mathrm{WF}(u)$ contains information about points in $\mathbb{R}^n$ at which u is singular, but also about directions in the momentum space (i.e. after the Fourier transform) in which $\widehat{u}(k)$ fails to decay sufficiently fast. In other words, $\mathrm{WF}(u)$ characterizes *singular directions* of u. For a pedagogical introduction to WF sets see [5]. Knowing the WF sets of distributions u_1, u_2 one can apply the criterion due to Hörmander [28] to check if the pointwise product of u_1, u_2 is well defined. This motivates using WF sets of functional derivatives $F^{(n)}(\varphi)$ to distinguish classes of "well-behaving" functionals. One such class is called *microcausal functionals* $\mathscr{F}_{\mu c}$. For the precise definition and for possible modifications of this notion see [4, 41]. For the purpose of this paper, it is enough to know that $\mathscr{F}_{\mathrm{loc}} \subset \mathscr{F}_{\mu c}$ and that some important algebraic structures are well defined on this space.

3 The S-Matrix and Time-Ordered Products

In the next step we introduce the S-matrix. Since we work perturbatively, the S-matrix is understood as a formal power series in the coupling constant λ and a Laurent series in $\hbar$, with coefficients in smooth functionals. First we introduce the time-ordered products.

Definition 4 Time ordered products are multilinear maps $\mathscr{T}^n : \mathscr{F}_{\mathrm{loc}}^{\otimes n} \to \mathscr{F}_{\mu c}[[\hbar]]$, $n \in \mathbb{N}$, satisfying:

---

[1]Prime always denotes the topological dual, so $\mathscr{E}'(\mathbb{M}^n)$ is the space of continuous linear maps from $\mathscr{E}(\mathbb{M}^n)$ to $\mathbb{R}$ and similarly, $\mathscr{E}'(\mathbb{M}^n, \mathbb{C})$ is the space of continuous linear maps to $\mathbb{C}$. $\mathscr{E}(\mathbb{M}^n)$ is always understood as equipped with its natural Fréchet topology. It is a standard result in functional analysis that the dual of the space of smooth functions is exactly the space of distributions with compact support.

1. **Causal factorisation property**

$$\mathcal{T}^n(F_1,\ldots,F_n)=\mathcal{T}^k(F_1,\ldots,F_k)\star\mathcal{T}^{n-k}(F_{k+1},\ldots,F_n)\,,$$

if the supports $\operatorname{supp} F_i$, $i=1,\ldots,k$ of the first k entries do not intersect the past of the supports $\operatorname{supp} F_j$, $j=k+1,\ldots,n$ of the last $n-k$ entries. Here $\star$ is the operator product of the quantum theory defined by

$$(F\star G)(\varphi)\doteq e^{\hbar\left\langle\Delta_+,\frac{\delta^2}{\delta\varphi\delta\varphi'}\right\rangle}F(\varphi)G(\varphi')|_{\varphi'=\varphi}\,,$$

where Δ_+ is the Wightman 2-point function.
2. $\mathcal{T}^0=1$, $\mathcal{T}^1=\mathrm{id}$.
3. **Symmetry**: For a purely bosonic theory $\mathcal{T}^n$s are symmetric in their arguments. If the fermions are present, $\mathcal{T}^n$s are graded-symmetric.
4. **Field independence**: $\mathcal{T}^n(F_1,\ldots,F_n)$, as a functional on $\mathcal{E}$, depends on φ only via the functional derivatives of $F_1,\ldots,F_n$, i.e.

$$\frac{\delta}{\delta\varphi}\mathcal{T}^n(F_1,\ldots,F_n)=\sum_{i=1}^n\mathcal{T}^n\left(F_1,\ldots,\frac{\delta F_i}{\delta\varphi},\ldots,F_n\right)$$

5. **φ-Locality**: $\mathcal{T}^n(F_1,\ldots,F_n)=\mathcal{T}^n(F_1^{[N]},\ldots,F_n^{[N]})+\mathcal{O}(\hbar^N)$, where $F_i^{[N]}$ is the Taylor series expansion of the functional F_i up to the N-th order.
6. **Poincaré invariance**: Let $\alpha\in\mathcal{P}_+^\uparrow$ (the proper ortochronous Poincaré group). We define $\sigma_\alpha(\varphi)(x)\doteq\varphi(\alpha^{-1}x)$ for $\varphi\in\mathcal{E}$, $x\in\mathbb{M}$ and define the action of $\alpha\in\mathcal{P}_+^\uparrow$ on functionals using $\sigma_\alpha(F)\doteq F(\sigma_\alpha(\varphi))$. We require $\sigma_\alpha\circ\mathcal{T}^n\circ(\sigma_\alpha^{-1})^{\otimes n}=\mathcal{T}^n$.

We refer to these conditions as the Epstein-Glaser (EG) axioms.

Definition 5 The formal S-matrix is a map from $\mathcal{F}_{\mathrm{loc}}$ to $\mathcal{F}_{\mu c}[[\lambda]]((\hbar))$ defined as

$$S(\lambda F)=\sum_{n=0}^\infty\frac{(\lambda i)^n}{n!\hbar^n}\mathcal{T}_n(F^{\otimes n})\,,\tag{3}$$

With $\mathcal{T}^n$s satisfying the EG axioms. Let $(\mathcal{F}_{\mathrm{loc}})^{\otimes n}_{\mathrm{pds}}$ denote the subset of $\mathcal{F}_{\mathrm{loc}}^{\otimes n}$ consisting of functionals with pairwise disjoint supports. On such functionals one can define the n-fold time-ordered product to be

$$\mathcal{T}^n(F_1,\ldots,F_n)=m\circ e^{\hbar\sum_{i<j}D_{\mathrm{F}}^{ij}}(F_1\otimes\cdots\otimes F_n)\,,\tag{4}$$

where $D_{\mathrm{F}}^{ij}\doteq\langle\Delta^{\mathrm{F}},\frac{\delta^2}{\delta\varphi_i\delta\varphi_j}\rangle$, m denotes the pointwise multiplication and Δ^{F} is the Feynman propagator of the free scalar field theory on $\mathbb{M}$. Unfortunately, this definition doesn't trivially extend to arbitrary local functionals, due to singularities of the Feynman propagator. Instead, one has to use more sophisticated analytical tools, which we will review in the next section. We will refer to (4) as the *non-renormalized* n-fold time-ordered product and the problem of extending $\mathscr{T}^n$ to arbitrary local functional is referred to as *the renormalization problem.*

To organize the combinatorics present in the construction of time-ordered products, it is convenient to write them in terms of Feynman graphs. To see how this comes about, we use the identity

$$e^{\hbar\sum_{i<j} D_{\mathrm{F}}^{ij}} = \prod_{i<j}\sum_{l_{ij}=0}^{\infty}\frac{\left(\hbar D_{\mathrm{F}}^{ij}\right)^{l_{ij}}}{l_{ij}!} \tag{5}$$

to obtain the expansion

$$\mathscr{T}^n = \sum_{\Gamma\in\mathscr{G}_n}\mathscr{T}^\Gamma ,$$

where $\mathscr{G}_n$ is the set of all graphs with n vertices and no tadpoles (i.e. no loops in the graph-theoretic sense). Let $E(\Gamma)$ denote the set of edges and $V(\Gamma)$ the set of vertices of the graph Γ. Contributions from particular graphs are given by

$$\mathscr{T}^\Gamma = \frac{1}{\mathrm{Sym}(\Gamma)} m\circ\langle t^\Gamma, \delta_\Gamma\rangle , \tag{6}$$

with

$$\delta_\Gamma = \frac{\delta^{2\,|E(\Gamma)|}}{\prod_{i\in V(\Gamma)}\prod_{e:i\in\partial e}\delta\varphi_i(x_{e,i})}$$

and

$$t^\Gamma = \prod_{e\in E(\Gamma)}\hbar\Delta^{\mathrm{F}}(x_{e,i}, i\in\partial e) \tag{7}$$

The symmetry factor Sym is the number of possible permutations of lines joining the same two vertices, $\mathrm{Sym}(\Gamma)=\prod_{i<j} l_{ij}!$.

Note that the map δ_Γ applied to $F\in\mathscr{F}_{\mathrm{loc}}^{\otimes n}$ yields, at any n-tuple of field configurations $(\varphi_1,\ldots,\varphi_n)$, a compactly supported distribution in the variables $x_{e,i}$, $i\in\partial e$, $e\in E(\Gamma)$ with support on the partial diagonal

$$\mathrm{Diag}_\Gamma = \{x_{e,i}=x_{f,i}, i\in\partial e\cap\partial f, e, f\in E(\Gamma)\}\subset\mathbb{M}^{2|E(\Gamma)|} .$$

This partial diagonal can be parametrized using the centre of mass coordinates

$$z_v \doteq \frac{1}{\mathrm{valence}(v)}\sum_{e:v\in\partial e} x_{e,v} ,$$

assigned to each vertex. The remaining relative coordinates are $x_{e,v}^{\mathrm{rel}} = x_{e,v}-z_v$, where $v\in V(\Gamma)$, $e\in E(\Gamma)$ and $v\in\partial e$. Obviously, we have $\sum_{e|v\in\partial e} x_{e,v}^{\mathrm{rel}}=0$ for all

$v \in V(\Gamma)$, so in fact Diag_Γ is parametrized by $|V(\Gamma)|-1$ independent variables. In this parametrization $\delta_\Gamma F$ can be written as a finite sum

$$\delta_\Gamma F = \sum_\beta f^\beta \partial_\beta \delta_{\mathrm{rel}}\,,$$

where $\beta \in \mathbb{N}_0^{D(|V(\Gamma)|-1)}$, each $f^\beta(\varphi_1,\ldots,\varphi_n)$ is a test function on Diag_Γ and δ_{rel} is the Dirac delta distribution in relative coordinates, i.e. $\delta_{\mathrm{rel}}(g) = g(0,\ldots,0)$, where g is a function of $(x^{\mathrm{rel}}_{e,v}, v \in V(\Gamma), e \in E(\Gamma))$.

Let Y_Γ denote the vector space spanned by derivatives of the Dirac delta distributions $\partial_\beta\delta_{\mathrm{rel}}$, where $\beta \in \mathbb{N}_0^{D(|V(\Gamma)|-1)}$ and let $\mathscr{D}(\mathrm{Diag}_\Gamma, Y_\Gamma)$ denote the graded space of test functions on Diag_Γ with values in Y_Γ. With this notation we have $\delta_\Gamma F \in \mathscr{D}(\mathrm{Diag}_\Gamma, Y_\Gamma)$ and if $F \in (\mathscr{F}_{\mathrm{loc}})^{\otimes n}_{\mathrm{pds}}$, then $\delta_\Gamma F$ is supported on $\mathrm{Diag}_\Gamma \setminus \mathrm{DIAG}$, where DIAG is the large diagonal:

$$\mathrm{DIAG} = \left\{ z \in \mathrm{Diag}_\Gamma | \, \exists v, w \in V(\Gamma), v \neq w : z_v = z_w \right\}.$$

We can therefore write (6) in the form

$$\frac{1}{\mathrm{Sym}(\Gamma)}\langle t^\Gamma, \delta_\Gamma\rangle = \sum_{\mathrm{finite}} \left\langle f^\beta \partial_\beta \delta_{\mathrm{rel}}, t^\Gamma \right\rangle$$

where t^Γ is written in terms of centre of mass and relative coordinates. To see that this expression is well defined, note that we can move all the partial derivatives ∂_β to t^Γ by formal partial integration. Then the contraction with δ_{rel} is just the pullback through the diagonal map $\rho_\Gamma\colon \mathrm{Diag}_\Gamma \to \mathbb{M}^{2|E(\Gamma)|}$ by

$$(\rho_\Gamma(z))_{e,v} = z_v \quad \text{if } v \in \partial e\,.$$

The pullback ρ^*_Γ of each $t^\Gamma_\beta \doteq \partial_\beta t^\Gamma$ is a well defined distribution on $\mathrm{Diag}_\Gamma \backslash \mathrm{DIAG}$, so (6) makes sense if $F \in (\mathscr{F}_{\mathrm{loc}})^{\otimes n}_{\mathrm{pds}}$.

The renormalization problem to extend $\mathscr{T}^n$'s to maps on the full $\mathscr{F}^{\otimes n}_{\mathrm{loc}}$ is now reduced to extending distributions $\rho^*_\Gamma t^\Gamma_\beta$ to the diagonal.

In this and the next section we will consider the simplest situation, where the free theory is the free massless scalar field and the possible interactions are local functionals $F_1,\ldots,F_n$ that depend on the field itself but not on its derivatives. Without the loss of generality, we can assume them to be monomials, i.e. of the form

$$F(\varphi) = \int f(x)\varphi(x)^l d^D x\,,$$

where $f \in \mathscr{D}$, $l \in \mathbb{N}$. Such a functional can be graphically represented as a vertex of valence l, decorated by the test function f.

The distributions we need to extend are then $u^\Gamma = \rho_\Gamma^* t^\Gamma$, where t^Γ is given by (7). We can write the explicit expression for u^Γ using the following rules:

1. Choose a vertex of Γ and label it as $x_0 = 0$. Label the remaining vertices with variables $x_1, \dots, x_n$, where $n = |V(\Gamma)| - 1$.
2. Assign the Feynman propagator $\Delta^{\mathrm{F}}(x_i, x_j)$ to each edge $e \in E(\Gamma)$, where $x_i, x_j \in \partial e$.

Because of the translational symmetry, the Feynman propagator $\Delta^{\mathrm{F}}(x, y)$ depends only on the difference $x - y$. Explicitly, it is given by

$$\Delta^{\mathrm{F}}(x, y) = (-1)^{\frac{D}{2}-1} \frac{\boldsymbol{\Gamma}(\frac{D}{2} - 1)}{4\pi^{\frac{D}{2}}} \lim_{\varepsilon \to 0^+} \frac{1}{((x-y)^2 - i\varepsilon)^{\frac{D}{2}-1}} \equiv \frac{k_D}{((x-y)^2 - i0)^{\frac{D}{2}-1}},$$

where $(x - y)^2 \doteq \eta(x - y, x - y)$ is the square with respect to the Minkowski metric and $\boldsymbol{\Gamma}$ denotes the Gamma function. We use the bold symbol to distinguish this from the notation we use for graphs. It follows now that

$$u^\Gamma(x_1, \dots, x_{n-1}) = \frac{k_D^{|E(\Gamma)|}}{\prod_{e \in E(\Gamma)} ((x_{s(e)} - x_{f(e)})^2 - i0)^{\frac{D}{2}-1}}, \tag{8}$$

where $\{x_{s(e)}, x_{f(e)}\} = \partial e$ is the pair of vertices that constitute the boundary of an edge e and the order of these vertices is irrelevant.

Example 1 Consider the following examples:

1. For the fish graph: $u^\Gamma(x) = \frac{k_D^2}{(x^2 - i0)^{D-2}}$,
2. For the triangle graph:

$$u^\Gamma(x, y) = \frac{k_D^3}{(x^2 - i0)^{\frac{D}{2}-1}(y^2 - i0)^{\frac{D}{2}-1}((x-y)^2 - i0)^{\frac{D}{2}-1}}.$$

We have seen how to reduce the renormalization problem to extension of distributions. The construction of $\mathcal{T}^n$s proceeds inductively. Given renormalized time-ordered products of order $k < n$, we can use the causal factorisation property to fix the time-ordered products at order n up to the thin diagonal D_n (see (2)). On the level of graphs it means that all the distributions u^γ corresponding to proper subgraphs $\gamma \subset \Gamma$ have been constructed and substituted into u^Γ. The renormalization problem for u^Γ is now the extension of a distribution defined everywhere outside the *thin diagonal of the graph* Γ understood as the subset of Diag_Γ with all the variables equal. Because of the translation symmetry, this is in fact extension problem for a distribution defined everywhere outside the origin.

4 Distributional Residues and Periods

The framework of pAQFT is different from the one of Connes and Kreimer in two fundamental ways: one works in position rather than momentum space and the metric of the underlying spacetime has Lorentzian rather than Euclidean signature. The latter is the reason for invoking Epstein-Glaser causal approach to renormalization, as outlined in the previous section.

There has been a lot of work done concerning periods in position space approach to renormalization. The most recent comprehensive review has been given in [37], while for historical remarks on the development of the subject, it is worth to look up [47]. A very detailed analysis of renormalization of Feynman integrals and its relation to periods and motives has been done in the series of papers [14, 15]. However, the computations performed in these works are done in Euclidean signature. Another noteworthy work, focusing on relations between Epstein-Glaser renormalization and "wonderful compactifications" is [3].

There are some serious technical difficulties arising when changing the signature to Lorentzian. In the present paper we show how some standard methods used in Euclidean setting can, nevertheless, be applied also to the Lorentzian case.

Before coming to the main result of this paper, let us recall some basic facts about the problem of extension of almost homogeneous distributions [4, 7, 23, 31, 37, 45].

Definition 6 We say that a distribution $u \in \mathscr{D}'(\mathbb{R}^N \setminus \{0\})$ scales almost homogeneously, if $(\rho \frac{d}{d\rho})^{k+1} \rho^\alpha u(\rho .) = 0$ for some $k \in \mathbb{N}_0$ (called scaling order), $\alpha \in \mathbb{R}$ (called scaling degree).

The almost homogeneous scaling relation can also be written in terms of the Euler operator $\mathscr{E} = \sum_{i=1}^d x^i \frac{\partial}{\partial x^i}$, namely a distribution with scaling degree α and order k satisfies

$$(\mathscr{E} + \alpha)^{k+1} u = 0 \,,$$

while $(\mathscr{E} + \alpha)^k u \neq 0$.

Example 2 For a graph Γ with n vertices the distribution $u \equiv u^\Gamma$ that we need to extend belongs to $\mathscr{D}'(\mathbb{R}^N \setminus \{0\})$, where $N = (n-1)D$ and D is the dimension of $\mathbb{M}$.

The following result was proven in [26, Proposition 1] (see also [37, section 4.4]):

Proposition 1 *Let u be a (Lorentz invariant) almost homogeneously scaling distribution with degree $\alpha = N + \mathbb{N}_0$, then there exists a non-unique (Lorentz invariant) extension $\bar{u} \in \mathscr{D}'(\mathbb{R}^N)$ of u and*

$$\left(\rho \frac{d}{d\rho}\right)^{k+1} \rho^\alpha \bar{u}(\rho .)\Big|_{\rho=1} = (\mathscr{E} + \alpha)^{k+1} \bar{u} = \sum_{|\beta| = \alpha - N} c_\beta \partial^\beta \delta \,,$$

where $\beta \in \mathbb{N}_0^N$ is a multiindex.

In the proof of the above proposition provided in [26], the coefficients c^α are computed by integrating certain (closed) distributional forms over a closed codimension 1 surface enclosing the origin. We will now review the construction of these forms and it will become clear that these do not depend on the choice of the extension. Moreover, their closeness is the reason why c^αs do not depend on the choice of the integration surface and hence the homogeneous differential operator

$$\sum_{|\beta|=\alpha-N} c_\beta \partial^\beta \tag{9}$$

doesn't depend on the choice of the extension $\bar{u}$. This fact has also been highlighted in the discussion following formula (4.21) in [37, section 4.4]).

We will call (9) the residue of u and denote it by $\mathrm{Res}(u)$, so that

$$(\mathscr{E}+\alpha)^{k+1}\bar{u} = \mathrm{Res}(u)\delta\,.$$

Coefficients of the differential operator $\mathrm{Res}(u)$ can be explicitly computed using the construction of $\bar{u}$ proposed in [26, eq. (186)] and [37, Theorem 4.8]. Let us outline the main ideas behind this construction. First, note that the almost homogeneous scaling implies that the distributional kernel of u can be written as [26, eq. (172)], [37, eq. (3.12)]

$$u(rx) = \sum_{m=0}^{k} r^{-l}\frac{(\log r)^m}{m!} v_m(x) \quad r>0\,, \tag{10}$$

where $v_m = (\mathscr{E}+\alpha)^m u$. Let $\langle u, f\rangle$ denote the dual pairing between the distribution u and the test function $f \in \mathscr{D}(\mathbb{R}^N \setminus \{0\})$. This pairing is usually realized as the integral

$$\langle u, f\rangle = \int_{\mathbb{M}^N} u(x)f(x)d^N x\,. \tag{11}$$

We rewrite this integral using the representation (10). First, choose a compact $N-1$ dimensional hypersurface around the origin, homoeomorphic to the (Euclidean) sphere S^{N-1} that intersects each orbit of the scaling transformation $x \mapsto \mu x$ exactly once. Note that the map $\mathbb{R}_+ \times \Sigma \ni (r, \hat{x}) \mapsto r\hat{x} \in \mathbb{R}^N \setminus \{0\}$ is a diffeomorphism, since the surface Σ is transverse to the orbits of dilations in $\mathbb{R}^N$.

Using microlocal analysis techniques [28] one can show that distributions v_m appearing in (10) have well defined restrictions to Σ (see [26], Section 3.3, after eq. (173)). Denote points on Σ by $\hat{x}$ and write the restriction of v_m as $v_m(\hat{x})$. Next, define for $r>0$ the following space

$$\Sigma_r \doteq \{r\hat{x} \in \mathbb{R}^N | \hat{x} \in \Sigma\}\,.$$

Denote the natural inclusion of Σ_r into $\mathbb{R}^N$ by i_r. One introduces a $(N-1)$-form Ω on $\mathbb{R}^N$ by

$$\Omega(x) = \sum_{a=1}^{N} (-1)^{a-1} x_a dx_1 \wedge \cdots \wedge \widehat{dx_a} \wedge \cdots \wedge dx_N \, ,$$

where x_a are components of $x \in \mathbb{R}^N$. The caret symbol $\widehat{\ }$ means that the corresponding factor is omitted. We can now write

$$d^N x = \frac{dr}{r} \wedge i_r^* \Omega \, .$$

Let $\rho_\Sigma : \mathbb{R}^N \setminus \{0\} \to \mathbb{R}_+$ denote the smooth function defined by the condition

$$\frac{x}{\rho_\Sigma(x)} \in \Sigma \, .$$

We obtain a measure on Σ by setting

$$d\sigma(\hat{x}) = \rho_\Sigma(x)^{-N} \Omega(x) \, ,$$

and express the pairing (11) as

$$\langle u, f \rangle = \int_{-\infty}^{\infty} \sum_{m=0}^{k} \theta(r) r^{N-1-l} \frac{(\log r)^m}{m!} \left(\int_\Sigma v_m(\hat{x}) f(r\hat{x}) d\sigma(\hat{x}) \right) dr \, , \tag{12}$$

where θ denotes the Heaviside step function. Denote $F(r) \doteq \int_\Sigma v_m(\hat{x}) f(r\hat{x}) d\sigma(\hat{x})$. Formula (12) makes sense, since the support of the test function f is bounded away from the origin in $\mathbb{R}^N$ and hence $F(r)$ is a test function on $\mathbb{R}_+$ (i.e. smooth compactly supported), whose support is bounded away from $r = 0$. If we want f to be an arbitrary test function, then $F(r)$ vanishes for sufficiently large r, but does not vanish near $r = 0$ [26, discussion following eq. (184)].

The renormalization problem has therefore been reduced to extension of the distribution $\theta(r) r^{N-1-l} (\log r)^m$ on $\mathbb{R}$. This is done by various methods, see for example [22, 24, 25, 37, 43]. The idea that we are going to follow here (proposed by [23] based on the ideas of [21, 39]) is to consider first the extension of the distribution $\theta(r) r^{N-1-l+\varepsilon} (\log r)^m$ for a complex, non integer $N - 1 - l + \varepsilon$. If we require the almost homogeneous scaling, then the extension exists and is unique. Next, we expand the resulting extended distribution in ε and subtract the pole part.

Let us come back to our original extension problem for $u \in \mathscr{D}'(\mathbb{R}^N \setminus \{0\})$. It is well known in the literature on differential renormalization (see e.g. [26, eq. (186)] or [37, Thm. 4.8]) that an extension $\bar{u}$ of an almost homogeneously scaling distribution u of order k and degree α to an everywhere-defined distribution can be obtained by setting

$$\langle \bar{u}, f \rangle \doteq \lim_{\varepsilon \to 0} \left(\int_0^\infty \int_\Sigma \overline{r^\varepsilon u(r\hat{x})}^{\text{uhe}} f(r\hat{x}) d\sigma(\hat{x}) dr \right.$$
$$\left. - \sum_{m=0}^{k} \frac{(-1)^{m+\alpha-N}}{\varepsilon^{m+1}} \sum_{|\beta|=\alpha-N} \frac{1}{\beta!} \int_0^\infty \int_\Sigma v_m(\hat{x}) f(r\hat{x}) \partial^\beta \delta(r\hat{x}) d\sigma(\hat{x}) dr \right),$$

where $\overline{\cdot}^{\text{uhe}}$ denotes the *unique almost homogeneous extension*, $\beta \in \mathbb{N}_0^N$ is a multiindex, $\beta! \equiv \beta_1! \dots \beta_N!$ and $\partial^\beta \doteq \partial_{x_1}^{\beta_1} \dots \partial_{x_N}^{\beta_N}$.

We are now ready to compute the almost homogeneous scaling violation for the extension $\bar{u}$. The coefficients c_β of $\mathrm{Res}(u)$ in formula (9) are obtained from (see e.g. [26, eq. (92)])

$$c_\beta \doteq (-1)^{\alpha-N} \frac{1}{\beta!} \int_\Sigma \hat{x}^\beta v_m(\hat{x}) d\sigma(\hat{x})$$

that manifestly doesn't depend on the choice of the extension, but only on u. Note that c_β does not depend on the choice of Σ because the integrand is a (distributional) closed form (see [26, eq. (210)] for the proof of closedness).

As a special case we can consider a distribution with scaling degree $\alpha = N$ and scaling order 0. In this case the residue is given in terms of a complex number

$$\mathrm{Res}(u) = c_0 = \int_\Sigma u(\hat{x}) d\sigma(\hat{x}) . \tag{13}$$

Definition 7 For a graph Γ with n vertices and no derivatives decorating the edges, the scaling degree of the distribution u^Γ is given by the formula

$$\alpha_\Gamma = (D-2)|E(\Gamma)| .$$

Definition 8 We define the divergence degree of a graph Γ by

$$\omega_\Gamma = \alpha_\Gamma - (|V(\Gamma)| - 1)D .$$

A graph Γ is called *superficially divergent* if $\omega_\Gamma \geq 0$.

Hence graphs with $\alpha_\Gamma = N$ are characterized by the condition

$$(D-2)|E(\Gamma)| = (|V(\Gamma)| - 1)D . \tag{14}$$

Note that the loop number of a graph (the first Betti number) is given by $h_1 = |E(\Gamma)| - |V(\Gamma)| + 1$, so the above condition can be also expressed as

$$|E(\Gamma)| = \frac{D}{2} h_1 .$$

In four dimensions ($D = 4$) this reduces to $|E(\Gamma)| = 2h_1$. If Γ satisfies (14) and has no superficially divergent subgraphs (here a subgraph $\gamma \subset \Gamma$ is specified by choosing a subset of vertices of Γ and taking all the edges connecting these), then it has scaling degree $\alpha_\Gamma = N$ (so the divergence degree vanishes) and scaling order $k_\Gamma = 0$. Such graphs coincide with *primitive graphs* in the Connes-Kreimer approach, if we restrict to $D = 4$ and fix the interaction.

Remark 1 The class of primitive graphs in the Epstein-Glaser Hopf algebra [18, 25, 34, 38] differs from the class of primitive graphs in the Connes-Kreimer approach. As an example consider the two vertex graph, which has $|E(\Gamma)| = 4$ and $h_1 = 3$. This graph is primitive in the Epstein-Glaser Hopf algebra, but not primitive in the Connes-Kreimer approach.

Consider a graph Γ with $|E(\Gamma)| = \frac{D}{2}h_1$ and no superficially divergent subgraphs. Let Δ be the simplex defined by $\sum_{e\in E(\Gamma)} \alpha_e = 1$ and $\alpha_e > 0$. We introduce the measure $\mu(\boldsymbol{\alpha}) \doteq \delta(1 - \sum_{e\in E(\Gamma)} \alpha_e) \prod_{e\in E(\Gamma)} \alpha_e^{\frac{D}{2}-2} d\alpha_e$ on Δ. Let

$$\hat{\Psi}_\Gamma(\boldsymbol{\alpha}) = \sum_{\substack{T \text{ spanning} \\ \text{tree}}} \prod_{e\in T} \alpha_e$$

be the dual graph polynomial (see e.g. [6, 9, 33, 49]). We define

$$P_\Gamma \doteq \int_\Delta \frac{\mu(\boldsymbol{\alpha})}{(\hat{\Psi}_\Gamma(\boldsymbol{\alpha}))^{D/2}} . \tag{15}$$

If P_Γ converges absolutely, then it defines a *real period* of the graph Γ in the sense of Definition 36 of [11].

It was shown in [6] that, in $D = 4$, under assumptions on Γ stated above, P_Γ indeed converges absolutely. For explicit computations of these periods in Euclidean ϕ^4 theory in 4 dimensions, see for example [42].

It is highly plausible that this result can also be generalized to other dimensions, e.g. $D = 6$. For an elementary argument, first note that potential singularities of the integrand lie on $C \doteq X_\Gamma \cap \partial\Delta$, the intersection of the hyper-surface $X_\Gamma \doteq \{\boldsymbol{\alpha} \in \mathbb{R}^{|E(\Gamma)|} | \hat{\Psi}_\Gamma(\boldsymbol{\alpha}) = 0\}$ with the boundary $\partial\Delta$. If C is just a collection of points, one can split the integration region into small neighborhoods of these points and the rest. For each such neighborhood one parametrizes the integral using spherical coordinates around the point and examines the behaviour of the integrand as the radius r approaches 0. One can now observe that for each such integral, extra factors of α_e contribute $r^{(|E(\Gamma)|-1)(\frac{D}{2}-2)}$, the integration measure contributes $r^{|E(\Gamma)|-2}$, while the denominator contributes $r^{-(|V(\Gamma)|-2)\frac{D}{2}}$. The last assertion follows from the fact that $\hat{\Psi}_\Gamma$ is a degree $|V(\Gamma)| - 1$ polynomial and because we are integrating over the simplex, the dominant contribution comes from degree $|V(\Gamma)| - 2$ terms. Since $|V(\Gamma)| - 1 = \frac{D-2}{2}|E(\Gamma)|$, the integrand can be bounded by a constant, as $r \to 0$. We perform these estimates explicitly in Example 4.

In proposition 2 we show how periods defined by (15) appear in distributional residues in Lorentzian signature. Before we do that, it is worth to recall a few facts concerning graph polynomials (see [9, 13] for a more comprehensive review).

Definition 9 ([44, 48]) The generic graph Laplacian (or Kirchhoff matrix) is the $|V(\Gamma)| \times |V(\Gamma)|$ matrix defined by

$$\mathscr{L}_{ij}(\boldsymbol{\alpha}) = \begin{cases} \sum\limits_{\substack{e\in E(\Gamma)\\ v_i,v_j\in\partial e}} -\alpha_e & \text{if } i \neq j, \\ \sum\limits_{\substack{e\in E(\Gamma)\\ v_i\in\partial e}} \alpha_e & \text{if } i = j, \end{cases}$$

for all $v_i, v_j \in V(\Gamma)$. A sum over the empty set is set to be zero.

Theorem 1 (tree-matrix theorem in [48], thm. VI.29) *Let Γ be a graph with N edges, all of them labelled by the set $\{\alpha_1, \ldots, \alpha_N\}$ and let v_i be an arbitrary vertex of Γ. Let $\mathscr{L}_\Gamma(\boldsymbol{\alpha})$ be the generic Laplacian and $\hat{\Psi}_\Gamma$ the dual graph polynomial. Then we have*

$$\hat{\Psi}_\Gamma = \mathrm{Det}(\mathscr{L}_\Gamma(\boldsymbol{\alpha})[v_i])\,,$$

where the notation $\mathscr{L}_\Gamma(\boldsymbol{\alpha})[v_i]$ means the (i,i) minor of the matrix $\mathscr{L}_\Gamma(\boldsymbol{\alpha})$.

We are now ready to prove our main result of this section.

Proposition 2 *Let Γ be a graph with $|E(\Gamma)| = \frac{D}{2}h_1$ and such that every proper subgraph γ satisfies $|E(\Gamma)| > 2h_1$. If P_Γ converges absolutely, then the distributional residue* $\mathrm{Res}\, u^\Gamma$ *is given by*

$$\mathrm{Res}\, u_\Gamma = c_0 = \frac{2i^{(2D-1)(|V|-1)}}{(4\pi)^{|E(\Gamma)|}} P_\Gamma\,.$$

Proof First recall that the integral (13) doesn't depend on the choice of Σ. The simplest choice is the unit Euclidean sphere in $\mathbb{R}^{Dn}$, where $n = |V(\Gamma)| - 1$. Denote

$$X \equiv (x_1^0, \ldots, x_n^0, \ldots, x_1^{D-1}, \ldots, x_n^{D-1})$$

Using the formula (8) we obtain

$$c_0 = (-1)^{(\frac{D}{2}-1)|E(\Gamma)|} \left(\frac{\Gamma(\frac{D}{2}-1)}{4\pi^{\frac{D}{2}}}\right)^{|E(\Gamma)|} \lim_{\varepsilon\to 0^+} \int_\Sigma \frac{d\sigma(X)}{\prod_{e\in E(\Gamma)}((x_{s(e)} - x_{f(e)})^2 - i\varepsilon)^{\frac{D}{2}-1}}\,,$$

Denote $b_e \equiv (x_{s(e)} - x_{f(e)})^2 - i\varepsilon$, $e \in E(\Gamma)$. We have $\Re(ib_e) = \varepsilon > 0$, so we can use the well known Schwinger trick to write

$$\frac{1}{\prod_{e\in E(\Gamma)} b_e^{\frac{D}{2}-1}}$$
$$= \frac{\boldsymbol{\Gamma}((\frac{D}{2}-1)|E(\Gamma)|)}{(\boldsymbol{\Gamma}(\frac{D}{2}-1))^{|E(\Gamma)|}} \int_0^1 \cdots \int_0^1 \frac{\delta(1-\sum_{e\in E(\Gamma)}\alpha_e)}{(\sum_{e\in E(\Gamma)}\alpha_e b_e)^{(\frac{D}{2}-1)|E(\Gamma)|}} \prod_{e\in E(\Gamma)} \alpha_e^{\frac{D}{2}-1} d\alpha_e$$
$$\equiv \frac{\boldsymbol{\Gamma}((\frac{D}{2}-1)|E(\Gamma)|)}{(\boldsymbol{\Gamma}(\frac{D}{2}-1))^{|E(\Gamma)|}} \int_\Delta \frac{\mu(\boldsymbol{\alpha})}{(\sum_{e\in E(\Gamma)}\alpha_e b_e)^{(\frac{D}{2}-1)|E(\Gamma)|}},$$

where $k = |E(\Gamma)|$. Now we want to perform a change of variables to put the quadratic form $B \equiv \sum_{e\in E(\Gamma)}\alpha_e b_e$ into its normal form. We write $B = X^T M X$, where M is a block diagonal matrix of the form

$$M = \begin{pmatrix} N & 0 & 0 & \dots & 0 \\ 0 & -N & 0 & \dots & 0 \\ 0 & 0 & -N & \dots & 0 \\ \vdots & & & & \vdots \\ 0 & 0 & 0 & \dots & -N \end{pmatrix}.$$

Each block is a $(|V(\Gamma)|-1)$-dimensional symmetric positive semidefinite matrix (as $\alpha_e \geq 0$, $\forall e \in E(\Gamma)$), which is in fact the $(0,0)$ minor of the generic graph Laplacian $\mathscr{L}_\Gamma(\boldsymbol{\alpha})$ introduced in Definition 9. We can find a non-singular matrix Λ such that

$$\Lambda^T N \Lambda = \mathbb{1}.$$

The argument proceeds now exactly the same as in [6, 8]. Defining

$$S \doteq \begin{pmatrix} \Lambda & 0 & 0 & \dots & 0 \\ 0 & \Lambda & 0 & \dots & 0 \\ 0 & 0 & \Lambda & \dots & 0 \\ \vdots & & & & \vdots \\ 0 & 0 & 0 & \dots & \Lambda \end{pmatrix},$$

we obtain

$$S^T M S \doteq \begin{pmatrix} \mathbb{1} & 0 & 0 & \dots & 0 \\ 0 & -\mathbb{1} & 0 & \dots & 0 \\ 0 & 0 & -\mathbb{1} & \dots & 0 \\ \vdots & & & & \vdots \\ 0 & 0 & 0 & \dots & -\mathbb{1} \end{pmatrix} \equiv \Xi, \tag{16}$$

This suggests a change of variables $X \mapsto S^{-1}X$ that puts the quadratic form B into the normal form. In order to perform this change of variables we only need to ensure that in the following formula the order of integration can be interchanged:

$$\int_{\Sigma}\int_{\Delta} \frac{1}{(\sum_{e\in E(\Gamma)} \alpha_e b_e)^{(\frac{D}{2}-1)|E(\Gamma)|}}\, \mu(\boldsymbol{\alpha}) d\sigma \tag{17}$$

For this, note that

$$|\sum_{e\in E(\Gamma)} \alpha_e b_e|^2 \geq |\sum_{e\in E(\Gamma)} \alpha_e|^2 \varepsilon^2 .$$

Since on the simplex Δ we have $\sum_{e\in E(\Gamma)} \alpha_e = 1$, we can conclude that the integrand in (17) is uniformly bounded by $\frac{1}{\varepsilon^2}$ and as long as $\varepsilon > 0$, we can interchange the order of integration and perform the desired change of variables $X \mapsto S^{-1}X$. The Jacobian for this change of variables is

$$\text{Det}\, S = (\text{Det}\, \Lambda)^D = (\text{Det}\, N)^{-D/2} ,$$

since $(\text{Det}\, \Lambda)^2\, \text{Det}\, N = 1$. It follows now from the tree-matrix Theorem 1 that

$$\text{Det}\, N = \hat{\Psi}_\Gamma(\boldsymbol{\alpha}) .$$

It is now also explicitly seen that the result doesn't depend on the choice of the vertex to which we assigned 0 in our Feynman rules, as the tree-matrix theorem gives the same result for any choice of the minor $\mathscr{L}_\Gamma[v_i]$, $v_i \in V(\Gamma)$.

We can now rewrite c_0 as

$$\begin{aligned}
c_0 = &\boldsymbol{\Gamma}\left(|E(\Gamma)|(\tfrac{D}{2}-1)\right)\left(\frac{(-1)^{(\frac{D}{2}-1)}}{4\pi^{\frac{D}{2}}}\right)^{|E(\Gamma)|} \\
&\times \lim_{\varepsilon\to 0^+} \int_\Sigma \frac{d\sigma}{(X^T \Xi X - i\varepsilon)^{|E(\Gamma)|(\frac{D}{2}-1)}} \int_\Delta \hat{\Psi}_\Gamma^{-D/2}(\boldsymbol{\alpha})\mu(\boldsymbol{\alpha}) = \\
&\boldsymbol{\Gamma}\left(|E(\Gamma)|(\tfrac{D}{2}-1)\right)\left(\frac{(-1)^{(\frac{D}{2}-1)}}{4\pi^{\frac{D}{2}}}\right)^{|E(\Gamma)|} P_\Gamma \lim_{\varepsilon\to 0^+} \int_\Sigma \frac{d\sigma}{(X^T \Xi X - i\varepsilon)^{|E(\Gamma)|(\frac{D}{2}-1)}} ,
\end{aligned} \tag{18}$$

where Ξ is a diagonal metric given in (16).

The remaining integral in (18) is easy to evaluate. It is the residue of the distribution

$$t(X) = \frac{1}{(X^T \Xi X - i0)^{(\frac{D}{2}-1)|E(\Gamma)|}}$$

on the indefinite product space $(\mathbb{R}^{(D-2)|E(\Gamma)|}, \Xi)$, with divergence degree 0 and scaling order 0. Now we use formula [4, Appendix C, formula after eq. (102)]:

$$\text{Res}\, t = i^s |S^{d-1}| ,$$

where d is the total dimension of the indefinite product space (in our case $d = (|V(\Gamma)| - 1)D = |E(\Gamma)|(D-2)$) and s is the number of minus signs in the signature of Ξ (in our case $s = (|V(\Gamma)| - 1)(D-1)$) and $|S^{d-1}|$ is the volume of the unit sphere in d dimensions. We obtain

$$\operatorname{Res} t = i^{(D-1)(|V(\Gamma)|-1)} \left|S^{|E(\Gamma)|(D-2)-1}\right| .$$

With this result and the formula for the volume of the unit sphere in d dimensions

$$|S^{d-1}| = \frac{2\pi^{n/2}}{\boldsymbol{\Gamma}(\frac{n}{2})} ,$$

we arrive at

$$c_0 = i^{(2D-1)(|V(\Gamma)|-1)} \frac{2}{(4\pi)^{|E(\Gamma)|}} P_\Gamma .$$

In particular, for $D = 4$ we have

$$c_0 = (-i)^{(|V(\Gamma)|-1)} \frac{2}{(4\pi)^{|E(\Gamma)|}} \int_\Delta |\hat{\Psi}_\Gamma|^{-2} \Omega(\alpha) ,$$

where $\Omega(\alpha)$ is the standard measure on the simplex.

Example 3 The simplest example is the fish graph in 4 dimensions:

The scaling degree and the scaling order vanish, so from Proposition 2 we obtain

$$c_0 = -i \frac{2}{(4\pi)^2} P_\Gamma .$$

Here $\hat{\Psi}_\Gamma = \alpha_1 + \alpha_2$, so $P_\Gamma = 1$ and hence $c_0 = \frac{-i}{8\pi^2}$.

Example 4 Following [4], consider the triangle in 6 dimensions:

Proposition 2 implies that

$$c_0 = -\frac{2}{(4\pi)^3} P_\Gamma ,$$

if P_Γ converges. Since $\hat{\Psi}(\boldsymbol{\alpha}) = \alpha_1\alpha_2 + \alpha_1\alpha_3 + \alpha_2\alpha_3$, we have

$$P_\Gamma = \int_\Delta \frac{\alpha_1\alpha_2\alpha_3\delta(1 - \alpha_1 - \alpha_2 - \alpha_3) d\alpha_1 d\alpha_2 d\alpha_3}{(\alpha_1\alpha_2 + \alpha_2\alpha_3 + \alpha_1\alpha_3)^3} .$$

To see that this integral is absolutely convergent, note that singularities of the integrand appear only in the "corners" of the simplex. Using the symmetry of the problem, we pick the $\alpha_3 = 1$ and consider the integral I_ε of the same integrand as above, but over a small neighborhood of the point $(\alpha_1, \alpha_2, 1)$ on the simplex Δ. Using polar coordinates $\alpha_1 = r\cos\theta$, $\alpha_2 = r\sin\theta$, this integral takes the form

$$I_\varepsilon = \int_0^\varepsilon \int_0^{\pi/2} \frac{r^3 \sin^2\theta \cos\theta (1 - r\sqrt{2}\sin(\theta + \frac{\pi}{4}))}{r^3(\frac{1}{2}r\sin 2\theta + \sqrt{2}\sin(\theta + \frac{\pi}{4}) - 2r\sin^2(\theta + \frac{\pi}{4}))^3} d\theta dr$$

Since $\sin(\theta + \frac{\pi}{4})$ does not vanish in the interval $[0, \frac{\pi}{2}]$, the integrand can be bounded by a constant when $r \to 0$, so I_ε is absolutely convergent and so is P_Γ.

Following [4, example on p. 39] we evaluate this integral by integrating out α_3 and then changing the variables to λ, κ, so that $\alpha_1 = \lambda\kappa$ and $\alpha_2 = (1-\lambda)\kappa$. We obtain

$$P_\Gamma = \int_0^1 \int_0^1 \frac{\lambda(1-\lambda)\kappa^2(1-\kappa)}{(\lambda(1-\lambda)\kappa^2 + \kappa(1-\kappa))^3} d\kappa d\lambda = \frac{1}{2},$$

so

$$c_0 = -\frac{1}{2^6\pi^3}.$$

Example 5 The final example is the well known "wheel with three spokes" graph in 4 dimensions:

This one also satisfies the assumptions of Proposition 2, so using the general formula we obtain

$$c_0 = \frac{i}{2^{11}\pi^6} P_\Gamma = \frac{3i}{2^{10}\pi^6}\zeta(3),$$

where we used the well-known value $P_\Gamma = 6\zeta(3)$ (see e.g. [10]).

Proposition 2 allows to reduce the problem of computing a large class of distributional residues to the problem of evaluating periods arising from graph polynomials, of the form discussed in [1, 6, 11, 42], so can be used to easily translate the existing results and apply them to theories in Lorentzian signature.

Let us come back to the general case. Let Γ be a graph with $\omega_\Gamma \geq 0$. If it contains proper subgraphs with $\omega_\gamma \geq 0$, then one has to renormalize these first and substitute the result to the expression for t^Γ. If overlapping divergences are present, a partition of unity might be required. However, there are convincing arguments that this step can be avoided; compare the Example 4.16 in [18] (using the partition of unity) with example 5.3 of [24] (without the partition of unity). A distribution constructed this way is denoted by $\tilde{u}^\Gamma$ and it was shown in [29] that the property of almost homogeneous scaling is preserved in the recursive procedure of renormalization of proper subgraphs. Hence $\tilde{u}^\Gamma$ is an almost homogeneously scaling distribution and the general formula for its residue is

$$\mathrm{Res}(\tilde{u}^\Gamma) = \sum_{|\beta|=\alpha-N} c_\beta \partial^\beta ,$$

where

$$c_\beta \doteq (-1)^{\alpha-N} \frac{1}{\beta!} \int_\Sigma \hat{x}^\beta (\mathscr{E}+\alpha)^k \tilde{u}^\Gamma(\hat{x}) d\sigma(\hat{x}) , \tag{19}$$

If a graph is EG primitive, then $k=0$, $\tilde{u}^\Gamma = u^\Gamma$ and the residue is uniquely determined by the graph. Residues for EG primitive graphs which are not CK primitive can be obtained by using the fact that coefficients c_β are Lorentz invariant. This implies that integrals (19) can be reduced to scalar integrals multiplying appropriate powers of $\eta_{\mu\nu}$.

We believe that a result generalizing Proposition 2 can be established also in this case and we will address it in future work.

Example 6 Consider the sunset diagram in 4 dimensions:

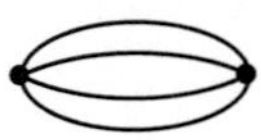

We have $m=0$ and $\alpha=8$. This implies that $|\beta|=4$ so we need to compute

$$c_{\mu\nu\alpha\beta} = \frac{1}{(2\pi)^8 4!} \int_\Sigma \frac{x^\mu x^\nu x^\alpha x^\beta}{(x^2-i0)^4} d\sigma(x) .$$

The Lorenz invariance and the symmetry of the problem imply that

$$\begin{aligned}(2\pi)^8 c_{\mu\nu\alpha\beta} &= \frac{1}{4!24}(\eta_{\alpha\beta}\eta_{\mu\nu}+\eta_{\mu\beta}\eta_{\nu\alpha}+\eta_{\mu\alpha}\eta_{\nu\beta}) \int_\sigma \frac{(x^2)^2}{(x^2-i0)^4} d\sigma(x) \\ &= \frac{1}{2^6 3^2}(\eta_{\alpha\beta}\eta_{\mu\nu}+\eta_{\mu\beta}\eta_{\nu\alpha}+\eta_{\mu\alpha}\eta_{\nu\beta}) \int_\sigma \frac{d\sigma(x)}{(x^2-i0)^2} \\ &= -\frac{i\pi^2}{2^5 3^2}(\eta_{\alpha\beta}\eta_{\mu\nu}+\eta_{\mu\beta}\eta_{\nu\alpha}+\eta_{\mu\alpha}\eta_{\nu\beta}) \end{aligned} \tag{20}$$

Hence

$$\mathrm{Res}(u_\Gamma) = -\frac{i}{2^{13} 3 \pi^6} \Box^2 .$$

In fact there is a different, more direct, way to obtain residues for all the "sunset" type diagrams with arbitrary number of lines. For details see [37, section 5.2] or [4, Appendix C]. The general formula is

$$\mathrm{Res}\left(\frac{1}{(X^2 - i0)^{\frac{d}{2}+l}}\right) = c_l \Box^l \,,$$

where

$$c_l = i^s |S^{d-1}| \frac{\Gamma(\frac{d}{2})}{2^{2l} l! \Gamma(\frac{d}{2}+l)}$$

and $X \in \mathbb{R}^d$ with the diagonal metric of the form $\mathrm{diag}(1, \ldots, 1 \underbrace{-\mathbb{1}, \ldots, -\mathbb{1}}_{s})$. The Example 6 is then the special case of this formula with $d = 4$, $s = 3$ and $l = 2$.

5 Renormalization Group Flow

In [4] the breaking of the homogeneous scaling is shown to relate to the definition of the β-function. In this section we review the main ideas of that argument.

In the first step we generalize the discussion from the previous sections from the massless to the massive scalar field. For studying the scaling properties, it is crucial to work with time-ordered products that are smooth in mass.[2] This is, unfortunately, not the case if we use the standard Feynman propagator Δ^{F}. To rectify this, we replace in our framework the 2-point function Δ^+ with a Hadamard 2-point function H and the Feynman propagator Δ^{F} with a corresponding modified Feynman propagator H^{F}. Crucially, H and H^{F} are smooth in mass. The choice of these objects is unique up to a parameter $M > 0$ with the dimension of mass. Explicit formula for H^{F}_M was derived in [4] and it reads:

$$H^{\mathrm{F}}_M(x) = \frac{m^{D-2}}{(2\pi)^{\frac{D}{2}}\, y^{D-2}} \left(K_{\frac{D}{2}-1}(y) + (-1)^{\frac{D}{2}} \log\frac{M}{m}\, I_{\frac{D}{2}-1}(y)\right) , \qquad (21)$$

where $y \doteq \sqrt{-m^2(x^2 - i0)}$ and K, I are modified Bessel's functions. In 4 dimensions this amounts to

$$\begin{aligned} H^{\mathrm{F}}_M(x) = &\frac{-1}{4\pi^2(x^2 - i0)} \\ &+ \log(-M^2(x^2 - i0))\, m^2 f(m^2x^2) + m^2 F(m^2x^2) \,, \end{aligned}$$

[2] The usual physical argument for the 2-point functions not being smooth at $m^2 = 0$ is that it should not be possible to go smoothly to models with imaginary mass. However, the smoothness in mass is crucial for renormalization on curved spacetimes, as argued in [4, 30–32]. Another approach was proposed in [19], where the "usual" 2-point function can be used and the smoothness in mass is replaced by the smoothen of appropriately rescaled time-ordered products.

while in 6 dimensions

$$H^{\mathrm{F}}_{M}(x) = \frac{1}{4\pi^3(x^2 - i0)^2} + \frac{m^2 f(m^2x^2)}{\pi\,(x^2 - i0)} + \frac{1}{\pi}\left(\log(-M^2(x^2 - i0))\, m^4 f'(m^2x^2) + m^4 F'(m^2x^2)\right)\,,$$

where f and F are real-valued analytic functions. f and f' can be expressed in terms of the Bessel functions J_1 and J_2, respectively, namely

$$f(z) \doteq \frac{1}{8\pi^2\sqrt{z}}\, J_1(\sqrt{z})\,, \quad f(0) = \frac{1}{2^4\,\pi^2}\,, \quad f'(z) = \frac{-1}{16\,\pi^2\,z}\, J_2(\sqrt{z})\,;$$

and F is given by a power series

$$F(z) \doteq -\frac{1}{4\pi}\sum_{k=0}^{\infty}\{\psi(k+1) + \psi(k+2)\}\frac{(-z/4)^k}{k!(k+1)!}\,, \quad F(0) = \frac{2\,C-1}{4\pi}\,,$$

where C is Euler's constant and the Psi-function is related to the Gamma-function by $\psi(x) \doteq \boldsymbol{\Gamma}'(x)\,/\,\boldsymbol{\Gamma}(x)$.

The non-uniqueness of H and H^{F} forces one to use a bit more abstract construction to define the observables and time-ordered product.

Definition 10 For a mass m we define a family of algebras $\mathfrak{A}(m)_M \doteq (\mathscr{F}_{\mu c}[[\hbar]], \star_H)$, labeled by $M > 0$, where $H \equiv H^m_M$ and $\star_H$ is defined by

$$(F \star_H G)(\varphi) \doteq e^{\hbar\left\langle H, \frac{\delta^2}{\delta\varphi\delta\varphi'}\right\rangle} F(\varphi)G(\varphi')|_{\varphi'=\varphi}$$

Different choices of the Hadamard 2-point function for a given mass m differ by a smooth function, i.e. $H^m_{M_1} - H^m_{M_2}$ is smooth. This allows to define a homomorphism

$$\alpha^m_{M_1M_2} \doteq e^{\hbar\left\langle H^m_{M_1} - H^m_{M_2}, \frac{\delta^2}{\delta\varphi^2}\right\rangle}\,,$$

between the algebras $\mathfrak{A}(m)_{M_1}$ and $\mathfrak{A}(m)_{M_2}$. We are now ready to define the algebra of observables for a fixed mass.

Definition 11 $\mathfrak{A}(m)$, the algebra of observables for mass m consists of families $A = (A_M)_{M>0}$, where $A_M \in \mathfrak{A}(m)_M$ and we have $A_{M_1} = \alpha^m_{M_1M_2}(A_{M_2})$.

We can identify abstract elements of the algebra $\mathfrak{A}(m)$ with concrete functionals in $\mathscr{F}_{\mu c}[[\hbar]]$. For $A \in \mathfrak{A}(m)$ denote

$$A_M \doteq \alpha_H(A)\,,$$

where $\alpha_H \equiv e^{\left\langle \hbar H, \frac{\delta^2}{\delta\varphi^2}\right\rangle}$ and $H \equiv H^m_M$ is the appropriate Hadamard 2-point function. A_M defined this way is now a functional in $\mathscr{F}_{\mu c}[[\hbar]]$. Conversely, let $F \in \mathscr{F}_{\mu c}$. We denote by $\alpha_H^{-1} F$ the element of $\mathfrak{A}(m)$ such that $(\alpha_H^{-1} F)_M = F$, where $H \equiv H^m_M$, as above. The rationale behind this notation is explained in [4] and further clarified in [41]. Let $\mathfrak{A}_{\rm loc}(m)$ denote the subspace of $\mathfrak{A}(m)$ arising from local functionals.

Now, following [4], we want to combine algebras corresponding to different masses in a common algebraic structure.

Definition 12 We define the following bundle of algebras

$$\mathscr{B} = \bigsqcup_{m^2 \in \mathbb{R}} \mathfrak{A}(m) \, .$$

Let $A = (A^m)_{m^2 \in \mathbb{R}}$ be a section of $\mathscr{B}$. We fix $M > 0$ and define a function from $\mathbb{R}_+$ to $\mathscr{F}_{\mu c}[[\hbar]]$ by

$$m^2 \mapsto \alpha_M(A)(m) \doteq \alpha_H(A^m) \, , \qquad \text{where } H \equiv H^m_M \, .$$

Definition 13 A section A of $\mathscr{B}$ is called smooth if $\alpha_M(A)$ is smooth for some (and hence all) $M > 0$. The space of smooth sections of $\mathscr{B}$ is denoted by $\mathfrak{A}$. Similarly, $\mathfrak{A}_{\rm loc}$ denotes the space of smooth sections of $\mathscr{B}$ such that $A(m) \in \mathfrak{A}_{\rm loc}(m)$ for all m.

$\mathfrak{A}$ is equipped with a non-commutative product defined as follows:

$$(A \star B)^m_M \doteq A^m_M \star_H B^m_M \, ,$$

where $H \equiv H^m_M$. The n-fold time-ordered product $\mathscr{T}^n$ is a map from $\mathfrak{A}_{\rm loc}$ to $\mathfrak{A}$ defined by

$$\mathscr{T}^n(A_1, \ldots, A_n)(m) \doteq \alpha_H^{-1} \circ \mathscr{T}^n_H(\alpha_H A_1 \ldots, \alpha_H A_n) \, ,$$

where $H \equiv H^m_M$ is a Hadamard 2-point function for mass m and maps $\mathscr{T}^n_H$: $\mathscr{F}_{\rm loc}[[\hbar]] \to \mathscr{F}_{\mu c}[[\hbar]]$ satisfy axioms from Definition 4 with Δ_+ replaced by H.

The S-matrix is now a map from $\mathfrak{A}_{\rm loc}$ to $\mathfrak{A}$ defined by

$$S(A) \doteq \sum_{n=0}^{\infty} \frac{1}{n!} \mathscr{T}^n(A^{\otimes n}) \, .$$

Axioms for time-ordered products can be conveniently formulated on the level of S-matrices.

S 1. Causality $S(A + B) = S(A) \star S(B)$ if $\operatorname{supp}(A^m)$ is later than $\operatorname{supp}(B^m)$ for all $m^2 \in \mathbb{R}_+$.[3]

---

[3] We define $\operatorname{supp} A^m \doteq \operatorname{supp}(\alpha_H(A))$, where $H \equiv H^m_M$ and this definition is independent of the choice of M.

S 2. $S(0) = 1$, $S^{(1)}(0) = \mathrm{id}$,
S 3. **φ-Locality**: $\alpha_M \circ S(A)(\varphi_0) = \alpha_M \circ S \circ \alpha_M^{-1} \left(\alpha_M(A)^{[N]}_{\varphi_0}\right)(\varphi_0) + O(\hbar^{N+1})$, where

$$\alpha_M(A)^{[N]}_{\varphi_0}(\varphi) = \sum_{n=0}^{N} \frac{1}{n!} \left\langle \frac{\delta^n \alpha_M(A)}{\delta\varphi^n}(\varphi_0), (\varphi - \varphi_0)^{\otimes n} \right\rangle$$

is the Taylor expansion up to order N. The dependence on mass m is kept implicit in all these formulas.
S 4. **Field independence**: S doesn't explicitly depends on field configurations.

In Epstein-Glaser renormalization the freedom in defining the renormalized S-matrix is controlled by the Stückelberg-Petermann renormalization group.

Definition 14 The Stückelberg-Petermann renormalization group $\mathscr{R}$ is defined as the group of maps $Z : \mathfrak{A}_{\mathrm{loc}} \to \mathfrak{A}_{\mathrm{loc}}$ with the following properties:

Z 1. $Z(0) = 0$,
Z 2. $Z^{(1)}(0) = \mathrm{id}$,
Z 3. $Z = \mathrm{id} + \mathscr{O}(\hbar)$,
Z 4. $Z(F + G + H) = Z(F + G) + Z(G + H) - Z(G)$, if $\mathrm{supp}\, F \cap \mathrm{supp}\, G = \emptyset$,
Z 5. $\frac{\delta Z}{\delta\varphi} = 0$.

Note that constructing Z's can be reduced to constructing maps $Z_H : \mathscr{F}_{\mathrm{loc}}[[\hbar]] \to \mathscr{F}_{\mathrm{loc}}[[\hbar]]$ which control the freedom in constructing $\mathscr{T}_H^n$, so the abstract formalism reviewed in the present section can be related to the more concrete description presented in Sects. 1–3. We have

$$Z = \alpha_H^{-1} \circ Z_H \circ \alpha_H .$$

The fundamental result in the Epstein-Glaser approach to renormalization is *the Main Theorem of Renormalization* [4, 17, 40, 46].

Theorem 2 *Given two S-matrices S and $\widehat{S}$ satisfying conditions* **S 1– S 5**, *there exists a unique $Z \in \mathscr{R}$ such that*

$$\widehat{S} = S \circ Z . \tag{22}$$

Conversely, given an S-matrix S satisfying the mentioned conditions and a $Z \in \mathscr{R}$, Eq. (22) defines a new S-matrix $\widehat{S}$ satisfying **S 1– S 5**.

Let us now discuss symmetries. Again, we follow closely [4]. Let G be a subgroup of the automorphism group of $\mathfrak{A}$. Assume that it has a well defined action on $\mathscr{S}$, the space of S-matrices, by

$$S \mapsto g \circ S \circ g^{-1} ,$$

where $S \in \mathscr{S}$, $g \in G$. Since $g \circ S \circ g^{-1} \in \mathscr{S}$, it follows from the Main Theorem of Renormalization that there exists an element $Z(g) \in \mathscr{R}$ such that

$$g \circ S \circ g^{-1} = S \circ Z(g) .$$

We obtain a cocycle in $\mathscr{R}$,

$$Z(gh) = Z(g) g Z(h) g^{-1} . \tag{23}$$

The cocycle can be trivialized, i.e. is a coboundary, if there exists an element $Z \in \mathscr{R}$ such that

$$Z(g) = Z g Z^{-1} g^{-1} \qquad \forall g \in G . \tag{24}$$

If this is the case, then

$$g \circ S \circ g^{-1} = S \circ Z g Z^{-1} g^{-1} .$$

Hence

$$g \circ S \circ Z \circ g^{-1} = S \circ Z ,$$

so the S-matrix $S \circ Z$ is G-invariant.

The non-triviality of the cocycle corresponds to the existence of anomalies. One of the most prominent examples where the cocycle cannot be trivialized is the action of the scaling transformations.

The scaling transformation is defined first on the level of field configurations $\varphi \in \mathscr{E}$ as

$$(\sigma_\rho \varphi)(x) = \rho^{\frac{2-D}{2}} \varphi(\rho^{-1} x) , \tag{25}$$

where D is the dimension of $\mathbb{M}$. This induces the action on functionals by the pullback $\sigma_\rho(F)(\varphi) \doteq F(\sigma_\rho(\varphi))$ and finally, the action on $\mathfrak{A}$ can be defined by

$$\sigma_\rho(A)^m = \sigma_\rho(A^{\rho^{-1} m}) .$$

Let now

$$\sigma_\rho \circ S \circ \sigma_\rho^{-1} = S \circ Z(\rho) . \tag{26}$$

$Z(\rho)$ is called the *Gell-Mann Low cocycle* and it satisfies the cocycle condition

$$Z(\rho_1 \rho_2) = Z(\rho_1) \sigma_{\rho_1} Z(\rho_2) \sigma_{\rho_1}^{-1} . \tag{27}$$

Typically this cocycle cannot be trivialized. The generator of this cocycle, denoted by B is related to the β-function known from the physics literature. Following [4] we define

$$B \doteq \rho \frac{d}{d\rho} Z(\rho) \Big|_{\rho=1} , \tag{28}$$

The physical β-function can be obtained from B after one corrects for the "wave function renormalization" and "mass renormalization" (see [4, section 6.4] for details).

To find B we differentiate (26) and obtain

$$\rho\frac{d}{d\rho}(\sigma_\rho\circ S\circ\sigma_\rho^{-1})(V)\Big|_{\rho=1}=\rho\frac{d}{d\rho}(S\circ Z(\rho))(V)\Big|_{\rho=1}=\left\langle S^{(1)}(V),B(V)\right\rangle\,,$$

Note that $\left\langle S^{(1)}(V),.\right\rangle$ is invertible in the sense of formal power series so

$$B(V)=S^{(1)}(V)^{-1}\circ\rho\frac{d}{d\rho}(\sigma_\rho\circ S\circ\sigma_\rho^{-1})(V)\Big|_{\rho=1}$$

To compute B, first we write it in terms of its Taylor expansion:

$$B(V)=\sum_{n=0}^{\infty}\frac{1}{n!}\left\langle B^{(n)}(0),V^{\otimes n}\right\rangle\,, \tag{29}$$

where

$$\left\langle B^{(n)}(0),V^{\otimes n}\right\rangle=\frac{d^n}{d\lambda^n}B(\lambda V)\Big|_{\lambda=0}=\rho\frac{d}{d\rho}\,\frac{d^n}{d\lambda^n}\,Z(\rho)(\lambda V)\Big|_{\lambda=0,\rho=1}$$

Denote $B^{(n)}(0)\equiv B^{(n)}$. The computation of $B^{(n)}$ amounts to summing up the scaling violations of distributional extensions appearing at order n in construction of time-ordered products. To see that lower orders do not contribute, we use the fact that

$$Z(\rho)^{(n)}(0)=\sigma_\rho\circ S^{(n)}(0)\circ\sigma_\rho^{-1}-(S\circ Z_{n-1}(\rho))^{(n)}(0)\,, \tag{30}$$

where Z_n is an element of $\mathscr{R}$ defined in terms of its Taylor expansion as

$$Z_n^{(k)}(0)\doteq\begin{cases}Z^{(k)}(0)\,, & k\le n\,,\\ 0\,, & k>n\,.\end{cases} \tag{31}$$

The proof of (30) is provided in [4] and relies on the proof of the Main Theorem of Renormalization (Theorem 4.1 in [4]). We expand $Z(\rho)^{(n)}(0)$ in terms of Feynman graphs:

$$Z(\rho)^{(n)}(0)=\sum_{\Gamma\in\mathcal{G}_n}Z(\rho)^{\Gamma}\,.$$

where the sum is over all graphs with n vertices. Similarly for $S^{(n)}(0)$ and $B^{(n)}(0)$. We can rewrite (30) as

$$Z(\rho)^{\Gamma}=\sigma_\rho\circ\mathscr{T}^{\Gamma}\circ\sigma_\rho^{-1}-\sum_{P\in\mathrm{Part}'(V(\Gamma))}\mathscr{T}^{\Gamma_P}\circ\bigotimes_{I\in P}Z(\rho)^{\Gamma_I}\,, \tag{32}$$

where $\mathrm{Part}'(V(\Gamma))$ denotes the set of partitions of the vertex set $V(\Gamma)$, excluding the partition with n elements; Γ_P is the graph with vertex set $V(\Gamma_P) = V(\Gamma)$, with all lines connecting different index sets of the partition P, and Γ_I is the graph with vertex set $V(\Gamma_I) = I$ and all lines of Γ which connect two vertices in I. Differentiating (32) with respect ot ρ gives

$$B^\Gamma = \rho\frac{d}{d\rho}(\sigma_\rho \circ \mathscr{T}^\Gamma \circ \sigma_\rho^{-1})\Big|_{\rho=1} - \sum_{P\in \mathrm{Part}'(V(\Gamma))} \mathscr{T}^{\Gamma_P} \circ \bigotimes_{I\in P} B^{\Gamma_I}\,, \tag{33}$$

Note that B^Γ is an operator on $\mathscr{F}_{\mathrm{loc}}^{\otimes n}[[\hbar]]$.

It is now clear that the second term in (32) subtracts contributions from scaling violations corresponding to renormalization of all proper subgraphs of Γ. Hence the only contributions to B^Γ arise from scaling violations resulting from extending distributions defined everywhere outside the thin diagonal of the graph Γ.

For performing computations we need to express $V \in \mathfrak{A}$ in terms of a concrete functional in $\mathscr{F}_{\mathrm{loc}}$. Let's take $V = \alpha_M^{-1} F$ for some $F \in \mathscr{F}_{\mathrm{loc}}$. In the computation of B we have to take into account that α_M, does not commute with the scaling transformations. Define

$$S_M \doteq \alpha_M \circ S \circ \alpha_M^{-1}$$

and

$$B_M \doteq \alpha_M \circ B \circ \alpha_M^{-1}$$

We obtain

$$\begin{aligned}\rho\frac{\partial}{\partial\rho}(\sigma_\rho \circ S_M \circ \sigma_\rho^{-1})(F) - M\frac{\partial}{\partial M} S_M(F)\Big|_{\rho=1} &= \rho\frac{d}{d\rho}(\sigma_\rho \circ S_{\rho^{-1}M} \circ \sigma_\rho^{-1})(F)\Big|_{\rho=1}\\ &= \left\langle S_M^{(1)}(F), B_M(F)\right\rangle .\end{aligned}$$

for $V \in \mathscr{F}_{loc}$. The expression for $-M\frac{\partial}{\partial M} S_M$ was derived in [4] and is given by

$$M\frac{\partial}{\partial M} S_M^{(n)} = 2\hbar\; S_M^{(n)} \circ \sum_{i\neq j} D_v^{ij}\,,$$

where $D_v^{ij} \doteq \frac{1}{2}\left\langle v, \frac{\delta^2}{\delta\varphi_i\delta\varphi_j}\right\rangle$ is a functional differential operator on $\mathscr{F}_{\mathrm{loc}}^{\otimes n}$ and $v \doteq \frac{1}{2}M\frac{d}{dM}H_M^m$.

Again, B_M can be written in terms of its Taylor expansion and $B_M^{(n)}(0)$ is expressed as a sum over graphs with n vertices. Finally, note that due to the field independence of S and Z, we have

$$\frac{\delta^n}{\delta\varphi^n} \circ B_M(F) = \sum_{P\in\mathrm{Part}(n)} B_M^{(|P|)} \circ \bigotimes_{I\in P} F^{|I|}\,.$$

It follows that the Taylor expansion of $B_M(F)$ around $\varphi = 0$ is determined by the values of $B_M^{(k)}(F^{(n_1)} \otimes \cdots \otimes F^{(n_k)})$ at $\varphi = 0$, where $n_1 + \cdots + n_k = n$. We will see now that this allows to express everything in terms of connected graphs.

Let $F \in \mathscr{F}_{\mathrm{loc}}$. Without loss of generality we can assume F to be monomial, i.e. of the form

$$F(\varphi) = \int_{\mathbb{M}} f(x) p(j_x(\varphi)) d^D x\,, \tag{34}$$

where $f \in \mathscr{D}$ and p is a monomial function on the jet space and $j_x(\varphi)$ is a finite order jet of φ at point x. Graphically, we can represent F as a vertex, decorated by f with one external leg for each factor of φ, some of them carrying derivatives. For example $\int_{\mathbb{M}} f(x)\varphi^4(x) d^D x$ is

Given a monomial p on the jet space, define the set of Wick submonomials W_p as the set of all monomials that are factors of p. For example, for $\varphi^4(x)$, the set of Wick submonomials consists of $\varphi^4(x)$, $\varphi^3(x)$, $\varphi^2(x)$, $\varphi(x)$, 1. To indicate derivatives, we put lines across edges, e.g. $p(j_x(\varphi)) = \partial_\mu \varphi \partial_\nu \varphi$ is

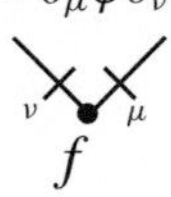

and after summing up over the index μ we obtain $\partial_\mu \varphi \partial^\mu \varphi \equiv (\partial\varphi)^2$ represented for simplicity by

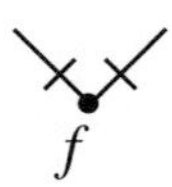

The Taylor expansion induces a coproduct

$$p(j_x(\varphi + \psi)) = \Delta(p)(j_x(\varphi) \otimes j_x(\psi))\,,$$

which can be written explicitly as

$$\Delta(p) = \sum_{q \in W_p} \mathrm{Sym}(q)\, p/q \otimes q\,,$$

where p/q is the graph obtained by removing the edges corresponding to q and $\mathrm{Sym}(q)$ is the number of ways in which graph q can be embedded into graph p. For the local functional F in (34) we obtain

$$F(\varphi + \psi) = \int_{\mathbb{M}} f(x) \Delta p(j_x(\varphi) \otimes j_x(\psi)) d^D x\,.$$

Using Sweedler's notation:

$$\Delta p = \sum_p p_{(1)} \otimes p_{(2)} \,.$$

By a small abuse of notation, we define a functional $F_{(1)}(\varphi) \doteq \int_{\mathbb{M}} f(x) p_{(1)}(j_x(\varphi)) d^D x$, while $F_{(2)}(\varphi)(x)$ is a smooth function defined by $x \mapsto p_{(2)}(j_x(\varphi))$. Using this notation:

$$B_M^{(n)}(F_1, \dots, F_n)(\varphi) = \sum_{F_1,\dots,F_n} \left\langle B_M^{(n)}(F_{1(1)}, \dots, F_{n(1)})(0), F_{1(2)}, \dots, F_{n(2)} \right\rangle .$$

Here $B_M^{(n)}(F_{1(1)}, \dots, F_{n(1)})(0)$ is a distribution, which we can write as

$$B_M^{(n)}(F_{1(1)}, \dots, F_{n(1)})(0)(x_1, \dots, x_n) = f_1(x_1) \dots f_n(x_n) \sum_\Gamma b^\Gamma(x_1, \dots, x_n) \,,$$

where the sum runs over connected graphs Γ with vertices representing $p_{1(1)}, \dots, p_{n(1)}$. Distributions b^Γ are given by

$$b^\Gamma = \rho \frac{d}{d\rho} \sigma_\rho(\overline{u}^\Gamma) \Big|_{\rho=1} \,,$$

where $\overline{u}^\Gamma$ is the extension to the total diagonal of the distribution $\tilde{u}^\Gamma$ constructed as in Sect. 4, where all the proper subgraphs have been renormalized. Hence

$$B_M^{(n)}(F_1, \dots, F_n)(\varphi) = \sum_{F_1,\dots,F_n} \sum_\Gamma \left\langle (f_1 \otimes \dots \otimes f_n) \cdot b^\Gamma, F_{1(2)}, \dots, F_{n(2)} \right\rangle . \quad (35)$$

If Γ is EG primitive, then $\tilde{u}^\Gamma = u^\Gamma$ and u^Γ scales homogeneously. In this case

$$b^\Gamma = \operatorname{Res} u^\Gamma \,.$$

This result provides a link between Kontsevich-Zagier periods appearing in Proposition 2 and physical quantities computed in the pAQFT framework. However, the class of distributional residues relevant for the computation of B is larger than the ones discussed in Sect. 4, since here we need to replace D_{F} with H^{F} given by the formula (21). To give an idea of how the computation proceeds at low loop orders, we review the example of φ^4 in 4 dimensions discussed in [4], but in contrast to [4] we use the Feynman graphs notation to make it easier to follow.

Example 7 Consider the functional

$$F(\varphi) = \lambda \int_{\mathbb{M}} f(x) \varphi^4(x) d^4 x \,.$$

The corresponding element of $\mathfrak{A}$ is

$$V = \alpha_M^{-1} F\,,$$

i.e.

$$V(m)_M = \lambda\, \alpha^{-1}_{H^m_M} \int_{\mathbb{M}} f(x)\varphi^4(x) d^4x\,.$$

We are interested in finding B_M for the QFT model with this interaction. First note that the orbit of the renormalization group is spanned by 1 and functionals of the form $\int_{\mathbb{M}} f_1(x)\varphi^4(x)d^4x$, $\int_{\mathbb{M}} f_2(x)\varphi^2(x)d^4x$, $\int_{\mathbb{M}} f_3(x)(\partial\varphi)^2(x)d^4x$, where $f_1, f_2, f_3 \in \mathscr{D}$. Hence, we need to determine B_M only on such functionals. Graphically we represent them as decorated vertices:

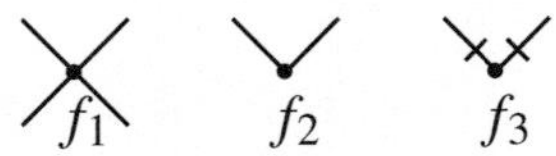

Let us now compute $B_M^{(2)}$ on these functionals. We have

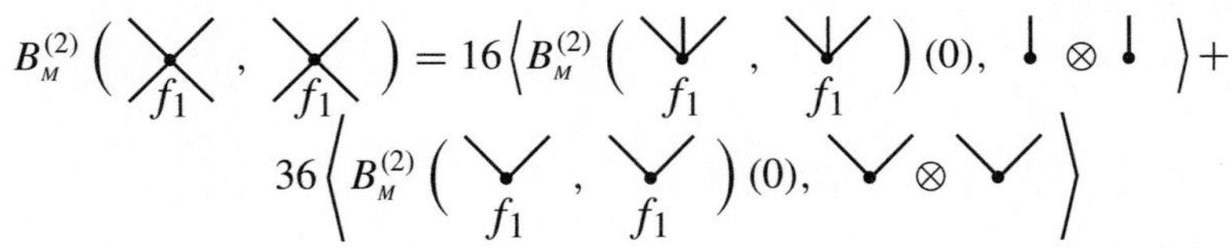

+ constant and linear terms, (36)

since the co-product acts as:

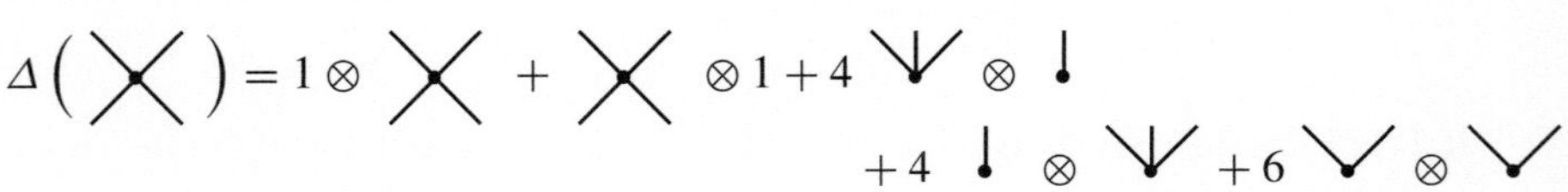

It follows from (36) now the graphs contributing to $B_M^{(2)}$ are

Hence, neglecting constant and linear terms:

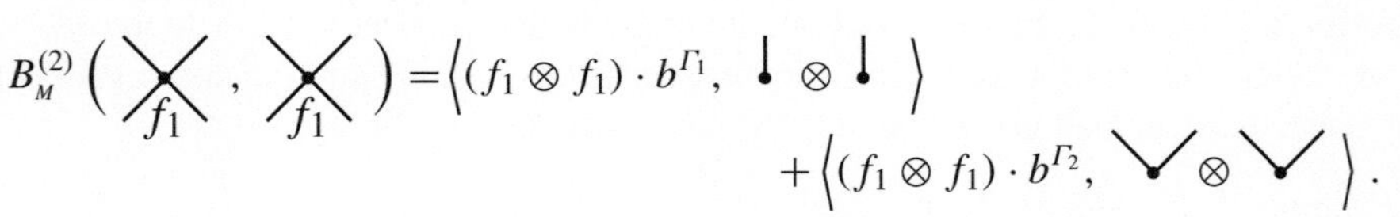

A similar reasoning leads to

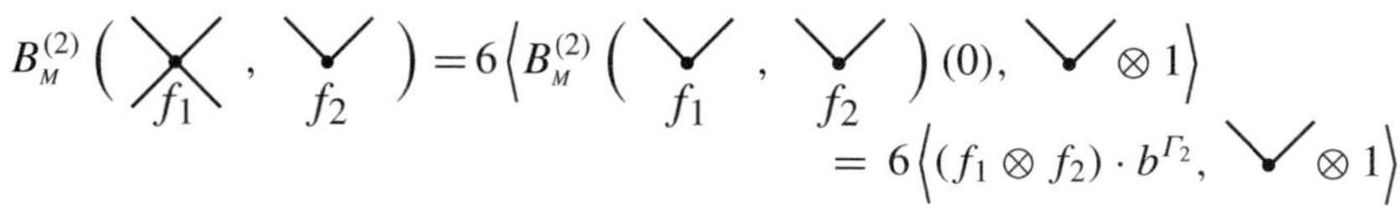

and

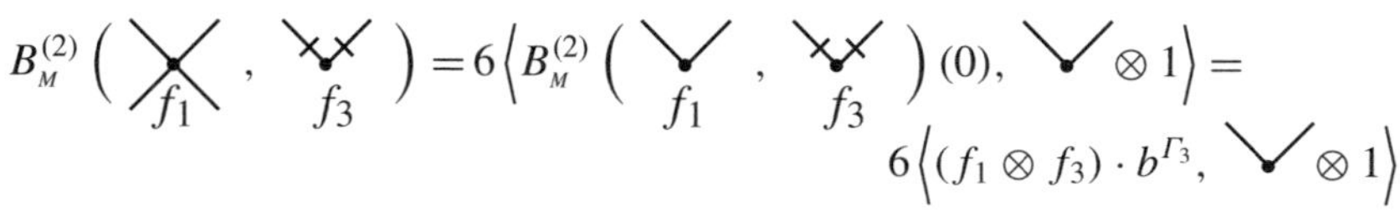

In the latter case there is a new graph appearing, namely

$\Gamma_3 =$  .

Calculating $B_M^{(2)}$ is now reduced to finding the residues: $\operatorname{Res} u^{\Gamma i}$, $i = 1, 2, 3$. The (rather lengthy) computation can be found in Sect. 7.2 of [4].

From the point of view of Kontsevich-Zagier periods, one gets some more interesting numbers in calculating higher orders of B. In particular, the wheel with three spokes appears as a contribution to

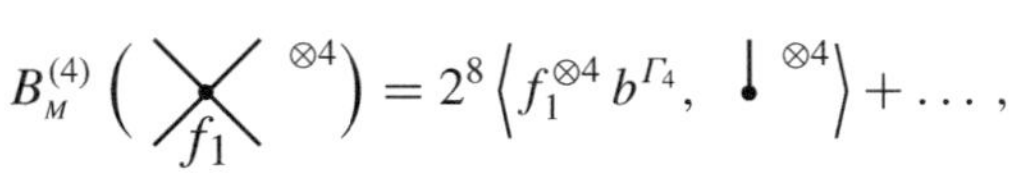

where

$\Gamma_4 =$ 

and $b^{\Gamma_4} = \operatorname{Res} u^{\Gamma_4}$.

6 Conclusion

In this paper we reviewed some important algebraic structures appearing in perturbative Algebraic Quantum Field Theory (pAQFT) on Minkowski spacetime [4] and we have shown how these relate to periods, usually investigated in a different context in Euclidean QFT in momentum space. The approach we advocate here provides a natural interpretation of these periods both in the mathematical and physical context. Mathematically, these correspond to distributional residues and are therefore intrinsic characterizations of scaling properties of certain class of distributions. Physically, they are relevant in computing the β-function. Note that, in our approach, the later characterization is independent of any regularization scheme. In fact, regularization is not needed at all and there is no need to recur to ill defined divergent expressions. Instead, the whole analysis is centered around the singularity structure of distributions that arise from taking powers of the Feynman propagator.

The main result of this paper is that distributional residues in pAQFT, corresponding to CK primitive graphs, are up to a factor that we compute, the same as Feynman periods in the CK framework (as conjectured in [4]). The remaining EG primitive graphs, which are not CK primitive, also give rise to multiples of the same periods.

For the future research it would be worth investigating the distributional residues arising in pAQFT on other Lorentzian manifolds. Some interesting results have already been obtained for de Sitter spacetime in [27]. All the fundamental structures of pAQFT presented in this paper generalize easily to curved spacetimes. The only difference is the form of the Feynman propagator (or rather the "Feynman-like" propagator H^{F}). The hope is that looking at more general propagators, one would obtain a richer structure of residues and some new structures would appear, which are not present in the Minkowski spacetime context (and would not be periods anymore).

In recent work, [16] investigated the dependence of Feynman amplitudes on variations of the metric in Riemannian setting and shows that integrals of non divergent Feynman amplitudes associated to closed graphs are functions on the moduli space of Riemannian metrics. It would be interesting to extend that work to the Lorentzian setting.

Acknowledgements I would like to thank ICMAT (Madrid) for hospitality and financial support. The ideas presented in this paper were developed during my stay in Madrid in 2014 as part of the "Research Trimester on Multiple Zeta Values, Multiple Polylogarithms, and Quantum Field Theory".

References

1. Aluffi, P., Marcolli, M.: Parametric Feynman integrals and determinant hypersurfaces. Adv. Theor. Math. Phys. **14**, 911–963 (2010)
2. Bastiani, A.: Applications différentiables et variétés différentiables de dimension infinie. J. d'Anal. math. **13**(1), 1–114 (1964)
3. Bergbauer, C., Brunetti, R., Kreimer, D., Renormalization and resolution of singularities. arXiv:hep-th/0908.0633
4. Brunetti, R., Dütsch, M., Fredenhagen, K.: Perturbative algebraic quantum field theory and the renormalization groups. Adv. Theor. Math. Phys. **13**(5), 1541–1599 (2009)
5. Brouder, C., Dang, N.V., Hélein, F.: A smooth introduction to the wavefront set. J. Phys. A **47**(44) 443001 (2014)
6. Bloch, S., Esnault, H., Kreimer, D.: On motives associated to graph polynomials. Commun. Math. Phys. **267**(1), 181–225 (2006)
7. Brunetti, R., Fredenhagen, K.: Microlocal analysis and interacting quantum field theories. Commun. Math. Phys. **208**(3), 623–661 (2000)
8. Bloch, S.: Motives associated to graphs. Jpn. J. Math. **2**(1), 165–196 (2007)
9. Bogner, C.: Mathematical aspects of feynman integrals. Ph.D. thesis, Mainz
10. Brown, F.: The massless higher-loop two-point function. Commun. Math. Phys. **287**(3), 925–958 (2009)
11. Brown, F.: On the periods of some Feynman integrals (2009). arXiv:math.AGsps0910.0114]

12. Bogner, C., Weinzierl, S.: Periods and Feynman integrals. J. Math. Phys. **50**(4), 042302 (2009)
13. Bogner, C., Weinzierl, S.: Feynman graph polynomials. Int. J. Mod. Phys. A **25**(13), 2585–2618 (2010)
14. Ceyhan, O., Marcolli, M.: Feynman integrals and motives of configuration spaces. Commun. Math. Phys. **313**, 35–70 (2012)
15. Ceyhan, O., Marcolli, M.: Algebraic renormalization and Feynman integrals in configuration spaces. Adv. Theor. Math. Phys. **18** (2013). https://doi.org/10.4310/ATMP.2014.v18.n2.a5
16. Dang, N.V.: Wick squares of the gaussian free field and Riemannian rigidity. http://math.univ-lyon1.fr/homes-www/dang/Riemann_invariants_GFF_rigidity.pdf
17. Dütsch, M., Fredenhagen, K.: Causal perturbation theory in terms of retarded products, and a proof of the action ward identity. Rev. Math. Phys. **16**(10), 1291–1348 (2004)
18. Dütsch, M., Fredenhagen, K., Keller, K.J., Rejzner, K.: Dimensional regularization in position space and a forest formula for Epstein-Glaser renormalization. J. Math. Phys. **55**(12), 122303 (2014)
19. Dütsch, M.: The scaling and mass expansion. Ann. Henri Poincare **16**, 163–188 (2015)
20. Epstein, H., Glaser, V.: The role of locality in perturbation theory. AHP **19**(3), 211–295 (1973)
21. Estrada, R., Kanwal, R.P.: Regularization, pseudofunction, and Hadamard finite part. J. Math. Anal. Appl. **141**(1), 195–207 (1989)
22. Freedman, D.Z., Johnson, K., Latorre, J.: Differential regularization and renormalization: a new method of calculation in quantum field theory. Nucl. Phys. B **371**(1–2), 353–414 (1992)
23. Gracia-Bondía, J.M.: Improved Epstein-Glaser renormalization in coordinate space I Euclidean framework. Math. Phys., Anal. Geom. **6**(1), 59–88 (2003)
24. Gracia-Bondía, J.M., Gutiérrez, H., Várilly, J.C.: Improved Epstein-Glaser renormalization in x-space versus differential renormalization. Nuclear Physics B **886**, 824–869 (2014)
25. Gracia-Bondía, J.M., Lazzarini, S.: Connes-Kreimer-Epstein-Glaser renormalization (2000). arXiv:hep-th/0006106
26. Hollands, S.: Renormalized quantum Yang-Mills fields in curved spacetime. Rev. Math. Phys. **20**, 1033–1172 (2008). arXiv:grspsqcsps705.3340v3
27. Hollands, S.: Correlators, Feynman diagrams, and quantum no-hair in De-Sitter spacetime. Commun. Math. Phys. **319**, 1–68 (2013)
28. Hörmander, L.: The analysis of the linear partial differential operators I: Distribution theory and Fourier analysis. Classics in Mathematics. Springer, Berlin (2003)
29. Hollands, S., Wald, R.M.: Local Wick polynomials and time ordered products of quantum fields in curved spacetime. Commun. Math. Phys. **223**(2), 289–326 (2001)
30. Hollands, S., Wald, R.M.: Existence of local covariant time ordered products of quantum fields in curved spacetime. Commun. Math. Phys. **231**(2), 309–345 (2002)
31. Hollands, S., Wald, R.M.: On the renormalization group in curved spacetime. Commun. Math. Phys. **237**, 123–160 (2002)
32. Hollands, S., Wald, R.M.: Axiomatic quantum field theory in curved spacetime. pp. 1–44
33. Itzykson, C., Zuber, J.-B.:Quantum field theory. Courier Corporation, 2006
34. Keller, K.J.: Dimensional regularization in position space and a forest formula for regularized epstein-glaser renormalization. Ph.D. thesis, University of Hamburg (2010). arXiv:math-ph/1006.2148
35. Kontsevich, M., Zagier, D.: Periods, 2001. In: Engquis, B., Schmid, W. (Eds.) Mathematics unlimited - 2001 and beyond, pp. 771–808 (2001)
36. Neeb, K.H.: Monastir summer school. Infinite-dimensional Lie groups, TU Darmstadt Preprint **2433** (2006)
37. Nikolov, N.M., Stora, R., Todorov, I.: Renormalization of massless Feynman amplitudes in configuration space. Rev. Math. Phys. **26**, 1430002 (2014)
38. Pinter, G.: The Hopf algebra structure of Connes and Kreimer in Epstein-Glaser renormalization. Prog. Theor. Phys. **54**, 227–233 (2000)
39. Prange, D.: Epstein-Glaser renormalization and differential renormalization. J. Phys. A Math. Gen. **32**(11), 2225 (1999)

40. Popineau, G., Stora, R.: A pedagogical remark on the main theorem of perturbative renormalization theory. Nucl. Phys. B **912**, 70–78 (2016)
41. Rejzner, K.: Perturbative Algebraic Quantum Field Theory. An introduction for Mathematicians, Mathematical Physics Studies, Springer (2016)
42. Schnetz, O.: Quantum periods: A census of φ^4-transcendentals. Commun. Num. Theor. Phys. **4**, 1–48 (2010)
43. Speer, E.R.: On the structure of analytic renormalization. Commun. Math. Phys. **23**(1), 23–36 (1971)
44. Stanley, R.P.: Spanning trees and a conjecture of Kontsevich. Ann. Comb. **2**(4), 351–363 (1998)
45. Steinmann, O.: Perturbation expansions in axiomatic field theory, vol. 11. Springer, Berlin (1971)
46. Stora, R.: Pedagogical experiments in renormalized perturbation theory, contribution to the conference . In: Theory of Renormalization and Regularization. Hesselberg, Germany (2002). February
47. Todorov, I.: Relativistic causality and position space renormalization. Nucl. Phys. B **912**, 79–87 (2016)
48. Tutte, W.: Graph theory, encyclopedia of mathematics, vol. 21. Cambridge University Press, Cambridge (1984)
49. Weinzierl, S.: The art of computing loop integrals. In: Binder, I., Kreimer, D. (eds.) Proceedings of the Workshop on Renormalization and Universality in Mathematical Physics. Providence, Am. Math. Soc. pp. 345–395 (2007)

Symmetril Moulds, Generic Group Schemes, Resummation of MZVs

Claudia Malvenuto and Frédéric Patras

Abstract The present article deals with various generating series and group schemes (not necessarily affine ones) associated with MZVs. Our developments are motivated by Ecalle's mould calculus approach to the latter. We propose in particular a Hopf algebra–type encoding of symmetril moulds and introduce a new resummation process for MZVs.

Keywords Multiple zeta values · Mould calculus · Quasi-shuffle

1 Introduction

Motivated by the study of multiple zeta values (MZVs), Jean Ecalle has introduced various combinatorial notions such as the ones of "symmetral moulds", "symmetrel moulds", "symmetril moulds" or "symmetrul moulds" [4, 7]. The first two are well-understood classical objects: they are nothing but characters on the shuffle algebra, resp. the quasi-shuffle algebra over the integers, both isomorphic to the algebra **QSym** of Quasi-symmetric functions. These two notions are closely related to the interpretation of properly regularized MZVs as real points of two prounipotent affine group schemes (associated respectively to the integral and power series representations of MZVs), whose interactions through double shuffle relations has given rise to the modern approach to MZVs (by Zagier, Deligne, Ihara, Racinet, Brown, Furusho and many others) [3, 10, 12, 18].

Although fairly natural from the point of view of MZVs (the resummation of MZVs into suitable generating series gives rise to a symmetril mould), the notion of

C. Malvenuto (✉)
Dipartimento di Matematica, Sapienza Università di Roma, Piazzale A. Moro, 5, 00198 Rome, Italy
e-mail: claudia@mat.uniroma1.it

F. Patras
Université Côte d'Azur UMR 7351 CNRS, Parc Valrose, 06108 Nice Cedex 02, France
e-mail: patras@unice.fr

J. I. Burgos Gil et al. (eds.), *Periods in Quantum Field Theory and Arithmetic*, Springer Proceedings in Mathematics & Statistics 314,
https://doi.org/10.1007/978-3-030-37031-2_14

symmetrility is more intriguing and harder to account for using classical combinatorial Hopf algebraic tools.

The aim of this article is accordingly threefold. We first show that Ecalle's mould calculus can be interpreted globally, beyond the cases of symmetral and symmetrel moulds, as a rephrasing of the theory of MZVs into the framework of prounipotent groups. However, these are not necessarily associated to affine group schemes (that is, to groups whose elements are characters on suitable Hopf algebras), at least in our interpretation and indeed, to account for symmetrility we introduce a new class of functors from commutative algebras to groups referred to as generic group schemes (because the elements of these groups are characters on suitably defined "generic" Hopf algebras). Second, we focus on this notion of symmetrility, develop systematic foundations for the notion and prove structure theorems for the corresponding algebraic structures. Third, we interpret Ecalle's resummation of MZVs by means of formal power series as the result of a properly defined Hopf algebra morphism. This construction is reminiscent in many aspects of the resummation of the various Green's functions in the functional calculus approach to quantum field theory or statistical physics, see e.g. [17]. This approach leads us to introduce a new resummation process, different from Ecalle's. The new process is more complex combinatorially but more natural from the group and Lie theoretical point of view: indeed, it encodes MZVs into new generating series that behave according to the usual combinatorics of tensor bialgebras and their dual shuffle bialgebras.

In the process, we introduce various Hopf algebraic structures that, besides being motivated by the mould calculus approach to MZVs, seem to be interesting on their own from a combinatorial algebra point of view.

We refer the readers not acquainted with classical arguments on the theory of MZVs to Cartier's Bourbaki seminar [2] that provides a short and mostly self contained treatment of the key notions.

2 Hopf Algebras

We recall first briefly the definition of a Hopf algebra and related notions. The reader is referred to [2] for details. All the maps we will consider between vector spaces will be assumed to be linear, unless otherwise stated explicitly. We will be mostly interested in graded or filtered connected Hopf algebras, and restrict therefore our presentation to that case.

Let $H = \bigoplus_{n\in\mathbf{N}} H_n$ be a graded vector space over a field k of characteristic zero. We will always assume that the H_n are finite dimensional. We write $H_{\leq n} := \bigoplus\limits_{m\leq n} H_m$, $H_{\geq n} := \bigoplus\limits_{m\geq n} H_m$ and $H^+ := \bigoplus_{n\in\mathbf{N}^*} H_n$. The graded vector space H is said to be connected if $H_0 \cong k$.

An associative and unital product $\mu : H \otimes H \to H$ on H (also written $h \cdot h' := \mu(h \otimes h')$) with unit map $\eta : k \to H_0 \subset H$ (so that for any $h \in H$ and $\eta(1) =:$

$\mathbf{1} \in H_0, \mathbf{1} \cdot h = h \cdot \mathbf{1} = h$) makes H a graded (resp. filtered) algebra if, for any integers n, m, $\mu(H_n \otimes H_m) \subset H_{n+m}$ (resp. $\mu(H_n \otimes H_m) \subset H_{\leq n+m}$).

Dualizing, a coassociative and counital coproduct $\Delta : H \to H \otimes H$ on H (also written using the abusive but useful Sweedler notation $\Delta(h) = h^{(1)} \otimes h^{(2)}$) with counit map $\nu : H \to k$ (with ν the null map on H^+) makes H a graded coalgebra if, for any integer n, $\Delta(H_n) \subset \bigoplus_{p+q=n} H_p \otimes H_q =: (H \otimes H)_n$. The coproduct (resp. the coalgebra H) is cocommutative if for any $h \in H$, $h^{(1)} \otimes h^{(2)} = h^{(2)} \otimes h^{(1)}$.

Recall that the category of associative unital algebras is monoidal: the tensor product of two associative unital algebras is a unital associative algebra. Assume that (μ,η) and (Δ,ν) equip H with the structure of an associative unital algebra and coassociative counital coalgebra: they equip H with the structure of a bialgebra if furthermore Δ and ν are maps of algebras (or equivalently μ and η are maps of coalgebras). The bialgebra H is called a Hopf algebra if furthermore there exists a endomorphism S of H (called the antipode) such that

$$\mu \circ (Id \otimes S) \circ \Delta = \mu \circ (S \otimes Id) \circ \Delta = \eta \circ \nu =: \varepsilon. \tag{1}$$

A bialgebra or a Hopf algebra is graded (resp. filtered) if it is a graded algebra and coalgebra (resp. a filtered algebra and a graded coalgebra). Graded and filtered connected bialgebras are automatically equipped with an antipode and are therefore Hopf algebras, and the two notions of Hopf algebras and bialgebras identify in that case, see e.g. [2] for the graded case, the filtered one being similar. This observation will apply to the bialgebras we will consider.

Example 1 The first example of a bialgebra occurring in the theory of MZVs is **QSym**, the quasi-shuffle bialgebra over the integers $\mathbf{N}^*$. The underlying graded vector space is the vector space over the sequences of integers (written as bracketed words) $[n_1 ... n_k]$. The bracket notation is assumed to behave multilinearly: for example, for two words $n_1 \dots n_k, m_1 \dots m_l$ and two scalars α, β

$$[\alpha\, n_1 \dots n_k + \beta\, m_1 \dots m_l] = \alpha[n_1 \dots n_k] + \beta[m_1 \dots m_l].$$

The words of length k span the degree k component of **QSym** (another graduation is obtained by defining the word $[n_1 ... n_k]$ to be of degree $n_1 + \cdots + n_k$). The graded coproduct is the deconcatenation coproduct:

$$\Delta([n_1 ... n_k]) := \sum_{i=0}^{k} [n_1 \dots n_i] \otimes [n_{i+1} \dots n_k].$$

The unital "quasi-shuffle" product $\uplus$ is the filtered product defined inductively by (the empty word identifies with the unit):

$$[n_1...n_k] \uplus [m_1 \dots m_l] := [n_1(n_2...n_k \uplus m_1 \dots m_l)] +$$
$$[m_1(n_1...n_k \uplus m_2 \dots m_l)] + [(n_1 + m_1)(n_2...n_k \uplus m_2 \dots m_l)].$$

For example,

$$[35] \uplus [1] = [3(5 \uplus 1) + 1(35) + 45] = [351] + [315] + [36] + [135] + [45].$$

Notice that, here and later on, we use in such formulas the shortcut notation $[3(5 \uplus 1)]$ for the concatenation of [3] with $[5 \uplus 1]$ (so that $[3(5 \uplus 1)] = [3(51 + 15 + 6)] = [351 + 315 + 36]$).

Algebra characters on **QSym** (i.e. unital multiplicative maps from **QSym** to a commutative unital algebra A) are called by Ecalle *symmetrel moulds*. The convolution product of linear morphisms from **QSym** to A,

$$f * g := m_A \circ (f \otimes g) \circ \Delta,$$

where m_A stands for the product in A, equips the set $G_{\mathbf{QSym}}(A)$ of A-valued characters with a group structure. Since **QSym** is a filtered connected commutative Hopf algebra, the corresponding functor $G_{\mathbf{QSym}}$ is (by Cartier's correspondence between group schemes and commutative Hopf algebras over a field of characteristic 0) a prounipotent affine group scheme. Properly regularized MZVs are real valued algebra characters on **QSym** and probably the most important example of elements in $G_{\mathbf{QSym}}(\mathbf{R})$ [2].

The quasi-shuffle bialgebra $\mathbf{QSh}(B)$ over an arbitrary commutative algebra $(B, \times)$ is defined similarly: the underlying vector space is $T(B) := \bigoplus\limits_{n \in \mathbf{N}} B^{\otimes n}$, the coproduct is the deconcatenation coproduct and the product is defined recursively by (we use a bracketed word notation for tensor products): $[b_1 \dots b_k] := b_1 \otimes ... \otimes b_k$,

$$[b_1 \dots b_k] \uplus [c_1 \dots c_l] := [b_1(b_2...b_k \uplus c_1 \dots c_l)] +$$
$$[c_1(b_1...b_k \uplus c_2 \dots c_l)] + [(b_1 \times c_1)(b_2...b_k \uplus c_2 \dots c_l)].$$

Example 2 The second example arises from the integral representation of MZVs. The corresponding graded vector space $T(x, y)$ is spanned by words in two variables x and y. The length of a word defines the grading. The coproduct is again the deconcatenation coproduct acting on words. The product $\sqcup\!\sqcup$ is the shuffle product, defined inductively on sequences by

$$a_1...a_k \sqcup\!\sqcup b_1 \dots b_l := a_1(a_2...a_k \sqcup\!\sqcup b_1 \dots b_l) + b_1(a_1...a_k \sqcup\!\sqcup b_2 \dots b_l).$$

The Hopf algebra $T(x, y)$ is called the shuffle bialgebra over the set $\{x, y\}$. Properly regularized MZVs are algebra characters on $T(x, y)$ (or on subalgebras thereof), but the regularization process fails to preserve simultaneously the shuffle and quasi-shuffle products [2].

Shuffle bialgebras over arbitrary sets X are defined similarly and denoted $\mathbf{Sh}(X)$ (see [8]). In the mould calculus terminology, a character on $\mathbf{Sh}(X)$ is called a *symmetral mould*. The shuffle bialgebra over $\mathbf{N}^*$, $\mathbf{Sh}(\mathbf{N}^*)$, is written simply $\mathbf{Sh}$ and will be called the *integer shuffle bialgebra*. It is isomorphic to $\mathbf{QSym}$ as a bialgebra [11].

Example 3 Rota-Baxter quasi-shuffle bialgebras. This third example departs from the two previous ones in that it is not a classical one but already illustrates a leading idea of mould calculus, namely: the application of fundamental identities of integral calculus to word-indexed formal power series. We refer e.g. to [5] and to the survey [6] for an overview of Rota–Baxter algebras and their relations to integral calculus and MZVs as well as for their general properties.

Let A be a commutative Rota-Baxter algebra of weight θ, that is a commutative algebra equipped with a linear endomorphism R such that

$$\forall x, y \in A, \;\; R(x)R(y) = R(R(x)y + xR(y) + \theta xy).$$

The term $R(x)y + xR(y) + \theta xy =: x *_R y$ defines a new commutative (and associative) product $*_R$ on A called the *double Rota-Baxter product*. We define the *double quasi-shuffle bialgebra* over a Rota–Baxter algebra A, $\mathbf{QSh}^R(A)$, as the bialgebra which identifies with $T(A) := \bigoplus\limits_{n\in\mathbf{N}} A^{\otimes n}$ as a vector space, equipped with the deconcatenation coproduct, and equipped with the following recursively defined product $⧢_R$:

$$x_1...x_k ⧢_R y_1...y_l := x_1(x_2...x_k ⧢_R y_1...y_l) + y_1(x_1...x_k ⧢_R y_2...y_l) + (x_1 *_R y_1)(x_2...x_k ⧢_R y_2...y_l).$$

The fact that $\mathbf{QSh}^R(A)$ is indeed a bialgebra follows from the general definition of the quasi-shuffle bialgebra over a commutative algebra A, see [9, 11].

Example 4 This fourth example (a particular case of the previous one) and the following one are the first concrete examples of the kind of Hopf algebraic structure showing up specifically in mould calculus. The definitions we introduce are inspired by the notion of *symmetrul mould* [7, p. 418] of which they aim at capturing the underlying combinatorial structure.

Let $\mathbb{R}[X]$ be equipped with the Riemann integral $R := \int_0^X$ viewed as a Rota–Baxter operator of weight zero. With the notation $a_i := X^{i-1}, \;\; i \in \mathbf{N}^*$ we get: $R(a_i) := \frac{a_{i+1}}{i}$ and

$$a_i *_R a_j = \frac{i+j}{ij} a_{i+j}.$$

This associative and commutative product gives rise to the following definition:

Definition 1 The bialgebra of quasi-symmetrul functions **QSul** is the quasi-shuffle bialgebra over the linear span of the integers $\mathbf{N}^*$ equipped with the product

$$[i] * [j] := \frac{i+j}{ij}[i+j].$$

Proposition 1 *The bialgebras* **QSym**, **Sh** *and* **QSul** *are isomorphic, the isomorphism ϕ from* **Sh** *to* **QSul** *is given by:*

$$\phi([n_1 \dots n_k]) := \sum_{\mu_1+\dots+\mu_i=k} \frac{(n_1+\dots+n_{\mu_1})\dots(n_{\mu_1+\dots+\mu_{i-1}+1}+\dots+n_{\mu_1+\dots+\mu_i})}{\mu_1!\dots\mu_i! n_1\dots n_k}.$$

$$[n_1+\dots+n_{\mu_1},\dots,n_{\mu_1+\dots+\mu_{i-1}+1}+\dots+n_{\mu_1+\dots+\mu_i}].$$

The proposition is an application of Hoffman's structure theorems for quasi-shuffle bialgebras [11]. It also follows from the combinatorial analysis of quasi-shuffle bialgebras understood as deformations of shuffle bialgebras in [9].

Example 5 The previous example gives the pattern for the notion of symmetrulity (and gives a hint for its analytic meaning). Let now V be a vector space with a distinguished basis $\mathcal{B} := (v_i)_{i\in I}$ and M a subsemigroup of $(\mathbb{R}^{>0}, +)$, the strictly positive real numbers. Let A be the linear span of $M \times \mathcal{B}$ whose elements (m, v) are represented $\binom{m}{v}$, to stick to the "bimould" calculus notation [7]. We set:

$$\binom{m_1}{v_1} * \binom{m_2}{v_2} := -\frac{1}{m_2}\binom{m_1+m_2}{v_1} - \frac{1}{m_1}\binom{m_1+m_2}{v_2}.$$

This product $*$ is associative and commutative.

Using the notation $\binom{m_1 \dots m_n}{v_1 \dots v_n}$ for the tensor product of the $\binom{m_i}{v_i}$ in $T(A)$, equipped with the deconcatenation coproduct, the following recursively defined product yields a bialgebra structure denoted **QSul**(M, V) on $T(A)$:

$$\binom{m_1 \dots m_n}{v_1 \dots v_n} ⊔⊔_{ul} \binom{p_1 \dots p_k}{w_1 \dots w_l} := \binom{m_1}{v_1}\left(\binom{m_2 \dots m_n}{v_2 \dots v_n} ⊔⊔_{ul} \binom{p_1 \dots p_k}{w_1 \dots w_l}\right)$$
$$+ \binom{p_1}{w_1}\left(\binom{m_1 \dots m_n}{v_1 \dots v_n} ⊔⊔_{ul} \binom{p_2 \dots p_k}{w_2 \dots w_l}\right)$$
$$-\left(\frac{1}{p_1}\binom{m_1+p_1}{v_1} + \frac{1}{m_1}\binom{m_1+p_1}{w_1}\right)\left(\binom{m_2 \dots m_n}{v_2 \dots v_n} ⊔⊔_{ul} \binom{p_2 \dots p_k}{w_2 \dots w_l}\right).$$

Definition 2 (*Corollary*) For B an arbitrary commutative algebra, denote the group of B-valued characters of **QSul**(M, V) by $G_{\mathbf{QSul}(M,V)}(B)$: then the functor $G_{\mathbf{QSul}(M,V)}$ is a prounipotent affine group scheme whose points are called symmetrul moulds [7].

Let us write **Sh**(M, V) for the shuffle bialgebra over A, and $G_{\mathbf{Sh}(M,V)}$ for the corresponding prounipotent affine group scheme, we have:

Theorem 1 *The prounipotent affine group schemes $G_{\mathbf{QSul}(M,V)}$ and $G_{\mathbf{Sh}(M,V)}$ are isomorphic. The isomorphism is induced by the bialgebra isomorphism ϕ between* **Sh**(M, V) *and* **QSul**(M, V) *defined by:*

$$\phi\binom{m_1 \dots m_n}{v_1 \dots v_n} := \sum_{\mu_1+\dots+\mu_i=n} \frac{(-1)^{n-i}}{\mu_1! \dots \mu_i!}.$$

$$\left(\sum_{i=1}^{\mu_1} \frac{1}{m_1 \dots m_{i-1} m_{i+1} \dots m_{\mu_1}} \binom{m_1+\dots+m_{\mu_1}}{v_i}\right) \dots$$

$$\left(\sum_{i=\mu_1+\dots+\mu_{i-1}+1}^{n} \frac{1}{m_{\mu_1+\dots+\mu_{i-1}+1} \dots m_{i-1} m_{i+1} \dots m_n} \binom{m_{\mu_1+\dots+\mu_{i-1}+1}+\dots+m_n}{v_i}\right).$$

This theorem follows once again from Hoffman's structure theorem for quasi-shuffle algebras by identification of the coefficients of the exponential isomorphism in the particular case under consideration.

In Ecalle's terminology, symmetrul moulds and symmetral moulds on $M \times V$ are canonically in bijection. Notice that whereas the definition of symmetrul moulds as characters on the Hopf algebra $\mathbf{QSul}(M, V)$ is essentially a group-theoretical interpretation of the definitions given in [7], the equivalence between the two notions of symmetrulity and symmetrality of Theorem 1 (and therefore also the precise formula for the isomorphism) is new at our best knowledge.

We do not insist further on the notion of symmetrulity that is relatively easy to handle group-theoretically as we just have seen, and will focus preferably on the one of symmetrility, whose signification for MZVs seems deeper, and for which a group-theoretical account is harder to obtain, since it does not seem possible to interpret symmetril moulds as elements of a prounipotent affine group scheme, but only as elements of a properly defined prounipotent group scheme.

3 Generic Bialgebras

Symmetril moulds, of which a formal definition will be given later, behave very much as characters on **QSym** or $T(x, y)$. There are even some conversion rules to move from one notion to the other, that we will explain later. Unfortunately, this notion of symmetrility fails to be accounted for by using a naive theory of characters on a suitable Hopf algebra. The aim of this section is to explain what has to be changed in the classical theory of Hopf algebras to make sense of the notion.

The constructions in this section are motivated by the two notions of twisted bialgebras (also called Hopf species) explored in [1, 14–16] and the one of constructions in the sense of Eilenberg and MacLane [13]. However, both the theory of constructions and vector species are too functorial to account for the very specific combinatorics of symmetrility, and we have to introduce for its proper understanding a different framework. In view of the similarities with the theory of constructions, we decided to keep however the terminology of "generic structures" used in [13].

Let X be a finite or countable alphabet, partitioned into subsets $X = \coprod_{i \in I} X_i$. We say that the partition is trivial if the X_i are singletons. A word over X (possibly empty) is said to be *generic* if it contains at most one letter in each X_i: for example, when $X = \{a, b, c\}$, abc is generic for the trivial partition $X = \{\{a\}, \{b\}, \{c\}\}$ but is not generic for the partition $X = \{\{a, b\}, \{c\}\}$. This means that no letter can appear twice. Similarly, tensor products of words are generic if they contain overall at most one letter in each X_i. Two generic tensor products of words, t, t', are said to be *in generic position* if $t \otimes t'$ is again a generic tensor product. Two linear combinations of generic tensor products of words $\sum_t \lambda_t t$, $\sum_{t'} \lambda_{t'} t'$ are in generic position if all the pairs (t, t') are. The underlying word $u(t)$ of a tensor product t of words is the word obtained by concatenating its components: observe that different tensors might have the same underlying word. For instance $u(x_1 \otimes x_2 x_4 \otimes x_3) = u(x_1 x_2 \otimes x_4 x_3) = x_1 x_2 x_4 x_3$, so that $u(t)$ is generic if and only if t is generic.

Definition 3 The category $\mathbf{Gen}_X$ of generic expressions over X is the smallest linear (i.e. such that Hom-sets are k-vector spaces) subcategory of the category of vector spaces:

- containing the null vector space,
- containing the one-dimensional vector spaces V_t generated by generic tensor products of words t,
- closed by direct sums (although this will not be the case in the examples we will consider, multiple copies of the V_t can be allowed, the following rules are applied to each of these copies),

and such that furthermore Hom-sets contain:

- for t, t' two generic tensors with $u(t) = u(t')$, the map from V_t to $V_{t'}$ induced by $f(t) := t'$,
- the maps induced by substitutions of the letters inside the blocks X_i,
- the maps obtained by erasing letters in the tensor products (e.g. the map induced by $f(x_1 \otimes x_2 x_4 \otimes x_3) := x_1 \otimes x_4$) (the example refers to the case where $X = \{x_i\}_{i \in \mathbf{N}^*}$, with the trivial partition).

Most importantly for our purposes, $\mathbf{Gen}_X$ is equipped with a symmetric monoidal category structure by the generic tensor product $\hat{\otimes}$ defined on the V_t by

$$V_t \hat{\otimes} V_{t'} := \begin{cases} V_{t \otimes t'} & \text{if } t \otimes t' \text{ is generic,} \\ 0 & \text{otherwise.} \end{cases}$$

The generic tensor product is extended to direct sums by the rule

$$(A \oplus B)\hat{\otimes}(C \oplus D) = A\hat{\otimes}C \oplus A\hat{\otimes}D \oplus B\hat{\otimes}C \oplus B\hat{\otimes}D.$$

Notice, for further use, the canonical embedding $A\hat{\otimes}B \hookrightarrow A \otimes B$.

The reader familiar with homological algebra will have recognized the main ingredients of the theory of constructions [13]. *Generic algebras, coalgebras, Hopf algebras, Lie algebras*, and so on, are, by definition, algebras, coalgebras, Hopf algebras, Lie algebras, and so on, in a given $\mathbf{Gen}_X$. For example, a generic algebra A without unit is an object of $\mathbf{Gen}_X$ equipped with an associative product map μ from $A\hat{\otimes}A$ to A. Notice that μ can be viewed alternatively as a partially defined product map on A (it is defined only on elements in $A \otimes A$ in generic position and linear combinations thereof).

We will study from now on only *standard generic bialgebras* H, by which we mean that $(H, \pi\,\Delta)$ is a generic bialgebra with product π and coproduct Δ such that:

- $H = \bigoplus\limits_{n\in\mathbf{N}} H_n$, where $H_0 = V_\emptyset$ is identified with the ground field k and $\emptyset$ behaves as a unit/counit for the product and the coproduct,
- the coproduct is graded,
- the product satisfies the filtering condition: $\forall k, l > 0,\quad \pi(H_k \otimes H_l) \subset \bigoplus_{0<n\leq k+l} H_n$.

These bialgebras behave as the analogous usual bialgebras (the same arguments and proofs apply, we refer e.g. to [2] for the classical case). In particular such a bialgebra is equipped with a convolution product of linear endomorphims: for arbitrary $f, g \in Hom_{\mathbf{Gen}_X}(H, H)$, $f * g := \pi \circ (f\hat{\otimes}g) \circ \Delta$. The projection u from H to H_0 orthogonally to the $H_i,\ i \geq 1$ is a unit for $*$. Convolution of linear forms on H is defined similarly.

The existence of an antipodal map, that is a convolution inverse S to the identity map I, follows from the identity

$$S = (u + (I - u))^{*\,-1} = u + \sum_{n>0}(-1)^n(I - u)^{*n}, \tag{2}$$

where the rightmost sum restricts to a finite sum when S is acting on a graded component H_n since the coproduct is graded. In particular, a standard generic bialgebra H is automatically a generic Hopf algebra.

Since $A\hat{\otimes}B \subset A \otimes B$, one can define morphisms from an algebra, bialgebra, Hopf algebra... in Gen_X to a classical algebra, bialgebra, Hopf algebra... We will call such morphisms *regularizing morphisms*. For example, a regularizing morphism between a standard generic bialgebra H equipped with the product μ and the coproduct Δ and a graded Hopf algebra H' equipped with the product μ' and the coproduct Δ' is a morphism of graded vector spaces f that maps the unit $\emptyset$ of H to the unit $1 \in H_0'$ of H' and such that, for any h, h' in generic position in H,

$$f(\mu(h\hat{\otimes}h')) = \mu'(f(h) \otimes f(h')),\ (f \otimes f) \circ \Delta(h) = \Delta'(f(h)).$$

Example 6 A first example of a standard generic bialgebra will look familiar to readers acquainted with the theory of free Lie algebras and Reutenauer's monograph [19]. Let $X = \{1, \ldots, n\}$ be equipped with the trivial partition. Then, let $T_k^g(X)$ be the linear span of generic words of length k, we have: $T^g(X) = \bigoplus_{k\in\mathbf{N}} T_k^g(X) = \bigoplus_{k\leq n} T_k^g(X)$; the highest order non trivial component of this direct sum, $T_n^g(X)$, is usually called the multilinear part of the tensor algebra over X in the literature. Concatenation of words defines a map from $T_k^g(X)\hat{\otimes}T_l^g(X)$ to $T_{k+l}^g(X)$ and a generic algebra structure on $T^g(X) = \bigoplus_{n\in\mathbf{N}} T_n^g(X)$. Similarly, the usual unshuffling of words Δ (the coproduct dual to the one introduced in example 2) defines, when restricted to generic words, a generic coalgebra structure, and, together with the concatenation product, a standard generic bialgebra structure on $T^g(X)$. The generic Lie algebra of primitive elements of $T^g(X)$ is defined as usual: $Prim(T^g(X)) := \{w \in T^g(X), \Delta(w) = w \otimes 1 + 1 \otimes w\}$, its highest order non trivial component with respect to the graduation by the length of words is simply the multilinear part of the usual free Lie algebra over X.

Dually, the shuffle product and the deconcatenation product (as in Example 2) define a (dual) standard generic bialgebra structure on $T^g(X)$, that will be named the generic shuffle bialgebra over X and denoted $\mathbf{Sh}^g(X)$. We write simply $\mathbf{Sh}^g$ for $\mathbf{Sh}^g(\mathbf{N}^*)$.

This example is particularly easy to understand: the embedding of $T^g(X)$ into the usual tensor algebra $T(X)$ over X is a regularizing morphism, and all our assertions are direct consequences of the behaviour of $T(X)$ as exposed e.g. in [19].

Definition 4 Let H be a standard generic bialgebra. For B an arbitrary commutative algebra, a B-valued character on H is, by definition, a unital multiplicative map from H to B, that is a map ϕ such that:

- $\phi(\emptyset) = 1$,
- for any h_1, h_2 in generic position, writing $h_1 \cdot h_2 := \pi(h_1\hat{\otimes}h_2)$

$$\phi(h_1 \cdot h_2) = \phi(h_1)\phi(h_2). \tag{3}$$

Proposition 2 *Let H be a standard generic bialgebra. The set $G_H(B)$ of B-valued characters is equipped with a group structure by the convolution product $*$. The corresponding functor G_H from commutative algebras over the reference ground field k to groups is called, by analogy with the classical case, a generic group scheme.*

Indeed, we have, for any $\phi, \phi' \in G_H(B)$, and any $h, h' \in H^+ := \bigoplus_{n>0} H_n$ in generic position:

$$\phi * \phi'(\emptyset) = \phi(\emptyset)\phi'(\emptyset) = 1,$$

$$\phi * \phi'(h \cdot h') = \phi(h^{(1)} \cdot h'^{(1)})\phi'(h^{(2)} \cdot h'^{(2)})$$

$$= \phi(h^{(1)})\phi'(h^{(2)})\phi(h'^{(1)})\phi'(h'^{(2)})$$

$$= \phi * \phi'(h) \cdot \phi * \phi'(h'),$$

where we used a Sweedler-type notation $\Delta(h) = h^{(1)}\hat{\otimes}h^{(2)}$.

Similarly, $\phi \circ S$ is the convolution inverse of ϕ since: $((\phi \circ S) * \phi)(\emptyset) = \phi(\emptyset)^2 = 1$ and, for h as above,

$$((\phi \circ S) * \phi)(h) = \phi(S(h^{(1)}))\phi(h^{(2)}) = \phi(S(h^{(1)}) \cdot h^{(2)}) = \phi \circ u(h) = 0.$$

Notice that, contrary to the classical case, the identity $\phi(S(h^{(1)}))\phi(h^{(2)}) = \phi(S(h^{(1)}) \cdot h^{(2)})$ is not straightforward, since identity Eq. 3 holds only under the assumption that h_1, h_2 are in generic position. Here, we can apply the identity because S, in view of Eq. 2, can be written on each graded component as a sum of convolution powers I^{*k} of the identity map. It is then enough to check that, given $h \in H^+$, $I^{*k}(h^{(1)}) \otimes h^{(2)}$ can be written as a linear combination of tensor products $w \otimes w'$, where w, w' are in generic position, which follows from the definition of the convolution product $*$ and the coassociativity of Δ.

4 Symmetril Moulds and Generic Group Schemes

We come now to the main examples of generic structures in view of the scope of the present article—symmetrility properties. This section aims at abstracting the key combinatorial features of symmetrility in order to study them and link them with classical combinatorial objects, such as quasi-symmetric functions. The next section will move forward by sticking closer to Ecalle's study of MZVs, linking symmetrility phenomena to the resummation of MZVs.

Definition 5 Let $X = \mathbf{N}^*$, equipped with the trivial partition. We define the *generic divided* quasi-shuffle bialgebra over $\mathbf{N}^*$, $\mathbf{QSh}^g_d$, as the generic bialgebra which identifies with $T^g(\mathbf{N}^*)$ as a vector space, equipped with the deconcatenation coproduct, and equipped with the following recursively defined product $\overline{\sqcup\!\sqcup}$ (elements of $T^g(\mathbf{N}^*)$ are written using a bracketed word notation):

$$[n_1...n_k]\overline{\sqcup\!\sqcup}[m_1...m_l] := [n_1(n_2...n_k\overline{\sqcup\!\sqcup}m_1...m_l)] + [m_1(n_1...n_k\overline{\sqcup\!\sqcup}m_2...m_l)] + \frac{1}{n_1 - m_1}\{[n_1(n_2...n_k\overline{\sqcup\!\sqcup}m_2...m_l)] - [m_1(n_2...n_k\overline{\sqcup\!\sqcup}m_2...m_l)]\}.$$

The elements of the groups $G_{\mathbf{QSh}^g_d}(B)$ are called symmetril moulds (over $\mathbf{N}^*$).

Proving that $\mathbf{QSh}^g_d$ is indeed a Hopf algebra in $\mathbf{Gen}_X$ is not entirely straightforward and is better stated at a more general level, by mimicking for generic structures the theory of quasi-shuffle algebras.

Definition 6 (*Proposition*) Let X be a partitioned alphabet and assume that $*$ equips $k\langle X\rangle$, the linear span of X, with the structure of a generic commutative algebra. Then, the generic quasi-shuffle bialgebra denoted $\mathbf{QSh}^g_*(X)$ over $(k\langle X\rangle, *)$ is, by definition, the generic bialgebra whose underlying generic coalgebra is $T^g(X)$ equipped with the deconcatenation coproduct Δ, and whose commutative product is defined inductively (for words satisfying the genericity conditions) by:

$$[n_1...n_k]\overline{\sqcup\!\sqcup}[m_1...m_l] := [n_1(n_2...n_k\overline{\sqcup\!\sqcup}m_1...m_l)] + [m_1(n_1...n_k\overline{\sqcup\!\sqcup}m_2...m_l)] + [(n_1 * m_1)(n_2...n_k\overline{\sqcup\!\sqcup}m_2...m_l)].$$

The fact that the product is well defined and sends two generic words in generic position on a linear combination of generic words follows from the very definition of the category of generic expressions.

The associativity of the product follows by induction on the total length $k + l + q$ from the identity of the expansion:

$$\begin{aligned}
&[(n_1...n_k\overline{\sqcup\!\sqcup}m_1...m_l)\overline{\sqcup\!\sqcup}p_1..p_q] = [n_1(n_2...n_k\overline{\sqcup\!\sqcup}m_1...m_l\overline{\sqcup\!\sqcup}p_1..p_q)]\\
&+ [p_1((n_1(n_2...n_k\overline{\sqcup\!\sqcup}m_1...m_l))\overline{\sqcup\!\sqcup}p_2..p_q)] + [(n_1 * p_1)(n_2...n_k\overline{\sqcup\!\sqcup}m_1...m_l\overline{\sqcup\!\sqcup}p_1..p_q)]\\
&+ [m_1(n_1...n_k\overline{\sqcup\!\sqcup}m_2...m_l\overline{\sqcup\!\sqcup}p_1..p_q)] + [p_1((m_1(n_1...n_k\overline{\sqcup\!\sqcup}m_2...m_l))\overline{\sqcup\!\sqcup}p_2..p_q)]\\
&+ [(m_1 * p_1)(n_1...n_k\overline{\sqcup\!\sqcup}m_2...m_l\overline{\sqcup\!\sqcup}p_2...p_q)] + [(n_1 * m_1)(n_2...n_k\overline{\sqcup\!\sqcup}m_2...m_l\overline{\sqcup\!\sqcup}p_1..p_q)]\\
&+ [p_1((n_1 * m_1)(n_2...n_k\overline{\sqcup\!\sqcup}m_2...m_l))\overline{\sqcup\!\sqcup}p_2..p_q)]\\
&+ [(n_1 * m_1 * p_1)(n_1...n_k\overline{\sqcup\!\sqcup}m_2...m_l\overline{\sqcup\!\sqcup}p_1...p_q)]\\
&= [n_1(n_2...n_k\overline{\sqcup\!\sqcup}m_1...m_l\overline{\sqcup\!\sqcup}p_1..p_q)] + [m_1(n_1...n_k\overline{\sqcup\!\sqcup}m_2...m_l\overline{\sqcup\!\sqcup}p_1..p_q)]\\
&+ [p_1(n_1...n_k\overline{\sqcup\!\sqcup}m_1...m_l\overline{\sqcup\!\sqcup}p_2...p_q)] + [(n_1 * m_1)(n_2...n_k\overline{\sqcup\!\sqcup}m_2...m_l\overline{\sqcup\!\sqcup}p_1..p_q)]\\
&+ [(n_1 * p_1)(n_2...n_k\overline{\sqcup\!\sqcup}m_1...m_l\overline{\sqcup\!\sqcup}p_2...p_q)] + [(m_1 * p_1)(n_1...n_k\overline{\sqcup\!\sqcup}m_2...m_l\overline{\sqcup\!\sqcup}p_2...p_q)]\\
&+ [(n_1 * m_1 * p_1)(n_1...n_k\overline{\sqcup\!\sqcup}m_2...m_l\overline{\sqcup\!\sqcup}p_2...p_q)],
\end{aligned}$$

with the same symmetric expansion in the n_i, m_i, p_i for

$$[n_1...n_k\overline{\sqcup\!\sqcup}(m_1...m_l\overline{\sqcup\!\sqcup}p_1...p_q)].$$

The compatibility of the deconcatenation coproduct with the product is obtained similarly and follows the same pattern as the proof that usual quasi-shuffle algebras over commutative algebras are indeed equipped with a Hopf algebra structure by the deconcatenation coproduct [9, 11], and is omitted.

We can now conclude that $\mathbf{QSh}_d^g$ is indeed a generic bialgebra from the Lemma:

Lemma 1 *The product* $*$ *defined by*

$$[n] * [m] := \frac{1}{n - m}([n] - [m])$$

equips $k < \mathbf{N}^* >$ *with the structure of a generic commutative algebra.*

Indeed, for distinct m, n, p,

$$[(n * m) * p] = \frac{1}{n - m}([n] - [m]) * [p] = \frac{1}{(n - m)(n - p)}[n]$$

$$+\frac{1}{(m-n)(m-p)}[m]+(\frac{1}{(n-m)(p-n)}+\frac{1}{(m-n)(m-p)})[p]$$

$$=\frac{1}{(n-m)(n-p)}[n]+\frac{1}{(m-n)(m-p)}[m]+\frac{1}{(p-n)(p-m)}[p]$$

which is equal to the same symmetric expression for $[n*(m*p)]$.

For later use, we also calculate iterated products in $k < \mathbf{N}^* >$.

Lemma 2 *For distinct* $n_1, \ldots, n_k \in N^*$ *we have*

$$[n_1]*\cdots*[n_k] := \sum_{i=1}^{k} \frac{[n_i]}{\prod_{j\neq i}(n_i-n_j)}$$

Let us assume that the Lemma holds for $k \leq p$ and prove it by induction. Since the product $*$ is commutative, it is enough to show that the coefficient of $[n_{p+1}]$ in $[n_1]*\cdots*[n_{p+1}]$ is given by $\frac{1}{\prod_{j\leq p}(n_{p+1}-n_j)}$. Equivalently, we have to show that $\alpha = 1$, where

$$\alpha = \sum_{i=1}^{p} \frac{\prod_{j\leq p}(n_{p+1}-n_j)}{(n_{p+1}-n_i)\prod_{j\neq i, j\leq p}(n_i-n_j)} = \sum_{i=1}^{p}\prod_{j\neq i, j\leq p}\frac{(n_{p+1}-n_j)}{(n_i-n_j)}.$$

Notice that the induction hypothesis amounts to assuming that the following two identities hold for arbitrary distinct integers $m_1, \ldots, m_p$ (the two identities are shown to be equivalent by multiplying the i-th term of the sum in the left hand side of the first identity by $(m_p-m_i)/(m_p-m_i)$)

$$\sum_{i=1}^{p-1}\prod_{j\neq i, j\leq p-1}\frac{(m_p-m_j)}{(m_i-m_j)} = 1, \sum_{i=1}^{p}\prod_{j\neq i, j\leq p}\frac{1}{(m_i-m_j)} = 0.$$

We get:

$$\alpha = \sum_{i=1}^{p-1}\left(\prod_{j\neq i, j\leq p}\frac{(n_{p+1}-n_j)}{(n_i-n_j)}\right)+\prod_{j\leq p-1}\frac{(n_{p+1}-n_j)}{(n_p-n_j)}$$

$$=\sum_{i=1}^{p-1}\left(\frac{\prod_{j\neq i, j\leq p-1}(n_{p+1}-n_j)}{\prod_{j\neq i, j\leq p}(n_i-n_j)}\right)\left((n_{p+1}-n_i)+(n_i-n_p)\right)+\prod_{j\leq p-1}\frac{(n_{p+1}-n_j)}{(n_p-n_j)}$$

$$= \left(\sum_{i=1}^{p-1} \left(\frac{\prod_{j\neq i, j\leq p-1} (n_{p+1} - n_j)}{\prod_{j\neq i, j\leq p} (n_i - n_j)} \right) (n_{p+1} - n_i) + \prod_{j\leq p-1} \frac{(n_{p+1} - n_j)}{(n_p - n_j)} \right)$$

$$+ \left(\sum_{i=1}^{p-1} \left(\frac{\prod_{j\neq i, j\leq p-1} (n_{p+1} - n_j)}{\prod_{j\neq i, j\leq p} (n_i - n_j)} \right) (n_i - n_p) \right)$$

$$= \sum_{i=1}^{p} \left(\frac{1}{\prod_{j\neq i, j\leq p} (n_i - n_j)} \right) \cdot \prod_{j\leq p-1} (n_{p+1} - n_j) + \sum_{i=1}^{p-1} \left(\prod_{j\neq i, j\leq p-1} \frac{(n_{p+1} - n_j)}{(n_i - n_j)} \right) = 0 + 1 = 1,$$

where the last identity follows from the induction hypothesis.

Theorem 2 *The following map ψ defines a linear embedding of $\mathbf{QSh}_d^g$ into $\mathbf{Sh}$ and is a regularizing bialgebra map.*

$$\psi([n_1 \dots n_k]) := \sum_{\mu_1+\cdots+\mu_i=k} \frac{(-1)^{k-i}}{\mu_1 \dots \mu_i} \left(\sum_{j=1}^{\mu_1} \frac{[n_j]}{\prod_{l\neq j, l\leq \mu_1} (n_j - n_l)} \right) \cdots$$

$$\cdots \left(\sum_{j=\mu_1+\cdots+\mu_{i-1}+1}^{k} \frac{[n_j]}{\prod_{l\neq j, \mu_1+\cdots+\mu_{i-1}+1\leq l\leq k} (n_j - n_l)} \right)$$

In particular, the product and coproduct maps on $\mathbf{QSh}_d^g$ are mapped to the product and coproduct on **Sh**.

The Theorem can be rephrased internally to the category $\mathbf{Gen}_{\mathbf{N}^*}$. This is because the image of ψ identifies with the subspace $T^g(\mathbf{N}^*)$ of **Sh** (the latter identifying with $T(\mathbf{N}^*)$ as a graded vector space).

Corollary 1 *The standard generic bialgebras $\mathbf{QSh}_d^g$ and $\mathbf{Sh}^g$ are isomorphic under ψ.*

The theorem is an extension to the generic case of the Hoffman isomorphism between shuffle and quasi-shuffle bialgebras. Following [9, 11], the proof of the isomorphism relies only on the combinatorics of partitions and on a suitable lift to formal power series of natural coalgebra endomorphisms of shuffle bialgebras (we refer to [9] for details). Let us show here that these arguments still hold in the generic framework.

Let $P(X) = \sum_{i=1}^{\infty} p_i X^i$ be a formal power series $X\mathbf{Q}[[X]]$. This power series induces a generic coalgebra endomorphism ϕ_P of $T^g(\mathbf{N}^*)$ equipped with the deconcatenation coproduct: on an arbitrary generic tensor $[n_1 \dots n_k] \in T^g(\mathbf{N}^*)$ the action is given by

$$\phi_P([n_1 \dots n_k]) = \sum_{j=1}^{k} \sum_{i_1+\dots+i_j=k} p_{i_1} \dots p_{i_j}([n_1] * \dots * [n_{i_1}]) \otimes \dots \otimes ([n_{i_1+\dots+i_{j-1}+1}] * \dots * [n_k]), \tag{4}$$

where we recall that $[n] * [m] := \frac{[n]-[m]}{n-m}$. When $p_1 \neq 0$, ϕ_P is bijective (by a triangularity argument), and a coalgebra automorphism of $T^g(\mathbf{N}^*)$.

Let us show now that, for arbitrary $P(X),\ Q(X) \in X\mathbf{Q}[[X]]$,

$$\phi_P \circ \phi_Q = \phi_{P \circ Q}, \tag{5}$$

where $(P \circ Q)(X) := P(Q(X))$. We have indeed, for an arbitrary sequence of distinct integers $n_1, \dots, n_k$:

$$\phi_P \circ \phi_Q(n_1 \dots n_k) =$$

$$= \phi_P(\sum_{j=1}^{k} \sum_{i_1+\dots+i_j=k} q_{i_1} \dots q_{i_k}(n_1 * \dots * n_{i_1}) \otimes \dots \otimes (n_{i_1+\dots+i_{j-1}+1} * \dots * n_k))$$

$$= \sum_{j=1}^{k} \sum_{l=1}^{j} \sum_{h_1+\dots+h_l=j} \sum_{i_1+\dots+i_j=k} p_{h_1} \dots p_{h_l} q_{i_1} \dots q_{i_k}(n_1 * \dots * n_{i_1+\dots+i_{h_1}}) \otimes$$

$$\dots \otimes (n_{i_1+\dots+i_{h_1+\dots+h_{l-1}}+1} * \dots * n_k)$$

$$= \phi_{P(Q)}(n_1 \dots n_k).$$

The proof of the theorem follows: $\psi = \phi_{log}$ has for inverse $\rho = \phi_{exp}$, which maps isomorphically $\mathbf{Sh}^g$ to $\mathbf{QSh}_d^g$ (Hoffman's combinatorial argument in the classical case in [11] applies *mutatis mutandis* when restricted to generic tensors).

5 Resummation of MZVs

In order to resum MZVs into formal power series equipped with interesting group-theoretical operations and structures, let us introduce first a formal analogue of the standard generic bialgebra $\mathbf{QSh}_d^g$ studied previously. Here, "formal" means that numbers and sequences of numbers are replaced by formal power series and words over an alphabet. Proofs of the properties and structure theorems are similar to the ones for $\mathbf{QSh}_d^g$ and are omitted. Our definitions and constructions are motivated by [7].

Definition 7 Let $V = \{v_i\}_{i \in \mathbf{N}^*}$, equipped with the trivial partition. We define the generic divided quasi-shuffle bialgebra over V, $\mathbf{QSh}_d^g(V)$, as the generic bialgebra defined over $k_V := k((V))$, the field of fractions of the ring of formal power series over V, which identifies with $T^g(V)$ as a vector space, equipped with the deconcatenation coproduct, and equipped with the following recursively defined product $\overline{\sqcup\!\sqcup}$ (elements of $T^g(V)$ are written using a bracketed word notation):

$$[v_{i_1}...v_{i_k}]\overline{\sqcup\!\sqcup}[v_{i_{k+1}}...v_{i_{k+l}}] := [v_{i_1}(v_{i_2}...v_{i_k}\overline{\sqcup\!\sqcup}v_{i_{k+1}}...v_{i_{k+l}})]$$
$$+ [v_{i_{k+1}}(v_{i_1}...v_{i_k}\overline{\sqcup\!\sqcup}v_{i_{k+2}}...v_{i_{k+l}})]$$
$$+ \frac{1}{v_{i_1} - v_{i_{k+1}}}\{[v_{i_1}(v_{i_2}...v_{i_k}\overline{\sqcup\!\sqcup}v_{i_{k+2}}...v_{i_{k+l}})] - [v_{i_{k+1}}(v_{i_2}...v_{i_k}\overline{\sqcup\!\sqcup}v_{i_{k+2}}...v_{i_{k+l}})]\},$$

where $[v_{i_1}...v_{i_k}]$ and $[v_{i_{k+1}}...v_{i_{k+l}}]$ are in generic position (so that $\frac{1}{v_{i_1}-v_{i_{k+1}}}$ is well-defined).

The elements of the groups $G_{\mathbf{QSh}_d^g(V)}(B)$, where B runs over algebras over k_V associated to the generic group scheme $G_{\mathbf{QSh}_d^g(V)}$ over k_V, are called symmetril moulds (over V).

Let us denote $\mathbf{Sh}_V^g$ the *generic shuffle bialgebra* (or g-shuffle bialgebra) over V with k_V as a field of coefficients. Corollary 1 generalizes to $\mathbf{QSh}_d^g(V)$ and $\mathbf{Sh}_V^g$: the two g-bialgebras are isomorphic under ψ_V:

$$\psi_V([v_1 \dots v_k]) := \sum_{\mu_1+\dots+\mu_i=k} \frac{(-1)^{k-i}}{\mu_1 \dots \mu_i} \left(\sum_{j=1}^{\mu_1} \frac{[v_j]}{\prod_{l\neq j, l\leq \mu_1} (v_j - v_l)} \right) \dots$$
$$\dots \left(\sum_{j=\mu_1+\dots+\mu_{i-1}+1}^{k} \frac{[v_j]}{\prod_{l\neq j, \mu_1+\dots+\mu_{i-1}+1\leq l\leq k} (v_j - v_l)} \right). \quad (6)$$

Let us denote now $\mathbf{QSym}_V$ the completion (with respect to the grading) of the bialgebra of quasi-symmetric functions over the base field k_V. Since properly regularized MZVs at positive values are characters on **QSym**, generating series for MZVs such as

$$\sum_{n_1,\dots,n_k\geq 1} v_1^{n_1-1} \dots v_k^{n_k-1} \zeta(n_1, \dots, n_k)$$

and the study of their algebraic structure can be lifted to $\mathbf{QSym}_V$. Let us show how this idea translates group-theoretically.

Theorem 3 *The following morphism γ is a regularizing bialgebra map from $\mathbf{QSh}_d^g(V)$ to $\mathbf{QSym}_V$:*

$$\gamma([v_{i_1} \dots v_{i_k}]) := \sum_{n_1,\dots,n_k\geq 1} v_{i_1}^{n_1-1} \dots v_{i_k}^{n_k-1} \cdot [n_1 \dots n_k]. \quad (7)$$

Notice first that γ is, by its very definition, multiplicative for the concatenation product:

$$\gamma\big([v_{i_1}\dots v_{i_k}]\big)=\gamma([v_{i_1}])\cdot\gamma([v_{i_2}\dots v_{i_k}])=\gamma([v_{i_1}])\cdot\gamma([v_{i_2}])\dots\gamma([v_{i_k}]),\qquad(8)$$

from which it follows that γ is a coalgebra map (recall that the later is induced on $\mathbf{QSh}_d^g(V)$ and $\mathbf{QSym}_V$ by deconcatenation).

Let us prove that, for any $v_{i_1}...v_{i_k}$, $v_{i_{k+1}}...v_{i_{k+l}}$ in generic position, we have

$$\gamma\Big([v_{i_1}\dots v_{i_k}]\overline{\sqcup\!\sqcup}[v_{i_{k+1}}\dots v_{i_{k+l}}]\Big)=\gamma\Big([v_{i_1}\dots v_{i_k}]\Big)\uplus\gamma\Big([v_{i_{k+1}}\dots v_{i_{k+l}}]\Big)$$

by induction on $k+l$. So, let v_{i_0} an element of V distinct from v_{i_1}, ..., $v_{i_{k+l}}$. We get from (7):

$$\gamma\Big([v_{i_0}\dots v_{i_k}]\overline{\sqcup\!\sqcup}[v_{i_{k+1}}\dots v_{i_{k+l}}]\Big)=\gamma\Big([v_{i_0}(v_{i_1}...v_{i_k}\overline{\sqcup\!\sqcup}v_{i_{k+1}}...v_{i_{k+l}})]\Big)$$
$$+\gamma\Big([v_{i_{k+1}}(v_{i_0}...v_{i_k}\overline{\sqcup\!\sqcup}v_{i_{k+2}}...v_{i_{k+l}})]\Big)$$
$$+\gamma\Big(\frac{1}{v_{i_0}-v_{i_{k+1}}}\{[v_{i_0}(v_{i_1}...v_{i_k}\overline{\sqcup\!\sqcup}v_{i_{k+2}}...v_{i_{k+l}})]-[v_{i_{k+1}}(v_{i_1}...v_{i_k}\overline{\sqcup\!\sqcup}v_{i_{k+2}}...v_{i_{k+l}})]\}\Big).$$

From Eq. 8 and the induction hypothesis, we get:

$$\gamma\Big([v_{i_0}(v_{i_1}...v_{i_k}\overline{\sqcup\!\sqcup}v_{i_{k+1}}...v_{i_{k+l}})]\Big)=\gamma\Big([v_{i_0}]\Big)\gamma\Big([v_{i_1}...v_{i_k}\overline{\sqcup\!\sqcup}v_{i_{k+1}}...v_{i_{k+l}})]\Big)$$
$$=\gamma\Big([v_{i_0}]\Big)\gamma\Big([v_{i_1}\dots v_{i_k}]\Big)\uplus\gamma\Big([v_{i_{k+1}}\dots v_{i_{k+l}}]\Big)$$

and similarly

$$\gamma\Big([v_{i_{k+1}}(v_{i_0}...v_{i_k}\overline{\sqcup\!\sqcup}v_{i_{k+2}}...v_{i_{k+l}})]\Big)=\gamma\Big([v_{i_{k+1}}]\Big)\Big(\gamma\Big([v_{i_0}\dots v_{i_k}]\Big)\uplus\gamma\Big([v_{i_{k+2}}\dots v_{i_{k+l}}]\Big)\Big).$$

At last,

$$\gamma\Big(\frac{1}{v_{i_0}-v_{i_{k+1}}}[v_{i_0}-v_{i_{k+1}}][v_{i_1}...v_{i_k}\overline{\sqcup\!\sqcup}v_{i_{k+2}}...v_{i_{k+l}}]\Big)=$$
$$\frac{1}{v_{i_0}-v_{i_{k+1}}}\gamma\Big([v_{i_0}-v_{i_{k+1}}]\Big)\gamma\Big([v_{i_1}...v_{i_k}\overline{\sqcup\!\sqcup}v_{i_{k+2}}...v_{i_{k+l}}]\Big)$$

and, in view of the recursive definition of $\uplus$, to conclude the proof it remains to show that

$$\frac{1}{v_{i_0}-v_{i_{k+1}}}\gamma\Big([v_{i_0}-v_{i_{k+1}}]\Big)=\gamma\Big([v_{i_0}]\Big)\odot\gamma\Big([v_{i_{k+1}}]\Big)$$

where, to avoid confusion with other already introduced symbols, $\odot$ denotes the product of bracketed integers induced by the addition: $[n]\odot[m]=[n+m]$.

Indeed, we have:

$$\gamma\Big([v_{i_0}-v_{i_{k+1}}]\Big)=\sum_{n\geq 1}(v_{i_0}^{n-1}-v_{i_{k+1}}^{n-1})[n]=\sum_{n\geq 2}(v_{i_0}^{n-1}-v_{i_{k+1}}^{n-1})[n],$$

and

$$(v_{i_0}-v_{i_{k+1}})\gamma([v_{i_0}])\odot\gamma([v_{i_{k+1}}])=(v_{i_0}-v_{i_{k+1}})\sum_{n,m\geq 1}v_{i_0}^{n-1}v_{i_{k+1}}^{m-1}[n+m]$$

$$=\sum_{p\geq 2}(v_{i_0}-v_{i_{k+1}})\Big(\sum_{n,m\geq 0,n+m=p-2}v_{i_0}^{n}v_{i_{k+1}}^{m}\Big)[p]=\sum_{p\geq 2}(v_{i_0}^{p-1}-v_{i_{k+1}}^{p-1})[p].$$

Corollary 2 *Let V be an infinite alphabet. The regularizing morphism γ induces, for any commutative algebra B over a base field k a group map from $G_{\mathbf{QSym}}(B)$ to $G_{\mathbf{QSh}_d^g(V)}(B\otimes_k k_V)$.*

In particular, regularized ζ functions, viewed as a real-valued characters on **QSym**, give rise to symmetril $\mathbb{R}((V))$-valued moulds over V. More generally, symmetrel moulds give rise to symmetril moulds by resummation [4, 7]—the very reason for the introduction of the latter.

Recall the definition of the Multiple Zeta Values (MZVs for short) associated to $(s_1,s_2,\ldots,s_r)$, where the s_i's are positive integers, and $s_1>1$:

$$\zeta(s_1s_2\ldots s_r):=\sum_{n_1>\cdots>n_r>0}\frac{1}{n_1^{s_1}\ldots n_r^{s_r}}.$$

For $\epsilon_i\in\mathbf{Q}/\mathbf{Z}$, the *modular MZVs* are defined as

$$\zeta\begin{pmatrix}\epsilon_1\ldots\epsilon_r\\ s_1\ldots s_r\end{pmatrix}:=\sum_{n_1>\cdots>n_r}\frac{e^{2\pi i n_1\epsilon_1}\ldots e^{2\pi i n_r\epsilon_r}}{n_1^{s_1}\ldots n_r^{s_r}}.$$

Observe that when $\epsilon_i=0$ for all i, then $\zeta\binom{0\ldots 0}{s_1\ldots s_r}=\zeta(s_1s_2\ldots s_r)$. Let us mention that, when dealing with modular MZVs, a "bimould" version of the previous construction has to be used. We only sketch the constructions in that case, they could be developed in more detail following the previous ones in this section.

Definition 8 Let $W:=\mathbf{Q}/\mathbf{Z}\times V$, with $V=\{v_i\}_{i\in\mathbf{N}^*}$, equipped with the partition $W=\coprod W_i$, $W_i:=\mathbf{Q}/\mathbf{Z}\times\{v_i\}$. We define the *generic divided quasi-shuffle bialgebra* over W, $\mathbf{QSh}_d^g(W)$, or g-divided quasi-shuffle algebra as the g-bialgebra defined over $k_V := k((V))$, which identifies with $T^g(W)$ as a vector space, equipped with the deconcatenation coproduct, and equipped with the following recursively defined product $\overline{\sqcup\!\sqcup}$ (elements of W are represented as column vector):

$$\begin{aligned}
\begin{pmatrix}\epsilon_1\dots\epsilon_r\\ v_{i_1}\dots v_{i_r}\end{pmatrix}\overline{\sqcup\!\sqcup}\begin{pmatrix}\epsilon_{r+1}\dots\epsilon_{r+s}\\ v_{i_{r+1}}\dots v_{i_{r+s}}\end{pmatrix} &:= \begin{pmatrix}\epsilon_1\\ v_{i_1}\end{pmatrix}\left(\begin{pmatrix}\epsilon_2\dots\epsilon_r\\ v_{i_2}\dots v_{i_r}\end{pmatrix}\overline{\sqcup\!\sqcup}\begin{pmatrix}\epsilon_{r+1}\dots\epsilon_{r+s}\\ v_{i_{r+1}}\dots v_{i_{r+s}}\end{pmatrix}\right)\\
&+\begin{pmatrix}\epsilon_{r+1}\\ v_{i_{r+1}}\end{pmatrix}\left(\begin{pmatrix}\epsilon_1\dots\epsilon_r\\ v_{i_1}\dots v_{i_r}\end{pmatrix}\overline{\sqcup\!\sqcup}\begin{pmatrix}\epsilon_{r+2}\dots\epsilon_{r+s}\\ v_{i_{r+2}}\dots v_{i_{r+s}}\end{pmatrix}\right)\\
&+\frac{1}{v_{i_1}-v_{i_{r+1}}}\begin{pmatrix}\epsilon_1+\epsilon_{r+1}\\ v_{i_1}\end{pmatrix}\left(\begin{pmatrix}\epsilon_2\dots\epsilon_r\\ v_{i_2}\dots v_{i_r}\end{pmatrix}\overline{\sqcup\!\sqcup}\begin{pmatrix}\epsilon_{r+2}\dots\epsilon_{r+s}\\ v_{i_{r+2}}\dots v_{i_{r+s}}\end{pmatrix}\right)\\
&-\frac{1}{v_{i_1}-v_{i_{r+1}}}\begin{pmatrix}\epsilon_1+\epsilon_{r+1}\\ v_{i_{r+1}}\end{pmatrix}\left(\begin{pmatrix}\epsilon_2\dots\epsilon_r\\ v_{i_2}\dots v_{i_r}\end{pmatrix}\overline{\sqcup\!\sqcup}\begin{pmatrix}\epsilon_{r+2}\dots\epsilon_{r+s}\\ v_{i_{r+2}}\dots v_{i_{r+s}}\end{pmatrix}\right)
\end{aligned}$$

where $\begin{pmatrix}\epsilon_1\dots\epsilon_r\\ v_{i_1}\dots v_{i_r}\end{pmatrix}$ and $\begin{pmatrix}\epsilon_{r+1}\dots\epsilon_{r+s}\\ v_{i_{r+1}}\dots v_{i_{r+s}}\end{pmatrix}$ are in generic position (so that $\frac{1}{v_{i_1}-v_{i_{k+1}}}$ is well-defined).

The elements of the groups $G_{\mathbf{QSh}_d^g(W)}(B)$, where B runs over algebras over k_V, associated to the generic group scheme $G_{\mathbf{QSh}_d^g(W)}$ over k_W are called symmetril moulds (over W).

Symmetril moulds over W can be used to resum modular MZVs by the same process that allows the resummation of usual MZVs by symmetril moulds over V, see [7].

6 A New Resummation Process

In this last section, we introduce a new resummation process for MZVs, based on Theorem 2. Contrary to Ecalle's resummation process, which maps a symmetrel mould (a character on the algebra of quasi-symmetric functions) to a symmetril mould, the new resummation is much more satisfactory in that it maps a symmetrel mould to a character on $\mathbf{Sh}_V^g$, so that calculus on MZVs and other characters on **QSym** can be interpreted in terms of the usual rules of Lie calculus (recall that the set of primitive elements in the dual of $\mathbf{Sh}_V^g$ is simply the multilinear part of the free Lie algebra over the integers, a well-known object whose study is even easier than the one of the usual free Lie algebra).

Theorem 4 *The inverse ρ_V of the standard g-bialgebra isomorphism ψ_V between $\mathbf{QSh}_d^g(V)$ and $\mathbf{Sh}_V^g$ is given by*

$$\rho_V([v_1\dots v_k]) := \sum_{\mu_1+\dots+\mu_i=k}\frac{1}{\mu_1!\dots\mu_i!}\left(\sum_{j=1}^{\mu_1}\frac{[v_j]}{\prod\limits_{l\neq j,l\leq\mu_1}(v_j-v_l)}\right)\dots$$

$$\cdots \left(\sum_{j=\mu_1+\cdots+\mu_{i-1}+1}^{k} \frac{[v_j]}{\prod\limits_{l\neq j, \mu_1+\cdots+\mu_{i-1}+1\leq l\leq k} (v_j - v_l)} \right).$$

The theorem follows by adapting to $T^g(V)$ the correspondence between formal power series in $X\mathbb{Q}[[X]]$ and generic coalgebra endomorphisms of $T^g(\mathbb{N}^*)$: with the same notation than the one used for $T^g(\mathbb{N}^*)$, each $P \in X\mathbb{Q}[[X]]$ defines a generic coalgebra endomorphism ϕ_P of $T^g(V)$. We have $\rho_V = \phi_{exp}$ and $\psi_V = \phi_{log}$, and the two morphisms are mutually inverse.

Corollary 3 *The morphism*

$$reg_V := \gamma \circ \rho_V$$

is a regularizing Hopf algebra morphism from $\mathbf{Sh}_V^g$ *to* $\mathbf{QSym}_V$*. It induces, for any commutative algebra* B *over the base field* k*, a group map from* $G_{\mathbf{QSym}}(B)$ *to* $G_{\mathbf{Sh}_V^g}(B \otimes_k k_V)$*.*

Naming generic symmetral moulds the characters on $\mathbf{Sh}_V^g$, we get that this last map resums symmetrel moulds (such as regularized MZVs at the positive integers) into generic symmetral moulds. As announced, this approach should provide a new way to investigate group-theoretically the properties of MZVs. Together with the study of the various combinatorial structures introduced in the present article, this will the object of further studies.

We conclude by illustrating the resummation process on low dimensional examples that show the behaviour of the map reg_V. We write ζ for a character on **QSym** (a symmetrel mould), having in mind the example of regularized multizetas. The morphism reg_V is given in low degrees by:

$$reg_V([v_1]) = \gamma([v_1]) = \sum_{n\geq 1} v_1^{n-1}[n],$$

$$\begin{aligned} reg_V([v_1, v_2]) =& \gamma([v_1, v_2] + \frac{1}{2}\frac{[v_1] - [v_2]}{v_1 - v_2}) \\ =& \sum_{n,m\geq 1} v_1^{n-1} v_2^{m-1}[n, m] + \frac{1}{2(v_1 - v_2)} \sum_{n\geq 1} (v_1^{n-1} - v_2^{n-1})[n] \\ =& \sum_{n,m\geq 1} v_1^{n-1} v_2^{m-1}[n, m] + \frac{1}{2} \sum_{\substack{n\geq 2 \\ p+q=n-2}} v_1^p v_2^q [n]. \end{aligned}$$

$$
\begin{aligned}
reg_V([v_1, v_2, v_3]) =&\gamma([v_1, v_2, v_3] + \frac{1}{2}\left(\frac{[v_1v_3]-[v_2v_3]}{v_1-v_2} + \frac{[v_1v_2]-[v_1v_3]}{v_2-v_3}\right) \\
&+ \frac{1}{6}\left(\frac{[v_1]}{(v_1-v_2)(v_1-v_3)} + \frac{[v_2]}{(v_2-v_1)(v_2-v_3)} + \frac{[v_3]}{(v_3-v_1)(v_3-v_2)}\right) \\
&= \sum_{n,m,p\geq 1} v_1^{n-1}v_2^{m-1}v_3^{p-1}[n,m,p] + \frac{1}{2}\Bigg(\sum_{\substack{n\geq 2, m\geq 1\\ p+q=n-2}} v_1^p v_2^q v_3^{m-1}[n,m] + \\
&\sum_{\substack{n\geq 1, m\geq 2\\ p+q=m-2}} v_1^{n-1}v_2^p v_3^q[n,m]\Bigg) + \frac{1}{6}\sum_{\substack{n\geq 3\\ p+q+r=n-3}} v_1^p v_2^q v_3^r[n],
\end{aligned}
$$

where we used the identity

$$
\begin{aligned}
&\frac{v_1^{n-1}}{(v_1-v_2)(v_1-v_3)} + \frac{v_2^{n-1}}{(v_2-v_1)(v_2-v_3)} + \frac{v_3^{n-1}}{(v_3-v_1)(v_3-v_2)} \\
&= \sum_{p+q+r=n-3} v_1^p v_2^q v_3^r.
\end{aligned}
$$

We get, for the ζ character:

$$
\begin{aligned}
\zeta \circ reg_V([v_1]⧢[v_2]) =&\zeta\Bigg(\sum_{n,m\geq 1} v_1^{n-1}v_2^{m-1}([n,m]+[m,n]) + \sum_{n\geq 2, p+q=n-2} v_1^p v_2^q[n]\Bigg) \\
&= \zeta\Bigg(\sum_{n,m\geq 1} v_1^{n-1}v_2^{m-1}[n] \uplus [m]\Bigg) \\
&= \zeta \circ reg_V([v_1]) \cdot \zeta \circ reg_V([v_2]).
\end{aligned}
$$

$$
\zeta \circ reg_V([v_1, v_2]⧢[v_3]) =\zeta \circ reg_V([v_1, v_2, v_3] + [v_1, v_3, v_2] + [v_3, v_1, v_2])
$$

$$
\begin{aligned}
=&\zeta\Bigg(\sum_{n,m,p\geq 1} v_1^{n-1}v_2^{m-1}v_3^{p-1}[n,m]⧢[p] \\
&+ \sum_{\substack{n\geq 2\\ p+q=n-2\\ m\geq 1}} \left((\frac{1}{2}v_1^p v_2^q v_3^{m-1} + v_1^p v_2^{m-1} v_3^q)[n,m] + (\frac{1}{2}v_1^p v_2^q v_3^{m-1} + v_1^{m-1} v_2^p v_3^q)[m,n]\right) \\
&+ \frac{1}{2}\sum_{\substack{n\geq 3\\ p+q+r=n-3}} v_1^p v_2^q v_3^r[n]\Bigg) \\
=&\zeta\Bigg(\Bigg(\sum_{n,m\geq 1} v_1^{n-1}v_2^{m-1}[n,m] + \frac{1}{2}\sum_{\substack{n\geq 2\\ p+q=n-2}} v_1^p v_2^q[n]\Bigg) \uplus \Bigg(\sum_{r\geq 1} v_3^{r-1}[r]\Bigg)\Bigg) \\
=&\zeta \circ reg_V([v_1, v_2]) \cdot \zeta \circ reg_V([v_3]).
\end{aligned}
$$

Acknowledgements The authors acknowledge support from ICMAT, Madrid, and from the grant CARMA, ANR-12-BS01-0017.

References

1. Aguiar, M., Mahajan, S.: Monoidal functors, species and Hopf algebras. CRM Monogr. Ser. **29** (2010)
2. Cartier, P.: A primer of Hopf algebras. Frontiers in Number Theory, Physics, and Geometry II, pp. 537–615. Springer, Berlin (2007)
3. Cartier, P.: Fonctions polylogarithmes, nombres polyzêtas et groupes prounipotents, *Séminaire Bourbaki*, Mars 2001, 53ème année, 2000–2001, no 885
4. Cresson, J.: Calcul moulien. Annales de la Faculté des Sciences de Toulouse. Mathématiques **18**(2), 307–395 (2009)
5. Ebrahimi-Fard, K., Guo, L.: Multiple zeta values and Rota-Baxter algebras. Integers **8**(2), 1553–1732 (2008)
6. Ebrahimi-Fard, K., Patras, F.: La structure combinatoire du calcul intégral. Gazette des Mathématiciens **138** (2013)
7. Ecalle, J.: ARI/GARI, la dimorphie et l'arithmétique des multizêtas: un premier bilan. Journal de Théorie des Nombres Bordeaux **15**(2), 411–478 (2003)
8. Foissy, L., Patras, F.: Natural endomorphisms of shuffle algebras. Int. J. Algebra Comput. **23**(4), 989–1009 (2013)
9. Foissy, L., Patras, F., Thibon, J.-Y.: Deformations of shuffles and quasi-shuffles. Annales de l'Institut Fourier (Grenoble) **66**(1), 209–237 (2016)
10. Furusho, H.: Double shuffle relation for associators. Ann. Math. **174**, 341–360 (2011)
11. Hoffman, M.E.: Quasi-shuffle products. J. Algebr. Comb. **11**(1), 49–68 (2000)
12. Ihara, K., Kaneko, M., Zagier, D.: Derivation and double shuffle relations for multiple zeta values. Compos. Math. **142**, 307–338 (2006)
13. Patras, F.: Generic algebras and iterated Hochschild homology. J. Pure Appl. Algebra **162**, 337–357 (2001)
14. Patras, F., Reutenauer, C.: On descent algebras and twisted bialgebras. Moscow Math. J. **4**(1), 199–216 (2004)
15. Patras, F., Schocker, M.: Twisted descent algebras and the Solomon-Tits algebra. Adv. Math. **199**(1), 151–184 (2006)
16. Patras, F., Schocker, M.: Trees, set compositions and the twisted descent algebra. J. Algebr. Comb. **28**, 3–23 (2008)
17. Peskin, M.E., Schroeder, D.V.: An Introduction to Quantum Field Theory. Westview (1995)
18. Racinet, G.: Doubles mélanges des polylogarithmes multiples aux racines de l'unité. Publications Mathématiques de l'Institut des Hautes Études Scientifiques **95**(1), 185–231 (2002)
19. Reutenauer, C.: Free Lie Algebras. Oxford University Press (1993)

Mould Theory and the Double Shuffle Lie Algebra Structure

Adriana Salerno and Leila Schneps

Abstract The real multiple zeta values $\zeta(k_1, \ldots, k_r)$ are known to form a $\mathbb{Q}$-algebra; they satisfy a pair of well-known families of algebraic relations called the double shuffle relations. In order to study the algebraic properties of multiple zeta values, one can replace them by formal symbols $Z(k_1, \ldots, k_r)$ subject only to the double shuffle relations. These form a graded Hopf algebra over $\mathbb{Q}$, and quotienting this algebra by products, one obtains a vector space. A complicated theorem due to G. Racinet proves that this vector space carries the structure of a Lie coalgebra; in fact Racinet proved that the dual of this space is a Lie algebra, known as the double shuffle Lie algebra $\mathfrak{ds}$. J. Ecalle developed a new theory to explore combinatorial and algebraic properties of the formal multiple zeta values. His theory is sketched out in some publications. However, because of the depth and complexity of the theory, Ecalle did not include proofs of many of the most important assertions, and indeed, even some interesting results are not always stated explicitly. The purpose of the present paper is to show how Racinet's theorem follows in a simple and natural way from Ecalle's theory. This necessitates an introduction to the theory itself, which we have pared down to only the strictly necessary notions and results.

Keywords Mould · Double shuffle · Multiple zeta values · Lie algebra · Dimorphy · Flexions

1 Introduction

In his doctoral thesis from 2000, Georges Racinet ([10], see also [11]) proved a remarkable theorem using astute combinatorial and algebraic reasoning. His proof

A. Salerno (✉)
Bates College, 3 Andrews Rd, Lewiston 04240, USA
e-mail: asalerno@bates.edu

L. Schneps
Institute de Mathématiques de Jussieu, 4 place Jussieu, Case 247, 75252 Paris Cedex, France
e-mail: leila@math.jussieu.fr

J. I. Burgos Gil et al. (eds.), *Periods in Quantum Field Theory and Arithmetic*, Springer Proceedings in Mathematics & Statistics 314,
https://doi.org/10.1007/978-3-030-37031-2_15

was later somewhat simplified and streamlined by Furusho [8], but it remains really difficult to grasp the essential key that makes it work. The purpose of this article is to show how Ecalle's theory of moulds yields a very different and natural proof of the same result. The only difficulty is to enter into the universe of moulds and learn its language; the theory is equipped with a sort of standard all-purpose "toolbox" of objects and identities which, once acquired, serve to prove all kinds of results, in particular the one we consider in this paper. Therefore, the goal of this article is not only to present the mould-theoretic proof of Racinet's theorem, but also to provide an initiation into mould theory in general. Ecalle's seminal article on the subject is [6], and a detailed introduction with complete proofs can be found in [12]; the latter text will be referred to here for some basic lemmas.

We begin by recalling the definitions necessary to state Racinet's theorem.

Definition 1 Let u, v be two monomials in x and y. Then the commutative *shuffle product* $\mathrm{sh}(u,v)$ is defined recursively by $\mathrm{sh}(u,v) = \{\{u\}\}$ if $v = 1$ and $\{\{v\}\}$ if $u = 1$, where $\{\{\cdot\}\}$ denotes a multiset, i.e. an unordered list with possible repetitions; otherwise, writing $u = Xu'$ and $v = Yv'$ where $X, Y \in \{x, y\}$ represents the first letter of the word, we have the recursive rule

$$\mathrm{sh}(Xu, Yv) = \{\{X \cdot \mathrm{sh}(u, Yv)\}\} \cup \{\{Y \cdot \mathrm{sh}(Xu, v)\}\}, \tag{1}$$

where $\cup$ denotes the union of the two multisets which preserves repetitions and $X \cdot \mathrm{sh}(u, v)$ means we multiply every member in the multiset $\mathrm{sh}(u, v)$ on the left by X.

For example,

$$\begin{aligned}
\mathrm{sh}(xy, x) &= \{\{x \cdot \mathrm{sh}(y, x)\}\} \cup \{\{x \cdot \mathrm{sh}(xy, 1)\}\} \\
&= \{\{x \cdot \{\{yx, xy\}\}\ \}\} \cup \{\{x \cdot \{\{xy\}\}\ \}\} \\
&= \{\{xyx, xxy\}\} \cup \{\{xxy\}\} \\
&= \{\{xyx, xxy, xxy\}\}
\end{aligned}$$

If u, v are two words ending in y, we can write them uniquely as words in the letters $y_i = x^{i-1}y$. The *stuffle product* of u, v is defined by $\mathrm{st}(u, v) = \{\{u\}\}$ if $v = 1$ and $\{\{v\}\}$ if $u = 1$, and

$$\mathrm{st}(y_i u, y_j v) = \{\{y_i \cdot \mathrm{st}(u, y_j v)\}\} \cup \{\{y_j \cdot \mathrm{st}(y_i u, v\})\} \cup \{\{y_{i+j} \cdot \mathrm{st}(u, v)\}\}, \tag{2}$$

where y_i and y_j are respectively the first letters of the words u and v written in the y_j.

For example,

$$\begin{aligned}\mathrm{st}(y_1y_2, y_1) &= \{\{y_1 \cdot \mathrm{st}(y_2, y_1)\}\} \cup \{\{y_1 \cdot \mathrm{st}((y_1y_2, 1)\}\} \cup \{\{y_2 \cdot \mathrm{st}(y_2, 1)\}\} \\ &= \{\{y_1y_2y_1, y_1y_1y_2, y_1y_3\}\} \cup \{\{y_1y_1y_2\}\} \cup \{\{y_2y_2\}\} \\ &= \{\{y_1y_2y_1, y_1y_1y_2, y_1y_3, y_1y_1y_2, y_2y_2\}\}\end{aligned}$$

Definition 2 The *double shuffle space* $\mathfrak{ds}$ is the space of polynomials $f \in \mathbb{Q}\langle x, y\rangle$, the polynomial ring on two non-commutative variables x and y, of degree ≥ 3 that satisfy the following two properties:

1. The coefficients of f satisfy the *shuffle relations*
$$\sum_{w \in \mathrm{sh}(u,v)} (f|w) = 0, \tag{3}$$
where u, v are words in x, y and $\mathrm{sh}(u, v)$ is the set of words obtained by shuffling them. This condition is equivalent to the assertion that f lies in the free Lie algebra $\mathrm{Lie}[x, y]$, a fact that is easy to see by using the characterization of Lie polynomials in the non-commutative polynomial ring $\mathbb{Q}\langle x, y\rangle$ as those that are "Lie-like" under the coproduct Δ defined by $\Delta(x) = x \otimes 1 + 1 \otimes x$ and $\Delta(y) = y \otimes 1 + 1 \otimes y$, i.e. such that $\Delta(f) = f \otimes 1 + 1 \otimes f$ ([13, Chap. 3, Theorem 5.4]). Indeed, when the property of being Lie-like under Δ is expressed explicitly on the coefficients of f it is nothing other than the shuffle relations (3).
2. Let $f_* = \pi_y(f) + f_{\mathrm{corr}}$, where $\pi_y(f)$ is the projection of f onto just the words ending in y, and
$$f_{\mathrm{corr}} = \sum_{n\geq 1} \frac{(-1)^{n-1}}{n}(f|x^{n-1}y)y^n. \tag{4}$$
Considering f_* as being rewritten in the variables $y_i = x^{i-1}y$, the coefficients of f_* satisfy the *stuffle relations*:
$$\sum_{w \in \mathrm{st}(u,v)} (f_*|w) = 0, \tag{5}$$
where u and v are words in the y_i.

The double shuffle space $\mathfrak{ds}$ is the one defined by Racinet in [10] (which he denoted $\mathfrak{dm}$, for the French term "double mélange"). It should not be confused with the bigraded space Dsh studied in [9]. The space Dsh is a linearized version of $\mathfrak{ds}$, which has also been the subject of a great deal of study, but is more often denoted $\mathfrak{ls}$ (cf. for example [3]).

For every $f \in \mathrm{Lie}[x, y]$, define a derivation D_f of $\mathrm{Lie}[x, y]$ by setting it to be

$$D_f(x) = 0, \quad D_f(y) = [y, f]$$

on the generators. Define the *Poisson (or Ihara) bracket* on (the underlying vector space of) $\mathrm{Lie}[x, y]$ by

$$\{f, g\} = [f, g] + D_f(g) - D_g(f). \tag{6}$$

This definition corresponds naturally to the Lie bracket on the space of derivations of Lie$[x, y]$; indeed, it is easy to check that

$$[D_f, D_g] = D_f \circ D_g - D_g \circ D_f = D_{\{f,g\}}. \tag{7}$$

Theorem 1 (Racinet) *The double shuffle space* $\mathfrak{ds}$ *is a Lie algebra under the Poisson bracket.*

The goal of this paper is to give the mould-theoretic proof of this result, which first necessitates rephrasing the relevant definitions in terms of moulds. The paper is organized as follows. In Sect. 2, we give basic definitions from mould theory that will be used throughout the rest of the paper, and in Sect. 3 we define dimorphy and consider the main dimorphic subspaces related to double shuffle. In Sect. 4 we give the dictionary between mould theory and the double shuffle situation. In Sect. 5 we give some of the definitions and basic results on the group aspect of mould theory. In Sect. 6 we describe the special mould *pal* that lies at the heart of much of mould theory, and introduce Ecalle's fundamental identity. The final Sect. 7 contains the simple and elegant proof of the mould version of Racinet's theorem. Sections 2, 3, 5 and 6 can serve as a short introduction to the basics of mould theory; a much more complete version with full proofs and details is given in [12], which is cited for some results. Every mould-theory definition in this paper is due to Ecalle, as are all of the statements, although some of these are not made explicitly in his papers, but used as assumptions. Our contribution has been firstly to provide complete proofs of many statements which are either nowhere proved in his articles or proved by arguments that are difficult to understand (at least by us), secondly to pick a path through the dense forest of his results that leads most directly to the desired theorem, and thirdly, to give the dictionary that identifies the final result with Racinet's theorem above.

In order to preserve the expository flow leading to the proof of the main theorem, we have chosen to consign the longer and more technical proofs to appendices or, for those that already appear in [12], to simply give the reference.

2 Definitions for Mould Theory

This section constitutes what could be called the "first drawer" of the mould toolbox, with only the essential definitions of moulds, some operators on moulds, and some mould symmetries. We work over a base field K, and let $u_1, u_2, \ldots$ be a countable set of indeterminates, and $v_1, v_2, \ldots$ another. The definitions below arise from Ecalle's papers (see especially [6], and are also developed at length in [5, 12]).

Moulds. A *mould* in the variables u_i is a family $A = (A_r)_{r \geq 0}$ of functions of the u_i, where each A_r is a function of $u_1, \ldots, u_r$. We call A_r the *depth* r component of the mould. In this paper we let $K = \mathbb{Q}$, and in fact we consider only rational-function valued moulds, i.e. we have $A_r(u_1, \ldots, u_r) \in \mathbb{Q}(u_1, \ldots, u_r)$ for $r \geq 0$. Note that $A_0(\emptyset)$ is a constant. We often drop the index r when the context is clear, and write $A(u_1, \ldots, u_r)$. Moulds can be added and multiplied by scalars componentwise, so the set of moulds forms a vector space. A mould in the v_i is defined identically for the variables v_i.

Let ARI (resp. $\overline{\text{ARI}}$) denote the space of moulds in the u_i (resp. in the v_i) such that $A_0(\emptyset) = 0$.[1] These two vector spaces are obviously isomorphic, but they will be equipped with very different Lie algebra structures. We use superscripts on ARI to denote the type of moulds we are dealing with; in particular ARI^{pol} denotes the space of polynomial-valued moulds, and ARI^{rat} denotes the space of rational-function moulds.

Operators on moulds. We will use the following operators on moulds in ARI:

$$\text{neg}(A)(u_1, \ldots, u_r) = A(-u_1, \ldots, -u_r) \tag{8}$$

$$\text{push}(A)(u_1, \ldots, u_r) = A(-u_1 - \cdots - u_r, u_1, \ldots, u_{r-1}) \tag{9}$$

$$\text{mantar}(A)(u_1, \ldots, u_r) = (-1)^{r-1} A(u_r, \ldots, u_1) \tag{10}$$

We also introduce the swap, which is a map from ARI to $\overline{\text{ARI}}$ given by

$$\text{swap}(A)(u_1, \ldots, u_r) = A(v_r, v_{r-1} - v_r, v_{r-2} - v_{r-1}, \ldots, v_1 - v_2), \tag{11}$$

and its inverse, also called swap, from $\overline{\text{ARI}}$ to ARI:

$$\text{swap}(A)(v_1, \ldots, v_r) = A(u_1 + \cdots + u_r, u_1 + \cdots + u_{r-1}, \ldots, u_1 + u_2, u_1). \tag{12}$$

Thanks to this formulation, which is not ambiguous since to know which swap is being used it suffices to check whether swap is being applied to a mould in ARI or one in $\overline{\text{ARI}}$, we can treat swap like an involution: $\text{swap} \circ \text{swap} = \text{id}$.

Let us now introduce some notation necessary for the Lie algebra structures on ARI and $\overline{\text{ARI}}$.

Flexions. Let $\mathbf{w} = (u_1, \cdots, u_r)$. For every possible way of cutting the word $\mathbf{w}$ into three (possibly empty) subwords $\mathbf{w} = \mathbf{abc}$ with

$$\mathbf{a} = (u_1, \ldots, u_k), \quad \mathbf{b} = (u_{k+1}, \ldots, u_{k+l}), \quad \mathbf{c} = (u_{k+l+1}, \ldots, u_r),$$

set

$$\begin{cases} \mathbf{a}\rceil = (u_1, u_2, \cdots, u_k + u_{k+1} + \cdots + u_{k+l}) & \text{if } \mathbf{b} \neq \emptyset, \text{ otherwise } \mathbf{a}\rceil = \mathbf{a} \\ \lceil\mathbf{c} = (u_{k+1} + \cdots + u_{k+l+1}, u_{k+l+2}, \cdots, u_r) & \text{if } \mathbf{b} \neq \emptyset, \text{ otherwise. } \lceil\mathbf{c} = \mathbf{c}. \end{cases}$$

---

[1] Ecalle works with *bimoulds*, which are moulds that are simultaneously in the variables u_i and v_i. However, while bimoulds are well-adapted to the study of certain more complex objects such as multizeta values colored by roots of unity, they do not arise naturally in the context of the simple multizeta values used here, and we found that using moulds in only the u_i or only the v_i made the proofs and the notation considerably simpler.

If now $\mathbf{w} = (v_1, \ldots, v_r)$ is a word in the v_i, then for every decomposition $\mathbf{w} = \mathbf{abc}$ with

$$\mathbf{a} = (v_1, \ldots, v_k), \quad \mathbf{b} = (v_{k+1}, \ldots, v_{k+l}), \quad \mathbf{c} = (v_{k+l+1}, \ldots, v_r),$$

we set

$$\begin{cases} \mathbf{b}\rfloor = (v_{k+1} - v_{k+l+1}, v_{k+2} - v_{k+l+1}, \ldots, v_{k+l} - v_{k+l+1}) & \text{if } \mathbf{c} \neq \emptyset, \text{ otherwise } \mathbf{b}\rfloor = \mathbf{b} \\ \lfloor\mathbf{b} = (v_{k+1} - v_k, v_{k+2} - v_k, \ldots, v_{k+l} - v_k) & \text{if } \mathbf{a} \neq \emptyset, \text{ otherwise } \lfloor\mathbf{b} = \mathbf{b}. \end{cases}$$

Operators on pairs of moulds. For $A, B \in \mathrm{ARI}$ or $A, B \in \overline{\mathrm{ARI}}$, we set

$$\mathrm{mu}(A, B)(\mathbf{w}) = \sum_{\mathbf{w}=\mathbf{ab}} A(\mathbf{a})B(\mathbf{b}) \tag{13}$$

$$\mathrm{lu}(A, B) = \mathrm{mu}(A, B) - \mathrm{mu}(B, A) \tag{14}$$

For any mould $B \in \mathrm{ARI}$, we define two operators on ARI, $\mathrm{amit}(B)$ and $\mathrm{anit}(B)$, defined by

$$\begin{aligned} \big(\mathrm{amit}(B) \cdot A\big)(\mathbf{w}) &= \sum_{\substack{\mathbf{w}=\mathbf{abc} \\ \mathbf{b},\mathbf{c}\neq\emptyset}} A(\mathbf{a}\lceil\mathbf{c})B(\mathbf{b}) \\ \big(\mathrm{anit}(B) \cdot A\big)(\mathbf{w}) &= \sum_{\substack{\mathbf{w}=\mathbf{abc} \\ \mathbf{a},\mathbf{b}\neq\emptyset}} A(\mathbf{a}\rceil\mathbf{c})B(\mathbf{b}) \end{aligned} \tag{15}$$

For any mould $B \in \mathrm{ARI}$, the operators $\mathrm{amit}(B)$ and $\mathrm{anit}(B)$ are derivations of ARI for the lu-bracket (see [12, Proposition 2.2.1]). We define a third derivation, $\mathrm{arit}(B)$, by

$$\big(\mathrm{arit}(B) \cdot A\big)(\mathbf{w}) = \mathrm{amit}(B) \cdot A - \mathrm{anit}(B) \cdot A. \tag{16}$$

If $B \in \overline{\mathrm{ARI}}$ we have derivations of $\overline{\mathrm{ARI}}$ given by

$$\begin{aligned} \big(\mathrm{amit}(B) \cdot A\big)(\mathbf{w}) &= \sum_{\substack{\mathbf{w}=\mathbf{abc} \\ \mathbf{b},\mathbf{c}\neq\emptyset}} A(\mathbf{ac})B(\mathbf{b}\rfloor) \\ \big(\mathrm{anit}(B) \cdot A\big)(\mathbf{w}) &= \sum_{\substack{\mathbf{w}=\mathbf{abc} \\ \mathbf{a},\mathbf{b}\neq\emptyset}} A(\mathbf{ac})B(\lfloor\mathbf{b}), \end{aligned} \tag{17}$$

and again we define the derivation $\mathrm{arit}(B)$ as in (16).

Finally, for $A, B \in \mathrm{ARI}$ or $A, B \in \overline{\mathrm{ARI}}$, we set

$$\mathrm{ari}(A, B) = \mathrm{arit}(B) \cdot A + \mathrm{lu}(A, B) - \mathrm{arit}(A) \cdot B. \tag{18}$$

Remark. The condition $\mathbf{b} \neq \emptyset$ in the definitions of *amit* and *anit* above are not necessary in (15) and (17), since we are assuming that $B \in \text{ARI}$, so it has the property that $B(\emptyset) = 0$; this means that including decompositions with $\mathbf{b} = \emptyset$ in the sum would not actually change the values. However, we chose to reproduce Ecalle's definition, which also applies to moulds with non-zero value in depth 0, so as to make it easier to consult his articles and recognize the same definitions.

Since arit is a derivation for lu, the ari-operator is easily shown to be a Lie bracket. Note that although we use the same notation ari for the Lie brackets on both ARI and $\overline{\text{ARI}}$, they are two different Lie brackets on two different spaces. Indeed, while some formulas and properties (such as mu, or alternality, see (19) below) are written identically for ARI and $\overline{\text{ARI}}$, others, in particular all those that use flexions, are very different, since the definitions of upper flexions (on the u_i) and lower flexions (on the v_i) are very different. This can be seen in the following examples.

Examples. We give a few of the expressions above explicitly in low depth. The moulds $\text{amit}(B) \cdot A$ and $\text{amit}(B) \cdot A$ are all zero in depth 1. Let $A, B \in \text{ARI}$ and let us compute the mould $\text{amit}(B) \cdot A$ in depth 2. The only possible decomposition of $\mathbf{w} = (u_1, u_2)$ as $\mathbf{abc}$ with $\mathbf{b}, \mathbf{c} \neq \emptyset$ is $\mathbf{abc} = (\emptyset)(u_1)(u_2)$, so using the upper flexions as in (15), we have $\lceil \mathbf{c} = (u_1 + u_2)$ and

$$\big(\text{amit}(B) \cdot A\big)(u_1, u_2) = A(u_1 + u_2)B(u_1).$$

(Note that if we don't include the condition $\mathbf{b} \neq \emptyset$ in the sum, we would also consider the decomposition $\mathbf{abc} = (u_1)(\emptyset)(u_2)$ so we would add on the term $A(u_1, u_2)B(\emptyset)$, but as pointed out in the remark above, this term is zero since $B \in \text{ARI}$.)

Now let us compute the mould $\text{anit}(B) \cdot A$ in depth 3. Let $\mathbf{w} = (u_1, u_2, u_3)$. The decompositions $\mathbf{w} = \mathbf{abc}$ with $\mathbf{a}, \mathbf{b} \neq \emptyset$ are given by $(u_1)(u_2)(u_3)$, $(u_1, u_2)(u_3)(\emptyset)$ and $(u_1)(u_2, u_3)(\emptyset)$, so

$$\big(\text{anit}(B) \cdot A\big)(u_1, u_2, u_3) = \\ A(u_1 + u_2, u_3)B(u_2) + A(u_1, u_2 + u_3)B(u_3) + A(u_1 + u_2 + u_3)B(u_2, u_3).$$

If $A, B \in \overline{\text{ARI}}$, we again compute $\text{amit}(B) \cdot A$ in depth 2 and $\text{anit}(B) \cdot A$ in depth 3, but now using the lower flexions of (17); we obtain the expressions

$$\big(\text{amit}(B) \cdot A\big)(v_1, v_2) = A(v_2)B(v_1 - v_2),$$

$$\big(\text{anit}(B) \cdot A\big)(v_1, v_2, v_3) = \\ A(v_1, v_3)B(v_2 - v_1) + A(v_1, v_2)B(v_3 - v_2) + A(v_1)B(v_2 - v_1, v_3 - v_1).$$

Symmetries. A mould in ARI (resp. $\overline{\text{ARI}}$) is said to be *alternal* if for all words $\mathbf{u}, \mathbf{v}$ in the u_i (resp. v_i),

$$\sum_{\mathbf{w}\in \mathrm{sh}(\mathbf{u},\mathbf{v})} A(\mathbf{w}) = 0. \tag{19}$$

The relations in (19) are known as the *alternality* relations, and they are identical for moulds in ARI and $\overline{\mathrm{ARI}}$. Let us now define the *alternility relations*, which are only applicable to moulds in $\overline{\mathrm{ARI}}$. Just as the alternality conditions are the mould equivalent of the shuffle relations, the alternility conditions are the mould equivalent of the stuffle relations, translated in terms of the alphabet $\{v_1, v_2, \ldots\}$ as follows. Let $Y_1 = (y_{i_1}, \ldots, y_{i_r})$ and $Y_2 = (y_{j_1}, \ldots, y_{j_s})$ be two sequences; for example, we consider $Y_1 = (y_i, y_j)$ and $Y_2 = (y_k, y_l)$. Let w be a word in the stuffle product $\mathrm{st}\big(Y_1, Y_2\big)$, which in our example is the 13-element multiset

$$\begin{aligned}\{\{(y_i, y_j, y_k, y_l), (y_i, y_k, y_j, y_l), (y_i, y_k, y_l, y_j), (y_k, y_i, y_j, y_l), (y_k, y_i, y_l, y_j),\\ (y_k, y_l, y_i, y_j), (y_i, y_{j+k}, y_l), (y_{i+k}, y_j, y_l), (y_i, y_k, y_{j+l}), (y_{i+k}, y_l, y_j),\\ (y_k, y_i, y_{j+l}), (y_k, y_{i+l}, y_j), (y_{i+k}, y_{j+l})\}\}.\end{aligned} \tag{20}$$

To each such word we associate an alternility term for the mould A, given by associating the tuple (v_1, v_2, v_3, v_4) to the ordered tuple (y_i, y_j, y_k, y_l) and taking

$$\frac{1}{(v_i - v_j)}\big(A(\ldots, v_i, \ldots) - A(\ldots, v_j, \ldots)\big) \tag{21}$$

for each contraction occurring in the word w. For instance in our example we have the six alternility terms

$$\begin{gathered}A(v_1, v_2, v_3, v_4), A(v_1, v_3, v_2, v_4), A(v_1, v_3, v_4, v_2), A(v_3, v_1, v_2, v_4),\\ A(v_3, v_1, v_4, v_2), A(v_3, v_4, v_1, v_2)\end{gathered} \tag{22}$$

corresponding to the first six words in (20), the six terms

$$\begin{aligned}&\frac{1}{(v_2 - v_3)}\big(A(v_1, v_2, v_4) - A(v_1, v_3, v_4)\big), \quad \frac{1}{(v_1 - v_3)}\big(A(v_1, v_2, v_4) - A(v_3, v_2, v_4)\big),\\ &\frac{1}{(v_2 - v_4)}\big(A(v_1, v_3, v_2) - A(v_1, v_3, v_4)\big), \quad \frac{1}{(v_1 - v_3)}\big(A(v_1, v_4, v_2) - A(v_3, v_4, v_2)\big),\\ &\frac{1}{(v_2 - v_4)}\big(A(v_3, v_1, v_2) - A(v_3, v_1, v_4)\big), \quad \frac{1}{(v_1 - v_4)}\big(A(v_3, v_1, v_2) - A(v_3, v_4, v_2)\big)\end{aligned} \tag{23}$$

corresponding to the next six words, and the final term

$$\frac{1}{(v_1 - v_3)(v_2 - v_4)}\big(A(v_1, v_2) - A(v_3, v_2) - A(v_1, v_4) + A(v_3, v_4)\big) \tag{24}$$

corresponding to the final word with the double contraction.

Let us write $A_{\mathbf{w}}$ for the alternility term of A associated to a word $\mathbf{w}$ in the stuffle product $\mathrm{st}(Y_1, Y_2)$; note that the alternility terms (for example those in (22), (23) and (24) associated to the words $\mathbf{w}$ in the list (20)) are not all terms of the form $A(\mathbf{w})$ or even linear combinations of such terms (due to the denominators). However, the alternility terms $A_{\mathbf{w}}$ are all polynomials in the v_i, since the zeros of the denominators all correspond to zeros of the numerator.

The *alternility relation* associated to the pair (Y_1, Y_2) on A is the sum of the alternility terms associated to words in the stuffle of Y_1 and Y_2; it is given by

$$\sum_{\mathbf{w}\in \mathrm{st}(Y_1,Y_2)} A_{\mathbf{w}} = 0. \tag{25}$$

Let $A_{r,s}$ denote the left-hand side of (25). Note that indeed, $A_{r,s}$ does not depend on the actual sequences Y_1 and Y_2, but merely on the number of letters in Y_1 and in Y_2. For example when $r = s = 2$, the alternility sum $A_{2,2}$ is given by the sum of the terms (22)–(24) above. Furthermore, like for the shuffle, we may assume that $r \le s$ by symmetry. Thus we have the following definition: a mould in $\overline{\mathrm{ARI}}$ is said to be *alternil* if it satisfies the alternility relation $A_{r,s} = 0$ for all pairs of integers $1 \le r \le s$.

3 Lie Subalgebras of ARI

In this section, we show that the spaces of moulds satisfying certain important symmetry properties are closed under the ari-bracket. In particular, we introduce the following *dimorphic spaces* investigated by Ecalle, where the term dimorphy refers to the double description of a mould by a symmetry property on it and another one on its swap.

Definition 3 Let ARI_{al} denote the set of alternal moulds. Let $\mathrm{ARI}_{al/al}$ (resp. $\mathrm{ARI}_{al/il}$) denote the set of alternal moulds with alternal (resp. alternil) swap. Let ARI_{al*al} (resp. ARI_{al*il}) denote the set of alternal moulds whose swap is alternal (resp. alternil) up to addition of a constant-valued mould. Finally, let $\mathrm{ARI}_{\underline{al/al}}$ (resp. $\mathrm{ARI}_{\underline{al*al}}$, $\mathrm{ARI}_{\underline{al/il}}$, $\mathrm{ARI}_{\underline{al*il}}$) denote the subspace of $\mathrm{ARI}_{al/al}$ (resp. ARI_{al*al}, $\mathrm{ARI}_{al/il}$, ARI_{al*il}) consisting of moulds A such that A_1 is an even function, i.e. $A(-u_1) = A(u_1)$.

The first main theorem of this paper is the following result, which is used constantly in Ecalle's work although no explicit proof appears to have been written down, and the proof is by no means as easy as one might imagine.

Theorem 2 *The subspace* $\mathrm{ARI}_{al} \subset \mathrm{ARI}$ *of alternal moulds forms a Lie algebra under the* ari*-bracket, as does the subspace* $\overline{\mathrm{ARI}}_{al}$ *of* $\overline{\mathrm{ARI}}$.

The full proof is given in Appendix A. The idea is as follows: if $C = \mathrm{ari}(A, B)$, then by (18) it is enough to show separately that if A and B are alternal then $\mathrm{lu}(A, B)$

is alternal and $\mathrm{arit}(B) \cdot A$ is alternal. This is done via a combinatorial manipulation that is fairly straightforward for lu but actually quite complicated for arit.

We next have a simple but important result on polynomial-valued moulds.

Proposition 1 *The subspace* ARI^{pol} *of polynomial-valued moulds in* ARI *forms a Lie algebra under the* ari-*bracket.*

Proof This follows immediately from the definitions of mu, arit and ari in (13)–(18), as all the operations and flexions there are polynomial. □

Now we give another key theorem, the first main result concerning dimorphy. This result, again, is used repeatedly by Ecalle but we were not able to find a complete proof in his papers, so we have reconstructed one here (see also [12, Sect. 2.5]).

Theorem 3 *The subspaces* $\mathrm{ARI}_{\underline{al}/\underline{al}}$ *and* $\mathrm{ARI}_{\underline{al}*\underline{al}}$ *form Lie algebras under the* ari-*bracket.*

The proof is based on the following two propositions.

Proposition 2 *If* $A \in \mathrm{ARI}_{\underline{al}*\underline{al}}$, *then* A *is* neg-*invariant and* push-*invariant.*

The proof of this proposition is deferred to Appendix B.

Proposition 3 *If* A *and* B *are both* push-*invariant moulds, then*

$$\mathrm{swap}\Big(\mathrm{ari}\big(\mathrm{swap}(A), \mathrm{swap}(B)\big)\Big) = \mathrm{ari}(A, B), \tag{26}$$

Proof Explicit computation using the flexions shows that for all moulds $A, B \in \mathrm{ARI}$ we have the general formula:

$$\mathrm{swap}\big(\mathrm{ari}(\mathrm{swap}(A), \mathrm{swap}(B))\big) = \mathrm{axit}\big(B, -\mathrm{push}(B)\big) \cdot A - \mathrm{axit}\big(A, -\mathrm{push}(A)\big) \cdot B + \mathrm{lu}(A, B), \tag{27}$$

where here ari is the Lie bracket on $\overline{\mathrm{ARI}}$, and axit is the operator on ARI defined for a general pair of moulds $B, C \in \mathrm{ARI}$ by the formula

$$\mathrm{axit}(B, C) \cdot A = \mathrm{amit}(B) \cdot A + \mathrm{anit}(C) \cdot A.$$

(See [12, Sect. 4.1] for complete details of this flexion computation.) Comparing with (16) shows that $\mathrm{arit}(B) = \mathrm{axit}(B, -B)$. Thus if A and B are push-invariant, (27) reduces to

$$\mathrm{swap}\Big(\mathrm{ari}\big(\mathrm{swap}(A), \mathrm{swap}(B)\big)\Big) = \mathrm{arit}(B) \cdot A - \mathrm{arit}(A) \cdot B + \mathrm{lu}(A, B),$$

which is exactly $\mathrm{ari}(A, B)$ by (18). □

Proof (*Theorem* 3) Using these two propositions, the proof becomes reasonably easy. We first consider the case where $A, B \in \mathrm{ARI}_{\underline{al}/\underline{al}}$. In particular A and B are alternal. Set $C = \mathrm{ari}(A, B)$. The mould C is alternal by Theorem 2. By Proposition 2, we know that A and B are push-invariant, so by Proposition 3 we have $\mathrm{swap}(C) = \mathrm{swap}\big(\mathrm{ari}(A, B)\big) = \mathrm{ari}\big(\mathrm{swap}(A), \mathrm{swap}(B)\big)$. But this is also alternal by Theorem 2, so $C \in \mathrm{ARI}_{al/al}$. Furthermore, it follows directly from the defining formula for the ari-bracket, which is additive in the mould depths, that if C is an ari-bracket of two moulds in ARI, i.e. with constant term equal to 0, we must have $C(u_1) = 0$, so $C \in \mathrm{ARI}_{\underline{al}/\underline{al}}$.

Now we consider the more general situation where $A, B \in \mathrm{ARI}_{\underline{al}*\underline{al}}$. Let A_0, B_0 be the constant-valued moulds such that $\mathrm{swap}(A) + A_0$ and $\mathrm{swap}(B) + B_0$ are alternal. From the definitions (13)–(16), we see that for any constant-valued mould M_0, we have $\mathrm{arit}(M_0) \cdot M = 0$. Indeed if M_0 is constant-valued, say with constant value c_r in depth r, then

$$\big(\mathrm{arit}(M_0) \cdot M\big)(\mathbf{w}) = \sum_{\substack{\mathbf{w}=\mathbf{abc}\\ \mathbf{b},\mathbf{c}\neq 0}} M(\mathbf{a}\lceil\mathbf{c})M_0(\mathbf{b}) - \sum_{\substack{\mathbf{w}=\mathbf{abc}\\ \mathbf{a},\mathbf{b}\neq 0}} M(\mathbf{a}\rceil\mathbf{c})M_0(\mathbf{b}).$$

Writing $\mathbf{w} = \mathbf{abc} = (u_1, \ldots, u_i)(u_{i+1}, \ldots, u_{i+j})(u_{i+j+1}, \ldots, u_r)$, we can rewrite this as

$$\sum_{i=0}^{r-2}\sum_{j=1}^{r-1} c_j M(u_1, \ldots, u_i, u_{i+1} + \cdots + u_{i+j+1}, u_{i+j+2}, \ldots, u_r)$$
$$- \sum_{i=1}^{r-1}\sum_{j=1}^{r-1} c_j M(u_1, \ldots, u_{i-1}, u_i + \cdots + u_{i+j}, u_{i+j+1}, \ldots, u_r).$$

But by renumbering i as $i + 1$ in the first sum shows that these two sums are in fact equal, so their difference is zero. An analogous computation shows that $\mathrm{arit}(M) \cdot M_0 = \mathrm{lu}(M, M_0)$; thus by (18), we have $\mathrm{ari}(M, M_0) = 0$. Thus we find that

$$\mathrm{ari}(A + A_0, B + B_0) = \mathrm{ari}(A, B) + \mathrm{ari}(A, B_0) + \mathrm{ari}(A_0, B) + \mathrm{ari}(A_0, B_0) = \mathrm{ari}(A, B). \tag{28}$$

Now, A and B are push-invariant by Proposition 2, and constant-valued moulds are always push-invariant, so $A + A_0$ and $B + B_0$ are also push-invariant; thus we have

$$\begin{aligned} \mathrm{swap}(C) &= \mathrm{swap}\big(\mathrm{ari}(A, B)\big) \\ &= \mathrm{swap}\big(\mathrm{ari}(A + A_0, B + B_0)\big) \quad \text{by (28)} \\ &= \mathrm{ari}\big(\mathrm{swap}(A + A_0), \mathrm{swap}(B + B_0)\big) \quad \text{by (26)}. \end{aligned}$$

But since swap preserves constant-valued moulds, we have $\mathrm{swap}(A + A_0) = \mathrm{swap}(A) + A_0$ and $\mathrm{swap}(B + B_0) = \mathrm{swap}(B) + B_0$. These two moulds are alternal by hypothesis, so by Theorem 2, their ari-bracket is alternal, i.e. $\mathrm{swap}(C)$ is alternal. Since as above we have $C(u_1) = 0$, we find that in fact C is not just in $\mathrm{ARI}_{\underline{al}*\underline{al}}$ but in $\mathrm{ARI}_{\underline{al}/\underline{al}}$. This completes the proof of Theorem 3. □

We will see in the next section that the double shuffle space $\mathfrak{ds}$ defined in Sect. 1 is isomorphic to the space of polynomial-valued moulds $\mathrm{ARI}^{pol}_{\underline{al}*\underline{il}}$, with the alternality property translating shuffle and the alternility property translating stuffle. Thus dimorphy is closely connected to double shuffle, but much more general, since the symmetry properties of alternality or alternility on itself or its swap can hold for any mould, not just polynomial ones.

4 Dictionary with the Lie Algebra and Double Shuffle Framework

Let $C_i = \mathrm{ad}(x)^{i-1}y \in \mathbb{Q}\langle x, y\rangle$, where $\mathrm{ad}(x)y = [x, y]$. By Lazard elimination (see [2, Proposition 10a]), the subring $\mathbb{Q}\langle C_1, C_2, \ldots\rangle$, which we denote simply by $\mathbb{Q}\langle C\rangle$, is free on the C_i. Let $\mathbb{Q}_0\langle C\rangle$ denote the subspace of polynomials in the C_i with constant term equal to 0. Define a linear map

$$\begin{aligned} ma : \mathbb{Q}_0\langle C\rangle &\overset{\sim}{\rightarrow} \mathrm{ARI}^{pol} \\ C_{a_1} \cdots C_{a_r} &\mapsto A_{a_1,\ldots,a_r} \end{aligned} \tag{29}$$

where $A_{a_1,\ldots,a_r}$ is the polynomial mould concentrated in depth r defined by

$$A_{a_1,\ldots,a_r}(u_1, \ldots, u_r) = (-1)^{a_1+\cdots+a_r-r} u_1^{a_1-1} \cdots u_r^{a_r-1}. \tag{30}$$

This map ma is trivially invertible and thus an isomorphism of vector spaces. Let $\mathrm{Lie}[C]$ denote the free Lie algebra $\mathrm{Lie}[C_1, C_2, \ldots]$ on the C_i. Note that, again by Lazard elimination, we can write $\mathrm{Lie}[x, y] = \mathbb{Q}x \oplus \mathrm{Lie}[C]$. Since by its definition, all elements of the double shuffle space $\mathfrak{ds} \subset \mathrm{Lie}[x, y]$ are polynomials of degree ≥ 3, we have

$$\mathfrak{ds} \subset \mathrm{Lie}[C] \subset \mathbb{Q}_0\langle C\rangle.$$

Definition 4 Let $\mathcal{MT}_0$ denote the Lie algebra whose underlying space is the space of polynomials $\mathbb{Q}_0\langle C\rangle$, equipped with the Poisson bracket (6), and let $\mathfrak{mt}$ denote the subspace of Lie polynomials in the C_i, i.e. the vector space $\mathrm{Lie}[C]$ equipped with the Poisson bracket. Observe that $\mathfrak{mt}$ is closed under the Poisson bracket since if f, g are Lie then so are $D_f(g)$, $D_g(f)$ and $[f, g]$, so $\mathfrak{mt}$ is a Lie algebra. The letters "M-T" stand for *twisted Magnus* (cf. [10]).

Let $\mathcal{MT}$ denote the universal enveloping algebra of $\mathfrak{mt}$. It is isomorphic as a vector space to $\mathbb{Q}\langle C\rangle$, and like all universal enveloping algebras, it is equipped with a pre-Lie[2] law $\odot$. In the special case where $g \in \mathfrak{mt}$, the pre-Lie law on $\mathcal{MT}$ reduces to the expression $f \odot g = fg - D_g(f)$, so that we have $f \odot g - g \odot f = \{f, g\}$ as befits the pre-Lie law of a universal enveloping algebra.

Let us also define the twisted Magnus group as the exponential $MT = \exp^{\odot}(\mathfrak{mt})$, where

$$\exp^{\odot}(f) = \sum_{n\geq 0} \frac{1}{n!} f^{\odot n}.$$

Note that

$$f^{\odot n} = f^{\odot(n-1)} \odot f = f^n - D_f(f^{\odot n}),$$

which gives an explicit recursive expression for $f^{\odot n}$.

Theorem 4 (Racinet) *The linear isomorphism (29) is a Lie algebra isomorphism*

$$ma : \mathcal{MT}_0 \xrightarrow{\sim} \mathrm{ARI}^{pol}, \tag{31}$$

and it restricts to a Lie algebra isomorphism of the Lie subalgebras

$$ma : \mathfrak{mt} \xrightarrow{\sim} \mathrm{ARI}_{al}^{pol}. \tag{32}$$

Proof In view of the fact that ma is invertible as a linear map, the isomorphism (31) follows from the following identity relating the Poisson bracket and the ari-bracket on polynomial-valued moulds, which was proven by Racinet in his thesis ([10, Appendix A], see also [12, Corollary 3.3.4]):

$$ma\big(\{f, g\}\big) = \mathrm{ari}\big(ma(f), ma(g)\big). \tag{33}$$

The isomorphism (32), identifying Lie polynomials with alternal polynomial moulds, follows from a standard argument that we indicate briefly, as it is merely an adaptation to Lie$[C]$ of the similar argument following the definition of the shuffle relations in (3). Let Δ denote the standard cobracket on $\mathbb{Q}\langle C\rangle$ defined by $\Delta(C_i) = C_i \otimes 1 + 1 \otimes C_i$. Then the Lie subspace Lie$[C]$ of the polynomial algebra $\mathbb{Q}\langle C\rangle$ is the space of primitive elements for Δ, i.e. elements $f \in$ Lie$[C]$ satisfying $\Delta(f) = f \otimes 1 + 1 \otimes f$. This condition on f is given explicitly on the coefficients of f by the family of shuffle relations

$$\sum_{D \in \mathrm{sh}(C_{a_1} \cdots C_{a_r}, C_{b_1} \cdots C_{b_s})} (f|D) = 0,$$

where $(f|D)$ denotes the coefficient in the polynomial f of the monomial D in the C_i. But these conditions are exactly equivalent to the alternality relations

[2] A pre-Lie law must satisfy the defining relation $\big((f \odot g) \odot h\big) - \big(f \odot (g \odot h)\big) = \big((f \odot h) \odot g\big) - \big(f \odot (h \odot g)\big)$.

$$\sum_{D\in \mathrm{sh}((a_1,\dots,a_r),(b_1,\dots,b_s))} ma(f)(D) = 0,$$

proving (32). □

Theorem 5 *The linear isomorphism (32) restricts to a linear isomorphism of the subspaces*

$$ma : \mathfrak{ds} \xrightarrow{\sim} \mathrm{ARI}^{pol}_{\underline{al}*\underline{il}}. \tag{34}$$

Proof By (32), since $\mathfrak{ds} \subset \mathfrak{mt}$, we have $ma : \mathfrak{ds} \hookrightarrow \mathrm{ARI}^{pol}_{\underline{al}}$. If an element $f \in \mathfrak{ds}$ has a depth 1 component, i.e. if the coefficient of $x^{n-1}y$ in f is non-zero, then n is odd; this is a simple consequence of solving the depth 2 stuffle relations (see [4, Theorem 2.30 (i)] for details). Thus, if the mould $ma(f)$ has a depth 1 component, it will be an even function, since by the definition of ma the degree of $ma(f)(u_1)$ is equal to the degree of f minus 1. This shows that ma maps $\mathfrak{ds}$ to moulds that are even in depth 1, i.e.

$$ma : \mathfrak{ds} \hookrightarrow \mathrm{ARI}^{pol}_{\underline{al}}.$$

It remains only to show that if $f \in \mathfrak{ds}$ then $\mathrm{swap}\big(ma(f)\big)$ is alternil up to addition of a constant mould, i.e. that the stuffle conditions (5) imply the alternility of $\mathrm{swap}\big(ma(f)\big)$.

By additivity, we may assume that f is of homogeneous degree n. Let C be the constant mould concentrated in depth n given by

$$C(u_1,\dots,u_n) = \frac{(-1)^{n-1}}{n}(f|x^{n-1}y),$$

and let $A = \mathrm{swap}\big(ma(f)\big) + C$. Ecalle showed (see [10, Appendix A] or [12, (3.2.6)] for full details) that we have the following explicit expression for $\mathrm{swap}\big(ma(f)\big)$. If for $r \geq 1$ we write the depth r part of f_* as

$$(f_*)^r = \sum_{\mathbf{a}=(a_1,\dots,a_r)} c_{\mathbf{a}}\, y_{a_1}\cdots y_{a_r}, \tag{35}$$

then $\mathrm{swap}\big(ma(f)\big)$ is given by

$$\mathrm{swap}\big(ma(f)\big)(v_1,\dots,v_r) = \sum_{\mathbf{a}=(a_1,\dots,a_r)} c_{\mathbf{a}}\, v_1^{a_1-1}\cdots v_r^{a_r-1} \tag{36}$$

Note that since f is homogeneous of degree n, the associated mould

$$A = \mathrm{swap}\big(ma(f)\big) + C$$

is concentrated in depths $\leq n$. We will use this close relation between the polynomial f_* and the mould A to show that the stuffle relations (5) on f_* are equivalent to the alternility of A.

For any pair of integers $1 \leq r \leq s$, let $A_{r,s}$ denote the alternility sum associated to the mould A as in (25). By definition, A is alternil if and only if $A_{r,s} = 0$ for all pairs $1 \leq r \leq s$. Recall from Sect. 2 that the alternility sum $A_{r,s}$ is a polynomial in $v_1, \ldots, v_{r+s}$ obtained by summing up polynomial terms in one-to-one correspondence with the terms of the stuffle of two sequences of lengths r and s. By construction, the coefficient of a monomial $w = v_1^{b_1-1} \cdots v_{r+s}^{b_{r+s}-1}$ in the alternility term corresponding to a given stuffle term is equal to the coefficient in f_* of the stuffle term itself. This follows from a direct calculation obtained by expanding the alternility terms; for example, the alternility term corresponding to the stuffle term (y_i, y_{j+k}, y_l) in (20) is given by

$$\frac{1}{v_2 - v_2}\big(A(v_1, v_2, v_4) - A(v_1, v_3, v_4)\big)$$

(see (22)), whose polynomial expansion is given by

$$\sum_{\mathbf{a}=(a_1,a_2,a_3)} c_{\mathbf{a}} v_1^{a_1-1}\Big(\sum_{m=0}^{a_2-2} v_2^m v_3^{a_2-2-m}\Big) v_4^{a_3-1},$$

and the coefficient of the monomial $v_1^{i-1} v_2^{j-1} v_3^{k-1} v_4^{l-1}$ in this alternility term corresponds to $a_1 - 1 = i - 1$, $m = j - 1$, $a_2 - 2 - m = k - 1$ and $a_3 - 1 = l - 1$, i.e. $a_1 = i$, $a_2 = j + k$, $a_3 = l$, so it is equal to $c_{i,j+k,l}$ which is exactly the coefficient $(f_*|y_i y_{j+k} y_l)$ in (35). The alternility sum is equal to zero if and only if the coefficient of each monomial in $v_1, \ldots, v_{r+s}$ is equal to zero, which is thus equivalent to the full set of stuffle relations on f_*. $\square$

In view of (33) and (34), a mould-theoretic proof of Racinet's theorem consists in proving that $\mathrm{ARI}_{\underline{al*il}}^{pol}$ is a Lie algebra under the ari-bracket. To prove this mould-theoretic version, we need to make use of the Lie group GARI associated to ARI, defined in the next section. In Sect. 6 we give the necessary results from Ecalle's theory, and the theorem is proved in Sect. 7.

5 The Group GARI

In this section we introduce several notions on the group GARI of moulds with constant term 1, which are group analogs of the Lie notions introduced in Sect. 2. To move from the Lie algebra ARI to the associated group GARI, Ecalle introduces a pre-Lie law on ARI, defined as follows:

$$\mathrm{preari}(A,B) = \mathrm{arit}(B)\cdot A + \mathrm{mu}(A,B), \tag{37}$$

where arit and mu are as defined in (16) and (13). Indeed, if $A, B \in \mathrm{ARI}$ then $\mathrm{preari}(A, B)$ also lies in ARI, and it is straightforward to check that preari satisfies the defining condition of pre-Lie laws given in Sect. 4. Using preari, Ecalle defines an exponential map on ARI in the standard way:

$$\exp_{\mathrm{ari}}(A) = \sum_{n\geq 0} \frac{1}{n!}\,\mathrm{preari}(\underbrace{A,\ldots,A}_{n}), \tag{38}$$

where

$$\mathrm{preari}(\underbrace{A,\ldots,A}_{n}) = \mathrm{preari}(\mathrm{preari}(\underbrace{A,\ldots,A}_{n-1}),A).$$

This map is the exponential isomorphism $\exp_{\mathrm{ari}} : \mathrm{ARI} \to \mathrm{GARI}$, where GARI is nothing other than the group of all moulds with constant term equal to 1, equipped with the multiplication law, denoted gari, that comes as always from the Campbell-Hausdorff law $\mathrm{ch}(\cdot,\cdot)$ on ARI:

$$\mathrm{gari}\big(\exp_{\mathrm{ari}}(A), \exp_{\mathrm{ari}}(B)\big) = \exp_{\mathrm{ari}}(\mathrm{ch}(A,B)). \tag{39}$$

The gari-inverse of a mould $B \in \mathrm{GARI}$ is denoted $\mathrm{inv}_{\mathrm{gari}}(B)$. The inverse isomorphism of $\exp_{\mathrm{ari}}$ is denoted by $\log_{\mathrm{ari}}$.

Like all Lie algebras, ARI is equipped with an action of the associated group GARI, namely the standard adjoint action, denoted $\mathrm{Ad}_{\mathrm{ari}}$ (Ecalle denotes it simply adari, but we have modified it to stress the fact that it represents the adjoint action of the group $GARI$ on ARI):

$$\begin{aligned}\mathrm{Ad}_{\mathrm{ari}}(A)\cdot B &= \mathrm{gari}\big(\mathrm{preari}(A,B), \mathrm{inv}_{\mathrm{gari}}(A)\big)\\ &= \frac{d}{dt}|_{t=0}\ \mathrm{gari}\big(A, \exp_{\mathrm{ari}}(tB), \mathrm{inv}_{\mathrm{gari}}(A)\big)\\ &= B + \mathrm{ari}\big(\log_{\mathrm{ari}}(A), B\big) + \frac{1}{2}\,\mathrm{ari}\big(\log_{\mathrm{ari}}(A), \mathrm{ari}\big(\log_{\mathrm{ari}}(A), B\big)\big) + \cdots\end{aligned} \tag{40}$$

Finally, to any mould $A \in \mathrm{GARI}$ (i.e. any mould in the u_i with constant term 1), Ecalle associates an automorphism $\mathrm{ganit}(A)$ of the ring of all moulds in the u_i under the mu-multiplication which is just the exponential of the derivation $\mathrm{anit}\big(\log_{\mathrm{ari}}(A)\big)$.

The analogous objects exist for moulds in the v_i. If preari denotes the pre-Lie law on $\overline{\mathrm{ARI}}$ given by (37) (but for the derivation arit of $\overline{\mathrm{ARI}}$), then the formula (38) defines an analogous exponential isomorphism $\overline{\mathrm{ARI}} \to \overline{\mathrm{GARI}}$, where $\overline{\mathrm{GARI}}$ consists of all moulds in the variables v_i with constant term 1 and multiplication determined by (39) (note that this definition depends on that of arit, so just as the Lie bracket ari is different for ARI and $\overline{\mathrm{ARI}}$, the multiplication is different for GARI and $\overline{\mathrm{GARI}}$).

As above, we let the automorphism $\text{ganit}(A)$ of $\overline{\text{GARI}}$ associated to each $A \in \overline{\text{GARI}}$ be defined as the exponential of the derivation $\text{anit}\big(\log_{\text{ari}}(A)\big)$ of $\overline{\text{ARI}}$.

Definition 5 A mould $A \in \text{GARI}$ (resp. $\overline{\text{GARI}}$) is *symmetral* if for all words $\mathbf{u}$, $\mathbf{v}$ in the u_i (resp. in the v_i), we have

$$\sum_{\mathbf{w}\in \text{sh}(\mathbf{u},\mathbf{v})} A(\mathbf{w}) = A(\mathbf{u})A(\mathbf{v}). \tag{41}$$

Following Ecalle, we write GARI_{as} (resp. $\overline{\text{GARI}}_{as}$) for the set of symmetral moulds in GARI (resp. $\overline{\text{GARI}}$). The property of *symmetrality* is the group equivalent of alternality; in particular,

$$A \in \text{ARI}_{al} \ \ (\text{resp. } \overline{\text{ARI}}_{al}) \ \Leftrightarrow \ \exp_{\text{ari}}(A) \in \text{GARI}_{as} \ \ (\text{resp. } \overline{\text{GARI}}_{as}). \tag{42}$$

Remark Let MT denote the *twisted Magnus group* of power series in $\mathbb{Q}\langle\langle C_1, C_2, \ldots\rangle\rangle$ with constant term 1, identified with the exponential of the twisted Magnus Lie algebra $\mathfrak{mt}$ defined by

$$\exp^{\odot}(f) = \sum_{n\geq 0} \frac{1}{n!} f^{\odot n}$$

for $f \in \mathfrak{mt}$, where $\odot$ is the pre-Lie law

$$f \odot g = fg + D_f(g) \tag{43}$$

defined for $f, g \in \mathfrak{mt}$ (see Sect. 4). The group MT is equipped with the twisted Magnus multiplication

$$\big(f \odot g\big)(x, y) = f(x, gyg^{-1})g(x, y). \tag{44}$$

Notice that it makes sense to use the same symbol $\odot$ for (43) and (44), because in fact $\odot$ is the multiplication on the completion of the universal enveloping algebra of $\mathfrak{mt}$, and (43) and (44) merely represent the particular expressions that it takes on two elements of $\mathfrak{mt}$ resp. two elements of MT.

The multiplication (44) corresponds to the gari-multiplication in the sense that the map ma defined in (29) yields a group isomorphism $MT \xrightarrow{\sim} \text{GARI}^{pol}$. If $g \in MT$, then the automorphism $\text{ganit}\big(ma(g)\big)$ is the GARI-version of the automorphism of MT given by mapping $x \mapsto x$ and $y \mapsto yg$.

The fact of having non-polynomial moulds in GARI gives enormously useful possibilities of expanding the familiar symmetries and operations (derivations, shuffle and stuffle relations etc.) to a broader situation. In particular, the next section contains some of Ecalle's most important results in mould multizeta theory, which make use of moulds with denominators and have no analog within the usual polynomial framework.

6 The Mould Pair *pal*/*pil* and Ecalle's Fundamental Identity

In this section we enter into the "second drawer" of Ecalle's powerful toolbox, with the mould pair *pal*/*pil* and Ecalle's fundamental identity.

Definition 6 Let *dupal* be the mould defined explicitly by the following formulas: $dupal(\emptyset) = 0$ and for $r \geq 1$,

$$dupal(u_1, \ldots, u_r) = \frac{B_r}{r!} \frac{1}{u_1 \cdots u_r} \left(\sum_{i=0}^{r} (-1)^i \binom{r-1}{i} u_{i+1} \right), \tag{45}$$

where B_r denotes the r-th Bernoulli number.

This mould is actually quite similar to a power series often studied in classical situations. Indeed, if we define dar to be the mould operator defined by

$$\text{dar} \cdot A(u_1, \ldots, u_r) = u_1 \cdots u_r \, A(u_1, \ldots, u_r),$$

then dar $\cdot dupal$ is a polynomial-valued mould, so it is the image of a power series under ma; explicitly

$$\text{dar} \cdot dupal = ma\Big(x - \frac{\text{ad}(-y)}{\exp(\text{ad}(-y)) - 1}(x)\Big).$$

Ecalle gave several equivalent definitions of the key mould *pal*, but the most recent one (see [7]) appears to be the simplest and most convenient. If we define dur to be the mould operator defined by

$$\text{dur} \cdot A(u_1, \ldots, u_r) = (u_1 + \cdots + u_r) \, A(u_1, \ldots, u_r),$$

then the mould *pal* is defined recursively by

$$\text{dur} \cdot pal = \text{mu}(pal, dupal). \tag{46}$$

Calculating the first few terms of *pal* explicitly, we find that

$$\begin{aligned} pal(\emptyset) &= 1 \\ pal(u_1) &= \frac{1}{2u_1} \\ pal(u_1, u_2) &= \frac{u_1 + 2u_2}{12u_1u_2(u_1 + v_2)} \\ pal(u_1, u_2, u_3) &= \frac{-1}{24u_1u_3(u_1 + u_2)}. \end{aligned}$$

Let $pil = \mathrm{swap}(pal)$. The most important result concerning pal, necessary for the proof of Ecalle's fundamental identity below, is the following.

Theorem 6 *The moulds pal and pil are symmetral.*

In [6, Sect. 4.2], the mould pil (called $\mathfrak{ess}$) is given an independent definition which makes it easy to prove that it is symmetral. Similarly, it is not too hard to prove that pal is symmetral using the definition (46). The real difficulty is to prove that pil (as defined in [6]) is the swap of pal (as defined in (46)). Ecalle sketched beautiful proofs of these two facts in [7], and the details are fully written out in [12, Sects. 4.2, 4.3].

Before proceeding to the fundamental identity, we need a useful result in which a very simple v-mould is used to give what amounts to an equivalent definition of alternility.[3]

Proposition 4 *Let pic be the v-mould defined by* $\mathrm{pic}(v_1,\ldots,v_r) = 1/v_1\cdots v_r$. *Then for any alternal mould $A \in \overline{\mathrm{ARI}}$, the mould* $\mathrm{ganit}(pic)\cdot A$ *is alternil.*

Proof The proof is deferred to Appendix C. □

We now come to Ecalle's fundamental identity.
Ecalle's fundamental identity: For any push-invariant mould A, we have

$$\mathrm{swap}\big(\mathrm{Ad}_{\mathrm{ari}}(pal)\cdot A\big) = \mathrm{ganit}(pic)\cdot\big(\mathrm{Ad}_{\mathrm{ari}}(pil)\cdot\mathrm{swap}(A)\big). \tag{47}$$

The proof of this fundamental identity actually follows as a consequence of (27) and a more general fundamental identity, similar but taking place in the group GARI and valid for all moulds. It is given in full detail in [12, Theorem 4.5.2].

7 The Main Theorem

In this section we give Ecalle's main theorem on dimorphy, which shows how the mould pal transforms moulds with the double symmetry $\underline{al}*\underline{al}$ to moulds that are $\underline{al}*\underline{il}$. We then show how Racinet's theorem follows directly from this. We first need a useful lemma.

Lemma 1 *If C is a constant-valued mould, then*

$$\mathrm{ganit}(pic)\cdot\mathrm{Ad}_{\mathrm{ari}}(pil)\cdot C = C. \tag{48}$$

[3] This is just one example of a general identity valid for *flexion units*, see [6, p. 64] where Ecalle explains the notion of alternality twisted by a flexion unit and asserts that alternility is merely alternality twisted by the flexion unit $1/v_1$.

Proof [1, Corollary 4.43] We apply the fundamental identity (47) in the case where $A = \text{swap}(A) = C$ is a constant-valued mould, obtaining

$$\text{swap}\big(\text{Ad}_{\text{ari}}(pal)\cdot C\big) = \text{ganit}(pic)\cdot \text{Ad}_{\text{ari}}(pil)\cdot C.$$

So it is enough to show that the left-hand side of this is equal to C, i.e. that $\text{Ad}_{\text{ari}}(pal)\cdot C = C$, since a constant mould is equal to its own swap. As we saw just before (28), the definitions (13)–(16) imply that $\text{ari}(A, C) = 0$ for all $A \in \text{ARI}$. Now, by (40) we see that $\text{Ad}_{\text{ari}}(pal)\cdot C$ is a linear combination of iterated ari-brackets of $\log_{\text{ari}}(pal)$ with C, but since $pal \in \text{GARI}$, $\log_{\text{ari}}(pal) \in \text{ARI}$, so $\text{ari}(\log_{\text{ari}}(pal), C) = 0$, i.e. all the bracketed terms in (40) are 0. Thus $\text{Ad}_{\text{ari}}(pal)\cdot C = C$. This concludes the proof. □

We can now state the main theorem on moulds.

Theorem 7 *The action of the operator* $\text{Ad}_{\text{ari}}(pal)$ *on the Lie subalgebra* $\text{ARI}_{\underline{al}*\underline{al}} \subset \text{ARI}$ *yields a Lie isomorphism of subspaces*

$$\text{Ad}_{\text{ari}}(pal) : \text{ARI}_{\underline{al}*\underline{al}} \xrightarrow{\sim} \text{ARI}_{\underline{al}*\underline{il}}. \tag{49}$$

Thus in particular $\text{ARI}_{\underline{al}*\underline{il}}$ *forms a Lie algebra under the* ari-*bracket.*

Proof The proof we give appears not to have been published anywhere by Ecalle, but we learned its outline from him through a personal communication to the second author, for which we are grateful.

Note first that $\text{Ad}_{\text{ari}}(pal)$ preserves the depth 1 component of moulds in ARI, so if A is even in depth 1 then so is $\text{Ad}_{\text{ari}}(pal)\cdot A$. We first consider the case where $A \in \text{ARI}_{\underline{al}/\underline{al}}$, i.e. *swap*$(A)$ is alternal without addition of a constant correction. By (42), the mould $\text{Ad}_{\text{ari}}(pal)\cdot A$ is alternal, since pal is symmetral by Theorem 6. By Proposition 2, A is push-invariant, so Ecalle's fundamental identity (47) holds. Since $A \in \text{ARI}_{\underline{al}/\underline{al}}$, swap$(A)$ is alternal, and by Theorem 6, pil is alternal; thus by (42), $\text{Ad}_{\text{ari}}(pil)\cdot \text{swap}(A)$ is alternal. Then by Proposition 4, $\text{ganit}(pic)\cdot \text{Ad}_{\text{ari}}(pil)\cdot \text{swap}(A)$ is alternil, and finally by (47), $\text{swap}\big(\text{Ad}_{\text{ari}}(pal)\cdot A\big)$ is alternil, which proves that $\text{Ad}_{\text{ari}}(pal)\cdot A \in \text{ARI}_{\underline{al}/\underline{il}}$ as desired.

We now consider the general case where $A \in \text{ARI}_{\underline{al}*\underline{al}}$. Let C be the constant-valued mould such that $\text{swap}(A) + C$ is alternal. As above, we have that $\text{Ad}_{\text{ari}}(pal)\cdot A$ is alternal, so to conclude the proof of the theorem it remains only to show that its swap is alternil up to addition of a constant mould, and we will show that this constant mould is exactly C. As before, since $\text{swap}(A) + C \in \overline{\text{ARI}}$ is alternal, the mould

$$\text{Ad}_{\text{ari}}(pil)\cdot\big(\text{swap}(A) + C\big) = \text{Ad}_{\text{ari}}(pil)\cdot \text{swap}(A) + \text{Ad}_{\text{ari}}(pil)\cdot C$$

is also alternal. Thus by Proposition 4, applying $\text{ganit}(pic)$ to it yields the alternil mould

$$\text{ganit}(pic) \cdot \text{Ad}_{\text{ari}}(pil) \cdot \text{swap}(A) + \text{ganit}(pic) \cdot \text{Ad}_{\text{ari}}(pil) \cdot C.$$

By Lemma 1, this is equal to

$$\text{ganit}(pic) \cdot \text{Ad}_{\text{ari}}(pil) \cdot \text{swap}(A) + C, \tag{50}$$

which is thus alternil. Now, since A is push-invariant by Proposition 2, we can apply (47) and find that (50) is equal to

$$\text{swap}\big(\text{Ad}_{\text{ari}}(pal) \cdot A\big) + C,$$

which is thus also alternil. Therefore $\text{swap}\big(\text{Ad}_{\text{ari}}(pal) \cdot A\big)$ is alternil up to a constant, which precisely means that $\text{Ad}_{\text{ari}}(pal) \cdot A \in \text{ARI}_{\underline{al}*\underline{il}}$ as claimed. Since $\text{Ad}_{\text{ari}}(pal)$ is invertible (with inverse $\text{Ad}_{\text{ari}}\big(\text{inv}_{\text{gari}}(pal)\big)$) and by the analogous arguments this inverse takes $\text{ARI}_{\underline{al}*\underline{il}}$ to $\text{ARI}_{\underline{al}*\underline{al}}$, this proves that (49) is an isomorphism. □

Corollary 1 $\text{ARI}^{pol}_{\underline{al}*\underline{il}}$ *forms a Lie algebra under the* ari*-bracket.*

Proof By Proposition 1, ARI^{pol} is a Lie algebra under the ari-bracket, so since $\text{ARI}_{\underline{al}*\underline{il}}$ is as well by Theorem 7, their intersection also forms a Lie algebra. □

In view of (33) and (34), this corollary is equivalent to Racinet's theorem that $\mathfrak{ds}$ is a Lie algebra under the Poisson bracket.

Acknowledgements The authors wish to thank the ICMAT and the Women in Number Theory group for hosting wonderful workshops, the referees for the careful reading of this paper and the many improvements that resulted, and Jean Ecalle for his guidance. Salerno was partially supported by the NSF-AWM Mentoring Travel grant.

Appendix A

Proof of Theorem 2. We cut it into two separate results as explained in the main text.

Proposition 5 *If A, B are alternal moulds then $C = \text{lu}(A, B)$ is alternal.*

Proof We have

$$C(\mathbf{w}) = \text{lu}(A, B)(\mathbf{w}) = \sum_{\mathbf{w}=\mathbf{ab}} \big(A(\mathbf{a})B(\mathbf{b}) - B(\mathbf{a})A(\mathbf{b})\big),$$

so we need to show that the following sum vanishes:

$$\begin{aligned} \sum_{\mathbf{w}\in \text{sh}(\mathbf{u},\mathbf{v})} C(\mathbf{w}) &= \sum_{\mathbf{w}\in \text{sh}(\mathbf{u},\mathbf{v})} \text{lu}(A, B)(\mathbf{w}) \\ &= \sum_{\mathbf{w}\in \text{sh}(\mathbf{u},\mathbf{v})} \sum_{\mathbf{w}=\mathbf{ab}} \big(A(\mathbf{a})B(\mathbf{b}) - B(\mathbf{a})A(\mathbf{b})\big). \end{aligned} \tag{51}$$

This sum breaks into three pieces: the terms where **a** contain letters from both **u** and **v**, the case where **a** contains only letters from **u** or from **v** but **b** contains letters from both, and finally the cases $\mathbf{a} = \mathbf{u}$, $\mathbf{b} = \mathbf{v}$ and $\mathbf{a} = \mathbf{v}$, $\mathbf{b} = \mathbf{u}$.

The first type of terms add up to zero because we can break up the sum into smaller sums where **a** lies in the shuffle of the first i letters of **u** and j letters of **b**, and these terms already sum to zero since A and B are alternal.

The second type of term adds up to zero for the same reason, because even though **a** may contain only letters from one of **u** and **v**, **b** must contain letters from both and therefore the same reasoning holds.

The third type of term yields $A(\mathbf{u})B(\mathbf{v}) - B(\mathbf{u})A(\mathbf{v})$ when $\mathbf{a} = \mathbf{u}, \mathbf{b} = \mathbf{v}$ and $A(\mathbf{v})B(\mathbf{u}) - B(\mathbf{v})A(\mathbf{u})$ when $\mathbf{a} = \mathbf{v}$, $\mathbf{b} = \mathbf{u}$, which cancel out. Thus the sum (51) adds up to zero. □

Proposition 6 *If A and B are alternal moulds in* ARI*, then* $C = \mathrm{arit}(B) \cdot A$ *is alternal.*

Proof By definition, C is alternal if

$$\sum_{\mathbf{w}=\mathrm{sh}(\mathbf{x},\mathbf{y})} C(\mathbf{w}) = 0,$$

for all pairs of non-trivial words **x**, **y**.

Pick an arbitrary pair of non-trivial words **x**, **y**, of appropriate length (that is, so that their lengths add up to the length of A plus the length of B). We will be shuffling **x** and **y** together, and the resulting word is then broken up into three parts (all possible ones) in order to compute the flexions. Thus, if we break up $\mathbf{w} = \mathbf{abc}$, **a** must be a shuffle of some parts at the beginning of each word **x**, **y**, **b** must come from shuffling their middles, and **c** can only come from shuffling the last parts. Then we can rewrite this computation as follows:

$$\begin{aligned}
\sum_{\mathbf{w}=\mathrm{sh}(\mathbf{x},\mathbf{y})} \mathrm{arit}(B)\cdot A(\mathbf{w}) &= \sum_{\mathbf{w}=\mathrm{sh}(\mathbf{x},\mathbf{y})} \left(\sum_{\substack{\mathbf{w}=\mathbf{abc}\\ \mathbf{c}\neq\emptyset}} A(\mathbf{a}\lceil\mathbf{c})B(\mathbf{b}) - \sum_{\substack{\mathbf{w}=\mathbf{abc}\\ \mathbf{a}\neq\emptyset}} A(\mathbf{a}\rceil\mathbf{c})B(\mathbf{b}) \right) \\
&= \sum_{\substack{\mathbf{x}=\mathbf{x}_1\mathbf{x}_2\mathbf{x}_3\\ \mathbf{y}=\mathbf{y}_1\mathbf{y}_2\mathbf{y}_3,\, \mathbf{x}_3\mathbf{y}_3\neq\emptyset}} \quad \sum_{\substack{\mathbf{a}=\mathrm{sh}(\mathbf{x}_1,\mathbf{y}_1)\\ \mathbf{b}=\mathrm{sh}(\mathbf{x}_2,\mathbf{y}_2),\mathbf{c}=\mathrm{sh}(\mathbf{x}_3,\mathbf{y}_3)}} A(\mathbf{a}\lceil\mathbf{c})B(\mathbf{b}) \\
&- \sum_{\substack{\mathbf{x}=\mathbf{x}_1\mathbf{x}_2\mathbf{x}_3\\ \mathbf{y}=\mathbf{y}_1\mathbf{y}_2\mathbf{y}_3,\, \mathbf{x}_1\mathbf{y}_1\neq\emptyset}} \quad \sum_{\substack{\mathbf{a}=\mathrm{sh}(\mathbf{x}_1,\mathbf{y}_1)\\ \mathbf{b}=\mathrm{sh}(\mathbf{x}_2,\mathbf{y}_2),\mathbf{c}=\mathrm{sh}(\mathbf{x}_3,\mathbf{y}_3)}} A(\mathbf{a}\rceil\mathbf{c})B(\mathbf{b}).
\end{aligned}$$

Now for a fixed splitting of each **x** and **y** into three parts, we have the following possibilities.

Case I. Both $\mathbf{x}_2 = \mathbf{y}_2 = \emptyset$. Then $B(\emptyset) = 0$ so we are done.

Case II. Both $\mathbf{x}_2$ and $\mathbf{y}_2$ are nonempty. The trick here is that because of the flexion operations, no matter how $\mathbf{b} = \mathrm{sh}(\mathbf{x}_2, \mathbf{y}_2)$ is shuffled, the part being added together with the last letter in **a** and the first letter in **c** remains the same. Thus, if we further fix a particular **a** and **c**, we get that

$$\sum_{\mathbf{b}=\mathrm{sh}(\mathbf{x}_2,\mathbf{y}_2)} A(\mathbf{a}\lceil\mathbf{c})B(\mathbf{b}) = A(\mathbf{a}\lceil\mathbf{c}) \sum_{\mathbf{b}=\mathrm{sh}(\mathbf{x}_2,\mathbf{y}_2)} B(\mathbf{b}) = 0$$

and

$$\sum_{\mathbf{b}=\mathrm{sh}(\mathbf{x}_2,\mathbf{y}_2)} A(\mathbf{a}\rceil\mathbf{c})B(\mathbf{b}) = A(\mathbf{a}\rceil\mathbf{c}) \sum_{\mathbf{b}=\mathrm{sh}(\mathbf{x}_2,\mathbf{y}_2)} B(\mathbf{b}) = 0,$$

by alternality of B. And thus,

$$\sum_{\substack{\mathbf{a}=\mathrm{sh}(\mathbf{x}_1,\mathbf{y}_1)\\ \mathbf{c}=\mathrm{sh}(\mathbf{x}_3,\mathbf{y}_3)}} \sum_{\mathbf{b}=\mathrm{sh}(\mathbf{x}_2,\mathbf{y}_2)} A(\mathbf{a}\lceil\mathbf{c})B(\mathbf{b}) = 0$$

and

$$\sum_{\substack{\mathbf{a}=\mathrm{sh}(\mathbf{x}_1,\mathbf{y}_1)\\ \mathbf{c}=\mathrm{sh}(\mathbf{x}_3,\mathbf{y}_3)}} \sum_{\mathbf{b}=\mathrm{sh}(\mathbf{x}_2,\mathbf{y}_2)} A(\mathbf{a}\rceil\mathbf{c})B(\mathbf{b}) = 0.$$

Case III. Either $\mathbf{x}_2 = \emptyset$ or $\mathbf{y}_2 = \emptyset$, but not both. Without loss of generality, assume $\mathbf{x}_2 = \emptyset$. Then we have

$$\sum_{\substack{\mathbf{a}=\mathrm{sh}(\mathbf{x}_1,\mathbf{y}_1)\\ \mathbf{b}=\mathbf{y}_2,\mathbf{c}=\mathrm{sh}(\mathbf{x}_3,\mathbf{y}_3)}} A(\mathbf{a}\lceil\mathbf{c})B(\mathbf{b}) = B(\mathbf{y}_2) \sum_{\substack{\mathbf{a}=\mathrm{sh}(\mathbf{x}_1,\mathbf{y}_1)\\ \mathbf{c}=\mathrm{sh}(\mathbf{x}_3,\mathbf{y}_3)}} A(\mathbf{a}\lceil\mathbf{c}).$$

And similarly,

$$\sum_{\substack{\mathbf{a}=\mathrm{sh}(\mathbf{x}_1,\mathbf{y}_1)\\ \mathbf{b}=\mathbf{y}_2,\mathbf{c}=\mathrm{sh}(\mathbf{x}_3,\mathbf{y}_3)}} A(\mathbf{a}\rceil\mathbf{c})B(\mathbf{b}) = B(\mathbf{y}_2) \sum_{\substack{\mathbf{a}=\mathrm{sh}(\mathbf{x}_1,\mathbf{y}_1)\\ \mathbf{c}=\mathrm{sh}(\mathbf{x}_3,\mathbf{y}_3)}} A(\mathbf{a}\rceil\mathbf{c}).$$

Recall that by definition

$$\mathrm{sh}(\mathbf{x}_1,\mathbf{y}_1) = \mathrm{sh}(\mathbf{x}_1',\mathbf{y}_1)(\text{last letter in } \mathbf{x}_1) + \mathrm{sh}(\mathbf{x}_1,\mathbf{y}_1')(\text{last letter in } \mathbf{y}_1)$$

and

$$\mathrm{sh}(\mathbf{x}_3,\mathbf{y}_3) = (\text{first letter in } \mathbf{x}_3)\,\mathrm{sh}(\mathbf{x}_3',\mathbf{y}_3) + (\text{first letter in } \mathbf{y}_3)\,\mathrm{sh}(\mathbf{x}_3,\mathbf{y}_3').$$

Thus,

$$\mathbf{a}\lceil\mathbf{c} = \mathrm{sh}(\mathbf{x}_1,\mathbf{y}_1)(\text{sum of letters in } \mathbf{y}_2 \text{ plus first letter in } \mathbf{x}_3)\,\mathrm{sh}(\mathbf{x}_3',\mathbf{y}_3) \quad (52)$$

or

$$\mathbf{a}\lceil\mathbf{c} = \mathrm{sh}(\mathbf{x_1}, \mathbf{y_1})(\text{sum of letters in } \mathbf{y_2} \text{ plus first letter in } \mathbf{y_3})\,\mathrm{sh}(\mathbf{x_3}, \mathbf{y}'_3) \tag{53}$$

and

$$\mathbf{a}\rceil\mathbf{c} = \mathrm{sh}(\mathbf{x}'_1, \mathbf{y_1})(\text{sum of letters in } \mathbf{y_2} \text{ plus last letter in } \mathbf{x_1})\,\mathrm{sh}(\mathbf{x_3}, \mathbf{y_3}) \tag{54}$$

or

$$\mathbf{a}\rceil\mathbf{c} = \mathrm{sh}(\mathbf{x_1}, \mathbf{y}'_1)(\text{sum of letters in } \mathbf{y_2} \text{ plus last letter in } \mathbf{y_1})\,\mathrm{sh}(\mathbf{x_3}, \mathbf{y_3}). \tag{55}$$

Recall that, since $\mathbf{x_2}$ is assumed to be empty, then for a given $\mathbf{x_1}, \mathbf{x_3}$, we can let $\overline{\mathbf{x_1}}, \overline{\mathbf{x_3}}$ be so that $\overline{\mathbf{x_1}}$ is $\mathbf{x_1}$ with an additional letter given by the first letter of $\mathbf{x_3}$ and $\overline{\mathbf{x_3}}$ is defined in the logical way. That means that Eqs. (52) and (54) are exactly the same. Thus, we get direct cancellation for all possible choices of $\mathbf{x_1}, \mathbf{x_3}$ (this is compatible with the restrictions on nonemptiness given by the definition).

We cannot do the same for (53) and (55), since $\mathbf{y_2}$ is assumed to be nonempty. For these, notice that if we keep $\mathbf{y}$ fixed and sum over all possible partitions of $\mathbf{x} = \mathbf{x_1}\mathbf{x_2}\mathbf{x_3}$ where $\mathbf{x_2} = \emptyset$, and $\mathbf{x_3} \neq \emptyset$ we get that each

$$\mathbf{a}\lceil\mathbf{c} = \mathrm{sh}(\mathbf{x_1}, \mathbf{y_1})(\text{sum of letters in } \mathbf{y_2} \text{ plus first letter in } \mathbf{y_3})\,\mathrm{sh}(\mathbf{x_3}, \mathbf{y}'_3)$$

could be seen as a term in the shuffle $\mathrm{sh}(\mathbf{x}, \mathbf{y_1}\lceil\mathbf{y_3})$. To see this, suppose that

$$\mathbf{x} = u_1 \cdots u_k | u_{k+1} \cdots u_l = \mathbf{x_1}|\mathbf{x_3}$$

and that

$$\mathbf{y} = u_{l+1} \cdots u_{l+i} | u_{l+i+1} \cdots u_{l+j} | u_{l+j+1} \cdots u_n = \mathbf{y_1}|\mathbf{y_2}|\mathbf{y_3}.$$

Then

$$\begin{aligned}\mathbf{a}\lceil\mathbf{c} = \mathrm{sh}((u_1 \cdots u_k), (u_{l+1} \cdots u_{l+i}))(u_{l+i+1} + \cdots + u_{l+j} + u_{l+j+1})\\ \cdot\, \mathrm{sh}((u_{k+1} \cdots u_l), (u_{l+j+2} \cdots u_n)).\end{aligned}$$

And so if we allow the k to shift from 1 to l, this is essentially the shuffling of the words $u_1 \cdots u_l = \mathbf{x}$ and $u_{l+1} \cdots u_{l+i}(u_{l+i+1} + \cdots + u_{l+j} + u_{l+j+1})u_{l+j+2} \cdots u_n = \mathbf{y_1}\lceil\mathbf{y_3}$. Thus we have

$$\sum_{\substack{\mathbf{x}=\mathbf{x_1}\mathbf{x_3} \\ \mathbf{x_3}\neq\emptyset}} \sum_{\substack{\mathbf{a}=\mathrm{sh}(\mathbf{x_1},\mathbf{y_1}) \\ \mathbf{b}=\mathbf{y_2}, \mathbf{c}=y_{\mathrm{first}}\,\mathrm{sh}(\mathbf{x_3},\mathbf{y}'_3)}} A(\mathbf{a}\lceil\mathbf{c}) = \sum_{\mathbf{w}=\mathrm{sh}(\mathbf{x},\mathbf{y_1}\lceil\mathbf{y_3})} A(\mathbf{w}) = 0$$

by alternality of A.

A similar argument holds for the terms corresponding to the other flexion (the terms corresponding to Eq. (55)).

Putting all of these cases together, we see that indeed, C is alternal. □

Proposition 7 *If A and B are alternal moulds in $\overline{\mathrm{ARI}}$, then $C = \mathrm{arit}(B)\cdot A$ is alternal.*

Proof As with the proof for ARI_{al}, we have to show that

$$\sum_{\mathbf{w}=\mathrm{sh}(\mathbf{x},\mathbf{y})} C(\mathbf{w}) = 0,$$

for all pairs of non-trivial words $\mathbf{x}$, $\mathbf{y}$. Again, this can be rewritten as follows:

$$\begin{aligned}\sum_{\mathbf{w}=\mathrm{sh}(\mathbf{x},\mathbf{y})} \mathrm{arit}(B)\cdot A(\mathbf{w}) &= \sum_{\mathbf{w}=\mathrm{sh}(\mathbf{x},\mathbf{y})} \left(\sum_{\substack{\mathbf{w}=\mathbf{abc}\\ \mathbf{c}\neq\emptyset}} A(\mathbf{ac})B(\mathbf{b}\rfloor) - \sum_{\substack{\mathbf{w}=\mathbf{abc}\\ \mathbf{a}\neq\emptyset}} A(\mathbf{ac})B(\lfloor\mathbf{b}) \right)\\ &= \sum_{\substack{\mathbf{x}=\mathbf{x}_1\mathbf{x}_2\mathbf{x}_3\\ \mathbf{y}=\mathbf{y}_1\mathbf{y}_2\mathbf{y}_3,\mathbf{x}_3\mathbf{y}_3\neq\emptyset}} \sum_{\substack{\mathbf{a}=\mathrm{sh}(\mathbf{x}_1,\mathbf{y}_1)\\ \mathbf{b}=\mathrm{sh}(\mathbf{x}_2,\mathbf{y}_2),\mathbf{c}=\mathrm{sh}(\mathbf{x}_3,\mathbf{y}_3)}} A(\mathbf{ac})B(\mathbf{b}\rfloor)\\ &- \sum_{\substack{\mathbf{x}=\mathbf{x}_1\mathbf{x}_2\mathbf{x}_3\\ \mathbf{y}=\mathbf{y}_1\mathbf{y}_2\mathbf{y}_3,\mathbf{x}_1\mathbf{y}_1\neq\emptyset}} \sum_{\substack{\mathbf{a}=\mathrm{sh}(\mathbf{x}_1,\mathbf{y}_1)\\ \mathbf{b}=\mathrm{sh}(\mathbf{x}_2,\mathbf{y}_2),\mathbf{c}=\mathrm{sh}(\mathbf{x}_3,\mathbf{y}_3)}} A(\mathbf{ac})B(\lfloor\mathbf{b})\end{aligned}$$

Again, for a fixed splitting of each $\mathbf{x}$ and $\mathbf{y}$ into three parts, we have the following possibilities.

Case I. Both $\mathbf{x}_2 = \mathbf{y}_2 = \emptyset$. Then $B(\emptyset) = 0$ so we are done.

Case II. Both $\mathbf{x}_2$ and $\mathbf{y}_2$ are nonempty.

Here, no matter how $\mathbf{b} = \mathrm{sh}(\mathbf{x}_2, \mathbf{y}_2)$ is shuffled, the part being subtracted from $\mathbf{b}$, which is either the last letter in $\mathbf{a}$ or the first letter in $\mathbf{c}$, remains the same if we fix a particular $\mathbf{a}$ and $\mathbf{c}$. Thus, we get that

$$\mathbf{b}\rfloor_i = \mathrm{sh}(\mathbf{x}_2, \mathbf{y}_2)_i - \text{first letter in } \mathbf{c} = \mathrm{sh}((\mathbf{x}_{2_\mathbf{k}} - \text{first letter in } \mathbf{c}), (\mathbf{y}_{2_\mathbf{k}} - \text{first letter in } \mathbf{c}))_i$$

and

$$\lfloor\mathbf{b}_i = \mathrm{sh}(\mathbf{x}_2, \mathbf{y}_2)_i - \text{last letter in } \mathbf{a} = \mathrm{sh}((\mathbf{x}_{2_\mathbf{k}} - \text{last letter in } \mathbf{a}), (\mathbf{y}_{2_\mathbf{k}} - \text{last letter in } \mathbf{a}))_i.$$

Thus,

$$\sum_{\mathbf{b}=\mathrm{sh}(\mathbf{x}_2,\mathbf{y}_2)} A(\mathbf{ac})B(\mathbf{b}\rfloor) = A(\mathbf{ac}) \sum_{\mathbf{b}=\mathrm{sh}(\mathbf{x}_2,\mathbf{y}_2)} B(\mathbf{b}\rfloor) = 0$$

and

$$\sum_{\mathbf{b}=\mathrm{sh}(\mathbf{x}_2,\mathbf{y}_2)} A(\mathbf{ac})B(\lfloor\mathbf{b}) = A(\mathbf{ac}) \sum_{\mathbf{b}=\mathrm{sh}(\mathbf{x}_2,\mathbf{y}_2)} B(\lfloor\mathbf{b}) = 0,$$

by alternality of B. And thus,

$$\sum_{\substack{\mathbf{a}=\mathrm{sh}(\mathbf{x_1},\mathbf{y_1})\\ \mathbf{c}=\mathrm{sh}(\mathbf{x_3},\mathbf{y_3})}} \sum_{\mathbf{b}=\mathrm{sh}(\mathbf{x_2},\mathbf{y_2})} A(\mathbf{ac})B(\mathbf{b}\rfloor) = 0$$

and

$$\sum_{\substack{\mathbf{a}=\mathrm{sh}(\mathbf{x_1},\mathbf{y_1})\\ \mathbf{c}=\mathrm{sh}(\mathbf{x_3},\mathbf{y_3})}} \sum_{\mathbf{b}=\mathrm{sh}(\mathbf{x_2},\mathbf{y_2})} A(\mathbf{ac})B(\lfloor\mathbf{b}) = 0.$$

Case III. Either $\mathbf{x_2} = \emptyset$ or $\mathbf{y_2} = \emptyset$, but not both. Without loss of generality, assume $\mathbf{x_2} = \emptyset$.

Recall, again, that by definition

$$\mathrm{sh}(\mathbf{x_1}, \mathbf{y_1}) = \mathrm{sh}(\mathbf{x}_1', \mathbf{y_1})(\text{last letter in } \mathbf{x_1}) + \mathrm{sh}(\mathbf{x_1}, \mathbf{y}_1')(\text{last letter in } \mathbf{y_1})$$

and

$$\mathrm{sh}(\mathbf{x_3}, \mathbf{y_3}) = (\text{first letter in } \mathbf{x_3})\, \mathrm{sh}(\mathbf{x}_3', \mathbf{y_3}) + (\text{first letter in } \mathbf{y_3})\, \mathrm{sh}(\mathbf{x_3}, \mathbf{y}_3').$$

Since $\mathbf{x_2} = \emptyset$, we can see that

$$\mathbf{b}\rfloor_i = \mathbf{y_2}_{\mathrm{i}} - \text{first letter in } \mathbf{c}$$

and

$$\lfloor\mathbf{b}_i = \mathbf{y_2}_{\mathrm{i}} - \text{last letter in } \mathbf{a}.$$

For a given $\mathbf{x_1}, \mathbf{x_3}$, we can let $\overline{\mathbf{x_1}}, \overline{\mathbf{x_3}}$ be so that $\overline{\mathbf{x_1}}$ is $\mathbf{x_1}$ with an additional letter given by the first letter of $\mathbf{x_3}$ and $\overline{\mathbf{x_3}}$ is defined in the logical way. That means that

$$A(\mathrm{sh}(\overline{\mathbf{x_1}}', \mathbf{y_1})(\text{last letter in } \overline{\mathbf{x_1}})\, \mathrm{sh}(\overline{\mathbf{x_3}}, \mathbf{y_3}))B(\lfloor\mathbf{b})$$

and

$$A(\mathrm{sh}(\mathbf{x_1}, \mathbf{y_1})(\text{first letter in } \mathbf{x_3})\, \mathrm{sh}(\mathbf{x}_3', \mathbf{y_3}))B(\mathbf{b}\rfloor)$$

are identical (for each fixed shuffling).

Thus, we get direct cancellation for all possible choices of $\mathbf{x_1}, \mathbf{x_3}$ (this is compatible with the restrictions on nonemptiness given by the definition).

The only terms that have not cancelled out are the ones coming from the second term in the shuffle equations above. Now, suppose that

$$\mathbf{x} = v_1 \cdots v_k | v_{k+1} \cdots v_l = \mathbf{x_1}|\mathbf{x_3}$$

and that

$$\mathbf{y} = v_{l+1} \cdots v_{l+i} | v_{l+i+1} \cdots v_{l+j} | v_{l+j+1} \cdots v_n = \mathbf{y_1}|\mathbf{y_2}|\mathbf{y_3},$$

and fix this splitting of $\mathbf{y}$. Then

$$\mathbf{ac} = \mathrm{sh}(v_1 \cdots v_k, v_{l+1} \cdots v_{l+i})v_{l+j+1}\,\mathrm{sh}(v_{k+1} \cdots v_l, v_{l+j+2} \cdots v_n).$$

And so if we allow the k to shift from 1 to l, this is essentially the shuffling of the words $v_1 \cdots v_l = \mathbf{x}$ and $v_{l+1} \cdots v_{l+i}, v_{l+j+1}, v_{l+j+2} \cdots v_n = \mathbf{y}_1\mathbf{y}_3$. Notice that this shuffling fixes $\mathbf{b}\rfloor$, since

$$\mathbf{b}\rfloor = (v_{l+i+1} - v_{l+j+1}, \ldots, v_{l+j} - v_{l+j+1}).$$

Thus we have

$$\sum_{\substack{\mathbf{x}=\mathbf{x}_1\mathbf{x}_3 \\ \mathbf{x}_3\neq\emptyset}} \sum_{\substack{\mathbf{a}=\mathrm{sh}(\mathbf{x}_1,\mathbf{y}_1),\mathbf{b}=\mathbf{y}_2 \\ \mathbf{c}=y_{\text{first}}\,\mathrm{sh}(\mathbf{x}_3,\mathbf{y}_3')}} A(\mathbf{ac})B(\mathbf{b}\rfloor) = B(\mathbf{b}\rfloor) \sum_{w=\mathrm{sh}(\mathbf{x},\mathbf{y}_1\mathbf{y}_3)} A(w) = 0$$

by alternality of A.

A similar argument holds for the terms corresponding to the other flexion. Combining all the cases, we see that indeed, C is alternal. □

Appendix B

Proof of Proposition 2. By additivity, we may assume that A is concentrated in a fixed depth d, meaning that $A(u_1, \ldots, u_r) = 0$ for all $r \neq d$. We use the following two lemmas.

Lemma 2 *If $A \in \mathrm{ARI}_{al}$, then*

$$A(u_1, \ldots, u_r) = (-1)^{r-1} A(u_r, \ldots, u_1);$$

in other words, A is mantar*-invariant. Similarly, if $A \in \overline{\mathrm{ARI}}_{al}$ then again A is* mantar*-invariant.*

Proof We give the argument for ARI; the result in $\overline{\mathrm{ARI}}$ comes from the identical computation with u_i replaced by v_i. We first show that the sum of shuffle relations

$$\mathrm{sh}\big((1), (2, \ldots, r)\big) - \mathrm{sh}\big((2, 1), (3, \ldots, r)\big) + \mathrm{sh}\big((3, 2, 1), (4, \ldots, r)\big) + \cdots$$

$$+ (-1)^{r-2}\,\mathrm{sh}\big((r-1, \ldots, 2, 1), (r)\big) = (1, \ldots, r) + (-1)^r (r, \ldots, 1).$$

Indeed, using the recursive formula for shuffle, we can write the above sum with two terms for each shuffle, as

$$\begin{aligned}(1,\ldots,r)&+2\cdot \mathrm{sh}\big((1),(3,\ldots,r)\big)\\ &-2\cdot \mathrm{sh}\big((1),(3,\ldots,r)\big)-3\cdot \mathrm{sh}\big((2,1),(4,\ldots,r)\big)\\ &+3\cdot \mathrm{sh}\big((2,1),(4,\ldots,r)\big)+4\cdot \mathrm{sh}\big((3,2,1),(5,\ldots,r)\big)\\ &+\cdots+(-1)^{r-3}(r-1)\cdot \mathrm{sh}\big((r-2,\ldots,1),(r)\big)\\ &+(-1)^{r-2}(r-1)\cdot \mathrm{sh}\big((r-2,\ldots,1),(r)\big)+(-1)^{r-2}(r,r-1,\ldots,1)\\ &=(1,\ldots,r)+(-1)^r(r,\ldots,1).\end{aligned}$$

Using this, we conclude that if A satisfies the shuffle relations, then

$$A(u_1,\ldots,u_r)+(-1)^{r-1}A(u_r,\ldots,u_1)=0,$$

which is the desired result. □

Lemma 3 *If* $A\in \mathrm{ARI}_{\underline{al}*\underline{al}}$, *then* A *is* neg ∘ push*-invariant.*

Proof We first consider the case where $A\in \mathrm{ARI}_{\underline{al}/\underline{al}}$. Using the easily verified identity

$$\mathrm{neg}\circ \mathrm{push}=\mathrm{mantar}\circ \mathrm{swap}\circ \mathrm{mantar}\circ \mathrm{swap}, \tag{56}$$

and the fact that by Lemma 2, if $A\in \mathrm{ARI}_{\underline{al}/\underline{al}}$, then both A and $\mathrm{swap}(A)$ are mantar-invariant, we have

$$\begin{aligned}\mathrm{neg}\circ \mathrm{push}(A)(u_1,\ldots,u_r)&=\mathrm{mantar}\circ \mathrm{swap}\circ \mathrm{mantar}\circ \mathrm{swap}(A)(u_1,\ldots,u_r)\\ &=\mathrm{mantar}\circ \mathrm{swap}\circ \mathrm{swap}(A)(u_1,\ldots,u_r)\\ &=\mathrm{mantar}(A)(u_1,\ldots,u_r)\\ &=A(u_1,\ldots,u_r),\end{aligned} \tag{57}$$

so A is neg ∘ push-invariant.

Now suppose that $A\in \mathrm{ARI}_{\underline{al}*\underline{al}}$, so A is alternal and $\mathrm{swap}(A)+A_0$ is alternal for some constant mould A_0. By additivity, we may assume that A is concentrated in depth r. First suppose that r is odd. Then $\mathrm{mantar}(A_0)(v_1,\ldots,v_r)=(-1)^{r-1}A_0(v_r,\ldots,v_1)$, so since A_0 is a constant mould, it is mantar-invariant. But $\mathrm{swap}(A)+A_0$ is alternal, so it is also mantar-invariant by Lemma 2; thus $\mathrm{swap}(A)$ is mantar-invariant, and the identity neg ∘ push = mantar ∘ swap ∘ mantar ∘ swap shows that A is neg ∘ push-invariant as in (57).

Finally, we assume that A is concentrated in even depth r. Here we have $\mathrm{mantar}(A_0)=-A_0$, so we cannot use the argument above; indeed $\mathrm{swap}(A)+A_0$ is mantar-invariant, but

$$\mathrm{mantar}(\mathrm{swap}(A))=\mathrm{swap}(A)+2A_0. \tag{58}$$

Instead, we note that if A is alternal then so is $\mathrm{neg}(A)=A$. Thus we can write A as a sum of an even and an odd function of the u_i via the formula

$$A = \frac{1}{2}(A + \text{neg}(A)) + \frac{1}{2}(A - \text{neg}(A)). \tag{59}$$

So it is enough to prove the desired result for all moulds concentrated in even depth r such that either $\text{neg}(A) = A$ (even functions) or $\text{neg}(A) = -A$ (odd functions). First suppose that A is even. Then since neg commutes with push and push is of odd order $r + 1$ and neg is of order 2, we have

$$(\text{neg} \circ \text{push})^{r+1}(A) = \text{neg}(A) = A. \tag{60}$$

However, we also have

$$\begin{aligned}
\text{neg} \circ \text{push}(A) &= \text{mantar} \circ \text{swap} \circ \text{mantar} \circ \text{swap}(A) \\
&= \text{mantar} \circ \text{swap}\big(\text{swap}(A) + 2A_0\big) \text{ by (58)} \\
&= \text{mantar}\big(A + 2A_0\big) \\
&= A - 2A_0.
\end{aligned}$$

Thus $(\text{neg} \circ \text{push})^{r+1}(A) = A - 2(r+1)A_0$, and this is equal to A by (60), so $A_0 = 0$; thus in fact $A \in \text{ARI}_{\underline{al}/\underline{al}}$ and that case is already proven.

Finally, if A is odd, i.e. $\text{neg}(A) = -A$, the same argument as above gives $A - 2(r+1)A_0 = -A$, so $A = (r+1)A_0$, so A is a constant-valued mould concentrated in depth r, but this contradicts the assumption that A is alternal since constant moulds are not alternal, unless $A = A_0 = 0$. Note that this argument shows that all moulds in $\text{ARI}_{\underline{al}*\underline{al}}$ that are not in $\text{ARI}_{\underline{al}/\underline{al}}$ must be concentrated in odd depths. □

We can now complete the proof of Proposition 2.[4] Because $A = \text{neg} \circ \text{push}(A)$, we have $\text{neg}(A) = \text{push}(A)$, so in fact we only need to show that $\text{neg}(A) = A$. As before, we may assume that A is concentrated in depth r. If $r = 1$, then A is an even function by assumption. If r is even, then as before we have $A = (\text{neg} \circ \text{push})^{2s+1}(A) = \text{neg}(A)$. Finally, assume $r = 2s + 1$ is odd. Since we can write A as a sum of an even and an odd part as in (59), we may assume that $\text{neg}(A) = -A$. Then, since A is alternal, using the shuffle $\text{sh}\big((u_1, \ldots, u_{2s})(u_{2s+1})\big)$, we have

$$\sum_{i=0}^{2s} A(u_1, \ldots, u_i, u_{2s+1}, u_{i+1}, \ldots, u_{2s}) = 0.$$

Making the variable change $u_0 \leftrightarrow u_{2s+1}$ gives

$$\sum_{i=0}^{2s} A(u_1, \ldots, u_i, u_0, u_{i+1}, \ldots, u_{2s}) = 0. \tag{61}$$

[4]Ecalle states this result in [6, Sect. 2.4] and there is also a proof in [7, Sect. 12], but we were not able to follow the argument, so we have provided this alternative proof.

Now consider the shuffle relation $\mathrm{sh}((u_1)(u_2,\ldots,u_{2s+1}))$, which gives

$$\sum_{i=1}^{2s+1} A(u_2,\ldots,u_i,u_1,u_{i+1},\ldots,u_{2s+1}) = 0. \tag{62}$$

Set $u_0 = -u_1 - \cdots - u_{2s+1}$. Since neg $\circ$ push acts like the identity on A, we can apply it to each term of (62) to obtain

$$\sum_{i=1}^{2s} -A(u_0,u_2,\ldots,u_i,u_1,u_{i+1},\ldots,u_{2s}) - A(u_0,u_2,\ldots,u_{2s},u_{2s+1}).$$

We apply neg $\circ$ push again to the final term of this sum in order to get the u_{2s+1} to disappear, obtaining

$$\sum_{i=1}^{2s} -A(u_0,u_2,\ldots,u_i,u_1,u_{i+1},\ldots,u_{2s}) + A(u_1,u_0,u_2,\ldots,u_{2s-1},u_{2s}) = 0.$$

Making the variable change $u_0 \leftrightarrow u_1$ in this identity yields

$$\sum_{i=1}^{2s} -A(u_1,u_2,\ldots,u_i,u_0,u_{i+1},\ldots,u_{2s}) + A(u_0,u_1,u_2,\ldots,u_{2s-1},u_{2s}) = 0. \tag{63}$$

Finally, adding (61) and (63) yields $2A(u_0,u_1,\ldots,u_{2s}) = 0$, so $A = 0$. This concludes the proof that $\mathrm{neg}(A) = A$ for all $A \in \mathrm{ARI}_{\underline{al}*\underline{al}}$, and thus, by Lemma 3, that $\mathrm{push}(A) = A$. This concludes the proof of Proposition 2. □

Appendix C

We follow Ecalle's more general construction of *twisted alternality* from [6, pp. 57–64]. Let $\mathbf{e} \in \overline{\mathrm{ARI}}$ be a *flexion unit*, which is a mould concentrated in depth 1 satisfying

$$\mathbf{e}(v_1) = -\mathbf{e}(-v_1)$$

and

$$\mathbf{e}(v_1)\mathbf{e}(v_2) = \mathbf{e}(v_1 - v_2)\mathbf{e}(v_2) + \mathbf{e}(v_1)\mathbf{e}(v_2 - v_1).$$

Associate to $\mathbf{e}$ the mould $\mathbf{ez} \in \overline{\mathrm{GARI}}$ defined by

$$\mathbf{ez}(v_1,\ldots,v_r) = \mathbf{e}(v_1)\cdots\mathbf{e}(v_r).$$

Then a mould $A \in \overline{\text{ARI}}$ is said to be **e**-*alternal* if $A = \text{ganit}(\mathbf{ez}) \cdot B$ where $B \in \overline{\text{ARI}}$ is alternal. The conditions for **e**-alternality can be written out using the explicit expression for ganit, using flexions, computed by Ecalle [6, (2.36)]:

$$\big(\text{ganit}(B) \cdot A\big)(\mathbf{w}) = \sum A(\mathbf{b}^1 \cdots \mathbf{b}^s) B(\lfloor \mathbf{c}^1) \cdots A(\lfloor \mathbf{c}^s), \tag{64}$$

where the sum runs over the decompositions of the word $\mathbf{w} = (u_1, \ldots, u_r)$ $(r \geq 1)$ as

$$\mathbf{w} = \mathbf{b}^1 \mathbf{c}^1 \cdots \mathbf{b}^s \mathbf{c}^s, \quad (s \geq 1)$$

where all $\mathbf{b}^i$ and $\mathbf{c}^i$ are non-empty words except possibly for $\mathbf{c}^s$. For example in small depths, setting $C = \text{ganit}(B) \cdot A$, we have

$$\begin{aligned} C(v_1) &= A(v_1) \\ C(v_1, v_2) &= A(v_1, v_2) + A(v_1)B(v_2 - v_1) \\ C(v_1, v_2, v_3) &= A(v_1, v_2, v_3) + A(v_1, v_2)B(v_3 - v_2) \\ &\quad + A(v_1)B(v_2 - v_1, v_3 - v_1) + A(v_1, v_3)B(v_2 - v_1). \end{aligned}$$

Using the expression (64) for $\text{ganit}(B) \cdot A$, the **e**-alternality relations can be written explicitly as follows. Let $Y_1 = (y_1, \ldots, y_r)$ and $Y_2 = (y_{r+1}, \ldots, y_{r+s})$. Then for each word in the stuffle set $\text{st}(Y_1, Y_2)$, we construct the associated **e**-*alternality term*, with an expression of the form

$$\big(C(\ldots, v_i, \ldots) - C(\ldots, v_j)\big)\mathbf{e}(v_i - v_j)$$

corresponding to each contraction (cf. (21)). For example, taking $Y_1 = (y_i, y_j)$ and $Y_2 = (y_k, y_l)$, the stuffle set $\text{st}(Y_1, Y_2)$ is given in (20), and the corresponding 13 **e**-alternality terms are, first of all the six shuffle terms

$$C(v_1, v_2, v_3, v_4), C(v_1, v_3, v_2, v_4), C(v_1, v_3, v_4, v_2), C(v_3, v_1, v_2, v_4),$$

$$C(v_3, v_1, v_4, v_2), C(v_3, v_4, v_1, v_2)$$

(cf. (22)), then the six terms with a single contraction

$$\big(C(v_1, v_2, v_4) - C(v_1, v_3, v_4)\big)\mathbf{e}(v_2 - v_3), \quad \big(C(v_1, v_2, v_4) - C(v_3, v_2, v_4)\big)\mathbf{e}(v_1 - v_3),$$

$$\big(C(v_1, v_3, v_2) - C(v_1, v_3, v_4)\big)\mathbf{e}(v_2 - v_4), \quad \big(C(v_1, v_4, v_2) - C(v_3, v_4, v_2)\big)\mathbf{e}(v_1 - v_3),$$

$$\big(C(v_3, v_1, v_2) - C(v_3, v_1, v_4)\big)\mathbf{e}(v_2 - v_4), \quad \big(C(v_3, v_1, v_2) - C(v_3, v_4, v_2)\big)\mathbf{e}(v_1 - v_4)$$

(cf. (23)), and finally the single term with two contractions,

$$\big(C(v_1, v_2) - C(v_3, v_2) - C(v_1, v_4) + C(v_2, v_4)\big)\mathbf{e}(v_1 - v_3)\mathbf{e}(v_2 - v_4).$$

The **e**-*alternality sum* $C_{r,s}$ is defined to be the sum of all the **e**-alternality terms corresponding to words in the stuffle set $\mathrm{st}(Y_1, Y_2)$; this sum is independent of the actual sequences Y_1, Y_2, depending only on their lengths r, s. The mould C is said to satisfy the **e**-*alternality relations* if $C_{r,s} = 0$ for all $1 \leq r \leq s$. Comparing with (22)–(24) we see that the notion of alternality is nothing but the special case of **e**-alternality for the flexion unit $\mathbf{e}(v_1) = 1/v_1$. The associated mould **ez** is thus equal to pic, so we find that $\mathrm{ganit}(pic) \cdot A$ is alternil if A is alternal.

References

1. Baumard, S.: Aspects modulaires et elliptiques des relations entre multizêtas. Ph.D. dissertation. Paris, France (2014)
2. Bourbaki, N.: Éléments de Mathématique. Groupes et Algèbres de Lie, Chapitres 2 et 3, 2nd ed. Springer (2006)
3. Brown, F.: Depth-graded motivic multiple zeta values (2013) arXiv:1301.3053
4. Carr, S.: Multizeta Values: lie algebras and periods on $M_{0,n}$. Ph.D. dissertation. Paris, France. http://math.unice.fr/~brunov/GdT/Carr.pdf (2008)
5. Cresson, J.: Calcul Moulien. Ann. Fac. Sci. Toulouse Math **XVIII**(2), 307–395 (2009)
6. Ecalle, J.: The flexion structure and dimorphy: flexion units, singulators, generators, and the enumeration of multizeta irreducibles. In: Costin, O., Fauvet, F., Menous, F., Sauzin, D. (eds.) Asymptotics in Dynamics, Geometry and PDEs; Generalized Borel Summation, vol. II, pp. 27–211. Edizioni della Normale, Pisa (2011)
7. Ecalle, J.: Eupolars and their bialternality grid. Acta Math. Vietnam. **40**, 545 (2015). https://doi.org/10.1007/s40306-015-0152-x
8. Furusho, H.: Double shuffle relation for associators. Ann. Math. **174**(1), 341–360 (2011)
9. Ihara, K., Kaneko, M., Zagier, D.: Derivation and double shuffle relations for multiple zeta values. Comp. Math. **142**(2), 307–338 (2006)
10. Racinet, G.: Séries génératrices non-commutatives de polyzêtas et associateurs de Drinfel'd. Ph.D. dissertation. Paris, France (2000)
11. Racinet, G.: Doubles mélanges des polylogarithmes multiples aux racines de l'unité. Publ. Math. Inst. Hautes Etudes Sci. **95**, 185–231 (2002)
12. Schneps, L.: ARI, GARI, Zig and Zag: an introduction to Ecalle's theory of multizeta values (2014). arXiv:1507.01534
13. Serre, J.P.: Lie Algebras and Lie Groups. Springer Lecture Notes, vol. 1500. Springer, Heidelberg (1965)

On Some Tree-Indexed Series with One and Two Parameters

F. Chapoton

Abstract There is a rich algebraic setting involving free pre-Lie algebras and the combinatorics of rooted trees. In this context, one can consider the analog of formal power series, called tree-indexed series. Several interesting such series are known, including one called Ω and its more recent one-parameter and two-parameters generalizations. This survey article explains how one can compute their coefficients using Ehrhart polynomials of lattice polytopes.

Keywords Order polytope · Rooted tree · Pre-Lie algebra · Ehrhart polynomial · q-Bernoulli number

Let us start by recalling how tree-indexed series appear when considering ordinary differential equations. This is very classical, but maybe not so well-known.

Let $V : \mathbb{R}^D \to \mathbb{R}^D$ be a smooth vector field and X be a curve $\mathbb{R} \to \mathbb{R}^D$. Let t be the coordinate on $\mathbb{R}$ (time) and let $x_1, \ldots, x_D$ be coordinates on $\mathbb{R}^D$.

The ordinary differential equation

$$\frac{d}{dt}X(t) = V(X(t)), \tag{1}$$

with initial condition $X(0) = X_0$, describes the movement of a material point X whose velocity is given by the vector field V.

When the vector field is linear, the solution is given by the exponential of the associated matrix. Otherwise, one can try to look for a solution as a formal power series in t. The result is

$$X(t) - X_0 = V(X_0)t + (V \triangleright V)(X_0)\frac{t^2}{2} + \cdots + (V \triangleright (V \triangleright (V \ldots)))(X_0)\frac{t^n}{n!} + \ldots \tag{2}$$

F. Chapoton (✉)
Institut de Recherche Mathématique Avancée, UMR 7501, Université de Strasbourg et CNRS, 7 rue René Descartes, 67000 Strasbourg, France
e-mail: chapoton@unistra.fr

J. I. Burgos Gil et al. (eds.), *Periods in Quantum Field Theory and Arithmetic*, Springer Proceedings in Mathematics & Statistics 314,
https://doi.org/10.1007/978-3-030-37031-2_16

Here $\triangleright$ is a bilinear operation on smooth vector fields defined by

$$V \triangleright W = \sum_{i,j} V_i (\partial_i W_j) \partial_j \tag{3}$$

where $V = \sum_i V_i \partial_i$ and $W = \sum_j W_j \partial_j$ in the chosen coordinates on $\mathbb{R}^D$.

This is half of the Lie bracket of vector fields, and $[V, W] = V \triangleright W - W \triangleright V$. The operation $\triangleright$ is modified by a general change-of-coordinates, but preserved under affine change-of-coordinates. It has the following property:

$$a \triangleright (b \triangleright c) - (a \triangleright b) \triangleright c = b \triangleright (a \triangleright c) - (b \triangleright a) \triangleright c, \tag{4}$$

as can be checked by an easy computation. This property is the definition of a pre-Lie product.

Let us now spend some time to describe free pre-Lie algebras.

Free associative algebras are very well known and can be easily described: they have bases indexed by words, and the associative product is given by concatenation of words. It turns out that free pre-Lie algebras also have a very nice explicit description, just slightly more complicated: they have bases indexed by decorated rooted trees, and the pre-Lie product is given by grafting of rooted trees.

A rooted tree is a finite connected graph without cycles and with a distinguished vertex called the root. A decoration of a rooted tree T by a set E is a map from the vertices of T to E. Here is an example, with $E = \{\heartsuit, \diamondsuit, \spadesuit, \clubsuit\}$:

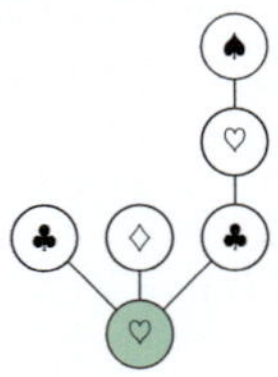

Note that trees will be drawn growing up, with their root at the bottom.

Consider a set of generators E and let $\mathrm{PL}(E)$ be the vector space over $\mathbb{Q}$ spanned by (isomorphism classes of) rooted trees decorated by E. The pre-Lie product is defined on the basis of $\mathrm{PL}(E)$ by

$$S \triangleright T = \sum_{v \in T} S \curvearrowright_v T, \tag{5}$$

where $S \curvearrowright_v T$ is the rooted tree obtained from the disjoint union of the rooted trees S and T by adding an edge between v and the root of S, and taking the root of T as root.

Here is an example with $E = \{\heartsuit, \clubsuit\}$:

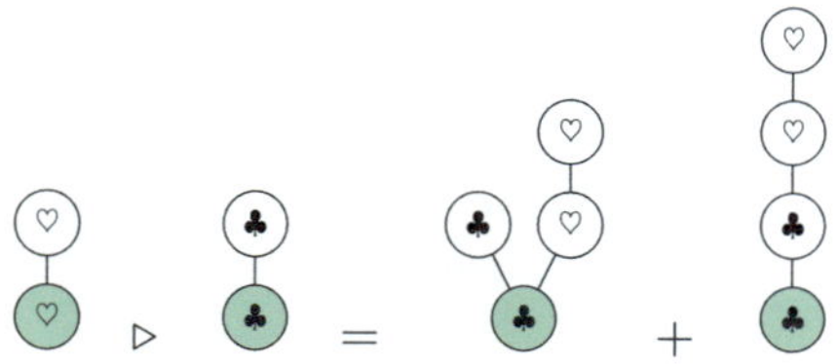

Let us now go back to the flow of vector fields. One can lift the formal power series solution (2) (letting also $t = 1$) to the element

$$A = \bigcirc + (\bigcirc \triangleright \bigcirc)\frac{1}{2} + \cdots + (\bigcirc \triangleright (\bigcirc \triangleright (\bigcirc \ldots)))\frac{1}{n!} + \cdots \tag{6}$$

which belongs to the completion of the free pre-Lie algebra on one generator $\bigcirc$ with respect to its natural grading by the number of vertices. If a finite sum of (undecorated) rooted trees is like a polynomial in one variable, then this is like a formal power series in one variable.

Using the definition of $\triangleright$ on rooted trees, one can compute the first few terms:

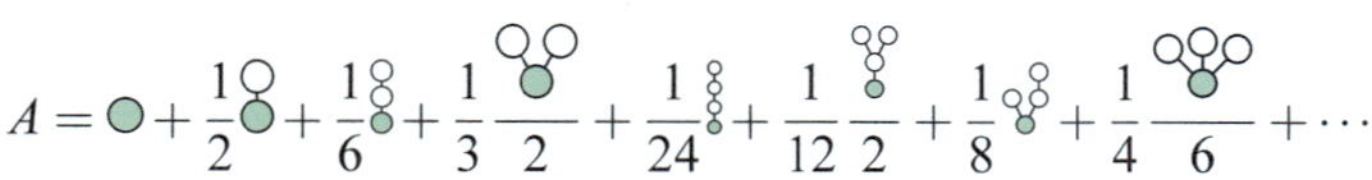

(7)

Because of its origin, this tree-indexed series could be called the "exact solution of the generic flow equation of vector fields". Note that this is completely independent of the ambient dimension D of vector fields.

Note also that one has written the coefficients of A in a specific way, by separating out (under every tree) a factor corresponding to the number of automorphisms of the tree (sometimes called a symmetry factor). Then the remaining coefficients have a very simple expression as the inverse of the tree-factorials.[1] All the coefficients are positive rational numbers.

All this story is tightly connected with several important theories in numerical analysis and mathematical physics. First, it is very close with John Butcher's theory of composition of Runge-Kutta methods, which is very well-known in numerical analysis under the keywords of B-series or Butcher series [4]. This setting has been recently generalized to numerical analysis on manifolds, see for instance [20, 21, 24]. In a similar direction, usage of pre-Lie products has been pioneered by Agracev and Gamkrelidze under a different name in [1], where the pre-Lie logarithm and the pre-Lie exponential were introduced. In relation with mathematical physics, the work of Connes and Kreimer on Hopf algebras and renormalisation of quantum field theories [11–13], at least in its initial stage, used an Hopf algebra of decorated rooted

[1]For more on the tree-factorial, see for example [3, 17].

trees, which is the dual of the universal enveloping algebra of a free pre-Lie algebra (also known as a Grossman-Larson Hopf algebra [16]).

Let PL be the free pre-Lie algebra on one generator (spanned by undecorated rooted trees), and let $\widehat{\mathrm{PL}}$ be its completion with respect to its graduation.

Let us now explain briefly why the space $\widehat{\mathrm{PL}}$ is a monoid, just like the space of formal power series with no constant term in one variable is a monoid under composition.

Every element B of $\widehat{\mathrm{PL}}$ can be considered as a linear operator acting on $\widehat{\mathrm{PL}}$ by

$$C \mapsto \phi_B(C), \tag{8}$$

for any $C \in \widehat{\mathrm{PL}}$, where ϕ_B is the unique morphism of completed pre-Lie algebras from $\widehat{\mathrm{PL}}$ to $\widehat{\mathrm{PL}}$ that maps the generator ● to B. Existence and uniqueness of ϕ_B are given by the freeness of PL as a pre-Lie algebra.

One can then show that the composition of two such operators is again of the same kind, and this defines an associative (but linear only with respect to its left argument) product $\circ$ on $\widehat{\mathrm{PL}}$, that it is natural to call the composition of tree-indexed series.

Let us now introduce our main tree-indexed series of interest: let Ω be the inverse of A for the composition law of tree-indexed series.

Using this definition, the first few terms can be computed and are given by

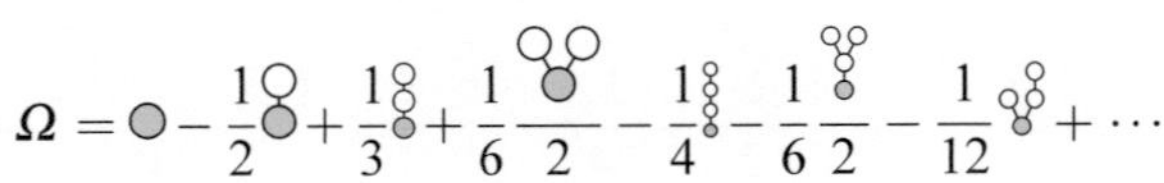
(9)

This series is also used in numerical analysis under the name of "backward error analysis" element. It can also be considered as providing a "pre-Lie Magnus expansion", similar to the more classical Magnus expansion, see [15]. The tree-indexed series A and Ω are the concrete expansions of the pre-Lie exponential and logarithm that play a crucial rôle in the general theory of pre-Lie algebras, see for example [23] for more details on this.

The story we want to tell here has started with the aim to understand the coefficients of Ω, which are complicated-looking rational numbers with signs, including Bernoulli numbers and other interesting numbers. This has lead to consider a one parameter version Ω_x with coefficients in the polynomial ring $\mathbb{Q}[x]$. Then, motivated by a connection found with the study of Lie idempotents, the author has introduced a q-analogue of Ω, called Ω_q, with coefficients in the field of rational functions in q.

Let us now explain how one can give an expression for the coefficients of Ω.

The aim is therefore to describe a procedure starting from a rooted tree T and giving a rational number Ω_T such that

$$\Omega = \sum_T \Omega_T \frac{T}{\# \operatorname{Aut} T}, \tag{10}$$

where Aut T is the automorphism group of the rooted tree T. This group is defined, after choosing a bijection between the set of vertices of T and the set of integers in the interval $[1, n]$, as the subgroup of the symmetric group S_n preserving the root and the adjacency of vertices.

This result that we are going to explain now has been obtained by Wright and Zhao in [30] and can be described using several steps:

$$\text{rooted tree} \xrightarrow{\textcircled{1}} \text{partial order} \xrightarrow{\textcircled{2}} \text{polytope} \xrightarrow{\textcircled{3}} \text{polynomial} \xrightarrow{\textcircled{4}} \text{rational number}$$

Let us present these steps in order.

(1) First, a rooted tree defines a partial order on the set of its vertices, where $x \leq y$ if and only if there is a path from the root up to y going through x. This partial order has the root as unique minimal element, and maximal elements are the leaves.

For example, the partial order associated with the tree 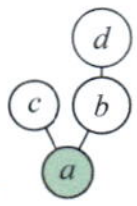 is described by the relations $a \leq c$ and $a \leq b \leq d$. Let us keep this tree as a running example.

(2) Next, given a partial order $(P, \leq)$, one can consider its order polytope in the space $\mathbb{R}^P$ (with coordinates x_i for $i \in P$). This is defined by the inequalities

$$\forall i \in P \quad 0 \leq x_i \leq 1, \tag{11}$$

$$\forall i, j \in P \quad i \leq j \implies x_i \leq x_j. \tag{12}$$

For the running example, this gives the polytope described by the inequalities $0 \leq x_a \leq x_c \leq 1$ and $0 \leq x_a \leq x_b \leq x_d \leq 1$.

The order polytope was introduced by Stanley [27]. It has some nice general properties. First, it is a lattice polytope, which means that all its vertices are in $\mathbb{Z}^P$ (and in fact in $\{0, 1\}^P$). Moreover, it is an empty lattice polytope, which means that it has no other points in $\mathbb{Z}^P$ than its vertices.

Some new results on order polytopes have been obtained recently in [26].

In the example, there are seven vertices:

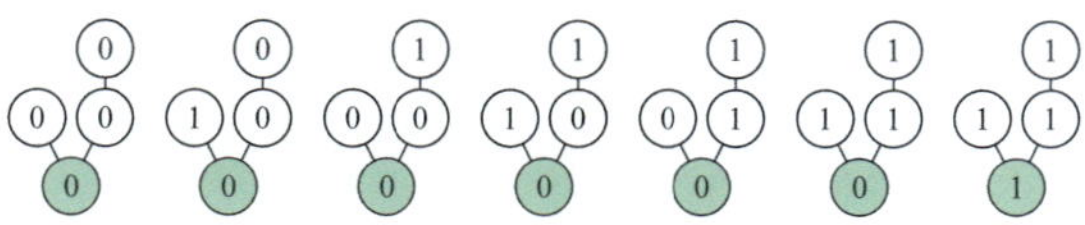

(3) Now, from a lattice polytope Q, one can define a polynomial, called the Ehrhart polynomial E_Q. Its defining property is that, for every integer $n \geq 0$, the value $E_Q(n)$ is the number of lattice points in the dilated polytope nQ (where all coordinates of vertices have been multiplied by n).

For the running example of rooted tree, one can compute (by interpolation)

$$E_T(n) = \frac{(n+1)(n+2)(n+3)(3n+4)}{24}, \tag{13}$$

where one uses the convenient shortcut notation E_T for the Ehrhart polynomial of the order polytope of the poset associated with T.

The Ehrhart polynomial has a wonderful general property called reciprocity, namely its value $E_Q(-n)$ at negative integers is (up to sign) the number of interior lattice points (points not belonging to any proper face) in the dilated polytope nQ. In particular, if the polytope Q is an empty lattice polytope, then

$$E_Q(-1) = 0. \tag{14}$$

Because the order polytope of a poset is an empty lattice polytope, one therefore knows that the Ehrhart polynomial of the order polytope of the poset of a rooted tree vanishes at -1.

A general reference for the theory of Ehrhart polynomials is [2].

(4) The last step is to go from a polynomial Z vanishing at -1 to the rational number $Z'(-1)$, which is also the value of the polynomial $Z/(1+x)$ at $x = -1$.

In our favorite example of rooted tree, from the Ehrhart polynomial displayed in (13), one gets the number $\frac{1}{12}$.

Theorem 1 (Wright-Zhao) *The coefficient Ω_T in the tree-indexed series Ω is given (up to sign) by the derivative at -1 of the Ehrhart polynomial $E_T(n)$ (defined using the order polytope of the poset of the rooted tree T).*

This is the promised explicit description of the coefficients of Ω. Given the last step of this description, it is natural to introduce the 1-parameter tree-indexed series Ω_x defined by

$$\Omega_x = \sum_T E_T(x) \frac{T}{\# \operatorname{Aut} T}. \tag{15}$$

This is like a generating series of the Ehrhart polynomials of trees. Note that every coefficient is divisible by $1 + x$, because all order polytopes are empty lattice polytopes.

Now let us turn to another one-parameter deformation of Ω, with a very different origin.

Let

$$\mathrm{FQSYM} = \oplus_{n \geq 0} \mathbb{Q} S_n \tag{16}$$

be the direct sum of group rings of symmetric groups. This has a natural structure of graded Hopf algebra, introduced by Malvenuto and Reutenauer [22]. It is also known as the algebra of free-quasi-symmetric functions [14]. The associative product is defined on the basis by

$$\pi * \sigma = \pi \shuffle \operatorname{shift}(\sigma), \tag{17}$$

where $\shuffle$ is the classical shuffle product of words and “shift” is the minimal shift on indices of σ that makes them disjoint from the indices of π. For example, the product of the permutation 12 in S_2 with itself is given by

$$12 * 12 = 12 \shuffle 34 = 1234 + 1324 + 1342 + 3124 + 3142 + 3412. \tag{18}$$

In this algebra, one can also cut the product $*$ into two halves: let $\pi \prec \sigma$ be the sum of words in $\shuffle$ where the first letter comes from π and let $\pi \succ \sigma$ be the sum of words in $\shuffle$ where the first letter comes from σ. For example:

$$12 \prec 12 = 1234 + 1324 + 1342, \tag{19}$$

$$12 \succ 12 = 3124 + 3142 + 3412. \tag{20}$$

Then obviously $\pi * \sigma = \pi \prec \sigma + \pi \succ \sigma$. These two products define on FQSYM a structure of dendriform algebra, as defined in [19].

It is then rather easy to check that the operation

$$\pi \triangleright \sigma = \pi \succ \sigma - \sigma \prec \pi \tag{21}$$

defines a pre-Lie product of FQSYM.[2]

By the universal property of the free pre-Lie algebra on one generator PL, there is a unique morphism of pre-Lie algebras ψ from PL to FQSYM that maps the generator ○ to the permutation 1 in S_1. This morphism can be extended into a morphism (still denoted by ψ) from the completion $\widehat{\mathrm{PL}}$ of PL to the completion of FQSYM. This gives a way to map tree-indexed series to infinite linear combinations of permutations.

It happens that the image of Ω by ψ has a nice explicit expression.

Theorem 2 *There holds*

$$\psi(\Omega) = \sum_{n \geq 1} \frac{(-1)^n}{n} \sum_{\sigma \in S_n} \frac{(-1)^{d(\sigma)}}{\binom{n-1}{d(\sigma)}} \sigma, \tag{22}$$

where $d(\sigma)$ is the number of descents of the permutation σ.

On the other hand, a one-parameter deformation of this formula has appeared in the study of Lie idempotents [18], namely

$$\sum_{n \geq 1} \frac{(-1)^n}{[n]_q} \sum_{\sigma \in S_n} \frac{(-1)^{d(\sigma)} q^{\operatorname{maj}(\sigma) - \binom{d(\sigma)+1}{2}}}{\begin{bmatrix} n-1 \\ d(\sigma) \end{bmatrix}_q} \sigma, \tag{23}$$

[2]In fact, the same formula gives a pre-Lie algebra for every dendriform algebra.

where q is a formal parameter, $[n]_q = 1 + q + q^2 + \cdots + q^{n-1}$, $[n]_q! = [1]_q[2]_q \ldots [n]_q$, and

$$\begin{bmatrix} n \\ d \end{bmatrix}_q = \frac{[n]!_q}{[d]!_q[n-d]!_q} \tag{24}$$

are the usual q-integers, q-factorials and q-binomial coefficients. In (23), one can see $\mathrm{maj}(\sigma)$, which is the major index, defined as the sum of the positions of the descents.

One can then prove (see [6]) that this deformation is still in the image of ψ.

Theorem 3 *There exists a tree-indexed series Ω_q, with coefficients in the field of fractions $\mathbb{Q}(q)$, such that $\psi(\Omega_q)$ is equal to* (23)*. Moreover, one can evaluate this series at $q = 1$, and the result is Ω.*

The first few terms of Ω_q are

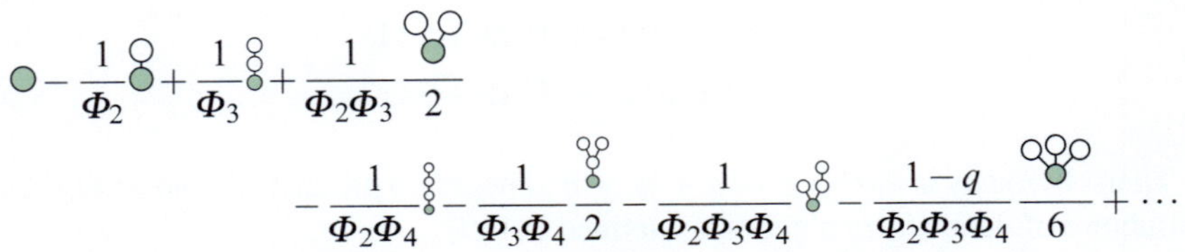

where, for every integer d, $\Phi_d(q)$ is the cyclotomic polynomial of order d.

Now the main question is: can we understand the coefficients of Ω_q in the same way as explained before in Theorem 1 for Ω ? The answer is yes, with just a single new ingredient, which is a q-analog of the Ehrhart polynomial, as defined in [9].

Recall that a tree T has an associated poset and order polytope Q_T.

Proposition 1 *There exists a unique polynomial $E_{q,T}$ in $\mathbb{Q}(q)[x]$ such that, for every integer $n \geq 0$, there holds*

$$E_{q,T}([n]_q) = \sum_{z \in nQ_T} q^{\sum z}, \tag{25}$$

where $\sum z$ is the sum of the coordinates of z.

Moreover, there is still a kind of reciprocity.

Proposition 2 *For every integer $n \geq 1$, there holds*

$$E_{q,T}([-n]_q) = (-1)^d \sum_{z \in \text{ interior of } nQ_T} q^{\sum z}, \tag{26}$$

where d is the dimension of the ambient space.

For example, consider the tree ○. The polytope Q_T is just the segment $[0, 1]$ in $\mathbb{R}$. Then

$$\sum_{z \in nQ_T} q^{\sum z} = 1 + q + \cdots + q^n = 1 + q[n]_q. \tag{27}$$

Therefore the q-Ehrhart polynomial is $1 + qx$. Evaluated at the q-integer $[-1]_q = -1/q$, it gives 0, as there are no interior points in Q_T. The same vanishing at $[-1]_q$ holds for all empty lattice polytopes, hence in particular for order polytopes.

Theorem 4 *The coefficient $\Omega_{q,T}$ of the rooted tree T in Ω_q is the value at $[-1]_q$ (i.e. $-1/q$) of the polynomial $E_{q,T}(x)/(1+qx)$.*

This is an explicit description of the coefficients of Ω_q. Note that this reduces to Theorem 1 when q tends to 1. This is also essentially the derivative at $[-1]_q$.

Just as in the case of Ω, it is natural to introduce a generating series for the q-Ehrhart polynomials:

$$\Omega_{q,x} = \sum_T E_{q,T}(x) \frac{T}{\# \operatorname{Aut} T}. \tag{28}$$

The first few terms of $\Omega_{q,x}$ are given by

$$(1+qx)\bigcirc + \frac{(1+qx)(1+q+q^2x)}{\Phi_2}$$ 

$$+ \frac{(1+qx)(1+q+q^2x)(1+q+q^2+q^3x)}{\Phi_2\Phi_3}$$ 

$$+ \frac{(1+qx)(1+q+q^2x)(1+q+q^2+q^2x+q^3x)}{\Phi_2\Phi_3} \frac{}{2} + \cdots$$ 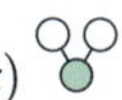

All the series A, Ω, Ω_q and $\Omega_{q,x}$ can be characterized by some functional equations, using the completed free pre-Lie algebra on one generator and the universal enveloping algebra of its Lie algebra.

The tree-indexed A can be defined concisely by the equality

$$A = \bigcirc \curvearrowleft \left(\frac{\exp(\bigcirc) - 1}{\bigcirc} \right), \tag{29}$$

where the fraction $\frac{\exp(\bigcirc)-1}{\bigcirc}$ has to be understood as an element of the (completed) enveloping algebra of PL, and $\curvearrowleft$ is the action deduced from the pre-Lie product.

Composing this functional equation by Ω (using the monoid structure on tree-indexed series and its compatibility with the other algebraic structures), one gets

$$\bigcirc = A \circ \Omega = \Omega \curvearrowleft \left(\frac{\exp(\Omega) - 1}{\Omega} \right). \tag{30}$$

Inverting then the element of the completed enveloping algebra, one gets

$$\Omega = \bigcirc \curvearrowleft \left(\frac{\Omega}{\exp(\Omega) - 1} \right), \tag{31}$$

which has Ω as unique solution.

With a little more work, one can show that Ω is characterized by

$$\Omega \curvearrowleft (\exp(\Omega) - 1) = \bigcirc \curvearrowleft \Omega. \tag{32}$$

A similar-looking equation can then be given for the q-analog Ω_q:

$$\Omega_q[q] \curvearrowleft (\exp(\Omega) - 1) + \Omega_q[q] - \Omega_q = \bigcirc \curvearrowleft \Omega_q + (q-1)\bigcirc, \tag{33}$$

where $\Omega_q[q]$ is obtained from Ω_q by multiplying by q^{n-1} the coefficients of trees with n vertices, for all n.

There is a similar functional equation for $\Omega_{q,x}$, that the interested reader can find in [8].

Let us now return to the coefficients of Ω and Ω_q.

The coefficients of corollas (trees made of one root and some vertices attached to the root) in Ω are interesting and well-known numbers, namely the Bernoulli numbers, classically defined as the coefficients B_n in the exponential generating series

$$\frac{z}{\exp(z) - 1} = \sum_{n \geq 0} B_n \frac{z^n}{n!}.$$

This can be easily proved using the morphism of monoids from $(\widehat{\mathrm{PL}}, \circ)$ to the monoid $(\mathbb{Q}[[z]], \times)$ which maps the corolla with n leaves to z^n.

It follows that the coefficients of corollas in Ω_q must be q-analogues of the Bernoulli numbers, In fact, these q-analogues were introduced by Carlitz already in the 1950s [5]. They have been related to the values of a q-analogue of the zeta function in [7]. They can be defined as follows

$$q(q\beta + 1)^n - \beta^n = \begin{cases} q - 1 & \text{si } n = 0, \\ 1 & \text{si } n = 1, \\ 0 & \text{si } n > 1, \end{cases} \tag{34}$$

where one has to expand the binomial and then replace β^k by β_k.

These numbers lead to another interesting point of view on the coefficients of Ω_q. Instead of obtaining them by a derivative at $[-1]_q$, one can get them using the linear form Ψ_q defined by

$$\Psi_q(x^n) = \beta_n. \tag{35}$$

By reference to the classical umbral calculus, the image of a polynomial in x by the linear form Ψ_q can be called its q-umbra.

The description of the coefficients in Ω_q can be rephrased using q-umbra. More precisely, let T be the rooted tree obtained by grafting several rooted trees $T_1, \ldots, T_k$ on a common root. Then the coefficient of T in Ω_q is given by

$$\Psi_q\left(\prod_{i=1}^{k} E_{T_i}\right), \tag{36}$$

using the q-Ehrhart polynomials attached to the rooted trees $T_1, \ldots, T_k$. For more details, see [8].

Let us now go back to the classical case $q = 1$.

It is worth remarking that the coefficients of the double corollas[3] in Ω have appeared in the works of Ramanujan, as coefficients of an asymptotic expansion involving triangular numbers [28, 29]. By the same umbral expression (36), but taken at $q = 1$, they can be obtained as the values $\Psi\left(\binom{x+2}{2}^n\right)$, where Ψ is the linear form on polynomials in x that maps x^n to the Bernoulli number B_n.

Let us end this article by some more results and questions about the coefficients of Ω.

Computing the first few terms of Ω, one observes that the corolla with three leaves has zero coefficient. More generally, this is also true for corollas with k leaves when $k \geq 3$ is odd, because the Bernoulli numbers B_k vanish.

One can therefore ask the following question: is it possible to describe all trees that have zero coefficients in Ω ?

It has been proved in [25] that if the number n is composite (not a prime number) then there is at least one tree with n vertices which has zero coefficient in Ω. This is proved by exhibiting an explicit family of trees with vanishing coefficients.

On the other hand, it has been checked up to $n = 19$ included that if n is a prime number, then no tree with n vertices has zero coefficient in Ω. For $n = 19$, this verification already involved 4 688 676 trees. One may wonder if this continues to hold for greater prime numbers.

Let us end with a last statement. Define the multi-corolla M_ℓ^p as the rooted tree obtained by grafting p linear trees[4] of size ℓ on a common root. It has therefore $p\ell + 1$ vertices.

Theorem 5 *If $p \geq 3$ is odd and ℓ is odd, then the coefficient of M_ℓ^p in Ω is zero.*

When $\ell = 1$, this is true by the classical vanishing of Bernoulli numbers, which are the coefficients of corollas in Ω. This theorem is just a very special case of the results in [10], which involve the notion of Gorenstein polytope.

---

[3] A *double corolla* is obtained by grafting several copies of the rooted tree with 2 vertices on a common root.

[4] A *linear tree* is a rooted tree which is a path graph (no forking vertex), with the root at one end.

References

1. Agračev, A.A., Gamkrelidze, R.V.: Chronological algebras and nonstationary vector fields. In: Problems in Geometry, Vol. 11 (Russian), pp. 135–176, 243. Akad. Nauk SSSR, Vsesoyuz. Inst. Nauchn. i Tekhn. Informatsii, Moscow (1980)
2. Beck, M., Robins, S.: Computing the continuous discretely. Undergraduate Texts in Mathematics. Springer, New York (2007). Integer-point enumeration in polyhedra
3. Brouder, C.: Runge-Kutta methods and renormalization. Eur. Phys. J. C **12**, 521–534 (2000). https://doi.org/10.1007/s100529900235
4. Butcher, J.C.: An algebraic theory of integration methods. Math. Comp. **26**, 79–106 (1972)
5. Carlitz, L.: q-Bernoulli numbers and polynomials. Duke Math. J. **15**, 987–1000 (1948)
6. Chapoton, F.: A rooted-trees q-series lifting a one-parameter family of Lie idempotents. Algebra Number Theory **3**(6), 611–636 (2009). https://doi.org/10.2140/ant.2009.3.611
7. Chapoton, F.: Fractions de Bernoulli-Carlitz et opérateurs q-zeta. J. Théor. Nombres Bordeaux **22**(3), 575–581 (2010)
8. Chapoton, F.: Sur une série en arbres à deux paramètres. Séminaire Lotharingien de Combinatoire **70** (2013)
9. Chapoton, F.: q-analogues of Ehrhart polynomials. Proc. Edinb. Math. Soc. (2) **59**(2), 339–358 (2016). https://doi.org/10.1017/S0013091515000243
10. Chapoton, F., Essouabri, D.: q-Ehrhart polynomials of Gorenstein polytopes, Bernoulli umbra and related Dirichlet series. Mosc. J. Comb. Number Theory **5**(4), 13–38 (2015)
11. Connes, A., Kreimer, D.: Hopf algebras, renormalization and noncommutative geometry. Comm. Math. Phys. **199**(1), 203–242 (1998). https://doi.org/10.1007/s002200050499
12. Connes, A., Kreimer, D.: Renormalization in quantum field theory and the Riemann-Hilbert problem. I. The Hopf algebra structure of graphs and the main theorem. Comm. Math. Phys. **210**(1), 249–273 (2000). https://doi.org/10.1007/s002200050779
13. Connes, A., Kreimer, D.: Renormalization in quantum field theory and the Riemann-Hilbert problem. II. The β-function, diffeomorphisms and the renormalization group. Comm. Math. Phys. **216**(1), 215–241 (2001). https://doi.org/10.1007/PL00005547
14. Duchamp, G., Hivert, F., Thibon, J.Y.: Noncommutative symmetric functions. VI. Free quasi-symmetric functions and related algebras. Internat. J. Algebra Comput. **12**(5), 671–717 (2002). https://doi.org/10.1142/S0218196702001139
15. Ebrahimi-Fard, K., Manchon, D.: A Magnus- and Fer-type formula in dendriform algebras. Found. Comput. Math. **9**(3), 295–316 (2009). https://doi.org/10.1007/s10208-008-9023-3
16. Grossman, R., Larson, R.G.: Hopf-algebraic structure of families of trees. J. Algebra **126**(1), 184–210 (1989). https://doi.org/10.1016/0021-8693(89)90328-1
17. Krajewski, T., Martinetti, P.: Wilsonian renormalization, differential equations and Hopf algebras. Combinatorics Phys. Contemp. Math. **539**, 187–236. Amer. Math. Soc., Providence, RI (2011). https://doi.org/10.1090/conm/539/10635
18. Krob, D., Leclerc, B., Thibon, J.Y.: Noncommutative symmetric functions. II. Transformations of alphabets. Internat. J. Algebra Comput. **7**(2), 181–264 (1997). https://doi.org/10.1142/S0218196797000113
19. Loday, J.L.: Dialgebras. In: Dialgebras and related operads, Lecture Notes in Math., vol. 1763, pp. 7–66. Springer, Berlin (2001). https://doi.org/10.1007/3-540-45328-82
20. Lundervold, A., Munthe-Kaas, H.: Hopf algebras of formal diffeomorphisms and numerical integration on manifolds. Combinatorics Phys. Contemp. Math. **539**, 295–324. Amer. Math. Soc., Providence, RI (2011). https://doi.org/10.1090/conm/539/10641
21. Lundervold, A., Munthe-Kaas, H.: Backward error analysis and the substitution law for Lie group integrators. Found. Comput. Math. **13**(2), 161–186 (2013). URL http://dx.doi.org.docelec.univ-lyon1.fr/10.1007/s10208-012-9130-z
22. Malvenuto, C., Reutenauer, C.: Duality between quasi-symmetric functions and the Solomon descent algebra. J. Algebra **177**(3), 967–982 (1995). https://doi.org/10.1006/jabr.1995.1336

23. Manchon, D.: A short survey on pre-Lie algebras. In: Noncommutative geometry and physics: renormalisation, motives, index theory, ESI Lect. Math. Phys., pp. 89–102. Eur. Math. Soc., Zürich (2011). https://doi.org/10.4171/008-1/3
24. Munthe-Kaas, H.Z., Lundervold, A.: On post-Lie algebras, Lie-Butcher series and moving frames. Found. Comput. Math. **13**(4), 583–613 (2013). https://doi.org/10.1007/s10208-013-9167-7
25. Oger, B.: Etudes de séries particulières dans le groupe des séries en arbres. mémoire de Master 2, ENS Lyon (2010)
26. Santos, F., Stump, C., Welker, V.: Noncrossing sets and a Grassmann associahedron. Forum Math. Sigma **5**, e5, 49 (2017). https://doi.org/10.1017/fms.2017.1
27. Stanley, R.P.: Two poset polytopes. Discrete Comput. Geom. **1**(1), 9–23 (1986). https://doi.org/10.1007/BF02187680
28. Villarino, M.B.: Ramanujan's harmonic number expansion (2005). arXiv.org:math/0511335
29. Villarino, M.B.: Ramanujan's harmonic number expansion into negative powers of a triangular number. JIPAM. J. Inequal. Pure Appl. Math. **9**(3), Article 89, 12 (2008)
30. Wright, D., Zhao, W.: D-log and formal flow for analytic isomorphisms of n-space. Trans. Amer. Math. Soc. **355**(8), 3117–3141 (electronic) (2003). https://doi.org/10.1090/S0002-9947-03-03295-1

Evaluating Generating Functions for Periodic Multiple Polylogarithms via Rational Chen–Fliess Series

Kurusch Ebrahimi-Fard, W. Steven Gray and Dominique Manchon

Abstract The goal of the paper is to give a systematic way to numerically evaluate the generating function of a periodic multiple polylogarithm using a Chen–Fliess series with a rational generating series. The idea is to realize the corresponding Chen–Fliess series as a bilinear dynamical system. A standard form for such a realization is given. The method is also generalized to the case where the multiple polylogarithm has non-periodic components. This allows one, for instance, to numerically validate the Hoffman conjecture. Finally, a setting in terms of dendriform algebras is provided.

Keywords Chen–Fliess series · Dendriform algebra · Hoffman conjecture · Multiple polylogarithms · Rational formal power series

1 Introduction

Given any vector $\mathbf{s} = (s_1, s_2, \ldots, s_l) \in \mathbb{N}^l$ with $s_1 \geq 2$ and $s_i \geq 1$ for $i \geq 2$, the associated *multiple polylogarithm* (MPL) of *depth* l and *weight* $|\mathbf{s}| := \sum_{i=1}^{l} s_i$ is taken to be

$$\mathrm{Li}_{\mathbf{s}}(t) := \sum_{k_1 > k_2 > \cdots > k_l \geq 1} \frac{t^{k_1}}{k_1^{s_1} k_2^{s_2} \cdots k_l^{s_l}}, \quad |t| \leq 1, \tag{1}$$

Kurusch Ebrahimi-Fard: (on leave from UHA, Mulhouse, France).

K. Ebrahimi-Fard
Department of Mathematical Sciences, Norwegian University of Science and Technology (NTNU), 7491 Trondheim, Norway
e-mail: kurusch.ebrahimi-fard@ntnu.no; kurusch.ebrahimi-fard@uha.fr

W. Gray (✉)
Old Dominion University, Norfolk, VA 23529, USA
e-mail: sgray@odu.edu

D. Manchon
LMBP, CNRS - Université Clermont-Auvergne, CS 60026, 63178 Aubière Cedex, France
e-mail: Dominique.Manchon@uca.fr

J. I. Burgos Gil et al. (eds.), *Periods in Quantum Field Theory and Arithmetic*, Springer Proceedings in Mathematics & Statistics 314,
https://doi.org/10.1007/978-3-030-37031-2_17

whereupon the *multiple zeta value* (MVZ) of depth l and weight $|\mathbf{s}|$ is the value of (1) at $t = 1$, namely,

$$\zeta(\mathbf{s}) := \mathrm{Li}_{\mathbf{s}}(1).$$

Any such vector $\mathbf{s}$ will be referred to as *admissible*. The MPL in (1) can be represented in terms of iterated Chen integrals with respect to the 1-forms $\omega_j^{(1)} := dt_j/(1-t_j)$ and $\omega_j^{(0)} := dt_j/t_j$. Indeed, using the standard notation, $|\mathbf{s}_{(j)}| := s_1 + \cdots + s_j$, $j \in \{1, \ldots, l\}$, one can show that

$$\mathrm{Li}_{\mathbf{s}}(t) = \int_0^t \left(\prod_{j=1}^{|\mathbf{s}_{(1)}|-1} \omega_j^{(0)} \right) \omega_{|\mathbf{s}_{(1)}|}^{(1)} \cdots \left(\prod_{j=|\mathbf{s}_{(l-1)}|+1}^{|\mathbf{s}_{(l)}|-1} \omega_j^{(0)} \right) \omega_{|\mathbf{s}_{(l)}|}^{(1)}. \tag{2}$$

For instance,

$$\mathrm{Li}_{(2,1,1)}(t) = \int_0^t \frac{dt_1}{t_1} \int_0^{t_1} \frac{dt_2}{1-t_2} \int_0^{t_2} \frac{dt_3}{1-t_3} \int_0^{t_3} \frac{dt_4}{1-t_4} = \sum_{k_1>k_2>k_3\geq 1} \frac{t^{k_1}}{k_1^2 k_2 k_3}.$$

An MPL of depth l is said to be *periodic* if it can be written in the form $\mathrm{Li}_{\{\mathbf{s}\}^n}(t)$, where $\{\mathbf{s}\}^n$ denotes the n-tuple $(\mathbf{s}, \mathbf{s}, \ldots, \mathbf{s}) \in \mathbb{N}^{nl}$, $n \geq 0$ with $\mathrm{Li}_{\{\mathbf{s}\}^0}(t) := 1$.[1] In this case, the sequence $(\mathrm{Li}_{\{\mathbf{s}\}^n}(t))_{n\in\mathbb{N}_0}$ has the generating function

$$\mathscr{L}_{\mathbf{s}}(t, \theta) := \sum_{n=0}^{\infty} \mathrm{Li}_{\{\mathbf{s}\}^n}(t) \left(\theta^{|\mathbf{s}|}\right)^n. \tag{3}$$

In general, the integral representation (2) implies that $\mathscr{L}_{\mathbf{s}}$ will satisfy a linear ordinary differential equation in t whose solution can be written in terms of a hypergeometric function [1, 4, 5, 28–31]. For example, when $l = 1$ and $\mathbf{s} = (s)$, it follows that

$$\left(\left((1-t)\frac{d}{dt} \right) \left(t\frac{d}{dt} \right)^{s-1} - \theta^s \right) \mathscr{L}_{\mathbf{s}}(t, \theta) = 0, \tag{4}$$

and its solution is the Euler–Gauss hypergeometric function

$$\mathscr{L}_{(s)}(t, \theta) = {}_sF_{s-1} \left(\begin{matrix} -\omega\theta, -\omega^3\theta, \ldots, -\omega^{2s-1}\theta \\ 1, 1, \ldots, 1 \end{matrix} \middle| \, t \right),$$

where $\omega = \mathrm{e}^{\pi i/s}$, a primitive s-th root of -1 [4]. By expanding this solution into a hypergeometric series and equating like powers of θ with those in (3), it is possible to show, for example, when $s = 2$ that

[1]Following other authors, $\{\mathbf{s}\}^n = \{(s_1, s_2, \ldots, s_l)\}^n$ will be written more concisely as $\{s_1, s_2, \ldots, s_l\}^n$.

$$\zeta(\{2\}^n) = \frac{\pi^{2n}}{(2n+1)!}, \quad n \geq 1. \tag{5}$$

In a similar manner it can be shown that

$$\zeta(\{3,1\}^n) = \frac{2\pi^{4n}}{(4n+2)!}, \quad n \geq 1.$$

This method has yielded a plethora of such MZV identities [3, 4, 6, 32]. The most general case is treated in [31], where it is shown that $\mathscr{L}_{\mathbf{s}}$ satisfies the linear differential equation of Fuchs type

$$(P_{\mathbf{s}} - \theta^{|\mathbf{s}|})\mathscr{L}_{\mathbf{s}}(t,\theta) = 0, \tag{6}$$

where for $\mathbf{s} = (s_1, s_2, \ldots, s_l) \in \mathbb{N}^l$

$$P_{\mathbf{s}} := P_{s_l} P_{s_{l-1}} \cdots P_{s_1}$$

and

$$P_{s_i} := \left((1-t)\frac{d}{dt}\right)\left(t\frac{d}{dt}\right)^{s_i - 1}.$$

(The conventions in [31] are to use $-\theta$ in place of θ and t in place of $1-t$.) In [31] and related work [28–30], the authors develop WKB type asymptotic expansions of these hypergeometric solutions.

The ultimate goal of the present paper is to provide a numerical scheme for estimating $\mathscr{L}_{\mathbf{s}}(t,\theta)$ by in essence mapping the $|\mathbf{s}|$-order linear differential equation (6) to a *system* of $|\mathbf{s}|$ first-order bilinear differential equations which can be solved by standard tools found in software packages like MatLab. Specifically, it will be shown how to construct a dynamical system of the form

$$\dot{z} = N_0 z\, u_0 + N_1 z\, u_1, \quad z(0) = z_0 \tag{7a}$$

$$y = Cz, \tag{7b}$$

which when simulated over the interval $(0, 1)$ has the property that $y(t) = \mathscr{L}_{\mathbf{s}}(t,\theta)$ for any value of θ and $t \in (0, 1)$. In this case, the matrices N_0 and N_1 will depend on θ, and the initial condition z_0 and the input functions u_0, u_1 must be suitably chosen. Such a technique could be useful for either disproving certain conjectures involving MZVs or providing additional evidence for the truthfulness of other conjectures. For example, one could validate with a certain level of (numerical) confidence a conjecture of the form

$$\zeta(\{\mathbf{s}_a\}^n) = b^n \zeta(\{\mathbf{s}_b\}^n), \quad n, b \in \mathbb{N},$$

where $\mathbf{s}_a \in \mathbb{N}^{l_a}$, $\mathbf{s}_b \in \mathbb{N}^{l_b}$ with $|\mathbf{s}_a| = |\mathbf{s}_b|$. Take as a specific example the known identity

$$\zeta(\{4\}^n) = 4^n \zeta(\{3, 1\}^n) \tag{8}$$

for all $n \geq 1$, so that $\mathbf{s}_a = (4)$, $\mathbf{s}_b = (3, 1)$ and $b = 4$ [4]. Note that for $n = 1$ the identity follows immediately from double shuffle relations for MZVs [22]. On the level of generating functions it is evident that

$$\mathscr{L}_{(4)}(1, \theta) = \sum_{n=0}^{\infty} \mathrm{Li}_{\{4\}^n}(1) \left(\theta^4\right)^n = \sum_{n=0}^{\infty} \zeta(\{4\}^n)\, \theta^{4n}$$

$$\mathscr{L}_{(3,1)}(1, \sqrt{2}\theta) = \sum_{n=0}^{\infty} \mathrm{Li}_{\{3,1\}^n}(1) \left((\sqrt{2}\theta)^4\right)^n = \sum_{n=0}^{\infty} 4^n \zeta(\{3, 1\}^n)\, \theta^{4n}.$$

Therefore, identity (8) implies that

$$\mathscr{L}_{(4)}(1, \theta) - \mathscr{L}_{(3,1)}(1, \sqrt{2}\theta) = 0, \quad \forall \theta \in \mathbb{R}, \tag{9}$$

a claim that can be tested empirically if these generating functions can be accurately evaluated. The method can also be generalized to address the conjecture of Hoffman that

$$\zeta(\{2\}^n, 2, 2, 2) + 2\zeta(\{2\}^n, 3, 3) = \zeta(2, 1, \{2\}^n, 3), \tag{10}$$

for all integers $n > 0$, which has only been proved for $n \leq 8$ [6]. The idea here is to admit *non-periodic* components in the generating function calculation. For example, $(\{2\}^n, 3, 3)$ can be viewed as having the periodic component $\{2\}^n$ and the non-periodic component $(3, 3)$. In the general case, say when $\mathbf{s}_n := (\mathbf{s}_a, \{\mathbf{s}_b\}^n, \mathbf{s}_c)$, $n \geq 0$, the generating function is defined analogously as

$$\mathscr{L}_{(\mathbf{s}_a, \{\mathbf{s}_b\}, \mathbf{s}_c)}(t, \theta) := \sum_{n=0}^{\infty} \mathrm{Li}_{\mathbf{s}_n}(t) \left(\theta^{|\mathbf{s}_b|}\right)^n.$$

Therefore, relation (10), if true, would imply that

$$\mathscr{L}_{(\{2\},2,2,2)}(1, \theta) + 2\mathscr{L}_{(\{2\},3,3)}(1, \theta) - \mathscr{L}_{(2,1,\{2\},3)}(1, \theta) = 0, \quad \forall \theta \in \mathbb{R}. \tag{11}$$

The basic approach to estimating $\mathscr{L}_{\mathbf{s}}(t, \theta)$ is to map a periodic multiple polylogarithm to a rational series and then to employ well known concepts from control theory to produce bilinear state space realization (7) of the corresponding rational Chen–Fliess series [2, 16, 17]. The periodic nature of the MPL always ensures that these realizations have a certain built-in recursion/feedback structure. The technique will first be described in general, and then it will be demonstrated by empirically verifying the identities (5), (8), and (10). It should be noted that the connection between polylogarithms and differential equations with singularities at $\{0, 1, \infty\}$ has been well studied by a number of researchers, especially, [8, 10, 19, 20]. (See, in particular, [20, Chapter 4] and the references therein.) In addition, rational series of the type suitable for representing periodic multiple polylogarithms have appeared in [8, 20]. In this regard, the main contribution here is to customize these results specifically and explicitly for periodic multiple polylogarithms and then to actually apply them to problems like the Hoffman conjecture (10).

The paper is organized as follows. In the next section, a brief summary of rational Chen–Fliess series is given to establish the notation and the basic concepts to be employed. Then the general method for evaluating a generating function of a periodic multiple polylogarithm is given in the subsequent section, which also contains in Sect. 3.3 a short digression regarding another way of looking at periodic MPLs in terms of dendriform algebra along the lines of reference [15]. This is followed by several examples in Sect. 4. In particular, the last example shows that the Hoffman conjecture (10) has a high likelihood of being true. The final section gives the paper's conclusions.

2 Preliminaries

2.1 Chen–Fliess Series

A finite nonempty set of noncommuting symbols $X = \{x_0, x_1, \dots, x_m\}$ is called an *alphabet*. Each element of X is called a *letter*, and any finite sequence of letters from X, $\eta = x_{i_1} \cdots x_{i_k}$, is called a *word* over X. The *length* of word η, denoted $|\eta|$, is the number of letters in η. The set of all words with fixed length k is denoted by X^k. The set of all words including the empty word, $\emptyset$, is designated by X^*. It forms a monoid under catenation. The set $\eta X^* \xi \subseteq X^*$ is the set of all words with prefix η and suffix ξ. Any mapping $c : X^* \to \mathbb{R}^\ell$ is called a *formal power series*. The value of c at $\eta \in X^*$ is written as $(c, \eta) \in \mathbb{R}^\ell$ and called the *coefficient* of the word η in the series c. Typically, c is represented as the formal sum $c = \sum_{\eta \in X^*} (c, \eta)\eta$. If the *constant term* $(c, \emptyset) = 0$ then c is said to be *proper*. The collection of all formal power series over the alphabet X is denoted by $\mathbb{R}^\ell\langle\langle X \rangle\rangle$. The subset of polynomials is written as $\mathbb{R}^\ell\langle X \rangle$. Each set forms an associative $\mathbb{R}$-algebra under the catenation product.

Definition 1 Given $\xi \in X^*$, the corresponding ***left-shift operator*** $\xi^{-1} : X^* \to \mathbb{R}\langle X \rangle$ is defined:

$$\eta \mapsto \xi^{-1}(\eta) := \begin{cases} \eta' : \eta = \xi\eta' \\ 0 : \text{otherwise}. \end{cases}$$

It is extended linearly to $\mathbb{R}^\ell\langle\langle X \rangle\rangle$.

One can formally associate with any series $c \in \mathbb{R}^\ell\langle\langle X \rangle\rangle$ a causal m-input, ℓ-output operator, F_c, in the following manner. Let $t_0 < t_1$ be fixed, and consider a class of locally integrable functions $u = (u_1, \dots, u_m) \in L^m_{1,loc}[t_0, t_1)$ modulo almost-everywhere equality with respect to the Lebesgue measure. For any compact subset

$\Omega = [t_0, s] \subset [t_0, t_1)$, the usual L_1 norm restricted to Ω, denoted here by $\|\cdot\|_{1,\Omega}$, provides a family of seminorms on $L^m_{1,loc}[t_0, t_1)$. Define inductively for each word $\eta \in X^*$ and $u \in L^m_{1,loc}(\Omega)$ an iterated integral by setting $E_\emptyset[u] = 1$ and letting

$$E_{x_i\bar{\eta}}[u](t) := \int_{t_0}^{t} u_i(\tau) E_{\bar{\eta}}[u](\tau)\, d\tau, \tag{12}$$

where $x_i \in X$, $\bar{\eta} \in X^*$, $t \in \Omega$, and $u_0 = 1$. The input-output operator corresponding to the series $c \in \mathbb{R}^\ell\langle\langle X\rangle\rangle$ is the *Fliess operator* or *Chen–Fliess series*

$$F_c[u](t) = \sum_{\eta \in X^*} (c, \eta)\, E_\eta[u](t) \tag{13}$$

[17]. If there exist real numbers $K_c, M_c > 0$ such that the coefficients of the generating series $c = \sum_{\eta\in X^*}(c, \eta)\eta \in \mathbb{R}^\ell\langle\langle X\rangle\rangle$ satisfy the growth bound

$$|(c, \eta)| \le K_c M_c^{|\eta|}\, |\eta|!, \quad \forall \eta \in X^*, \tag{14}$$

then the series (13) converges absolutely and uniformly for every $t \in \Omega$ provided the measure of Ω and $\|u\|_{1,\Omega}$ are sufficiently small [18].

In the case of polylogarithms, it is sufficient to consider the single-input, single-output case $m = \ell = 1$ and to set $t_0 = 0$ and $t_1 = 1$. The convergence situation, however, is a bit different: the underlying iterated integrals (12) involve the locally integrable function $u_1(t) = 1/(1-t)$ on $[0, 1)$, but the function u_0 is now given by $u_0(t) = 1/t$, which is locally integrable only on $(0, 1)$. The growth condition (14) is *not* sufficient to ensure the convergence of a Chen-Fliess series. Even rationality of the generating series c is not sufficient as it can be shown using results from [20, Theorem 4.3.4], for example, that $F_c[\mathrm{Li}_0](t)$ with $c = \sum_{k\ge 0} x_0^k x_1$ and $\mathrm{Li}_0(t) := t/(1-t)$ is divergent. Therefore, the convergence of (13) will have to be addressed for the specific case of interest in the context of polylogarithms.

2.2 Bilinear Realizations of Rational Chen–Fliess Series

A series $c \in \mathbb{R}\langle\langle X\rangle\rangle$ is called *invertible* if there exists a series $c^{-1} \in \mathbb{R}\langle\langle X\rangle\rangle$ such that $cc^{-1} = c^{-1}c = 1$.[2] In the event that c is not proper, i.e., the coefficient $(c, \emptyset)$ is nonzero, it is always possible to write

$$c = (c, \emptyset)(1 - c'),$$

where $c' \in \mathbb{R}\langle\langle X\rangle\rangle$ is proper. It then follows that

[2]The polynomial $1\emptyset$ is abbreviated throughout as 1.

$$c^{-1} = \frac{1}{(c, \emptyset)}(1 - c')^{-1} = \frac{1}{(c, \emptyset)}(c')^*,$$

where the Kleene star of c' is defined by

$$(c')^* := \sum_{i=0}^{\infty} (c')^i.$$

In fact, $c \in \mathbb{R}\langle\langle X \rangle\rangle$ is invertible *if and only if* c is not proper. Now let S be a subalgebra of the $\mathbb{R}$-algebra $\mathbb{R}\langle\langle X \rangle\rangle$ with the catenation product. S is said to be *rationally closed* when every invertible $c \in S$ has $c^{-1} \in S$ (or equivalently, every proper $c' \in S$ has $(c')^* \in S$). The *rational closure* of any subset $E \subset \mathbb{R}\langle\langle X \rangle\rangle$ is the smallest rationally closed subalgebra of $\mathbb{R}\langle\langle X \rangle\rangle$ containing E.

Definition 2 A series $c \in \mathbb{R}\langle\langle X \rangle\rangle$ is ***rational*** if it belongs to the rational closure of $\mathbb{R}\langle X \rangle$.

Rational series have appeared in a number of different contexts including automata theory [26], control theory [17], formal language theory [25], and polylogarithms [20]. The monograph [2] provides a concise introduction to the area. Of particular importance is an alternative characterization of rationality using the following concept.

Definition 3 A ***linear representation*** of a series $c \in \mathbb{R}\langle\langle X \rangle\rangle$ is any triple (μ, γ, λ), where

$$\mu : X^* \to \mathbb{R}^{n \times n}$$

is a monoid morphism, and the vectors $\gamma, \lambda^T \in \mathbb{R}^{n \times 1}$ are such that each coefficient

$$(c, \eta) = \lambda \mu(\eta) \gamma, \quad \forall \eta \in X^*.$$

The integer n is the dimension of the representation.

Definition 4 A series $c \in \mathbb{R}\langle\langle X \rangle\rangle$ is called ***recognizable*** if it has a linear representation.

Theorem 1 [26] *A formal power series is rational if and only if it is recognizable.*

Returning to (13), Chen–Fliess series F_c is said to be rational when its generating series $c \in \mathbb{R}\langle\langle X \rangle\rangle$ is rational. The state space realization (7) is said to *realize* F_c *on some admissible input set* $\mathscr{U}$ when (7a) has a well defined solution, $z(t)$, on the interval $[t_0, t_0 + T]$ for every $T > 0$ with input $u \in \mathscr{U}$ and output

$$y(t) = F_c[u](t) = C(z(t)), \quad t \in [t_0, t_0 + T].$$

Identify with any linear representation (μ, γ, λ) of the series $c \in \mathbb{R}\langle\langle X \rangle\rangle$ the bilinear system

$$(N_0, N_1, z_0, C) := (\mu(x_0), \mu(x_1), \gamma, \lambda).$$

The following result is well known.

Theorem 2 [17, 18] *The statements below are equivalent for a given* $c \in \mathbb{R}\langle\langle X \rangle\rangle$*:*

i (μ, γ, λ) *is a linear representation of* c.
ii The bilinear system (N_0, N_1, z_0, C) *realizes* F_c *on the extended space* $L_{\mathfrak{p},e}(t_0)$ *for any* $\mathfrak{p} \geq 1$.

3 Evaluating Periodic Multiple Polylogarithms

It is first necessary to associate a periodic MPL and its generating function to a rational series. Elements of this idea have appeared in numerous places. The approach taken here is most closely related to the one presented in [21]. The next step is then to find the bilinear realization of the rational Chen–Fliess series in terms of its linear representation (see Theorem 4). The case when non-periodic components are present works similarly but is slightly more complicated (see Theorem 5). Recall that throughout $m = 1$, so that the underlying alphabet is $X := \{x_0, x_1\}$.

3.1 Periodic Multiple Polylogarithms

Given any admissible vector $\mathbf{s} \in \mathbb{N}^l$, there is an associated word $\eta_{\mathbf{s}} \in x_0X^*x_1$ of length $|\mathbf{s}|$

$$\eta_{\mathbf{s}} = x_0^{s_1-1}x_1x_0^{s_2-1}x_1 \cdots x_0^{s_l-1}x_1.$$

In which case, $c_{\mathbf{s}} := (\theta^{|\mathbf{s}|}\eta_{\mathbf{s}})^* = \sum_{n\geq 0} \left(\theta^{|\mathbf{s}|}\eta_{\mathbf{s}}\right)^n$ is a rational series satisfying the identity

$$1 + (\theta^{|\mathbf{s}|}\eta_{\mathbf{s}})c_{\mathbf{s}} = c_{\mathbf{s}}. \tag{15}$$

The idea is to now relate the generating function of the sequence $(\mathrm{Li}_{\{\mathbf{s}\}^n}(t))_{n>0}$ to the Chen–Fliess series with generating series $c_{\mathbf{s}}$. Recall that for any word $x_i\xi' \in X^*$ the iterated integral is defined inductively by

$$E_{x_i\xi'}[u](t) = \int_0^t u_i(\tau)E_{\xi'}[u](\tau)\,d\tau,$$

where $x_i \in X, \xi' \in X^*$. Assume here that the letters x_0 and x_1 correspond to the inputs $u_0(t) := 1/t$ and $u_1(t) := 1/(1-t)$, respectively, and $E_\emptyset := 1$. For the formal power series $c_{\mathbf{s}} \in \mathbb{R}\langle\langle X \rangle\rangle$, the corresponding Chen–Fliess series is then taken to be

$$F_{c_{\mathbf{s}}}[u] = \sum_{\xi \in X^*} (c_{\mathbf{s}}, \xi) E_\xi[u].$$

Comparing this to the classical definition (13), the factor $1/t$ can be extracted from u_0 and u_1 so that each integral can be viewed instead as integration with respect to the Haar measure. That is,

$$E_{x_i \xi'}[u](t) = \int_0^t \bar{u}_i(\tau) E_{\xi'}[u](\tau) \frac{d\tau}{\tau},$$

where $\bar{u}_0(t) := 1$ and $\bar{u}_1(t) = t u_1(t)$. The following theorem is central to the paper.

Theorem 3 *For any admissible vector* $\mathbf{s} \in \mathbb{N}^l$,

$$\mathscr{L}_{\mathbf{s}}(t, \theta) = F_{c_{\mathbf{s}}}[\mathrm{Li}_0](t), \quad t \in [0, 1),$$

where $\mathrm{Li}_0(t) := t/(1-t)$*, and the defining series for* $F_{c_{\mathbf{s}}}[\mathrm{Li}_0](t)$ *converges absolutely for any fixed* $t \in [0, 1)$ *provided* $\theta \in \mathbb{R}$ *is sufficiently small.*

Proof First observe that since $c_{\mathbf{s}} = \sum_{n \geq 0} \left(\theta^{|\mathbf{s}|} \eta_{\mathbf{s}}\right)^n$, it follows directly that

$$F_{c_{\mathbf{s}}}[u](t) = \sum_{n=0}^{\infty} F_{\left(\theta^{|\mathbf{s}|} \eta_{\mathbf{s}}\right)^n}[u](t) = \sum_{n=0}^{\infty} E_{\eta_{\mathbf{s}}^n}[u](t) \left(\theta^{|\mathbf{s}|}\right)^n.$$

Comparing this against the definition

$$\mathscr{L}_{\mathbf{s}}(t, \theta) = \sum_{n=0}^{\infty} \mathrm{Li}_{\{\mathbf{s}\}^n}(t) \left(\theta^{|\mathbf{s}|}\right)^n,$$

it is evident that one only needs to verify the identity

$$E_{\eta_{\mathbf{s}}^n}[\mathrm{Li}_0](t) = \mathrm{Li}_{\{\mathbf{s}\}^n}(t), \quad n \geq 0. \tag{16}$$

But this is clear from (2), i.e., for any admissible vector $\mathbf{s} \in \mathbb{N}^l$

$$\mathrm{Li}_{\mathbf{s}}(t) = \int_0^t u_i(\tau) \mathrm{Li}_{\mathbf{s}'}(\tau) \, d\tau,$$

where $\eta_{\mathbf{s}} = x_i \eta_{\mathbf{s}'}$,

$$u_i(t) = \begin{cases} \frac{1}{t} & : i = 0 \\ \frac{t}{1-t} \frac{1}{t} & : i = 1, \end{cases}$$

and $\mathrm{Li}_\emptyset(t) = 1$ [32]. Therefore, it follows directly that $\mathrm{Li}_{\mathbf{s}}(t) = E_{\eta_{\mathbf{s}}}[\mathrm{Li}_0](t)$, from which (16) also follows. To prove the convergence claim, it is sufficient to consider the special case where $\eta_{\mathbf{s}} = x_0^{s_1 - 1} x_1$ so that $c_{\mathbf{s}} = (\theta^{s_1} x_0^{s_1 - 1} x_1)^*$. The general case then follows similarly. Clearly, for any $t \in [0, 1)$

$$E_{x_1}[\mathrm{Li}_0](t, 0) = \ln\left(\frac{1}{1-t}\right) = \sum_{k=1}^{\infty} \frac{t^k}{k}.$$

Hence, for any $s_1 \geq 1$

$$E_{x_0^{s_1-1}x_1}[\mathrm{Li}_0](t, 0) = \sum_{k=1}^{\infty} \frac{t^k}{k^{s_1}} = \mathrm{Li}_{(s_1)}(t) < \infty,$$

and similarly, for any $n \geq 1$

$$E_{\left(x_0^{s_1-1}x_1\right)^n}[\mathrm{Li}_0](t, 0) = \sum_{k_1,k_2,\ldots,k_n=1}^{\infty} \frac{t^{k_1+k_2+\cdots+k_n}}{k_1^{s_1}(k_1+k_2)^{s_1}\cdots(k_1+k_2+\cdots+k_n)^{s_1}}.$$

The convergence claim for the series

$$F_{\left(\theta^{s_1}x_0^{s_1-1}x_1\right)^*}[\mathrm{Li}_0](t) = \sum_{n=0}^{\infty} E_{\left(x_0^{s_1-1}x_1\right)^n}[\mathrm{Li}_0](t, 0)\,\theta^{s_1 n}$$

can be verified by the ratio test. Observe

$$\begin{aligned}
&\frac{E_{\left(x_0^{s_1-1}x_1\right)^{n+1}}[\mathrm{Li}_0](t, 0)\,|\theta|^{s_1(n+1)}}{E_{\left(x_0^{s_1-1}x_1\right)^n}[\mathrm{Li}_0](t, 0)\,|\theta|^{s_1 n}} \\
&\quad = \frac{\sum_{k_1,k_2,\ldots,k_{n+1}=1}^{\infty} \frac{t^{k_1+k_2+\cdots+k_{n+1}}}{k_1^{s_1}(k_1+k_2)^{s_1}\cdots(k_1+k_2+\cdots+k_{n+1})^{s_1}}}{\sum_{k_1,k_2,\ldots,k_n=1}^{\infty} \frac{t^{k_1+k_2+\cdots+k_n}}{k_1^{s_1}(k_1+k_2)^{s_1}\cdots(k_1+k_2+\cdots+k_n)^{s_1}}}|\theta|^{s_1} \\
&\quad = \sum_{k_1=1}^{\infty} \frac{t^{k_1}}{k_1^{s_1}} \frac{\sum_{k_2,k_3,\ldots,k_{n+1}=1}^{\infty} \frac{t^{k_2+k_3+\cdots+k_{n+1}}}{(k_1+k_2)^{s_1}(k_1+k_2+k_3)^{s_1}\cdots(k_1+k_2+\cdots+k_{n+1})^{s_1}}}{\sum_{k_2,k_3,\ldots,k_{n+1}=1}^{\infty} \frac{t^{k_2+k_3+\cdots+k_{n+1}}}{k_2^{s_1}(k_2+k_3)^{s_1}\cdots(k_2+k_3+\cdots+k_{n+1})^{s_1}}}|\theta|^{s_1} \\
&\quad < \mathrm{Li}_{(s_1)}(t)|\theta|^{s_1},
\end{aligned}$$

so that ratio is less than one when $|\theta| < (1/\mathrm{Li}_{(s_1)}(t))^{1/s_1}$. □

The key idea now is to apply Theorem 2 and the rational nature of the series c_s in order to build a bilinear realization of the mapping $u \mapsto y = F_{c_s}[u]$ (see [23, 24]) so that $\mathscr{L}_s(t, \theta)$ can be evaluated by numerical simulation of a dynamical system. In principle, one could attempt to ensure that any such realization is minimal in dimension or even canonical in some sense [7, 9, 11, 27]. There is also the potential for lower dimensional realizations to exist if systems other than bilinear realizations are considered. But in the present context these issues are not really essential. In addition, the realizations considered here are in the same general class as those described in [8, 20] for realizing classes of hypergeometric functions and polylogarithms using rational generating series. But in this work they are customized specifically for periodic multiple polylogarithms.

Theorem 4 *For any admissible* $\mathbf{s} \in \mathbb{N}^l$, $\mathscr{L}_{\mathbf{s}}(t,\theta) = F_{c_{\mathbf{s}}}[\mathrm{Li}_0](t)$ *has the bilinear realization*

$$(N_0, N_1, z_0, C) := \big(\mu(x_0), \mu(x_1), \gamma, \lambda\big),$$

where

$$N_0 = \mathrm{diag}\big(N_0(s_1), N_0(s_2), \ldots, N_0(s_l)\big) \tag{17a}$$

$$N_1 = I^{+}_{|\mathbf{s}|} - N_0 + \theta^{|s|} e_{|\mathbf{s}|} e_1^T \tag{17b}$$

with $N_0(s_i) \in \mathbb{R}^{s_i \times s_i}$ *and* $I^{+}_{|\mathbf{s}|} \in \mathbb{R}^{|\mathbf{s}| \times |\mathbf{s}|}$ *being matrices of zeros except for a super diagonal of ones,* e_i *is an elementary vector with a one in the* i*-th position, and* $z_0 = C^T = e_1 \in \mathbb{R}^{|\mathbf{s}| \times 1}$.

Proof First recall Definition 1 describing the left-shift operator on X^*, i.e., for any $x_i \in X, x_i^{-1}(\cdot)$ is defined by $x_i^{-1}(x_i\eta) = \eta$ with $\eta \in X^*$ and zero otherwise. In which case, $(x_i\xi)^{-1}(\cdot) = \xi^{-1}x_i^{-1}(\cdot)$ for any $\xi \in X^*$. Now assign the first state of the realization to be

$$z_1(t) = F_{c_{\mathbf{s}}}[u](t) = 1 + F_{(\theta^{|\mathbf{s}|}\eta_{\mathbf{s}})c_{\mathbf{s}}}[u](t).$$

In light of the integral representation (2) of MPLs, differentiate z_1 exactly s_1 times so that the input $u_1(t) := \bar{u}_1(t)/t$ appears. Assign a new state at each step along the way. Specifically,

$$\dot{z}_1(t) = \frac{1}{t} F_{\theta^{|\mathbf{s}|} x_0^{-1}(\eta_{\mathbf{s}}) c_{\mathbf{s}}}[u](t) =: z_2(t)\frac{1}{t}$$

$$\vdots$$

$$\dot{z}_{s_1-1}(t) = \frac{1}{t} F_{\theta^{|\mathbf{s}|}(x_0^{s_1-1})^{-1}(\eta_{\mathbf{s}})c_{\mathbf{s}}}[u](t) =: z_{s_1}(t)\frac{1}{t}$$

$$\dot{z}_{s_1}(t) = \bar{u}_1(t)\frac{1}{t} F_{\theta^{|\mathbf{s}|}(x_0^{s_1-1}x_1)^{-1}(\eta_{\mathbf{s}})c_{\mathbf{s}}}[u](t) =: z_{s_1+1}(t)\bar{u}_1(t)\frac{1}{t}.$$

This produces the first s_1 rows of the matrices in (17) since when $l > 1$

$$\begin{bmatrix} \dot{z}_1(t) \\ \vdots \\ \dot{z}_{s_1-1}(t) \\ \dot{z}_{s_1}(t) \end{bmatrix} = I^+_{s_1\times(s_1+1)} \begin{bmatrix} z_1(t) \\ \vdots \\ z_{s_1}(t) \\ z_{s_1+1}(t)\bar{u}_1(t) \end{bmatrix} \frac{1}{t}$$

$$= \left[\, N_0(s_1) \,\middle|\, 0 \,\right] \begin{bmatrix} z_1(t) \\ \vdots \\ z_{s_1}(t) \\ \hline z_{s_1+1}(t) \end{bmatrix} \frac{1}{t} + \left[\, \mathbf{0}_{s_1} \,\middle|\, e_{s_1} \,\right] \begin{bmatrix} z_1(t) \\ \vdots \\ z_{s_1}(t) \\ \hline z_{s_1+1}(t) \end{bmatrix} \bar{u}_1(t)\frac{1}{t}.$$

Both $\left[\, N_0(s_1) \,\middle|\, 0 \,\right]$ and $\left[\, \mathbf{0}_{s_1} \,\middle|\, e_{s_1} \,\right]$ denote matrices in $\mathbb{R}^{s_1\times(s_1+1)}$. The pattern is exactly repeated until the final state, then the periodicity of $c_{\mathbf{s}}$ comes into play. Namely,

$$\dot{z}_{|\mathbf{s}|}(t) = \theta^{|\mathbf{s}|}\bar{u}_1(t)\frac{1}{t}F_{(\eta_{\mathbf{s}})^{-1}(\eta_{\mathbf{s}})c_{\mathbf{s}}}[u](t) =: \theta^{|\mathbf{s}|}z_1(t)\bar{u}_1(t)\frac{1}{t},$$

which gives the final rows of N_0 and N_1 in (17). □

It is worth pointing out that the validity of (6) is obvious in the present setting. Namely, (6) follows from the fact that (15) implies $\eta_{\mathbf{s}}^{-1}(c_{\mathbf{s}}) - \theta^{|\mathbf{s}|}c_{\mathbf{s}} = 0$, and thus, Theorem 3 gives

$$(P_{\mathbf{s}} - \theta^{|\mathbf{s}|})\mathscr{L}_{\mathbf{s}}(t,\theta) = (P_{\mathbf{s}} - \theta^{|\mathbf{s}|})F_{c_{\mathbf{s}}}[\mathrm{Li}_0](t) = F_{\eta_{\mathbf{s}}^{-1}(c_{\mathbf{s}})-\theta^{|\mathbf{s}|}c_{\mathbf{s}}}[\mathrm{Li}_0](t) = F_{0\cdot c_{\mathbf{s}}}[\mathrm{Li}_0](t) = 0.$$

3.2 Periodic Multiple Polylogarithms with Non-periodic Components

The non-periodic case requires a generalization of the basic set-up. The following lemma links this class of generating functions to the corresponding set of rational Fliess operators.

Lemma 1 *For any admissible* $\mathbf{s} := (\mathbf{s}_a, \{\mathbf{s}_b\}, \mathbf{s}_c)$

$$\mathscr{L}_{\mathbf{s}}(t,\theta) = F_{c_{\mathbf{s}}}[\mathrm{Li}_0](t), \quad t \in [0, 1), \quad \theta \in \mathbb{R},$$

where $c_{\mathbf{s}} := \eta_{\mathbf{s}_a}\left(\theta^{|\mathbf{s}_b|}\eta_{\mathbf{s}_b}\right)^{*}\eta_{\mathbf{s}_c}$.

Proof Similar to the periodic case, $c_{\mathbf{s}} = \sum_{n\geq 0}\eta_{\mathbf{s}_a}\left(\theta^{|\mathbf{s}_b|}\eta_{\mathbf{s}_b}\right)^n\eta_{\mathbf{s}_c}$, and therefore,

$$F_{c_{\mathbf{s}}}[u](t) = \sum_{n=0}^{\infty} F_{\eta_{\mathbf{s}_a}\left(\theta^{|\mathbf{s}_b|}\eta_{\mathbf{s}_b}\right)^n\eta_{\mathbf{s}_c}}[u](t) = \sum_{n=0}^{\infty} E_{\eta_{\mathbf{s}_a}\eta_{\mathbf{s}_b}^n\eta_{\mathbf{s}_c}}[u](t)\left(\theta^{|\mathbf{s}_b|}\right)^n.$$

The same argument used for proving (16) now shows that $E_{\eta_{s_a}\eta_{s_b}^n\eta_{s_c}}[\mathrm{Li}_0](t) = \mathrm{Li}_{\mathbf{s}_n}(t)$, $n \geq 0$. In which case, $F_{c_\mathbf{s}}[\mathrm{Li}_0](t) = \mathscr{L}_\mathbf{s}(t, \theta)$ as claimed. □

The required generalization of Theorem 4 is a bit more complicated. A simple example is given first to motivate the general approach.

Example 1 Consider the periodic MPL with non-periodic components specified by $\mathbf{s} = (2, 1, \{2\}, 3)$ as appearing in (11). In this case, $c_\mathbf{s} = \sum_{n\geq 0} x_0x_1^2(\theta^2 x_0x_1)^n x_0^2x_1 = x_0x_1^2\bar{c}$, where $\bar{c} = x_0^2x_1 + \theta^2 x_0x_1\bar{c}$. Assign the first state of the realization to be

$$z_1(t) = F_{c_\mathbf{s}}[u](t) = F_{x_0x_1^2\bar{c}}[u](t).$$

The strategy here is to differentiate z_1 exactly $|\eta_{\mathbf{s}_a}| = |x_0x_1^2| = 3$ times, assigning new states along the way, in order to remove the prefix $x_0x_1^2$ and isolate $\bar{c}$. At which point, the identity $\bar{c} = x_0^2x_1 + \theta^2 x_0x_1\bar{c}$ is used and the process is continued. This will yield a certain block diagonal structure for N_0 and an upper triangular form for N_1. As will be shown shortly, this structure is completely general but possibly redundant. Specifically,

$$\begin{aligned}
\dot{z}_1(t) &= \frac{1}{t}F_{x_1^2\bar{c}}[u](t) =: z_2(t)\frac{1}{t} \\
\dot{z}_2(t) &= \frac{1}{t}\bar{u}_1(t)F_{x_1\bar{c}}[u](t) =: z_3(t)\bar{u}_1(t)\frac{1}{t} \\
\dot{z}_3(t) &= \frac{1}{t}\bar{u}_1(t)F_{\bar{c}}[u](t) = \frac{1}{t}\bar{u}_1(t)F_{x_0^2x_1+\theta^2x_0x_1\bar{c}}[u](t) =: z_4(t)\bar{u}_1(t)\frac{1}{t} \\
\dot{z}_4(t) &= \frac{1}{t}F_{x_0x_1+\theta^2x_1\bar{c}}[u](t) =: z_5(t)\frac{1}{t} \\
\dot{z}_5(t) &= \frac{1}{t}F_{x_1}[u](t) + \frac{\theta^2}{t}\bar{u}_1(t)F_{\bar{c}}[u](t) =: z_6(t)\frac{1}{t} + \theta^2 z_4(t)\bar{u}_1(t)\frac{1}{t} \\
\dot{z}_6(t) &= \bar{u}_1(t)\frac{1}{t}.
\end{aligned}$$

The corresponding realization at this point has the form

$$\begin{aligned}
\dot{z} &= \tilde{N}_0 z\bar{u}_0 + \tilde{N}_1 z\bar{u}_1 + B_1\bar{u}_1, \quad z(0) = \tilde{z}_0 \\
y &= \tilde{C}z,
\end{aligned}$$

which does not have the form of a bilinear realization as defined in (7) since the state equation for z_6 does not depend on z, and thus, the term $B_1\bar{u}_1$ with $B_1 = e_6$ appears. Nevertheless, a permutation of the canonical embedding of Brockett (see [7, Theorem 1]), namely,

$$N_0 = \begin{bmatrix} \tilde{N}_0 & 0 \\ 0 & 0 \end{bmatrix}, \quad N_1 = \begin{bmatrix} \tilde{N}_1 & B_1 \\ 0 & 0 \end{bmatrix}, \quad z_0 = \begin{bmatrix} \tilde{z}_0 \\ 1 \end{bmatrix}, \quad C^T = \begin{bmatrix} \tilde{C}^T \\ 0 \end{bmatrix}, \tag{18}$$

renders an input-output equivalent bilinear realization of the desired form, albeit at the cost of increasing the dimension of the system by one. In this case,

$$N_0 = \left[\begin{array}{ccc|ccc|c} 0&1&0&0&0&0&0\\ 0&0&0&0&0&0&0\\ 0&0&0&0&0&0&0\\ \hline 0&0&0&0&1&0&0\\ 0&0&0&0&0&1&0\\ 0&0&0&0&0&0&0\\ \hline 0&0&0&0&0&0&0 \end{array}\right],\quad N_1 = \left[\begin{array}{ccc|ccc|c} 0&0&0&0&0&0&0\\ 0&0&1&0&0&0&0\\ 0&0&0&1&0&0&0\\ \hline 0&0&0&0&0&0&0\\ 0&0&0&\theta^2&0&0&0\\ 0&0&0&0&0&0&1\\ \hline 0&0&0&0&0&0&0 \end{array}\right],\quad z(0) = \left[\begin{array}{c} 1\\0\\0\\ \hline 0\\0\\0\\ \hline 1 \end{array}\right],\quad C^T = \left[\begin{array}{c} 1\\0\\0\\ \hline 0\\0\\0\\ \hline 0 \end{array}\right].$$

Theorem 5 *Consider any admissible* $\mathbf{s} := (\mathbf{s}_a, \{\mathbf{s}_b\}, \mathbf{s}_c)$ *with* $\eta_{\mathbf{s}_a} := x_{i_1} \cdots x_{i_k}$, $k = j_{|\mathbf{s}_a|}$, *and* $|\mathbf{s}_c| > 0$. *Then* $\mathscr{L}_{\mathbf{s}}(t, \theta) = F_{c_{\mathbf{s}}}[\mathrm{Li}_0](t)$ *has the bilinear realization* (N_0, N_1, z_0, C), *where*

$$N_0 = \mathrm{diag}(N_0(\mathbf{s}_a), N_0(\mathbf{s}_b, \mathbf{s}_c), 0), \quad N_1 = \begin{bmatrix} N_1(\mathbf{s}_a) & E_{|\mathbf{s}_a|1} \\ 0 & N_1(\mathbf{s}_b, \mathbf{s}_c) \end{bmatrix}$$

with $N_i(\mathbf{s}_a) \in \mathbb{R}^{|\mathbf{s}_a| \times |\mathbf{s}_a|}$ *being a matrix of zeros and ones depending only on* $\mathbf{s}_a$, $E_{|\mathbf{s}_a|1}$ *is the elementary matrix with a one in position* $(|\mathbf{s}_a|, 1)$, *and* $N_i(\mathbf{s}_b, \mathbf{s}_c) \in \mathbb{R}^{s_{bc} \times s_{bc}}$ *is a matrix of zeros, ones, and the entry* $\theta^{|\mathbf{s}_b|}$. *(Its dimension* s_{bc} *and exact structure depend only on* $\mathbf{s}_b$ *and* $\mathbf{s}_c$.*) Finally,* $z_0 = e_1 + e_{|\mathbf{s}_a|+s_{bc}} \in \mathbb{R}^{(|\mathbf{s}_a|+s_{bc})\times 1}$ *and* $C = e_1 \in \mathbb{R}^{1\times(|\mathbf{s}_a|+s_{bc})}$.

Proof Following Example 1, assign the first state of the realization to be

$$z_1(t) = F_{c_{\mathbf{s}}}[u](t) = F_{\eta_{\mathbf{s}_a}\bar{c}}[u](t),$$

where $\bar{c} := \eta_{\mathbf{s}_c} + \theta^{|\eta_{\mathbf{s}_b}|}\eta_{\mathbf{s}_b}\bar{c}$, and differentiate z_1 until the series $\bar{c}$ appears in isolation. Observe

$$\dot{z}_1(t) = \sum_{i=0}^{1} \bar{u}_i(t)\frac{1}{t}F_{x_i^{-1}(\eta_{\mathbf{s}_a})\bar{c}}[u](t) =: e_2^T z(t)\bar{u}_{i_1}(t)\frac{1}{t}.$$

So the first row of N_{i_1} is e_2^T, where x_{i_1} is the first letter of $\eta_{\mathbf{s}_a}$, and the first row of the other realization matrix contains all zeroes. Continuing in this way,

$$\dot{z}_k(t) = \sum_{i=0}^{1} \bar{u}_i(t)\frac{1}{t}F_{\eta_{\mathbf{s}_a}^{-1}(\eta_{\mathbf{s}_a})\bar{c}}[u](t) =: e_{k+1}^T z(t)\bar{u}_{i_k}(t)\frac{1}{t}.$$

Since in general $x_{i_k} = x_1$, the k-th row of N_1 is e_{k+1}^T, and the k-th row of the N_0 contains all zeroes. So far, this is in agreement with the proposed structure of the realization. Next observe that

$$\dot{z}_{k+1}(t) = \sum_{i=0}^{1} \bar{u}_i(t)\frac{1}{t} F_{x_i^{-1}(\bar{c})}[u](t)$$
$$= \sum_{i=0}^{1} \bar{u}_i(t)\frac{1}{t} \underbrace{F_{x_i^{-1}(\eta_{s_c})}[u](t)}_{=:z_{k+2}(t)} + \sum_{j=0}^{1} \bar{u}_j(t)\frac{1}{t} \underbrace{F_{x_j^{-1}(\eta_{s_b}\bar{c})}[u](t)}_{=:z_{k+3}(t)}.$$

In this way, new states are created until finally the term $F_{\bar{c}}[u](t) = z_{k+1}(t)$ reappears as it must. This produces an entry $\theta^{|s_b|}$ in N_1 and preserves the proposed structures of N_0 and N_1. But note, as in Example 1, that the process can continue to create new states, and the state $z_{k+1}(t)$ could reappear if η_{s_c} is a power of η_{s_b}, a possibility that has not been excluded. In addition, this realization could produce *copies* of the the first k states if η_{s_c} contains η_{s_a} as a factor. These copies will still preserve the desired structure, but this possibility points out that in general the final realization constructed by this process may not be minimal. Finally, the canonical embedding (18), which is always needed if $|s_c| > 0$, yields the final elements of the proposed structure. □

Clearly, when non-periodic components are present, giving a precise general form of the matrices N_0 and N_1 is not as simple as in the purely periodic case.

3.3 The Dendriform Setting

It is shown in this section that the generating function $\mathscr{L}_{\mathbf{s}}(t, \theta)$ defined in (3), more precisely its t-derivative, is a solution of a higher-order linear dendriform equation in the sense of [15]. The case with non-periodic components can also be considered from that perspective. This provides a purely algebraic setting for the problem and also motivates an interesting generalization in the context of the theory of linear dendriform equations.

Recall that MPLs satisfy shuffle product identities, which are derived from integration by parts for the iterated integrals in (2). For instance,

$$\mathrm{Li}_{(2)}(t)\mathrm{Li}_{(2)}(t) = 4\mathrm{Li}_{(3,1)}(t) + 2\mathrm{Li}_{(2,2)}(t).$$

In slightly more abstract terms this can be formulated using the notion of a dendriform algebra. The reader is referred to [15] for full details. Examples of dendriform algebras include the shuffle algebra as well as associative Rota–Baxter algebras. Indeed, for any $t_0 < t_1$, the space $L_{1,loc}[t_0, t_1)$ is naturally endowed with such a structure consisting of two products:

$$f \succ g := I(f)g \quad (19a)$$
$$f \prec g := fI(g), \quad (19b)$$

where I is the Riemann integral operator defined by $I(f)(t,t_0) := \int_{t_0}^{t} f(s)\,ds$—a Rota–Baxter map of weight zero. It is easily seen to satisfy the axioms of a *dendriform algebra*

$$\begin{aligned} f \succ (g \succ h) &= (f * g) \succ h \\ (f \succ g) \prec h &= f \succ (g \prec h) \\ (f \prec g) \prec h &= f \prec (g * h), \end{aligned}$$

where

$$f * g := f \succ g + f \prec g$$

is an associative product. The example (19) above moreover verifies the extra commutativity property $f \succ g = g \prec f$, making it a commutative dendriform or Zinbiel algebra[3]

$$(f \prec g) \prec h = f \prec (g \prec h + h \prec g).$$

This is another way of saying that Chen's iterated integrals define a shuffle product, which gives rise to the shuffle algebra of MPLs. For more details, including a link between general, i.e., not necessarily commutative, dendriform algebras and Fliess operators, see [13–15].

In the following, the focus is on the commutative dendriform algebra $(C[t_0,t_1), \succ, \prec)$, where $C[t_0,t_1)$ stands for the linear subspace of *continuous* (hence locally integrable!) functions on $[t_0,t_1)$. The linear operator $R_g^{\succ} : C[t_0,t_1) \to C[t_0,t_1)$ is defined for $g \in C[t_0,t_1)$ by right multiplication using (19a)

$$R_g^{\succ}(f) := f \succ g.$$

Now add the distribution $\delta = \delta_{t_0}$ to the dendriform algebra $C[t_0,t_1)$. In view of the identity $I(\delta) = 1$ on the interval $[t_0,t_1]$, it follows that $R_f^{\succ}(\delta) = \delta \succ f = f$ for any $f \in C[t_0,t_1)$. Consider next the specific functions $u_0(t) = 1/t$ and $u_1(t) = 1/(1-t)$ which appeared above (with $t_0 = 0$ and $t_1 = 1$ here), and the corresponding linear operators $R_{u_0}^{\succ}$ and $R_{u_1}^{\succ}$. Although u_0 is not locally integrable on $[0,1)$, the space $C[t_0,t_1)$ is invariant under $R_{u_0}^{\succ}$. For any word $w = x_0^{s_1-1}x_1 \cdots x_0^{s_l-1}x_1 \in x_0X^*x_1$, the linear operator $R_w^{\succ}$ is defined as the composition of the linear operators associated to its letters, namely,

$$R_w^{\succ} = (R_{u_0}^{\succ})^{s_1-1} R_{u_1}^{\succ} \cdots (R_{u_0}^{\succ})^{s_l-1} R_{u_1}^{\succ}$$

for $w = w_1 \cdots w_{|\mathbf{s}|} = x_0^{s_1-1}x_1 \cdots x_0^{s_l-1}x_1$. Using the shorthand notation $R_w^{\succ} = R_{\mathbf{s}}^{\succ}$ with $\mathbf{s} = (s_1,\ldots,s_l)$, the multiple polylogarithm $\mathrm{Li}_{\mathbf{s}}$ obviously satisfies

---

[3]The space of continuous maps on $[t_0,t_1]$ with values in the algebra $\mathscr{M}_n(\mathbb{R})$ is also a dendriform algebra, with $\prec$ and $\succ$ defined the same way. But it is Zinbiel only for $n = 1$.

$$\frac{d}{dt}\,\mathrm{Li}_{\mathbf{s}} = R_{\mathbf{s}}^{\succ}(\delta). \tag{21}$$

From (21) it follows immediately that

$$\frac{d}{dt}\mathscr{L}_{\mathbf{s}}(t,\theta) = \sum_{k=0}^{\infty} \theta^{k|\mathbf{s}|}(R_{\mathbf{s}}^{\succ})^{k}(\delta),$$

which in turn yields

$$\frac{d}{dt}\mathscr{L}_{s}(t,\theta) = \delta + \theta^{|\mathbf{s}|} R_{\mathbf{s}}^{\succ}\Big(\frac{d}{dt}\mathscr{L}_{s}(t,\theta)\Big). \tag{22}$$

Equation (22) is a dendriform equation of degree $(|\mathbf{s}|, 0)$ in the sense of [15, Section 7]. The general form of the latter is

$$X = a_{00} + \sum_{q=1}^{|\mathbf{s}|} \theta^{q} \sum_{j=1}^{q} (\cdots (X \succ a_{q1}) \succ a_{q1} \cdots) \succ a_{qq} \tag{23}$$

with $a_{00} := \delta, a_{qj} = 0$ for $q < |\mathbf{s}|$ and $a_{|\mathbf{s}|j} := \widetilde{w}_j$, matching the notations of equation (46) in Ref. [15]. The general solution X of (23) is the first coefficient of a vector Y of length $|\mathbf{s}|$ whose coefficients (discarding the first one) are given by $\theta^{j} R^{\succ}_{\widetilde{w}_1 \cdots w_j}(X)$ for $j = 1, \ldots, |\mathbf{s}| - 1$. This vector satisfies the following matrix dendriform equation of degree $(1, 0)$:

$$Y = (\delta, \underbrace{0, \ldots, 0}_{|\mathbf{s}|-1}) + \theta Y \succ N, \tag{24}$$

where the matrix[4] N is given by:

$$N = \begin{bmatrix} 0 & \widetilde{w}_1 & 0 & 0 & \cdots & 0 \\ 0 & 0 & \widetilde{w}_2 & 0 & \cdots & 0 \\ 0 & 0 & 0 & \widetilde{w}_3 & \cdots & 0 \\ \vdots & \vdots & \ddots & \ddots & \vdots & \vdots \\ 0 & 0 & 0 & 0 & \cdots & \widetilde{w}_{|\mathbf{s}|-1} \\ \widetilde{w}_{|\mathbf{s}|} & 0 & 0 & 0 & \cdots & 0 \end{bmatrix}.$$

First, observe that the $|\mathbf{s}|$-fold product $(\cdots (N \succ N) \succ \cdots) \succ N$ yields a diagonal matrix with the entry $\frac{d}{dt}\,\mathrm{Li}_{\mathbf{s}}(t)$ in the position $(1, 1)$. Second, matrix N splits into $N = N_0 u_0 + N_1 u_1$ with N_0, N_1 as in (17). Equation (24) essentially corresponds to the integral equation deduced from (7) giving the state $z(t)$.

[4]The size of the matrix can be reduced from $1 + |\mathbf{s}|(|\mathbf{s}| - 1)/2$ to $|\mathbf{s}|$ by eliminating rows and columns of zeroes due to the particular form of (22) compared to equation (46) in [15].

The case with non-periodic components can also be handled in this setting. Observe

$$\frac{d}{dt}\mathcal{L}_{\mathbf{s}_a\{\mathbf{s}_b\}\mathbf{s}_c} = R^{\succ}_{\mathbf{s}_a}\left(\frac{d}{dt}\mathcal{L}_{\{\mathbf{s}_b\}\mathbf{s}_c}\right),$$

and the term $X' = \frac{d}{dt}\mathcal{L}_{\{\mathbf{s}_b\}\mathbf{s}_c}$ satisfies the dendriform equation

$$X' = R^{\succ}_{\mathbf{s}_c}(\delta) + \theta^{|\mathbf{s}_b|}R^{\succ}_{\mathbf{s}_b}(X'). \tag{25}$$

Equation (25) is again a dendriform equation of degree $(|\mathbf{s}_b|, 0)$ with $a_{00} = R^{\succ}_{\mathbf{s}_c}(\delta)$, $a_{qj} = 0$ for $q < |\mathbf{s}_b|$ and $a_{|\mathbf{s}_b|j} = w_j$ using the notation in [15]. The general solution X' of (25) is the first coefficient of a vector Y' of length $|\mathbf{s}_b|$ whose coefficients (discarding the first one) are given by $\theta^j R^{\succ}_{w_1\cdots w_j}(X')$ for $j = 1, \ldots, |\mathbf{s}_b| - 1$. This vector satisfies the following matrix dendriform equation of degree $(1, 0)$

$$Y' = (R^{\succ}_{\mathbf{s}_c}(\delta), \underbrace{0, \ldots, 0}_{|\mathbf{s}_b|-1}) + \theta Y' \succ M',$$

where the matrix M' is given by:

$$M' = \begin{bmatrix} 0 & \widetilde{w}_1 & 0 & 0 & \cdots & 0 \\ 0 & 0 & \widetilde{w}_2 & 0 & \cdots & 0 \\ 0 & 0 & 0 & \widetilde{w}_3 & \cdots & 0 \\ \vdots & \vdots & \ddots & \ddots & \vdots & \vdots \\ 0 & 0 & 0 & 0 & \cdots & \widetilde{w}_{|\mathbf{s}_b|-1} \\ \widetilde{w}_{|\mathbf{s}_b|} & 0 & 0 & 0 & \cdots & 0 \end{bmatrix}.$$

One can ask the question whether the term $X = \frac{d}{dt}\mathcal{L}_{\mathbf{s}_a\{\mathbf{s}_b\}\mathbf{s}_c}$ itself is a solution of a dendriform equation. In fact, a closer look reveals that the theory of linear dendriform equations presented in [15] has not been sufficiently developed to embrace this more complex setting. In the light of Theorem 5, it is clear that the results in [15] should be adapted in order to address this question. Such a step, however, is beyond the scope of this paper and will thus be postponed to another work. It is worth mentioning that the matrix N needed in the linear dendriform equation

$$Y' = (0, \delta, 0, 0, 0, 0, 0) + \theta Y' \succ N$$

to match the result from Example 1 has the form

$$N = \begin{bmatrix} 0 & 0 & 0 & 0 & 0 & 0 & 0 \\ 0 & 0 & \widetilde{w}_1 & 0 & 0 & 0 & 0 \\ 0 & 0 & 0 & \widetilde{w}_2 & 0 & 0 & 0 \\ 0 & 0 & 0 & 0 & \widetilde{w}_3 & 0 & 0 \\ 0 & 0 & 0 & 0 & 0 & \widetilde{w}_4 & 0 \\ 0 & 0 & 0 & 0 & \widetilde{w}_6 & 0 & \widetilde{w}_5 \\ \widetilde{w}_7 & 0 & 0 & 0 & 0 & 0 & 0 \end{bmatrix},$$

which reflects the canonical embedding of Brockett. The first component of the vector Y' contains the solution. As indicated earlier, a proper derivation of this result in the context of general dendriform algebras, i.e., extending the results in [15], lies outside the scope of the present paper.

4 Examples

In this section, three examples of the method described above are given corresponding to the generating functions behind the identities (5), (8), and (10).

Example 2 Consider the generating function $\mathscr{L}_{(2)}(t, \theta)$. This example is simple enough that a bilinear realization can be identified directly from (4). For any fixed θ define the first state variable to be $z_1(t) = \mathscr{L}_{(2)}(t, \theta)$, and the second state variable to be $z_2(t) = t\, d\mathscr{L}_{(2)}(t, \theta)/dt$. In which case,

$$\dot{z}_1(t) = z_2(t)\frac{1}{t}, \quad z_1(0) = 1 \tag{26a}$$

$$\dot{z}_2(t) = \theta^2\, z_1(t)\frac{t}{1-t}\frac{1}{t}, \quad z_2(0) = 0 \tag{26b}$$

$$y(t) = z_1(t). \tag{26c}$$

Thereupon, system (26) assumes the form of a bilinear system as given by (17), where the inputs are set to be $\bar{u}_0(t) = 1$ and $\bar{u}_1(t) = \mathrm{Li}_0(t) = t/(1-t)$, i.e.,

$$N_0 = N_0(2) = \begin{bmatrix} 0 & 1 \\ 0 & 0 \end{bmatrix}, \quad N_1 = N_1(2) = \begin{bmatrix} 0 & 0 \\ \theta^2 & 0 \end{bmatrix}, \quad z(0) = C^T = \begin{bmatrix} 1 \\ 0 \end{bmatrix}$$

(recall the $1/t$ factors in (26) are absorbed into Haar integrators). A simulation diagram for this realization suitable for MatLab's Simulink simulation software is shown Fig. 1. Setting $\theta = 1$ and using Simulink's default integration routine `ode45` (Dormand-Prince method [12]) with a variable step size lower bounded by 10^{-8}, Fig. 2 was generated showing $\mathscr{L}_{(2)}(t, 1) = F_{(x_0x_1)^*}[\mathrm{Li}_0](t)$ as a function of t. In particular, it was found numerically that $\mathscr{L}_{(2)}(1, 1) \approx 3.6695$, which compares favorably to the theoretical value derived from (5):

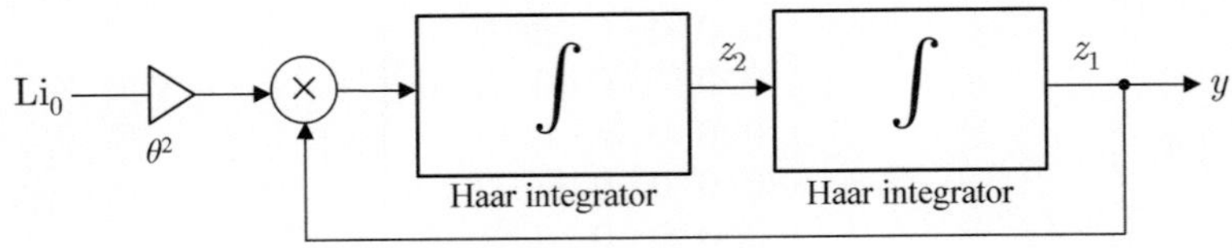

Fig. 1 Unity feedback system realizing $\mathscr{L}_{(2)}(t, 1)$

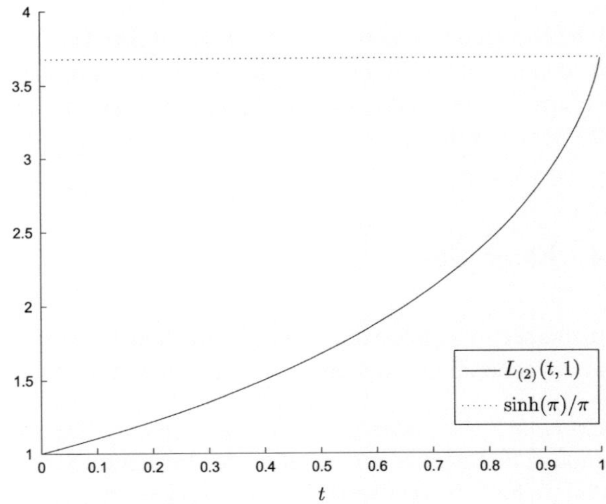

Fig. 2 Plot of $\mathscr{L}_{(2)}(t, 1)$ versus t

$$\mathscr{L}_{(2)}(1, 1) = \sum_{n=0}^{\infty} \zeta(\{2\}^n) = \sum_{n=0}^{\infty} \frac{\pi^{2n+1}}{(2n+1)^n} = \frac{\sinh(\pi)}{\pi} = 3.6761.$$

Better estimates can be found by more carefully addressing the singularities at the boundary conditions $t = 0$ and $t = 1$ in the Haar integrators.

Example 3 In order to validate (8), the identity (9) is checked numerically. Since the generating functions $\mathscr{L}_{(4)}$ and $\mathscr{L}_{(3,1)}$ are periodic, Theorem 4 applies. For $\mathbf{s} = (4)$ the corresponding bilinear realization is

$$N_0 = N_0(4) = \begin{bmatrix} 0\,1\,0\,0 \\ 0\,0\,1\,0 \\ 0\,0\,0\,1 \\ 0\,0\,0\,0 \end{bmatrix}, \quad N_1 = N_1(4) = \begin{bmatrix} 0\ 0\,0\,0 \\ 0\ 0\,0\,0 \\ 0\ 0\,0\,0 \\ \theta^4\,0\,0\,0 \end{bmatrix}, \quad z(0) = C^T = \begin{bmatrix} 1 \\ 0 \\ 0 \\ 0 \end{bmatrix}.$$

For $\mathbf{s} = (3, 1)$ the bilinear realization is

$$N_0 = \left[\begin{array}{c|c} 0\,1\,0 & 0 \\ 0\,0\,1 & 0 \\ 0\,0\,0 & 0 \\ \hline 0\,0\,0 & 0 \end{array}\right], \quad N_1 = \left[\begin{array}{c|c} 0 & 0\,0\,0 \\ 0 & 0\,0\,0 \\ 0 & 0\,0\,1 \\ \hline \theta^4 & 0\,0\,0 \end{array}\right], \quad z(0) = C^T = \begin{bmatrix} 1 \\ 0 \\ 0 \\ 0 \end{bmatrix}.$$

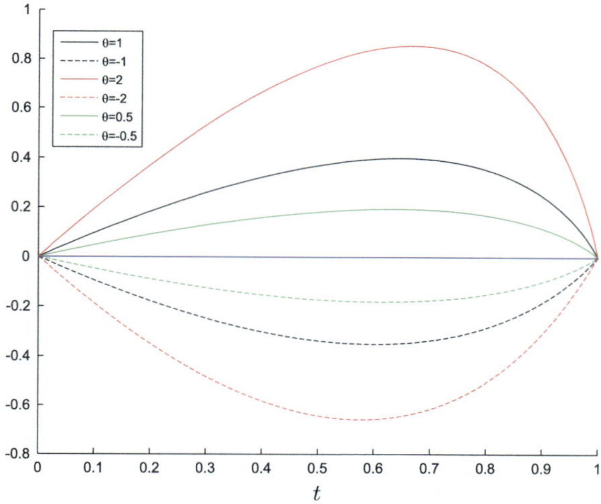

Fig. 3 Plot of $\mathscr{L}_{(4)}(t,\theta) - \mathscr{L}_{(3,1)}(t,\sqrt{2}\theta)$ versus t for different values of θ

These two dynamical systems were simulated using Haar integrators in Simulink and the difference (9) was computed as a function of t as shown in Fig. 3. As expected, this difference is very close to zero when $t = 1$ no matter how the parameter θ is selected. This is pretty convincing numerical evidence supporting (8), which as discussed in the introduction is known to be true.

Example 4 Now the method is applied to the generating functions behind the Hoffman conjecture (10). In this case, each multiple polylogarithm has non-periodic components, so Theorem 5 has to be applied three times. The realization for $\mathscr{L}_{(2,1,\{2\},3)}(t,\theta)$ was presented in Example 1. Following a similar approach, the realization for $\mathscr{L}_{(\{2\},2,2,2)}(t,\theta)$ and $\mathscr{L}_{(\{2\},3,3)}(t,\theta)$ are, respectively,

$$N_0 = \left[\begin{array}{cc|cccc|c} 0&1&0&0&0&0&0\\ 0&0&0&0&0&0&0\\ \hline 0&0&0&1&0&0&0\\ 0&0&0&0&0&0&0\\ 0&0&0&0&0&1&0\\ 0&0&0&0&0&0&0\\ \hline 0&0&0&0&0&0&0 \end{array}\right],\; N_1 = \left[\begin{array}{cc|cccc|c} 0&0&0&0&0&0&0\\ \theta^2&0&1&0&0&0&0\\ \hline 0&0&0&0&0&0&0\\ 0&0&0&0&1&0&0\\ 0&0&0&0&0&0&0\\ 0&0&0&0&0&0&1\\ \hline 0&0&0&0&0&0&0 \end{array}\right],\; z(0) = \left[\begin{array}{c} 1\\ 0\\ \hline 0\\ 0\\ 0\\ 0\\ \hline 1 \end{array}\right],\; C^T = \left[\begin{array}{c} 1\\ 0\\ \hline 0\\ 0\\ 0\\ 0\\ \hline 0 \end{array}\right]$$

and

$$N_0 = \left[\begin{array}{cc|cccc|c} 0&1&0&0&0&0&0\\ 0&0&1&0&0&0&0\\ \hline 0&0&0&0&0&0&0\\ 0&0&0&0&1&0&0\\ 0&0&0&0&0&1&0\\ 0&0&0&0&0&0&0\\ \hline 0&0&0&0&0&0&0 \end{array}\right],\; N_1 = \left[\begin{array}{cc|cccc|c} 0&0&0&0&0&0&0\\ \theta^2&0&0&0&0&0&0\\ \hline 0&0&1&0&0&0&0\\ 0&0&0&0&0&0&0\\ 0&0&0&0&0&0&0\\ 0&0&0&0&0&0&1\\ \hline 0&0&0&0&0&0&0 \end{array}\right],\; z(0) = \left[\begin{array}{c} 1\\ 0\\ \hline 0\\ 0\\ 0\\ 0\\ \hline 1 \end{array}\right],\; C^T = \left[\begin{array}{c} 1\\ 0\\ \hline 0\\ 0\\ 0\\ 0\\ \hline 0 \end{array}\right].$$

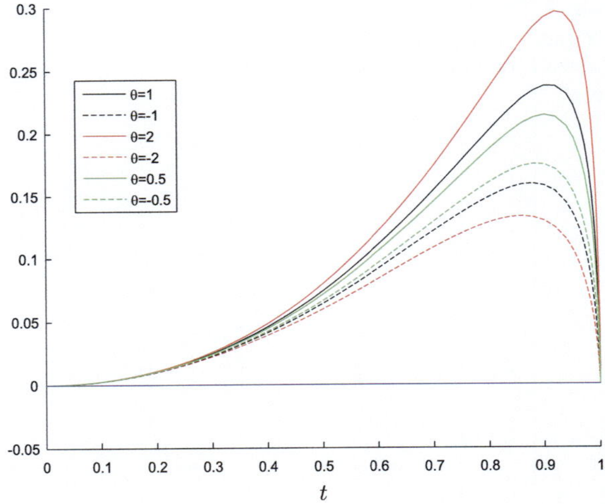

Fig. 4 Plot of $\mathscr{L}_{(\{2\},2,2,2)}(t,\theta) + 2\mathscr{L}_{(\{2\},3,3)}(t,\theta) - \mathscr{L}_{(2,1,\{2\},3)}(t,\theta)$ versus t for different values of θ

These dynamical systems were simulated to estimate numerically the left-hand side of (11) as shown in Fig. 4. As in the previous example, the case where $t = 1$ is of primary interest. This value is again very close to zero for every choice of θ tested. It is highly likely therefore that the Hoffman conjecture is true.

5 Conclusions

A systematic way was given to numerically evaluate the generating function of periodic multiple polylogarithm using Chen–Fliess series with rational generating series. The method involved mapping the corresponding Chen–Fliess series to a bilinear dynamical system, which could then be simulated numerically using Haar integration. A standard form for such a realization was given, and the method was generalized to the case where the multiple polylogarithm could have non-periodic components. The method was also described in the setting of dendriform algebras. Finally, the technique was used to numerically validate the Hoffman conjecture.

Acknowledgements The second author was supported by grant SEV-2011-0087 from the Severo Ochoa Excellence Program at the Instituto de Ciencias Matemáticas in Madrid, Spain. This research was also supported by a grant from the BBVA Foundation.

References

1. Aoki, T., Kombu, Y., Ohno, Y.: A generating function for sums of multiple zeta values and its applications. Proc. Amer. Math. Soc. **136**, 387–395 (2008)
2. Berstel, J., Reutenauer, C.: Noncommutative Rational Series with Applications. Cambridge University Press, Cambridge, UK (2010)
3. Borwein, J.M., Bradley, D.M., Broadhurst, D.J., Lisonek, P.: Combinatorial aspects of multiple zeta values. Electron. J. Combin. **5** R38, 12 (1998)
4. Borwein, J.M., Bradley, D.M., Broadhurst, D.J., Lisonek, P.: Special values of multidimensional polylogarithms. Trans. Amer. Math. Soc. **353**, 907–941 (2001)
5. Bowman, D., Bradley, D.M.: Multiple polylogarithms: a brief survey. In: Berndt, B.C., Ono, K. (eds.) Q-series with Applications to Combinatorics, Number Theory, and Physics: A Conference on Q-series with Applications to Combinatorics, Number Theory, and Physics, pp. 71–92. AMS, Providence, RI (2001)
6. Borwein, J.M., Zudilin, W.: Math honours: multiple zeta values, available at https://carma.newcastle.edu.au/MZVs/mzv.pdf
7. Brockett, R.W.: On the algebraic structure of bilinear systems. In: Mohler, R., Ruberti, R. (eds.) Theory and Applications of Variable Structure Systems, pp. 153–168. Academic Press, New York (1972)
8. Costermans, C., Minh, H.N.: Some results à l'Abel obtained by use of techniques à la Hopf. In: Proceeding of the Workshop on Global Integrability of Field Theories and Applications, Daresbury UK, pp. 63–83 (2006)
9. D'Alessandro, P., Isidori, A., Ruberti, A.: Realization and structure theory of bilinear dynamical systems. SIAM J. Control **12**, 517–535 (1974)
10. Deneufchâtel, M., Duchamp, G.H.E., Minh, V.H.N., Solomon, A.I.: Independence of hyperlogarithms over function fields via algebraic combinatorics. In: Winkler, F. (ed.) Algebraic Infomatics, Lecture Notes in Computer Science, vol. 6742, pp. 127–139. Springer, Berlin, Heidelberg (2011)
11. Dorissen, H.T.: Canonical forms for bilinear systems. Syst. Control Lett. **13**, 153–160 (1989)
12. Dormand, J.R., Prince, P.J.: A family of embedded Runge-Kutta formulae. J. Comput. Appl. Math. **6**, 19–26 (1980)
13. Duffaut Espinosa, L.A., Gray, W.S., Ebrahimi-Fard, K.: Dendriform-tree setting for fully noncommutative Fliess operators. In: Proceedings 53rd IEEE Conference on Decision and Control, Los Angeles, CA, pp. 4814–4819 (2014), arXiv:1409.0059
14. Duffaut Espinosa, L.A., Gray, W.S., Ebrahimi-Fard, K.: Dendriform-tree setting for fully noncommutative Fliess operators. IMA J. Math. Control Inform. **35**, 491–521 (2018)
15. Ebrahimi-Fard, K., Manchon, D.: Dendriform equations. J. Algebra **322**, 4053–4079 (2009)
16. Elliott, D.L.: Bilinear Control Systems. Springer, Dordrecht (2009)
17. Fliess, M.: Fonctionnelles causales non linéaires et indéterminées non commutatives. Bull. Soc. Math. France **109**, 3–40 (1981)
18. Gray, W.S., Wang, Y.: Fliess operators on L_p spaces: convergence and continuity. Syst. Control Lett. **46**, 67–74 (2002)
19. Minh, H.N.: Finite polyzêtas, Poly-Bernoulli numbers, identities of polyzêtas and noncommutative rational power series. In: Proceedings of WORDS'03, 4th International Conference on Combinatorics on Words, Turku, Finland, pp. 232–250 (2003)
20. Minh, V.H.N.: Calcul Symbolique Non Commutatif. Presses Académiques Francophones, Saarbrücken, Germany (2014)
21. Hoseaux, V., Jacob, G., Oussous, N.E., Petitot, M.: A complete Maple package for noncommutative rational power series. In: Li, Z., Sit, W.Y. (eds.) Computer Mathematics: Proceedings of the Sixth Asian Symposium, Beijing, China. World Scientific, pp. 174–188 (2003)
22. Ihara, K., Kaneko, M., Zagier, D.: Derivation and double shuffle relations for multiple zeta values. Compositio Math. **142**, 307–338 (2006)
23. Isidori, A.: Nonlinear Control Systems, 3rd edn. Springer, London (1995)

24. Jacob, G.: Réalisation des systèmes réguliers (ou bilinéaires) et séries génératrices non commutatives. In: Landau, I.D. (ed.) Outils et Modèles Mathématiques pour L'automatique, L'analyse de Systèmes et Le Traitement du Signal. CNRS-RCP 567, CNRS, Paris, pp. 325–357 (1981)
25. Nivat, M.: Transductions des languages de Chomsky. Ann. Inst. Fourier **18**, 339–455 (1968)
26. Schützenberger, M.P.: On the definition of a family of automata. Inform. Control **4**, 245–270 (1961)
27. Sussmann, H.J.: Minimal realizations and canonical forms for bilinear systems. J. Franklin Inst. **301**, 593–604 (1976)
28. Zakrzewski, M., Żoładek, H.: Linear differential equations and multiple zeta-values. I. Zeta(2). Fund. Math. **210**, 207–242 (2010)
29. Zakrzewski, M., Żoładek, H.: Linear differential equations and multiple zeta-values. II. A generalization of the WKB method. J. Math. Anal. Appl. **383**, 55–70 (2011)
30. Zakrzewski, M., Żoładek, H.: Linear differential equations and multiple zeta-values. III. Zeta(3). J. Math. Phys. **53** 013507 (2012)
31. Żoładek, H.: Note on multiple zeta-values. Bul. Acad. Ştiinţe Repub. Mold. Mat. **43**, 78–82 (2003)
32. Zudilin, W.: Algebraic relations for multiple zeta values. Russian Math. Surveys **58**, 1–29 (2003)

Arborified Multiple Zeta Values

Dominique Manchon

Abstract We describe some particular finite sums of multiple zeta values which arise from J. Ecalle's "arborification", a process which can be described as a surjective Hopf algebra morphism from the Hopf algebra of decorated rooted forests onto a Hopf algebra of shuffles or quasi-shuffles. This formalism holds for both the iterated sum picture and the iterated integral picture. It involves a decoration of the forests by the positive integers in the first case, by only two colours in the second case.

Keywords Multiple zeta values · Rooted trees · Hopf algebras · Shuffle · Quasi-shuffle · Arborification

1 Introduction

Multiple zeta values are defined by the following nested sums:

$$\zeta(n_1, \ldots, n_r) := \sum_{k_1 > k_2 > \cdots > k_r \geq 1} \frac{1}{k_1^{n_1} \cdots k_r^{n_r}}, \tag{1}$$

where the n_j's are positive integers. The nested sum (1) converges as long as $n_1 \geq 2$. The integer r is the *depth*, whereas the sum $p := n_1 + \cdots n_r$ is the *weight*. Although the multiple zeta values of depth one and two were already known by L. Euler, the full set of multiple zeta values first appears in 1981 in a preprint of Jean Ecalle under the name "moule $\zeta_<^\bullet$", in the context of resurgence theory in complex analysis [9, Page 429], together with its companion $\zeta_\leq^\bullet$ now known as the set of multiple star zeta values. The systematic study begins a decade later with the works of Hoffman [15] and Zagier [26]. It has been remarked by Kontsevich ([26], (see also the intriguing precursory Remark 4 on Page 431 in [9]) that multiple zeta values admit another representation by iterated integrals, namely:

D. Manchon (✉)
C.N.R.S. UMR 6620, 3 place Vasarély, BP 80026, 63178 Aubière, France
e-mail: Dominique.Manchon@uca.fr

J. I. Burgos Gil et al. (eds.), *Periods in Quantum Field Theory and Arithmetic*, Springer Proceedings in Mathematics & Statistics 314,
https://doi.org/10.1007/978-3-030-37031-2_18

$$\zeta(n_1,\ldots,n_r) = \int\cdots\int_{0\le u_p\le\cdots\le u_1\le 1} \frac{du_1}{\varphi_1(u_1)}\cdots\frac{du_p}{\varphi_p(u_p)}, \tag{2}$$

with $\varphi_j(u) = 1 - u$ if $j \in \{n_1, n_1 + n_2, n_1 + n_2 + n_3, \ldots, p\}$ and $\varphi_j(u) = u$ otherwise. For later use we set:

$$f_0(u) := u, \qquad f_1(u) := 1 - u.$$

Iterated integral representation (2) is the starting point to the modern approach in terms of mixed Tate motives over $\mathbb{Z}$, already outlined in [26] and widely developed in the literature since then [1–3, 7, 24]. Multiple zeta values verify a lot of polynomial relations with integer coefficients: the representation (1) by nested sums leads to *quasi-shuffle relations*, whereas representation (2) by iterated integrals leads to *shuffle relations*. A third family of relations, the *regularization relations*, comes from a subtle interplay between the first two families, involving divergent multiple zeta sums $\zeta(1, n_2 \ldots n_r)$. A representative example of each family (in the order above) is given by:

$$\zeta(2,3) + \zeta(3,2) + \zeta(5) = \zeta(2)\zeta(3), \tag{3}$$
$$\zeta(2,3) + 3\zeta(3,2) + 6\zeta(4,1) = \zeta(2)\zeta(3), \tag{4}$$
$$\zeta(2,1) = \zeta(3). \tag{5}$$

It is conjectured that these three families include all possible polynomial relations between multiple zeta values. Note that the rationality of the quotient $\dfrac{\zeta(2k)}{\pi^{2k}}$, proved by L. Euler, does not yield supplementary polynomial identities. As an example, $\zeta(2) = \dfrac{\pi^2}{6}$ and $\zeta(4) = \dfrac{\pi^4}{90}$ yield $2\zeta(2)^2 = 5\zeta(4)$, a relation which can also be deduced from quasi-shuffle, shuffle and regularization relations.

It is convenient to write multiple zeta values in terms of words. In view of representations (1) and (2), this can be done in two different ways. We consider the two alphabets:

$$X := \{x_0, x_1\}, \qquad Y := \{y_1, y_2, y_3, \ldots\}, \tag{6}$$

and we denote by X^* (resp. Y^*) the set of words with letters in X (resp. Y). The vector space $\mathbb{Q}\langle X\rangle$ freely generated by X^* is a commutative algebra for the *shuffle product*, which is defined by:

$$(v_1\cdots v_p) ш (v_{p+1}\cdots v_{p+q}) := \sum_{\sigma\in \mathrm{Sh}(p,q)} v_{\sigma_1^{-1}}\cdots v_{\sigma_{p+q}^{-1}} \tag{7}$$

with $v_j \in X$, $j \in \{1, \ldots, p+q\}$. Here, $\mathrm{Sh}(p,q)$ is the set of (p,q)-*shuffles*, i.e. permutations σ of $\{1, \ldots, p+q\}$ such that $\sigma_1 < \cdots < \sigma_p$ and $\sigma_{p+1} < \cdots < \sigma_{p+q}$. The vector space $\mathbb{Q}\langle Y\rangle$ freely generated by Y^* is a commutative algebra for the *quasi-shuffle product*, which is defined as follows: a (p,q)-*quasi-shuffle of type* r

is a surjection $\sigma : \{1, \ldots, p+q\} \twoheadrightarrow \{1, \ldots, p+q-r\}$ such that $\sigma_1 < \cdots \sigma_p$ and $\sigma_{p+1} < \cdots < \sigma_{p+q}$. Denoting by $\mathrm{Qsh}(p, q; r)$ the set of (p, q)-quasi-shuffles of type r, the formula for the quasi-shuffle product ⧢̄ is:

$$(w_1 \cdots w_p) \,\overline{⧢}\, (w_{p+1} \cdots w_{p+q}) := \sum_{r\geq 0} \sum_{\sigma\in \mathrm{Qsh}(p,q;r)} w_1^\sigma \cdots w_{p+q-r}^\sigma \tag{8}$$

with $w_j \in Y$, $j \in \{1, \ldots, p+q\}$, and where w_j^σ is the *internal product* of the letters in the set $\sigma^{-1}(\{j\})$, which contains one or two elements. The internal product is defined by $[y_k y_l] := y_{k+l}$.

We denote by Y^*_{conv} the submonoid of words $w = w_1 \cdots w_r$ with $w_1 \neq y_1$, and we set $X^*_{\mathrm{conv}} = x_0 X^* x_1$. An injective monoid morphism is given by changing letter y_n into the word $x_0^{n-1}x_1$, namely:

$$\begin{aligned} \mathfrak{s} : Y^* &\longrightarrow X^* \\ y_{n_1} \cdots y_{n_r} &\longmapsto x_0^{n_1-1}x_1 \cdots x_0^{n_r-1}x_1, \end{aligned}$$

and restricts to a monoid isomorphism from Y^*_{conv} onto X^*_{conv}. As notation suggests, Y^*_{conv} and X^*_{conv} are two convenient ways to symbolize convergent multiple zeta values through representations (1) and (2) respectively. The following notations are commonly adopted:

$$\zeta_{\overline{⧢}} (y_{n_1} \cdots y_{n_r}) := \zeta(n_1, \ldots n_r), \tag{9}$$

$$\zeta_{⧢} (x_{e_1} \cdots x_{e_p}) := \int \cdots \int_{0\leq u_p \leq \cdots \leq u_1 \leq 1} \frac{du_1}{f_{e_1}(u_1)} \cdots \frac{du_p}{f_{e_p}(u_p)}, \tag{10}$$

and extended to finite linear combinations of convergent words by linearity. In particular we have:

$$\zeta(n_1, \ldots, n_r) = \zeta_{⧢} (x_0 x_1^{n_1} \cdots x_0 x_1^{n_r}), \tag{11}$$

hence the relation:

$$\zeta_{\overline{⧢}} = \zeta_{⧢} \circ \mathfrak{s}$$

is obviously verified. The quasi-shuffle relations then write:

$$\zeta_{\overline{⧢}} (w \,\overline{⧢}\, w') = \zeta_{\overline{⧢}} (w)\zeta_{\overline{⧢}} (w') \tag{12}$$

for any $w, w' \in Y^*_{\mathrm{conv}}$, whereas the shuffle relations write:

$$\zeta_{⧢} (v \,⧢\, v') = \zeta_{⧢} (v)\zeta_{⧢} (v') \tag{13}$$

for any $v, v' \in X^*_{\mathrm{conv}}$. By assigning an indeterminate value θ to $\zeta(1)$ and setting $\zeta_{\overline{⧢}} (y_1) = \zeta_{⧢} (x_1) = \theta$, it is possible to extend $\zeta_{\overline{⧢}}$, resp. $\zeta_{⧢}$, to all words in Y^*, resp. to $X^* x_1$, such that (12), resp. (13), still holds. It is also possible to extend $\zeta_{⧢}$ to a map defined on X^* by assigning an indeterminate value θ' to $\zeta_{ш}(x_0)$, such that (13) is still valid. We will stick to $\theta' = \theta$ for symmetry reasons, reflecting the following formal equality between two infinite quantities:

$$\int_0^1 \frac{dt}{t} = \int_0^1 \frac{dt}{1-t}.$$

It is easy to show that for any word $v \in X^*$ or $w \in Y^*$, the expressions $\zeta_{ш}(v)$ and $\zeta_{ш\!\!\!-}(w)$ are polynomial with respect to θ. It is no longer true that extended $\zeta_{ш\!\!\!-}$ coincides with extended $\zeta_{ш} \circ \mathfrak{s}$, but the defect can be explicitly written:

Theorem 1 (Boutet de Monvel and Zagier [26]) *There exists an infinite-order invertible differential operator* $\rho : \mathbb{R}[\theta] \to \mathbb{R}[\theta]$ *such that*

$$\zeta_{ш} \circ \mathfrak{s} = \rho \circ \zeta_{ш\!\!\!-}. \tag{14}$$

The operator ρ *is explicitly given by the series:*

$$\rho = \exp\left(\sum_{n\geq 2} \frac{(-1)^n \zeta(n)}{n} \left(\frac{d}{d\theta}\right)^n\right). \tag{15}$$

In particular, $\rho(1) = 1$, $\rho(\theta) = \theta$, and more generally $\rho(P) - P$ is a polynomial of degree $\leq d-2$ if P is of degree d, hence ρ is invertible. A proof of Theorem 1 can be read in numerous references, e.g. [5, 17, 21]. Any word $w \in Y^*_{\mathrm{conv}}$ gives rise to Hoffman's regularization relation:

$$\zeta_{ш}\big(x_1 \,ш\, \mathfrak{s}(w) - \mathfrak{s}(y_1 \,ш\!\!\!-\, w)\big) = 0, \tag{16}$$

which is a direct consequence of Theorem 1. The linear combination of words involved above is convergent, hence (16) is a relation between convergent multiple zeta values, although divergent ones have been used to establish it. The simplest regularization relation (5) is nothing but (16) applied to the word $w = y_2$.

Rooted trees can enrich the picture in two ways: first of all, considering a rooted tree t with set of vertices $\mathcal{V}(t)$ and decoration $n_v \in \mathbb{Z}_{>0}$, $v \in \mathcal{V}(t)$, we define the associated *contracted arborified multiple zeta value* by:

$$\zeta^{\mathcal{T}}(t) := \sum_{k\in D_t} \prod_{v\in\mathcal{V}(t)} \frac{1}{k_v^{n_v}}, \tag{17}$$

where D_t is made of those maps $v \mapsto k_v \in \mathbb{Z}_{>0}$ such that $k_v < k_w$ if and only if there is a path from the root to w through v. The sum (17) is convergent as long as $n_v \geq 2$ if v is a leaf of t. The definition is multiplicatively extended to rooted forests. A similar definition can be introduced starting from the integral representation (2): considering a rooted tree τ with set of vertices $\mathcal{V}(\tau)$ and decoration $e_v \in \{0, 1\}$, $v \in \mathcal{V}(\tau)$, we define the associated *arborified multiple zeta value* by:

$$\zeta^T(\tau) := \int_{u \in \Delta_\tau} \prod_{v \in \mathcal{V}(\tau)} \frac{du_v}{f_{e_v}(u_v)}, \tag{18}$$

where $\Delta_\tau \subset [0, 1]^{|\mathcal{V}(\tau)|}$ is made of those maps $v \mapsto u_v \in [0, 1]$ such that $u_v \leq u_w$ if and only if there is a path from the root to w through v. The integral (18) is convergent as long as $e_v = 1$ if v is the root of τ and $e_v = 0$ if v is a leaf of τ. A multiplicative extension to two-coloured rooted forests will also be considered. A further extension of multiple zeta values to more general finite posets than rooted forests, in this non-contracted form, recently appeared in a paper by Yamamoto [25], see also [18]. We give a brief account of these "posetified" multiple zeta values in Sect. 6.

Arborified and contracted arborified multiple zeta values are finite linear combinations of ordinary ones. For example we have :

$\zeta^T($ 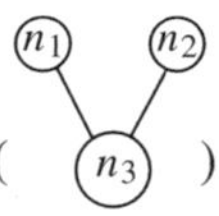 $) = \zeta(n_1, n_2, n_3) + \zeta(n_2, n_1, n_3) + \zeta(n_1 + n_2, n_3)$

and, choosing white for colour 0 and black for colour 1:

$\zeta^T\Big($  $\Big) = 2\zeta(3, 1) + \zeta(2, 2),$

$\zeta^T($  $) = 3\zeta(4).$

The terminology comes from J. Ecalle's *arborification*, a transformation which admits a "simple" and a "contracting" version [10, 11]. This transformation is best understood in terms of a canonical surjective morphism from Butcher-Connes-Kreimer Hopf algebra of rooted forests onto a corresponding shuffle Hopf algebra (quasi-shuffle Hopf algebra for the contracting arborification) [13].

The paper is organized as follows: after a reminder on shuffle and quasi-shuffle Hopf algebras, we describe the two versions of arborification in some detail, and we describe a possible transformation from contracted arborified to arborified multiple zeta values, which can be seen as an arborified version of the map $\mathfrak{s}$ from words in Y^* into words in X^*. A more natural version of this arborified $\mathfrak{s}$ with respect to the tree structures is still to be found. We finally give in Sect. 6 an account of the more general poset multiple zeta values in both simple and contracting versions, and we interpret the restricted sum formula of [12, 23] in terms of simple poset multiple zeta values.

2 Shuffle and Quasi-Shuffle Hopf Algebras

Let V be any commutative algebra on a base field k of characteristic zero. The product on V will be denoted by $(a, b) \mapsto [ab]$. This algebra is not supposed to be unital: in particular any vector space can be considered as a commutative algebra with trivial product $(a, b) \mapsto [ab] = 0$. The associated *quasi-shuffle Hopf algebra* is $\big(T(V), ⧢\!\!\!-\,, \Delta\big)$, where $\big(T(V), \Delta\big)$ is the tensor coalgebra:

$$T(V) = \bigoplus_{k \geq 0} V^{\otimes k}.$$

The decomposable elements of $V^{\otimes k}$ will be denoted by $v_1 \cdots v_k$ with $v_j \in V$. The coproduct Δ is the deconcatenation coproduct:

$$\Delta(v_1 \cdots v_k) := \sum_{r=0}^{k} v_1 \cdots v_r \otimes v_{r+1} \cdots v_k. \tag{19}$$

The quasi-shuffle product ⧢̄ is given for any $v_1, \ldots v_{p+q}$ by:

$$(v_1 \cdots v_p) \,\overline{⧢}\, (v_{p+1} \cdots v_{p+q}) := \sum_{r \geq 0} \sum_{\sigma \in \mathrm{Qsh}(p,q;r)} v_1^{\sigma} \cdots v_{p+q-r}^{\sigma} \tag{20}$$

with $v_j \in Y$, $j \in \{1, \ldots, p+q\}$, and where v_j^{σ} is the internal product of the letters in the set $\sigma^{-1}(\{j\})$, which contains one or two elements. Note that if the internal product vanishes, only ordinary shuffles (i.e. quasi-shuffles of type $r = 0$) do contribute to the quasi-shuffle product, which specializes to the shuffle product ⧢ in this case. The tensor coalgebra endowed with the quasi-shuffle product ⧢̄ is a Hopf algebra which, remarkably enough, does not depend on the particular choice of the internal product [16]. An explicit Hopf algebra isomorphism exp from $\big(T(V), \overline{⧢}, \Delta\big)$ onto $\big(T(V), ⧢, \Delta\big)$ is given in [16]. Although we won't use it, let us recall its expression: let $\mathcal{P}(k)$ be the set of compositions of the integer k, i.e. the set of sequences $I = (i_1, \ldots, i_r)$ of positive integers such that $i_1 + \cdots + i_r = k$. For any $u = v_1 \ldots v_k \in T(V)$ and any composition $I = (i_1, \ldots, i_r)$ of k we set:

$$I[u] := [v_1 \ldots v_{i_1}].[v_{i_1+1} \cdots v_{i_1+i_2}] \ldots [v_{i_1+\cdots+i_{r-1}+1} \cdots v_k].$$

Then:

$$\exp u = \sum_{I=(i_1,\ldots,i_r) \in \mathcal{P}(k)} \frac{1}{i_1! \ldots i_r!} I[u].$$

Moreover ([16], lemma 2.4), the inverse log of exp is given by :

$$\log u = \sum_{I=(i_1,\ldots,i_r)\in\mathcal{P}(k)} \frac{(-1)^{k-r}}{i_1\ldots i_r} I[u].$$

For example for $v_1, v_2, v_3 \in V$ we have:

$$\begin{aligned}
\exp v_1 &= v_1 \quad , \log v_1 = v_1,\\
\exp(v_1v_2) &= v_1v_2 + \frac{1}{2}[v_1v_2], \log(v_1v_2) = v_1v_2 - \frac{1}{2}[v_1v_2],\\
\exp(v_1v_2v_3) &= v_1v_2v_3 + \frac{1}{2}([v_1v_2]v_3 + v_1[v_2v_3]) + \frac{1}{6}[v_1v_2v_3],\\
\log(v_1v_2v_3) &= v_1v_2v_3 - \frac{1}{2}([v_1v_2]v_3 + v_1[v_2v_3]) + \frac{1}{3}[v_1v_2v_3].
\end{aligned}$$

Going back to the notations of the introduction, $\mathbb{Q}\langle Y\rangle$ is the quasi-shuffle Hopf algebra associated to the algebra $tk[t]$ of polynomials without constant terms, whereas $\mathbb{Q}\langle X\rangle$ is the shuffle Hopf algebra associated with the two-dimensional vector space spanned by X.

3 The Butcher-Connes-Kreimer Hopf Algebra of Decorated Rooted Trees

Let $\mathcal{D}$ be a set. A *rooted tree* is an oriented (non planar) graph with a finite number of vertices, among which one is distinguished and called the *root*, such that any vertex admits exactly one incoming edge, except the root which has no incoming edges. A *$\mathcal{D}$-decorated rooted tree* is a rooted tree t together with a map from its set of vertices $\mathcal{V}(t)$ into $\mathcal{D}$. Here is the list of (non-decorated) rooted trees up to five vertices:

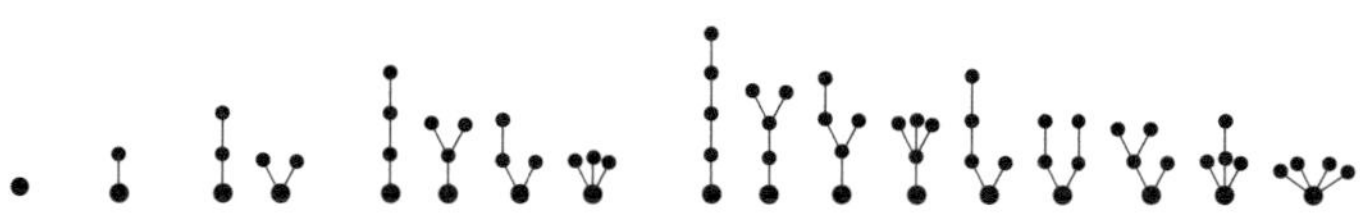

A *$\mathcal{D}$-decorated rooted forest* is a finite collection of $\mathcal{D}$-decorated rooted trees, with possible repetitions. The empty set is the forest containing no trees, and is denoted by $\mathbf{1}$. For any $d \in \mathcal{D}$, the *grafting operator* B_+^d takes any forest and changes it into a tree by grafting all components onto a common root decorated by d, with the convention $B_+^d(\mathbf{1}) = \bullet_d$.

Let $\mathcal{T}^{\mathcal{D}}$ denote the set of nonempty rooted trees and let $\mathcal{H}_{\mathrm{BCK}}^{\mathcal{D}} = k[\mathcal{T}^{\mathcal{D}}]$ be the free commutative unital algebra generated by elements of $\mathcal{T}^{\mathcal{D}}$. We identify a product of trees with the forest containing these trees. Therefore the vector space underlying

$\mathcal{H}_{\rm BCK}^{\mathcal{D}}$ is the linear span of rooted forests. This algebra is a graded and connected Hopf algebra, called the *Hopf algebra of $\mathcal{D}$-decorated rooted trees*, with the following structure: the grading is given by the number of vertices, and the coproduct on a rooted forest u is described as follows [14, 20]: the set $\mathcal{V}(u)$ of vertices of a forest u is endowed with a partial order defined by $x \le y$ if and only if there is a path from a root to y passing through x. Any subset W of $\mathcal{V}(u)$ defines a *subforest* $u_{|_W}$ of u in an obvious manner, i.e. by keeping the edges of u which link two elements of W. The coproduct is then defined by:

$$\Delta(u) = \sum_{\substack{V \sqcup W = \mathcal{V}(u) \\ W < V}} u_{|_V} \otimes u_{|_W}. \tag{21}$$

Here the notation $W < V$ means that $y < x$ for any vertex x in V and any vertex y in W such that x and y are comparable. Such a couple (V, W) is also called an *admissible cut*, with *crown* (or pruning) $u_{|_V}$ and *trunk* $u_{|_W}$. We have for example:

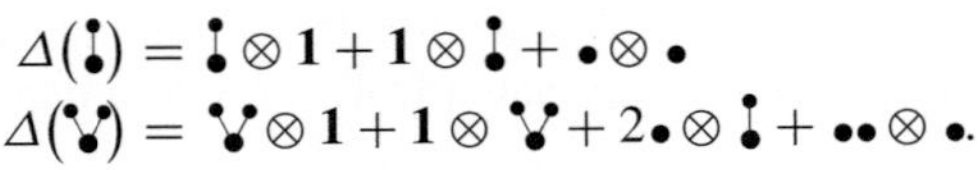

The counit is $\varepsilon(\mathbf{1}) = 1$ and $\varepsilon(u) = 0$ for any non-empty forest u. The coassociativity of the coproduct is easily checked using the following formula for the iterated coproduct:

$$\widetilde{\Delta}^{n-1}(u) = \sum_{\substack{V_1 \sqcup \cdots \sqcup V_n = \mathcal{V}(u) \\ V_n < \cdots < V_1}} u_{|_{V_1}} \otimes \cdots \otimes u_{|_{V_n}}.$$

The notation $V_n < \cdots < V_1$ is to be understood as $V_i < V_j$ for any $i > j$, with $i, j \in \{1, \ldots, n\}$.

This Hopf algebra first appeared in the work of Dür in 1986 [8]. Its dual algebra appears in [10] (Page 81 therein). It has been rediscovered and intensively studied by Kreimer in 1998 [19], as the Hopf algebra describing the combinatorial part of the BPHZ renormalization procedure of Feynman graphs in a scalar φ^3 quantum field theory. Its group of characters:

$$G_{\rm BCK}^{\mathcal{D}} = \mathrm{Hom}_{\rm alg}(\mathcal{H}_{\rm BCK}^{\mathcal{D}}, k) \tag{22}$$

is known as the *Butcher group* and plays a key role in approximation methods in numerical analysis [4]. Connes and Kreimer also proved in [6] that the operators B_+^d satisfy the property

$$\Delta\big(B_+^d(t_1 \cdots t_n)\big) = B_+^d(t_1 \cdots t_n) \otimes \mathbf{1} + (\mathrm{Id} \otimes B_+^d) \circ \Delta(t_1 \cdots t_n), \tag{23}$$

for any $t_1, \ldots, t_n \in \mathcal{T}$. This means that B_+^d is a 1-cocycle in the Hochschild cohomology of $\mathcal{H}_{\rm BCK}^{\mathcal{D}}$ with values in $\mathcal{H}_{\rm BCK}^{\mathcal{D}}$.

4 Simple and Contracting Arborification

The Hopf algebra of decorated rooted forests enjoys the following universal property (see e.g. [14]): let $\mathcal{D}$ be a set, let $\mathcal{H}$ be a graded Hopf algebra, and, for any $d \in \mathcal{D}$, let $L^d : \mathcal{H} \to \mathcal{H}$ be a Hochschild one-cocycle, i.e. a linear map such that:

$$\Delta\big(L^d(x)\big) = L^d(x) \otimes \mathbf{1}_{\mathcal{H}} + (\mathrm{Id} \otimes L^d) \circ \Delta(x). \tag{24}$$

Then there exists a unique Hopf algebra morphism $\Phi : \mathcal{H}_{\mathrm{BCK}}^{\mathcal{D}} \to \mathcal{H}$ such that:

$$\Phi \circ B_+^d = L^d \circ \Phi \tag{25}$$

for any $d \in \mathcal{D}$. Now let V be a commutative algebra, let $\big(T(V), ⧢, \Delta\big)$ be the corresponding quasi-shuffle Hopf algebra, let $(e_d)_{d\in\mathcal{D}}$ be a linear basis of V, and let $L^d : T(V) \to T(V)$ the right concatenation by e_d, defined by:

$$L^d(v_1 \ldots v_k) := v_1 \ldots v_k e_d. \tag{26}$$

One can easily check, due to the particular form of the deconcatenation coproduct, that L^d verifies the one-cocycle condition (24). The *contracting arborification* of the quasi-shuffle Hopf algebra above is the unique Hopf algebra morphism

$$\mathfrak{a}_V : \mathcal{H}_{\mathrm{BCK}}^{\mathcal{D}} \twoheadrightarrow \big(T(V), ⧢, \Delta\big) \tag{27}$$

such that $\mathfrak{a}_V \circ B_+^d = L^d \circ \mathfrak{a}_V$ for any $d \in \mathcal{D}$. The map $\mathfrak{a}_V$ sends any decorated forest to the sum of all its linear extensions, taking contractions into account (see Example (30) below). It is obviously surjective, since the word $w = e_{d_1} \cdots e_{d_r}$ can be obtained as the image of the ladder $\ell_Y(w)$ with r vertices decorated by $d_1, \ldots d_r$ from top to bottom. This map is invariant under linear base changes. For the shuffle algebra (i.e. when the internal product on V is set to zero), the corresponding Hopf algebra morphism $\mathfrak{a}_V$ is called *simple arborification*, and the corresponding section will be denoted by ℓ_X (see Examples (31) and (32) below).

Let us apply this construction to multiple zeta values (the base field k being the field $\mathbb{Q}$ of rational numbers): we denote by $\mathfrak{a}_X$ (resp. $\mathfrak{a}_Y$) the simple (resp. contracting) arborification from $\mathcal{H}_{\mathrm{BCK}}^X$ onto $\mathbb{Q}\langle X\rangle$ (resp. from $\mathcal{H}_{\mathrm{BCK}}^Y$ onto $\mathbb{Q}\langle Y\rangle$). The maps $\zeta_{ш}$ and $\zeta_{⧢}$ defined in the introduction are characters of the (Hopf) algebras $\mathbb{Q}\langle X\rangle$ and $\mathbb{Q}\langle Y\rangle$ respectively, with values in the algebra $\mathbb{R}[\theta]$. The simple and contracted arborified multiple zeta values are then respectively given by:

$$\begin{aligned} \zeta_{ш}^T : \mathcal{H}_{\mathrm{BCK}}^X &\longrightarrow \mathbb{R}[\theta] \\ \tau &\longmapsto \zeta_{ш}^T(\tau) = \zeta_{ш} \circ \mathfrak{a}_X(\tau). \end{aligned} \tag{28}$$

and:

$$\begin{aligned}\zeta_{\overline{ш}}^{\overline{T}} : \mathcal{H}_{\mathrm{BCK}}^{Y} &\longrightarrow \mathbb{R}[\theta] \\ t &\longmapsto \zeta_{\overline{ш}}^{\overline{T}}(t) = \zeta_{\overline{ш}} \circ \mathfrak{a}_Y(t).\end{aligned} \tag{29}$$

They are obviously characters of $\mathcal{H}_{\mathrm{BCK}}^{X}$ and $\mathcal{H}_{\mathrm{BCK}}^{Y}$ respectively, and respectively coincide with the maps ζ^T and $\zeta^{\overline{T}}$ defined in the introduction. This last statement comes from the fact that, for any X-decorated forest τ, the domain Δ_τ can be decomposed in a union of simplices the same way $\mathfrak{a}_X(\tau)$ is decomposed as the sum of its linear extensions, and similarly with contracting arborification $\mathfrak{a}_Y$ for any Y-decorated forest t, taking diagonals in D_t into account. Looking back at the examples given there we have:

$\mathfrak{a}_Y($ 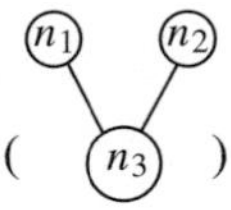 $) = y_{n_1} y_{n_2} y_{n_3} + y_{n_2} y_{n_1} y_{n_3} + y_{n_1+n_2} y_{n_3}$ (30)

and

$\mathfrak{a}_X($  $) = 2x_0x_0x_1x_1 + x_0x_1x_0x_1,$ (31)

$\mathfrak{a}_X($  $) = 3x_0x_0x_0x_1.$ (32)

5 Arborification of the Map $\mathfrak{s}$

We are looking for a map $\mathfrak{s}^T$ which makes the following diagram commutative:

$$\begin{array}{ccc} \mathcal{H}_{\mathrm{BCK}}^{Y} & \xrightarrow{\mathfrak{s}^T} & \mathcal{H}_{\mathrm{BCK}}^{X} \\ \downarrow{\scriptstyle \mathfrak{a}_Y} & & \downarrow{\scriptstyle \mathfrak{a}_X} \\ \mathbb{Q}\langle Y\rangle & \xrightarrow{\mathfrak{s}} & \mathbb{Q}\langle X\rangle \end{array}$$

An obvious answer to this problem is given by:

$$\mathfrak{s}^T = \ell_X \circ \mathfrak{s} \circ \mathfrak{a}_Y,$$

where ℓ_X is the section of $\mathfrak{a}_X$ described in the previous section. It has the drawback of completely destroying the geometry of trees: indeed, any Y-decorated forest is mapped on a linear combination of X-decorated ladders. We are then looking for a more natural map with respect to the tree structures, which makes the diagram above commute, or at least the outer square of the diagram below:

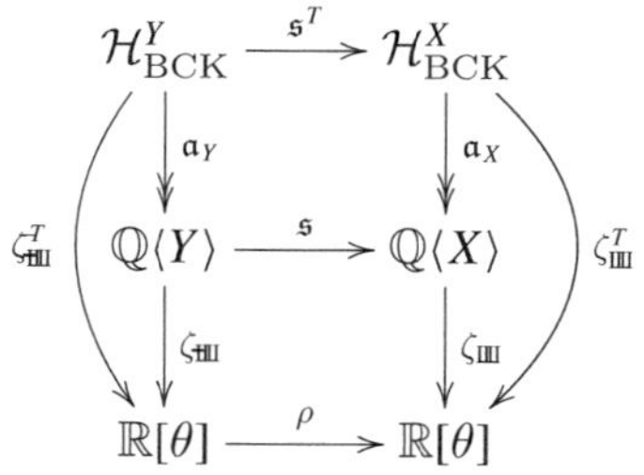

This interesting problem remains open.

6 Poset Multiple Zeta Values

A rooted forest is nothing but a particular finite poset in which each non-minimal element (i.e. each vertex different from a root) x has a unique predecessor, i.e. there exists a unique $y < x$ such that for any z with $y \leq z \leq x$, one has $z = x$ or $z = y$. It turns out that most of the concepts previously defined still make sense without this last condition. First of all, identities (17) and (18) define real numbers for any finite poset t (resp. τ) respectively decorated by Y and X, respectively named *contracted poset multiple zeta values* and *simple poset multiple zeta value*. Connected (non-decorated) posets up to four vertices are given by:

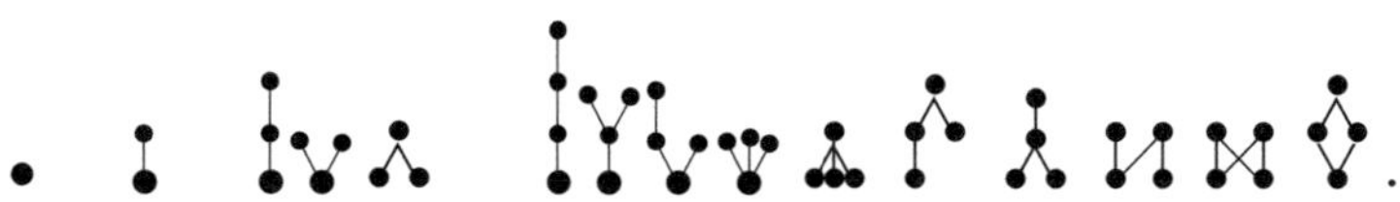

We have for example:

$\zeta^{T}($ 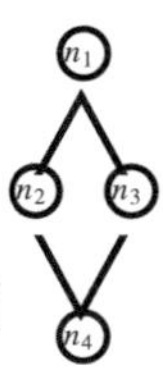 $) = \zeta(n_1, n_2, n_3, n_4) + \zeta(n_1, n_3, n_2, n_4) + \zeta(n_1, n_2 + n_3, n_4)$

and

$\zeta^{T}($  $) = \zeta(3, 1) + \zeta(2, 2).$

Next, for any set D, the linear span of isomorphism classes of D-decorated posets is a graded connected commutative Hopf algebra $\mathcal{H}_P^D$. The product is given by disjoint union, and the coproduct is still given by Formula (21). It is well-known that $\mathcal{H}_P^D$ is a commutative incidence Hopf algebra: see [22, Paragraph 16], taking for $\mathcal{F}$ the family of all finite posets with the notations therein. The forest Hopf algebra $\mathcal{H}_{BCK}^D$ is a Hopf subalgebra of $\mathcal{H}_P^D$. The simple arborification $\mathfrak{a}_X : \mathcal{H}_{BCK}^X \to (T(X), ш, \Delta)$ extends to a surjective Hopf algebra morphism $\mathfrak{p}_X : \mathcal{H}_P^X \to (T(X), ш, \Delta)$ and, similarly, the contracting arborification $\mathfrak{a}_Y : \mathcal{H}_{BCK}^Y \to (T(Y), ш\!\!\!-, \Delta)$ extends to a surjective Hopf algebra morphism $\mathfrak{p}_Y : \mathcal{H}_P^Y \to (T(Y), ш\!\!\!-, \Delta)$.

The "posetization" map $\mathfrak{p}_X$ and its contracting version $\mathfrak{p}_Y$ still map a poset on the sum of all its linear extensions, moreover taking contraction terms into account in the case of $\mathfrak{p}_Y$. The fact that both are Hopf algebra morphisms can be checked by a routine computation.

The canonical involution ι on the set of finite posets is given by reversing the order: for example,

$\iota($  $) =$  .

The duality involution σ on the set of X-decorated posets is given by both applying ι and switching the two colours, i.e. exchanging 0 and 1. The duality relations for multiple zeta values extends to poset multiple zeta values as follows:

$$\zeta^T(\tau) = \zeta^T \circ \sigma(\tau). \tag{33}$$

Poset multiple zeta values recently appeared (in the simple form only) in a paper by Yamamoto [25], as well as in another paper of the same author together with Kaneko [18]. Let us mention that the restricted sum formula of [12], (see [23], formula (2) therein) can be understood as an equality between two poset multiple zeta values (in the simple version) involving "kite-shaped" posets, namely:

$$\zeta^T(A_{a,b,c}) = \zeta^T(B_{a,b,c}), \tag{34}$$

where a, b, c are three non-negative integers, and where $A_{a,b,c}$ and $B_{a,b,c}$ are defined as follows:

- $A_{a,b,c}$ has a unique white maximum linked to two ladders, the first made of c white vertices, the second made of b black vertices. Both join to a black ladder (the tail, pointing downwards) of length $a + 1$.
- $B_{a,b,c}$ has a unique black minimum linked to two ladders, the first made of b white vertices, the second made of a black vertices. Both join to a white ladder (the tail, pointing upwards) of length $c + 1$.

Both posets defined above have total number of vertices equal to $a + b + c + 2$. From (33) and (34), we immediately get:

$$\zeta^T(A_{a,b,c}) = \zeta^T(A_{c,b,a}). \tag{35}$$

Finally, the question asked in Sect. 5 makes also sense in the poset context, replacing the two Hopf algebras $\mathcal{H}_{BCK}^X$ and $\mathcal{H}_{BCK}^Y$ respectively by $\mathcal{H}_P^X$ and $\mathcal{H}_P^Y$.

References

1. Brown, F.: On the decomposition of motivic multiple zeta values. Galois-Teichmüller theory and arithmetic geometry. Adv. Stud. Pure Math. **63**, 31–58 (2012)
2. Brown, F.: Mixed Tate motives over $\mathbb{Z}$. Ann. Math. **175**(1), 949–976 (2012)
3. Brown, F.: Depth-graded motivic multiple zeta values, preprint, arXiv:1301.3053 (2013)
4. Butcher, J.C.: An algebraic theory of integration methods. Math. Comp. **26**, 79–106 (1972)
5. Cartier, P.: Fonctions polylogarithmes, nombres polyzêtas et groupes pro-unipotents, Séminaire Bourbaki No 885. Astérisque **282**, 137–173 (2002)
6. Connes, A., Kreimer, D.: Hopf Algebras, Renormalization and Noncommutative Geometry. Comm. Math. Phys. **199**, 203–242 (1998)
7. Deligne, P., Goncharov, A.: Groupes fondamentaux motiviques de Tate mixtes. Ann. Sci. Ec. Norm. Sup. (4) **38**(1), 1–56 (2005)
8. Dür, A.: Möbius functions, incidence algebras and power series representations, Lect. Notes math. **1202**, Springer (1986)
9. Ecalle, J.: Les fonctions résurgentes Vol. 2, Publications Mathématiques d'Orsay (1981). Available at http://portail.mathdoc.fr/PMO/feuilleter.php?id=PMO_1981
10. Ecalle, J.: Singularités non abordables par la géométrie. Ann. Inst. Fourier **42**(1–2), 73–164 (1992)
11. Ecalle, J., Vallet, B.: The arborification-coarborification transform: analytic, combinatorial, and algebraic aspects. Ann. Fac. Sci. Toulouse **XIII**(4), 575–657 (2004)
12. Eie, M., Liaw, W., Ong, Y.L.: A restricted sum formula for multiple zeta values. J. Number Theory **129**, 908–921 (2009)
13. Fauvet, F., Menous, F.: Ecalle's arborification-coarborification transforms and the Connes-Kreimer Hopf algebra. Ann. Sci. Ec. Norm. **50**(1), 39–83 (2017)
14. Foissy, L.: Les algèbres de Hopf des arbres enracinés décorés I,II. Bull. Sci. Math. **126** , 193–239, 249–288 (2002)
15. Hoffman, M.E.: Multiple harmonic series. Pacific J. Math. **152**, 275–290 (1992)
16. Hoffman, M.E.: Quasi-shuffle products. J. Algebraic Combin. **11**, 49–68 (2000)
17. Ihara, K., Kaneko, M., Zagier, D.: Derivation and double shuffle relations for multiple zeta values. Comp. Math. **142**(2), 307–338 (2004)
18. Kaneko, M., Yamamoto, S.: A new integral-series identity of multiple zeta values and regularizations. Selecta Math. **24**(3), 2499–2521 (2018)
19. Kreimer, D.: On the Hopf algebra structure of perturbative quantum field theories. Adv. Theor. Math. Phys. **2**, 303–334 (1998)
20. Murua, A.: The Hopf algebra of rooted trees, free Lie algebras, and Lie series. Found. Computational Math. **6**, 387–426 (2006)
21. Racinet, G.: Doubles mélanges des polylogarithmes multiples aux racines de l'unité. Publ. Math. IHES **95**, 185–231 (2002)
22. Schmitt, W.: Incidence Hopf algebras. J. Pure Appl. Algebra **96**, 299–330 (1994)
23. Tanaka, T.: Restricted sum formula and derivation relation for multiple zeta values. arXiv:1303.0398 (2014)
24. Terasoma, T.: Mixed Tate motives and multiple zeta values. Invent. Math. **149**(2), 339–369 (2002)
25. Yamamoto, S.: Multiple zeta-star values and multiple integrals, preprint. arXiv:1405.6499 (2014)
26. Zagier, D.: Values of zeta functions and their applications. Proc. First Eur. Congress Math. **2**, 497–512, Birkhäuser, Boston (1994)

Lie Theory for Quasi-Shuffle Bialgebras

Loïc Foissy and Frédéric Patras

Abstract Many features of classical Lie theory generalize to the broader context of algebras over Hopf operads. However, this idea remains largely to be developed systematically. Quasi-shuffle algebras provide for example an interesting illustration of these phenomena, but have not been investigated from this point of view. The notion of quasi-shuffle algebras can be traced back to the beginnings of the theory of Rota-Baxter algebras, but was developed systematically only recently, starting essentially with Hoffman's work, that was motivated by multizeta values (MZVs) and featured their bialgebra structure. Many partial results on the fine structure of quasi-shuffle bialgebras have been obtained since then but, besides the fact that each of these articles features a particular point of view, they fail to develop systematically a complete theory. This article builds on these various results and develops the analog theory, for quasi-shuffle algebras, of the theory of descent algebras and their relations to free Lie algebras for classical enveloping algebras.

Keywords Combinatorial Hopf algebra · Quasi-shuffle algebra · Shuffle algebra · Planar tree · Dendriform algebra · Tridendriform algebra

L. Foissy (✉)
LMPA Joseph Liouville–Université du Littoral Côte d'opale, Centre Universitaire de la Mi-Voix, 50, rue Ferdinand Buisson, CS 80699, 62228 Calais Cedex, France
e-mail: foissy@lmpa.univ-littoral.fr

F. Patras
UMR 7351 CNRS–Université de Nice Parc Valrose, 06108 Nice Cedex 02, France
e-mail: patras@unice.fr

J. I. Burgos Gil et al. (eds.), *Periods in Quantum Field Theory and Arithmetic*, Springer Proceedings in Mathematics & Statistics 314,
https://doi.org/10.1007/978-3-030-37031-2_19

1 Introduction

Enveloping algebras of Lie algebras are known to be a fundamental notion, for an impressive variety of reasons. Their bialgebra structure allows to make a natural bridge between Lie algebras and groups. As such they are a key tool in pure algebra, algebraic and differential geometry, and so on. Their combinatorial structure is interesting on its own and is the object of the theory of free Lie algebras. Applications thereof include the theory of differential equations, numerics, control theory... From the modern point of view, featured in Reutenauer's *Free Lie algebras* [38], the "right" point of view on enveloping algebras is provided by the descent algebra: most of their key properties can indeed be obtained and finely described using computations in symmetric group algebras relying on the statistics of descents of permutations. More recently, finer structures have emerged that refine this approach. Let us quote, among others, the Malvenuto-Reutenauer or free quasi-symmetric functions Hopf algebra [29] and its bidendriform structure [14].

Many features of classical Lie theory generalize to the broader context of algebras over Hopf operads [24]. However, this idea remains largely to be developed systematically. Quasi-shuffle algebras provide for example an interesting illustration of these phenomena, but have not been investigated from this point of view.

The notion of quasi-shuffle algebras was developed systematically only recently, starting essentially with Hoffman's work, that was motivated by multizeta values (MZVs) and featured their bialgebra structure [23]. The reason for the appearance of quasi-shuffle products in many application fields (classical and stochastic integration, summation processes, probability, renormalization...) is explained by the construction by Ebrahimi-Fard of a forgetful functor from Rota–Baxter algebras of non-zero weight to quasi-shuffle algebras [11]. Many partial results on the structure of quasi-shuffle bialgebras have been obtained during the last two decades [17, 28, 30–32], fine structure theorems have been obtained in [2], but, besides the fact that each of these articles features a particular point of view, they fail to develop systematically a complete combinatorial theory.

This article builds on these various results and develops the analog theory, for quasi-shuffle bialgebras, of the theory of descent algebras and their relations to free Lie algebras for classical enveloping algebras.

The plan is as follows. Sections 2 and 3 recall the fundamental definitions. These are fairly standard ideas and materials, excepted for the fact that bialgebraic structures are introduced from the point of view of Hopf operads that will guide later developments.

The following section shows how the symmetrization process in the theory of twisted bialgebras (or Hopf species) can be adapted to define a noncommutative quasi-shuffle bialgebra structure on the operad of quasi-shuffle algebras (Theorem 1).

Section 5 deals with the algebraic structure of linear endomorphisms of quasi-shuffle bialgebras and studies from this point of view the structure of surjections. Section 6 deals with the projection on the primitives of quasi-shuffle bialgebras -the analog in the present setting of the canonical projection from an enveloping algebra to the Lie algebra of primitives. As in classical Lie theory, a structure theorem for quasi-shuffle algebras follows from the properties of this canonical projection.

Section 7 investigates the relations between the shuffle and quasi-shuffle operads when both are equipped with the Hopf algebra structure inherited from the Hopf operadic structure of their categories of algebras (as such they are isomorphic respectively to the Malvenuto-Reutenauer Hopf algebra, or Hopf algebra of free quasi-symmetric functions, and to the Hopf algebra of word quasi-symmetric functions). We recover in particular from the existence of a Hopf algebra morphism from the shuffle to the quasi-shuffle operad (Theorem 3) the exponential isomorphism relating shuffle and quasi-shuffle bialgebras. Section 8 studies coalgebra endomorphisms of quasi-shuffle bialgebras and classifies natural Hopf algebra endomorphisms and morphisms relating shuffle and quasi-shuffle bialgebras.

Section 9 studies coderivations. Quasi-shuffle bialgebras are considered classically as filtered objects (the product does not respect the tensor graduation), however the existence of a natural graded Hopf algebra structure can be deduced from the general properties of their coderivations.

Section 10 recalls briefly how the formalism of operads can be adapted to take into account graduations by using decorated operads. We detail then the case of quasi-shuffle algebras and conclude by initiating the study of the analog, in this context, of the classical descent algebra. Section 11 shows, using the bidendriform rigidity theorem, that the decorated quasi-shuffle operad is free as a noncommutative shuffle algebra.

Section 12 shows that the quasi-shuffle analog of the descent algebra, **QDesc**, is, up to a canonical isomorphism, a free noncommutative quasi-shuffle algebra over the integers (Theorem 6). The last section concludes by investigating the quasi-shuffle analog of the classical sequence of inclusions **Desc** $\subset$ **PBT** $\subset$ **Sh** of the descent algebra into the algebra of planar binary trees, resp. the operad of shuffle algebras. In the quasi-shuffle context, this sequence reads **Desc** $\subset$ **ST** $\subset$ **QSh**, where **ST** stands for the algebra of Schröder trees and **QSh** for the quasi-shuffle operad.

Terminology Following a suggestion by the referee, we include comments on the terminology. The behaviour of shuffle products was investigated by Eilenberg and MacLane in the early 50's [12]. They introduced the key idea of splitting shuffle products into two "half-shuffle products" and used the algebraic relations they satisfy to prove the associativity of shuffle products in topology. Soon after, and independently, Schützenberger axiomatized the shuffle products appearing in combinatorics and Lie algebra theory [42]. In control theory, shuffles and their relations appear in relation to products of iterated integrals under the name chronological products. The terminology is probably inspirated by the physicists' time-ordered products. The structure of the corresponding operad was implicit in Schützenberger's work as a consequence of his description of free shuffle algebras, it was introduced independently by Loday in the early 2000's [25]. Following a wit by the topologist J.-M. Lemaire, this operad of shuffle algebras is now often called operad of Zinbiel algebras (up to a few exceptions previous names such as "commutative dendriform algebras" do not seem to be used anymore). The wit is motivated by a Koszul duality phenomenon with the Bloh-Cuvier notion of Leibniz algebras. The operad encoding the axioms associated naturally to Hoffmann's quasi-shuffle algebras is called instead operad of commutative tridendriform algebras [28].

As far as the subject of the present article is concerned, quasi-shuffles are usually viewed as a deformation of shuffles (Hoffmann's isomorphism states for example that under relatively mild technical conditions quasi-shuffle bialgebras are isomorphic to shuffle bialgebras [17, 23]), and from this point of view the (weird and heavy) terminology commutative tridendriform algebras is not consistent with the one of Zinbiel algebras.

For that reason and other, historical and conceptual, ones we prefer to use the simple and coherent terminology promoted in articles such as [16, 17, 31] of "shuffle algebras" (resp. operad) and "quasi-shuffle algebras" (resp. operad) for algebras equipped with product operations satisfying the axioms obeyed by the various usual commutative shuffle and quasi-shuffle products that have appeared in the literature (resp. the corresponding operads). The reader familiar with the operadic terminology should therefore have in mind the dictionary:

- Shuffle algebra = Zinbiel algebra
- Quasi-shuffle algebra = commutative tridendriform algebra
- Noncommutative shuffle algebra = dendriform algebra
- Noncommutative quasi-shuffle algebra = tridendriform algebra.

Notations and conventions All the structures in the article (vector spaces, algebras, tensor products...) are defined over a field k. Algebraic theories and their categories (*Com*, *As*, *Sh*, *QSh* ...) are denoted in italic, as well as the corresponding free algebras over sets or vector spaces ($QSh(X)$, $Com(V)$ ...). Operads (of which we will study underlying algebra structures) and abbreviations of algebra names are written in bold (**QSh**, **NSh**, **Com**, **FQSym** ...).

2 Quasi-Shuffle Algebras

Quasi-shuffle algebras have mostly their origin in the theory of Rota-Baxter algebras and related objects such as MZVs (this because the summation operator of series is an example of a Rota–Baxter operator [10]). As we just mentioned, this is sometimes traced back to Cartier's construction of free commutative Rota-Baxter algebras [3]. They appeared independently in the study of adjunction phenomena in the theory of Hopf algebras. The relations defining quasi-shuffle algebras have also be written down in probability, in relation to semimartingales, but this does not seem to have given rise to a systematic algebraic approach. Recent developments really started with Hoffman's [23].

Another reason for the development of the theory lies in the theory of combinatorial Hopf algebras and, more specifically, into the developments originating in the theory of quasi-symmetric functions, the dual theory of noncommutative symmetric functions and other Hopf algebras such as the one of word quasi-symmetric functions. This line of thought is illustrated in [17, 30–32].

Still another approach originates in the work of Chapoton on the combinatorial and operadic properties of permutohedra and other polytopes (see e.g. [6, 7] and the introductions of [2, 32]). These phenomena lead to the axiomatic definition of noncommutative quasi-shuffle algebras (also known as dendriform trialgebras) in [28].

We follow here the Rota–Baxter approach to motivate the introduction of the axioms of quasi-shuffle algebras. This approach is the one underlying at the moment most of the applications of the theory and the motivations for its development. Rota–Baxter algebras encode for example classical integration, summation operations (as in the theory of MZVs), but also renormalization phenomena in quantum field theory, statistical physics and dynamical systems (see the survey article [10]). As explained below, any commutative Rota–Baxter algebra of weight non zero gives automatically rise to a quasi-shuffle algebra.

Definition 1 A Rota–Baxter (RB) algebra of weight θ is an associative algebra A equipped with a linear endomorphism R such that

$$\forall x, y \in A,\ R(x)R(y) = R(R(x)y + xR(y) + \theta xy).$$

It is a commutative Rota–Baxter algebra if it is commutative as an algebra.

Setting $R' := R/\theta$ when $\theta \neq 0$, one gets that the pair (A, R') is a Rota–Baxter algebra of weight 1. This implies that, in practice, there are only two interesting cases to be studied abstractly: the weight 0 and weight 1 (or equivalently any other non zero weight). The others can be deduced easily from the weight 1 case. Similar observations apply for one-parameter variants of the notion of quasi-shuffle algebras.

A classical example of a Rota–Baxter operator of weight 1 is the summation operator acting on sequences $(f(n))_{n\in\mathbf{N}}$ of elements of an associative algebra $\mathcal{A}$

$$R(f)(n) := \sum_{i=0}^{n-1} f(i).$$

This general property of summation operators applies in particular to MZVs. Recall that the latter are defined for k positive integers $n_1, \ldots, n_k \in \mathbb{N}^*$, $n_1 > 1$, by

$$\zeta(n_1, \ldots, n_k) := \sum_{m_1 > \cdots > m_k > 0} \frac{1}{m_1^{n_1} \cdots m_k^{n_k}}.$$

The Rota–Baxter property of summation operators translates then into the identity

$$\zeta(p)\zeta(q) = \zeta(p, q) + \zeta(q, p) + \zeta(p + q).$$

From now on in this article, *RB algebra* will stand for *RB algebra of weight 1*. When other RB algebras will be considered, their weight will be mentioned explicitly.

An important property of RB algebras, whose proof is left to the reader, is the existence of an associative product, the RB double product $\star$, defined by:

$$x \star y := R(x)y + xR(y) + xy \tag{1}$$

so that: $R(x)R(y) = R(x \star y)$. If one sets, in a RB algebra, $x \prec y := xR(y)$, $x \succ y := R(x)y$, one gets immediately relations such as

$$(x \cdot y) \prec z = xyR(z) = x \cdot (y \prec z),$$

$$(x \prec y) \prec z = xR(y)R(z) = x \prec (y \star z),$$

and so on. In the commutative case, $x \prec y = y \succ x$, and all relations between the products $\prec, \succ, \cdot$ and $\star := \prec + \succ + \cdot$ follow from these two. In the noncommutative case, the relations duplicate and one has furthermore $(x \succ y) \prec z = R(x)yR(z) = x \succ (y \prec z)$. These observations give rise to the axioms of quasi-shuffle algebras and noncommutative quasi-shuffle algebras.

From now on, "commutative algebra" without other precision means commutative and associative algebra; "product" on a vector space A means a bilinear product, that is a linear map from $A \otimes A$ to A.

Definition 2 A quasi-shuffle (QSh) algebra A is a nonunital commutative algebra (with product written $\bullet$) equipped with another product $\prec$ such that

$$(x \prec y) \prec z = x \prec (y \star z) \tag{2}$$
$$(x \bullet y) \prec z = x \bullet (y \prec z). \tag{3}$$

where $x \star y := x \prec y + y \prec x + x \bullet y$. We also set for further use $x \succ y := y \prec x$. As the RB double product in a commutative RB algebra, the product $\star$ is automatically associative and commutative and defines another commutative algebra structure on A.

Recall, for further use, that *shuffle algebras* correspond to weight 0 commutative RB algebras, that is quasi-shuffle algebras with a null product $\bullet = 0$. Equivalently:

Definition 3 A shuffle (Sh) algebra is a vector space equipped with a product $\prec$ satisfying (2) with $x \star y := x \prec y + y \prec x$.

It is sometimes convenient to equip quasi-shuffle algebras with a unit. The phenomenon is exactly similar to the case of shuffle algebras [42]: given a quasi-shuffle algebra, one sets $B := k \oplus A$, and the products $\prec, \bullet$ have a partial extension to B defined by, for $x \in A$:

$$1 \bullet x = x \bullet 1 := 0, \; 1 \prec x := 0, \; x \prec 1 := x.$$

The products $1 \prec 1$ and $1 \bullet 1$ cannot be defined consistently, but one sets $1 \star 1 := 1$, making B a unital commutative algebra for $\star$.

The categories of quasi-shuffle and of unital quasi-shuffle algebras are clearly equivalent (under the operation of adding or removing a copy of the ground field).

Definition 4 A noncommutative quasi-shuffle algebra (NQSh algebra) is a nonunital associative algebra (with product written $\bullet$) equipped with two other products $\prec, \succ$ such that, for all $x, y, z \in A$:

$$(x \prec y) \prec z = x \prec (y \star z) \tag{4}$$

$$(x \succ y) \prec z = x \succ (y \prec z) \tag{5}$$

$$(x \star y) \succ z = x \succ (y \succ z) \tag{6}$$

$$(x \prec y) \bullet z = x \bullet (y \succ z) \tag{7}$$

$$(x \succ y) \bullet z = x \succ (y \bullet z) \tag{8}$$

$$(x \bullet y) \prec z = x \bullet (y \prec z). \tag{9}$$

where $x \star y := x \prec y + x \succ y + x \bullet y$.

As the RB double product, the product $\star$ is automatically associative and equips A with another associative algebra structure. Indeed, the associativity relation

$$(x \bullet y) \bullet z = x \bullet (y \bullet z) \tag{10}$$

and (4)+ ... +(9) imply the associativity of $\star$:

$$(x \star y) \star z = x \star (y \star z). \tag{11}$$

If A is furthermore a quasi-shuffle algebra, then the product $\star$ is commutative.

One can show that these properties are equivalent to the associativity of the double product $\star$ in a Rota-Baxter algebra (this is because the free NQSh algebras embed into the corresponding free Rota–Baxter algebras).

Noncommutative shuffle algebras correspond to weight 0 RB algebras, that is NQSh algebras with a null product $\bullet = 0$. Equivalently:

Definition 5 A noncommutative shuffle (NSh) algebra is a vector space equipped with two products $\prec, \succ$ satisfying (4, 5, 6) with $x \star y := x \prec y + y \prec x$.

The most classical example of such a structure is provided by the topologists' shuffle product and its splitting into two "half-shuffles", an idea going back to [12].

As in the commutative case, it is sometimes convenient to equip NQSh algebras with a unit. Given a NQSh algebra, one sets $B := k \oplus A$, and the products $\prec, \succ, \bullet$ have a partial extension to B defined by, for $x \in A$:

$$1 \bullet x = x \bullet 1 := 0,\ 1 \prec x := 0,\ x \prec 1 := x,\ 1 \succ x := x,\ x \succ 1 := 0.$$

The products $1 \prec 1$, $1 \succ 1$ and $1 \bullet 1$ cannot be defined consistently, but one sets $1 * 1 := 1$, making B a unital commutative algebra for $*$.

The categories of NQSh and unital NQSh algebras are clearly equivalent.

The following Lemma encodes the previously described relations between RB algebras and quasi-shuffle algebras:

Lemma 1 *The identities $x \prec y := xR(y)$, $x \succ y := R(x)y$, $x \bullet y := xy$ induce a forgetful functor from RB algebras to NQSh algebras, resp. from commutative RB algebras to QSh algebras.*

Remark 1 Let A be a NQSh algebra.

1. If A is a commutative algebra (for the product $\bullet$) and if for $x, y \in A$: $x \prec y = y \succ x$, we say that A is commutative as a NQSh algebra. Then, $(A, \bullet, \prec)$ is a quasi-shuffle algebra.
2. We put $\preceq = \prec + \bullet$. Then (4)+(7)+(9)+(10), (5)+(9) and (6) give:

$$(x \preceq y) \preceq z = x \preceq (y \preceq z + y \succ z), \tag{12}$$

$$(x \succ y) \preceq y = x \succ (y \preceq z), \tag{13}$$

$$(x \preceq y + x \succ y) \succ z = x \succ (y \succ z). \tag{14}$$

These are the axioms that define a noncommutative shuffle algebra structure $(A, \preceq, \succ)$ on A. Similarly, if $\succeq = \succ + \bullet$, then $(A, \prec, \succeq)$ is a noncommutative shuffle algebra.

Example 1 (Hoffman, [23]) Let V be an associative, non unitary algebra. The product of $v, w \in V$ is denoted by $v.w$. The augmentation ideal $T^+(V) = \bigoplus_{n \in \mathbb{N}^*} V^{\otimes n}$ of the tensor algebra $T(V) = \bigoplus_{n \in \mathbb{N}^*} T_n(V) = \bigoplus_{n \in \mathbb{N}^*} V^{\otimes n}$ (resp. $T(V)$) is given a unique (resp. unital) NQSh algebra structure by induction on the length of tensors such that for all $a, b \in V$, for all $v, w \in T(V)$:

$$av \prec bw = a(v \uplus bw), \quad av \succ bw = b(av \uplus w), \quad av \bullet bw = (a.b)(v \uplus w), \tag{15}$$

where $\uplus = \prec + \succ + \bullet$ is called the quasi-shuffle product on $T(V)$ (by definition: $\forall v \in T(V), 1 \uplus v = v = v \uplus 1$).

Definition 6 The NQSh algebra $(T^+(V), \prec, \succ, \bullet)$ is called the *tensor quasi-shuffle algebra* associated to V. It is quasi-shuffle algebra if, and only if, $(V, .)$ is commutative (and then is called simply the *quasi-shuffle* algebra associated to V).

Here are examples of products in $T^+(V)$. Let $a, b, c \in V$.

$$\begin{aligned} a \prec b &= ab, & a \succ b &= ba, & a \bullet b &= a.b,\\ a \prec bc &= abc, & a \succ bc &= bac + bca + b(a.c), & a \bullet bc &= (a.b)c,\\ ab \prec c &= abc + acb + a(b.c), & ab \succ c &= cab, & ab \bullet c &= (a.c)b. \end{aligned}$$

In particular, the restriction of $\bullet$ to V is the product of V. If the product of V is zero, we obtain for $⊎$ the usual shuffle product $\sqcup\!\sqcup$.

A useful observation, to which we will refer as "Schützenberger's trick" (see [42]) is that, in $T^+(V)$, for $v_1, \ldots, v_n \in V$,

$$v_1 \ldots v_n = v_1 \prec (v_2 \prec \ldots (v_{n-1} \prec v_n) \ldots)). \tag{16}$$

3 Quasi-Shuffle Bialgebras

We recall that graded connected and more generally conilpotent bialgebras are automatically equipped with an antipode [5], so that the two notions of bialgebras and Hopf algebras identify when these conditions are satisfied—this will be most often the case in the present article.

Quasi-shuffle bialgebras are particular deformations of shuffle bialgebras associated to the exponential and logarithm maps. They were first introduced by Hoffman in [23] and studied further in [2, 17, 26]. The existence of a natural isomorphism between the two categories of bialgebras is known as Hoffman's isomorphism [23] and has been studied in depth in [17].

We introduce here a theoretical approach to their definition, namely through the categorical notion of Hopf operad, see [24]. The underlying ideas are elementary and deserve probably to be better known. We avoid using the categorical or operadic langage and present them simply (abstract definitions and further references on the subject are given in [24]).

Let us consider categories of binary algebras, that is algebras defined by one or several binary products satisfying homogeneous multilinear relations (i.e. algebras over binary operads). For example, commutative algebras are algebras equipped with a binary product $\cdot$ satisfying the relations $x \cdot (y \cdot z) = (x \cdot y) \cdot z$ and $x \cdot y = y \cdot x$, and so on. Multilinear means that letters should not be repeated in the defining relations: for example, n-nilpotent algebras defined by a binary product with $x^n = 0,\ n > 1$ are excluded.

The category of algebras will be said *non-symmetric* if in the defining relations the letters x, y, z... always appear in the same order. For example, the category *Com* of commutative algebras is not non-symmetric because of the relation $x \cdot y = y \cdot x$, whereas *As*, the one of associative algebras ($x \cdot (y \cdot z) = (x \cdot y) \cdot z$) is.

Notice that the categories *Sh*, *QSh* of shuffle and quasi-shuffle algebras are not non-symmetric (respectively because of the relation $x \star y = x \prec y + y \prec x$ and because of the commutativity of the $\bullet$ product) and are equipped with a forgetful functor to *Com*. The categories *NSh*, *NQSh* of noncommutative shuffle and quasi-shuffle algebras are non-symmetric (in their defining relations the letters x, y, z are not permuted) and are equipped with a forgetful functor to *As*.

Definition 7 Let C be a category of binary algebras. The category is said Hopfian if tensor products of algebras in C are naturally equipped with the structure of an algebra in C (i.e. the tensor product can be defined internally to C).

Classical examples of Hopfian categories are *Com* and *As*.

Definition 8 A bialgebra in a Hopfian category of algebras C (or C-bialgebra) is an algebra A in C equipped with a coassociative morphism to $A \otimes A$ in C.

Equivalently, it is a coalgebra in the tensor category of C-algebras.

Further requirements can be made in the definition of bialgebras, for example when algebras have units. When $C = Com$ or As, we recover the usual definition of bialgebras.

Proposition 1 *A category of binary algebras equipped with a forgetful functor to Com is Hopfian. In particular, Pois, Sh, QSh are Hopfian.*

Here *Po* is stands for the category of Poisson algebras, studied in [24] from this point of view.

Indeed, let C be a category of binary algebras equipped with a forgetful functor to *Com*. We write $\mu_1, \dots, \mu_n$ the various binary products on $A, B \in C$ and $\cdot$ the commutative product (which may be one of the μ_i, or be induced by these products as the $\star$ product is induced by the $\prec, \succ$ and $\bullet$ products in the case of shuffle and quasi-shuffle algebras). Notice that a given category may be equipped with several distinct forgetful functors to *Com*: the quasi-shuffle algebras carry, for example, two commutative products ($\bullet$ and $\star$).

The Proposition follows by defining properly the C-algebra structure on the tensor products $A \otimes B$:

$$\mu_i(a \otimes b, a' \otimes b') := \mu_i(a, a') \otimes b \cdot b'.$$

The new products μ_i on $A \otimes B$ clearly satisfy the same relations as the corresponding products on A, which concludes the proof. Notice that one could also define a "right-sided" structure by $\mu_i(a \otimes b, a' \otimes b') := a \cdot a' \otimes \mu_i(b, b')$.

A bialgebra (without a unit) in the category of quasi-shuffle algebras is a bialgebra in the Hopfian category *QSh*, where the Hopfian structure is induced by the $\star$ product. Concretely, it is a quasi-shuffle algebra A equipped with a coassociative map Δ in *QSh* to $A \otimes A$, where the latter is equipped with a quasi-shuffle algebra structure by:

$$(a \otimes b) \prec (a' \otimes b') = (a \prec a') \otimes (b \star b'), \tag{17}$$

$$(a \otimes b) \bullet (a' \otimes b') = (a \bullet a') \otimes (b \star b'). \tag{18}$$

The same process defines the notion of shuffle bialgebra (without a unit), e.g. by taking a null $\bullet$ product in the definition.

Using Sweedler's shortcut notation $\Delta(a) =: a^{(1)} \otimes a^{(2)}$, one has:

$$\Delta(a \prec b) = a^{(1)} \prec b^{(1)} \otimes a^{(2)} \star b^{(2)}, \tag{19}$$

$$\Delta(a \bullet b) = a^{(1)} \bullet b^{(1)} \otimes a^{(2)} \star b^{(2)}. \tag{20}$$

In the unital case, $B = k \oplus A$, one requires furthermore that Δ be a counital coproduct (with $\Delta(1) = 1 \otimes 1$) and, since $1 \prec 1$ and $1 \bullet 1$ are not defined, sets:

$$(1 \otimes b) \prec (1 \otimes b') = 1 \otimes (b \prec b'),$$

$$(1 \otimes b) \bullet (1 \otimes b') = 1 \otimes (b \bullet b').$$

Since unital quasi-shuffle and shuffle bialgebras are more important for applications, we call them simply quasi-shuffle bialgebras and shuffle bialgebras. In this situation it is convenient to introduce the reduced coproduct on A,

$$\tilde{\Delta}(a) := \Delta(a) - a \otimes 1 - 1 \otimes a.$$

Concretely, we get:

Definition 9 The unital QSh algebra $k \oplus A$ equipped with a counital coassociative coproduct Δ is a quasi-shuffle bialgebra if and only if for all $x, y \in A$ (we introduce for the reduced coproduct the Sweedler-type notation $\tilde{\Delta}(x) = x' \otimes x''$):

$$\tilde{\Delta}(x \prec y) = x' \prec y' \otimes x'' \star y'' + x' \otimes x'' \star y + x \prec y' \otimes y'' + x' \prec y \otimes x'' + x \otimes y, \tag{21}$$

$$\tilde{\Delta}(x \bullet y) = x' \bullet y' \otimes x'' \star y'' + x' \bullet y \otimes x'' + x \bullet y' \otimes y''. \tag{22}$$

The same constructions and arguments hold in the non-symmetric context. We do not repeat them and only state the conclusions.

Proposition 2 *A non-symmetric category of binary algebras equipped with a forgetful functor to As is Hopfian. In particular, NSh and NQSh are Hopfian.*

A bialgebra (without a unit) in the category of noncommutative quasi-shuffle (NQSh) algebras is a bialgebra in the Hopfian category *NQSh*, where the Hopfian structure is induced by the $\star$ product. Concretely, it is a NQSh algebra A equipped with a coassociative map Δ in *NQSh* to $A \otimes A$, where the latter is equipped with a NQSh algebra structure by:

$$(a \otimes b) \prec (a' \otimes b') = (a \prec a') \otimes (b \star b'), \tag{23}$$

$$(a \otimes b) \succ (a' \otimes b') = (a \succ a') \otimes (b \star b'), \tag{24}$$

$$(a \otimes b) \bullet (a' \otimes b') = (a \bullet a') \otimes (b \star b'). \tag{25}$$

The same process defines the notion of NSh (or dendriform) bialgebra (without a unit), e.g. by taking a null $\bullet$ product in the definition.

Recall that setting $\preceq := \prec + \bullet$ defines a forgetful functor from NQSh to NSh algebras. The same definition yields a forgetful functor from NQSh to NSh bialgebras.

In the unital case, one requires furthermore that Δ be a counital coproduct (with $\Delta(1) = 1 \otimes 1$) and sets

$$(1 \otimes b) \prec (1 \otimes b') = 1 \otimes (b \prec b'),$$

and similarly for $\succ$ and $\bullet$. Since this case is more important for applications, we call simply NQSh and NSh bialgebras the ones with a unit.

Definition 10 The unital NQSh algebra $k \oplus A$ equipped with counital coassociative coproduct Δ is a NQSh bialgebra if and only if for all $x, y \in A$:

$$\tilde{\Delta}(x \prec y) = x' \prec y' \otimes x'' \star y'' + x' \otimes x'' \star y + x \prec y' \otimes y'' + x' \prec y \otimes x'' + x \otimes y, \tag{26}$$

$$\tilde{\Delta}(x \succ y) = x' \succ y' \otimes x'' \star y'' + y' \otimes x \star y'' + x \succ y' \otimes y'' + x' \succ y \otimes x'' + y \otimes x, \tag{27}$$

$$\tilde{\Delta}(x \bullet y) = x' \bullet y' \otimes x'' \star y'' + x' \bullet y \otimes x'' + x \bullet y' \otimes y''. \tag{28}$$

Recall, for later use, that a NQSh bialgebra $k \oplus A$ is *connected* if the reduced coproduct is locally conilpotent:

$$A = \bigcup_{n \geq 0} Ker(\tilde{\Delta}^{(n)}),$$

where $\tilde{\Delta}^{(n)}$ is the iterated coproduct of order n ($Ker(\tilde{\Delta}$, the set of primitive elements, is also denoted $Prim(A)$) and similarly for the other unital bialgebras we will consider.

The reason for the importance of the unital case comes from Hoffman's:

Example 2 Let V be an associative, non unitary algebra. With the deconcatenation coproduct Δ, defined by:

$$\Delta(x_1 \dots x_n) = \sum_{i=0}^{n} x_1 \dots x_i \otimes x_{i+1} \dots x_n,$$

the tensor quasi-shuffle algebra $T(V)$ is a NQSh bialgebra. When V is commutative, it is a quasi-shuffle bialgebra.

4 Lie Theory for Quasi-Shuffle Bialgebras

The structural part of Lie theory, as developed for example in Bourbaki's *Groupes et Algèbres de Lie* [1] and Reutenauer's monograph on free Lie algebras [38], is largely concerned with the structure of enveloping algebras and cocommutative Hopf algebras. It was shown in [24] that many phenomena that might seem characteristic of Lie theory do actually generalize to other families of bialgebras -precisely the ones studied in the previous section, that is the ones associated with Hopfian categories of algebras equiped with a forgetful functor to *Com* or *As*.

The most natural way to study these questions is by working with twisted algebras over operads—algebras in the category of **S**-modules (families of representations of all the symmetric groups $\mathfrak{S}_n,\ n \geq 0$) or, equivalently, of functors from finite sets to vector spaces. However, doing so systematically requires the introduction of many terms and preliminary definitions (see [24]), and we prefer to follow here a more direct approach inspired by the theory of combinatorial Hopf algebras. The structures we are going to introduce are reminiscent of the Malvenuto–Reutenauer Hopf algebra [29], whose construction can be deduced from the Hopfian structure of *As*, see [35–37] and [24, Example 2.3.4]. The same process will allow us to contruct a combinatorial Hopf algebra structure on the operad **QSh** of quasi-shuffle algebras.

Recall that an algebraic theory such as the ones we have been studying (associative, commutative, quasi-shuffle, NQSh... algebras) is entirely characterized by the behaviour of the corresponding free algebra functor F: an analytic functor described by a sequence of symmetric group representation $\mathbf{F}_n$ (i.e. a **S**-module) so that, for a vector space V, $F(V) = \bigoplus\limits_n \mathbf{F}_n \otimes_{\mathfrak{S}_n} V^{\otimes n}$. Composition of operations for F-algebras are encoded by natural transformations from $F \circ F$ to F. By a standard process, this defines a monad, and F-algebras are the algebras over this monad. The direct sum $\mathbf{F} = \bigoplus\limits_n \mathbf{F}_n$ equipped with the previous (multilinear) composition law is called an operad, and F-algebras are algebras over this operad. Conversely, the $\mathbf{F}_n$ are most easily described as the multilinear part of the free F-algebras $F(X_n)$ over the vector space spanned by a finite set with n elements, $X_n := \{x_1, \ldots, x_n\}$. Here, multilinear means that $\mathbf{F}_n$ is the intersection of the n eigenspaces associated to the eigenvalue λ of the n operations induced on $F(X_n)$ by the map that scales x_i by λ (and acts as the identity on the $x_j,\ j \neq i$).

Let X be a finite set, and let us anticipate on the next Lemma and write $QSh(X) := T^+(k[X]^+)$ for the quasi-shuffle algebra associated to $k[X]^+$, the (non unital, commutative) algebra of polynomials without constant term over X. For I a multiset over X, we write x_I the associated monomial (e.g. if $I = \{x_1, x_3, x_3\}$, $x_I = x_1x_3^2$). The tensors $x_{I_1} \ldots x_{I_n} = x_{I_1} \otimes \cdots \otimes x_{I_n}$ form a basis of $QSh(X)$.

There are several ways to show that $QSh(X)$ is the free quasi-shuffle algebra over X: the property can be deduced from the classical constructions of commutative Rota-Baxter algebras by Cartier [3] or Rota [39, 40] (indeed the tensor product $x_{I_1} \ldots x_{I_n}$ corresponds to the Rota–Baxter monomial $x_{I_1} R(x_{I_2} R(x_{I_3} \ldots R(x_{I_n}) \ldots)))$ in the free RB algebra over X). It can be deduced from the construction of the free shuffle algebra over X by standard filtration/graduation arguments. It can also be deduced from a Schur functor argument [26]. The simplest proof is but the one due to Schützenberger for shuffle algebras that applies almost without change to quasi-shuffle algebras [42, p. 1–19].

Lemma 2 *The quasi-shuffle algebra $QSh(X)$ is the (unique up to isomorphism) free quasi-shuffle algebra over X.*

Proof Indeed, let A be an arbitrary quasi-shuffle algebra generated by X. Then, one checks easily by a recursion using the defining relations of quasi-shuffle algebras that every $a \in A$ is a finite sum of "normed terms", that is terms of the form

$$x_{I_1} \prec (x_{I_2} \prec (x_{I_3} \cdots \prec x_{I_n}) \dots).$$

But, if $A = QSh(X)$, by the Schützenberger's trick, $x_{I_1} \prec (x_{I_2} \prec (x_{I_3} \cdots \prec x_{I_n}) \dots)$ $= x_{I_1} \dots x_{I_n}$; the result follows from the fact that these terms form a basis of $QSh(X)$. □

Corollary 1 *The component* $\mathbf{QSh}_n$ *of the operad* $\mathbf{QSh}$ *identifies therefore with the linear span of tensors $x_{I_1} \dots x_{I_k}$, where $I_1 \amalg \cdots \amalg I_k = [n]$.*

Let us introduce useful notations. We write $x_{\mathcal{I}} := x_{I_1} \dots x_{I_k}$, where $\mathcal{I}$ denotes an arbitrary ordered sequence of disjoint subsets of $\mathbf{N}^*$, $I_1, \dots, I_k$, and set $|\mathcal{I}| := |I_1| + \cdots + |I_k|$. Recall that the standardization map associated to a subset $I = \{i_1, \dots, i_n\}$ of $\mathbf{N}^*$, where $i_1 < \cdots < i_n$ is the map st from I to $[n]$ defined by: $st(i_k) := k$. The *standardization* of $\mathcal{I}$ is then the ordered sequence $st(\mathcal{I}) := st(I_1, \dots, I_k)$, where st is the standardization map associated to the subset $I_1 \amalg \cdots \amalg I_k$ of the integers. We also set $st(x_{\mathcal{I}}) := x_{st(\mathcal{I})}$. For example, if $\mathcal{I} = \{2, 6\}, \{5, 9\}$, $st(\mathcal{I}) = \{1, 3\}, \{2, 4\}$ and $st(x_{\mathcal{I}}) = x_1x_3 \otimes x_2x_4$. The shift by k of a subset $I = \{i_1, \dots, i_n\}$ (or a sequence of subsets, and so on...) of $\mathbf{N}^*$, written $I + k$, is defined by $I + k := \{i_1 + k, \dots, i_n + k\}$.

Theorem 1 *The operad* $\mathbf{QSh}$ *of quasi-shuffle algebras inherits from the Hopfian structure of its category of algebras a NQSh bialgebra structure whose product operations are defined by:*

$$x_{\mathcal{I}} \prec x_{\mathcal{J}} := x_{\mathcal{I}} \prec_f x_{\mathcal{J}+n},$$

$$x_{\mathcal{I}} \succ x_{\mathcal{J}} := x_{\mathcal{I}} \succ_f x_{\mathcal{J}+n},$$

$$x_{\mathcal{I}} \bullet x_{\mathcal{J}} := x_{\mathcal{I}} \bullet_f x_{\mathcal{J}+n},$$

where $\mathcal{I}$ and $\mathcal{J}$ run over ordered partitions of $[n]$ and $[m]$; the coproduct is defined by:

$$\Delta(x) := (st \otimes st) \circ \Delta_f(x),$$

where, on the right-hand sides, $\prec_f, \succ_f, \bullet_f, \Delta_f$ stand for the corresponding operations on $QSh(\mathbf{N}^*)$ (where, as usual, $x \prec_f y =: y \succ_f x$).

The link with the Hopfian structure of the category of quasi-shuffle algebras refers to [24, Theorem 2.3.3]: any connected Hopf operad is a twisted Hopf algebra over this operad. The Theorem 1 can be thought of as a reformulation of this general result in terms of NQSh bialgebras.

The fact that **QSh** is a NQSh algebra follows immediately from the fact that $QSh(\mathbf{N}^*)$ is a NQSh algebra for $\prec_f, \succ_f, \bullet_f$, together with the fact that the category of NQSh algebras is non-symmetric. The coalgebraic properties and their compatibility with the NQSh algebra structure are less obvious and follow from the following Lemma (itself a direct consequence of the definitions):

Lemma 3 *Let $\mathcal{I} = I_1, \ldots, I_k$ and $\mathcal{J} = J_1, \ldots, J_l$ be two ordered sequence of disjoint subsets of $\mathbf{N}^*$ that for any $n \in \mathcal{I}_p,\ \ p \leq k$ and any $m \in \mathcal{J}_q,\ \ q \leq l$ we have $n < m$. Then:*

$$st(x_{\mathcal{I}} \prec_f x_{\mathcal{J}}) = x_{st(\mathcal{I})} \prec_f x_{st(\mathcal{J})+|\mathcal{I}|} = x_{st(\mathcal{I})} \prec x_{st(\mathcal{J})},$$

$$st(x_{\mathcal{I}} \succ_f x_{\mathcal{J}}) = x_{st(\mathcal{I})} \succ_f x_{st(\mathcal{J})+|\mathcal{I}|} = x_{st(\mathcal{I})} \succ x_{st(\mathcal{J})},$$

$$st(x_{\mathcal{I}} \bullet_f x_{\mathcal{J}}) = x_{st(\mathcal{I})} \bullet_f x_{st(\mathcal{J})+|\mathcal{I}|} = x_{st(\mathcal{I})} \bullet x_{st(\mathcal{J})}.$$

The Hopf algebra **QSh** is naturally isomorphic with **WQSym**, the Chapoton-Hivert Hopf algebra of word quasi-symmetric functions, that has been studied in [17, 31], also in relation to quasi-shuffle algebras, but from a different point of view.

Let us conclude this section by some insights on the "Lie theoretic" structure underlying the previous constructions on **QSh** (where "Lie theoretic" refers concretely to the behaviour of the functor of primitive elements in a class of bialgebras associated to an Hopfian category with a forgetful functor to *As* or *Com*). Recall that there is a forgetful functor from quasi-shuffle algebras to commutative algebras defined by keeping only the $\bullet$ product. Dually, the operad **Com** embeds into the operad **QSh**: $\mathbf{Com}_n$ is the vector space of dimension 1 generated by the monomial $x_1 \ldots x_n$, and through the embedding into **QSh** this monomial is sent to the monomial (a tensor of length 1) $x_1^{\bullet n} := x_1 \bullet \cdots \bullet x_1$ in **QSh** viewed as a NQSh algebra. Let us write slightly abusively **Com** for the image of **Com** in **QSh**, we have, by definition of the coproduct on **QSh**:

Theorem 2 *The operad* **Com** *embeds into the primitive part of the operad* **QSh** *viewed as a NQSh bialgebra. Moreover, the primitive part of* **QSh** *is stable under the $\bullet$ product.*

Proof Only the last sentence needs to be proved. It follows from the relations:

$$1 \bullet x = x \bullet 1 = 0$$

for $x \in \mathbf{QSh}_n,\ n \geq 1$. □

From the point of view of **S**-modules, the Theorem should be understood in the light of [24, Theorem 2.4.2]: for **P** a connected Hopf operad, the space of primitive elements of the twisted Hopf P-algebra **P** is a sub-operad of **P**.

As usual in categories of algebras a forgetful functor such as the one from *QSh* to *Com* induced by $\bullet$ has a left adjoint, see e.g. [19] for the general case and [26] for quasi-shuffle algebras. This left adjoint, written U (by analogy with the case of classical enveloping algebras: $U(A) \in QSh$ for $A \in Com$ equipped with a product written $\cdot$) is, up to a canonical isomorphism, the quotient of the free quasi-shuffle over the vector space A by the relations $a \bullet b = a \cdot b$. When the initial category is Hopfian, such a forgetful functor to a category of algebras over a naturally defined sub-operad arises from the properties of the tensor product of algebras in the initial category, see [24, Theorem 2.4.2 and Sect. 3.1.2]—this is exactly what happens with the pair (As, Lie) in the classical situation where the left adjoint is the usual enveloping algebra functor, and here for the pair (QSh, Com).

Lemma 4 (Quasi-shuffle PBW theorem) *The left adjoint U of the forgetful functor from QSh to Com, or "quasi-shuffle enveloping algebra" functor from Com to QSh, is (up to isomorphism) Hoffman's quasi-shuffle algebra functor T^+.*

Proof An elementary proof follows once again from (a variant of) Schützenberger's construction of the free shuffle algebra. Notice first that $T^+(A)$ is generated by A as a quasi-shuffle algebra, and that, in it, the relations $a \bullet b = a \cdot b$ hold. Moreover, choosing a basis $(a_i)_{i \in I}$ of A, the tensors $a_{i_1} \dots a_{i_n} = a_{i_1} \prec (a_{i_2} \prec \dots \prec a_{i_n}) \dots)$ form a basis of $T^+(A)$. On the other hand, by the definition of the left adjoint $U(A)$ as a quotient of $Sh(A)$ by the relations $a \bullet b = a \cdot b$, using the defining relations of quasi-shuffle algebras, any term in $U(A)$ can be written recursively as a sum of terms in "normed form" $a_{i_1} \prec (a_{i_2} \prec \dots (a_{i_{n-1}} \prec a_{i_n}) \dots)$. The Lemma follows. □

Notice that the existence of a basis of $T^+(A)$ of tensors $a_{i_1} \dots a_{i_n} = a_{i_1} \prec (a_{i_2} \prec \dots \prec a_{i_n}) \dots)$ is the analog, for quasi-shuffle enveloping algebras, of the Poincaré-Birkhoff-Witt (PBW) basis for usual enveloping algebras.

5 Endomorphism Algebras

We follow once again the analogy with the familiar notion of usual enveloping algebras and connected cocommutative Hopf algebras and study, in this section the analogs of the convolution product of their linear endomorphisms. Surjections happen to play, for quasi-shuffle algebras $T(A)$ associated to commutative algebras A, the role played by bijections in classical Lie theory, see [29] and [17, 31].

Proposition 3 *Let A be a coassociative (non necessarily counitary) coalgebra with coproduct $\tilde{\Delta} : A \longrightarrow A \otimes A$, and B be a NQSh algebra. The space of linear morphisms $Lin(A, B)$ is given a NQSh algebra structure in the following way: for all $f, g \in Lin(A, B)$,*

$$f \prec g = \prec \circ (f \otimes g) \circ \tilde{\Delta}, \quad f \succ g = \succ \circ (f \otimes g) \circ \tilde{\Delta}, \quad f \bullet g = \bullet \circ (f \otimes g) \circ \tilde{\Delta}. \tag{29}$$

Proof The construction follows easily from the fact that *NQSh* is non-symmetric and from the coassociativity of the coproduct. As an example, let us prove (5) using Sweedler's notation for $\tilde{\Delta}$. Let $f, g, h \in Lin(A, B)$. For all $x \in A$,

$$\begin{aligned}(f \succ g) \prec h(x) &= (f \succ g)(x') \prec h(x'')\\ &= (f((x')') \succ g((x')'')) \prec h(x'')\\ &= f(x') \succ (g((x'')') \prec h((x'')''))\\ &= f(x') \succ (g \prec h)(x'')\\ &= f \succ (g \prec h)(x).\end{aligned}$$

So $(f \succ g) \prec h = f \succ (g \prec h)$. □

Remark 2 The induced product $\star$ on $Lin(A, B)$ is the usual convolution product.

Corollary 2 *The set of linear endomorphisms of A, where $k \oplus A$ is a NQSh bialgebra, is naturally equiped with the structure of a NQSh algebra.*

Let us turn now to the quasi-shuffle analog of the Malvenuto-Reutenauer non-commutative shuffle algebra of permutations. The appearance of a noncommutative shuffle algebra of permutations in Lie theory in [29] can be understood operadically by noticing that the linear span of the n-th symmetric group $\mathfrak{S}_n$ is $\mathbf{As}_n$, the n-th component of the operad of associative algebras. The same reason explain why surjections appear naturally in the study of quasi-shuffle algebras: ordered partitions of initial subsets of the integers (say $\{2, 4\}, \{5\}, \{1, 3\}$) parametrize a natural basis of $\mathbf{QSh}_n$, and such ordered partitions are canonically in bijection with surjections (here, the surjection s from [5] to [3] defined by $s(2) = s(4) = 1,\ s(5) = 2,\ s(1) = s(3) = 3$). Let us show how the NQSh algebra structure of **QSh** can be recovered from the point of view of the structure of NQSh algebras of linear endomorphisms. In the process, we also give explicit combinatorial formulas for the corresponding structure maps $\prec, \succ, \bullet$. We also point out that composition of endomorphisms leads to a new product on **QSh** (such a product is usually called "internal product" in the theory of combinatorial Hopf algebras, we follow the use, see [18, 31]).

Recall that a word $n_1 \dots n_k$ over the integers is called packed if the underlying set $S = \{n_1, \dots, n_k\}$ is an initial subset of $\mathbb{N}^*$, that is, $S = [m]$ for a certain m. For later use, recall also that any word $n_1 \dots n_k$ over the integers can be packed: $pack(n_1 \dots n_k) = m_1 \dots m_k$ is the unique packed word preserving the natural order of letters ($m_i < m_j \Leftrightarrow n_i < n_j, m_i = m_j \Leftrightarrow n_i = n_j$, e.g. $pack(6353) = 3121$).

Let $n \geq 0$. We denote by $\mathbb{S}urj_n$ the set of maps $\sigma : [n] := \{1, \dots, n\} \longrightarrow \mathbb{N}^*$ such that $\sigma(\{1, \dots, n\}) = \{1, \dots, k\}$ for a certain k. The corresponding elements in $\mathbf{QSh}_n$ are the ordered partitions $\sigma^{-1}(\{1\}), \dots, \sigma^{-1}(\{k\})$ of $[n]$. The integer k is the maximum of σ and denoted by $max(\sigma)$. The element $\sigma \in \mathbb{S}urj_n$ will be represented by the packed word $(\sigma(1) \dots \sigma(n))$. We identify in this way elements of $\mathbb{S}urj_n$ with packed words of length n.

We assume that V is an associative, commutative algebra and work with the quasi-shuffle algebra $T^+(V)$. Let $\sigma \in \mathbb{S}urj_n$, $n \geq 1$. We define $F_\sigma \in End_k(T(V))$ in the following way: for all $x_1, \ldots, x_l \in V$,

$$F_\sigma(x_1 \ldots x_l) = \begin{cases} \left(\prod_{\sigma(i)=1} x_i\right) \cdots \left(\prod_{\sigma(i)=max(\sigma)} x_i\right) & \text{if } l = n, \\ 0 \text{ otherwise.} \end{cases}$$

Note that in each parenthesis, the product is the product of V. For example, if x, y, $z \in V$,

$$\begin{array}{lll} F_{(123)}(xyz) = xyz & F_{(132)}(xyz) = xzy & F_{(213)}(xyz) = yxz \\ F_{(231)}(xyz) = zxy & F_{(312)}(xyz) = yzx & F_{(321)}(xyz) = zyx \\ F_{(122)}(xyz) = x(y.z) & F_{(212)}(xyz) = y(x.z) & F_{(221)}(xyz) = z(x.y) \\ F_{(112)}(xyz) = (x.y)z & F_{(121)}(xyz) = (x.z)y & F_{(211)}(xyz) = (y.z)x \\ & F_{(111)}(xyz) = x.y.z. & \end{array}$$

We also define F_1, where 1 is the empty word, by $F_1(x_1 \ldots x_n) = \varepsilon(x_1 \ldots x_n)1$, where ε is the augmentation map from $T(V)$ to k (with kernel $T^+(V)$).

Notations. Let $k, l \geq 0$.

1. a. We denote by $QSh_{k,l}$ the set of (k, l) quasi-shuffles, that is to say elements $\sigma \in \mathbb{S}urj_{k+l}$ such that $\sigma(1) < \ldots < \sigma(k)$ and $\sigma(k+1) < \ldots < \sigma(k+l)$.
   b. $QSh^{\prec}_{k,l}$ is the set of (k, l) quasi-shuffles σ such that $\sigma^{-1}(\{1\}) = \{1\}$.
   c. $QSh^{\succ}_{k,l}$ is the set of (k, l) quasi-shuffles σ such that $\sigma^{-1}(\{1\}) = \{k+1\}$.
   d. $QSh^{\bullet}_{k,l}$ is the set of (k, l) quasi-shuffles σ such that $\sigma^{-1}(\{1\}) = \{1, k+1\}$.
   Note that $QSh_{k,l} = QSh^{\prec}_{k,l} \sqcup QSh^{\succ}_{k,l} \sqcup QSh^{\bullet}_{k,l}$.
2. If $\sigma \in \mathbb{S}urj_k$ and $\tau \in \mathbb{S}urj_l$, $\sigma \otimes \tau$ is the element of $\mathbb{S}urj_{k+l}$ represented by the packed word $\sigma\tau[\max(\sigma)]$, where $[k]$ denotes the translation by k ($312[5] = 867$).

The subspace of $End_K(T(V))$ generated by the maps F_σ is stable under composition and the products:

Proposition 4 *Let $\sigma \in \mathbb{S}urj_k$ and $\tau \in \mathbb{S}urj_l$.*

1. If $max(\tau) = k$, then $F_\sigma \circ F_\tau = F_{\sigma\circ\tau}$. Otherwise, this composition is equal to 0.
2.

$$F_\sigma \prec F_\tau = \sum_{\zeta \in QSh^{\prec}_{k,l}} F_{\zeta\circ(\sigma\otimes\tau)}, \qquad F_\sigma \succ F_\tau = \sum_{\zeta \in QSh^{\succ}_{k,l}} F_{\zeta\circ(\sigma\otimes\tau)},$$

$$F_\sigma \bullet F_\tau = \sum_{\zeta \in QSh^{\bullet}_{k,l}} F_{\zeta\circ(\sigma\otimes\tau)}, \qquad F_\sigma \uplus F_\tau = \sum_{\zeta \in QSh_{k,l}} F_{\zeta\circ(\sigma\otimes\tau)}.$$

The same formulas describe the structure of the operad **QSh** *as a NQSh algebra (i.e., in* **QSh**, *using the identification between surjections and ordered partitions, $\sigma \prec \tau = \sum_{\zeta \in QSh^{\prec}_{k,l}} \zeta \circ (\sigma \otimes \tau)$, and so on).*

Proof The proof of 1 and 2 follows by direct computations. The identification with the corresponding formulas for **QSh** follows from the identities, for all $x_1, \ldots, x_{k+l} \in V$, in the quasi-shuffle algebra $T^+(V)$:

$$x_1 \ldots x_k \prec x_{k+1} \ldots x_{k+l} = \sum_{\zeta \in QSh^{\prec}_{k,l}} F_\zeta(x_1 \ldots x_{k+l}),$$

$$x_1 \ldots x_k \succ x_{k+1} \ldots x_{k+l} = \sum_{\zeta \in QSh^{\succ}_{k,l}} F_\zeta(x_1 \ldots x_{k+l}),$$

$$x_1 \ldots x_k \bullet x_{k+1} \ldots x_{k+l} = \sum_{\zeta \in QSh^{\bullet}_{k,l}} F_\zeta(x_1 \ldots x_{k+l}),$$

$$x_1 \ldots x_k \uplus x_{k+1} \ldots x_{k+l} = \sum_{\zeta \in QSh_{k,l}} F_\zeta(x_1 \ldots x_{k+l}).$$

Moreover:

$$x_1 \ldots x_k \sqcup\!\sqcup x_{k+1} \ldots x_{k+l} = \sum_{\zeta \in Sh_{k,l}} F_\zeta(x_1 \ldots x_{k+l}),$$

where $Sh_{k,l}$ is the set of (k, l)-shuffles, that is to say $\mathfrak{S}_{k+l} \cap QSh_{k,l}$. □

Remark 3 1. $F_{(1\ldots n)}$ is the projection on the space of words of length n. Consequently:

$$Id = \sum_{n=0}^{\infty} F_{(1\ldots n)}.$$

2. In general, this action of packed words is not faithful. For example, if A is a trivial algebra, then for any $\sigma \in \mathbb{S}urj_k \setminus \mathfrak{S}_k$, $F_\sigma = 0$.
3. Here is an example where the action is faithful. Let $A = K[X_i \mid i \geq 1]_+$. Let us assume that $\sum a_\sigma F_\sigma = 0$. Acting on the word $X_1 \ldots X_k$, we obtain:

$$\sum_{\sigma \in \mathbb{S}urj_k} a_\sigma \left(\prod_{\sigma(i)=1} X_i \right) \ldots \left(\prod_{\sigma(i)=max(\sigma)} X_i \right) = 0.$$

As the X_i are algebraically independent, the words appearing in this sum are linearly independent, so for all σ, $a_\sigma = 0$.

6 Canonical Projections on Primitives

This section studies the analog, for quasi-shuffle bialgebras, of the canonical projection from a connected cocommutative Hopf algebra to its primitive part—the logarithm of the identity (see e.g. [33, 34, 38]). See also [2] where this particular topic and related ones are addressed in a more general setting.

Recall that a coalgebra C with a coassociative coproduct $\tilde{\Delta}$ is *connected* if and only if the coproduct il locally conilpotent (for $c \in C$ there exists $n \in \mathbf{N}^*$ such that $\tilde{\Delta}^{(n)}(c) = 0$).

Proposition 5 *Let A be a coassociative, non counitary, coalgebra with a locally conilpotent coproduct*

$$\tilde{\Delta} : A \longrightarrow A \otimes A, \quad A = \bigcup_{n \geq 0} Ker(\tilde{\Delta}^{(n)}),$$

and let B be a NQSh algebra. Then, for any $f \in Lin(A, B)$, there exists a unique map $\pi_f \in Lin(A, B)$, such that

$$f = \pi_f + \pi_f \prec f.$$

Proof For all $n \geq 1$, we put $F_n = Ker(\tilde{\Delta}^{(n)})$: this defines the coradical filtration of A. In particular, $F_1 =: Prim(A)$. Moreover, if $n \geq 1$:

$$\tilde{\Delta}(F_n) \subseteq F_{n-1} \otimes F_{n-1}.$$

Let us choose for all n a subspace E_n of A such that $F_n = F_{n-1} \oplus E_n$. In particular, $E_1 = F_1 = Prim(A)$. Then, A is the direct sum of the E_n's and for all n:

$$\tilde{\Delta}(E_n) \subseteq \bigoplus_{i,j<n} E_i \otimes E_j.$$

Existence. We inductively define a map $\pi_f : E_n \longrightarrow B$ for all $n \geq 1$ in the following way:

- For all $a \in E_1$, $\pi_f(a) = f(a)$.
- If $a \in E_n$, as $\tilde{\Delta}(a) \in \bigoplus_{i+j<n} E_i \otimes E_j$, $(\pi_f \otimes f) \circ \tilde{\Delta}(a)$ is already defined. We then put:
$$\pi_f(a) = f(a) - \prec \circ (\pi_f \otimes f) \circ \tilde{\Delta}(a) = f(a) - (\pi_f \prec f)(a)$$

Unicity. Let μ_f such that $f = \mu_f + (\mu_f \prec f)$. For all $a \in E_1$, $f(a) = \mu_f(a) + 0$, so $\mu_f(a) = \pi_f(a)$. Let us assume that for all $k < n$, $\mu_f(a) = \pi_f(a)$ if $a \in E_k$. Let $a \in E_n$. Then:

$$a = \mu_f(a) + \mu_f(a') \prec a'' = \mu_f(a) + \pi_f(a') \prec a'' = \mu_f(a) + a - \pi_f(a),$$

so $\mu_f(a) = \pi_f(a)$. Hence, $\mu_f = \pi_f$. □

Proposition 6 *When $A = B = T^+(V)$ and $f = Id$, the map $\pi := \pi_f$ defined in Proposition 5 is equal to the projection $F_{(1)}$.*

Proof First, observe that, as $QSh^{\prec}_{1,k} = \{(1, \ldots, k)\}$, for all packed words $(a_1 \ldots a_k)$, $F_{(1)} \prec F_{(a_1 \ldots a_k)} = F_{(1(a_1+1)\ldots(a_k+1))}$. Hence, in A:

$$\begin{aligned} F_{(1)} + F_{(1)} \prec Id_A &= F_{(1)} + \sum_{n=1}^{\infty} F_{(1)} \prec F_{(1\ldots n)} = F_{(1)} + \sum_{n=1}^{\infty} F_{(1\ldots n+1)} \\ &= \sum_{n=1}^{\infty} F_{(1\ldots n)} = Id_A. \end{aligned}$$

By unicity in Proposition 5, $\pi_f = F_{(1)}$. □

More generally, we have:

Proposition 7 *Let A be a non unital, connected NQSh bialgebra, and π the unique solution to*

$$Id_A = \pi + \pi \prec Id_A,$$

then π is a projection on $Prim(A)$, and for all $x \in Prim(A)$, $y \in A$, $\pi(x \prec y) = 0$.

Proof Let us prove that for all $a \in E_n$, $\pi(a) \in Prim(A)$ by induction on n. As $E_1 = Prim(A)$, this is obvious if $n = 1$. Let us assume the result for all $k < n$. Let $a \in E_n$. Then $\pi(a) = a - \pi(a') \prec a''$. By the induction hypothesis, we can assume that $\pi(a') \in Prim(A)$, so:

$$\begin{aligned} \tilde{\Delta}(\pi(a)) &= a' \otimes a'' - \pi(a') \prec a'' \otimes a''' - \pi(a') \otimes a'' \\ &= (a' - (\pi \prec Id)(a') - \pi(a')) \otimes a'' = 0. \end{aligned}$$

Hence, for all $a \in A$, $\pi(a) \in Prim(a)$. So π that, by its very definition, acts as the identity on $Prim(A)$, is a projection on $Prim(A)$.

Let $x \in Prim(A)$ and $y \in E_n$, let us prove that $\pi(x \prec y) = 0$ by induction on n. If $n = 1$, then $y \in Prim(A)$, so $\tilde{\Delta}(x \prec y) = x \otimes y$, and $\pi(x \prec y) = x \prec y - \pi(x) \prec y = x \prec y - x \prec y = 0$. Let us assume the result at all rank $< n$. We have:

$$\tilde{\Delta}(x \prec y) = x \prec y' \otimes y'' + x \otimes y.$$

By the induction hypothesis, we can assume that $\pi(x \prec y') = 0$, so $\pi(x \prec y) = x \prec y - 0 - \pi(x) \prec y = x \prec y - x \prec y = 0$. □

Remark 4 For all $x, y \in Prim(A)$:

$$\pi(x \prec y) = 0, \qquad \pi(x \succ y) = x \succ y - y \prec x, \qquad \pi(x \bullet y) = x \bullet y.$$

Proposition 8 *Let A be a nonunital, connected quasi-shuffle bialgebra. Then $Prim(A)$ is stable under $\bullet$ and the following map is an isomorphism of quasi-shuffle bialgebras:*

$$\theta : \begin{cases} T^+(Prim(A)) \longrightarrow A \\ \qquad a_1 \dots a_k \longrightarrow a_1 \prec (a_2 \prec (\dots \prec a_k)\dots). \end{cases}$$

Proof Let $a_1, \dots, a_k \in Prim(A)$. An easy induction on k proves that:

$$\tilde{\Delta}(\theta(a_1 \otimes \dots \otimes a_k)) = \sum_{i=1}^{k-1} \theta(a_1 \otimes \dots \otimes a_i) \otimes \theta(a_{i+1} \otimes \dots \otimes a_k).$$

So θ is a coalgebra morphism.

From this coalgebra morphism property and the identity $\pi(x \prec y) = 0$ for $x \in Prim(A)$, we get for $a_1, \dots, a_k \in Prim(A)$, $(Id_A \otimes \pi) \circ \tilde{\Delta}(\theta(a_1 \otimes \dots \otimes a_k)) = \theta(a_1 \otimes \dots \otimes a_{k-1}) \otimes \theta(a_k)$. Since θ is the identity on its restriction to $Prim(A)$, its injectivity follows by induction.

Let $a = a_1 \dots a_k$ and $b = b_1 \dots b_l \in T^+(Prim(A))$. Let us prove by induction on $k + l$ that:

$$\theta(a \prec b) = \theta(a) \prec \theta(b), \quad \theta(a \succ b) = \theta(a) \succ \theta(b), \quad \theta(a \bullet b) = \theta(a) \bullet \theta(b).$$

If $k = 1$, then $a \prec b_1 \dots b_l = ab_1 \dots b_l$, so $\theta(a \prec b) = a \prec \theta(b) = \theta(a) \prec \theta(b)$. If $l = 1$, then $a \succ b = ba$, so $\theta(a \succ b) = b \prec \theta(a) = \theta(b) \prec \theta(a) = \theta(a) \succ \theta(b)$. If $k = l = 1$, , $x \bullet y = \pi(x \bullet y) \in Prim(A)$, so $\theta(a \bullet b) = a \bullet b = \theta(a) \bullet \theta(b)$. All these remarks give the results for $k + l \leq 2$. Let us assume the result at all ranks $< k + l$. If $k = 1$, we already proved that $\theta(a \prec b) = \theta(a) \prec \theta(b)$. If $k \geq 2$, $a \prec b = a_1(a_2 \dots a_k \uplus b)$. By the induction hypothesis applied to $a_2 \dots a_k$ and b:

$$\theta(a \prec b) = a_1 \prec (\theta(a_2 \dots a_k) \star \theta(b)) = (a_1 \prec \theta(a_2 \dots a_k)) \prec \theta(b) = \theta(a) \prec \theta(b).$$

Using the commutativity of $T^+(Prim(A))$ and A, we obtain $\theta(a \succ b) = \theta(a) \succ \theta(b)$. If $l > 1$, $a \bullet b = a \bullet (b_1 \prec b_2 \dots b_l) = (a \bullet b_1) \prec b_2 \dots b_l$. Moreover, $a \bullet b_1$ is a linear span of words of length $\leq k + 1$, so, by the preceding computation and the induction hypothesis:

$$\theta(a \bullet b) = \theta(a \bullet b_1) \prec \theta(b_2 \dots b_l).$$

The induction hypothesis holds for a and b_1, so:

$$\theta(a \bullet b) = (\theta(a) \bullet \theta(b_1)) \prec \bullet(b_2 \dots b_l) = \theta(a) \bullet (b_1 \prec \theta(b_2 \dots b_l)) = \theta(a) \bullet \theta(b).$$

If $l = 1$, then $k > 1$ and we conclude with the commutativity of $\bullet$.

Let us now prove that $Prim(A)$ generates A as a quasi-shuffle algebra. Let A' be the quasi-shuffle subalgebra of A generated by $Prim(A)$. Let $a \in E_n$, let us prove that $x \in A'$ by induction on n. As $E_1 = Prim(A)$, this is obvious if $n = 1$. Let us assume the result for all ranks $< n$. Then $a = \pi(a) + \pi(a') \prec a''$. By the induction hypothesis, $a'' \in A'$. Moreover, $\pi(a)$ and $\pi(a') \in Prim(A)$, so $a \in A'$.

As a conclusion, θ is a morphism of quasi-shuffle algebras, whose image contains $Prim(A)$, which generates A, so θ is surjective. □

7 Relating the Shuffle and Quasi-Shuffle Operads

A fundamental theorem of the theory of quasi-shuffle algebras relates quasi-shuffle bialgebras and shuffle bialgebras and, under some hypothesis (combinatorial and graduation hypothesis on the generators in Hoffman's original version of the theorem [23]), shows that the two categories of bialgebras are isomorphic. This result allows to understand quasi-shuffle bialgebras as deformations of shuffle bialgebras and, as such, can be extended to other deformations of the shuffle product than the one induced by Hoffman's exponential map, see [17]. We will come back to this line of arguments in the next section.

Here, we stick to the relations between shuffle and quasi-shuffle algebras and show that Hoffman's theorem can be better understood and refined in the light of an Hopf algebra morphism relating the shuffle and quasi-shuffle operads.

Let us notice first that the same construction that allows to define a NQSh algebra structure on the operad **QSh** allows, *mutatis mutandis*, to define a noncommutative shuffle algebra structure on **Sh**, the operad of shuffle algebras. A natural basis of the latter operad is given by permutations (the result goes back to Schützenberger, who showed that the tensor algebra over a vector space V is a model of the free shuffle algebra over V [42]). Let us stick here to the underlying Hopf algebra structures.

Recall first that the set of packed words (or surjections, or ordered partitions of initial subsets of the integers) $\mathbb{S}urj$ is a basis of **QSh**. As a Hopf algebra, **QSh** is isomorphic to **WQSym**, the Hopf algebra of word symmetric functions, see e.g. [17] for references on the subject. This Hopf algebra structure is obtained as follows. For all $\sigma \in \mathbb{S}urj_k$, $\tau \in \mathbb{S}urj_l$:

$$\sigma \star \tau = \sum_{\zeta \in QSh_{k,l}} \zeta \circ (\sigma \otimes \tau).$$

For all $\sigma \in \mathbb{S}urj_n$:

$$\Delta(\sigma) = \sum_{k=0}^{max(\sigma)} \sigma_{|\{1,\ldots,k\}} \otimes Pack(\sigma_{|\{k+1,\ldots,max(\sigma)\}}),$$

where for all $I \subseteq \{1, \ldots, max(\sigma)\}$, $\sigma_{|I}$ is the packed word obtained by keeping only the letters of σ which belong to I.

On the other hand, the set of permutations is a basis of the operad **Sh**. As a Hopf algebra, the latter identifies with the Malvenuto-Reutenauer Hopf algebra [29] and with the Hopf algebra of free quasi-symmetric functions **FQSym**. Its Hopf structure is obtained as follows. For all $\sigma \in \mathfrak{S}_k$, $\tau \in \mathfrak{S}_l$:

$$\sigma \star \tau = \sum_{\zeta \in Sh_{k,l}} \zeta \circ (\sigma \otimes \tau).$$

For all $\sigma \in \mathfrak{S}_n$:

$$\Delta(\sigma) = \sum_{k=0}^{max(\sigma)} \sigma_{|\{1,\ldots,k\}} \otimes Pack(\sigma_{|\{k+1,\ldots,max(\sigma)\}}).$$

There is an obvious surjective Hopf algebra morphism Ξ from **QSh** to **Sh**, sending a packed word σ to itself if σ is a permutation, and to 0 otherwise. From an operadic point of view, this maps amounts to put to zero the $\bullet$ product. There is however another, non operadic, transformation, relating the two structures.

We use the following notations:

1. Let $\sigma \in \mathfrak{S}_n$ and $\tau \in \mathbb{S}urj_n$. We shall say that $\tau \propto \sigma$ if:
$$\forall 1 \leq i, j \leq n, (\sigma(i) \leq \sigma(j) \Longrightarrow \tau(i) \leq \tau(j)).$$
2. Let $\tau \in \mathbb{S}urj_n$. We put $\tau! = \prod_{i=1}^{max(\tau)} |\tau^{-1}(\{i\})|!$.

Theorem 3 *Consider the following map:*

$$\Phi : \begin{cases} \mathbf{Sh} \longrightarrow \mathbf{QSh} \\ \sigma \in \mathfrak{S}_n \longrightarrow \sum_{\tau \propto \sigma} \frac{\tau}{\tau!}. \end{cases}$$

Then Φ is an injective Hopf algebra morphism. Moreover it is equivariant: for all $\sigma, \tau \in \mathfrak{S}_n$,

$$\Phi(\sigma \circ \tau) = \Phi(\sigma) \circ \tau.$$

Proof Let $\sigma, \tau \in \mathfrak{S}_n$. Then $\tau \propto \sigma$ if, and only if, $\sigma = \tau$. So, for all $\sigma \in \mathfrak{S}_n$:

$$\Phi(\sigma) = \sigma + \text{linear span of packed words which are not permutations.}$$

So $\Xi \circ \Phi = Id_{\mathbf{Sh}}$, and Φ is injective.

Let $\tau \in \mathbb{S}urj_n$ and $\sigma \in \mathfrak{S}_n$. Then $\tau \propto \sigma$ if, and only if, $\tau \circ \sigma^{-1} \propto I_n$. Moreover, $|\tau \circ \sigma^{-1}|! = \tau!$, as σ is a bijection. Hence:

$$\Phi(\sigma) = \sum_{\tau \propto \sigma} \frac{\tau}{\tau!} = \sum_{\rho \propto I_n} \frac{\rho \circ \sigma}{\rho!} = \Phi(I_n) \circ \sigma.$$

More generally, if $\sigma, \tau \in \mathfrak{S}_n$, $\Phi(\sigma \circ \tau) = \Phi(I_n) \circ (\sigma \circ \tau) = (\Phi(I_n) \circ \sigma) \circ \tau = \Phi(\sigma) \circ \tau$.

Let $\sigma_1 \in \mathfrak{S}_{n_1}$ and $\sigma_2 \in \mathfrak{S}_{n_2}$.

$$\Phi(\sigma_1) \star \Phi(\sigma_2) = \sum_{\substack{\tau_1 \propto \sigma_1, \tau_2 \propto \sigma_2 \\ \zeta \in QSh(max(\tau_1), max(\tau_2))}} \frac{\zeta \circ (\tau_1 \otimes \tau_2)}{\tau_1! \tau_2!}.$$

Let S be the set of elements $\sigma \in \mathbb{S}urj_{n_1+n_2}$ such that:

- For all $1 \leq i, j \leq n_1$, $\sigma_1(i) \leq \sigma_1(j) \Longrightarrow \sigma(i) \leq \sigma(j)$.
- For all $1 \leq i, j \leq n_2$, $\sigma_2(i) \leq \sigma_2(j) \Longrightarrow \sigma(i + n_1) \leq \sigma(j + n_2)$.

Let $\tau_1 \propto \sigma_1$, $\tau_2 \propto \sigma_2$ and $\zeta \in QSh(max(\tau_1), max(\tau_2))$. As ζ is increasing on $\{1, \ldots, max(\tau_1)\}$ and $\{max(\tau_1) + 1, \ldots, max(\tau_1) + max(\tau_2)\}$, $\zeta \circ (\tau_1 \otimes \tau_2) \in S$. Conversely, if $\sigma \in S$, there exists a unique $\tau_1 \in \mathbb{S}urj_{n_1}$, $\tau_2 \in \mathbb{S}urj_{n_2}$ and $\zeta \in QSh_{max(\tau_1),max(\tau_2)}$ such that $\sigma = \zeta \circ (\tau_1 \otimes \tau_2)$: in particular, $\tau_1 = Pack(\sigma(1) \ldots \sigma(n_1))$ and $\tau_2 = Pack(\sigma(n_1 + 1) \ldots \sigma(n_1 + n_2))$. As $\sigma \in S$ and $\zeta \in QSh_{max(\tau_1),max(\tau_2)}$, $\tau_1 \propto \sigma_1$ and $\tau_2 \propto \sigma_2$. Hence:

$$\Phi(\sigma_1) \star \Phi(\sigma_2) = \sum_{\sigma \in S} \frac{\sigma}{Pack(\sigma(1) \ldots \sigma(n_1))! Pack(\sigma(n_1 + 1) \ldots \sigma(n_1 + n_2))!}.$$

On the other hand:

$$\Phi(\sigma_1 \star \sigma_2) = \sum_{\substack{\zeta \in Sh(n_1,n_2) \\ \tau \propto \zeta \circ (\sigma_1 \otimes \sigma_2)}} \frac{\tau}{\tau!}.$$

Let $\zeta \in Sh(n_1, n_2)$ and $\tau \propto \zeta \circ (\sigma_1 \otimes \sigma_2)$. If $1 \leq i, j \leq n_1$ and $\sigma_1(i) \leq \sigma_1(j)$, then:

$$\zeta \circ (\sigma_1 \otimes \sigma_2)(i) = \zeta(\sigma_1(i)) \leq \zeta(\sigma_1(j)) = \zeta \circ (\sigma_1 \otimes \sigma_2)(i),$$

so $\tau(i) \leq \tau(j)$. If $1 \leq i, j \leq n_2$ and $\sigma_2(i) \leq \sigma_2(j)$, then:

$$\zeta \circ (\sigma_1 \otimes \sigma_2)(i + n_1) = \zeta(\sigma_2(i) + max(\sigma_1)) \leq \zeta(\sigma_2(j) + max(\sigma_1)) = \zeta \circ (\sigma_1 \otimes \sigma_2)(j + n_1),$$

so $\tau(i + n_1) \leq \tau(j + n_2)$. Hence, $\tau \in S$ and finally:

$$\Phi(\sigma_1 \star \sigma_2) = \sum_{\tau \in S} \frac{\tau}{\tau!} \sharp\{\zeta \in Sh(n_1, n_2) \mid \tau \propto \zeta \circ (\sigma_1 \otimes \sigma_2)\}.$$

Let $\tau \in S$. We put $\tau_1 = (\tau(1) \ldots \tau(n_1))$ and $\tau_2 = (\tau(n_1 + 1) \ldots \tau(n_1 + n_2))$. Let $\zeta \in Sh(n_1, n_2)$, such that $\tau \propto \zeta \circ (\sigma_1 \otimes \sigma_2)$. For all $1 \leq i \leq max(\tau)$, $\zeta(\tau^{-1}(\{i\})) = I_i$ is entirely determined and does not depend on ζ. By the increasing conditions on ζ, the determination of such a ζ consists of choosing for all $1 \leq i \leq max(\tau)$ a bijective map ζ_i from $\tau^{-1}(\{i\})$ to I_i, such that ζ_i is increasing on $\tau^{-1}(\{i\}) \cap \{1, \ldots, n_1\} =$

$\tau_1^{-1}(\{i\})$ and on $\tau^{-1}(\{i\}) \cap \{n_1+1, \ldots, n_1+n_2\} = \tau_2^{-1}(\{i\})$. Hence, the number of possibilities for ζ is:

$$\begin{aligned}
&\prod_{i=1}^{max(\tau)} \frac{|\tau^{-1}(i)|!}{|\tau_1^{-1}(\{i\})|!|\tau_2^{-1}(\{i\})|!} \\
&= \frac{\prod_{i=1}^{max(\tau)} |\tau^{-1}(\{i\})|!}{\prod_{i=1}^{max(\tau_1)} |\tau_1^{-1}(\{i\})|! \prod_{i=1}^{max(\tau_2)} |\tau_2^{-1}(\{i\})|!} \\
&= \frac{\prod_{i=1}^{max(\tau)} |\tau^{-1}(\{i\})|!}{\prod_{i=1}^{max(Pack(\tau_1))} |Pack(\tau_1)^{-1}(\{i\})|! \prod_{i=1}^{max(Pack(\tau_2))} |Pack(\tau_2)^{-1}(\{i\})|!} \\
&= \frac{\tau!}{Pack(\tau_1)!Pack(\tau_2)!}.
\end{aligned}$$

Hence:

$$\begin{aligned}
\Phi(\sigma_1 \star \sigma_2) &= \sum_{\tau \in S} \frac{\tau}{\tau!} \frac{\tau!}{Pack(\tau(1)\ldots\tau(n_1))!Pack(\tau(n_1+1)\ldots\tau(n_1+n_2))!} \\
&= \Phi(\sigma_1) \star \Phi(\sigma_2).
\end{aligned}$$

So Φ is an algebra morphism.

Let $\sigma \in \mathfrak{S}_n$.

$$\begin{aligned}
&\Delta(\Phi(\sigma)) \\
&= \sum_{\tau \propto \sigma} \sum_{k=0}^{max(\tau)} \frac{1}{\tau!} \tau_{|\{1,\ldots,k\}} \otimes Pack(\tau_{|\{k+1,\ldots,max(\tau)\}} \\
&= \sum_{\tau \propto \sigma} \sum_{k=0}^{max(\tau)} \frac{1}{\tau_{|\{1,\ldots,k\}}!Pack(\tau_{|\{k+1,\ldots,max(\tau)\}}!} \tau_{|\{1,\ldots,k\}} \otimes Pack(\tau_{|\{k+1,\ldots,max(\tau)\}} \\
&= \sum_{k=0}^{n} \sum_{\substack{\tau_1 \propto \sigma_{|\{1,\ldots,k\}} \\ \tau_2 \propto Pack(\sigma_{|\{k+1,\ldots,n\}})}} \frac{\tau_1}{\tau_1!} \otimes \frac{\tau_2}{\tau_2!} \\
&= (\Phi \otimes \Phi) \circ \Delta(\sigma).
\end{aligned}$$

Hence, Φ is a coalgebra morphism. □

Example 3

$$
\begin{aligned}
\Phi((1)) &= (1),\\
\Phi((12)) &= (12) + \frac{1}{2}(11),\\
\Phi((123)) &= (123) + \frac{1}{2}(112) + \frac{1}{2}(122) + \frac{1}{6}(111),\\
\Phi((1234)) &= (1234) + \frac{1}{2}(1123) + +\frac{1}{2}(1223) + +\frac{1}{2}(1233)\\
&\quad + \frac{1}{4}(1122) + \frac{1}{6}(1112) + \frac{1}{6}(1222) + \frac{1}{24}(1111).
\end{aligned}
$$

More generally:

$$
\Phi((1\dots n)) = \sum_{k=1}^{n} \sum_{i_1+\dots+i_k=n} \frac{1}{i_1!\dots i_k!}(1^{i_1}\dots k^{i_k}).
$$

Remark 5 The map Φ is not a morphism of NSh algebras from $(\mathbf{Sh}, \prec, \succ)$ to $(\mathbf{QSh}, \preceq, \succ)$, nor to $(\mathbf{QSh}, \prec, \succeq)$. Indeed:

$$
\begin{aligned}
\Phi((1) \prec (1)) &= (12) + \frac{1}{2}(11),\\
\Phi((1)) \prec \Phi((1)) &= (12),\\
\Phi((1)) \preceq \Phi((1)) &= (12) + (11).
\end{aligned}
$$

We extend the map $\sigma \longrightarrow F_\sigma$ into a linear map from $\mathbf{QSh}$ to $End(T(V))$. By Proposition 4, F is an algebra morphism.

Corollary 3 (Exponential isomorphism) *Le us consider the following linear map:*

$$
\phi : \begin{cases} T(V) \longrightarrow T(V)\\ x_1 \dots x_n \longrightarrow F_{\Phi(I_n)}(x_1 \dots x_n). \end{cases}
$$

Then ϕ is a Hopf algebra isomorphism from $(T(V), ⧢, \Delta)$ to $(T(V), \uplus, \Delta)$.

Proof Let $x_1, \dots, x_{k+l} \in V$.

$$\begin{aligned}
\phi(x_1 \dots x_k \sqcup x_{k+1} \dots x_{k+l}) &= \sum_{\zeta \in Sh(k,l)} F_{\Phi(I_{k+l})} \circ F_\zeta(x_1 \dots x_{k+l}) \\
&= \sum_{\zeta \in Sh(k,l)} F_{\Phi(I_{k+l}) \circ \zeta}(x_1 \dots x_{k+l}) \\
&= \sum_{\zeta \in Sh(k,l)} F_{\Phi(\zeta)}(x_1 \dots x_{k+l}) \\
&= F_{\Phi(I_k \star I_l)}(x_1 \dots x_{k+l}) \\
&= F_{\Phi(I_k) \star \Phi(I_l)}(x_1 \dots x_{k+l}) \\
&= F_{\Phi(I_k)} \uplus F_{\Phi(I_l)}(x_1 \dots x_{k+l}) \\
&= \sum_{i=0}^{k+l} F_{\Phi(I_k)}(x_1 \dots x_i) \uplus F_{\Phi(I_l)}(x_{i+1} \dots x_{k+l}) \\
&= F_{\Phi(I_k)}(x_1 \dots x_k) \uplus F_{\Phi(I_l)}(x_{k+1} \dots x_{k+l}) \\
&= \phi(x_1 \dots x_k) \uplus \phi(x_{k+1} \dots x_l).
\end{aligned}$$

So ϕ is an algebra morphism.

For any packed words $\sigma \in \mathbb{S}urj_k$, $\tau \in \mathbb{S}urj_l$ and all $x_1, \dots, x_n \in V$ we define $G_{\sigma \otimes \tau}$ by:

$$G_{\sigma \otimes \tau}(x_1 \dots x_n) = F_\sigma(x_1 \dots x_k) \otimes F_\tau(x_{k+1} \dots x_n)$$

is $k + l = n$ and $= 0$ else. Then, for all increasing packed word σ, for all $x \in T(V)$:

$$\Delta(F_\sigma(x)) = G_{\Delta(\sigma)}(x).$$

Hence, if $x_1, \dots, x_n \in V$:

$$\begin{aligned}
\Delta \circ \phi(x_1 \dots x_n) &= G_{\Delta(\Phi(I_n))}(x_1 \dots x_n) \\
&= G_{(\Phi \otimes \Phi) \circ \Delta(I_n)}(x_1 \dots x_n) \\
&= \sum_{k=0}^{n} G_{\Phi(I_k) \otimes \Phi(I_{n-k})}(x_1 \dots x_n) \\
&= \sum_{k=0}^{n} F_{\Phi(I_k)}(x_1 \dots x_k) \otimes F_{\Phi(I_{n-k})}(x_{k+1} \dots x_n) \\
&= \sum_{k=0}^{n} \phi(x_1 \dots x_k) \otimes \phi(x_{k+1} \dots x_n) \\
&= (\phi \otimes \phi) \circ \Delta(x_1 \dots x_n).
\end{aligned}$$

So ϕ is a coalgebra morphism.

As the unique bijection appearing in $\Phi(I_n)$ is I_n, for all word $x_1 \dots x_n$:

$$\phi(x_1 \dots x_n) = x_1 \dots x_n + \text{linear span of words of length} < n.$$

So ϕ is a bijection. □

Example 4 Let $x_1, x_2, x_3, x_4 \in V$.

$$\begin{aligned}
\phi(x_1) &= x_1, \\
\phi(x_1x_2) &= x_1x_2 + \frac{1}{2}x_1.x_2, \\
\phi(x_1x_2x_3) &= x_1x_2x_3 + \frac{1}{2}(x_1.x_2)x_3 + \frac{1}{2}x_1(x_2.x_3) + \frac{1}{6}x_1.x_2.x_3, \\
\phi(x_1x_2x_3x_4) &= x_1x_2x_3x_4 + \frac{1}{2}(x_1.x_2)x_3x_4 + \frac{1}{2}x_1(x_2.x_3)x_4 \\
&\quad + \frac{1}{2}x_1x_2(x_3.x_4) + \frac{1}{4}(x_1.x_2)(x_3.x_4) + \frac{1}{6}(x_1.x_2.x_3)x_4 \\
&\quad + \frac{1}{6}x_1(x_2.x_3.x_4) + \frac{1}{24}x_1.x_2.x_3.x_4.
\end{aligned}$$

More generally, for all $x_1, \dots, x_n \in V$:

$$\phi(x_1 \dots x_n) = \sum_{k=1}^{n} \sum_{i_1+\dots+i_k=n} \frac{1}{i_1! \dots i_k!} F_{(1^{i_1} \dots k^{i_k})}(x_1 \dots x_n).$$

Remark 6 1. This isomorphism is the morphism denoted by exp and obtained in the graded case by Hoffman in [23].
2. If V is a trivial algebra, then $\phi = Id_{T(V)}$.
3. This morphism is not a NSh algebra morphism, except if V is a trivial algebra. In fact, except if the product of V is zero, the NSh algebras $(T(V), \preceq, \succ)$ and $(T(V), \prec, \succeq)$ are not commutative, so cannot be isomorphic to a shuffle algebra.

8 Coalgebra and Hopf Algebra Endomorphisms

In the previous section, we studied the links between shuffle and quasi-shuffle operads and obtained as a corollary the exponential isomorphism of Corollary 3 between the shuffle and quasi-shuffle Hopf algebra structures on $T(V)$. This section aims at classifying all such possible (natural, i.e. functorial in commutative algebras V) morphisms. We refer to our [17] for applications of natural coalgebra endomorphisms to the study of deformations of shuffle bialgebras.

Recall that we defined π as the unique linear endomorphism of the quasi-shuffle bialgebra $T^+(V)$ such that $\pi + \pi \prec Id_{T^+(V)} = Id_{T^+(V)}$. By Proposition 6, it is equal to $F_{(1)}$, so is the canonical projection on V. This construction generalizes as follows.

Hereafter, we work in the unital setting and write ε for the canonical projection from $T(V)$ to the scalars (the augmentation map). It behaves as a unit w.r.t. the NQSh products on $End(T^+(V))$: for $g \in End(T^+(V))$, $\varepsilon \prec g = 0$, $g \prec \varepsilon = g$.

Proposition 9 *Let $f : T(V) \longrightarrow V$ be a linear map such that $f(1) = 0$. There exists a unique coalgebra endomorphism ψ of $T(V)$ such that $\pi \circ \psi = f$. This coalgebra endomorphism is the unique linear endomorphism of $T(V)$ such that $\varepsilon + f \prec \psi = \psi$.*

Proof *First step.* Let us prove the unicity of the coalgebra morphism ψ such that $\pi \circ \psi = f$. Let ψ_1, ψ_2 be two (non zero) coalgebra endomorphisms such that $\pi \circ \psi_1 = \pi \circ \psi_2$. Let us prove that for all $x_1, \ldots, x_n \in V$, $\psi_1(x_1 \ldots x_n) = \psi_2(x_1 \ldots x_n)$ by induction on n. If $n = 1$, as $\psi_1(1)$ and $\psi_2(1)$ are both nonzero group-like elements, they are both equal to 1. Let us assume the result at all rank $< n$. Then:

$$\begin{aligned}
\Delta \circ \psi_1(x_1 \ldots x_n) &= (\psi_1 \otimes \psi_1) \circ \Delta(x_1 \ldots x_n) \\
&= \psi_1(x_1 \ldots x_n) \otimes 1 + 1 \otimes \psi_1(x_1 \ldots x_n) \\
&\quad + \sum_{i=1}^{n-1} \psi_1(x_1 \ldots x_i) \otimes \psi_1(x_{i+1} \ldots x_n), \\
\Delta \circ \psi_2(x_1 \ldots x_n) &= \psi_2(x_1 \ldots x_n) \otimes 1 + 1 \otimes \psi_2(x_1 \ldots x_n) \\
&\quad + \sum_{i=1}^{n-1} \psi_2(x_1 \ldots x_i) \otimes \psi_2(x_{i+1} \ldots x_n).
\end{aligned}$$

Applying the induction hypothesis, for all $i \leq 1 \leq n-1$, $\psi_1(x_1 \ldots x_i) = \psi_2(x_1 \ldots x_i)$ and $\psi_1(x_{i+1} \ldots x_n) = \psi_2(x_{i+1} \ldots x_n)$. Consequently, $\psi_1(x_1 \ldots x_n) - \psi_2(x_1 \ldots x_n)$ is primitive, so belongs to V and:

$$\psi_1(x_1 \ldots x_n) - \psi_2(x_1 \ldots x_n) = \pi \circ \psi_1(x_1 \ldots x_n) - \pi \circ \psi_2(x_1 \ldots x_n) = 0.$$

Second step. Let us prove the existence of a (necessarily unique) endomorphism ψ such that $\psi = \varepsilon + f \prec \psi$. We construct $\psi(x_1 \ldots x_n)$ for all $x_1, \ldots, x_n \in V$ by induction on n in the following way: $\psi(1) = 1$ and, if $n \geq 1$:

$$\psi(x_1 \ldots x_n) := f(x_1 \ldots f_n) + \sum_{i=1}^{n-1} f(x_1 \ldots x_i) \prec \psi(x_{i+1} \ldots x_n).$$

Then $(\varepsilon + f \prec \psi)(1) = \varepsilon(1) = 1 = \psi(1)$. If $n \geq 1$:

$$\begin{aligned}
&(\varepsilon + f \prec \psi)(x_1 \dots x_n)\\
&= \varepsilon(x_1 \dots x_n) + f(x_1 \dots x_n) + \sum_{i=1}^{n-1} f(x_1 \dots x_i) \prec \psi(x_{i+1} \dots x_n)\\
&= 0 + f(x_1 \dots x_n) + \sum_{i=1}^{n-1} f(x_1 \dots x_i) \prec \psi(x_{i+1} \dots x_n)\\
&= \psi(x_1 \dots x_n).
\end{aligned}$$

Hence, $\varepsilon + f \prec \psi = \psi$.

Third step. Let ψ such that $\varepsilon + f \prec \psi = \psi$. Let us prove that $\Delta \circ \psi(x_1 \dots x_n) = (\psi \otimes \psi) \circ \Delta(x_1 \dots x_n)$ by induction on n. If $n = 0$, then $\psi(1) = \varepsilon(1) + f(1) = 1 + 0 = 1$, so $\Delta \circ \psi(1) = (\psi \otimes \psi) \circ \Delta(1) = 1 \otimes 1$. If $n \geq 1$, we put $x = x_1 \dots x_n$, $\Delta(x) = x \otimes 1 + 1 \otimes x + x' \otimes x''$. The induction hypothesis holds for x''. Moreover:

$$\psi(x) = \varepsilon(x) + f(x) + f(x') \prec \psi(x'') = f(x) + f(x') \prec \psi(x'').$$

As $f(x), f(x') \in V$ are primitive:

$$\begin{aligned}
\tilde{\Delta} \circ \psi(x) &= f(x') \otimes \psi(x'') + f(x') \prec \psi(x'')' \otimes \psi(x'')'\\
&= f(x') \otimes \psi(x'') + f(x') \prec \psi(x'') \otimes \psi(x''')\\
&= \psi(x') \otimes \psi(x'')\\
&= (\psi \otimes \psi) \circ \tilde{\Delta}(x).
\end{aligned}$$

As $\psi(1) = 1$, we deduce that $\Delta \circ \psi(x) = (\psi \otimes \psi) \circ \Delta(x)$. So ψ is a coalgebra morphism. Moreover, $\pi \circ \psi(1) = \pi(1) = 0 = f(1)$. If $\varepsilon(x) = 0$:

$$\pi \circ \psi(x) = \pi \circ f(x) + \pi(f(x') \prec f(x'')) = f(x),$$

as $f(x), (x') \in V$ (so $f(x') \prec f(x'')$ is a linear span of words of length ≥ 2, so vanishes under the action of π). Hence, $\pi \circ \psi = f$. □

Proposition 10 *Let $A = \sum_{n \geq 1} a_n X^n$ be a formal series without constant term. Let f_A be the linear map from $T(V)$ to V defined by $f_A(x_1 \dots x_n) = a_n x_1 \bullet \dots \bullet x_n$ and let ϕ_A be the unique coalgebra endomorphism of $T(V)$ such that $\pi \circ \phi_A = f_A$. For all $x_1, \dots, x_n \in V$:*

$$\phi_A(x_1 \dots x_n) = \sum_{k=1}^{n} \sum_{i_1+\dots+i_k=n} a_{i_1} \dots a_{i_k} F_{(1^{i_1} \dots k^{i_k})}(x_1 \dots x_n). \tag{30}$$

Proof Note that $f_A(x_1 \dots x_n) = a_n F_{(1^n)}(x_1 \dots x_n)$. Let ϕ be the morphism defined by the second member of (30). Then $(\varepsilon + f_A \prec \phi)(1) = 1 + f_A(1) = 1 = \phi(1)$. If $n \geq 1$:

$$\begin{aligned}
&(\varepsilon + f_A \prec \phi)(x_1 \dots x_n)\\
&= f_A(x_1 \dots x_n) + \sum_{i=1}^{n-1} f_A(x_1 \dots x_i) \prec \phi(x_{i+1} \dots x_n)\\
&= a_n F_{(1^n)}(x_1 \dots x_n)\\
&+ \sum_{i=1}^{n-1} \sum_{k=2}^{n} \sum_{i_2+\dots+i_k=n-i} a_i a_{i_2} \dots a_{i_k} F_{(1^i)} \prec F_{(1^{i_2} \dots (k-1)^{i_k})}(x_1 \dots x_n)\\
&= a_n F_{(1^n)}(x_1 \dots x_n)\\
&+ \sum_{i=1}^{n-1} \sum_{k=2}^{n} \sum_{i+i_2+\dots+i_k=n} a_i a_{i_2} \dots a_{i_k} \prec F_{(1^i 2^{i_2} \dots k^{i_k})}(x_1 \dots x_n)\\
&= \phi(x_1 \dots x_n).
\end{aligned}$$

By unicity in Proposition 9, $\phi = \phi_A$. □

Remark 7 The morphism ϕ defined in corollary 3 is $\phi_{exp(X)-1}$.

Proposition 11 *$\phi_X = Id$ and for all formal series A, B without constant terms, $\phi_A \circ \phi_B = \phi_{A \circ B}$.*

Proof For all $x_1, \dots, x_n \in V$, $\pi \circ Id(x_1 \dots x_n) = \delta_{1,n} x_1 \dots x_n = f_X(x_1 \dots x_n)$. By unicity in Proposition 9, $\phi_X = Id$. Moreover:

$$\begin{aligned}
&\pi \circ \phi_A \circ \phi_B(x_1 \dots x_n)\\
&= f_A\left(\sum_{k=1}^{n} \sum_{i_1+\dots+i_k=n} b_{i_1} \dots b_{i_k} (x_1 \bullet \dots \bullet x_{i_1}) \dots (x_{i_1+\dots+i_{k-1}+1} \bullet \dots \bullet x_{1+\dots+i_k})\right)\\
&= \sum_{k=1}^{n} \sum_{i_1+\dots+i_k=n} a_k b_{i_1} \dots b_{i_k} x_1 \bullet \dots \bullet x_n\\
&= f_{A \circ B}(x_1 \dots x_n).
\end{aligned}$$

By unicity in Proposition 9, $\phi_A \circ \phi_B = \phi_{A \circ B}$. □

So the set of all ϕ_A, where A is a formal series such that $A(0) = 0$ and $A'(0) \neq 1$, is a subgroup of the group of coalgebra isomorphisms of $T(V)$, isomorphic to the group of formal diffeomorphisms of the line.

Corollary 4 *The inverse of the isomorphism ϕ defined in corollary 3 is $\phi_{ln(1+X)}$:*

$$\phi^{-1}(x_1\dots x_n)=\sum_{k=1}^{n}\sum_{i_1+\dots+i_k=n}\frac{(-1)^{n+k}}{i_1\dots i_k}F_{(1^{i_1}\dots k^{i_k})}(x_1\dots x_n).$$

Proposition 12 *Let $A\in K[[X]]^+$.*

1. *$\phi_A:(T(V),\sqcup\!\sqcup,\Delta)\longrightarrow(T(V),\sqcup\!\sqcup,\Delta)$ is a Hopf algebra morphism for any commutative algebra V if, and only if, $A=aX$ for a certain $a\in K$.*
2. *$\phi_A:(T(V),\sqcup\!\sqcup,\Delta)\longrightarrow(T(V),\uplus,\Delta)$ is a Hopf algebra morphism for any commutative algebra V if, and only if, $A=exp(aX)-1$ for a certain $a\in K$.*
3. *$\phi_A:(T(V),\uplus,\Delta)\longrightarrow(T(V),\uplus,\Delta)$ is a Hopf algebra morphism for any commutative algebra V if, and only if, $A=(1+X)^a-1$ for a certain $a\in K$.*
4. *$\phi_A:(T(V),\uplus,\Delta)\longrightarrow(T(V),\sqcup\!\sqcup,\Delta)$ is a Hopf algebra morphism for any commutative algebra V if, and only if, $A=a\ln(1+X)$ for a certain $a\in K$.*

Proof First, note that for any $x_1,\dots,x_k\in V$:

$$\pi\circ\phi_A(x_1\dots x_k)=a_kF_{(1\dots 1)}(x_1\dots x_k).$$

Consequently, for any commutative algebra V, for any $x,x_1,\dots,x_k\in V$, $k\geq 1$:

$$\begin{aligned}\pi\circ\phi_A(x\sqcup\!\sqcup x_1\dots x_k)&=\pi(xx_1\dots x_{k+1}+\dots+x_1\dots x_{k+1}x)\\&=(k+1)a_{k+1}x.x_1\cdot\ \dots\ \cdot x_k,\\ \pi(\phi_A(x)\sqcup\!\sqcup\phi_A(x_1\dots x_k))&=0,\\ \pi(\phi_A(x)\uplus\phi_A(x_1\dots x_k))&=a_1a_kx.x_1\cdot\dots\ \cdot x_k.\end{aligned}$$

1. We assume that ϕ_A is an algebra morphism for any V for the shuffle product. Let us choose an algebra V and elements $x,x_1,\dots,x_k\in V$ such that $x.x_1\cdot\ \dots\ \cdot x_k\neq 0$ in V. As $\phi(x\sqcup\!\sqcup x_1\dots x_k)=\phi(x)\sqcup\!\sqcup\phi(x_1\dots x_k)$, applying π, we deduce that for all $k\geq 1$, $(k+1)a_{k+1}=0$, so $a_{k+1}=0$. Hence, $A=a_1X$. Conversely, for any $x_1,\dots,x_k\in V$, $\phi_{aX}(x_1\dots x_k)=a_1^kx_1\dots x_k$, so ϕ_{aX} is an endomorphism of the Hopf algebra $(T(V),\sqcup\!\sqcup,\Delta)$.

2. We already proved that $\phi_{exp(X)-1}$ is a Hopf algebra morphism from $(T(V),\sqcup\!\sqcup,\Delta)$ to $(T(V),\uplus,\Delta)$. By composition:

$$\phi_{exp(aX)-1}=\phi_{exp(X)-1}\circ\phi_{aX}:(T(V),\sqcup\!\sqcup,\Delta)\longrightarrow(T(V),\sqcup\!\sqcup,\Delta)\longrightarrow(T(V),\uplus,\Delta)$$

is a Hopf algebra morphism.

We assume that ϕ_A is an algebra morphism for any V from the shuffle product to the quasi-shuffle product. Let us choose an algebra V, and $x,x_1,\dots,x_k\in V$, such that $x.x_1\cdot\ \dots\ \cdot x_k\neq 0$ in V. As $\phi(x\sqcup\!\sqcup x_1\dots x_k)=\phi(x)\uplus\phi(x_1\dots x_k)$, applying π, we deduce that for all $k\geq 1$, $(k+1)a_{k+1}=a_1a_k$, so $a_k=\frac{a_1^k}{k!}$ for all $k\geq 1$. Hence, $A=exp(a_1X)-1$.

3. The following conditions are equivalent:

- For any V, $\phi_A : (T(V), \uplus, \Delta) \longrightarrow (T(V), \uplus, \Delta)$ is a Hopf algebra morphism.
- For any V, $\phi_{\ln(1+X)} \circ \phi_A \circ \phi_{exp(X)-1} : (T(V), \sqcup\!\sqcup, \Delta) \longrightarrow (T(V), \sqcup\!\sqcup, \Delta)$ is a Hopf algebra morphism. For any V, $\phi_{\ln(1+X)\circ A\circ(exp(X)-1)} : (T(V), \sqcup\!\sqcup, \Delta) \longrightarrow (T(V), \sqcup\!\sqcup, \Delta)$ is a Hopf algebra morphism.
- There exists $a \in K$, $\ln(1 + X) \circ A \circ (exp(X) - 1) = aX$.
- There exists $a \in K$, $A = (1 + X)^a - 1$.

4. Similar proof. □

Remark 8 The Proposition 12 classifies actually all the Hopf algebra endomorphisms and morphisms relating shuffle and quasi-shuffle algebras $T(V)$, that are natural (i.e. functorial) in V. This naturality property follows formally from the study of nonlinear Schur-Weyl duality in [17, 31].

9 Coderivations and Graduations

The present section complements the previous one that studied coalgebra endomorphisms. We aim at investigating here coderivations of quasi-shuffle bialgebras. As an application we recover the existence of a natural graded structure on the Hopf algebras $(T(V), \uplus, \Delta)$ [17].

Notations. Let A be a NQSh algebra, $f \in End_K(A)$ and $v \in A$. We define:

$$f \prec v : \begin{cases} A \longrightarrow A \\ x \longrightarrow f(x) \prec v, \end{cases} \qquad v \prec f : \begin{cases} A \longrightarrow A \\ x \longrightarrow v \prec f(x). \end{cases}$$

Proposition 13 *Let $f : T(V) \longrightarrow V$ be a linear map. There exists a unique coderivation D of $T(V)$ such that $\pi \circ D = f$. Moreover, D is the unique linear endomorphism of $T(V)$ such that $D = f + \pi \prec D + f \prec Id$.*

Proof *First step*. Let us prove that the unicity of the coderivation D such that $\pi \circ D = f$. The result is classical [20] and elementary, we include its proof for completeness sake. Let D_1 and D_2 be two coderivations such that $\pi \circ D_1 = \pi \circ D_2$. Let us prove that $D_1(x_1 \dots x_n) = D_2(x_1 \dots x_n)$ by induction on n.

$$\Delta \circ D_1(1) = (D_1 \otimes Id + Id \otimes D_1)(1 \otimes 1) = D_1(1) \otimes 1 + 1 \otimes D_1(1),$$

so $D_1(1) \in Prim(T(V)) = V$. Similarly, $D_2(1) \in V$. Hence, $D_1(1) = \pi \circ D_1(1) = \pi \circ D_2(1) = D_2(1)$. Let us assume the result at all ranks $< n$. If $p = 1$ or 2:

$$\Delta \circ D_p(x_1 \dots x_n) = \sum_{i=0}^{n} D_p(x_1 \dots x_i) \otimes x_{i+1} \dots x_n + \sum_{i=0}^{n} x_1 \dots x_i \otimes D_p(x_{i+1} \dots x_n).$$

Applying the induction hypothesis at all ranks $< k$, we obtain by subtraction:

$$\Delta \circ (D_1 - D_2)(x_1 \dots x_n) = (D_1 - D_2)(x_1 \dots x_n) \otimes 1 + 1 \otimes (D_1 - D_2)(x_1 \dots x_n).$$

So $(D_1 - D_2)(x_1 \dots x_n) \in V$. Applying π:

$$(D_1 - D_2)(x_1 \dots x_n) = \pi \circ (D_1 - D_2)(x_1 \dots x_n) = 0.$$

So $D_1(x_1 \dots x_n) = D_2(x_1 \dots x_n)$.

Second step. Let us prove the existence of a map D such that $D = f + \pi \prec D + f \prec Id$. We define $D(x_1 \dots x_n)$ by induction on n by $D(1) = f(1)$ and:

$$D(x_1 \dots x_n) = x_1 \prec D(x_2 \dots x_n) + \sum_{i=0}^{n-1} f(x_1 \dots x_i) \prec x_{i+1} \dots x_n + f(x_1 \dots x_n).$$

Then $(f + \pi \prec D + f \prec Id)(1) = f(1) = D(1)$. If $n \geq 1$:

$$\begin{aligned}
&(f + \pi \prec D + f \prec Id)(x_1 \dots x_n)\\
&= f(x_1 \dots x_n) + \sum_{i=1}^{n} \pi(x_1 \dots x_i) \prec D(x_{i+1} \dots x_n)\\
&+ \sum_{i=0}^{n-1} f(x_1 \dots x_i) \prec x_{i+1} \dots x_n\\
&= f(x_1 \dots x_n) + x_1 \prec D(x_2 \dots x_n) + \sum_{i=0}^{n-1} f(x_1 \dots x_i) \prec x_{i+1} \dots x_n\\
&= D(x_1 \dots x_n).
\end{aligned}$$

So $D = f + \pi \prec D + f \prec Id$.

Last step. Let D be such that $D = f + \pi \prec D + f \prec Id$. Let us prove that $\Delta \circ D(x_1 \dots x_n) = (D \otimes Id + Id \otimes D) \circ \Delta(x_1 \dots x_n)$ by induction on n. If $n = 0$:

$$\begin{aligned}
\Delta \circ D(1) &= \Delta(f(1))\\
&= f(1) \otimes 1 + 1 \otimes f(1)\\
&= D(1) \otimes 1 + 1 \otimes D(1)\\
&= (D \otimes Id + Id \otimes D)(1 \otimes 1).
\end{aligned}$$

Let us assume the result at all ranks $< n$.

$$
\begin{aligned}
D(x_1 \dots x_n) &= (f + \pi \prec D + f \prec Id)(x_1 \dots x_n) \\
&= \sum_{i=1}^{n} \pi(x_1 \dots x_i) \prec D(x_{i+1} \dots x_n) + \sum_{i=0}^{n-1} f(x_1 \dots x_i) \prec x_{i+1} \dots x_n \\
&+ f(x_1 \dots x_n) \\
&= x_1 D(x_2 \dots x_n) + \sum_{i=0}^{n} f(x_1 \dots x_i) x_{i+1} \dots x_n .
\end{aligned}
$$

Hence:

$$
\begin{aligned}
&\Delta \circ D(x_1 \dots x_n)) \\
&= \sum_{j=1}^{n} x_1 D(x_2 \dots x_j) \otimes x_{j+1} \dots x_n \\
&+ \sum_{j=1}^{n} x_1 \dots x_j \otimes D(x_{j+1} \dots x_n) + 1 \otimes x_1 D(x_2 \dots x_n) \\
&+ \sum_{i=0}^{n} \sum_{j=i}^{n} f(x_1 \dots x_i) x_{i+1} \dots x_j \otimes x_{j+1} \dots x_n \\
&+ \sum_{i=0}^{n} 1 \otimes f(x_1 \dots x_i) x_{i+1} \dots x_n \\
&= \sum_{j=1}^{n} x_1 D(x_2 \dots x_j) \otimes x_{j+1} \dots x_n \\
&+ \sum_{j=1}^{n} x_1 \dots x_j \otimes D(x_{j+1} \dots x_n) + 1 \otimes x_1 D(x_2 \dots x_n) \\
&+ \sum_{j=1}^{n} \sum_{i=1}^{j} f(x_1 \dots x_i) x_{i+1} \dots x_j \otimes x_{j+1} \dots x_n \\
&+ f(1) \otimes x_1 \dots x_n + \sum_{i=0}^{n} 1 \otimes f(x_1 \dots x_i) x_{i+1} \dots x_n \\
&= \sum_{j=0}^{n} D(x_1 \dots x_j) \otimes x_{j+1} \dots x_n + \sum_{j=1}^{n} x_1 \dots x_j \otimes D(x_{j+1} \dots x_n) \\
&= (D \otimes Id + Id \otimes D) \circ \Delta(x_1 \dots x_n).
\end{aligned}
$$

Moreover, $\pi \circ D(1) = \pi \circ f(1) = f(1)$; if $n \geq 1$:

$$\pi \circ D(x_1 \dots x_n) = \pi(x_1 D(x_2 \dots x_n)) + \sum_{i=0}^{n} \pi(f(x_1 \dots x_i) x_{i+1} \dots x_n)$$
$$= 0 + f(x_1 \dots x_n).$$

So $\pi \circ D = f$. □

Proposition 14 *Let $A = \sum_{n \geq 1} a_n X^n$ be a formal series without constant term. Let D_A be the unique coderivation of $T(V)$ such that $\pi \circ \phi_A = f_A$. For all $x_1, \dots, x_n \in V$:*

$$D_A(x_1 \dots x_n) = \sum_{i=1}^{n} a_i \sum_{j=1}^{n-i+1} F_{(12\dots j-1 j^i j+1 \dots n-i+1)}(x_1 \dots x_n). \tag{31}$$

Proof Let D be the linear endomorphism defined by the right side of (31). As $f_A(1) = 0$, we get by induction on n:

$$(f + \pi \prec D + f \prec Id)(x_1 \dots x_n)$$
$$= f(x_1 \dots x_n) + x_1 D(x_2 \dots x_n) + \sum_{i=1}^{n-1} f(x_1 \dots x_i) x_{i+1} \dots x_n$$
$$= x_1 D(x_2 \dots x_n) + \sum_{i=1}^{n} f(x_1 \dots x_i) x_{i+1} \dots x_n$$
$$= \sum_{i=1}^{n-1} a_i \sum_{j=2}^{n-i+1} F_{(12\dots j-1 j^i j+1 \dots n-i+1)}(x_1 \dots x_n) + \sum_{i=1}^{n} a_i F_{(1^i 2 \dots n-i+1)}(x_1 \dots x_n)$$
$$= \sum_{i=1}^{n} a_i \sum_{j=1}^{n-i+1} F_{(12\dots j-1 j^i j+1 \dots n-i+1)}(x_1 \dots x_n)$$
$$= D(x_1 \dots x_n).$$

Moreover, $\pi \circ D(x_1 \dots x_n) = a_n x_1 \bullet \dots \bullet x_n = f_A(x_1 \dots x_n)$. The unicity in Proposition 13 implies that $D = D_A$. □

Corollary 5 *For all word $x_1 \dots x_n$, $D_X(x_1 \dots x_n) = n x_1 \dots x_n$.*

Proof Indeed, $D_X(x_1 \dots x_n) = \sum_{j=1}^{n} F_{(12\dots j-1 j^1 j+1 \dots n)}(x_1 \dots x_n) = n x_1 \dots x_n$. □

Remark 9 Let A and B be two formal series and $\lambda \in K$. As $D_A + \lambda D_B$ is a coderivation and $\pi \circ (D_A + \lambda D_B) = f_A + \lambda f_B = f_{A+\lambda B}$:

$$D_A + \lambda D_B = D_{A+\lambda B}.$$

Moreover, the group of coalgebra automorphims of $T(V)$ acts on the space of coderivations of $T(V)$ by conjugacy. Let us precise this action if we work only with automorphisms and coderivations associated to formal series.

Proposition 15 *Let A, B be two formal series without constant terms, such that $A'(0) \neq 0$. Then:*

$$\phi_A^{-1} \circ D_B \circ \phi_A = D_{\frac{B \circ A}{A'}}.$$

Proof By linearity and continuity of the action, it is enough to prove this formula if $B = X^p$. We denote by C the inverse of A for the composition.

$$\begin{aligned}
&\pi \circ \phi_A^{-1} \circ D_{X^p} \circ \phi_A(x_1 \ldots x_n)\\
&= f_C \circ D_{X_p}\left(\sum_{k=1}^{n} \sum_{i_1+\ldots+i_k=n} a_{i_1} \ldots a_{i_k} F_{(1^{i_1} \ldots k^{i_k})}(x_1 \ldots x_n)\right)\\
&= \sum_{k=p-1}^{n} \sum_{i_1+\ldots+i_k=n} (k-p-1) c_{k-p+1} a_{i_1} \ldots a_{i_k} x_1 \bullet \ldots \bullet x_n.
\end{aligned}$$

So $\pi \circ \phi_{A^{-1}} \circ D_{X^p} \circ \phi_A$ is the linear map associated to the formal series:

$$\begin{aligned}
\left(\sum_{k=p-1}^{\infty} (k-p+1) c_{k-p+1} X^k\right) \circ A &= \left(\sum_{i=0}^{\infty} i a_i X^{i-1+p}\right) \circ A\\
&= (X^p C') \circ A\\
&= A^p C' \circ A\\
&= \frac{A^p}{A'}.
\end{aligned}$$

Hence, $\phi_{A^{-1}} \circ D_{X^p} \circ \phi_A = D_{\frac{A^p}{A'}}$. □

Corollary 6 *The eigenspaces of the coderivation $D_{(1+X)ln(1+X)}$ give a gradation of the Hopf algebra $(T(V), \uplus, \Delta)$.*

Proof Let $D = \phi \circ D_X \circ \phi^{-1}$. As $\phi = \phi_{exp(X)-1}$:

$$D = \phi_{ln(1+X)}^{-1} \circ D_X \circ \phi_{ln(1+X)} = D_{(1+X)ln(1+X)}.$$

As D_X is a derivation of the algebra $(T(V), \sqcup\!\sqcup)$ and ϕ is an algebra isomorphism from $(T(V), \sqcup\!\sqcup)$ to $(T(V), \uplus)$, D is is a derivation of the algebra $(T(V), \uplus)$. As it is conjugated to D_X, its eigenvalues are the elements of $\mathbb{N}$. □

Remark 10 As $(1+X)ln(1+X) = 1 + \sum_{k=2}^{\infty} \frac{(-1)^k}{k(k-1)} X^k$:

$$D_{(1+X)ln(1+X)}(x_1 \dots x_n)$$
$$= nx_1 \dots x_n + \sum_{i=2}^{n} \sum_{j=1}^{n-i+1} \frac{(-1)^i}{i(i-1)} x_1 \dots x_{j-1}(x_j \bullet \dots \bullet x_{j+i-1})x_{j+i} \dots x_n.$$

The gradation of $A = (T(V), \uplus)$ is given by:

$$A_n = Vect\left(\sum_{k=1}^{n} \sum_{i_1+\ldots+i_k=n} \frac{1}{i_1! \ldots i_k!} \left(\prod_{i=1}^{i_1} x_i\right) \cdots \left(\prod_{i=i_1+\ldots+i_{k-1}+1}^{i_1+\ldots+i_k} x_i\right), \atop x_1, \ldots, x_n \in V \right).$$

10 Decorated Operads and Graded Structures

In many applications, algebras over operads carry a natural graduation. This is because geometrical objects (polynomial vector fields, spaces, differential forms…), but also combinatorial and algebraic ones carry often a graduation (or a dimension, a cardinal…) that is better taken into account in the associated algebra structures. As far as quasi-shuffle algebras are concerned, they often naturally carry a graduation in their application domains : think to quasi-symmetric functions and multizeta values (MZVs) [4]; Ecalle's mould calculus and dynamical systems [13]; iterated integrals of Itô type in stochastic calculus [8, 9].

Here, we recall briefly how the formalism of operads can be adapted to take into account graduations [41]. We detail then the case of quasi-shuffle algebras and conclude by studying the analogue, in this context, of the classical descent algebra of a graded commutative or cocommutative Hopf algebra [34].

In this section, we denote by $A = \bigoplus_{n \in \mathbf{N}} A_n$ (where $A_0 = k$, the ground field), a graded, connected, quasi-shuffle bialgebra. By graded we mean that all the structure maps ($\prec$, $\bullet$, Δ) are graded maps. Then $Prim(A) = V = \bigoplus_{n \in \mathbf{N}^*} V_n$ is an associative, commutative graded algebra for the product $\bullet$ and we can identify A and the quasi-shuffle bialgebra $T(V)$ as graded quasi-shuffle algebras. Be aware however that the graduation of $T(V)$ is not the tensor length: for example, for $v_1 \in V_{n_1}, \ldots, v_k \in V_{n_k}$, the degree of the tensor $v_1 \ldots v_k \in V^{\otimes k}$ is now $n_1 + \cdots + n_k$.

It is an easy exercise to adapt the definition of operads to the graded case: whereas the component $\mathbf{F}_n$ of an operad identifies with the set of multilinear elements in the n letters $x_1, \ldots, x_n$ in the free algebra $F(X_n)$, $X_n := \{x_1, \ldots, x_n\}$, the corresponding component of the associated graded operad $\mathbf{F}_n^d$ is obtained by allowing the x_is to be decorated by integers (corresponding to degrees). Each sequence $(d_1, \ldots, d_n)$ of decorations gives then rise to a component of the associated decorated operad, isomorphic to $\mathbf{F}_n$ and corresponding to n-ary operations that act on a sequence $(a_1, \dots, a_n)$ of elements of a $\mathbf{F}$-graded algebra as the corresponding element of $\mathbf{F}_n$ would when $deg(a_i) = d_i$, and as the null map else, see [41] for details. We call $\mathbf{F}^d = \cup_n \mathbf{F}_n^d$ the (integer-)decorated operad associated to $\mathcal{F}$-algebras.

The decorated operad $\mathbf{QSh}^d$ is then spanned by decorated packed words, where:

Definition 11 A *decorated packed word* of length k is a pair (σ, d), where σ is a packed word of length k and d is a map from $\{1, \dots, k\}$ into $\mathbb{N}^*$. We denote it by $\begin{pmatrix} \sigma(1) \dots \sigma(k) \\ d(1) \dots d(k) \end{pmatrix}$.

Notation. Let $(\sigma, d) = \begin{pmatrix} \sigma(1) \dots \sigma(k) \\ d(1) \dots d(k) \end{pmatrix}$ be a decorated packed word. Let m be the maximum of σ. We define $F_{(\sigma,d)} \in End_k(A)$ in the following way: for all $x_1, \dots, x_l \in V$, homogeneous,

$$F_{(\sigma,d)}(x_1 \dots x_l) = \begin{cases} \left(\prod\limits_{\sigma(i)=1} x_i\right) \cdots \left(\prod\limits_{\sigma(i)=m} x_i\right) & \begin{array}{r} \text{if } k = l \text{ and} \\ deg(x_1) = d(1), \\ \vdots \\ deg(x_k) = d(k), \end{array} \\ 0 \text{ otherwise.} \end{cases}$$

Note that in each parenthesis, the product is the product $\bullet$ of V. For example, if $x, y, z \in V$ are homogeneous,

$$F_{\begin{pmatrix} 2\ 1\ 2 \\ a\ b\ c \end{pmatrix}}(xyz) = y(x \bullet z)$$

if $deg(x) = a$, $deg(y) = b$, and $deg(z) = c$, and 0 otherwise.

The subspace of $End_k(A)$ generated by these maps is stable under composition and the noncommutative quasi-shuffle products:

Proposition 16 *Let*

$$(\sigma, d) = \begin{pmatrix} \sigma(1) \dots \sigma(k) \\ d(1) \dots d(k) \end{pmatrix} \text{ and } (\tau, e) = \begin{pmatrix} \tau(1) \dots \tau(l) \\ e(1) \dots e(l) \end{pmatrix}$$

be two decorated packed words. $max(\tau) = k$ *and for all* $1 \leq j \leq k$, $\sum\limits_{\tau(i)=j} e(i) = d(j)$, *then:*

$$F_{(\sigma,d)} \circ F_{(\tau,e)} = F_{\begin{pmatrix} \sigma\circ\tau(1) \dots \sigma\circ\tau(l) \\ e(1) \quad \dots \quad e(l) \end{pmatrix}}.$$

Otherwise, this composition is equal to 0. *Moreover:*

$$\begin{aligned}
&F_{(\sigma,d)} \prec F_{(\tau,e)}\\
&= \sum_{\substack{Pack(u(1)...u(k))=\sigma,\\ Pack(u(k+1)...u(k+l))=\tau,\\ min(u(1)...u(k))<min(u(k+1)...u(k+l))}} F_{\begin{pmatrix} u(1) & \ldots & u(k) & u(k+1) & \ldots & u(k+l)\\ d(1) & \ldots & d(k) & e(1) & \ldots & e(l)\end{pmatrix}},\\
&F_{(\sigma,d)} \succ F_{(\tau,e)}\\
&= \sum_{\substack{Pack(u(1)...u(k))=\sigma,\\ Pack(u(k+1)...u(k+l))=\tau,\\ min(u(1)...u(k))>min(u(k+1)...u(k+l))}} F_{\begin{pmatrix} u(1) & \ldots & u(k) & u(k+1) & \ldots & u(k+l)\\ d(1) & \ldots & d(k) & e(1) & \ldots & e(l)\end{pmatrix}},\\
&F_{(\sigma,d)} \bullet F_{(\tau,e)}\\
&= \sum_{\substack{Pack(u(1)...u(k))=\sigma,\\ Pack(u(k+1)...u(k+l))=\tau,\\ min(u(1)...u(k))=min(u(k+1)...u(k+l))}} F_{\begin{pmatrix} u(1) & \ldots & u(k) & u(k+1) & \ldots & u(k+l)\\ d(1) & \ldots & d(k) & e(1) & \ldots & e(l)\end{pmatrix}}.
\end{aligned}$$

Proof Direct computations. □

Remark 11 1. For all packed word $(\sigma(1)\ldots\sigma(n))$:

$$F_{(\sigma(1)\ldots\sigma(n))} = \sum_{d(1),\ldots,d(n)\geq 1} F_{\begin{pmatrix}\sigma(1) & \ldots & \sigma(n)\\ d(1) & \ldots & d(n)\end{pmatrix}}.$$

2. In general, this action of decorated packed words is not faithful. For example, if $V = K[X]_+$, where X is homogeneous of degree n, then $F_{\begin{pmatrix}1\,2\\1\,1\end{pmatrix}} = F_{\begin{pmatrix}2\,1\\1\,1\end{pmatrix}}$. Indeed, both sends the word XX on itself and all the other words on 0.
3. Here is an example where the action is faithful. Let $V = K[X_i \mid i \geq 1]_+$, where X_i is homogeneous of degree 1 for all i. Let us assume that $\sum a_{(\sigma,d)} F_{(\sigma,d)} = 0$. Acting on the word $(X_1^{a_1})\ldots(X_k^{a_k})$, we obtain:

$$\sum_{length(\sigma)=k} a_{\begin{pmatrix}\sigma(1) & \ldots & \sigma(k)\\ a_1 & \ldots & a_k\end{pmatrix}} \left(\prod_{\sigma(i)=1} X_i^{a_i}\right)\cdots\left(\prod_{\sigma(i)=max(\sigma)} X_i^{a_i}\right) = 0.$$

As the X_i are algebraically independent, the words appearing in this sum are linearly independent, so for all (σ, d), $a_{(\sigma,d)} = 0$.

Notations.

1. For all $n \geq 1$, we put:

$$p_n = \sum_{k=1}^{n} \sum_{d(1)+\ldots+d(k)=n} F_{\begin{pmatrix} 1 & \ldots & k \\ d(1) & \ldots & d(k) \end{pmatrix}}.$$

The map p_n is the projection on the space of words of degree n, so $\sum_{n \geq 1} p_n = Id_A$.

2. For all $n \geq 1$, we put:

$$q_n = F_{\begin{pmatrix} 1 \\ n \end{pmatrix}}.$$

The map q_n is the projection on the space of letters of degree n, so, by Proposition 6, $q = \sum_{n \geq 1} q_n = F_{(1)}$ is the projection π of Proposition 5. It is not difficult to deduce, in the same way as proposition 12 of [16], the following result:

Theorem 4 *The NQSh subalgebra $QDesc(A)$ of $End_K(A)$ generated by the homogeneous components p_n of Id_A is also generated by the homogeneous components q_n of the projection on $Prim(A)$ of Proposition 5. Moreover, for all $n \geq 1$:*

$$q_n = \sum_{k=1}^{n} (-1)^{k+1} \sum_{a_1+\ldots+a_k=n} p_{a_1} \prec (p_{a_2} \uplus \ldots \uplus p_{a_k}).$$

Remark 12 This result is the quasi-shuffle analog of the statement that the descent algebra of a graded connected cocommutative Hopf algebra H (the convolution subalgebra of $End(H)$ generated by the graded projections) is equivalently generated by the graded components of the convolution logarithm of the identity [34].

11 Structure of the Decorated Quasi-Shuffle Operad

In this section, we show that the decorated quasi-shuffle operad $\mathbf{QSh}^d$ is free as a NSh algebra using the bidendriform techniques developed in [14].

We denote by $\mathbf{QSh}^d_+$ the subspace of the decorated quasi-shuffle operad generated by nonempty decorated packed words. As for a well-chosen graded quasi-shuffle bialgebra A the action of packed words is faithful, we deduce that $\mathbf{QSh}^d_+$ inherits a NQSh algebra structure by:

$$(\sigma, d) \prec (\tau, e) = \sum_{\substack{Pack(u(1)...u(k))=\sigma,\\ Pack(u(k+1)...u(k+l))=\tau,\\ min(u(1)...u(k))<min(u(k+1)...u(k+l))}} \begin{pmatrix} u(1) & \dots & u(k) & u(k+1) & \dots & u(k+l) \\ d(1) & \dots & d(k) & e(1) & \dots & e(l) \end{pmatrix},$$

$$(\sigma, d) \succ (\tau, e) = \sum_{\substack{Pack(u(1)...u(k))=\sigma,\\ Pack(u(k+1)...u(k+l))=\tau,\\ min(u(1)...u(k))>min(u(k+1)...u(k+l))}} \begin{pmatrix} u(1) & \dots & u(k) & u(k+1) & \dots & u(k+l) \\ d(1) & \dots & d(k) & e(1) & \dots & e(l) \end{pmatrix},$$

$$(\sigma, d) \bullet (\tau, e) = \sum_{\substack{Pack(u(1)...u(k))=\sigma,\\ Pack(u(k+1)...u(k+l))=\tau,\\ min(u(1)...u(k))=min(u(k+1)...u(k+l))}} \begin{pmatrix} u(1) & \dots & u(k) & u(k+1) & \dots & u(k+l) \\ d(1) & \dots & d(k) & e(1) & \dots & e(l) \end{pmatrix}.$$

Notations. Let (σ, d) be a decorated packed word of length k and let $I \subseteq \{1, \dots, max(\sigma)\}$. We put $\sigma^{-1}(I) = \{i_1, \dots, i_l\}$, with $i_1 < \dots < i_l$. The decorated packed word $(\sigma, d)_{|I}$ is $(Pack(\sigma(i_1), \dots, \sigma(i_l)), (d(i_1), \dots, d(i_l)))$.

Definition 12 We define two coproducts on $\mathbf{QSh}_+^d$ in the following way: for all nonempty packed word (σ, d),

$$\Delta_\prec(\sigma, d) = \sum_{i=\sigma(1)}^{max(\sigma)-1} (\sigma, d)_{|\{1,\dots,i\}} \otimes (\sigma, d)_{|\{\{i+1,\dots,max(\sigma)\}},$$

$$\Delta_\succ(\sigma, d) = \sum_{i=1}^{\sigma(1)-1} (\sigma, d)_{|\{1,\dots,i\}} \otimes (\sigma, d)_{|\{\{i+1,\dots,max(\sigma)\}}.$$

Then $\mathbf{QSh}_+^d$ is a NSh coalgebra, that is to say:

$$(\Delta_\prec \otimes Id) \circ \Delta_\prec = (Id \otimes (\Delta_\prec + \Delta_\succ)) \circ \Delta_\prec, \tag{32}$$

$$(\Delta_\succ \otimes Id) \circ \Delta_\prec = (Id \otimes \Delta_\prec) \circ \Delta_\succ, \tag{33}$$

$$((\Delta_\prec + \Delta_\succ) \otimes Id) \circ \Delta_\succ = (Id \otimes \Delta_\succ) \circ \Delta_\succ. \tag{34}$$

For all $a, b \in \mathbf{QSh}_+^d$:

$$\Delta_\prec(a \prec b) = a'_\prec \prec b' \otimes a''_\prec \star b'' + a'_\prec \prec b \otimes a''_\prec + a'_\prec \otimes a''_\prec \star b + a \prec b' \otimes b'' + a \otimes b, \tag{35}$$

$$\Delta_\prec(a \succ b) = a'_\prec \succ b' \otimes a''_\prec \star b'' + a \succ b' \otimes b'' + a'_\prec \succ b \otimes a''_\prec, \tag{36}$$

$$\Delta_\prec(a \bullet b) = a'_\prec \bullet b' \otimes a''_\prec \star b'' + a'_\prec \bullet b \otimes a''_\prec + a \bullet b' \otimes b'', \tag{37}$$

$$\Delta_\succ(a \prec b) = a'_\succ \prec b' \otimes a''_\succ \star b'' + a'_\succ \prec b \otimes a''_\succ + a'_\succ \otimes a''_\succ \star b, \tag{38}$$

$$\Delta_\succ(a \succ b) = a'_\succ \succ b'' \otimes a''_\succ \star b'' + a'_\succ \succ b \otimes a''_\succ + b'_\succ \otimes a \star b'' + b \otimes a, \quad (39)$$
$$\Delta_\succ(a \bullet b) = a'_\succ \bullet b' \otimes a''_\succ \star b'' + a'_\succ \bullet b \otimes a''_\succ. \quad (40)$$

Proof Let (σ, d) be a decorated packed word. Then:

$$(\Delta_\prec \otimes Id) \circ \Delta_\prec(\sigma, d) = (Id \otimes (\Delta_\prec + \Delta_\succ)) \circ \Delta_\prec(\sigma, d)$$
$$= \sum_{\sigma(1)\leq i<j\leq max(\sigma)-1} (\sigma, d)_{|\{1,\ldots,i\}} \otimes (\sigma, d)_{|\{i+1,\ldots,j\}} \otimes (\sigma, d)_{|\{j+1,\ldots,max(\sigma)\}},$$

$$(\Delta_\succ \otimes Id) \circ \Delta_\prec(\sigma, d) = (Id \otimes \Delta_\prec) \circ \Delta_\succ(\sigma, d)$$
$$= \sum_{1\leq i<\sigma(1)\leq j\leq max(\sigma)-1} (\sigma, d)_{|\{1,\ldots,i\}} \otimes (\sigma, d)_{|\{i+1,\ldots,j\}} \otimes (\sigma, d)_{|\{j+1,\ldots,max(\sigma)\}},$$

$$((\Delta_\prec + \Delta_\succ) \otimes Id) \circ \Delta_\succ(\sigma, d) = (Id \otimes \Delta_\succ) \circ \Delta_\succ(\sigma, d)$$
$$= \sum_{1\leq i<j<\sigma(1)} (\sigma, d)_{|\{1,\ldots,i\}} \otimes (\sigma, d)_{|\{i+1,\ldots,j\}} \otimes (\sigma, d)_{|\{j+1,\ldots,max(\sigma)\}}.$$

Let us prove (35), for $a = (\sigma, d)$ and $b = (\tau, e)$ two decorated packed words of respective length k and l. We put:

$$a \otimes b = \begin{pmatrix} \sigma(1) \ldots \sigma(k) & \tau(1) + max(\sigma) & \ldots & \tau(l) + max(\tau) \\ d(1) \ldots d(k) & e(1) & \ldots & e(l) \end{pmatrix}.$$

Then $a \prec b$ is the sum of all decorated packed words obtained by quasi-shuffling in all possible ways the values of the letters in the first row of $a \otimes b$, in such a way that 1 occurs only in the first k columns; $\Delta_\prec(a \otimes b)$ is then given by separating the letters of the first row of these decorated packed words in such a way that the first letter appears in the left side. So at least one of the k first letters appears on the left side. This gives five possible cases:

1. All the k first letters are on the left and all the l last letters are on the right. Necessarily, this case comes from the decorated packed word $a \otimes b$, and this gives the term $a \otimes b$.
2. All the k first letters are on the left and at least one of the l last letters is on the left. This gives the term $a \prec b' \otimes b''$.
3. At least one of the k first letters is on the right and all the l last letters are on the left. This gives the term $a'_\prec \prec b \otimes a''_\prec$.
4. At least one of the k first letters is on the right and all the l last letters are on the right. This gives the term $a'_\prec \otimes a''_\prec \star b$.
5. At least one of the k first letters is on the right and there are some of the l last letters on both sides. This gives the term $a'_\prec \prec b' \otimes a''_\prec \star b''$.

Summing all these terms, we obtain (35). The other compatibilities can be proved similarly. □

Remark 13 We also obtain, by addition:

$$\begin{aligned}
\Delta_{\prec}(a \preceq b) &= a'_{\prec} \preceq b' \otimes a''_{\prec} \star b'' + a'_{\prec} \preceq b \otimes a''_{\prec} + a'_{\prec} \otimes a''_{\prec} \star b \\
&\quad + a \preceq b' \otimes b'' + a \otimes b, && (41)\\
\Delta_{\prec}(a \succeq b) &= a'_{\prec} \succeq b' \otimes a''_{\prec} \star b'' + a \succeq b' \otimes b'' + a'_{\prec} \succeq b \otimes a''_{\prec}, && (42)\\
\Delta_{\succ}(a \preceq b) &= a'_{\succ} \preceq b' \otimes a''_{\succ} \star b'' + a'_{\succ} \preceq b \otimes a''_{\succ} + a'_{\succ} \otimes a''_{\succ} \star b, && (43)\\
\Delta_{\succ}(a \succeq b) &= a'_{\succ} \succeq b'' \otimes a''_{\succ} \star b'' + a'_{\succ} \succeq b \otimes a''_{\succ} + b'_{\succ} \otimes a \star b'' + b \otimes a; && (44)\\
\tilde{\Delta}(a \prec b) &= a' \prec b' \otimes a'' \star b'' + a' \prec b \otimes a'' + a' \otimes a'' \star b \\
&\quad + a \prec b' \otimes b'' + a \otimes b, && (45)\\
\tilde{\Delta}(a \succ b) &= a' \succ b' \otimes a'' \star b'' + a' \succ b \otimes a'' + a \succ b' \otimes b'' \\
&\quad + b' \otimes a \star b'' + b \otimes a, && (46)\\
\tilde{\Delta}(a \bullet b) &= a' \bullet b' \otimes a'' \star b'' + a' \bullet b \otimes a'' + a \bullet b' \otimes b''; && (47)\\
\tilde{\Delta}(a \preceq b) &= a' \preceq b' \otimes a'' \star b'' + a' \preceq b \otimes a'' + a' \otimes a'' \star b \\
&\quad + a \preceq b' \otimes b'' + a \otimes b, && (48)\\
\tilde{\Delta}(a \succeq b) &= a' \succeq b' \otimes a'' \star b'' + a' \succeq b \otimes a'' + a \succeq b' \otimes b'' \\
&\quad + b' \otimes a \star b'' + b \otimes a. && (49)
\end{aligned}$$

Consequently, $(\mathbf{QSh}^d_+, \succ^{op}, \preceq^{op}, \Delta^{op}_{\succ}, \Delta^{op}_{\prec})$ and $(\mathbf{QSh}^d_+, \succeq^{op}, \prec^{op}, \Delta^{op}_{\succ}, \Delta^{op}_{\prec})$ are bidendriform bialgebras. By the bidendriform rigidity theorem of [14], we have:

Theorem 5 *$(\mathbf{QSh}^d_+, \preceq, \succ)$ and $(\mathbf{QSh}^d_+, \prec, \succeq)$ are free NSh algebras.*

Forgetting the decoration, we get back theorem 2.5 of [32], up to a permutation of maximum and minimum, and first and last letters.

Forgeting again the decorations, we obtain a NQSh algebra structure on $\mathbf{QSh}_+$ and a NSh coalgebra structure, with compatibilities (35)–(40). Let us describe, for completeness sake, the dual (half-)products and coproducts. The elements of the dual basis of packed words are denoted by N_u.

Proposition 17 *1. For all nonempty packed words σ, τ, of respective lengths k and l:*

$$N_\sigma \prec N_\tau = \sum_{\alpha \in Sh^{\prec}_{k,l}} N_{(\sigma\otimes\tau)\circ\alpha^{-1}}, \qquad N_\sigma \succ N_\tau = \sum_{\alpha \in Sh^{\succ}_{k,l}} N_{(\sigma\otimes\tau)\circ\alpha^{-1}}.$$

2. *For any nonempty packed word σ of length n, denoting by $f(\sigma)$ the index of the first appearance of* 1 *in σ and by $l(\sigma)$ the index of the last appearance of* 1 *in σ:*

$$\tilde{\Delta}_{\prec}(N_\sigma) = \sum_{k=l(\sigma)}^{n-1} N_{pack(\sigma(1)\ldots\sigma(k))} \otimes N_{pack(\sigma(k+1)\ldots\sigma(n))},$$

$$\tilde{\Delta}_{\succ}(N_\sigma) = \sum_{k=1}^{f(\sigma)-1} N_{pack(\sigma(1)\ldots\sigma(k))} \otimes N_{pack(\sigma(k+1)\ldots\sigma(n))},$$

$$\tilde{\Delta}_{\bullet}(N_\sigma) = \sum_{k=f(\sigma)}^{l(\sigma)-1} N_{pack(\sigma(1)\ldots\sigma(k))} \otimes N_{pack(\sigma(k+1)\ldots\sigma(n))}.$$

12 The Quasi-Shuffle Analog of the Descent Algebra

Recall that, given a graded NQSh bialgebra A, we introduced $QDesc(A)$, the quasi-shuffle analogue of the descent algebra defined as the NQSh subalgebra of $End(A)$ generated by the graded projections or, equivalently, by the graded components of the projection on $Prim(A)$. We write **QDesc** for the corresponding NQSh subalgebra of $\mathbf{QSh}^d$ (the subalgebra generated by the $\binom{1}{d}$).

Recall first some properties of NSh algebras.

Notations. Let $n \geq 1$.

1. a. Let $\mathbb{T}_{Sch}(n)$ be the set of Schröder trees of degree n, that is to say reduced planar rooted trees with $n+1$ leaves.
   b. For any set D, let $\mathbb{T}^D_{Sch}(n)$ be the set of reduced planar rooted trees t with $n+1$ leaves, such that the n spaces between the leaves of t are decorated by elements of D.
   c. $\mathbb{T}^D_{Sch} = \bigsqcup_{n\geq 1} \mathbb{T}^D_{Sch}(n)$.
2. Let $t_1, \ldots, t_k \in \mathbb{T}^{\mathbb{N}^*}_{Sch}$ and let $d_1, \ldots, d_{k-1} \in \mathbb{N}^*$. The element $t_1 \vee_{d_1} \ldots \vee_{d_{k-1}} t_k$ is obtained by grafting $t_1, \ldots, t_k$ on a common root; for all $1 \leq i \leq k$, the space between the right leaf of t_i and the left leaf of t_{i+1} is decorated by d_i.

Following [28], $\mathbb{T}^D_{Sch}$ is a basis of the free NQSh algebra generated by D, $NQSh(D)$. The three products are inductively defined: if $t = t_1 \vee_{d_1} \ldots \vee_{d_{k-1}} t_k$ and $t' = t'_1 \vee_{d'_1} \ldots \vee_{d'_{l-1}} t'_l \in \mathbb{T}_{Sch}(D)$, then

$$t \succ t' = (t \star t'_1) \vee_{d'_1} t'_2 \vee_{d'_2} \ldots \vee_{d'_{l-1}} t'_l,$$
$$t \prec t' = t_1 \vee_{d_1} \ldots \vee t_{k-1} \vee_{d_{k-2}} \ldots \vee_{d_{k-1}} (t_k \star t'),$$
$$t \bullet t' = t_1 \vee_{d_1} \ldots \vee_{d_{k-1}} (t_k \star t'_1) \vee_{d'_1} \ldots \vee_{d'_{l-1}} t'_l.$$

Sending any non binary tree to 0, we obtain the free NSh algebra $NSh(D)$ generated by D. A basis is given by the set of planar binary trees $\mathbb{T}_{bin}(D) \subseteq \mathbb{T}_{Sch}(D)$ whose spaces between the leaves are decorated by elements of D. The products are given in the following way: if $t = t_1 \vee_d t_2$ and $t' = t'_1 \vee_{d'} t'_2$, then:

$$t \succ t' = (t \star t'_1) \vee_{d'} t'_2,$$
$$t \prec t' = t_1 \vee_d (t_2 \star t').$$

We denote by $NQSh(1)$ and by $NSh(1)$ the free NQSh and the free NSh algebra on one generator. The set $\mathbb{T}_{Sch}$ is a basis of $NQSh(1)$, and $\mathbb{T}_{bin}$ is a basis of $NSh(1)$.

Example 5

$\mathbb{T}_{Sch}(0) = \mathbb{T}_{bin}(0) = \{ | \}$, $\mathbb{T}_{Sch}(1) = \mathbb{T}_{bin}(1) = \{ Y \}$,

$\mathbb{T}_{Sch}(2) =$ , $\mathbb{T}_{bin}(2) =$ 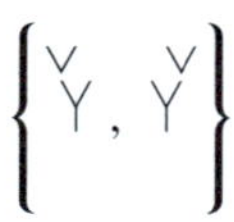,

$\mathbb{T}_{Sch}(3) =$ 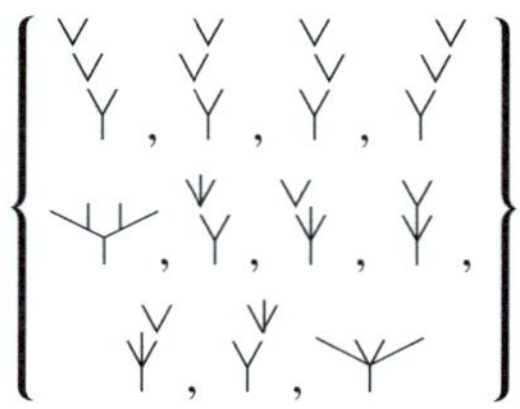, $\mathbb{T}_{bin}(3) =$ 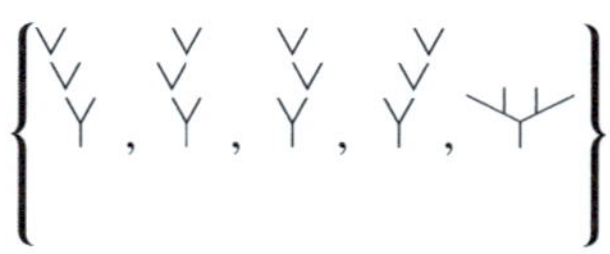.

We define now inductively a surjective map ϱ from the set of packed words decorated by D into $\mathbb{T}^D_{Sch}$ in the following way:

1. $\varrho(1) = |$.
2. If $w = (\sigma, d)$, let $\sigma^{-1}(1) = \{i_1, \ldots, i_k\}$, $i_1 < \ldots < i_k$. We put:

$$w_1 = Pack\begin{pmatrix} \sigma(1) \ldots \sigma(i_1 - 1) \\ d(1) \ldots d(i_1 - 1) \end{pmatrix},$$
$$w_2 = Pack\begin{pmatrix} \sigma(i_1 + 1) \ldots \sigma(i_2 - 1) \\ d(i_1 + 1) \ldots d(i_2 - 1) \end{pmatrix},$$
$$\vdots$$
$$w_{k+1} = Pack\begin{pmatrix} \sigma(i_k + 1) \ldots \sigma(n) \\ d(i_k + 1) \ldots d(n) \end{pmatrix}.$$

Then:

$$\varrho(\sigma, d) = \varrho(w_1) \vee_{d(i_1)} \ldots \vee_{d(i_k)} \varrho(w_{k+1}).$$

If $w = (\sigma, d)$ is a decorated packed word of length n, $\varrho(w)$ is an element of $\mathbb{T}^D_{Sch}(n)$ such that the spaces between the leaves are decorated from left to right by $d(1), \ldots, d(n)$. In particular $\varrho\binom{1}{d}$ is the tree Y d-decorated.

For any $t \in \mathbb{T}^{\mathbb{N}^*}_{Sch}$, we put:

$$\Omega(t) = \sum_{\sigma \in \mathbb{S}urj, \varrho(\sigma)=t} \sigma \in \mathbf{QSh}^d_+.$$

We extend $\Omega : NQSh(\mathbb{N}^*) \longrightarrow \mathbf{QSh}^d_+$ by linearity map. It is clearly injective.

Example 6

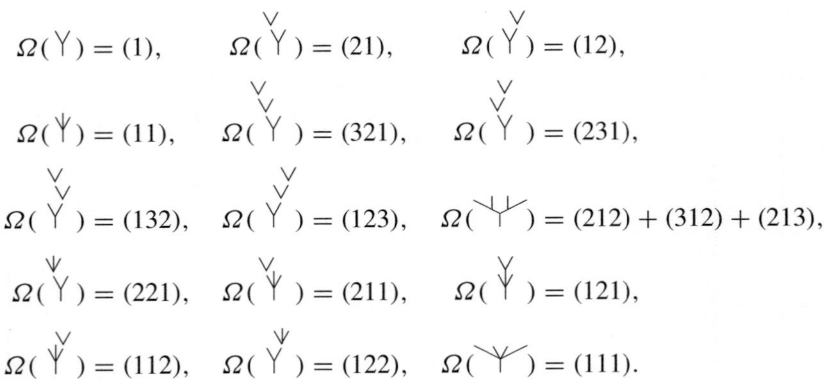

Theorem 6 *The map Ω is an injective morphism of NQSh algebras. Consequently,* **QDesc**, *the NQSh subalgebra of* $\mathbf{QSh}^d_+$ *generated by the elements* $\binom{1}{d}$, $d \geq 1$, *is free and isomorphic to* $NQSh(\mathbb{N}^*)$.

Proof Let $w = (\sigma, d)$ be a packed word of length n and let $i_1, \ldots, i_k$ be integers such that $i_1 + \ldots + i_k = n$. For all $d_1, \ldots, d_{k-1} \geq 1$, we put:

$$\begin{aligned} &ins^{d_1,\ldots,d_{k-1}}_{i_1,\ldots,i_k}(w) \\ &= \begin{pmatrix} \sigma(1)+1 & \ldots & \sigma(i_1)+1 & 1 & \ldots & 1 & \sigma(i_1+\ldots+i_{k-1}+1)+1 & \ldots & \sigma(n)+1 \\ d(1) & \ldots & d(i_1) & d_1 & \ldots & d_{k-1} & d(i_1+\ldots+i_{k-1}+1) & \ldots & d(n) \end{pmatrix}. \end{aligned}$$

It is not difficult to show that:

$$\Omega(t_1 \vee_{d_1} \ldots \vee_{d_{k-1}} t_k) = ins^{d_1,\ldots,d_{k-1}}_{|t_1|,\ldots,|t_k|}(\Omega(t_1) \star \ldots \star \Omega(t_k)).$$

Hence, if $t = t_1 \vee_{d_1} \ldots \vee_{d_{k-1}} t_k$ and $t' = t'_1 \vee_{d'_1} \ldots \vee_{d'_{l-1}} t'_l$:

$$\Omega(t) \succ \Omega(t') = ins^{d'_1,\ldots,d'_{l-1}}_{|t|+|t'_1|,\ldots,|t'_l|}(\Omega(t) \star \Omega(t'_1) \star \ldots \star \Omega(t'_l)),$$
$$\Omega(t) \prec \Omega(t') = ins^{d_1,\ldots,d_{k-1}}_{|t_1|,\ldots,|t_k|+|t|}(\Omega(t_1) \star \ldots \star \Omega(t_k) \star \Omega(t')),$$
$$\Omega(t) \bullet \Omega(t') = ins^{d_1,\ldots,d_{k-1},d'_1,\ldots,d'_{l-1}}_{[t_1|,\ldots,|t_k|+|t'_1|,\ldots,|t'_l|}(\Omega(t_1) \star \ldots \star \Omega(t_k) \star \Omega(t'_1) \star \ldots \star \Omega(t'_l)).$$

An induction on $m+n$ proves that for $t \in \mathbb{T}^{\mathbb{N}^*}_{Sch}(m)$, $t' \in \mathbb{T}^{\mathbb{N}^*}_{Sch}(n)$:

$$\Omega(t \succ t') = \Omega(t) \succ \Omega(t'), \quad \Omega(t \prec t') = \Omega(t) \prec \Omega(t'), \quad \Omega(t \bullet t') = \Omega(t) \bullet \Omega(t').$$

So Ω is an injective morphism of NQSh algebras. □

13 Lie Theory, Continued

In classical Lie theory, it has been realized progressively that many applications of the combinatorial part of the theory rely on the freeness of the Malvenuto-Reutenauer algebra of permutations (for us, the operad **Sh** or, equivalently, the algebra of free quasi-symmetric functions **FQSym**) as a noncommutative shuffle bialgebra (and more precisely, as a bidendriform bialgebra [14]). As such, **Sh** has two remarkable subalgebras. The first is **PBT**, the noncommutative shuffle sub-bialgebra freely generated as a noncommutative shuffle algebra by the identity permutation in $\mathfrak{S}_1$ (in particular **PBT** is isomorphic to $NSH(1)$, the free NQSh algebra on one generator). Its elements can be understood as linear combinations of planar binary trees (**PBT** can be constructed directly as a subspace of the direct sum of the symmetric group algebras is by using a construction going back to Viennot: a natural partition of the symmetric groups parametrized by planar binary trees), see [21, 22, 27]. The second, **Desc**, is known as the descent algebra [38], is isomorphic to **Sym**, the Hopf algebra of noncommutative symmetric functions, and is the sub Hopf algebra of **PBT** and **Sh** freely generated as an associative algebra by (all) the identity permutations using the convolution product $\star$. We get:

$$\mathbf{Desc} = \mathbf{Sym} \subset \mathbf{PBT} = NSH(1) \subset \mathbf{Sh} = \mathbf{FQSym}.$$

The situation is similar when moving to surjections, that is to **QSh**. As we already saw, the noncommutative quasi-shuffle sub-bialgebra freely generated by the identity permutation in $\mathfrak{S}_1$ (i.e. the packed word 1) is the free NQSh algebra on one generator, identified with **ST**, the linear span of Schröder trees. The sub Hopf algebra of **ST** and **QSh** freely generated as an associative algebra by (all) the identity permutations using the convolution product $\star$ is isomorphic (using e.g. that it is a free associative algebra over a countable set of generators) to **Desc**. We get:

$$\mathbf{Desc} = \mathbf{Sym} \subset \mathbf{ST} = NQSH(1) \subset \mathbf{QSh} = \mathbf{WQSym}.$$

The aim of the present and last section is to compare explicitely the two sequences of inclusions. The existence of a Hopf algebra map from **Sh** = **FQSym** to **QSh** = **WQSym** was obtained in [17, Corollary 18]. The existence of a map comparing the two copies of the descent algebra follows, a simple direct proof was given in [8, Lemma 7.1]. We aim here at refining these results and extend the constructions to planar and Schröder trees.

We start by showing how planar trees (**PBT**) can be embedded into Schröder trees (**ST**).

Definition 13 Let $t, t' \in \mathbb{T}_{Sch}$.

1. We denote by $R(t)$ the set of internal edges of t which are right, that is to say edges e such that:
   - both extremities of e are internal vertices.
   - e is the edge which is at most on the right among all the edges with the same origin as e.
2. Let $I \subseteq R(T)$. We denote by t/I the planar reduced tree obtained by contracting all the edges $e \in I$.
3. We shall say that $t' \leq t$ if there exists $I \subseteq R(t)$, such that $t' = t/I$.

Remark 14 If $I \subseteq R(t)$, then $R(t/I) = R(t) \setminus I$. Moreover, if $I, J \subseteq R(t)$ are disjoint, then $(t/I)/J = t/(I \sqcup J)$. This implies that $\leq$ is a partial order on $\mathbb{T}_{Sch}$.

Example 7 Here are the Hasse graphs of $\mathbb{T}_{Sch}(2)$ and $\mathbb{T}_{Sch}(3)$.

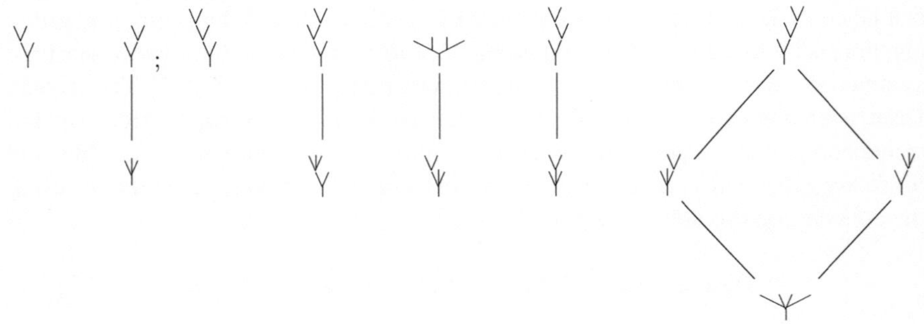

It is possible to prove the following points:

- For any $t \in \mathbb{T}_{Sch}$, there exists a unique $b(t) \in \mathbb{T}_{bin}$, such that $t \leq b(t)$. We denote by $I(t)$ the unique subset $I \subseteq R(b(t))$, such that $t = b(t)/I$.
- For any $t, t' \in \mathbb{T}_{Sch}$, $t \leq t'$ if, and only if, $b(t) = b(t')$ and $I(t) \supseteq I(t')$.

Theorem 7 *The following map is an injective morphism of bidendriform bialgebras:*

$$\psi : \begin{cases} (\mathbf{PBT}, \prec, \succ, \Delta_\prec, \Delta_\succ) \longrightarrow (\mathbf{ST}, \preceq, \succ, \Delta_\prec, \Delta_\succ) \\ \qquad\qquad t \in \mathbb{T}_{bin} \longrightarrow \displaystyle\sum_{t' \leq t} t'. \end{cases}$$

Proof By universal properties of free objects, there exists a unique morphism of noncommutative shuffle algebras ψ' from $(NSh(1) = \mathbf{PBT}, \prec, \succ)$ to $(NQSh(1) = \mathbf{ST}, \preceq, \succ)$, sending Y to Y. As Y is a primitive element (in the bidendriform sense) for both sides, ψ' is a morphism of bidendriform bialgebras. We shall prove that $\psi = \psi'$.

Let us show that for all $t_1, t_2 \in \mathbb{T}_{bin}$,

$$\psi'(t_1 \vee t_2) = \psi'(t_1) \succ \text{Y} \preceq \psi'(t_2),$$

$$\psi(t_1 \vee t_2) = \psi(t_1) \succ \text{Y} \preceq \psi(t_2).$$

The identity $\psi = \psi'$ will follow by induction.

The identity involving ψ' follows immediately from the identity, in $\mathbb{T}_{bin}$:

$$t_1 \vee t_2 = t_1 \succ \text{Y} \prec t_2.$$

Let us consider the action of ψ. We put $t = t_1 \vee t_2$. We first consider the case where $t_2 = \mid$. In this case, $R(t) = R(t_1)$ and for any $I \subseteq R(t_1)$, $t/I = (t_1/I) \vee \mid$. Hence:

$$\psi(t) = \sum_{I \subseteq R(t_1)} (t_1/I) \vee \mid = \left(\sum_{I \subseteq R(t_1)} t_1/I \right) \succ \text{Y} = \psi(t_1) \succ \text{Y} \preceq \mid.$$

We now consider the case where $t_2 \neq \mid$. Let r be the internal edge of t relating the root of t to the root of t_2. Then $R(t) = R(t_1) \sqcup R(t_2) \sqcup \{r\}$. Let $I_1 \subseteq R(t_1)$, $I_2 \subseteq R(t_2)$. Then:

$$t/I_1 \sqcup I_2 = (t_1/I_1) \vee (t_2/I_2) = (t_1/I_1) \succ \text{Y} \prec (t_2/I_2).$$

We put $t_2/i_2 = t_3 \vee \ldots \vee t_k$. Then:

$$\begin{aligned} t/I_1 \sqcup I_2 \sqcup \{r\} &= t_1/I_1 \vee t_3 \vee \ldots \vee t_k \\ &= (t_1/I_1 \vee \mid) \bullet (t_3 \vee \ldots \vee t_k) \\ &= ((t_1/I_1) \succ \text{Y}) \bullet (t_2/I_2) \\ &= (t_1/I_1) \succ \text{Y} \bullet (t_2/I_2). \end{aligned}$$

Hence:

$$\begin{aligned}\psi(t) &= \sum_{I_1\subseteq R(t_1), I_2\subseteq R(t_2)} (t_1/I_1) \succ \mathrm{Y} \prec (t_2/I_2) + (t_1/I_1) \succ \mathrm{Y} \bullet (t_2/I_2)\\ &= \sum_{I_1\subseteq R(t_1), I_2\subseteq R(t_2)} (t_1/I_1) \succ \mathrm{Y} \preceq (t_2/I_2)\\ &= \psi(t_1) \succ \mathrm{Y} \preceq \psi(t_2).\end{aligned}$$

So $\psi = \psi'$. As $\leq$ is an order, ψ is injective. □

We investigate now how the injection of **PBT** into **ST** behaves with respect to the respective embeddings into **Sh** and **QSh**. We consider the morphism:

$$\Omega : \begin{cases} \mathbf{ST} = NQSh(1) \longrightarrow \mathbf{QSh} \\ \qquad\qquad\quad t \longrightarrow \displaystyle\sum_{\sigma, \varrho(\sigma)=t} \sigma. \end{cases}$$

There exists a unique map from $\mathbf{PBT} = NSh(1)$ to $\mathcal{S}h$, denoted by Ω', making the following diagram commuting:

$$\begin{array}{ccc} \mathbf{ST} & \xrightarrow{\ \Omega\ } & \mathbf{QSh} \\ \downarrow & & \downarrow \\ \mathbf{PBT} & \xrightarrow[\ \Omega'\ ]{} & \mathcal{S}h \end{array}$$

where the vertical arrows are the canonical projection. For any $t \in \mathbb{T}_{bin}$:

$$\Omega'(t) = \sum_{\sigma\in\mathfrak{S}, \varrho(\sigma)=t} \sigma.$$

Example 8

$$\Omega'(\mathrm{Y}) = (1), \qquad \Omega'(\overset{\vee}{\mathrm{Y}}) = (21), \qquad \Omega'(\overset{\vee}{\mathrm{Y}}) = (12),$$

$$\Omega'(\overset{\overset{\vee}{\vee}}{\mathrm{Y}}) = (321), \qquad \Omega'(\overset{\overset{\vee}{\vee}}{\mathrm{Y}}) = (231), \qquad \Omega'(\overset{\overset{\vee}{\vee}}{\mathrm{Y}}) = (132),$$

$$\Omega'(\overset{\overset{\vee}{\vee}}{\mathrm{Y}}) = (123), \qquad \Omega'(\overset{\vee\ \vee}{\mathrm{Y}}) = (312) + (213).$$

Proposition 18 *[15] Let σ, τ be two packed words of the same length n. We shall say that $\sigma \leq \tau$ if:*

1. *If $i, j \in [n]$ and $\sigma(i) \leq \sigma(j)$, then $\tau(i) \leq \tau(j)$.*
2. *If $i, j \in [n]$, $i < j$ and $\sigma(i) > \sigma(j)$, then $\tau(i) > \tau(j)$.*

Then $\leq$ is a partial order. Moreover, the following map is a Hopf algebra morphism:

$$\Psi : \begin{cases} \mathbf{Sh} \longrightarrow \mathbf{QSh} \\ \sigma \longrightarrow \displaystyle\sum_{\tau \leq \sigma} \tau. \end{cases}$$

Here are the Hasse graphs of $Surj_2$ and $Surj_3$:

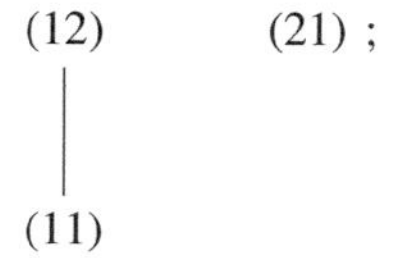

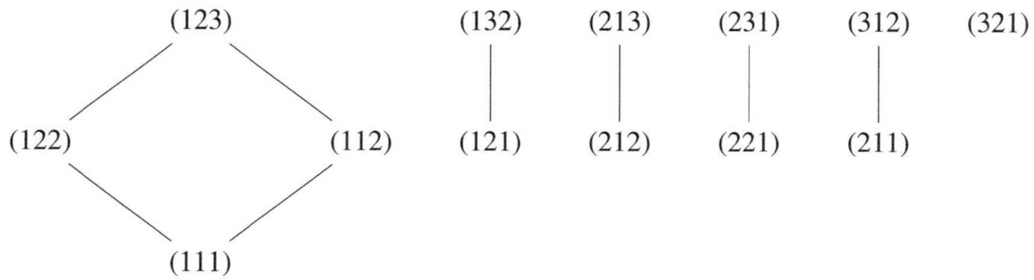

Lemma 5 *For any packed word σ, we put $\iota(\sigma) = \min\{i \mid \sigma(i) = 1\}$. If $\sigma \leq \tau$, then $\iota(\sigma) = \iota(\tau)$.*

Proof We put $i = \iota(\tau)$. For any j, $\tau(j) \geq \tau(i)$, so $\sigma(j) \geq \sigma(i)$ as $\sigma \leq \tau$. So $\sigma(i) = 1$, and by definition $\iota(\sigma) \leq i$. Let us assume that $j < i$. By definition of $\iota(\tau)$, $\tau(j) > \tau(i)$. As $\sigma \leq \tau$, $\sigma(j) > \sigma(i)$, so $\sigma(j) \neq 1$, and $\iota(\sigma) \neq j$. So $\iota(\sigma) = i$. □

Proposition 19 *The map $\varrho : \mathbb{S}urj \longrightarrow \mathbb{T}_{Sch}$ is a morphism of posets: for any packed words σ, τ,*

$$\sigma \leq \tau \Longrightarrow \varrho(\sigma) \leq \varrho(\tau).$$

We define a map $\omega : \mathbb{T}_{Sch} \longrightarrow \mathbb{S}urj$ by:

- $\omega(\mathsf{I}) = 1$,
- $\omega(t_1 \vee \ldots \vee t_k) = (\omega(t_1)[1])1 \ldots 1(\omega(t_k)[1])$.

Then $\varrho \circ \omega = Id_{\mathbb{T}_{Sch}}$, and ω is a morphism of posets: for any $t, t' \in \mathbb{T}_{Sch}$,

$$t \leq t' \Longrightarrow \omega(t) \leq \omega(t').$$

Proof Let us prove that ϱ is a morphism. Let σ, τ be two packed words, such that $\sigma \leq \tau$; let us prove that $\varrho(\sigma) \leq \varrho(\tau)$. We proceed by induction on the common length n of σ and τ. If $n = 0$ or 1, the result is obvious. Let us assume the result at all rank $< n$. As $\iota(\sigma) = \iota(\tau)$, we can write $\sigma = \sigma' 1 \sigma''$ and $\tau = \tau' 1 \tau''$, where σ' and τ' have the same length and do not contain any 1. By restriction, $Pack(\sigma') \leq Pack(\tau')$ and $Pack(\sigma'') \leq Pack(\tau'')$. By the induction hypothesis, $s_0 = \varrho(\sigma') \leq \varrho(\tau') = t_0$ and $s_1 \vee \ldots \vee s_k = \varrho(\sigma'') \leq \varrho(\tau'') = t_1 \vee \ldots \vee t_l$. Then:

$$\varrho(\sigma) = s_0 \vee s_1 \vee \ldots \vee \ldots s_k \leq t_0 \vee t_1 \vee \ldots \vee t_l = \varrho(\tau).$$

Let us now prove that ω is a morphism. Let $t, t' \in \mathbb{T}_{Sch}$, such that $t \leq t'$. By transitivity, we can assume that there exists $e \in R(t')$, such that $t = t'_{|e}$. Let us prove that $\omega(t) \leq \omega(t')$. We proceed by induction on the common degree n of t and t'. The result is obvious if $n = 0$ or 1. Let us assume the result at all ranks $< n$. We put $t' = t'_1 \vee \ldots \vee t'_k$. If e is an edge of t'_i, then $t = t'_1 \vee \ldots \vee (t'_i)_{|I} \vee \ldots \vee t'_k$. We put $\sigma'_j = \omega(t'_j)$ and $\sigma_j = \omega(t_j)$ for all j. If $j \neq i$, $\sigma'_j = \sigma_j$; by the induction hypothesis, $\sigma_i \leq \sigma'_i$. Then:

$$\begin{aligned}\omega(t) &= (\sigma_1[1])1 \ldots 1(\sigma_i[1])1 \ldots 1(\sigma_k[1]) \\ &\leq (\sigma_1[1])1 \ldots 1(\sigma'_i[1])1 \ldots 1(\sigma_k[1]) = \omega(t').\end{aligned}$$

If e is the edge relation the root of t to the root of t'_k, putting $t = t_1 \vee \ldots \vee t_k \vee \ldots \vee t_l$, then $t'_i = t_i$ if $i < k$ and $t'_k = t_k \vee \ldots \vee t_l$. Putting $\sigma_i = \omega(t_i)$, we obtain:

$$\omega(t) = (\sigma_1[1])1 \ldots 1(\sigma_k[1])1 \ldots 1(\sigma_l[1]),$$
$$\omega(t') = (\sigma_1[1])1 \ldots 1(\sigma_k[2])2 \ldots 2(\sigma_l[2]).$$

It is not difficult to prove that $\omega(t) \leq \omega(t')$. □

Remark 15 There are similar results for decorated packed words, replacing $NSh(1)$ and $NQSh(1)$ by $NSh(\mathbb{N}^{*n})$ and $NQSh(\mathbb{N}^*)$.

Example 9

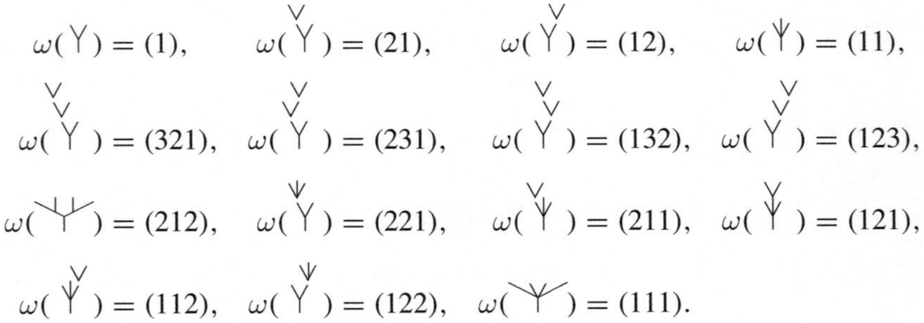

Proposition 20 *The map* Ψ *is a bidendriform bialgebra morphism from* $(\mathbf{Sh}, \prec, \succ, \Delta_\prec, \Delta_\succ)$ *to* $(\mathbf{QSh}, \preceq, \succ, \Delta_\prec, \Delta_\succ)$*. Moreover, the following diagram commutes:*

$$\begin{array}{ccc} \mathbf{PBT} & \xrightarrow{\psi} & \mathbf{ST} \\ {\scriptstyle \Omega'}\downarrow & & \downarrow{\scriptstyle \Omega} \\ \mathbf{Sh} & \xrightarrow[\Psi]{} & \mathbf{QSh} \end{array}$$

Proof Let σ be a packed word. We put:

$$A = \{(k, \tau) \mid \tau \leq \sigma, k \in [\max(\tau)]\},$$
$$B = \{(k, \tau', \tau'') \mid k \in [\max(\sigma)], \tau' \leq \sigma_{|[k]}, \tau'' \leq Pack(\sigma_{|[\max(\sigma)]\setminus[k]}).$$

As Ψ is a coalgebra morphism,

$$\begin{aligned} \Delta \circ \Psi(\sigma) &= \sum_{\tau \leq \sigma} \sum_{k=0}^{\max(\tau)} \tau_{|[k]} \otimes Pack(\tau_{|[\max(\tau)]\setminus[k]}) \\ &= \sum_{(k,\tau)\in A} \tau_{|[k]} \otimes Pack(\tau_{|[\max(\tau)]\setminus[k]}) \\ = (\Psi \otimes \Psi) \circ \Delta(\sigma) &= \sum_{k=0}^{\max(\sigma)} \sum_{\substack{\tau' \leq \sigma_{|[k]} \\ \tau'' \leq Pack(\sigma_{|[\max(\sigma)]\setminus[k]})}} \tau' \otimes \tau'' \\ &= \sum_{(l,\tau',\tau'')\in B} \tau' \otimes \tau''. \end{aligned}$$

Hence, there exists a bijection $F : A \longrightarrow B$, such that, if $F(k, \tau) = (l, \tau', \tau'')$, then:

- $\tau' = \tau_{|[k]}$ and $\tau'' = Pack(\tau_{|[\max(\tau)]\setminus[k]})$;
- l is the unique integer such that $\tau' \leq \sigma_{|[l]}$.

If $k \geq \tau(1)$, then the first letter of τ appears in $\tau_{|[k]}$, so the first letter of σ appears also in $\sigma_{|[l]}$. Consequently $l \geq \sigma(1)$. Similarly, if $l \geq \sigma(1)$, then $k \geq \tau(1)$. We obtain:

$$\begin{aligned} \Delta_\prec \circ \Psi(\sigma) &= \\ &= \sum_{(k,\tau)\in A, k\geq\tau(1)} \tau_{|[k]} \otimes Pack(\tau_{|[\max(\tau)]\setminus[k]}) \\ &= \sum_{(l,\tau',\tau'')\in B, l\geq\sigma(1)} \tau' \otimes \tau'' \\ &= (\Psi \otimes \Psi) \circ \Delta_\prec(\sigma) \end{aligned}$$

So Ψ is a morphism of dendriform coalgebras.

Let σ, τ be two permutations. We put:

$$C = \{(\alpha, \zeta) \mid \alpha \in Sh(\max(\sigma), \max(\tau)), \zeta \leq \alpha \circ (\sigma \otimes \tau)\},$$
$$D = \{(\beta, \sigma', \tau') \mid \sigma' \leq \sigma, \tau' \leq \tau, \beta \in QSh(\max(\sigma'), \max(\tau'))\},$$

Then:

$$\begin{aligned}
\Psi(\sigma \sqcup\!\sqcup \tau) &= \sum_{\alpha \in Sh(\max(\sigma),\max(\tau))} \sum_{\zeta \leq \alpha \circ (\sigma \otimes \tau)} \zeta \\
&= \sum_{(\alpha,\zeta) \in C} \zeta \\
= \Psi(\sigma) \uplus \Psi(\tau) &= \sum_{\substack{\sigma' \leq \sigma \\ \tau' \leq \tau}} \sum_{\beta \in QSh(\max(\sigma'),\max(\tau'))} \beta \circ (\sigma' \otimes \tau') \\
&= \sum_{(\beta,\sigma',\tau') \in D} \beta \circ (\sigma' \otimes \tau').
\end{aligned}$$

Hence, there exists a bijection $G : D \longrightarrow C$, such that if $G(\beta, \sigma', \tau') = (\alpha, \zeta)$, then:

1. $\zeta = \beta \circ (\sigma' \otimes \tau')$;
2. α is the unique $(\max(\sigma), max(\tau))$-shuffle such that $\zeta \leq \alpha \circ (\sigma \otimes \tau)$.

Let us assume that $\alpha(1) = 1$, and let us prove that $\beta(1) = 1$. Denoting by k the length of σ, 1 appears in the k first letters of $\zeta' = \alpha \circ (\sigma \otimes \tau)$. Let $i \in [k]$, such that $\zeta'(i) = 1$. For any j, $\zeta'(i) \leq \zeta'(j)$. As $\zeta \leq \zeta'$, $\zeta(i) \leq \zeta(j)$, so $\zeta(i) = 1$: 1 appears among the k first letters of ζ, so $\beta(1) = 1$.

Let us assume that $\alpha(1) \neq 1$. Then 1 does not appear in the first k letters of ζ'. Let $j > k$, such that $\zeta'(j) = 1$. For all $i \in [k]$, $\zeta'(i) > \zeta'(j)$ and $i < j$. As $\zeta \leq \zeta'$, $\zeta(i) > \zeta(j)$, so $\zeta(i) \neq 1$: 1 does not appear among the first k letters of ζ, so $\beta(1) \neq 1$. Finally, $\alpha(1) = 1$ if, and only if, $\beta(1) = 1$. Hence:

$$\Psi(\sigma \prec \tau) = \sum_{(\alpha,\zeta) \in C, \alpha(1)=1} \zeta = \sum_{(\beta,\sigma',\tau') \in D, \beta(1)=1} \beta \circ (\sigma' \otimes \tau') = \Psi(\sigma) \preceq \Psi(\tau).$$

By composition, $\Omega \circ \psi$ and $\Psi \circ \Omega$ are both noncommutative shuffle algebra morphisms, sending Y to (1), so, since **PBT** is a free NSh algebra, they are equal. □

Acknowledgements The authors were supported by the grant CARMA ANR-12-BS01-0017. We thank its participants and especially Jean-Christophe Novelli and Jean-Yves Thibon, for stimulating discussions on noncommutative symmetric functions and related structures. This article is, among others, a follow up of our joint works [17, 31]. We also thank the ICMAT Madrid for its hospitality.

References

1. Bourbaki, N.: Groupes et algèbres de Lie, Hermann, Paris (1968)
2. Burgunder, E., Ronco, M.: Tridendriform structure on combinatorial Hopf algebras. J. Algebra **324**(10), 2860–2883 (2010)
3. Cartier, P.: On the structure of free Baxter algebras. Adv. Math. **253**(9) (1972)
4. Cartier, P.: Fonctions polylogarithmes, nombres polyzêtas et groupes pro-unipotents. Séminaire Bourbaki **43**, 137–173 (2002)
5. Cartier, P.: A Primer of Hopf algebras, Frontiers in Number Theory, Physics, and Geometry II, pp. 537–615. Springer, Berlin, Heidelberg (2007)
6. Chapoton, F.: Algebres de Hopf des permutoèdres, associaèdres et hypercubes. Adv. Math. **150**, 264–275 (2000)
7. Chapoton, F.: Opérades différentielles graduées sur les simplexes et les permutoèdres. Bulletin de la Société mathématique de France **130**(2), 233–251 (2002)
8. Ebrahimi-Fard, K., Malham, S.J.A., Patras, F., Wiese, A.: Flows and stochastic taylor series in Itô calculus. J. Phys.: Math. Theor. **48**(49), 495202 (2015)
9. Ebrahimi-Fard, K., Malham, S.J.A., Patras, F., Wiese, A.: The exponential Lie series for continuous semimartingales. Proc. R. Soc. A **471**, The Royal Society, p. 20150429 (2015)
10. Ebrahimi-Fard, K., Patras, F.: La structure combinatoire du calcul intégral. Gazette des Mathématiciens **138** (2013)
11. Ebrahimi-Fard, K.: Loday-type algebras and the rota-baxter relation. Lett. Math. Phys. **61**(2), 139–147 (2002)
12. Eilenberg, S., Mac Lane, S.: On the groups $h(\pi, n)$. Ann. Math. **58**(55), 55–106 (1953)
13. Fauvet, F., Menous, F.: Ecalle's arborification-coarborification transforms and Connes-Kreimer Hopf algebra. Ann. Sci. Éc. Norm. Supér. (4) **50**(1), 39–83 (2017)
14. Foissy, L.: Bidendriform bialgebras, trees, and free quasi-symmetric functions. J. Pure Appl. Algebra **209**(2), 439–459 (2007)
15. Foissy, L., Malvenuto, C.: The Hopf algebra of finite topologies and t-partitions. J. Algebra **438**, 130–169 (2015)
16. Foissy, L., Patras, F.: Natural endomorphisms of shuffle algebras. Int. J. Algebra Comput. **23**(4), 989–1009 (2013)
17. Foissy, L., Patras, F., Thibon, J.-Y.: Deformations of shuffles and quasi-shuffles. Ann. Inst. Fourier **66**, 209–237 (2016)
18. Gelfand, I., Krob, D., Lascoux, A., Leclerc, B., Retakh, V.S., Thibon, J.-Y.: Noncommutative symmetric functions. Adv. Math. **112**(2), 218–348 (1994)
19. Getzler, E., Jones, J.D.S.: Operads, homotopy algebra and iterated integrals for double loop spaces. arXiv preprint [hep-th/9403055] (1994)
20. Getzler, E., Jones, J.D.S.S: a_∞-algebras and the cyclic bar complex. Illinois J. Math **34**(2), 256–283 (1990)
21. Hivert, F., Novelli, J.-C., Thibon, J.-Y.: The algebra of binary search trees. Theor. Comput. Sci. **339**(1), 129–165 (2005)
22. Hivert, F., Novelli, J.-C., Thibon, J.-Y.: Trees, functional equations, and combinatorial Hopf algebras. Eur. J. Combinatorics **29**(7), 1682–1695 (2008)
23. Hoffman, M.E.: Quasi-shuffle products. J. Algebraic Combin. **11**(1), 49–68 (2000)
24. Livernet, M., Patras, F.: Lie theory for Hopf operads. J. Algebra **319**, 4899–4920 (2008)
25. Loday, J.-L.: Dialgebras. Dialgebras and related operads, 7–66 (2001)
26. Loday, J.-L.: On the algebra of quasi-shuffles. manuscripta mathematica **123**(1), 79–93 (2007)
27. Loday, J.-L., Ronco, M.: Hopf algebra of the planar binary trees. Adv. Math. **139**(2), 293–309 (1998)
28. Loday, J.-L., Ronco, M.: Trialgebras and families of polytopes. Homotopy theory: relations with algebraic geometry, group cohomology, and algebraic K-theory. Contemp. Math., Amer. Math. Soc. **346**, 369–398 (2004)
29. Malvenuto, C., Reutenauer, C.: Duality between quasi-symmetrical functions and the solomon descent algebra. J. Algebra **177**(3), 967–982 (1995)

30. Manchon, D., Ebrahimi-Fard, K.: The tridendriform structure of a discrete magnus expansion. Discr. and Cont. Dynamical Systems **34**(3), 1021–1040 (2014)
31. Novelli, J.-C., Patras, F., Thibon, J.-Y.: Natural endomorphisms of quasi-shuffle Hopf algebras. Bull. Soc. Math. France **141**(1), 107–130 (2013)
32. Novelli, J.-C., Thibon, J.-Y.: Polynomial realizations of some trialgebras. In: Proceedings of Formal Power Series and Algebraic Combinatorics, San Diego, California (2006)
33. Patras, F.: La décomposition en poids des algèbres de Hopf. Annales de l'institut Fourier **43**(4), 1067–1087 (1993)
34. Patras, F.: L'algèbre des descentes d'une bigèbre graduée. J. Algebra **170**(2), 547–566 (1994)
35. Patras, F., Reutenauer, C.: On descent algebras and twisted bialgebras. Mosc. Math. J. **4**(1), 199–216 (2004)
36. Patras, F., Schocker, M.: Twisted descent algebras and the Solomon-Tits algebra. Adv. Math. **199**(1), 151–184 (2006)
37. Patras, F., Schocker, Ma.: Trees, set compositions and the twisted descent algebra. J. Algebraic Combinatorics **28**(1), 3–23 (2008)
38. Reutenauer, Ch.: Free Lie algebras. Oxford University Press (1993)
39. Rota, G.-C.: Baxter algebras and combinatorial identities. i, ii. Bull. Amer. Math. Soc. (75), 325, 330 (1969)
40. Rota, G.-C.: Fluctuation theory and Baxter algebras. Istituto Nazionale di Alta Matematica, (IX), 325, 330 (1972)
41. Saidi, A.: The pre-Lie operad as a deformation of NAP. J. Algebra Appl. **13**(01), 1350076 (2014)
42. Schützenberger, M.P.: Sur une propriété combinatoire des algèbres de Lie libres pouvant être utilisée dans un problème de mathématiques appliquées. Séminaire Dubreil–Jacotin Pisot (Algèbre et théorie des nombres) (1958/59)

Galois Action on Knots II: Proalgebraic String Links and Knots

Hidekazu Furusho

Abstract We discuss an action of the Grothendieck-Teichmüller proalgebraic group on the linear span of proalgebraic tangles, oriented tangles completed by a filtration of Vassiliev. The action yields a motivic structure on tangles. We derive distinguished properties of the action particularly on proalgebraic string links and on proalgebraic knots which can not be observed in the action on proalgebraic braids. By exploiting the properties, we explicitly calculate the inverse image of the trivial (the chordless) chord diagram under the Kontsevich isomorphism.

Keywords Proalgebraic tangles · Chord diagrams · Grothendieck-Teichmüller group · Associators

1 Introduction

This paper is a continuation of our previous paper [22], where the action of the absolute Galois group on profinite knots was constructed by an action of the Grothendieck-Teichmüller *profinite* group $\widehat{GT}$ there. While in this paper, the action of the motivic Galois group, which is the Galois group of the tannakian category of mixed Tate motives over Spec $\mathbb{Z}$, on the linear span of proalgebraic tangles is deduced from an action of the Grothendieck-Teichmüller *proalgebraic* group $GT(\mathbb{K})$ ($\mathbb{K}$: a field of characteristic 0) there.

Proalgebraic tangles (Definition 3.3) means the $\mathbb{K}$-linear span of oriented tangles completed by a filtration à la Vassiliev. *Proalgebraic n-string links* (resp. *proalgebraic knots*) are proalgebraic analogues of n-string links (a good example can be found in Fig. 5) (resp. knots) and they are subspaces of proalgebraic tangles spanned by them. In Sect. 3, we give a $GT(\mathbb{K})$-action on proalgebraic tangles by following a method indicated in [28]. The action is interpreted as an extension of the

H. Furusho (✉)
Graduate School of Mathematics, Nagoya University, Chikusa-ku, Furo-cho, Nagoya 464-8602, Japan
e-mail: furusho@math.nagoya-u.ac.jp

J. I. Burgos Gil et al. (eds.), *Periods in Quantum Field Theory and Arithmetic*, Springer Proceedings in Mathematics & Statistics 314,
https://doi.org/10.1007/978-3-030-37031-2_20

$GT(\mathbb{K})$-action on the proalgebraic braids (reviewed in Sect. 2) into the one on the proalgebraic tangles. In Sect. 4 we derive distinguished properties of the $GT(\mathbb{K})$-action on proalgebraic tangles which can not be observed in $GT(\mathbb{K})$-action on proalgebraic braids. Particularly

Theorem A (Theorem 4.14 and Proposition 4.19) *Let $GT_1(\mathbb{K})$ be the unipotent part of $GT(\mathbb{K})$. Then*

(1) The $GT_1(\mathbb{K})$-action on proalgebraic string links is given by an inner conjugation.

(2) The $GT_1(\mathbb{K})$-action on proalgebraic knots is trivial, which yields a non-trivial decomposition (68) *of each oriented knot.*

They are derived by Twistor Lemma (Lemmas 4.2 and 4.12), which can be seen as reformulations of [2] Theorem 7.5, [3] Theorem 2, [30] Theorem 8 and [32] Theorem 2.1 in our setting, though all of which originate from [16] Theorem A′. The action on proalgebraic n-string links in Theorem A (1) is faithful for $n \geqslant 3$ and is not faithful (actually it is trivial) for $n = 1$ (see Remark 4.16). However the case for $n = 2$ remains unsolved (Problem 4.17). Since $GT(\mathbb{K})$ contains the motivic Galois group, our $GT(\mathbb{K})$-action on proalgebraic tangles yields the Galois action there, which yields a structure of mixed Tate motives of Spec $\mathbb{Z}$ there. Particularly Theorem A (2) says that proalgebraic knots can be decomposed into infinite summations of Tate motives.

By exploiting the properties shown in the above theorem, we explicitly determine the inverse image of the unit e, the chordless chord diagram on the oriented circle, under Kontsevich isomorphism I in (52):

Theorem B (Theorem 4.22) *Let c_0 be the proalgebraic knot which is the infinite summation of (topological) knots given in Fig.* 15. *Put*

$$\gamma_0 := \circlearrowleft - c_0 + c_0 \sharp c_0 - c_0 \sharp c_0 \sharp c_0 + c_0 \sharp c_0 \sharp c_0 \sharp c_0 - \cdots \tag{1}$$

where $\circlearrowleft$ is the trivial knot and $\sharp$ is the product called the connected sum. Then Kontsevich (knot) invariant of γ_0 becomes trivial, i.e. $I(\gamma_0) = e$. Namely

$$I^{-1}(e) = \gamma_0.$$

We recall that the image $I(\circlearrowleft)$ of the unit $\circlearrowleft$, the trivial knot, under the Kontsevich isomorphism I was calculated in [8]. Hence the above theorem could be regarded as a calculation in an opposite direction to their calculation.

The contents of the paper go as follows: Sect. 2 is a review on Drinfeld's tools of Grothendieck-Teichmüller groups and their actions on braids, which serves for a good understanding of the actions of the groups on proalgebraic tangles given in Sect. 3. Main results will be shown in Sect. 4.

Convention. In this paper, $\mathbb{K}$ means a commutative field with characteristic 0. (Actually we may more generally assume that it is a commutative ring containing the rational number field $\mathbb{Q}$.) The symbols $\mathbb{C}$, $\mathbb{R}$, $\mathbb{Q}_p$, $\mathbb{Z}_p$, $\mathbb{Z}$ and $\widehat{\mathbb{Z}}$ stand for the complex number field, the real number field, the p-adic number field (p: a prime), the p-adic integer ring, the integer ring and its profinite completion respectively.

2 Proalgebraic Braids and Infinitesimal Braids

This is an expository section for non-experts on Drinfeld's works on the action of the proalgebraic Grothendieck-Teichmüller groups on proalgebraic braids and also on infinitesimal braids and also a short review on a relationship of the groups with the motivic Galois group.

2.1 *The GT-Action*

We recall in Definition 2.2 explicitly the definition of the Grothendieck-Teichmüller group $GT(\mathbb{K})$, a proalgebraic group introduced by Drinfeld [16], and explain its action on the proalgebraic braids $\widehat{\mathbb{K}[B_n]}$ for $n \geqslant 2$ in Proposition 2.5.

Notation 2.1 (1) Let B_n be the *Artin braid group* with n-strings ($n \geqslant 2$) with standard generators σ_i ($1 \leqslant i \leqslant n-1$) and defining relations $\sigma_i\sigma_{i+1}\sigma_i = \sigma_{i+1}\sigma_i\sigma_{i+1}$ and $\sigma_i\sigma_j = \sigma_j\sigma_i$ for $|i-j| > 1$. The generator σ_i in B_n is depicted as in Fig. 1. And for b and $b' \in B_n$, we draw the product $b \cdot b' \in B_n$ as in Fig. 2 (the order of product $b \cdot b'$ is chosen to combine the bottom endpoints of b with the top endpoints of b').

We denote the pure part of B_n by P_n, i.e. the kernel of the natural homomorphism $P_n \to \mathfrak{S}_n$, and call it by the *pure braid group*. For $1 \leqslant i < j \leqslant n$, special elements

$$x_{i,j} = x_{j,i} = (\sigma_{j-1} \cdots \sigma_{i+1})\sigma_i^2(\sigma_{j-1} \cdots \sigma_{i+1})^{-1}$$

form a generating set of P_n. For $1 \leqslant a \leqslant a+\alpha < b \leqslant b+\beta \leqslant n$, we define

$$\begin{aligned} x_{a\cdots a+\alpha,b\cdots b+\beta} := &(x_{a,b}x_{a,b+1}\cdots x_{a,b+\beta}) \cdot (x_{a+1,b}x_{a+1,b+1}\cdots x_{a+1,b+\beta}) \\ &\cdots (x_{a+\alpha,b}x_{a+\alpha,b+1}\cdots x_{a+\alpha,b+\beta}) \in P_n. \end{aligned}$$

They are drawn in Figs. 3 and 4.

We mean $\widehat{\mathbb{K}[B_n]}$ by the completion of the group algebra $\mathbb{K}[B_n]$ with respect to the two-sided ideal I generated by $\sigma_i - \sigma_i^{-1}$ for $1 \leqslant i \leqslant n-1$;

$$\widehat{\mathbb{K}[B_n]} := \varprojlim_N \mathbb{K}[B_n]/I^N$$

(cf. [22]). By abuse of notation, we denote the induced filtration on $\widehat{\mathbb{K}[B_n]}$ by the same symbol $\{I^n\}_{n \geqslant 0}$. It is checked that $\widehat{\mathbb{K}[B_n]}$ is a filtered Hopf algebra. We call its group-like part by the *proalgebraic braid group* and denote it by $B_n(\mathbb{K})$. It naturally admits a structure of a proalgebraic group over $\mathbb{K}$. We note that it is Hain's [23] relative completion of B_n with respect to the natural projection $B_n \to \mathfrak{S}_n$.

Fig. 1 σ_i

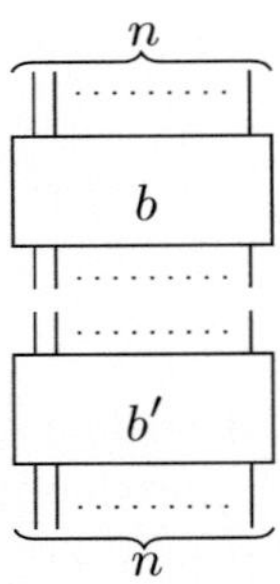

Fig. 2 $b \cdot b'$

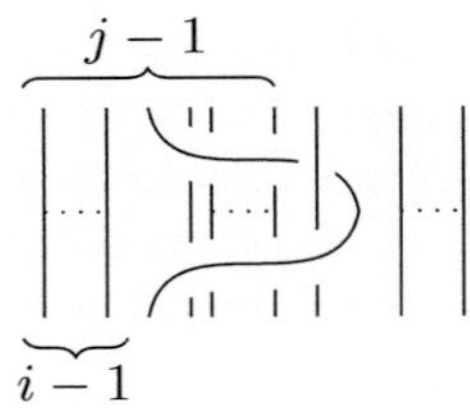

Fig. 3 x_{ij}

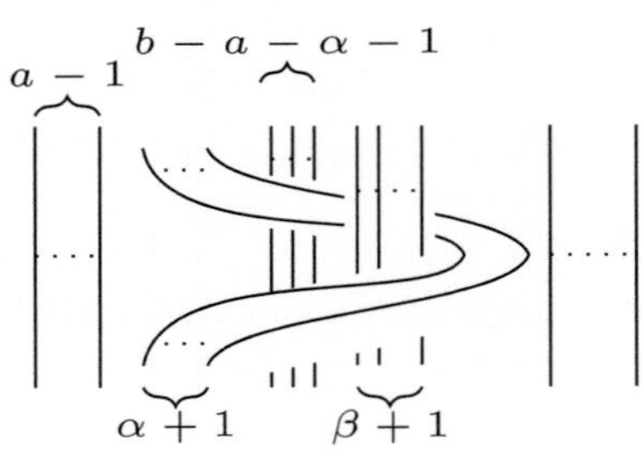

Fig. 4 $x_{a\cdots a+\alpha, b\cdots b+\beta}$

(2) Similarly we denote its pure part by $\widehat{\mathbb{K}[P_n]}$. Namely

$$\widehat{\mathbb{K}[P_n]} := \varprojlim_N \mathbb{K}[P_n]/I_0^N$$

with $I_0 = I \cap \mathbb{K}[P_n]$. It is also a filtered Hopf algebra. The *proalgebraic pure braid group* $P_n(\mathbb{K})$ means its group-like part. We note that it is a unipotent (Malcev) completion of P_n because I_0 forms an augmentation ideal of $\mathbb{K}[P_n]$.

(3) Let $F_2(\mathbb{K})$ be the prounipotent algebraic group over $\mathbb{K}$, the unipotent completion of the free group F_2 of rank 2 with two variables x and y, that is, the group-like part of the Hopf algebra $\widehat{\mathbb{K}[F_2]}$ completed by the augmentation ideal. Similarly to the convention in [22], for any $f \in F_2(\mathbb{K})$ and any homomorphism $\tau : F_2(\mathbb{K}) \to G(\mathbb{K})$ of proalgebraic groups sending $x \mapsto \alpha$ and $y \mapsto \beta$, the symbol $f(\alpha, \beta)$ stands for the image $\tau(f)$. Particularly for the (actually injective) homomorphism $F_2(\mathbb{K}) \to P_n(\mathbb{K})$ of proalgebraic groups sending $x \mapsto x_{a\cdots a+\alpha, b\cdots b+\beta}$ and $y \mapsto x_{b\cdots b+\beta, c\cdots c+\gamma}$ $(1 \leqslant a \leqslant a+\alpha < b \leqslant b+\beta < c \leqslant c+\gamma \leqslant n)$, the image of $f \in F_2(\mathbb{K})$ is denoted by $f_{a\cdots a+\alpha, b\cdots b+\beta, c\cdots c+\gamma}$.

The proalgebraic group $GT(\mathbb{K})$ is defined by Drinfeld [16] to be a subgroup of the automorphism (proalgebraic) group of $F_2(\mathbb{K})$.

Definition 2.2 ([16]) The *proalgebraic Grothendieck-Teichmüller group* $GT(\mathbb{K})$ is the proalgebraic group over $\mathbb{K}$ whose set of $\mathbb{K}$-values points forms a subgroup of $\mathrm{Aut} F_2(\mathbb{K})$ and is defined by

$$GT(\mathbb{K}) := \left\{ \sigma \in \mathrm{Aut} F_2(\mathbb{K}) \;\middle|\; \begin{array}{l} \sigma(x) = x^\lambda, \sigma(y) = f^{-1} y^\lambda f \\ \text{for some } (\lambda, f) \in \mathbb{K}^\times \times F_2(\mathbb{K}) \\ \text{satisfying the three relations below.} \end{array} \right\}$$

$$f(x, y) f(y, x) = 1 \quad \text{in } F_2(\mathbb{K}), \tag{2}$$

$$f(z, x) z^m f(y, z) y^m f(x, y) x^m = 1 \text{ in } F_2(\mathbb{K}) \text{ with } z = (xy)^{-1},\ m = \frac{\lambda - 1}{2}, \tag{3}$$

$$f_{1,2,34} f_{12,3,4} = f_{2,3,4} f_{1,23,4} f_{1,2,3} \quad \text{in } P_4(\mathbb{K}). \tag{4}$$

The powers x^λ, y^λ, x^m, y^m, z^m appearing in the above all make sense because $F_2(\mathbb{K})$ is the prounipotent completion of F_2. For $f_{1,2,34}$ etc., see Notation 2.1.

We remark that each $\sigma \in GT(\mathbb{K})$ determines a pair (λ, f) uniquely because the pentagon equation (4) implies that f belongs to the commutator of $F_2(\mathbb{K})$. By abuse of notation, we occasionally express the pair (λ, f) to represents σ and denote as $\sigma = (\lambda, f) \in GT(\mathbb{K})$.

The above set-theoretically defined $GT(\mathbb{K})$ forms indeed a proalgebraic group whose product is induced from that of $\mathrm{Aut} F_2(\mathbb{K})$ and is given by[1]

$$(\lambda_2, f_2) \circ (\lambda_1, f_1) := \left(\lambda_2\lambda_1, f_1(f_2 x^{\lambda_2} f_2^{-1}, y^{\lambda_2}) \cdot f_2\right) = \left(\lambda_2\lambda_1, f_2 \cdot f_1(x^{\lambda_2}, f_2^{-1} y^{\lambda_2} f_2)\right). \tag{5}$$

[1]For our purpose to make (6) not anti-homomorphic but homomorphic, we reverse the order of the product given in the original paper [16].

The first equality is the definition and the second equality can be easily verified. We denote the subgroup of $GT(\mathbb{K})$ with $\lambda = 1$ by $GT_1(\mathbb{K})$;

$$GT_1(\mathbb{K}) := \{\sigma = (\lambda, f) \in GT(\mathbb{K}) \mid \lambda = 1\}.$$

It is easily seen that it forms a proalgebraic unipotent subgroup of $GT(\mathbb{K})$.

Remark 2.3 In some literatures, (2), (3) and (4) are called *2-cycle, 3-cycle and 5-cycle relation* respectively. The author often calls (2) and (3) by *two hexagon equations* and (4) by *one pentagon equation* because they reflect the three axioms, two hexagon and one pentagon axioms, of braided monoidal (tensor) categories [25]. We remind that (4) represents

$$f(x_{12}, x_{23}x_{24})f(x_{13}x_{23}, x_{34}) = f(x_{23}, x_{34})f(x_{12}x_{13}, x_{24}x_{34})f(x_{12}, x_{23}) \quad \text{in } P_4(\mathbb{K}).$$

In several literatures such as [19, 24], the Eq. (4) is replaced by a different (more symmetric) formulation:

$$f(x_{12}^*, x_{23}^*)f(x_{34}^*, x_{45}^*)f(x_{51}^*, x_{12}^*)f(x_{23}^*, x_{34}^*)f(x_{45}^*, x_{51}^*) = 1 \quad \text{in } P_5^*(\mathbb{K})$$

where P_5^* is the pure sphere braid group with 5-strings.

The author actually showed that the pentagon equation implies two hexagon equations:

Proposition 2.4 ([21]) *Let $\mathbb{K}$ be an algebraically closed field of characteristic* 0. *For each $f \in F_2(\mathbb{K})$ satisfying* (4), *there always exists (actually unique up to signature) $\lambda \in \mathbb{K}$ such that the pair (λ, f) satisfies the two hexagon equations* (2) *and* (3).

The following Drinfeld's $GT(\mathbb{K})$-action on $\widehat{\mathbb{K}[B_n]}$ plays a fundamental role in our paper here.

Proposition 2.5 ([16]) *Let $n \geqslant 2$. There is a continuous $GT(\mathbb{K})$-action on the filtered Hopf algebra $\widehat{\mathbb{K}[B_n]}$*

$$\rho_n : GT(\mathbb{K}) \to \mathrm{Aut}\, \widehat{\mathbb{K}[B_n]} \tag{6}$$

which is induced by, for each $\sigma = (\lambda, f) \in GT(\mathbb{K})$,

$$\rho_n(\sigma) : \begin{cases} \sigma_1 & \mapsto \qquad \sigma_1^\lambda, \\ \sigma_i & \mapsto \quad f_{1\cdots i-1,i,i+1}^{-1} \; \sigma_i^\lambda f_{1\cdots i-1,i,i+1} \qquad (2 \leqslant i \leqslant n-1). \end{cases}$$

Here $\sigma_i^\lambda := \sigma_i \cdot (\sigma_i^2)^{\frac{\lambda-1}{2}} \in \widehat{\mathbb{K}[B_n]}$ and $f_{1\cdots i-1,i,i+1} = f(x_{1i}x_{2i}\cdots x_{i-1,i}, x_{i,i+1}) \in P_n(\mathbb{K})$ (see Notation 2.1). It is well defined because σ_i^2 belongs to the unipotent completion $P_n(\mathbb{K})$ and $\frac{\lambda-1}{2}$-th power of σ_i^2 makes sense in $P_n(\mathbb{K})$. We denote $\rho_n(\sigma)(b)$ simply by $\sigma(b)$ when there is no confusion.

We note that ρ_n is injective when $n \geqslant 3$.

Remark 2.6 For each prime l, there are natural homomorphisms $\widehat{B}_n \to B_n(\mathbb{Q}_l)$ and $\widehat{P}_n \to P_n(\mathbb{Q}_l)$. Hence we have

$$\widehat{B}_n \to \widehat{\mathbb{Q}_l[B_n]}.$$

By natural homomorphisms $\widehat{\mathbb{Z}} \to \mathbb{Q}_l$, $\widehat{F}_2 \to F_2(\mathbb{Q}_l)$ and $\widehat{P}_4 \to P_4(\mathbb{Q}_l)$, we obtain a continuous group homomorphism

$$\widehat{GT} \to GT(\mathbb{Q}_l) \tag{7}$$

from the profinite Grothendieck-Teichmüller group $\widehat{GT}$ (cf. [22]). By direct calculations, it can be verified that the above two homomorphisms are consistent with the $GT(\mathbb{Q}_l)$-action on $\widehat{\mathbb{Q}_l[B_n]}$ in Proposition 2.5 and the $\widehat{GT}$-action on $\widehat{B}_n$ (given in [16] and see also [22]).

In Sect. 4, we will extend the action on proalgebraic braids to the one on proalgebraic tangles and will show that actually the above action on proalgebraic pure braid groups is realized as an inner automorphism of proalgebraic string links (Corollary 4.15).

2.2 The GRT-Action

We recall explicitly Drinfeld's definitions of the *graded* Grothendieck-Teichmüller group $GRT(\mathbb{K})$ in Definition 2.9. We discuss their actions on the algebra $\widehat{U\mathfrak{b}_n}$ of infinitesimal braids for $n \geqslant 2$ in Proposition 2.12.

Notation 2.7 (1) Let $\mathfrak{p}_n$ be the *infinitesimal pure braid Lie algebra* with n-strings ($n \geqslant 2$) with standard generators t_{ij} ($1 \leqslant i, j \leqslant n$) and defining relations $t_{ii} = 0$, $t_{ij} = t_{ji}$, $[t_{ij}, t_{ik} + t_{jk}] = 0$ and $[t_{ij}, t_{kl}] = 0$ when i, j, k, l all differ. For $1 \leqslant a \leqslant a + \alpha < b \leqslant b + \beta \leqslant n$, we define

$$t_{a\cdots a+\alpha, b\cdots b+\beta} := \sum_{0 \leqslant i \leqslant \alpha, 0 \leqslant j \leqslant \beta} t_{a+i, b+j} \in \mathfrak{p}_n.$$

We denote by $U\mathfrak{p}_n$ its enveloping algebra and by $\widehat{U\mathfrak{p}_n}$ its completion with respect to its augmentation ideal.

(2) We denote $\mathfrak{S}_n$ to be the symmetric group acting on $\{1, 2, \ldots, n\}$ ($n \geqslant 1$) and $\tau_{i,i+1}$ in $\mathfrak{S}_n$ to be the transpose of i and $i+1$ ($1 \leqslant i \leqslant n-1$). The group $\mathfrak{S}_n$ acts $\widehat{U\mathfrak{p}_n}$ by $\tau \cdot t_{ij} = t_{\tau^{-1}(i), \tau^{-1}(j)}$ for $\tau \in \mathfrak{S}_n$ and $1 \leqslant i, j \leqslant n$. We may consider the crossed product (cf. [33] etc.)

$$\widehat{U\mathfrak{b}_n} := \mathbb{K}[\mathfrak{S}_n] * \widehat{U\mathfrak{p}_n}.$$

It is $\mathbb{K}[\mathfrak{S}_n] \otimes_{\mathbb{K}} \widehat{U\mathfrak{p}_n}$ as vector space with the product structure given by

$$(\tau_1 \otimes t_{i_1 j_1}) \cdot (\tau_2 \otimes t_{i_2 j_2}) := \tau_1\tau_2 \otimes (t_{\tau_2^{-1}(i_1)\tau_2^{-1}(j_1)} \cdot t_{i_2 j_2})$$

for $\tau_1, \tau_2 \in \mathfrak{S}_n$. By abuse of notation, occasionally τ indicate $\tau \otimes 1$ for $\tau \in \mathbb{K}[\mathfrak{S}_n]$ and t indicate $1 \otimes t$ for $t \in \widehat{U\mathfrak{p}_n}$ in this paper. Hence we have

$$\tau \cdot t_{ij} = t_{\tau(i)\tau(j)} \cdot \tau \quad (=\tau \otimes t_{ij}) \tag{8}$$

for $\tau \in \mathfrak{S}_n$. We occasionally depict $t_{ij} \in \widehat{U\mathfrak{p}_n}$ as the diagram with n vertical lines and a dotted horizontal line (called a *chord*) connecting i-th and j-th lines and $\tau \in \mathfrak{S}_n$ as the diagram connecting each i-th point on the bottom with $\tau(i)$-th point on the top by an interval. The order of the product $b \cdot b'$ is chosen to combine the bottom endpoints of b with the top endpoints of b'. By putting $\deg t_{ij} = 1$ $(1 \leqslant i, j \leqslant n)$ and $\deg \tau = 0$ $(\tau \in \mathfrak{S}_n)$, we can show that both $\widehat{U\mathfrak{p}_n}$ and $\widehat{U\mathfrak{b}_n}$ carry structures of graded Hopf algebras.

(3) Let $\mathfrak{f}_2$ be the free Lie algebra over $\mathbb{K}$ with two variables A and B and $\widehat{U\mathfrak{f}_2}$ be its completed Hopf algebra. Again similarly, for any $g \in \widehat{U\mathfrak{f}_2}$ and any algebra homomorphism $\tau : \widehat{U\mathfrak{f}_2} \to S$ sending $A \mapsto v$ and $B \mapsto w$, the symbol $g(v, w)$ stands for the image $\tau(g)$. Particularly for the (actually injective) homomorphism $\widehat{U\mathfrak{f}_2} \to \widehat{U\mathfrak{p}_n}$ sending $A \mapsto t_{a\cdots a+\alpha, b\cdots b+\beta}$ and $B \mapsto t_{b\cdots b+\beta, c\cdots c+\gamma}$ $(1 \leqslant a \leqslant a+\alpha < b \leqslant b+\beta < c \leqslant c+\gamma \leqslant n)$, the image of $g \in \widehat{U\mathfrak{f}_2}$ is denoted by $g_{a\cdots a+\alpha, b\cdots b+\beta, c\cdots c+\gamma} \in \widehat{U\mathfrak{p}_n}$.

We note that $\widehat{U\mathfrak{p}_n}$ is not algebraically generated by $t_{i,i+1}$ $(1 \leqslant i \leqslant n-1)$ while the following holds for $\widehat{U\mathfrak{b}_n}$.

Lemma 2.8 *The algebra $\widehat{U\mathfrak{b}_n}$ is algebraically generated by $t_{i,i+1}$ and $\tau_{i,i+1}$ for $1 \leqslant i \leqslant n-1$.*

Proof The elements $\tau_{i,i+1}$ generate $\mathfrak{S}_n$ and, by (8), any t_{kl} is obtained from $t_{i,i+1}$ and $\tau_{i,i+1}$ $(1 \leqslant i \leqslant n-1)$, which yields our claim.

The proalgebraic group $GRT_1(\mathbb{K})$ is defined by Drinfeld [16] to be a proalgebraic subgroup of the automorphism (proalgebraic) group of $\exp \mathfrak{f}_2$.

Definition 2.9 ([16]) The *proalgebraic graded Grothendieck-Teichmüller group* $GRT(\mathbb{K})$ is the subgroup of $\mathrm{Aut} \exp \mathfrak{f}_2$ defined by

$$GRT(\mathbb{K}) := \left\{ \sigma \in \mathrm{Aut}\exp\mathfrak{f}_2 \;\middle|\; \begin{array}{l} \sigma(e^A) = e^{A/c},\ \sigma(e^B) = g^{-1}e^{B/c}g \text{ for some } c \in \mathbb{K}^\times \\ \text{and } g \in \exp\mathfrak{f}_2 \text{ satisfying two hexagon equations} \\ \text{(9)-(10) and one pentagon equation (11) below.} \end{array} \right\}$$

$$g(A, B)g(B, A) = 1 \quad \text{in } \exp\mathfrak{f}_2, \tag{9}$$

$$g(C,A)g(B,C)g(A,B)=1 \quad \text{in } \exp\mathfrak{f}_2 \text{ with } C=-A-B, \tag{10}$$

$$g_{1,2,34}g_{12,3,4}=g_{2,3,4}g_{1,23,4}g_{1,2,3} \quad \text{in } \exp\mathfrak{p}_4. \tag{11}$$

Similarly to Definition 2.2, we remark that each $\sigma \in GRT(\mathbb{K})$ determines a pair (c, g) uniquely. By abuse of notation, we occasionally express the pair (c, g) to represents $\sigma \in GRT(\mathbb{K})$ and denote as $\sigma = (c, g) \in GRT(\mathbb{K})$.

The above set-theoretically defined $GRT(\mathbb{K})$ forms indeed a proalgebraic group whose product is induced from that of $\mathrm{Aut}\exp\mathfrak{f}_2(\mathbb{K})$ and is given by[2]

$$(c_2,g_2)\circ(c_1,g_1)=\left(c_2c_1,\ g_1\left(g_2\frac{A}{c_2}g_2^{-1},\ \frac{B}{c_2}\right)\cdot g_2\right)=\left(c_2c_1,\ g_2\cdot g_1\left(\frac{A}{c_2},\ g_2^{-1}\frac{B}{c_2}g_2\right)\right). \tag{12}$$

The first equality is the definition and the second equality can be easily verified. Notice the simple equality $(1, g)\circ(c, 1) = (c, g)$. We denote the subgroup of $GRT(\mathbb{K})$ with $c = 1$ by $GRT(\mathbb{K})_1$, i.e. $GRT_1(\mathbb{K}) := \{\sigma = (c, g) \in GRT(\mathbb{K}) \mid c = 1\}$. It is easily seen that it forms a proalgebraic unipotent subgroup of $GRT(\mathbb{K})$.

Remark 2.10 The symbol GRT stands for 'graded Grothendieck-Teichmüller group.' Indeed its grading on $GRT_1(\mathbb{K})$ is equipped by the action of $\mathbb{G}_m(\mathbb{K})$ $(=\mathbb{K}^\times)$ given by

$$(1, g(A,B)) \mapsto \left(1,\ g(\frac{A}{c},\frac{B}{c})\right) \tag{13}$$

for $c \in \mathbb{K}^\times$ and $g \in GRT_1(\mathbb{K})$, which is reformulated by $(1, g) \mapsto (c, 1)\circ(1, g)$. Thus we have, by the action,

$$GRT(\mathbb{K}) = \mathbb{K}^\times \ltimes GRT_1(\mathbb{K}).$$

Two specific elements of $GRT(\mathbb{K})$ are known.

Example 2.11 (1) The *p-adic Drinfeld associator* $\Phi^p_{\mathrm{KZ}}(A, B)$, a p-adic analogue of the Drinfeld (KZ-)associator (cf. Example 2.16) is a non-commutative formal power series whose coefficients are p-adic multiple zeta values [18]. It was constructed as a regularized holonomy of the p-adic KZ-equation and was shown in [20] by the results of [38] that it belongs to $GRT_1(\mathbb{K})$ with $\mathbb{K} = \mathbb{Q}_p$.

(2) The *p-adic Deligne associator* $\Phi^p_{\mathrm{De}}(A, B)$, a variant of the above $\Phi^p_{\mathrm{KZ}}(A, B)$ (cf. [20]) is shown in [38] to be in $GRT_1(\mathbb{K})$ with $\mathbb{K} = \mathbb{Q}_p$. It was in [20] shown that each of its coefficients is given by a certain polynomial combination of p-adic multiple zeta values.

The following $GRT(\mathbb{K})$-action on $\widehat{U\mathfrak{b}_n}$ was explicitly presented neither in Drinfeld's paper [16] nor Bar-Natan's paper [7], where they showed $GRT_1(\mathbb{K})$-action there.

[2]We remark again that, for our purpose, we reverse the order of the product given in [16].

Proposition 2.12 *Let $n \geqslant 2$. There is a continuous $GRT(\mathbb{K})$-action on the graded Hopf algebra $\widehat{U\mathfrak{b}_n}$*

$$\rho_n : GRT(\mathbb{K}) \to \operatorname{Aut} \widehat{U\mathfrak{b}_n} \tag{14}$$

which is induced by, for each $\sigma = (c, g) \in GRT(\mathbb{K})$,

$$\rho_n(\sigma) : \begin{cases} t_{1,2} \mapsto & \frac{t_{1,2}}{c}, \\ t_{i,i+1} \mapsto & g^{-1}_{1\cdots i-1,i,i+1} \ \frac{t_{i,i+1}}{c} \ g_{1\cdots i-1,i,i+1} \quad (2 \leqslant i \leqslant n-1), \\ \tau_{1,2} \mapsto & \tau_{1,2}, \\ \tau_{i,i+1} \mapsto & g^{-1}_{1\cdots i-1,i,i+1} \ \tau_{i,i+1} \ g_{1\cdots i-1,i,i+1} \quad (2 \leqslant i \leqslant n-1). \end{cases}$$

We recall that $\tau_{i,i+1}$ means the transpose of i and $i+1$ in $\mathfrak{S}_n$. We again note that ρ_n is injective when $n \geqslant 3$.

In Sect. 4 we will extend the above action on infinitesimal braids to the one on chord diagrams and will show that actually the above action on infinitesimal braids is realized as an inner automorphism of chord diagrams.

The associated Lie algebra $\mathfrak{grt}_1$ of GRT_1, which was independently introduced by Ihara [24] and called the stable derivation algebra, is equipped grading by the $\mathbb{G}_m$-action (13).

Conjecture 2.13 ([14, 16, 24]) *The graded Lie algebra $\mathfrak{grt}_1$ is freely generated by one element in each degree* $3, 5, 7, \ldots$.

Remark 2.14 By [10], we know that $\mathfrak{grt}_1$ contains such a free Lie subalgebra (see Remark 2.23 below).

2.3 Associators

We recall Drinfeld's definition of the associator set $M(\mathbb{K})$ in Definition 2.15 and its $(GRT(\mathbb{K}), GT(\mathbb{K}))$-bitorsor structure in Proposition 2.18. Then we will explain how associators give isomorphisms between $\widehat{\mathbb{K}[B_n]}$ and $\widehat{U\mathfrak{b}_n}$ in Proposition 2.19.

Definition 2.15 ([16]) The *associator set* $M(\mathbb{K})$ is the proalgebraic variety whose set of $\mathbb{K}$-valued points is given by

$$M(\mathbb{K}) := \left\{ p = (\mu, \varphi) \in \mathbb{K} \times \exp \mathfrak{f}_2 \;\middle|\; \mu \in \mathbb{K}^\times \text{ and } (\mu, \varphi) \text{ satisfies (9), (11) and (15).} \right\}$$

$$\exp\{\frac{\mu A}{2}\}\varphi(C, A) \exp\{\frac{\mu C}{2}\}\varphi(B, C) \exp\{\frac{\mu B}{2}\}\varphi(A, B) = 1 \quad \text{in } \exp \mathfrak{f}_2 \tag{15}$$

with $C = -A - B$. For each fixed $\mu_0 \in \mathbb{K}$, define the proalgebraic variety $M_{\mu_0}(\mathbb{K})$ by

$$M_{\mu_0}(\mathbb{K}) := \left\{\varphi \in \exp \mathfrak{f}_2 \;\middle|\; \varphi \text{ satisfies (9), (11) and (15) with } \mu = \mu_0\right\}.$$

Hence we have $M_0(\mathbb{K}) = GRT_1(\mathbb{K})$. Three examples of associators are known:

Example 2.16 (1) The *KZ-associator*, also known as the Drinfeld associator, $\Phi_{KZ}(A, B)$ is a non-commutative formal power series whose coefficients are multiple zeta values. It was constructed by Drinfeld [16] as a regularized holonomy of the KZ-equation and was shown by him that it belongs to $M_\mu(\mathbb{K})$ with $\mathbb{K} = \mathbb{C}$ and $\mu = \pm 2\pi\sqrt{-1}$. It is known to be expressed as follows:

$$\Phi_{KZ}(A, B) = 1 + \sum_{\substack{m,k_1,\dots,k_m \in \mathbb{N} \\ k_m > 1}} (-1)^m \zeta(k_1, \cdots, k_m) A^{k_m - 1} B \cdots A^{k_1 - 1} B \qquad (16)$$
$$+ \text{(regularized terms)}.$$

Here $\zeta(k_1, \cdots, k_m)$ is the *multiple zeta value* (MZV in short), the real number defined by the following power series

$$\zeta(k_1, \cdots, k_m) := \sum_{0 < n_1 < \cdots < n_m} \frac{1}{n_1^{k_1} \cdots n_m^{k_m}}$$

for m, $k_1, \dots, k_m \in \mathbb{N}(= \mathbb{Z}_{>0})$ with $k_m > 1$ (its convergent condition). All of the coefficients of Φ_{KZ} (including its regularized terms) are explicitly calculated in terms of MZV's in [17] Proposition 3.2.3 by Le-Murakami's method in [31].

(2) The *Deligne associator* $\Phi_{De}(A, B)$ [11] (denoted by $\Phi_{KZ}^{-}(A, B)$ in [20]) is a non-commutative formal power series in $M_\mu(\mathbb{K})$ with $\mathbb{K} = \mathbb{R}$ and $\mu = 1$ which is located in the 'middle' of $\Phi_{KZ}\left(\frac{1}{2\pi\sqrt{-1}}A, \frac{1}{2\pi\sqrt{-1}}B\right)$ and $\Phi_{KZ}\left(\frac{-1}{2\pi\sqrt{-1}}A, \frac{-1}{2\pi\sqrt{-1}}B\right)$. Its explicit relationship with the above $\Phi_{KZ}(A, B)$ is given in [20] Lemma 2.25.

(3) The *AT-associator* $\Phi_{AT}(A, B)$ is another associator. It was constructed by Alekseev and Torossian [1] as a holonomy of AT-connection, a certain non-holomorphic flat connection on a certain configuration space. Ševera and Willwacher [36] showed that it belongs to $M_\mu(\mathbb{K})$ with $\mathbb{K} = \mathbb{R}$ and $\mu = 1$. Rossi and Willwacher showed that $\Phi_{AT} \neq \Phi_{De}$ in [35].

We remark again that the author also in this setting showed that the pentagon equation implies two hexagon equations.

Proposition 2.17 ([22]) *Let $\mathbb{K}$ be an algebraically closed field of characteristic* 0. *For each $\varphi \in \exp \mathfrak{f}_2$ satisfying* (11), *there always exists (actually unique up to signature) $\mu \in \mathbb{K}$ such that the pair $p = (\mu, \varphi)$ satisfies the two hexagon equations* (9) *and* (15).

It was shown by Drinfeld that $GRT(\mathbb{K})$ acts freely and transitively on $M(\mathbb{K})$ from the left, $GT(\mathbb{K})$ acts freely and transitively on $M(\mathbb{K})$ from the right, and these two actions are commutative:

Proposition 2.18 ([16]) *The associator set $M(\mathbb{K})$ forms a $(GRT(\mathbb{K}), GT(\mathbb{K}))$-bitorsor by the left $GRT(\mathbb{K})$-action given by*

$$(c,g)\circ(\mu,\varphi) := \left(\frac{\mu}{c},\ \varphi\left(g\frac{A}{c}g^{-1},\ \frac{B}{c}\right)\cdot g\right) = \left(\frac{\mu}{c},\ g\cdot\varphi\left(\frac{A}{c},\ g^{-1}\frac{B}{c}g\right)\right)$$

for $(c,g)\in GRT(\mathbb{K})$ and $(\mu,\varphi)\in M(\mathbb{K})$ and the right $GT(\mathbb{K})$-action given by

$$(\mu,\varphi)\circ(\lambda,f) := \left(\lambda\mu,\ f(\varphi e^{\mu A}\varphi^{-1}, e^{\mu B})\cdot\varphi\right) = \left(\lambda\mu,\ \varphi\cdot f(e^{\mu A}, \varphi^{-1}e^{\mu B}\varphi)\right)$$

for $(\mu,\varphi)\in M(\mathbb{K})$ and $(\lambda,f)\in GT(\mathbb{K})$.

We must note again that we reverse the order of the product given in the paper [16] for our purpose. Drinfeld [16] showed that associators give an isomorphism between $\widehat{\mathbb{K}[B_n]}$ and $\widehat{U\mathfrak{b}_n}$.

Proposition 2.19 ([16]) *Let $n\geqslant 2$. The $(GRT(\mathbb{K}), GT(\mathbb{K}))$-bitorsor $M(\mathbb{K})$ is mapped to the $\left(\mathrm{Aut}(\widehat{U\mathfrak{b}_n}),\ \mathrm{Aut}(\widehat{\mathbb{K}[B_n]})\right)$-bitorsor $\mathrm{Isom}(\widehat{\mathbb{K}[B_n]},\ \widehat{U\mathfrak{b}_n})$ by the map*

$$\rho_n : M(\mathbb{K}) \to \mathrm{Isom}\left(\widehat{\mathbb{K}[B_n]},\ \widehat{U\mathfrak{b}_n}\right) \tag{17}$$

induced by, for each $p=(\mu,\varphi)$,

$$\rho_n(p) : \begin{cases} \sigma_1 \mapsto \tau_{1,2}\cdot\exp\left\{\frac{\mu t_{12}}{2}\right\}, \\ \sigma_i \mapsto \varphi^{-1}_{1\cdots i-1,i,i+1}\cdot\tau_{i,i+1}\cdot\exp\left\{\frac{\mu t_{i,i+1}}{2}\right\}\cdot\varphi_{1\cdots i-1,i,i+1} \\ \qquad (2\leqslant i\leqslant n-1). \end{cases}$$

It is a morphism as bitorsors, i.e. it is compatible with (6) *and* (14).

We again note that ρ_n is injective when $n\geqslant 3$.

2.4 The Motivic Galois Group

We briefly review the formulations of the motivic Galois groups and their torsor (consult also [4] as a nice exposition). We also review their relationship with the torsor of the Grothendieck-Teichmüller groups discussed in our previous subsections.

The triangulated category $DM(\mathbb{Q})_{\mathbb{Q}}$ of *mixed motives* over $\mathbb{Q}$ (a part of an idea of mixed motives is explained in [14] Sect. 1) was constructed by Hanamura, Levine and Voevodsky. *Tate motives* $\mathbb{Q}(n)$ ($n\in\mathbb{Z}$) are (Tate) objects of the category. Let $DMT(\mathbb{Q})_{\mathbb{Q}}$ be the triangulated sub-category of $DM(\mathbb{Q})_{\mathbb{Q}}$ generated by Tate motives $\mathbb{Q}(n)$ ($n\in\mathbb{Z}$). By the work of Levine a neutral tannakian $\mathbb{Q}$-category $MT(\mathbb{Q}) = MT(\mathbb{Q})_{\mathbb{Q}}$ of *mixed Tate motives over* $\mathbb{Q}$ is extracted by taking the heart with respect to a t-structure of $DMT(\mathbb{Q})_{\mathbb{Q}}$. Deligne and Goncharov [15] introduced the full subcategory $MT(\mathbb{Z}) = MT(\mathbb{Z})_{\mathbb{Q}}$ of *mixed Tate motives over* Spec $\mathbb{Z}$ inside of $MT(\mathbb{Q})_{\mathbb{Q}}$. The category $MT(\mathbb{Z})$ forms a neutral tannakian $\mathbb{Q}$-category and association of each object $M \in MT(\mathbb{Z})$ with the underlying $\mathbb{Q}$-linear space of its Betti and de Rham realizations give the fiber functor ω_{Be} and ω_{DR} respectively.

Definition 2.20 For $* \in \{\mathrm{Be}, \mathrm{DR}\}$, the *motivic Galois group* $\mathrm{Gal}_*^{\mathcal{M}}(\mathbb{Z})$ is defined to be the corresponding tannakian fundamental group of $MT(\mathbb{Z})$, that is, the pro-$\mathbb{Q}$-algebraic group defined by $\underline{\mathrm{Aut}}^{\otimes}(MT(\mathbb{Z}) : \omega_*)$.

For $*, *' \in \{\mathrm{Be}, \mathrm{DR}\}$, we denote the corresponding tannakian fundamental torsor $\underline{\mathrm{Isom}}^{\otimes}(MT(\mathbb{Z}) : \omega_*, \omega_{*'})$ by $\mathrm{Gal}_{*,*'}^{\mathcal{M}}(\mathbb{Z})$. This is a $(\mathrm{Gal}_{*'}^{\mathcal{M}}(\mathbb{Z}), \mathrm{Gal}_*^{\mathcal{M}}(\mathbb{Z}))$-bitorsor. We note that $\mathrm{Gal}_{*,*}^{\mathcal{M}}(\mathbb{Z}) = \mathrm{Gal}_*^{\mathcal{M}}(\mathbb{Z})$. By the fundamental theorem of tannakian category theory, each fiber functor ω_* induces an equivalence of categories

$$MT(\mathbb{Z}) \simeq \mathrm{Rep}\,\mathrm{Gal}_*^{\mathcal{M}}(\mathbb{Z}) \tag{18}$$

where the right hand side stands for the category of finite dimensional $\mathbb{Q}$-vector spaces equipped with $\mathrm{Gal}_*^{\mathcal{M}}(\mathbb{Z})$-action.

Remark 2.21 For $*, *' \in \{\mathrm{Be}, \mathrm{DR}\}$, the action of $\mathrm{Gal}_*^{\mathcal{M}}(\mathbb{Z})$ on $\omega_*(\mathbb{Q}(1)) \simeq \mathbb{Q}$ defines a surjection $\mathrm{Gal}_*^{\mathcal{M}}(\mathbb{Z}) \to \mathbb{G}_m$ and its kernel $\mathrm{Gal}_*^{\mathcal{M}}(\mathbb{Z})_1$ is the unipotent radical of $\mathrm{Gal}_*^{\mathcal{M}}(\mathbb{Z})$. For $* = \mathrm{DR}$, there is a natural splitting $\tau : \mathbf{G}_m \to \mathrm{Gal}_{\mathrm{DR}}^{\mathcal{M}}(\mathbb{Z})$ which gives a negative grading on its associated Lie algebra $\mathrm{Lie}\mathrm{Gal}_{\mathrm{DR}}^{\mathcal{M}}(\mathbb{Z})_1$. By the axiom on the structure of the category $MT(\mathbb{Z})$, it is known that the Lie algebra is the graded Lie algebra *freely* generated by one element in each degree $-3, -5, -7, \ldots$. (consult [14] Sect. 8 for the full story).

The *motivic fundamental group* $\pi_1^{\mathcal{M}}(\mathbb{P}^1\backslash\{0, 1, \infty\} : \overrightarrow{01})$ constructed in [15] Sect. 4 is a (pro-)object of $MT(\mathbb{Z})$. The KZ-associator (cf. Example 2.16) is essential in describing the Hodge realization of the motive (cf. [4, 15, 20]). Since its Betti and de Rham realization is given by $F_2(\mathbb{Q})$ and $\exp \mathfrak{f}_2$, the motivic Galois groups, $\mathrm{Gal}_{\mathrm{Be}}^{\mathcal{M}}(\mathbb{Z})$ and $\mathrm{Gal}_{\mathrm{DR}}^{\mathcal{M}}(\mathbb{Z})$, acts there respectively. The tannakian equivalence (18) induces a morphism of bitorsors

$$\Psi : \mathrm{Gal}_{\mathrm{Be},\mathrm{DR}}^{\mathcal{M}}(\mathbb{Z}) \to \mathrm{Isom}(F_2(\mathbb{Q}), \exp \mathfrak{f}_2)$$

from the $(\mathrm{Gal}_{\mathrm{DR}}^{\mathcal{M}}(\mathbb{Z}), \mathrm{Gal}_{\mathrm{Be}}^{\mathcal{M}}(\mathbb{Z}))$-bitorsor $\mathrm{Gal}_{\mathrm{Be},\mathrm{DR}}^{\mathcal{M}}(\mathbb{Z})$ to the $(\mathrm{Aut}\exp \mathfrak{f}_2, \mathrm{Aut} F_2(\mathbb{Q}))$-bitorsor $\mathrm{Isom}(F_2(\mathbb{Q}), \exp \mathfrak{f}_2)$. The following has been conjectured (Deligne-Ihara conjecture) and finally proved by Brown by using Zagier's relation on MZV's.

Theorem 2.22 ([10]) *The map Ψ is injective.*

It is a proalgebraic group analogue of the so-called Belyĭ's theorem [9] in the profinite group setting. The theorem says that all unramified mixed Tate motives over Spec $\mathbb{Z}$ are associated with MZV's.

Remark 2.23 We recall that our $(GRT(\mathbb{Q}), GT(\mathbb{Q}))$-bitorsor $M(\mathbb{Q})$ is naturally injected to the $(\mathrm{Aut}\exp\mathfrak{f}_2, \mathrm{Aut}F_2(\mathbb{Q}))$-bitorsor $\mathrm{Isom}(F_2(\mathbb{Q}), \exp\mathfrak{f}_2)$:

$$M(\mathbb{Q}) \hookrightarrow \mathrm{Isom}(F_2(\mathbb{Q}), \exp\mathfrak{f}_2).$$

As is explained in [4, 20], a certain geometric interpretations of the Grothendieck-Teichmüller groups shows that $\mathrm{Im}\Psi$ is injected in $M(\mathbb{Q})$ as bitorsors. Thus by the above theorem, $(\mathrm{Gal}^{\mathcal{M}}_{\mathrm{DR}}(\mathbb{Z}), \mathrm{Gal}^{\mathcal{M}}_{\mathrm{Be}}(\mathbb{Z}))$-bitorsor $\mathrm{Gal}^{\mathcal{M}}_{\mathrm{Be,DR}}(\mathbb{Z})$ is mapped injectively to $(GRT(\mathbb{Q}), GT(\mathbb{Q}))$-bitorsor $M(\mathbb{Q})$ as bitorsors:

$$\mathrm{Gal}^{\mathcal{M}}_{\mathrm{Be,DR}}(\mathbb{Z}) \hookrightarrow M(\mathbb{Q}). \tag{19}$$

The inclusion induces the one from $\mathrm{LieGal}^{\mathcal{M}}_{\mathrm{DR}}(\mathbb{Z})_1$ to $GRT(\mathbb{Q})$. By Remark 2.21 we get the claim in Remark 2.14. The $GT(\mathbb{Q})$-action on $\widehat{\mathbb{Q}[B_n]}$ given in (6) induces a $\mathrm{Gal}^{\mathcal{M}}_{\mathrm{Be}}(\mathbb{Z})$-action there and $GRT(\mathbb{Q})$-action on $\widehat{U\mathfrak{b}_n}$ given in (14) also induces a $\mathrm{Gal}^{\mathcal{M}}_{\mathrm{DR}}(\mathbb{Z})$-action there. Hence by the equivalence (18), $\widehat{\mathbb{Q}[B_n]}$ is the Betti realization of a certain mixed Tate (pro-)motive over Spec $\mathbb{Z}$, while whose de Rham realization is given by $\widehat{U\mathfrak{b}_n}$.

3 Proalgebraic Tangles and Chord Diagrams

We develop the actions of the Grothendieck-Teichmüller groups on proalgebraic braids and on infinitesimal braids explained in our previous section into the ones on proalgebraic tangles and on chord diagrams by following the method indicated in [28]. This section might be regarded as an extension of Bar-Natan's formalism [7] on a relationship of the Grothendieck-Teichmüller groups with proalgebraic braids into their relationship with proalgebraic tangles.

3.1 The GT-Action

In this subsection we give a review but with more detailed considerations on the last appendix of both [27, 28] where an interesting $GT(\mathbb{K})$-action on proalgebraic tangles and knots are briefly explained. Proalgebraic tangles and proalgebraic knots are recalled in Definition 3.3. They are shown in Proposition 3.7 to be described by proalgebraic pre-tangles (and knots) introduced by our ABC-construction in Definition 3.5. The $GT(\mathbb{K})$-action on proalgebraic tangles is explained in Definition 3.9 and Proposition 3.11. The induced $GT(\mathbb{K})$-action on proalgebraic knots is discussed in Proposition 3.13. In Proposition 3.17 we give a relationship of the $GT(\mathbb{K})$-action on the proalgebraic knots with the profinite $\widehat{GT}$-action on profinite knots which was constructed in our previous paper [22].

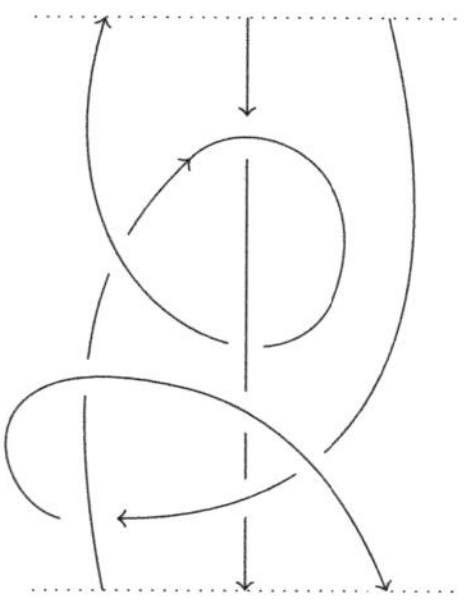

Fig. 5 A string link

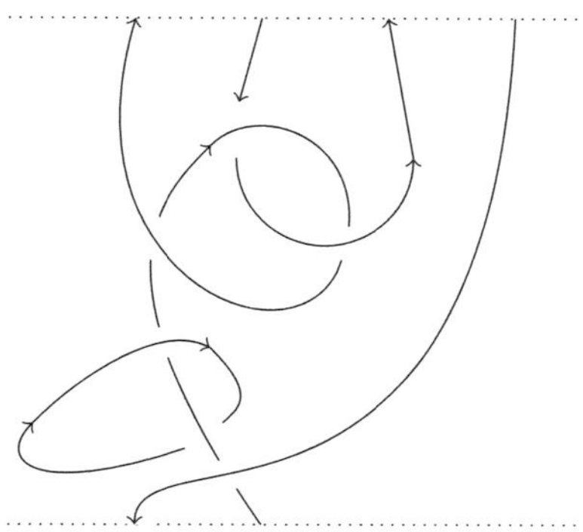

Fig. 6 A tangle in $\mathcal{T}_{\uparrow\downarrow\uparrow\downarrow,\downarrow\uparrow}$

Notation 3.1 Let $k, l \geqslant 0$. Let $\epsilon = (\epsilon_1, \ldots, \epsilon_k)$ and $\epsilon' = (\epsilon'_1, \ldots, \epsilon'_l)$ be sequences (including the empty sequence $\emptyset$) of symbols $\uparrow$ and $\downarrow$. An (oriented)[3] *tangle* of type (ϵ, ϵ') means a smooth embedded compact oriented one-dimensional real manifolds in $[0, 1] \times \mathbb{C}$ (hence it is a finite disjoint union of embedded one-dimensional intervals and circles), whose boundaries are $\{(1, 1), \ldots, (1, k), (0, 1) \ldots, (0, l)\}$ such that ϵ_i (resp. ϵ'_j) is $\uparrow$ or $\downarrow$ if the tangle is oriented upwards or downwards at $(1, i)$ (resp. at $(0, j)$) respectively. A *link* is a tangle of type $(\emptyset, \emptyset)$, i.e. $k = l = 0$, and a *knot* means a link with a single connected component. An *n-string link*,[4] a string link with n-components, means a tangle with $\epsilon = \epsilon'$ and $k = l = n$ which consists of n-intervals connecting $(0, i)$ with $(1, i)$ for each $1 \leqslant i \leqslant n$ and no circles (cf. Fig. 5).

We denote $\mathcal{T}$ to be the full set of isotopy classes of oriented tangles and $\mathcal{T}_{\epsilon,\epsilon'}$ to be its subset consisting of tangles with type (ϵ, ϵ'). Figure 6 might help the readers to have a good understanding of the definition. It is easily seen that there is a natural composition map

$$\cdot : \mathcal{T}_{\epsilon_1,\epsilon_2} \times \mathcal{T}_{\epsilon_2,\epsilon_3} \to \mathcal{T}_{\epsilon_1,\epsilon_3} \tag{20}$$

---

[3]We occasionally omit to mention it. Throughout the paper all tangles are assumed to be oriented.

[4]A string link is not a link in the sense of the previous sentence.

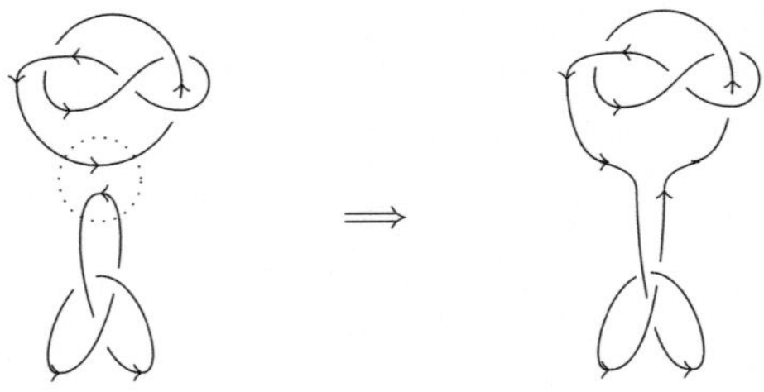

Fig. 7 Connected sum (knot sum)

for any sequences $\epsilon_1, \epsilon_2, \epsilon_3$. The set $\mathcal{SL}_\epsilon$ denotes the subspace of $\mathcal{T}_{\epsilon,\epsilon}$ consisting of string links. By the above composition $\mathcal{SL}_\epsilon$ forms a monoid for each ϵ. By putting on each i-th strand an orientation ϵ_i, the pure braid group P_n $(n > 1)$ may be regarded as a submonoid of $\mathcal{SL}_\epsilon$ with $\epsilon = (\epsilon_1, \ldots, \epsilon_n)$. By definition $\mathcal{T}_{\emptyset,\emptyset}$ is the set of isotopy classes of (oriented) links. We denote $\mathcal{K}$ to be its subspace consisting isotopy classes of (oriented) knots. The set $\mathcal{K}$ forms a monoid by the connected sum (the knot sum)

$$\sharp : \mathcal{K} \times \mathcal{K} \to \mathcal{K}. \tag{21}$$

It is a natural way to fuse two oriented knots, with an appropriate position of orientation, into one (an example is illustrated in Fig. 7). It can be done at any points. Our short caution is that the connected sum $\sharp$ in (21) is different from the composition in (20). (In fact a composition of two knots in not a knot but a link).

There is a fundamental identification between knots and long knots (string link with a single component).

Proposition 3.2 *Let $\epsilon = \uparrow$ or $\downarrow$. Then the set $\mathcal{SL}_\epsilon$ of long knots with the composition $\cdot$ is identified with the set $\mathcal{K}$ with the connected sum $\sharp$ by closing the two endpoints of each. Namely we have an identification of two monoids*

$$\mathrm{cl} : (\mathcal{SL}_\uparrow, \cdot) \simeq (\mathcal{K}, \sharp).$$

Proof The identification is simply obtained by combining the ends of long knots. Checking all compatibilities are easy to see.

For more on tangles, consult the standard textbook, say, [13].

Definition 3.3 ([22, 28]) (1) Let $\mathbb{K}[\mathcal{T}_{\epsilon,\epsilon'}]$ be the free $\mathbb{K}$-module of finite formal sums of elements of $\mathcal{T}_{\epsilon,\epsilon'}$. A singular oriented tangle, an 'oriented tangle' allowed to have a finite number of transversal double points (see [28] for detail), determines an element of $\mathbb{K}[\mathcal{T}_{\epsilon,\epsilon'}]$ by the singularization of each double point by the following relation

Let $\mathcal{T}_n$ ($n \geqslant 0$) be the $\mathbb{K}$-submodule of $\mathbb{K}[\mathcal{T}_{\epsilon,\epsilon'}]$ generated by all singular oriented tangles with type (ϵ, ϵ') and with n double points. The descending filtration $\{\mathcal{T}_n\}_{n\geqslant 0}$ is called the *singular filtration*. The topological $\mathbb{K}$-module $\widehat{\mathbb{K}[\mathcal{T}_{\epsilon,\epsilon'}]}$ of *proalgebraic tangles* of type (ϵ, ϵ') means its completion with respect to the singular filtration:

$$\widehat{\mathbb{K}[\mathcal{T}_{\epsilon,\epsilon'}]} := \varprojlim_N \mathbb{K}[\mathcal{T}_{\epsilon,\epsilon'}]/\mathcal{T}_N.$$

By abuse of notation, we denote the induced filtration on $\widehat{\mathbb{K}[\mathcal{T}_{\epsilon,\epsilon'}]}$ by the same symbol $\{\mathcal{T}_n\}_{n\geqslant 0}$. Note that there is a natural composition map

$$\cdot : \widehat{\mathbb{K}[\mathcal{T}_{\epsilon_1,\epsilon_2}]} \times \widehat{\mathbb{K}[\mathcal{T}_{\epsilon_2,\epsilon_3}]} \to \widehat{\mathbb{K}[\mathcal{T}_{\epsilon_1,\epsilon_3}]} \tag{22}$$

for any ϵ_1, ϵ_2 and ϵ_3. We denote the collection of $\widehat{\mathbb{K}[\mathcal{T}_{\epsilon,\epsilon'}]}$ for all ϵ and ϵ' by $\widehat{\mathcal{T}_{\mathbb{K}}}$.

(2) Let $\mathbb{K}[\mathcal{K}]$ be the $\mathbb{K}$-submodule of $\mathbb{K}[\mathcal{T}_{\emptyset,\emptyset}]$ generated by $\mathcal{K}$. By the product

$$\sharp : \widehat{\mathbb{K}[\mathcal{K}]} \times \widehat{\mathbb{K}[\mathcal{K}]} \to \widehat{\mathbb{K}[\mathcal{K}]} \tag{23}$$

induced by the connected sum $\sharp$ in (21) and the coproduct map $\Delta : \mathbb{K}[\mathcal{K}] \to \mathbb{K}[\mathcal{K}] \otimes_{\mathcal{K}} \mathbb{K}[\mathcal{K}]$ sending $k \mapsto k \otimes k$ and the augmentation map $\mathbb{K}[\mathcal{K}] \to \mathbb{K}$, it carries a structure of co-commutative and commutative bi-algebra. Put $\mathcal{K}_n := \mathcal{T}_n \cap \mathbb{K}[\mathcal{K}]$ ($n \geqslant 0$). Then $\mathcal{K}_n$ forms an ideal of $\mathbb{K}[\mathcal{K}]$ and the descending filtration $\{\mathcal{K}_n\}_{n\geqslant 0}$ is called the *singular knot filtration* (cf .loc .cit). The topological commutative $\mathbb{K}$-algebra $\widehat{\mathbb{K}[\mathcal{K}]}$ of *proalgebraic knots* means its completion with respect to the singular knot filtration:

$$\widehat{\mathbb{K}[\mathcal{K}]} := \varprojlim_N \mathbb{K}[\mathcal{K}]/\mathcal{K}_N.$$

It is a $\mathbb{K}$-linear subspace of $\widehat{\mathbb{K}[\mathcal{T}_{\emptyset\emptyset}]}$. Since each element γ in $\mathcal{K}$ is congruent to the unit $\circlearrowleft$ modulo $\mathcal{K}_1$ due to the finiteness property of the unknotting number (cf. [13]), the inverse of γ with respect to $\sharp$ always exists in $\widehat{\mathbb{K}[\mathcal{K}]}$. It defines the antipode map on $\widehat{\mathbb{K}[\mathcal{K}]}$, which yields a structure of co-commutative and commutative Hopf algebra there. Again by abuse of notation, we denote the induced filtration of $\widehat{\mathbb{K}[\mathcal{K}]}$ by the same symbol $\{\mathcal{K}_n\}_{n\geqslant 0}$, which is compatible with a structure of filtered Hopf algebra on $\widehat{\mathbb{K}[\mathcal{K}]}$.

(3) *Proalgebraic links* and *proalgebraic string links* can be defined in the same way. The subset $\widehat{\mathbb{K}[\mathcal{SL}_\epsilon]}$ of $\widehat{\mathbb{K}[\mathcal{T}_{\epsilon,\epsilon}]}$ consisting of proalgebraic string links forms a non-commutative $\mathbb{K}$-algebra by the composition map (22).

Here is a fundamental identification in our proalgebraic setting.

Lemma 3.4 *Let $\epsilon = \uparrow$ or $\downarrow$. There is an identification between two $\mathbb{K}$-algebras*

$$\mathrm{cl} : (\widehat{\mathbb{K}[\mathcal{SL}_\epsilon]}, \cdot) \simeq (\widehat{\mathbb{K}[\mathcal{K}]}, \sharp). \tag{24}$$

which is compatible with their filtrations.

Proof It is an immediate corollary of Proposition 3.2.

We give a piecewise construction of the above proalgebraic tangles by using proalgebraic pre-tangles introduced below.

Definition 3.5 (1) A *fundamental proalgebraic (oriented) tangle* means a vector belonging to an *ABC-space*, one of the following $\mathbb{K}$-linear spaces $A^{\epsilon}_{k,l}$, $\widehat{B^{\epsilon}_{\tau}}$ and $C^{\epsilon}_{k,l}$ for some k, l, ϵ, τ:

$$A^{\epsilon}_{k,l} := \mathbb{K}\cdot a^{\epsilon}_{k,l}, \text{ with } \epsilon = (\epsilon_i)_{i=1}^{k+l+1} \in \{\uparrow,\downarrow\}^k \times \{\curvearrowleft, \curvearrowright\} \times \{\uparrow,\downarrow\}^l \ (k,l = 0,1,2,\dots),$$

$$\widehat{B^{\epsilon}_{\tau}} := \widehat{\mathbb{K}[P_n]}\cdot\tau \text{ with } \epsilon = (\epsilon_i)_{i=1}^{n} \in \{\uparrow,\downarrow\}^n \text{ and } \tau \in \mathfrak{S}_n \ (n = 1,2,3,4,\dots),$$

$$C^{\epsilon}_{k,l} := \mathbb{K}\cdot c^{\epsilon}_{k,l}, \text{ with } \epsilon = (\epsilon_i)_{i=1}^{k+l+1} \in \{\uparrow,\downarrow\}^k \times \{\cup, \cup\} \times \{\uparrow,\downarrow\}^l \ (k,l = 0,1,2,\dots).$$

Here $\widehat{\mathbb{K}[P_n]}\cdot\tau$ stands for the coset of $\widehat{\mathbb{K}[B_n]} \big/ \widehat{\mathbb{K}[P_n]}$ corresponding to $\tau \in \mathfrak{S}_n = B_n/P_n$, the inverse image of $\mathbb{K}\cdot\sigma$ under the natural homomorphism $\widehat{\mathbb{K}[B_n]} \to \mathbb{K}[\mathfrak{S}_n]$. To stress that an element b belongs to $\widehat{B^{\epsilon}_{\tau}}$, we occasionally denote b^{ϵ} or (b, ϵ).

For each ABC-space V, its *source* $s(V)$ and *target* $t(V)$, which are sequences of $\uparrow$ and $\downarrow$, are defined in a completely same way to [22]; e.g. $s(\widehat{B^{\epsilon}_{\tau}}) = \epsilon$, $t(\widehat{B^{\epsilon}_{\tau}}) = \tau(\epsilon)$.

(2) A *proalgebraic pre-tangle* Γ of type (ϵ, ϵ') is a vector belonging to a $\mathbb{K}$-linear space which is a finite consistent (successively composable) $\mathbb{K}$-linear tensor product of ABC-spaces. Namely it is an element belonging to a $\mathbb{K}$-linear space $V_n \otimes \cdots V_2 \otimes V_1$ for some n such that $s(V_{i+1}) = t(V_i)$ for all $i = 1, 2, \dots, n-1$ and $s(V_1) = \epsilon$ and $t(V_n) = \epsilon'$. For our simplicity, we denote each of its element $\Gamma = \gamma_n \otimes \cdots \otimes \gamma_2 \otimes \gamma_1$ with $\gamma_i \in V_i$ by $\Gamma = \gamma_n \cdots \gamma_2 \cdot \gamma_1$. We also define $s(\Gamma) := s(V_1)$ and $t(\Gamma) := t(V_n)$. A *proalgebraic pre-link* Γ is a proalgebraic pre-tangle with $s(\Gamma) = t(\Gamma) = \emptyset$. Two proalgebraic pre-tangles $\Gamma = \gamma_n \cdots \gamma_2 \cdot \gamma_1$ and $\Gamma' = \gamma'_m \cdots \gamma'_2 \cdot \gamma'_1$ are called *composable* when $s(\Gamma) = t(\Gamma')$. Their *composition* $\Gamma \cdot \Gamma'$ is defined by $\gamma_n \cdots \gamma_2 \cdot \gamma_1 \cdot \gamma'_m \cdots \gamma'_2 \cdot \gamma'_1$.

For each ABC-space V, its *skeleton* $\mathbb{S}(V)$ is defined in a completely same way to [22]. For a proalgebraic pre-tangle $\Gamma = \gamma_n \cdots \gamma_2 \cdot \gamma_1$ with $\gamma_i \in V_i$, its *skeleton* $\mathbb{S}(\Gamma)$ stands for the graph of the compositions $\mathbb{S}(V_n) \cdots \mathbb{S}(V_2) \cdot \mathbb{S}(V_1)$ and its *connected components* mean the connected components of $\mathbb{S}(\Gamma)$ as graphs. A *proalgebraic pre-knot* is a proalgebraic pre-link with a single connected component. A *proalgebraic pre-string link* of type $\epsilon = (\epsilon_i)_{i=1}^{n}$ is a proalgebraic pre-tangle of type (ϵ, ϵ) whose connected components consist of n intervals connecting each i-th point on the bottom to the one on the top.

(3) Two proalgebraic pre-tangles are called *isotopic* when they are related by a finite number of the 6 moves replacing profinite tangles and profinite braids group $\widehat{B_n}$ by proalgebraic pre-tangles and proalgebraic braid algebras $\widehat{\mathbb{K}[B_n]}$ in (T1)–(T6) in [22] and $c \in \widehat{\mathbb{Z}}$ by $c \in \mathbb{K}$ in (T6). (N.B. We note that σ_i^c for $c \in \mathbb{K}$ makes sense in $\widehat{\mathbb{K}[B_n]}$ by the reason explained in [22] Proof of Proposition 2.29 (2).)

(4) We denote $\mathbb{K}[\mathcal{T}^{\mathrm{pre}}_{\epsilon,\epsilon'}]$ to be the $\mathbb{K}$-linear space which is the quotient of the $\mathbb{K}$-span of proalgebraic pre-tangles with type (ϵ, ϵ') divided by the equivalence linearly generated by the above isotopy. Note that there is a natural composition map

$$\cdot : \mathbb{K}[\mathcal{T}^{\mathrm{pre}}_{\epsilon_1,\epsilon_2}] \times \mathbb{K}[\mathcal{T}^{\mathrm{pre}}_{\epsilon_2,\epsilon_3}] \to \mathbb{K}[\mathcal{T}^{\mathrm{pre}}_{\epsilon_1,\epsilon_3}] \tag{25}$$

for any ϵ_1, ϵ_2 and ϵ_3. The subset $\mathbb{K}[\mathcal{SL}^{\mathrm{pre}}_{\epsilon}]$ of $\mathbb{K}[\mathcal{T}^{\mathrm{pre}}_{\epsilon,\epsilon}]$ consisting of proalgebraic pre-string links of type ϵ forms a non-commutative $\mathbb{K}$-algebra by the composition map.

The symbol $\mathbb{K}[\mathcal{K}^{\mathrm{pre}}]$ stands for the subspace of $\mathbb{K}[\mathcal{T}^{\mathrm{pre}}_{\emptyset,\emptyset}]$ generated by proalgebraic pre-knots. It can be proven in a same way to [22] that $\mathbb{K}[\mathcal{K}^{\mathrm{pre}}]$ inherits a structure of a commutative $\mathbb{K}$-algebra by the connected sum

$$\sharp : \mathbb{K}[\mathcal{K}^{\mathrm{pre}}] \times \mathbb{K}[\mathcal{K}^{\mathrm{pre}}] \to \mathbb{K}[\mathcal{K}^{\mathrm{pre}}]. \tag{26}$$

Here for any two proalgebraic knots $K_1 = \alpha_m \cdots \alpha_1$ and $K_2 = \beta_n \cdots \beta_1$ with $(\alpha_m, \alpha_1) = (\curvearrowleft, \circlearrowright)$ and $(\beta_n, \beta_1) = (\curvearrowleft, \circlearrowright)$ (we may assume such presentations by (T6)), their connected sum is defined by

$$K_1 \sharp K_2 := \alpha_m \cdots \alpha_2 \cdot \beta_{n-1} \cdots \beta_1.$$

Again we note that $\sharp$ is different from the above composition (25).

(5) For a $\mathbb{K}$-linear space V with $V = \widehat{B^{\epsilon}_{\tau}}$, we give its descending filtration $\{\mathcal{T}_N(V)\}_{N=0}^{\infty}$ of $\mathbb{K}$-linear subspaces by $\mathcal{T}_N(V) := I^N$ and for a $\mathbb{K}$-linear space V with $V = A^{\epsilon}_{k,l}$ or $C^{\epsilon}_{k,l}$ we give its filtration by $\mathcal{T}_0(V) = V$ and $\mathcal{T}_N(V) := \{0\}$ for $N > 0$. For a finite consistent tensor product $V = V_n \otimes \cdots \otimes V_1$ of ABC-spaces, we give its descending filtration $\{\mathcal{T}_N(V)\}_{N=0}^{\infty}$ with

$$\mathcal{T}_N(V) := \sum_{i_n+\cdots+i_1=N} \mathcal{T}_{i_n}(V_n) \otimes \cdots \otimes \mathcal{T}_{i_1}(V_1),$$

i.e. the $\mathbb{K}$-linear subspaces generated by the subspaces $\mathcal{T}_{i_n}(V_n) \otimes \cdots \otimes \mathcal{T}_{i_1}(V_1)$ with $i_n + \cdots + i_1 = N$. For any ϵ and ϵ', the collection of the filtrations of such consistent tensor product V with $s(V) = \epsilon'$ and $t(V) = \epsilon$ yields a filtration of $\mathbb{K}$-submodules of $\mathbb{K}[\mathcal{T}^{\mathrm{pre}}_{\epsilon,\epsilon'}]$, which we denote by $\{\mathcal{T}^{\mathrm{pre}}_N\}_{N\geqslant 0}$. They are compatible with the composition map (25). The filtration on $\mathbb{K}[\mathcal{K}^{\mathrm{pre}}]$ induced by the filtration of $\mathbb{K}[\mathcal{T}^{\mathrm{pre}}_{\emptyset,\emptyset}]$ is denoted by $\{\mathcal{K}^{\mathrm{pre}}_N\}_{N\geqslant 0}$. The filtration is compatible with its algebra structure given by (26).

Lemma 3.6 *(1) For any ϵ and ϵ' and any $N \geqslant 0$, there is an isomorphism of $\mathbb{K}$-linear spaces*

$$\mathbb{K}[\mathcal{T}_{\epsilon,\epsilon'}]/\mathcal{T}_N \simeq \mathbb{K}[\mathcal{T}^{\mathrm{pre}}_{\epsilon,\epsilon'}]/\mathcal{T}^{\mathrm{pre}}_N.$$

(2) For any ϵ and any $N \geqslant 0$, there is an isomorphism of non-commutative $\mathbb{K}$-algebras

$$\mathbb{K}[\mathcal{SL}_{\epsilon}]/\mathcal{T}_N \simeq \mathbb{K}[\mathcal{SL}^{\mathrm{pre}}_{\epsilon}]/\mathcal{T}^{\mathrm{pre}}_N.$$

(3) For any $N \geqslant 0$, there is an isomorphism of commutative $\mathbb{K}$-algebras

$$\mathbb{K}[\mathcal{K}]/\mathcal{K}_N \simeq \mathbb{K}[\mathcal{K}^{\mathrm{pre}}]/\mathcal{K}_N^{\mathrm{pre}}.$$

Proof (1) The map is obtained because the set $\mathcal{T}$ is described by the sequence of elements in the discrete sets A, B and C modulo the discrete version of our moves (T1)–(T6) (see [22]). Showing that it is an isomorphism is attained by the isomorphism

$$\mathbb{K}[B_n]/I^i \simeq \widehat{\mathbb{K}[B_n]}/I^i$$

for $i = 0, 1, 2, \ldots$.

(2) and (3) It is a direct consequence of (1).

Here are algebraic reformulations of proalgebraic tangles and knots.

Proposition 3.7 *(1) For each ϵ and ϵ', there is an identification of filtered K-linear spaces*

$$\varprojlim_n \mathbb{K}[\mathcal{T}_{\epsilon,\epsilon'}^{\mathrm{pre}}]/\mathcal{T}_n^{\mathrm{pre}} \simeq \widehat{\mathbb{K}[\mathcal{T}_{\epsilon,\epsilon'}]}.$$

(2) For each ϵ, there is an identification of filtered non-commutative $\mathbb{K}$-algebras

$$\varprojlim_n \mathbb{K}[\mathcal{SL}_{\epsilon}^{\mathrm{pre}}]/\mathcal{T}_n^{\mathrm{pre}} \simeq \widehat{\mathbb{K}[\mathcal{SL}_{\epsilon}]}.$$

(3) There is an identification of filtered commutative $\mathbb{K}$-algebras

$$\varprojlim_n \mathbb{K}[\mathcal{K}^{\mathrm{pre}}]/\mathcal{K}_n^{\mathrm{pre}} \simeq \widehat{\mathbb{K}[\mathcal{K}]}.$$

Proof It is immediate by Definition 3.3 and the above lemma.

Remark 3.8 By [5] Sect. 4.2, we have a natural inclusion $\widehat{\mathbb{K}[P_n]} \hookrightarrow \widehat{\mathbb{K}[\mathcal{SL}_{\uparrow^n}]}$.

Based on the identification, the action of the Grothendieck-Teichmüller group $GT(\mathbb{K})$ on proalgebraic tangles (in [27, 28] Appendix) can be explained as follows:

Definition 3.9 Let $\bar{\Gamma} \in \widehat{\mathbb{K}[\mathcal{T}_{\epsilon,\epsilon'}]}/\mathcal{T}_N$ with any sequences ϵ, ϵ' and $N \geqslant 0$. Let Γ be its representative in $\mathbb{K}[\mathcal{T}_{\epsilon,\epsilon'}^{\mathrm{pre}}]$ with a presentation $\Gamma = \gamma_m \cdots \gamma_2 \cdot \gamma_1$ (γ_j: fundamental proalgebraic tangle). For $\sigma = (\lambda, f) \in GT(\mathbb{K})$ with $\lambda \in \mathbb{K}^{\times}$ and $f \in F_2(\mathbb{K})$, we define

$$\sigma(\bar{\Gamma}) := \overline{\sigma(\gamma_m) \cdots \sigma(\gamma_2) \cdot \sigma(\gamma_1)} \in \widehat{\mathbb{K}[\mathcal{T}_{\epsilon,\epsilon'}]}/\mathcal{T}_N. \tag{27}$$

Here $\sigma(\gamma_j)$ is defined in a same way to [22] as follows.

(1) If $\gamma_j \in A_{k,l}^{\epsilon}$, we define

$$\sigma(\gamma_j) := \gamma_j \cdot (\nu_f)_{k+2}^{s(\gamma_j)} \cdot f_{1\cdots k,k+1,k+2}^{s(\gamma_j)}.$$

Here the middle term stands for the proalgebraic tangle whose source is $s(\gamma_j)$ which is obtained by putting the trivial braid with $k+1$-strands on the left of ν_f (see below) and the trivial braid with l-strands on its right.

(2) If $\gamma_j = (b_n, \epsilon) \in \widehat{B^\epsilon_\tau}$ with $b_n \in \widehat{\mathbb{K}[P_n]} \cdot \tau \subset \widehat{\mathbb{K}[B_n]}$, we define

$$\sigma(\gamma_j) := (\rho_n(\sigma)(b_n), \epsilon).$$

Here $\rho_n(\sigma)(b_n)$ is the image of b_n by the action $\sigma \in GT(\mathbb{K})$ on $\widehat{\mathbb{K}[B_n]}$ explained in Sect. 2.1.

(3) If $\gamma_j \in C^\epsilon_{k,l}$, we define

$$\sigma(\gamma_j) := f^{-1, t(\gamma_j)}_{1\cdots k, k+1, k+2} \cdot \gamma_j.$$

The symbol ν^ϵ_f ($\epsilon = \uparrow, \downarrow$) means the proalgebraic tangle in $\widehat{\mathbb{K}[\mathcal{T}_{\epsilon,\epsilon}]}$ which is a proalgebraic string link with a single strands such that $s(\mu^\epsilon_f) = t(\mu^\epsilon_f) = \epsilon$ and which is given by the inverse of Λ^ϵ_f with respect to the composition:

$$\nu^\epsilon_f := \{\Lambda^\epsilon_f\}^{-1}.$$

Here $\Lambda^\downarrow_f$ in $\widehat{\mathbb{K}[\mathcal{T}_{\epsilon,\epsilon}]}$ represents the proalgebraic string link with a single strands given by

$$\Lambda^\downarrow_f := a^{\downarrow\frown}_{1,0} \cdot f^{\downarrow\uparrow\downarrow} \cdot c^{\smile\downarrow}_{0,1}.$$

(cf. Fig. 8) and $\Lambda^\uparrow_f$ is the same one obtained by reversing its all arrows.

We note that the existence of the inverse of Λ^ϵ_f in $\widehat{\mathbb{K}[\mathcal{T}_{\epsilon,\epsilon}]}$ is immediate because Λ^ϵ_f is congruent to the trivial braid with a string modulo $\mathcal{T}_1$ and $\widehat{\mathbb{K}[\mathcal{T}_{\epsilon,\epsilon}]}$ is completed by the filtration $\{\mathcal{T}_n\}^\infty_{n=0}$.

Remark 3.10 (1) The inverse ν^ϵ_f does not look exist in $\mathbb{K}[\mathcal{T}^{\mathrm{pre}}_{\epsilon,\epsilon'}]$ generally. That is why we define $GT(\mathbb{K})$-action on $\widehat{\mathbb{K}[\mathcal{T}_{\epsilon,\epsilon'}]}$ below.

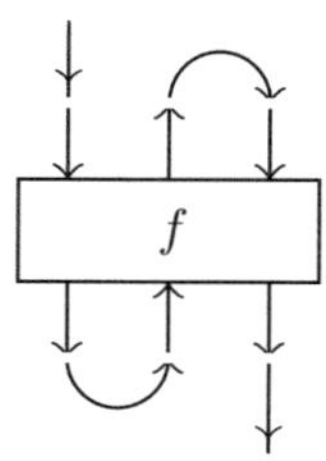

Fig. 8 $\Lambda^\downarrow_f$

(2) Our action (27) looks slightly different from the one given in [22]. One of the reasons is that we deal tangles in Definition 3.9 while we discuss knots in [22].

Proposition 3.11 ([27, 28]) *The Eq. (27) yields a well-defined $GT(\mathbb{K})$-action on $\widehat{\mathbb{K}[\mathcal{T}_{\epsilon,\epsilon'}]}/\mathcal{T}_N$ for any ϵ, ϵ' and any $N \geqslant 0$, and induces a well-defined $GT(\mathbb{K})$-action on $\widehat{\mathbb{K}[\mathcal{T}_{\epsilon,\epsilon'}]}$ such that the equality*

$$\sigma(\Gamma \cdot \Gamma') = \sigma(\Gamma) \cdot \sigma(\Gamma')$$

holds in $\widehat{\mathbb{K}[\mathcal{T}_{\epsilon_1,\epsilon_3}]}$ for two proalgebraic tangles $\Gamma \in \widehat{\mathbb{K}[\mathcal{T}_{\epsilon_1,\epsilon_2}]}$ and $\Gamma' \in \widehat{\mathbb{K}[\mathcal{T}_{\epsilon_2,\epsilon_3}]}$ for any ϵ_1, ϵ_2 and ϵ_3.

Proof Though its proof can be found in loc. cit where it is explained in terms of $\mathbb{K}$-linear braided monoidal categories, we can prove in a same way to the proof of [22] Theorem 2.38. by direct calculations.

We denote the above induced action by

$$\rho_{\epsilon,\epsilon'} : GT(\mathbb{K}) \to \text{Aut}\ \widehat{\mathbb{K}[\mathcal{T}_{\epsilon,\epsilon'}]}. \tag{28}$$

We may say that it is a generalization of the map (6) into the proalgebraic tangle case. As a consequence of the above proposition, we have

Proposition 3.12 *For each sequence ϵ, the subspace $\widehat{\mathbb{K}[\mathcal{SL}_\epsilon]}$ of proalgebraic string links is stable under the above $GT(\mathbb{K})$-action. The action is compatible with its non-commutative algebra structure whose product is given by the composition.*

We denote the above action by

$$\rho_\epsilon : GT(\mathbb{K}) \to \text{Aut}\ \widehat{\mathbb{K}[\mathcal{SL}_\epsilon]}. \tag{29}$$

We will see in Theorem 4.14 that this action restricted into $GT_1(\mathbb{K})$ is given by inner conjugation.

Particularly by restricting our action (28) into $\widehat{\mathbb{K}[\mathcal{T}_{\emptyset,\emptyset}]}$ we obtain the following.

Proposition 3.13 *The subspace $\widehat{\mathbb{K}[\mathcal{K}]}$ of proalgebraic knots is stable under the above $GT(\mathbb{K})$-action. The action there is not compatible with the connected sum $\sharp$ however the equality*

$$\sigma(K_1 \sharp K_2) \sharp \sigma(\acute{\bigcirc}) = \sigma(K_1) \sharp \sigma(K_2) \tag{30}$$

holds for any $\sigma \in GT(\mathbb{K})$ and any $K_1, K_2 \in \widehat{\mathbb{K}[\mathcal{K}]}$.

Proof Seeing that the subspace is stable can be verified by showing that connected components of skeletons of proalgebraic tangles are unchanged. The Eq. (30) can be proven in a same way to the proof of the Eq. (2.21) in [22].

We denote the above action by

$$\rho_0 : GT(\mathbb{K}) \to \mathrm{Aut}\ \widehat{\mathbb{K}[\mathcal{K}_0]}. \tag{31}$$

Remark 3.14 We will see in Proposition 4.19 that this action is given by $\mathbb{G}_m$-action. We will explicitly determine in Theorem 4.27 the subspace of $\widehat{\mathbb{K}[\mathcal{K}]}$ which is invariant under the above $GT(\mathbb{K})$-action.

We note that in the proalgebraic knot setting our action above is not compatible with the product structure (23) due to (30), while in the profinite knot setting the $\widehat{GT}$-action on the group of profinite knots is compatible with the product structure as shown in [22] Theorem 2.38 (4). In order to relate the $\widehat{GT}$-action on $\mathrm{Frac}\widehat{\mathcal{K}}$ constructed in [22] with the above $GT(\mathbb{K})$-action on $\widehat{\mathbb{K}[\mathcal{K}]}$, we introduce the proalgebraic group $\mathrm{Frac}\mathcal{K}(\mathbb{K})$ below.

Notation 3.15 Put $\mathcal{K}(\mathbb{K})$ to be the group-like part of $\widehat{\mathbb{K}[\mathcal{K}]}$, which carries a structure of proalgebraic group. It can be checked directly that $GT(\mathbb{K})$-action on $\widehat{\mathbb{K}[\mathcal{K}]}$ is compatible with its coproduct map and its antipode map. Hence we have $GT(\mathbb{K})$-action on $\mathcal{K}(\mathbb{K})$ though it is not compatible with its product structure of $\mathcal{K}(\mathbb{K})$ due to (30). We denote its group of fraction by $\mathrm{Frac}\mathcal{K}(\mathbb{K})$, which is the quotient space of $\mathcal{K}(\mathbb{K}) \times \mathcal{K}(\mathbb{K})$ by the equivalent relations $(r, s) \approx (r', s')$ if $r\sharp s'\sharp t = r'\sharp s\sharp t$ holds in $\mathcal{K}(\mathbb{K})$ for some $t \in \mathcal{K}(\mathbb{K})$.

Lemma 3.16 *The induced $GT(\mathbb{K})$-action on* $\mathrm{Frac}\mathcal{K}(\mathbb{K})$ *is compatible its group structure, i.e.*

$$\sigma(e) = e, \qquad \sigma(x\sharp y) = \sigma(x)\sharp\sigma(y), \qquad \sigma(\frac{1}{x}) = \frac{1}{\sigma(x)}$$

for any $\sigma \in GT(\mathbb{K})$ and $x, y \in$ $\mathrm{Frac}\mathcal{K}(\mathbb{K})$*. Here $e = \bigcirc/\bigcirc$.*

Proof Let $x = r_1/s_1$ and $y = r_2/s_2$. Then by (30) it is easy to see

$$\begin{aligned}
\sigma(x\sharp y) &= \sigma(\frac{r_1\sharp r_2}{s_1\sharp s_2}) = \frac{\sigma(r_1\sharp r_2)}{\sigma(s_1\sharp s_2)} = \frac{\sigma(r_1\sharp r_2)\sharp\sigma(\bigcirc)}{\sigma(s_1\sharp s_2)\sharp\sigma(\bigcirc)} = \frac{\sigma(r_1)\sharp\sigma(r_2)}{\sigma(s_1)\sharp\sigma(s_2)} \\
&= \frac{\sigma(r_1)}{\sigma(s_1)}\sharp\frac{\sigma(r_2)}{\sigma(s_2)} = \sigma(\frac{r_1}{s_1})\sharp\sigma(\frac{r_2}{s_2}) = \sigma(x)\sharp\sigma(y), \\
\sigma(\frac{1}{x})\sharp\sigma(x) &= \sigma(\frac{s_1}{r_1})\sharp\sigma(\frac{r_1}{s_1}) = \frac{\sigma(s_1)}{\sigma(r_1)}\sharp\frac{\sigma(r_1)}{\sigma(s_1)} = \frac{\sigma(s_1)\sharp\sigma(r_1)}{\sigma(r_1)\sharp\sigma(s_1)} = \frac{\sigma(r_1)\sharp\sigma(s_1)}{\sigma(r_1)\sharp\sigma(s_1)} = \frac{\bigcirc}{\bigcirc} = e.
\end{aligned}$$

In [22], the profinite $\widehat{GT}$-action on $\mathrm{Frac}\widehat{\mathcal{K}}$ was constructed. A relationship of the action with the above constructed $GT(\mathbb{K})$-action on $\mathcal{K}(\mathbb{K})$ is explicitly given below.

Proposition 3.17 *For each prime l, there is a natural group homomorphism*

$$\mathrm{Frac}\widehat{\mathcal{K}} \to \mathrm{Frac}\mathcal{K}(\mathbb{Q}_l). \tag{32}$$

The $\widehat{GT}$-action on Frac$\widehat{\mathcal{K}}$ *constructed in [22] is compatible with the above $GT(\mathbb{Q}_l)$-action on $\mathcal{K}(\mathbb{Q}_l)$ under the maps* (7) *and* (32).

Proof The map (32) is naturally induced by the map in [22] Proposition 2.30. It follows from our construction that the map is compatible with two actions.

3.2 The *GRT*-Action

In this subsection, we establish an 'infinitesimal' counterpart of the $GT(\mathbb{K})$-action on proalgebraic tangles given in the previous subsection. Infinitesimal tangles (and knots) are introduced in Definition 3.19 as an infinitesimal counterpart of proalgebraic tangles (and knots). (It will be shown in next subsection that this notion coincides with the notion of chord diagrams.) A consistent $GRT(\mathbb{K})$-action there is established in Proposition 3.23.

Definition 3.18 (1) A *fundamental infinitesimal tangle* means a vector belonging to one of the following $\mathbb{K}$-linear spaces (let us again call them ABC-spaces): $A^{\epsilon}_{k,l}$, $\widehat{IB^{\epsilon}_{\tau}}$ and $C^{\epsilon}_{k,l}$ for some k, l, ϵ, τ:

$$A^{\epsilon}_{k,l} := \mathbb{K}\cdot a^{\epsilon}_{k,l}, \text{ with } \epsilon = (\epsilon_i)_{i=1}^{k+l+1} \in \{\uparrow,\downarrow\}^k \times \{\frown\!\!\!\!\rightarrow, \leftarrow\!\!\!\!\frown\} \times \{\uparrow,\downarrow\}^l \; (k,l = 0,1,2,\dots),$$

$$\widehat{IB^{\epsilon}_{\tau}} := \widehat{U\mathfrak{p}_n}\cdot\tau \text{ with } \epsilon = (\epsilon_i)_{i=1}^{n} \in \{\uparrow,\downarrow\}^n \text{ and } \tau \in \mathfrak{S}_n \; (n = 1,2,3,4,\dots),$$

$$C^{\epsilon}_{k,l} := \mathbb{K}\cdot c^{\epsilon}_{k,l}, \text{ with } \epsilon = (\epsilon_i)_{i=1}^{k+l+1} \in \{\uparrow,\downarrow\}^k \times \{\smile\!\!\!\!\rightarrow, \leftarrow\!\!\!\!\smile\} \times \{\uparrow,\downarrow\}^l \; (k,l = 0,1,2,\dots).$$

Here $\widehat{U\mathfrak{p}_n}\cdot\tau$ stands for the coset of $\widehat{U\mathfrak{b}_n}\big/\widehat{U\mathfrak{p}_n}$ corresponding to $\tau \in \mathfrak{S}_n = B_n/P_n$. To stress that an element b belongs to $\widehat{IB^{\epsilon}_{\tau}}$, we occasionally denote b^{ϵ} or (b,ϵ).

For each above space V, its *source* $s(V)$ and *target* $t(V)$, are defined in a completely same way to Definition 3.5.

(2) An *infinitesimal pre-tangle* D of type (ϵ,ϵ') is defined in a same way to Definition 3.5. Namely it is $D = d_n \cdots d_2 \cdot d_1$ where each d_i is an infinitesimal fundamental tangle, a vector belonging to V_i, one of the above ABC-spaces, such that $s(V_{i+1}) = t(V_i)$ for all $i = 1,2,\dots,n-1$ and $s(V_1) = \epsilon$ and $t(V_n) = \epsilon'$. We also define $s(D) := s(V_1)$ and $t(D) := t(V_n)$. An *infinitesimal pre-link* D is an infinitesimal pre-tangle with $s(D) = t(D) = \emptyset$. Two infinitesimal pre-tangles $D = d_n \cdots d_2 \cdot d_1$ and $D' = d'_m \cdots d'_2 \cdot d'_1$ are called *composable* when $s(D) = t(D')$ and their *composition* $D \cdot D'$ is defined by $d_n \cdots d_2 \cdot d_1 \cdot d'_m \cdots d'_2 \cdot d'_1$.

For an infinitesimal pre-tangle D its *skeleton* $\mathbb{S}(D)$ and its *connected components* can be defined in the same way to Definition 3.5. An *infinitesimal pre-knot* is an infinitesimal pre-link with a single connected component. An *infinitesimal pre-string link* of type $\epsilon = (\epsilon_i)_{i=1}^n$ is an infinitesimal pre-tangle of type (ϵ,ϵ) consisting of n connected components whose each i-th component connect i-th point on the bottom to the one on the top.

(3) Two infinitesimal pre-tangles are called *isotopic* when they are related by a finite number of the 6 moves (IT1)–(IT6). Here (IT1)–(IT5) are the moves replacing profinite tangles and profinite braids group $\widehat{B_n}$ by infinitesimal pre-tangles and infinitesimal braid algebras $\widehat{U\mathfrak{b}_n}$ in (T1)–(T5) in [22] and (IT6) is an 'infinitesimal' variant of (T6), which is stated below.

(IT6): For $\alpha \in \mathbb{K}$, $c^{\epsilon}_{k,l} \in C^{\epsilon}_{k,l}$ and $t_{k+1,k+2} \cdot \tau \in \widehat{IB^{\epsilon'}_{\tau}}$ with $\tau = \tau_{k+1,k+2} \in \mathfrak{S}_{k+l+2}$ (the switch of $k+1$ and $k+2$) and $t(C^{\epsilon}_{k,l}) = \epsilon'$

$$\exp\{\alpha\, t_{k+1,k+2}\} \cdot \tau_{k+1,k+2} \cdot c^{\epsilon}_{k,l} = c^{\bar{\epsilon}}_{k,l}$$

where $\bar{\epsilon}$ is the sequence obtained by revering the $(k+1)$-st and $(k+2)$-nd arrows. And for $\alpha \in \mathbb{K}$, $a^{\epsilon}_{k,l} \in A^{\epsilon}_{k,l}$ and $\tau \cdot t_{k+1,k+2} \in \widehat{IB^{\epsilon'}_{\tau}}$ with $\tau = \tau_{k+1,k+2} \in \mathfrak{S}_{k+l+2}$ and $s(A^{\epsilon}_{k,l}) = \tau(\epsilon')$

$$a^{\epsilon}_{k,l} \cdot \tau_{k+1,k+2} \cdot \exp\{\alpha\, t_{k+1,k+2}\} = a^{\bar{\epsilon}}_{k,l}.$$

where $\bar{\epsilon}$ is the sequence obtained by revering the $(k+1)$-st and $(k+2)$-nd arrows. Figure 9 depicts the moves. (N.B. We note that $\exp\{\alpha\, t_{k+1,k+2}\}$ for $\alpha \in \mathbb{K}$ makes sense in $\widehat{U\mathfrak{p}}_{k+l+2}$.)

(4) We denote $\mathbb{K}[\mathcal{IT}^{\mathrm{pre}}_{\epsilon,\epsilon'}]$ to be the $\mathbb{K}$-linear space which is the quotient of the $\mathbb{K}$-span of infinitesimal pre-tangles with type (ϵ, ϵ') divided by the equivalence linearly generated by the above isotopy. Note that there is a natural composition map

$$\cdot : \mathbb{K}[\mathcal{IT}^{\mathrm{pre}}_{\epsilon_1,\epsilon_2}] \times \mathbb{K}[\mathcal{IT}^{\mathrm{pre}}_{\epsilon_2,\epsilon_3}] \to \mathbb{K}[\mathcal{IT}^{\mathrm{pre}}_{\epsilon_1,\epsilon_3}] \tag{33}$$

for any ϵ_1, ϵ_2 and ϵ_3. The subspace $\mathbb{K}[\mathcal{ISL}^{\mathrm{pre}}_{\epsilon}]$ of $\mathbb{K}[\mathcal{IT}^{\mathrm{pre}}_{\epsilon,\epsilon}]$ consisting of infinitesimal pre-string links forms a non-commutative $\mathbb{K}$-algebras by the composition (33).

The symbol $\mathbb{K}[\mathcal{IK}^{\mathrm{pre}}]$ stands for the subspace of $\mathbb{K}[\mathcal{IT}^{\mathrm{pre}}_{\emptyset,\emptyset}]$ generated by infinitesimal pre-knots. As in Definition 3.5, it inherits a structure of a commutative $\mathbb{K}$-algebra by the connected sum (which can be defined in the same way to [22])

$$\sharp : \mathbb{K}[\mathcal{IK}^{\mathrm{pre}}] \times \mathbb{K}[\mathcal{IK}^{\mathrm{pre}}] \to \mathbb{K}[\mathcal{IK}^{\mathrm{pre}}]. \tag{34}$$

Again we note that $\sharp$ is different from the above composition (33).

(5) For a $\mathbb{K}$-linear space V with $V = A^{\epsilon}_{k,l}$ or $C^{\epsilon}_{k,l}$ we give its descending filtration of $\mathbb{K}$-linear subspaces by $\mathcal{IT}_0(V) = V$ and $\mathcal{IT}_N(V) := \{0\}$ for $N > 0$, and when

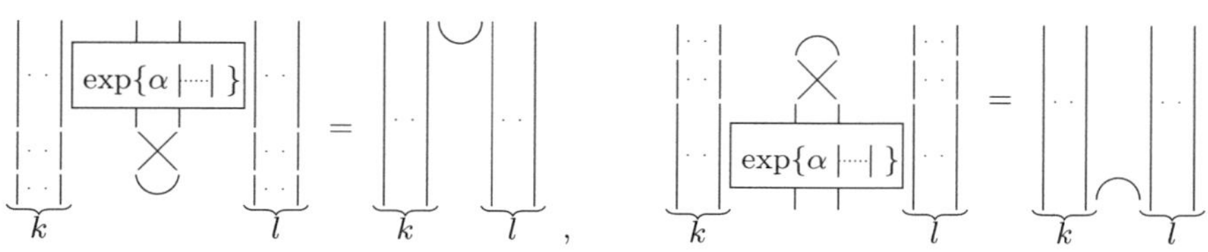

Fig. 9 (IT6)

$V = \widehat{IB^{\epsilon}_{\tau}}$, we give its descending filtration $\{\mathcal{IT}_N(V)\}_{N=0}^{\infty}$ such that $\mathcal{IT}_N(V)$ is the $\mathbb{K}$-linear subspace topologically generated by elements whose degrees are greater than or equal to N.

By the method indicated in Definition 3.5, for any sequences ϵ and ϵ', $\mathbb{K}[\mathcal{IT}^{\mathrm{pre}}_{\epsilon,\epsilon'}]$ is inherited a filtration of its $\mathbb{K}$-submodules which we denote by $\{\mathcal{IT}^{\mathrm{pre}}_N\}_{N\geqslant 0}$. They are compatible with the composition map (33). The filtration on $\mathbb{K}[\mathcal{IK}^{\mathrm{pre}}]$ induced by the filtration of $\mathbb{K}[\mathcal{IT}^{\mathrm{pre}}_{\emptyset,\emptyset}]$ is denoted by $\{\mathcal{IK}^{\mathrm{pre}}_N\}_{N\geqslant 0}$. The filtration is compatible with its algebra structure given by (34).

Definition 3.19 (1) An *infinitesimal tangle* with type (ϵ, ϵ') is an element of

$$\widehat{\mathbb{K}[\mathcal{IT}_{\epsilon,\epsilon'}]} := \varprojlim_n \mathbb{K}[\mathcal{IT}^{\mathrm{pre}}_{\epsilon,\epsilon'}]/\mathcal{IT}^{\mathrm{pre}}_n.$$

(2) An *infinitesimal string link* with type ϵ is an element of

$$\widehat{\mathbb{K}[\mathcal{ISL}_{\epsilon}]} := \varprojlim_n \mathbb{K}[\mathcal{ISL}^{\mathrm{pre}}_{\epsilon}]/\mathcal{IT}^{\mathrm{pre}}_n.$$

(3) An *infinitesimal knot* is an element of the following completed $\mathbb{K}$-algebra

$$\widehat{\mathbb{K}[\mathcal{IK}]} := \varprojlim_n \mathbb{K}[\mathcal{IK}^{\mathrm{pre}}]/\mathcal{IK}^{\mathrm{pre}}_n.$$

We note that $\widehat{\mathbb{K}[\mathcal{IT}_{\epsilon,\epsilon'}]}$ for ϵ, ϵ' is inherited a composition product $\cdot$ by (33) and the set $\widehat{\mathbb{K}[\mathcal{ISL}_{\epsilon}]}$ generally forms a non-commutative $\mathbb{K}$-algebra. The set $\widehat{\mathbb{K}[\mathcal{IK}]}$ is inherited a connected $\sharp$ by (34) and forms a commutative $\mathbb{K}$-algebra.

Remark 3.20 By [5] Sect. 4.2, we have a natural inclusion $\widehat{U\mathfrak{p}_n} \hookrightarrow \widehat{\mathbb{K}[\mathcal{ISL}_{\uparrow^n}]}$. Similarly for each $\epsilon = (\epsilon_i)_{i=1}^n$ with $\epsilon = \uparrow, \downarrow$, there is an inclusion $\widehat{U\mathfrak{p}_n} \hookrightarrow \widehat{\mathbb{K}[\mathcal{ISL}_{\epsilon}]}$. We denote the image of each element $h \in \widehat{U\mathfrak{p}_n}$ by h^{ϵ}.

As an analogue of Lemma 3.4, we have

Lemma 3.21 *Let $\epsilon = \uparrow$ or $\downarrow$. There is an identification of two algebras*

$$\mathrm{cl} : (\widehat{\mathbb{K}[\mathcal{ISL}_{\epsilon}]}, \cdot) \simeq (\widehat{\mathbb{K}[\mathcal{IK}]}, \sharp)$$

which is compatible with their filtrations.

Based on the identification, the action of the graded Grothendieck-Teichmüller group $GRT(\mathbb{K})$ on infinitesimal tangles is constructed as follows:

Definition 3.22 Let $\bar{D} \in \widehat{\mathbb{K}[\mathcal{IT}_{\epsilon,\epsilon'}]}/\mathcal{IT}_N$ with any sequences ϵ, ϵ' and $N \geqslant 0$. Let D be its representative in $\mathbb{K}[\mathcal{IT}^{\mathrm{pre}}_{\epsilon,\epsilon'}]$ with a presentation $D = d_m \cdots d_2 \cdot d_1$ (d_j:

fundamental infinitesimal tangle). For $\sigma = (c, g) \in GRT(\mathbb{K})$, hence $c \in \mathbb{K}^\times$ and $g \in \exp \mathfrak{f}_2$, we define

$$\sigma(\bar{D}) := \overline{\sigma(d_m) \cdots \sigma(d_2) \cdot \sigma(d_1)} \in \mathbb{K}[\widehat{\mathcal{IT}_{\epsilon,\epsilon'}}]/\mathcal{IT}_N. \tag{35}$$

with $\sigma(d_j)$ given below.

(1) when $d_j \in A^\epsilon_{k,l}$, we define

$$\sigma(d_j) := d_j \cdot (\nu_g)^{s(d_j)}_{k+2} \cdot g^{s(d_j)}_{1\cdots k,k+1,k+2}.$$

Here the middle term stands for the infinitesimal tangle whose source is $s(d_j)$ which is obtained by putting the trivial infinitesimal braid with $k+1$-strands on the left of ν_g (see below) and the trivial infinitesimal braid with l-strands on its right.

(2) when $d_j = (b_n, \epsilon) \in \widehat{IB^\epsilon_\tau}$ with $b_n \in \widehat{U\mathfrak{p}_n} \cdot \tau \subset \widehat{U\mathfrak{b}_n}$, we define

$$\sigma(d_j) := (\rho_n(\sigma)(b_n), \epsilon).$$

Here $\rho_n(\sigma)(b_n)$ is the image of b_n by the action $\sigma \in GRT(\mathbb{K})$ on $\widehat{U\mathfrak{b}_n}$ explained in Sect. 2.2.

(3) when $d_j \in C^\epsilon_{k,l}$, we define

$$\sigma(d_j) := g^{-1,t(d_j)}_{1\cdots k,k+1,k+2} \cdot d_j.$$

The symbol ν^ϵ_g ($\epsilon = \uparrow, \downarrow$) means the infinitesimal tangle in $\mathbb{K}[\widehat{\mathcal{IT}_{\epsilon,\epsilon}}]$ with a single connected component such that $s(\mu^\epsilon_g) = t(\mu^\epsilon_g) = \epsilon$ and which is given by the inverse of Λ^ϵ_g with respect to the composition:

$$\nu^\epsilon_g := \{\Lambda^\epsilon_g\}^{-1}.$$

Here $\Lambda^\downarrow_g$ in $\mathbb{K}[\widehat{\mathcal{IT}_{\epsilon,\epsilon}}]$ represents the infinitesimal string link with a single strands given by

$$\Lambda^\downarrow_g := a^{\downarrow\frown}_{1,0} \cdot g^{\downarrow\uparrow\downarrow} \cdot c^{\smile\downarrow}_{0,1}.$$

(which can be depicted as the picture in Fig. 8 replacing f by g) and $\Lambda^\uparrow_g$ is the same one obtained by reversing its all arrows.

We note that the existence of the inverse of Λ^ϵ_g in $\mathbb{K}[\widehat{\mathcal{IT}_{\epsilon,\epsilon}}]$ is immediate because Λ^ϵ_g is congruent to the unit in $\mathbb{K}[\widehat{\mathcal{IT}_{\epsilon,\epsilon}}]$ modulo $\mathcal{T}_1$ and $\mathbb{K}[\widehat{\mathcal{IT}_{\epsilon,\epsilon}}]$ is completed by the filtration $\{\mathcal{T}_n\}_{n=0}^\infty$.

Proposition 3.23 *(1). The Eq.* (35) *yields a well-defined* $GRT(\mathbb{K})$*-action on* $\mathbb{K}[\widehat{\mathcal{IT}_{\epsilon,\epsilon'}}]/\mathcal{IT}_N$ *for any* ϵ, ϵ' *and any* $N \geqslant 0$, *and induces a well-defined* $GRT(\mathbb{K})$*-action on* $\mathbb{K}[\widehat{\mathcal{IT}_{\epsilon,\epsilon'}}]$ *such that the equality*

$$\sigma(D \cdot D') = \sigma(D) \cdot \sigma(D')$$

holds in $\mathbb{K}[\widehat{\mathcal{IT}_{\epsilon_1,\epsilon_3}}]$ *for two infinitesimal tangles* $D \in \mathbb{K}[\widehat{\mathcal{IT}_{\epsilon_1,\epsilon_2}}]$ *and* $D' \in \mathbb{K}[\widehat{\mathcal{IT}_{\epsilon_2,\epsilon_3}}]$ *for any* ϵ_1, ϵ_2 *and* ϵ_3.

(2) For each ϵ, *the subspace* $\mathbb{K}[\widehat{\mathcal{ISL}_\epsilon}]$ *of infinitesimal string links with type* ϵ *is stable under the above* $GRT(\mathbb{K})$*-action. The action is compatible with its algebra structure whose product is given by the composition.*

(3) The subspace $\widehat{\mathbb{K}[\mathcal{IK}]}$ *of infinitesimal knots is stable under the above* $GRT(\mathbb{K})$*-action. The action there is not compatible with the connected sum* $\sharp$ *however the equality*

$$\sigma(D_1 \sharp D_2)\sharp\sigma(\acute{\bigcirc}) = \sigma(D_1)\sharp\sigma(D_2) \tag{36}$$

holds for any $\sigma \in GRT(\mathbb{K})$ *and any* $D_1, D_2 \in \widehat{\mathbb{K}[\mathcal{IK}]}$.

Proof Proof can be done in a completely same way to the proof of Proposition 3.11 and 3.13.

We denote the above induced actions respectively by

$$\rho_{\epsilon,\epsilon'} : GRT(\mathbb{K}) \to \mathrm{Aut}\ \mathbb{K}[\widehat{\mathcal{IT}_{\epsilon,\epsilon'}}], \tag{37}$$

$$\rho_\epsilon : GRT(\mathbb{K}) \to \mathrm{Aut}\ \mathbb{K}[\widehat{\mathcal{ISL}_\epsilon}], \tag{38}$$

$$\rho_0 : GRT(\mathbb{K}) \to \mathrm{Aut}\ \widehat{\mathbb{K}[\mathcal{IK}]}. \tag{39}$$

We may say that the map (37) is a generalization of the map (14) into the infinitesimal tangle case. We will see that the action (38) restricted into $GRT_1(\mathbb{K})$ is given by inner conjugation in Theorem 4.10 and that the action (39) is given by $\mathbb{G}_m$-action in Proposition 4.19.

3.3 Associators

In this subsection we reformulate the isomorphism given in [28] Theorem 2.4 (also shown in [6, 12, 28, 30, 34]) in our terminologies of proalgebraic and infinitesimal tangles. It is shown that each associator gives an isomorphism between the system of proalgebraic tangles and the system of infinitesimal tangles in Proposition 3.25. Proposition 3.28 shows an equivalence of the notion of infinitesimal tangles and the notion of chord diagrams.

Similarly to our previous subsections, such an isomorphism is constructed piecewise.

Definition 3.24 Let $\bar{\Gamma} \in \mathbb{K}[\widehat{\mathcal{T}_{\epsilon,\epsilon'}}]/\mathcal{T}_N$ with any sequences ϵ, ϵ' and $N \geqslant 0$. Let Γ be its representative in $\mathbb{K}[\mathcal{T}^{\mathrm{pre}}_{\epsilon,\epsilon'}]$ with a presentation $\Gamma = \gamma_m \cdots \gamma_2 \cdot \gamma_1$ (γ_j: fundamental proalgebraic tangle). For $p = (\mu, \varphi) \in M(\mathbb{K})$, hence $\mu \in \mathbb{K}^\times$ and $\varphi \in \exp \mathfrak{f}_2$, we define

$$p(\bar{\Gamma}) := \overline{p(\gamma_m)\cdots p(\gamma_2)\cdot p(\gamma_1)} \in \mathbb{K}[\widehat{\mathcal{IT}_{\epsilon,\epsilon'}}]/\mathcal{IT}_N. \tag{40}$$

with $p(\gamma_j)$ given below.

(1) when $\gamma_j \in A^{\epsilon}_{k,l}$, we define

$$p(\gamma_j) := \gamma_j \cdot (\nu_\varphi)^{s(\gamma_j)}_{k+2} \cdot \varphi^{s(\gamma_j)}_{1\cdots k,k+1,k+2}.$$

Here ν_φ is the infinitesimal string link defined in Definition 3.22.

(2) when $\gamma_j = (b_n, \epsilon) \in \widehat{B^{\epsilon}_{\tau}}$ with $b_n \in \widehat{\mathbb{K}[P_n]}\cdot\tau \subset \widehat{\mathbb{K}[B_n]}$, we define

$$p(\gamma_j) := (\rho_n(p)(b_n), \epsilon).$$

Here $\rho_n(p)(b_n) \in \widehat{U\mathfrak{b}_n}$ is the image of b_n by the map given in Proposition 2.19.

(3) when $\gamma_j \in C^{\epsilon}_{k,l}$, we define

$$p(\gamma_j) := \varphi^{-1,t(\gamma_j)}_{1\cdots k,k+1,k+2} \cdot \gamma_j.$$

As an analogue of Propositions 3.11 and 3.23 in this subsection, we have the following:

Proposition 3.25 *(1) For each $p = (\mu, \varphi) \in M(\mathbb{K})$, the Eq.* (40) *yields a well-defined isomorphism of $\mathbb{K}$-linear spaces*

$$\rho^N_{\epsilon,\epsilon'}(p) : \mathbb{K}[\widehat{\mathcal{T}_{\epsilon,\epsilon'}}]/\mathcal{T}_N \simeq \mathbb{K}[\widehat{\mathcal{IT}_{\epsilon,\epsilon'}}]/\mathcal{IT}_N$$

for any ϵ and ϵ' and $N \geqslant 0$.

(2) This induces an isomorphism of $\mathbb{K}$-linear spaces

$$\rho_{\epsilon,\epsilon'}(p) : \mathbb{K}[\widehat{\mathcal{T}_{\epsilon,\epsilon'}}] \simeq \mathbb{K}[\widehat{\mathcal{IT}_{\epsilon,\epsilon'}}] \tag{41}$$

such that the equality

$$\rho_{\epsilon_1,\epsilon_3}(p)(\Gamma\cdot\Gamma') = \rho_{\epsilon_1,\epsilon_2}(p)(\Gamma)\cdot\rho_{\epsilon_2,\epsilon_3}(p)(\Gamma')$$

holds in $\mathbb{K}[\widehat{\mathcal{IT}_{\epsilon_1,\epsilon_3}}]$ for two proalgebraic tangles $\Gamma \in \mathbb{K}[\widehat{\mathcal{T}_{\epsilon_1,\epsilon_2}}]$ and $\Gamma' \in \mathbb{K}[\widehat{\mathcal{T}_{\epsilon_2,\epsilon_3}}]$ for any ϵ_1, ϵ_2 and ϵ_3.

(3) Restriction of the map (41) *into proalgebraic string links of type ϵ yields an isomorphism*

$$\rho_\epsilon(p) : \mathbb{K}[\widehat{\mathcal{SL}_\epsilon}] \simeq \mathbb{K}[\widehat{\mathcal{ISL}_\epsilon}]. \tag{42}$$

It is compatible with non-commutative product structures of both algebras given by the compositions.

(4) Restriction of the map (41) *into proalgebraic knots yields an isomorphism*

$$\rho_0(p) : \widehat{\mathbb{K}[\mathcal{K}]} \simeq \widehat{\mathbb{K}[\mathcal{IK}]}. \tag{43}$$

It is not compatible with the commutative product structure given by the connected sum $\sharp$ however the equality

$$\rho_0(p)(K_1 \sharp K_2) \,\sharp\, \rho_0(p)(\circlearrowleft) = \rho_0(p)(K_1) \,\sharp\, \rho_0(p)(K_2) \tag{44}$$

holds in $\widehat{\mathbb{K}[\mathcal{IK}]}$ *for any* $K_1, K_2 \in \widehat{\mathbb{K}[\mathcal{K}]}$.

Proof (1) Firstly we have to show that Definition 3.24 makes sense, that is, the map $\rho^N_{\epsilon,\epsilon'}(p)$ is well-defined. It is enough to prove the equality $\rho^N_{\epsilon,\epsilon'}(p)(\bar{\Gamma}_1) = \rho^N_{\epsilon,\epsilon'}(p)(\bar{\Gamma}_2)$ for Γ_1 and $\Gamma_2 \in \widehat{\mathbb{K}[\mathcal{T}_{\epsilon,\epsilon'}]}$ when Γ_1 is obtained from Γ_2 by a single operation of one of the moves (T1)–(T6). This can be proved in a completely same way to the proof of [22] Theorem 2.38 (1).

Secondly we prove that the map $\rho^N_{\epsilon,\epsilon'}(p)$ is isomorphic. This is achieved by considering the $\mathbb{K}$-linear map

$$S_a : \mathcal{IT}_a \,/\, \mathcal{IT}_{a+1} \to \mathcal{T}_a \,/\, \mathcal{T}_{a+1}$$

for $a = 0, 1, 2, \ldots, N-1$. It is a map sending each $\Gamma = \gamma_m \cdots \gamma_1$ with each γ_i belonging to one V_i of the ABC-spaces to $S_a(\Gamma) := S_a(\gamma_m) \cdots S_a(\gamma_1)$ with $S_a(\gamma_i) = \gamma_i$ when $V_i = A^\epsilon_{k,l}$ or $C^\epsilon_{k,l}$ for some k, l, ϵ, and $S_a(\gamma_i) = \rho_n(p)^{-1}(\gamma_i)$ when $V_i = \widehat{IB^\tau_\epsilon}$ for some $\tau \in \mathfrak{S}_n$ (some $n \geqslant 1$) and ϵ. Here $\rho_n(p) : \widehat{\mathbb{K}[B_n]} \to \widehat{U\mathfrak{b}_n}$ is the isomorphism given by (17). The well-definedness of S_a can be checked directly. To see the compatibility for the move (IT6), we need to use the congruence $\rho_n(p)^{-1}(t_{k+1,k+2}) \equiv \sigma_{k+1} - \sigma^{-1}_{k+1} \pmod{I^1}$. By the construction, S_a is isomorphic. By checking that S_a gives an right inverse of

$$\rho^{a+1}_{\epsilon,\epsilon'}(p)\,\big|_{\mathcal{T}_a} : \mathcal{T}_a \,/\, \mathcal{T}_{a+1} \to \mathcal{IT}_a \,/\, \mathcal{IT}_{a+1},$$

inductively we get that $\rho^{a+1}_{\epsilon,\epsilon'}(p)$ is isomorphic.

(2) Since both filtration $\{\mathcal{T}_n\}_{n=0}^\infty$ and $\{\mathcal{IT}_n\}_{n=0}^\infty$ are compatible with the isomorphism $\rho^N_{\epsilon,\epsilon'}(p)$, the isomorphism (41) is obtained. Checking the equality of the composition is immediate to see.

(3) It can be proved by the same arguments of the proof of Proposition 3.12.

(4) The statements follow from the same arguments given in the proof of Proposition 3.13.

The above isomorphism (41) is compatible with both $GT(\mathbb{K})$-action on $\widehat{\mathbb{K}[\mathcal{T}_{\epsilon,\epsilon'}]}$ and $GRT(\mathbb{K})$-action on $\widehat{\mathbb{K}[\mathcal{IT}_{\epsilon,\epsilon'}]}$.

Proposition 3.26 *The induced maps*

$$\rho_{\epsilon,\epsilon'} : M(\mathbb{K}) \to \mathrm{Isom}\left(\widehat{\mathbb{K}[\mathcal{T}_{\epsilon,\epsilon'}]},\ \widehat{\mathbb{K}[\mathcal{IT}_{\epsilon,\epsilon'}]}\right), \tag{45}$$

$$\rho_\epsilon : M(\mathbb{K}) \to \mathrm{Isom}\left(\widehat{\mathbb{K}[\mathcal{SL}_\epsilon]},\ \widehat{\mathbb{K}[\mathcal{ISL}_\epsilon]}\right), \tag{46}$$

$$\rho_0 : M(\mathbb{K}) \to \mathrm{Isom}\left(\widehat{\mathbb{K}[\mathcal{K}]},\ \widehat{\mathbb{K}[\mathcal{IK}]}\right) \tag{47}$$

are all morphisms of bitorsors.

Proof It is derived from Proposition 2.19.

We may say that the map (45) is a generalization of the map (2.19) into proalgebraic tangles.

Next we will discuss a relation of infinitesimal tangles with chord diagrams.

Notation 3.27 ([27, 28] etc.) Let Γ be a tangle in $\mathcal{T}_{\epsilon,\epsilon'}$. A *chord diagram* on a curve Γ is a finite (possibly empty) set of unordered pair of points on $\Gamma \setminus \partial\Gamma$. A homeomorphism of chord diagrams means a homeomorphism of the underlying curves preserving their orientations and fixing their endpoints such that it preserves the distinguished pairs of points. In our picture, we draw a dashed line, called a chord, between the two points of a distinguished pair.

We denote $\mathcal{CD}^m_{\epsilon,\epsilon'}$ to be the $\mathbb{K}$-linear space generated by all homeomorphism classes of chord diagrams with m chords ($m \geqslant 0$) on tangles of type (ϵ, ϵ'), subject to the 4T-relation and the FI-relation. Here the *4T-relation* stands for the 4 terms relation defined by $D_1 - D_2 + D_3 - D_4 = 0$ where D_j are chord diagrams with four chords identical outside a ball in which they differ as illustrated in Fig. 10 and the *FI-relation* stands for the frame independent relation where we put $D = 0$ for any chord diagrams D with an isolated chord, a chord that does not intersect on any other one in their diagrams. We put $\widehat{\mathcal{CD}}_{\epsilon,\epsilon'} := \widehat{\bigoplus}_{m=0}^{\infty} \mathcal{CD}^m_{\epsilon,\epsilon'}$.

It is known that it is well-behaved under the composition

$$\cdot : \widehat{\mathcal{CD}}_{\epsilon_3,\epsilon_2} \times \widehat{\mathcal{CD}}_{\epsilon_2,\epsilon_1} \to \widehat{\mathcal{CD}}_{\epsilon_3,\epsilon_1}. \tag{48}$$

We denote the subspace of $\widehat{\mathcal{CD}}_{\epsilon,\epsilon'}$ consisting of chord diagrams whose underlying spaces are string links with type ϵ by $\widehat{\mathcal{CD}}(\epsilon)$. It forms a non-commutative $\mathbb{K}$-algebra by the composition map (48).

The subspace $\widehat{\mathcal{CD}}(\circlearrowleft)$ ($\subset \widehat{\mathcal{CD}}_{\emptyset,\emptyset}$) of chord diagrams whose underlying spaces are homeomorphic to the oriented circles forms a commutative algebra by the connected sum

$$\sharp : \widehat{\mathcal{CD}}(\circlearrowleft) \times \widehat{\mathcal{CD}}(\circlearrowleft) \to \widehat{\mathcal{CD}}(\circlearrowleft). \tag{49}$$

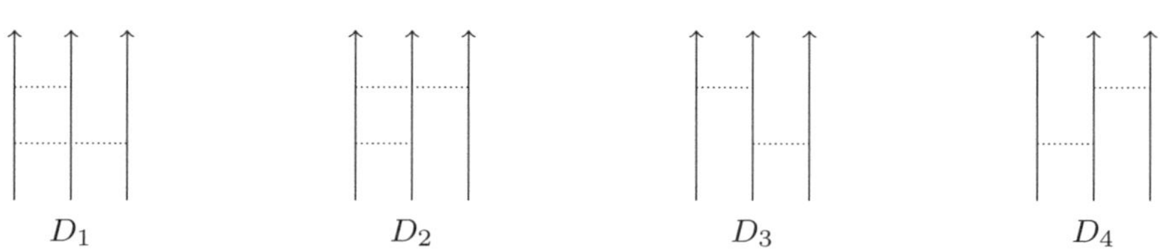

Fig. 10 4T-relation

We remind that the unit is given by the chordless chord diagram on the oriented circle $\circlearrowleft$.

Proposition 3.28 *(1) For each ϵ, ϵ', there is a natural identifications*

$$\mathbb{K}[\widehat{\mathcal{IT}_{\epsilon,\epsilon'}}] \simeq \widehat{\mathcal{CD}}_{\epsilon,\epsilon'}$$

which is compatible with the composition maps.

(2) For each ϵ, there is a natural identification of non-commutative graded $\mathbb{K}$-algebras:

$$\mathbb{K}[\widehat{\mathcal{ISL}_\epsilon}] \simeq \widehat{\mathcal{CD}}(\epsilon).$$

(3) There is a natural identification of commutative graded $\mathbb{K}$-algebras:

$$\widehat{\mathbb{K}[\mathcal{IK}]} \simeq \widehat{\mathcal{CD}}(\circlearrowleft).$$

Proof (1) By replacing each $\times$-part on the associated picture of each infinitesimal tangle D by $\times$ (actually we may replace it by $\times$ because both are equivalent modulo homeomorphisms of underlying tangles) and multiplying $(-1)^{D_\downarrow}$ to each D (which is necessary to keep 4T-relation), we obtain a well-defined $\mathbb{K}$-linear map

$$\mathbb{K}[\widehat{\mathcal{IT}_{\epsilon,\epsilon'}}] \to \widehat{\mathcal{CD}}_{\epsilon,\epsilon'}. \tag{50}$$

Here $D_\downarrow$ is the set of ends of chord on D which hit downward lines. Its composition with the map $\rho_{\epsilon,\epsilon'}(p)$ in (41) is the isomorphism

$$\mathbb{K}[\widehat{\mathcal{T}_{\epsilon,\epsilon'}}] \to \widehat{\mathcal{CD}}_{\epsilon,\epsilon'}.$$

given as the non-framed version of [28] Theorem 2.4. Since $\rho_{\epsilon,\epsilon'}(p)$ is isomorphic, the morphism in our claim should be isomorphic. Checking the compatibility with the composition map is immediate.

(2) It immediately follows from (1).

(3) It is obtained by a restriction of the above claim into the case $(\epsilon, \epsilon') = (\emptyset, \emptyset)$. It can be checked directly that the map is compatible with the connected sum.

Hereafter we identify $\mathbb{K}[\widehat{\mathcal{IT}_{\epsilon,\epsilon'}}]$ with $\widehat{\mathcal{CD}}_{\epsilon,\epsilon'}$, $\mathbb{K}[\widehat{\mathcal{ISL}_\epsilon}]$ with $\widehat{\mathcal{CD}}(\epsilon)$, and $\widehat{\mathbb{K}[\mathcal{IK}]}$ with $\widehat{\mathcal{CD}}(\circlearrowleft)$ by the map (50). Thus the identification given in Lemma 3.21 is reformulated as the identification below

$$(\widehat{\mathcal{CD}}(\epsilon), \cdot) \simeq (\widehat{\mathcal{CD}}(\circlearrowleft), \sharp) \tag{51}$$

for $\epsilon =\uparrow$ or $\downarrow$.

Remark 3.29 Kontsevich's isomorphism [26]

$$I : \widehat{\mathbb{C}[\mathcal{K}]} \simeq \widehat{\mathcal{CD}}(\circlearrowleft) \tag{52}$$

is given by specifying p of $\rho_0(p)$ of (43) to $p_{\mathrm{KZ}} = (1, \varphi_{\mathrm{KZ}}) \in M(\mathbb{C})$ with $\varphi_{\mathrm{KZ}} = \Phi_{\mathrm{KZ}}\left(\frac{1}{2\pi\sqrt{-1}}A, \frac{1}{2\pi\sqrt{-1}}B\right)$. We note that it is independent of the choice of φ (cf. [30]). For each oriented knot K, the image $I(K)$ is called the *Kontsevich invariant* of K.

Remark 3.30 By the same arguments to Remark 2.23, the proalgebraic tangles $\widehat{\mathbb{Q}[\mathcal{T}_{\epsilon,\epsilon'}]}$, the proalgebraic string links $\widehat{\mathbb{Q}[\mathcal{SL}_\epsilon]}$ and the proalgebraic knots $\widehat{\mathbb{K}[\mathcal{K}]}$ are regarded as Betti realizations of mixed Tate (pro-)motives over Spec $\mathbb{Z}$. And their corresponding de Rham realizations are given by the spaces $\widehat{\mathcal{CD}}_{\epsilon,\epsilon'}$, $\widehat{\mathcal{CD}}(\epsilon)$ and $\widehat{\mathcal{CD}}(\circlearrowleft)$ of chord diagrams located there respectively. In Remark 4.28, we will see that the proalgebraic knots $\widehat{\mathbb{K}[\mathcal{K}]}$ carries a structure of Tate (pro-)motives.

4 Main Results

We discuss and derive distinguished properties of the action of the Grothendieck-Teichmüller groups on proalgebraic tangles (constructed in Sect. 3.1) which can not be observed in the action on proalgebraic braids (discussed in Sect. 2.1). By exploiting the properties, we explicitly determine the proalgebraic knot whose Kontsevich invariant is the unit, the trivial chord diagram (Theorem 4.22).

4.1 Proalgebraic String Links

We restrict the previously constructed action of the Grothendieck-Teichmüller group $GT(\mathbb{K})$ on proalgebraic tangles into the action of its unipotent part $GT_1(\mathbb{K})$ on proalgebraic string links and show that it is simply described by an inner conjugation (Proposition 4.10 and Theorem 4.14). The proofs are based on Twistor Lemmas (Lemmas 4.2 and 4.12).

Notation 4.1 (1) For $n > 1$, $\epsilon_i : \widehat{\mathbb{K}[\mathcal{ISL}_{\uparrow^n}]} \to \widehat{\mathbb{K}[\mathcal{ISL}_{\uparrow^{n-1}}]}$ $(i = 1, 2 \ldots n)$ means the map sending an infinitesimal string link D to 0 if at least one chord of D has an endpoint on the i-th strand; otherwise $\epsilon_i(D)$ is obtained by removing the i-th strand.

(2) For $D \in \widehat{\mathbb{K}[\mathcal{ISL}_{\uparrow^n}]}$, we denote $D_{1,\ldots,n}$ (resp. $D_{2,\ldots,n+1}$) to be the element in $\widehat{\mathbb{K}[\mathcal{ISL}_{\uparrow^{n+1}}]}$ obtained by putting a chordless straight line on the right (resp. the left) of D and $D_{1,\ldots,i-1,i\ i+1,i+2,\ldots,n+1}$ $(i = 1, 2, \ldots, n)$ to be also the element in $\widehat{\mathbb{K}[\mathcal{ISL}_{\uparrow^{n+1}}]}$ obtained by doubling the i-th strand and taking the sum over all possible lifts of the chord endpoints of D from the i-th strand to one of the new two strands.

(3) For $D \in \mathbb{K}[\widehat{\mathcal{ISL}_{\uparrow^n}}]$ and $\tau \in \mathfrak{S}_n$, $D_{\tau(1),\dots,\tau(n)}$ denote the element in $\tau^{-1} \cdot D \cdot \tau$ where the product is taken as a product of infinitesimal tangles (recall the identification given in Proposition 3.28).

Hereafter we regard $\exp\mathfrak{f}_2$ and $U\mathfrak{p}_5$ to be the subspaces of $\mathbb{K}[\widehat{\mathcal{ISL}_{\uparrow\uparrow\uparrow}}]$ and $\mathbb{K}[\widehat{\mathcal{ISL}_{\uparrow\uparrow\uparrow\uparrow}}]$ respectively. The following lemma which is shown for $GRT(\mathbb{K})$ might be called as a reformulation of [30] Theorem 8 which is shown for a 'chord diagrammatic' analogue of $M_1(\mathbb{K})$.

Lemma 4.2 (Twistor Lemma) *Let $\sigma = (c, g) \in GRT(\mathbb{K})$, thus $c \in \mathbb{K}^\times$ and $g \in \exp\mathfrak{f}_2 \subset \mathbb{K}[\widehat{\mathcal{ISL}_{\uparrow\uparrow\uparrow}}]$. Then g is gauge equivalent to 1, namely, there exists $\Delta(\sigma) \in \mathbb{K}[\widehat{\mathcal{ISL}_{\uparrow\uparrow}}]^\times$ satisfying*

$$\epsilon_1(\Delta(\sigma)) = \epsilon_2(\Delta(\sigma)) = \uparrow \tag{53}$$

and the symmetric condition

$$\Delta(\sigma) = \Delta(\sigma)_{2,1} \tag{54}$$

such that

$$g = \Delta(\sigma)_{2,3} \cdot \Delta(\sigma)_{1,23} \cdot \Delta(\sigma)^{-1}_{12,3} \cdot \Delta(\sigma)^{-1}_{1,2} \tag{55}$$

holds in $\mathbb{K}[\widehat{\mathcal{ISL}_{\uparrow\uparrow\uparrow}}]^\times$.

Proof The proof can be done recursively in the same way to the proof of [30] Theorem 8, or rather, we may say that actually it is easier: Because we have $(1, g)$ and $(1, 1) \in GRT_1(\mathbb{K})$, both g and 1 satisfy the same relations (9)–(11). Our proof is obtained just by replacing Φ by g and Φ' by 1 in their proof.

Remark 4.3 Other variants of twistor lemma can be found in several literatures such as [32] Theorem 2.1 with twistors in $\mathrm{Aut}\widehat{F_2}$ for $\widehat{GT}_1$, [2] Theorem 7.5 with twistors in $\mathrm{TAut}\mathfrak{f}_2$ for KRV_3^0, and [3] Theorem 2 with twistors in $\mathrm{TAut}\mathfrak{f}_2$ for $M_1(\mathbb{K})$. Actually all of them are attributed to [16] Theorem A$'$.

The above $\Delta(\sigma)$ may not be uniquely chosen but it can be chosen independently from c by the construction. When we make such a choice, we occasionally denote $\Delta(g)$ instead of $\Delta(\sigma)$ by abuse of notations. We note that the first two terms on the right hand side of (55) commute each other, so do the last two terms.

Definition 4.4 We call such $\Delta(\sigma)$ in $\mathbb{K}[\widehat{\mathcal{ISL}_{\uparrow\uparrow}}]^\times$ a *twistor*[5] of $\sigma = (c, g) \in GRT(\mathbb{K})$. For a twisor $\Delta(\sigma)$, we put $\Delta(\sigma, \uparrow) := \uparrow \in \mathbb{K}[\widehat{\mathcal{ISL}_{\uparrow}}]^\times$ and

$$\Delta(\sigma, \uparrow^n) := \Delta(\sigma)_{12\cdots n-1,n} \cdots \Delta(\sigma)_{12,3} \cdot \Delta(\sigma)_{1,2} \in \mathbb{K}[\widehat{\mathcal{ISL}_{\uparrow^n}}]^\times \tag{56}$$

for $n \geqslant 2$. Here $\Delta(\sigma)_{1\cdots k,k+1}$ means the element in $\mathbb{K}[\widehat{\mathcal{ISL}_{\uparrow^n}}]$ obtained multi-doubling of the first strand of $\Delta(\sigma) \in \mathbb{K}[\widehat{\mathcal{ISL}_{\uparrow\uparrow}}]$ by k strands, summing up all

[5] We call this element twistor because it is related to Drinfeld's notion of twisting in [16].

possible lifts of chords and putting $n-k-1$ chordless line $\uparrow^{n-k-1}$ on its right. For any sequence $\epsilon = (\epsilon_i)_{i=1}^n$, we determine the element $\Delta(\sigma,\epsilon)$ in $\mathbb{K}[\widehat{\mathcal{ISL}_\epsilon}]$ by reversing corresponding all arrows of $\Delta(\sigma,\uparrow^n)$. We note that $\Delta(\sigma,\uparrow\uparrow)=\Delta(\sigma)$ and $g=\Delta(\sigma,\uparrow\uparrow\uparrow)_{3,2,1}\cdot\Delta(\sigma,\uparrow\uparrow\uparrow)^{-1}$ by (54).

An automorphism $\theta_{\epsilon,\epsilon'}^{\Delta(\sigma)}$ of $\mathbb{K}[\widehat{\mathcal{IT}_{\epsilon,\epsilon'}}]$ associated to a twistor $\Delta(\sigma)$ can be constructed piecewise as follows.

(1) when $D\in A_{k,l}^\epsilon$, we define

$$\theta_{\epsilon,\epsilon'}^{\Delta(\sigma)}(D) := D\cdot(\Delta(\sigma)_{k+1,k+2}^{-1})^{s(D)}\cdot(\nu_g)_{k+2}^{s(D)}\cdot g_{1\cdots k,k+1,k+2}^{s(D)}.$$

Here the term $(\Delta(\sigma)_{k+1,k+2}^{-1})^{s(D)}$ means the element in $\mathbb{K}[\widehat{\mathcal{ISL}_{s(D)}}]$ obtained by putting the trivial chordless diagram with k-strands on the left of $\Delta(\sigma)^{-1}$ and the trivial chordless diagram with l-strands on its right.

(2) when $D=(b_n,\epsilon)\in\widehat{IB_\tau^\epsilon}$, we define

$$\theta_{\epsilon,\epsilon'}^{\Delta(\sigma)}(D) := (\rho_n(\sigma)(b_n),\epsilon).$$

(3) when $D\in C_{k,l}^\epsilon$, we define

$$\theta_{\epsilon,\epsilon'}^{\Delta(\sigma)}(D) := g_{1\cdots k,k+1,k+2}^{-1,t(D)}\cdot(\Delta(\sigma)_{k+1,k+2})^{s(D)}\cdot D.$$

Here μ_g^ϵ is the one defined in Definition 3.22.

Lemma 4.5 *For any sequences ϵ,ϵ', any $\sigma\in GRT(\mathbb{K})$ and any twistor $\Delta(\sigma)$, the above construction determines a well-defined automorphism $\theta_{\epsilon,\epsilon'}^{\Delta(\sigma)}$ of $\mathbb{K}[\widehat{\mathcal{IT}_{\epsilon,\epsilon'}}]$ which is compatible with the composition map* (48).

Proof This can be verified by an almost same way to the proof of Proposition 3.23 except for compatibilities of (IT5) and (IT6).

- To check the compatibility of the first equality of (IT5), it is enough to show the equality illustrated in Fig. 11. By the identification of $\mathbb{K}[\widehat{\mathcal{ISL}_\uparrow}]$ with $\mathbb{K}[\widehat{\mathcal{IK}}]$ given in Lemma 3.21 (cf. (51)), showing the validity is deduced to showing the equality illustrated in Fig. 12. It is immediate to see because we have (IT6) and $\Delta(\sigma)_{2,1}=\Delta(\sigma)$ by (54).
  The compatibility of the second equality of (IT5) can be done in the same way.
- To check the compatibility of the first equality of (IT6), it is enough to show the equality illustrated in Fig. 13. Actually it is a consequence of (IT3), (IT5) and (IT6). The compatibility of the second equality of (IT6) can be done in the same way.

It is easily shown that the restriction of $\theta_{\epsilon,\epsilon}^{\Delta(\sigma)}$ into the algebra $\mathbb{K}[\widehat{\mathcal{ISL}_\epsilon}]$ of infinitesimal string links induces its automorphism, denoted by $\theta_\epsilon^{\Delta(\sigma)}$. The following explains its relationship with our previous automorphism $\rho_\epsilon(\sigma)$ in (37).

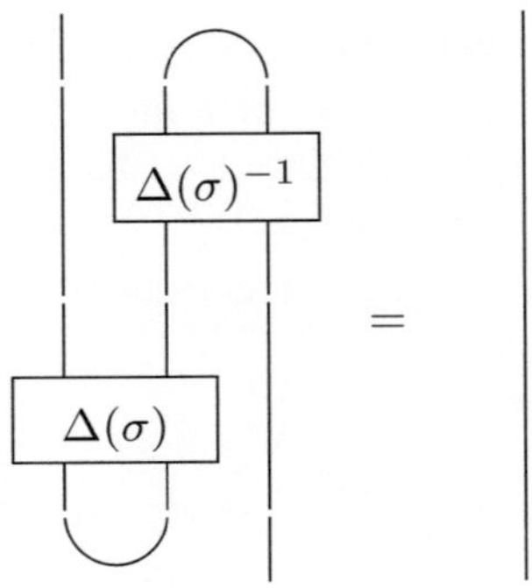

Fig. 11 The first equality of (IT5)

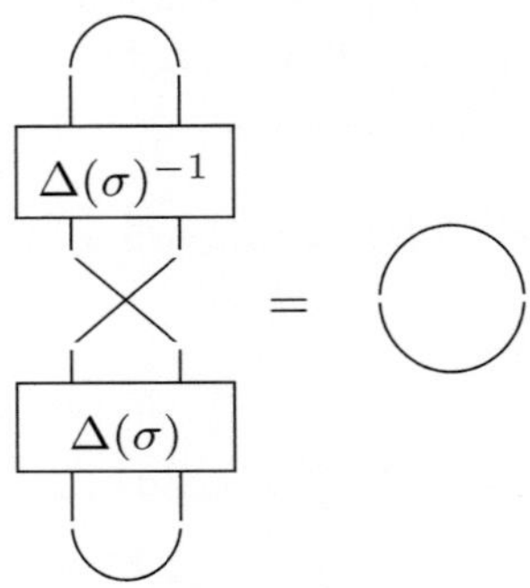

Fig. 12 Proof of Fig. 11

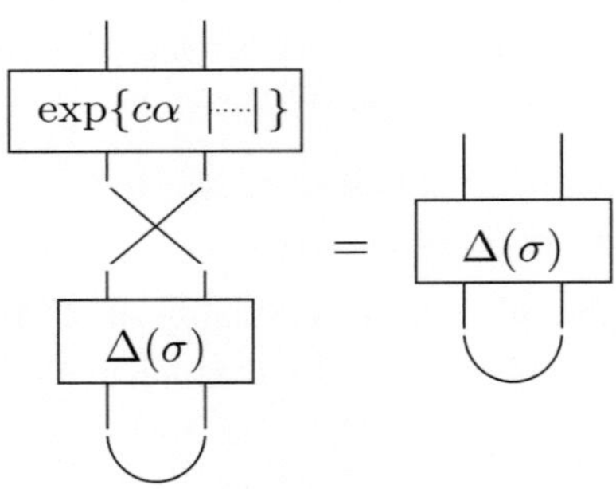

Fig. 13 The first equality of (IT6)

Proposition 4.6 *For any $\sigma \in GRT(\mathbb{K})$ and any sequence ϵ,*

$$\rho_\epsilon(\sigma)(D) = \theta_\epsilon^{\Delta(\sigma)}(D) \tag{57}$$

holds for all $D \in \mathbb{K}[\widehat{\mathcal{ISL}_\epsilon}]$.

Proof Differences between two actions $\theta_\epsilon^{\Delta(\sigma)}$ and $\rho_\epsilon(\sigma)$ are observed only on the action on $A_{k,l}^\epsilon$ and $C_{k,l}^\epsilon$. It is enough to show that equality of Fig. 12, which were proved in the above lemma.

When we restrict σ into the unipotent part $GRT_1(\mathbb{K})$, we obtain a relationship of $\theta^{\Delta(\sigma)}_{\epsilon,\epsilon'}$ with the identity map of $\mathbb{K}[\widehat{\mathcal{IT}_{\epsilon,\epsilon'}}]$ as follows.

Proposition 4.7 *For any $\sigma = (1, g) \in GRT_1(\mathbb{K})$ and any sequences ϵ and ϵ',*

$$\theta^{\Delta(\sigma)}_{\epsilon,\epsilon'}(D) = \Delta(\sigma, \epsilon) \cdot D \cdot \Delta(\sigma, \epsilon')^{-1}. \tag{58}$$

holds for any $D \in \mathbb{K}[\widehat{\mathcal{IT}_{\epsilon,\epsilon'}}]$.

Proof The equation can be checked piecewise.

- When $D \in A^{\epsilon}_{k,l}$, we may assume that $D = a^{\epsilon}_{k,l}$ with $s(D) = \epsilon_1$ and $t(D) = \epsilon_2$. Then

$$\begin{aligned}\theta^{\Delta(\sigma)}_{\epsilon_2,\epsilon_1}(a^{\epsilon}_{k,l}) &= a^{\epsilon}_{k,l} \cdot \Delta(\sigma)^{-1}_{k+1,k+2} \cdot (\mu_g)_{k+2} \cdot g_{1\cdots k,k+1,k+2} \\ &= a^{\epsilon}_{k,l} \cdot \Delta(\sigma)^{-1}_{k+1,k+2} \cdot (\mu_g)_{k+2} \\ &\quad \cdot \Delta(\sigma)_{k+1,k+2} \cdot \Delta(\sigma)_{1\cdots k,k+1\ k+2} \cdot \Delta(\sigma)^{-1}_{1\cdots k+1,k+2} \cdot \Delta(\sigma)^{-1}_{1\cdots k,k+1} \\ &= a^{\epsilon}_{k,l} \cdot (\mu_g)_{k+2} \cdot \Delta(\sigma)^{-1}_{1\cdots k+1,k+2} \cdot \Delta(\sigma)^{-1}_{1\cdots k,k+1}.\end{aligned}$$

  By Lemma 4.8,

$$\begin{aligned}&= a^{\epsilon}_{k,l} \cdot \Delta(\sigma)^{-1}_{1\cdots k+1,k+2} \cdot \Delta(\sigma)^{-1}_{1\cdots k,k+1} \\ &= \Delta(\sigma)_{1,2} \cdots \Delta(\sigma)_{1\cdots k+l-1,k+l} \cdot a^{\epsilon}_{k,l} \cdot \Delta(\sigma)^{-1}_{1\cdots k+l+1,k+l+2} \cdots \Delta(\sigma)^{-1}_{1,2} \\ &= \Delta(\sigma, \epsilon_2) \cdot a^{\epsilon}_{k,l} \cdot \Delta(\sigma, \epsilon_1)^{-1}.\end{aligned}$$

- When $D \in \widehat{IB}$, it is enough to to show the case when $D = \tau_{i,i+1}$ or $t_{i,i+1} \in U\mathfrak{b}_n$ $(1 \leqslant i \leqslant n-1)$. If $D = \tau_{i,i+1}$ with $s(D) = \epsilon_1$ and $t(D) = \epsilon_2$, by Proposition 2.12,

$$\begin{aligned}\theta^{\Delta(\sigma)}_{\epsilon_2,\epsilon_1}(\tau_{i,i+1}) &= g^{-1}_{1\cdots i-1,i,i+1} \cdot \tau_{i,i+1} \cdot g_{1\cdots i-1,i,i+1} \\ &= \Delta(\sigma)_{1\cdots i-1,i} \cdot \Delta(\sigma)_{1\cdots i,i+1} \cdot \Delta(\sigma)^{-1}_{1\cdots i-1,i\ i+1} \cdot \Delta(\sigma)^{-1}_{i,i+1} \\ &\quad \cdot \tau_{i,i+1} \cdot \Delta(\sigma)_{i,i+1} \cdot \Delta(\sigma)_{1\cdots i-1,i\ i+1} \cdot \Delta(\sigma)^{-1}_{1\cdots i,i+1} \cdot \Delta(\sigma)^{-1}_{1\cdots i-1,i}.\end{aligned}$$

  By (54),

$$\begin{aligned}&= \Delta(\sigma)_{1\cdots i-1,i} \cdot \Delta(\sigma)_{1\cdots i,i+1} \cdot \tau_{i,i+1} \cdot \Delta(\sigma)^{-1}_{1\cdots i,i+1} \cdot \Delta(\sigma)^{-1}_{1\cdots i-1,i} \\ &= \Delta(\sigma)_{1,2} \cdots \Delta(\sigma)_{1\cdots n-1,n} \cdot \tau_{i,i+1} \cdot \Delta(\sigma)^{-1}_{1\cdots n-1,n} \cdots \Delta(\sigma)^{-1}_{1,2} \\ &= \Delta(\sigma, \epsilon_2) \cdot \tau_{i,i+1} \cdot \Delta(\sigma, \epsilon_1)^{-1}.\end{aligned}$$

  If $D = t_{i,i+1}$ with $s(D) = \epsilon_1$ and $t(D) = \epsilon_2$, again by Proposition 2.12,

$$\begin{aligned}\theta^{\Delta(\sigma)}_{\epsilon_2,\epsilon_1}(t_{i,i+1}) &= g^{-1}_{1\cdots i-1,i,i+1} \cdot t_{i,i+1} \cdot g_{1\cdots i-1,i,i+1} \\ &= \Delta(\sigma)_{1\cdots i-1,i} \cdot \Delta(\sigma)_{1\cdots i,i+1} \cdot \Delta(\sigma)^{-1}_{1\cdots i-1,i\ i+1} \cdot \Delta(\sigma)^{-1}_{i,i+1} \\ &\quad \cdot t_{i,i+1} \cdot \Delta(\sigma)_{i,i+1} \cdot \Delta(\sigma)_{1\cdots i-1,i\ i+1} \cdot \Delta(\sigma)^{-1}_{1\cdots i,i+1} \cdot \Delta(\sigma)^{-1}_{1\cdots i-1,i}\end{aligned}$$

By Lemma 4.9,

$$
\begin{aligned}
=&\Delta(\sigma)_{1\cdots i-1,i}\cdot\Delta(\sigma)_{1\cdots i,i+1}\cdot t_{i,i+1}\cdot\Delta(\sigma)^{-1}_{1\cdots i,i+1}\cdot\Delta(\sigma)^{-1}_{1\cdots i-1,i}\\
=&\Delta(\sigma)_{1,2}\cdots\Delta(\sigma)_{1\cdots n-1,n}\cdot t_{i,i+1}\cdot\Delta(\sigma)^{-1}_{1\cdots n-1,n}\cdots\Delta(\sigma)^{-1}_{1,2}\\
=&\Delta(\sigma,\epsilon_2)\cdot t_{i,i+1}\cdot\Delta(\sigma,\epsilon_1)^{-1}.
\end{aligned}
$$

We note that we use $c=1$ for the second equality.

- When $D\in C^{\epsilon}_{k,l}$, we may assume that $D=c^{\epsilon}_{k,l}$ with $s(D)=\epsilon_1$ and $t(D)=\epsilon_2$. Then

$$
\begin{aligned}
\theta^{\Delta(\sigma)}_{\epsilon_2,\epsilon_1}(c^{\epsilon}_{k,l}) &= g^{-1}_{1\cdots k,k+1,k+2}\cdot\Delta(\sigma)_{k+1,k+2}\cdot c^{\epsilon}_{k,l}\\
&=\Delta(\sigma)_{1\cdots k,k+1}\cdot\Delta(\sigma)_{1\cdots k+1,k+2}\cdot\Delta(\sigma)^{-1}_{1\cdots k,k+1\ k+2}\cdot\Delta(\sigma)^{-1}_{k+1,k+2}\\
&\quad\cdot\Delta(\sigma)_{k+1,k+2}\cdot c^{\epsilon}_{k,l}\\
&=\Delta(\sigma)_{1\cdots k,k+1}\cdot\Delta(\sigma)_{1\cdots k+1,k+2}\cdot c^{\epsilon}_{k,l}\\
&=\Delta(\sigma)_{1,2}\cdots\Delta(\sigma)_{1\cdots k+l+1,k+l+2}\cdot c^{\epsilon}_{k,l}\cdot\Delta(\sigma)^{-1}_{1\cdots k+l,k+l}\cdots\Delta(\sigma)^{-1}_{1,2}\\
&=\Delta(\sigma,\epsilon_2)\cdot c^{\epsilon}_{k,l}\cdot\Delta(\sigma,\epsilon_1)^{-1}.
\end{aligned}
$$

Hence we get the equality (58) for any $D\in\mathbb{K}[\widehat{\mathcal{IT}_{\epsilon,\epsilon'}}]$.

The followings are required to prove the previous proposition.

Lemma 4.8 *For $\sigma=(c,g)\in GRT(\mathbb{K})$, the infinitesimal long knot $\mu_g\in\mathbb{K}[\widehat{\mathcal{ISL}_{\uparrow}}]$ is actually equal to the trivial chordless chord diagram on $\uparrow$.*

Proof It is enough to show that Λ_g is the trivial diagram $\downarrow$. Since $g=\Delta(\sigma)_{2,3}\cdot\Delta(\sigma)_{1,23}\cdot\Delta(\sigma)^{-1}_{12,3}\cdot\Delta(\sigma)^{-1}_{1,2}$, it is enough to show the equality depicted in Fig. 14. It can be proved in a same way to Fig. 12.

The following is proved in a topological way.

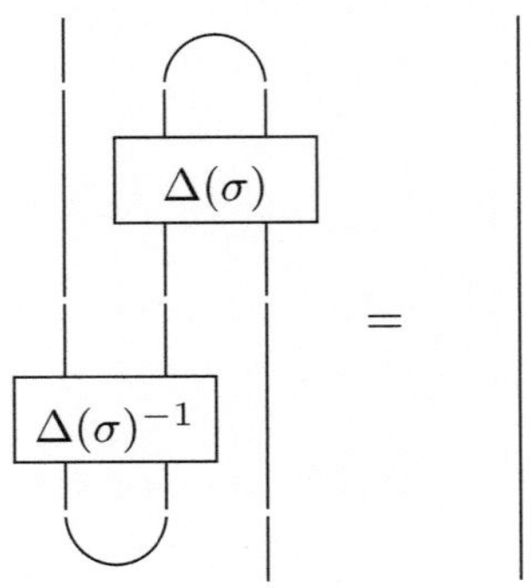

Fig. 14 Proof of Lemma 4.8

Lemma 4.9 *The element* t_{12} *lies in the center of* $\mathbb{K}[\widehat{\mathcal{ISL}_{\uparrow\uparrow}}]$. *Namely*

$$t_{12} \cdot D = D \cdot t_{12} \tag{59}$$

holds for any $D \in \mathbb{K}[\widehat{\mathcal{ISL}_{\uparrow\uparrow}}]$.

Proof First, for a discrete case, we have $T \cdot (\sigma_{1,2})^{2\alpha} = (\sigma_{1,2})^{2\alpha} \cdot T$ for any $T \in \mathcal{SL}(\uparrow\uparrow)$ and $\alpha \in \mathbb{Z}$. Whence we will have the same equality $T \cdot (\sigma_{1,2})^{2\alpha} = (\sigma_{1,2})^{2\alpha} \cdot T$ for any $T \in \mathbb{K}[\widehat{\mathcal{SL}_{\uparrow\uparrow}}]$ and $\alpha \in \mathbb{K}$. Then by the isomorphism (42), we have $D \cdot \exp\{\alpha t_{1,2}\} = \exp\{\alpha t_{1,2}\} \cdot D$ for any $D \in \mathbb{K}[\widehat{\mathcal{ISL}_{\uparrow\uparrow}}]$ and $\alpha \in \mathbb{K}$. Since the space $\mathbb{K}[\widehat{\mathcal{ISL}_{\uparrow\uparrow}}]$ is completed, we have the equality (59).

The proposition below states that our $GRT_1(\mathbb{K})$-action on $\mathbb{K}[\widehat{\mathcal{ISL}_{\epsilon}}]$ is described by an inner conjugation action of twistor.

Proposition 4.10 *Let* ϵ *be any sequence. The action* ρ_ϵ *of* (38) *restricted into the unipotent part* $GRT_1(\mathbb{K})$ *on the algebra* $\mathbb{K}[\widehat{\mathcal{ISL}_{\epsilon}}]$ *of infinitesimal string links is given by the inner conjugation of the twistor* $\Delta(\sigma, \epsilon)$, *i.e.*

$$\rho_\epsilon(\sigma)(D) = \Delta(\sigma, \epsilon) \cdot D \cdot \Delta(\sigma, \epsilon)^{-1} \tag{60}$$

holds for $\sigma = (1, g) \in GRT_1(\mathbb{K})$ *and* $D \in \mathbb{K}[\widehat{\mathcal{ISL}_{\epsilon}}]$.

Proof The formula is obtained by combining (4.6) and (4.7).

Next we discuss $GT_1(\mathbb{K})$-action on the algebra $\mathbb{K}[\widehat{\mathcal{SL}_{\epsilon}}]$ of proalgebraic string links.

Notation 4.11 (1) For $n > 1$, $\epsilon_i : \mathbb{K}[\widehat{\mathcal{SL}_{\uparrow^n}}] \to \mathbb{K}[\widehat{\mathcal{SL}_{\uparrow^{n-1}}}]$ $(i = 1, 2 \ldots n)$ means the map removing the i-th strand on each proalgebraic string links.

(2) For each $\Gamma \in \mathbb{K}[\widehat{\mathcal{SL}_{\uparrow^n}}]$, we denote $\Gamma_{1,\ldots,n}$ (resp. $\Gamma_{2,\ldots,n+1}$) to be the element in $\mathbb{K}[\widehat{\mathcal{SL}_{\uparrow^{n+1}}}]$ obtained by putting a straight line on the right (resp. the left) of Γ and $\Gamma_{1,\ldots,i-1,i\;i+1,i+2,\ldots,n+1}$ $(i = 1, 2, \ldots, n)$ to be also the element in $\mathbb{K}[\widehat{\mathcal{SL}_{\uparrow^{n+1}}}]$ obtained by *doubling* the i-th strand.

(3) Particularly for $\Gamma \in \mathbb{K}[\widehat{\mathcal{SL}_{\uparrow\uparrow}}]$, the symbol $\Gamma_{2,1}$ denotes the element $\sigma_{1,2}^{-1} \cdot \Gamma \cdot \sigma_{1,2}$ in $\Gamma \in \mathbb{K}[\widehat{\mathcal{SL}_{\uparrow\uparrow}}]$.

Hereafter we regard $F_2(\mathbb{K})$ and $P_5(\mathbb{K})$ to be the subspaces of $\mathbb{K}[\widehat{\mathcal{SL}_{\uparrow\uparrow\uparrow}}]$ and $\mathbb{K}[\widehat{\mathcal{SL}_{\uparrow\uparrow\uparrow\uparrow}}]$ respectively. The following lemma is a $GT(\mathbb{K})$-analogue of Twistor Lemma 4.2.

Lemma 4.12 (Twistor Lemma) *Let* $\sigma = (\lambda, f) \in GT(\mathbb{K})$, *thus* $\lambda \in \mathbb{K}^\times$ *and* $f \in F_2(\mathbb{K}) \subset \mathbb{K}[\widehat{\mathcal{SL}_{\uparrow\uparrow\uparrow}}]$. *Then* f *is gauge equivalent to* 1, *namely, there exists* $\varpi(\sigma) \in \mathbb{K}[\widehat{\mathcal{SL}_{\uparrow\uparrow}}]^\times$ *satisfying*

$$\epsilon_1(\varpi(\sigma)) = \epsilon_2(\varpi(\sigma)) = \uparrow \tag{61}$$

and the symmetric condition

$$\varpi(\sigma) = \varpi(\sigma)_{2,1} \tag{62}$$

such that

$$f = \varpi(\sigma)_{2,3} \cdot \varpi(\sigma)_{1,23} \cdot \varpi(\sigma)_{12,3}^{-1} \cdot \varpi(\sigma)_{1,2}^{-1} \tag{63}$$

holds in $\mathbb{K}[\widehat{\mathcal{SL}_{\uparrow\uparrow\uparrow}}]^\times$.

Proof Fix an element $p = (1, \varphi) \in M_1(\mathbb{K})$. Then by Proposition 2.18 it yields an isomorphism

$$r_p : GT_1(\mathbb{K}) \simeq GRT_1(\mathbb{K}).$$

By Proposition 3.25.(3), we have an isomorphism

$$\rho_{\uparrow\uparrow\uparrow}(p) : \mathbb{K}[\widehat{\mathcal{SL}_{\uparrow\uparrow\uparrow}}] \simeq \mathbb{K}[\widehat{\mathcal{ISL}_{\uparrow\uparrow\uparrow}}].$$

Put $\varpi(\sigma) := \rho_{\uparrow\uparrow\uparrow}(p)^{-1}\left(\Delta\left(r_p(\sigma)\right)\right)$. Then using Proposition 3.26 and Twistor Lemma 4.2, we can check the validities of (61)–(63) for $\varpi(\sigma)$ by direct calculations.

The above $\varpi(\sigma)$ may not be uniquely chosen and may depend on $\lambda \in \mathbb{K}^\times$ unlike the case for $GRT(\mathbb{K})$. Here we note again that the first two terms on the right hand side of (55) commute each other, so do the last two terms.

Definition 4.13 We call such $\varpi(\sigma)$ in $\mathbb{K}[\widehat{\mathcal{SL}_{\uparrow\uparrow}}]^\times$ a *twistor* of $\sigma = (\lambda, f) \in GT(\mathbb{K})$. For a twisor $\varpi(\sigma)$, we put $\varpi(\sigma, \uparrow) := \uparrow \in \mathbb{K}[\widehat{\mathcal{SL}_{\uparrow}}]^\times$ and

$$\varpi(\sigma, \uparrow^n) := \varpi(\sigma)_{12\cdots n-1,n} \cdots \varpi(\sigma)_{12,3} \cdot \varpi(\sigma)_{1,2} \in \mathbb{K}[\widehat{\mathcal{SL}_{\uparrow^n}}]^\times \tag{64}$$

for $n \geqslant 2$. Here $\varpi(\sigma)_{1\cdots k,k+1}$ means the element in $\mathbb{K}[\widehat{\mathcal{SL}_{\uparrow^n}}]$ obtained multi-doubling of the first strand of $\varpi(\sigma) \in \mathbb{K}[\widehat{\mathcal{SL}_{\uparrow\uparrow}}]$ by k strands, and putting $n-k-1$ straight lines $\uparrow^{n-k-1}$ on its right. For any sequence $\epsilon = (\epsilon_i)_{i=1}^n$, we determine the element $\varpi(\sigma, \epsilon)$ in $\mathbb{K}[\widehat{\mathcal{SL}_\epsilon}]^\times$ by reversing all corresponding arrows of $\varpi(\sigma, \uparrow^n)$. We note that $\varpi(\sigma, \uparrow\uparrow) = \varpi(\sigma)$ and $f = \varpi(\sigma, \uparrow\uparrow\uparrow)_{3,2,1} \cdot \varpi(\sigma, \uparrow\uparrow\uparrow)^{-1}$ by (62).

Our theorem in this subsection is to state that our $GT_1(\mathbb{K})$-action on $\mathbb{K}[\widehat{\mathcal{SL}_\epsilon}]$ is described by the inner conjugation action by the twistor.

Theorem 4.14 *Let ϵ be any sequence. The action*

$$\rho_\epsilon : GT_1(\mathbb{K}) \to \operatorname{Aut} \widehat{\mathbb{K}[\mathcal{SL}_\epsilon]}$$

induced by (29), *of the unipotent part* $GT_1(\mathbb{K})$ *on the algebra* $\widehat{\mathbb{K}[\mathcal{SL}_\epsilon]}$ *of proalgebraic string links is simply given by an inner conjugation of twistor* $\varpi(\sigma, \epsilon) \in \widehat{\mathbb{K}[\mathcal{SL}_\epsilon]}^\times$. *That is,*

$$\rho_\epsilon(\sigma)(\Gamma) = \varpi(\sigma, \epsilon) \cdot \Gamma \cdot \varpi(\sigma, \epsilon)^{-1} \tag{65}$$

holds for $\sigma = (1, f) \in GT_1(\mathbb{K})$ *and* $\Gamma \in \widehat{\mathbb{K}[\mathcal{SL}_\epsilon]}$.

Proof Let $p = (1, \varphi) \in M_1(\mathbb{K})$ be an element taken in the proof of Twistor Lemma 4.12. By Proposition 2.18 it yields an isomorphism $r_p : GT_1(\mathbb{K}) \simeq GRT_1(\mathbb{K})$. By Proposition 3.25.(3), we have an isomorphism

$$\rho_\epsilon(p) : \widehat{\mathbb{K}[\mathcal{SL}_{\uparrow^n}]} \simeq \widehat{\mathbb{K}[\mathcal{ISL}_{\uparrow^n}]}.$$

Our claim follows from Propositions 3.26 and 4.10. We may take $\varpi(\sigma, \epsilon)$ in the above (65). by $\rho_\epsilon(p)^{-1}\left(\Delta(r_p(\sigma), \epsilon)\right)$.

Here is a corollary of Theorem 4.14.

Corollary 4.15 *The restricted action of* (6) *into the unipotent part* $GT_1(\mathbb{K})$ *on proalgebraic pure braids, denoted by*

$$\rho_n : GT_1(\mathbb{K}) \to \operatorname{Aut} \widehat{\mathbb{K}[P_n]},$$

is simply given by an inner conjugation by the element $\varpi(\sigma, \uparrow^n) \in \widehat{\mathbb{K}[\mathcal{SL}_{\uparrow^n}]}$. *That is,*

$$\rho_n(\sigma)(x_{ij}) = \varpi(\sigma, \uparrow^n) \cdot x_{ij} \cdot \varpi(\sigma, \uparrow^n)^{-1} \tag{66}$$

holds for $1 \leqslant i, j \leqslant n$ *and* $\sigma \in GT_1(\mathbb{K})$.

We note that though $\varpi(\sigma, \uparrow^n)$ belongs to $\widehat{\mathbb{K}[\mathcal{SL}_{\uparrow^n}]}$ the right hand side of (66) belongs to its subspace $\widehat{\mathbb{K}[P_n]}$.

Remark 4.16 (1) Since the $GT(\mathbb{K})$-action on $\widehat{\mathbb{K}[P_n]}$ is faithful for $n \geqslant 3$, the action $\rho_\epsilon : GT_1(\mathbb{K}) \to \operatorname{Aut} \widehat{\mathbb{K}[\mathcal{SL}_\epsilon]}$ is faithful for $n \geqslant 3$, where n is the number of strings, i.e. the cardinality of ϵ, by Remark 3.8.

(2) When $n = 1$, the action is far from faithful. Actually the kernel of the action $\rho_\epsilon : GT_1(\mathbb{K}) \to \operatorname{Aut} \widehat{\mathbb{K}[\mathcal{SL}_\epsilon]}$ (with $\epsilon = \uparrow$ or $\downarrow$) on proalgebraic long knots is the unipotent part $GT_1(\mathbb{K})$ because $\varpi(\sigma, \epsilon)$ in Theorem 4.14 is trivial.

The rest case $n = 2$ looks unclear.

Problem 4.17 Is the action

$$\rho_\epsilon : GT(\mathbb{K}) \to \mathrm{Aut}\, \widehat{\mathbb{K}[\mathcal{SL}_\epsilon]}$$

(with $\epsilon = (\epsilon_1, \epsilon_2) \in \{\uparrow, \downarrow\}^2$) on proalgebraic 2-string links faithful?

4.2 Proalgebraic Knots

We discuss a non-trivial grading on proalgebraic knots induced by $GT(\mathbb{K})$-action. In [8] the image of the trivial knot, the unknot, under the Kontsevich isomorphism (52) is explicitly determined. In this subsection, we work in an opposite direction. That is, we explicitly calculate the inverse image γ_0 of the unit, the trivial (chordless) chord diagram, under the Kontsevich isomorphism. It is explicitly described as combinations of two-bridge knots (Theorem 4.22). We also show that the invariant space of proalgebraic knots under the $GT(\mathbb{K})$-action is one-dimensional generated by the element γ_0 (Theorem 4.27).

The following lemma is a knot analogue of Proposition 4.6.

Lemma 4.18 *For any $\sigma \in GRT(\mathbb{K})$*

$$\rho_0(\sigma)(D) = \theta_0^{\Delta(\sigma)}(D) \tag{67}$$

holds for all $D \in \widehat{\mathbb{K}[\mathcal{IK}]}$.

Proof It can be proven in a completely same way to the proof of Proposition 4.6.

We may say that the following is an analogue of Theorems 4.10 and 4.14 for proalgebraic knots.

Proposition 4.19 *(1) The $GRT(\mathbb{K})$-action ρ_0 constructed in* (39) *on the algebra $\widehat{\mathbb{K}[\mathcal{IK}]}$ of infinitesimal knots actually factors through $\mathbb{G}_m(\mathbb{K})(=\mathbb{K}^\times)$-action. Namely the kernel of the action is its unipotent part $GRT_1(\mathbb{K})$.*

(2) The $GT(\mathbb{K})$-action ρ_0 constructed in (31) *on the algebra $\widehat{\mathbb{K}[\mathcal{K}]}$ of proalgebraic knots actually factors through $\mathbb{G}_m(\mathbb{K})(=\mathbb{K}^\times)$-action. Namely the kernel of the action is its unipotent part $GT_1(\mathbb{K})$.*

Proof (1) It is obtained from a combination of (67) and (58) because we have $s(\acute{\bigcirc}) = t(\acute{\bigcirc}) = \emptyset$.

(2) The proof is almost same to the proof of Theorem 4.14. It can be derived from the above claim in this proposition and Proposition 3.26.

As a corollary of the above proposition, we obtain a non-trivial decomposition of knots below.

Corollary 4.20 *Each oriented knot K admits a canonical decomposition*

$$K = K_0 + K_1 + K_2 + \cdots \tag{68}$$

in $\widehat{\mathbb{K}[\mathcal{K}]}$ *such that*

$$\sigma(K_m) = \lambda^m \cdot K_m$$

holds for $\sigma = (\lambda, f) \in GT(\mathbb{K})$ *and* $m \geqslant 0$.

Proof It is a direct corollary of Proposition 4.19.(2). By the representation theory of $\mathbb{G}_m$, the completed vector space $\widehat{\mathbb{K}[\mathcal{K}]}$ is decomposed into the product of eigenspaces V_a $(a \in \mathbb{Z})$ where $\mathbb{G}_m$ acts as a multiplication of a-th power:

$$\widehat{\mathbb{K}[\mathcal{K}]} = \prod_{i \geqslant 0} V_i. \tag{69}$$

By our construction, the decomposition is compatible its filtration $\{\mathcal{K}\}_{N=0}^{\infty}$, i.e. $\prod_{i \geqslant N} V_i = \mathcal{K}_N$. We also have $\dim V_a < \infty$ for all $a \in \mathbb{Z}$ and actually $\dim V_a = 0$ for $a < 0$. Thus we get the claim.

Another Proof. Again by the representation theory of $\mathbb{G}_m$, Proposition 4.19(1) implies that the completed vector space $\widehat{\mathbb{K}[\mathcal{IK}]}$, hence $\widehat{\mathcal{CD}}(\circlearrowleft)$, is decomposed into the product of eigenspaces W_a $(a \in \mathbb{Z})$ where $\mathbb{G}_m$ acts as a multiplication of a-th power:

$$\widehat{\mathcal{CD}}(\circlearrowleft) = \prod_{i \geqslant 0} W_i. \tag{70}$$

Since the $\mathbb{G}_m$-action is compatible with the grading $\widehat{\mathcal{CD}}(\circlearrowleft) = \widehat{\oplus}_{m=0}^{\infty} \mathcal{CD}^m(\circlearrowleft)$ where $\mathcal{CD}^m(\circlearrowleft)$ is the $\mathbb{K}$-linear space spanned by the chord diagrams with m-chords, we have

$$W_{-m} = \mathcal{CD}^m(\circlearrowleft) \tag{71}$$

for $m \geqslant 0$. Thus $\dim W_a < \infty$ for all $a \in \mathbb{Z}$ and actually $\dim W_a = 0$ for $a > 0$. Let $p = (1, \varphi) \in M_1(\mathbb{K})$. Then we have an isomorphism

$$r_p : GT(\mathbb{K}) \simeq GRT(\mathbb{K})$$

by Proposition 2.18 and an isomorphism

$$\rho_0(p) : \widehat{\mathbb{K}[\mathcal{K}]} \simeq \widehat{\mathcal{CD}}(\circlearrowleft)$$

by Proposition 3.25.(4). By using Proposition 3.26, we can show that

$$V_a := \rho_0(p)^{-1}(W_{-a}) \tag{72}$$

is the eigenspace where $\mathbb{G}_m$ acts as a multiplication of a-th power and actually the space is invariant under any choice of $p = (1, \varphi) \in M_1(\mathbb{K})$.

We note that in the decomposition (69), we have $\dim V_m = \dim \mathcal{CD}^m(\bigcirc)$. for all $m \geqslant 0$.

Remark 4.21 Let $\iota_0 = (-1, 1) \in GT(\mathbb{K})$. Since it is an involution, the space $\widehat{\mathbb{K}[\mathcal{K}]}$ is divided into two eigenspaces V_+ and V_- where the action of ι_0 is given by the multiplication by 1 and -1 respectively. So an each oriented knot K is decomposed as

$$K = K_+ + K_- \in \widehat{\mathbb{K}[\mathcal{K}]} \quad \text{with} \quad K_\pm \in V_\pm.$$

Since $\iota_0(K)$ is nothing but the mirror image $\bar{K}$ of K, we have $K_+ = \frac{1}{2}(K + \bar{K})$ and $K_- = \frac{1}{2}(K - \bar{K})$. We note that in terms of the decomposition (68) they are expressed as

$$K_+ = \sum_{i \geqslant 0} K_{2i} \quad \text{and} \quad K_- = \sum_{i \geqslant 0} K_{2i+1}.$$

Our first theorem in this subsection is an explicit presentation of the proalgebraic knot whose Kontsevich invariant is trivial. That is, we explicitly calculate the inverse image of the unit, the trivial (chordless) chord diagram, under the Kontsevich isomorphism $I : \widehat{\mathbb{C}[\mathcal{K}]} \simeq \widehat{\mathcal{CD}}(\bigcirc)$ given in (52).

Theorem 4.22 *The inverse image $I^{-1}(e)$ of the unit $e \in \widehat{\mathcal{CD}}(\bigcirc)$ $(\simeq \widehat{\mathbb{K}[\mathcal{IK}]})$, under Kontsevich's isomorphism I is explicitly given by*

$$\gamma_0 := \bigcirc - c_0 + c_0 \sharp c_0 - c_0 \sharp c_0 \sharp c_0 + c_0 \sharp c_0 \sharp c_0 \sharp c_0 - \cdots \in \widehat{\mathbb{C}[\mathcal{K}]} \tag{73}$$

where $\sharp$ is the connected sum and $c_0 \in \widehat{\mathbb{C}[\mathcal{K}]}$ is given below (see also Fig. 15):

$$c_0 := \sum_{\substack{m,k_1,\ldots,k_m \in \mathbb{N} \\ k_m > 1}} (-1)^m \frac{\zeta^{\mathrm{inv}}(k_1,\ldots,k_m)}{(2\pi\sqrt{-1})^{k_1+\cdots+k_m}} \cdot \Big\{ a_{0,0}^{\curvearrowleft} \cdot a_{2,0}^{\downarrow\uparrow\curvearrowleft} \cdot (\log\sigma_2^2)^{k_m-1} \cdot (\log\sigma_3^2) \cdot$$
$$(\log\sigma_2^2)^{k_{m-1}-1} \cdot (\log\sigma_3^2) \cdots\cdots (\log\sigma_2^2)^{k_1-1} \cdot (\log\sigma_3^2) \cdot c_{1,1}^{\downarrow\curvearrowright\cup\uparrow} \cdot c_{0,0}^{\cup\curvearrowright} \Big\}. \tag{74}$$

Here we define

$$\log\sigma_i^2 = -\sum_{k=1}^{\infty} \frac{1}{k}\{1 - \sigma_i^2\}^k \in \widehat{\mathbb{C}[P_4]} \tag{75}$$

for $i = 2, 3$. And we define the *inversed MZV* $\zeta^{\mathrm{inv}}(k_1, \ldots, k_m)$ to be the coefficient of $A^{k_m-1}B \cdots A^{k_1-1}B$ multiplied by $(-1)^m$ in the *inversed KZ-associator* $\Phi_{\mathrm{KZ}}^{\mathrm{inv}}(A, B) \in$

$$c_0 = \sum_{\substack{m,k_1,\dots,k_m \in \mathbb{N} \\ k_m > 1}} \frac{(-1)^m \zeta^{\mathrm{inv}}(k_1,\dots,k_m)}{(2\pi\sqrt{-1})^{k_1+\cdots+k_m}} \cdot$$

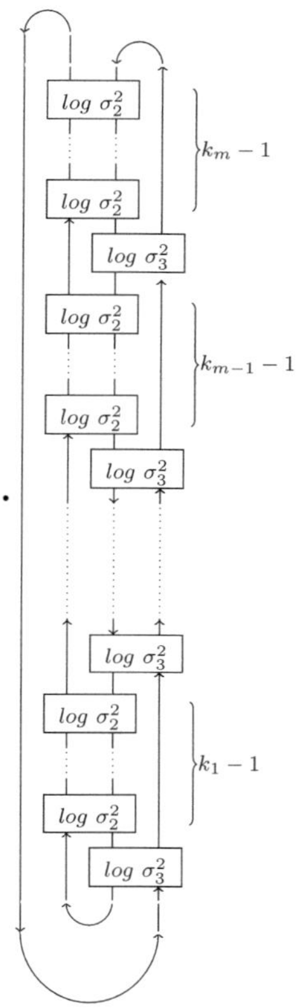

Fig. 15 c_0

$\mathbb{R}\langle\langle A, B\rangle\rangle$, which is the inverse of the KZ-associator $\Phi_{\mathrm{KZ}}(A, B)$ in (16) with respect to the multiplication (12). Namely

$$\Phi_{\mathrm{KZ}}^{\mathrm{inv}}(A, B) =: 1 + \sum_{\substack{m,k_1,\dots,k_m \in \mathbb{N} \\ k_m > 1}} (-1)^m \zeta^{\mathrm{inv}}(k_1, \cdots, k_m) A^{k_m - 1} B \cdots A^{k_1 - 1} B$$
$$+ \text{(other terms)} \tag{76}$$

where $\Phi_{\mathrm{KZ}}^{\mathrm{inv}}(A, B)$ is the series uniquely defined by

$$\Phi_{\mathrm{KZ}}^{\mathrm{inv}}\left(\Phi_{\mathrm{KZ}}(A, B) \cdot A \cdot \Phi_{\mathrm{KZ}}(A, B)^{-1}, B\right) = \Phi_{\mathrm{KZ}}(A, B)^{-1}. \tag{77}$$

The inversed MZV's with small depths are calculated in Example 4.24 and the first two terms of γ_0 is calculated in Example 4.25.

Proof By our construction, the degree 0-part of the image $I(\bigcirc)$ of the trivial knot (unknot) $\bigcirc$ under the Kontsevich isomorphism I given in (52) is the trivial (chordless) chord diagram e. Therefore on the decomposition $K = K_0 + K_1 + \cdots$ in (68) for $K = \bigcirc$, we have

$$K_0 = I^{-1}(e) \tag{78}$$

by (72). To calculate K_0, we take $\mathbb{K}$ to be the polynomial algebra $\mathbb{C}[T^{\pm}]$ generated by T and T^{-1} and take $p \in M_1(\mathbb{C})$ to be a specific element $p_{\mathrm{KZ}} = (1, \varphi_{\mathrm{KZ}}) \in M(\mathbb{C})$ with $\varphi_{\mathrm{KZ}} = \Phi_{\mathrm{KZ}}\left(\frac{1}{2\pi\sqrt{-1}}A, \frac{1}{2\pi\sqrt{-1}}B\right)$. We have $(T^{-1}, 1) \in GRT(\mathbb{K})$. By Proposition 2.18, we obtain a unique element $(T, f_T) \in GT(\mathbb{K}(T))$ satisfying

$$(T^{-1}, 1) \circ (1, \varphi_{\mathrm{KZ}}) = (1, \varphi_{\mathrm{KZ}}) \circ (T, f_T).$$

It can be read as

$$\varphi_{\mathrm{KZ}}(TA, TB) = f_T(\varphi_{\mathrm{KZ}} \cdot e^A \cdot \varphi_{\mathrm{KZ}}^{-1}, e^B) \cdot \varphi_{\mathrm{KZ}}.$$

We get that f_T belongs to $F_2(\mathbb{C}[T])$. So

$$f_T\left(\exp\{\varphi_{\mathrm{KZ}} \cdot A \cdot \varphi_{\mathrm{KZ}}^{-1}\}, \exp\{B\}\right) \equiv \varphi_{\mathrm{KZ}}^{-1} \pmod{T}.$$

By replacing A and B by $2\pi\sqrt{-1}A$ and $2\pi\sqrt{-1}B$ respectively, we obtain

$$f_T(e^{2\pi\sqrt{-1}A}, e^{2\pi\sqrt{-1}B}) \equiv \Phi_{\mathrm{KZ}}^{\mathrm{inv}} \pmod{T}.$$

It says that

$$f_T(\sigma_1^2, \sigma_2^2) \equiv \Phi_{\mathrm{KZ}}^{\mathrm{inv}}\left(\frac{\log \sigma_1^2}{2\pi\sqrt{-1}}, \frac{\log \sigma_2^2}{2\pi\sqrt{-1}}\right) \pmod{T},$$

where precisely it means that

$$\mathrm{red}\left(f_T(\sigma_1^2, \sigma_2^2)\right) = \mathrm{red}\left(\Phi_{\mathrm{KZ}}^{\mathrm{inv}}\left(\frac{\log \sigma_1^2}{2\pi\sqrt{-1}}, \frac{\log \sigma_2^2}{2\pi\sqrt{-1}}\right)\right)$$

with the reduction map $\mathrm{red} : F_2(\mathbb{C}[T]) \to F_2(\mathbb{C})$ caused by putting $T = 0$. Therefore by Definition 3.9

$$\begin{aligned}
\Lambda^{\uparrow}_{f_T} &= a^{\uparrow\curvearrowleft}_{1,0}\cdot f_T(\sigma_1^2,\sigma_2^2)\cdot c^{\smile\uparrow}_{0,1}\\
&\equiv a^{\uparrow\curvearrowleft}_{1,0}\cdot \Phi^{\mathrm{inv}}_{\mathrm{KZ}}\left(\frac{\log\sigma_1^2}{2\pi\sqrt{-1}},\frac{\log\sigma_2^2}{2\pi\sqrt{-1}}\right)\cdot c^{\smile\uparrow}_{0,1} \pmod T\\
&= a^{\uparrow\curvearrowleft}_{1,0}\cdot c^{\smile\uparrow}_{0,1} + \sum_{\substack{m,k_1,\dots,k_m\in\mathbb{N}\\ k_m>1}} (-1)^m\frac{\zeta^{\mathrm{inv}}(k_1,\dots,k_m)}{(2\pi\sqrt{-1})^{k_1+\cdots+k_m}}\cdot\Big\{a^{\uparrow\curvearrowleft}_{1,0}\cdot(\log\sigma_1^2)^{k_m-1}\cdot(\log\sigma_2^2)\\
&\qquad\cdot(\log\sigma_1^2)^{k_{m-1}-1}\cdot(\log\sigma_2^2)\cdot\cdots\cdot(\log\sigma_1^2)^{k_1-1}\cdot(\log\sigma_2^2)\cdot c^{\smile\uparrow}_{0,1}\Big\}.
\end{aligned}$$

The third equality follows from (76). Here we note that the non-admissible terms, the ‘other terms’ in (76), all vanish by (T6). By the isomorphism (24), we have

$$\mathrm{cl}(\Lambda^{\uparrow}_{f_T}) = a^{\curvearrowleft}_{0,0}\cdot(\downarrow\otimes\Lambda^{\uparrow}_{f_T})\cdot c^{\smile}_{0,0}\equiv \bigcirc + c_0 \pmod T$$

(for c_0, see Fig. 15). Whence, by taking inverse of both, we have

$$\mathrm{cl}(\nu^{\uparrow}_{f_T})\equiv\gamma_0 \pmod T \tag{79}$$

(for γ_0, see (1)). We note that the infinite summation on the right hand side of (73) converges because $c_0^{\sharp n}\in\mathcal{K}_n$ for $n\geqslant 1$. By Corollary 4.20, K_0 is obtained by evaluating $(T,f_T)(\bigcirc)$ at $T=0$, i.e.

$$(T,f_T)(\bigcirc)\equiv K_0 \pmod T. \tag{80}$$

Since $\bigcirc = a^{\curvearrowleft}_{0,0}\cdot c^{\smile}_{0,0}$, the action on $\bigcirc$ is calculated to be

$$(T,f_T)(\bigcirc) = a^{\curvearrowleft}_{0,0}\cdot(\downarrow\otimes\nu^{\uparrow}_{f_T})\cdot c^{\smile}_{0,0} = \mathrm{cl}(\nu^{\uparrow}_{f_T}). \tag{81}$$

By (79)–(81), we have

$$\gamma_0 = K_0 \tag{82}$$

because both γ_0 and K_0 are independent of T. Finally we obtain

$$\gamma_0 = I^{-1}(e)$$

by (78).

It might be worthy to mention that our c_0 is given by a linear combination of two bridge knots and our γ_0 is given by a linear combination of connected sums of two bridge knots.

Remark 4.23 We remind that the image $I(\bigcirc)$ of the unit $\bigcirc$, the trivial knot, under the Kontsevich isomorphism I was calculated in [8], which may be regarded as a calculation in an opposite direction to our theorem.

By the definition of inversed MZV's, (76) and (77), they are given by polynomial combinations of MZV's.

Example 4.24 (1) For $n > 1$,

$$\zeta^{\mathrm{inv}}(n) = -\zeta(n).$$

(2) For $a > 0$ and $b > 1$,

$$\zeta^{\mathrm{inv}}(a,b) = \zeta(a)\zeta(b) - \zeta(a,b) + \sum_{i=0}^{a-2}(-1)^i\binom{i+b-1}{i}\zeta(b+i)\zeta(a-i) \\ + (-1)^a\sum_{j=0}^{b-2}\binom{j+a-1}{j}\zeta(b-j)\zeta(a+j).$$

Here is a brief computation of our γ_0.

Example 4.25

$$\gamma_0 \equiv \bigcirc - \frac{1}{24}\{\bigcirc - 3_1^l\} \pmod{\mathcal{K}_4}.$$

Here 3_1^l stands for the left trefoil knot.

Proposition 4.26 $\gamma_0 \in \widehat{\mathbb{Q}[\mathcal{K}]}$.

Proof In the proof of above theorem, there is no specific reason to choose $(1, \varphi_{\mathrm{KZ}}) \in M_1(\mathbb{C})$. Take $(\mu, \varphi) \in M(\mathbb{K})$ to be any associator with an expansion

$$\varphi(A,B) = 1 + \sum_{\substack{m,k_1,\ldots,k_m\in\mathbb{N}\\ k_m>1}} (-1)^m c(k_1,\cdots,k_m)A^{k_m-1}B\cdots A^{k_1-1}B \\ + (\text{regularized terms}).$$

Then $(1, \varphi(A/\mu, B/\mu)) \in M_1(\mathbb{K})$. By using $(1, \varphi(A/\mu, B/\mu)) \in M_1(\mathbb{K})$ instead of $(1, \varphi_{\mathrm{KZ}}) \in M_1(\mathbb{C})$ in the proof of Theorem 4.22, we obtain an explicit formula of c_0, which is nothing but the replacement of $2\pi\sqrt{-1}$ by $\mu \in \mathbb{K}^\times$ and $\zeta(k_1,\ldots,k_m)$ by $c(k_1,\ldots,k_m)$ in (74). Thus our claim is obtained by particularly taking a rational associator, an element of $M(\mathbb{Q})$, whose existence is guaranteed in [16].

Our second theorem in this subsection is on the invariant subspace of $\widehat{\mathbb{K}[\mathcal{K}]}$ under our $GT(\mathbb{K})$-action (cf. (31)).

Theorem 4.27 *(1) The invariant subspace V_0 of $\widehat{\mathbb{K}[\mathcal{K}]}$ under our $GT(\mathbb{K})$-action is 1-dimensional and actually generated by γ_0.*

(2) On the decomposition $K = K_0 + K_1 + \cdots$ *given in* (68),

$$K_0 = \gamma_0$$

holds for any oriented knot K.

Proof (1) Since $\mathcal{CD}^0(\circlearrowleft)$ is 1-dimensional generated by the trivial (chordless) chord diagram e, the $\mathbb{K}$-linear space V_0 in (69) is 1-dimensional by (71) and (72). Thus it is enough to show that $\rho_0(p)^{-1}(e) = \gamma_0$ for a particular choice of $p = (1, \varphi) \in M_1(\mathbb{K})$. It is obtained by Theorem 4.22 because $\rho_0(p) = I$ for $p = p_{\mathrm{KZ}}$ (actually it holds for any $p \in M_1(\mathbb{K})$).

(2) By our construction, the degree 0-part of the image $I(K)$ is always equal to $e \in \mathcal{CD}^0(\circlearrowleft)$. So $I(K_0) = e$, by (72). Then by the arguments given above, we get $K_0 = \gamma_0$.

Though our knot γ_0 is not equal to the trivial knot, the unknot $\circlearrowleft$, we may say that it is the 'trivial' knot in a particular sense.

Remark 4.28 We saw in Remark 3.30 particularly that the space $\widehat{\mathbb{K}[\mathcal{K}]}$ of proalgebraic knots carries a structure of a mixed Tate (pro-)motive over Spec $\mathbb{Z}$. But Proposition 4.19 tells that it falls to infinite direct sum of Tate motives $\mathbb{Q}(n)$ for $n \geqslant 0$. Our γ_0 is the generator of the degree 0-part V_0. Finding a basis of the degree n-part V_n for another n, such as the basis which is dual to the Vassiliev invariants listed in [13] Table 3.2, is worthy to calculate.

Acknowledgements Part of the paper was written at Max Planck Institute for Mathematics. The author also thanks the institute for its hospitality. The referees' efforts to make this paper better is gratefully acknowledged. This work was supported by Grant-in-Aid for Young Scientists (A) 24684001.

References

1. Alekseev, A., Torossian, C.: Kontsevich deformation quantization and flat connections. Commun. Math. Phys. **300**(1), 47–64 (2010)
2. Alekseev, A., Torossian, C.: The Kashiwara-Vergne conjecture and Drinfeld's associators, Ann. Math. **175**(2), 415–463 (2012)
3. Alekseev, A., Enriquez, B., Torossian, C.: Drinfeld associators, braid groups and explicit solutions of the Kashiwara-Vergne equations. Publ. Math. Inst. Hautes Études Sci. **112**, 143–189 (2010)
4. André, Y.: Une introduction aux motifs (motifs purs, motifs mixtes, périodes), Panoramas et Synthèses, 17. Société Mathématique de France, Paris (2004)
5. Bar-Natan, D.: Vassiliev and quantum invariants of braids. In: Proceedings of Symposia in Applied Mathematics, vol. 51, 129–144. American Mathematical Society, Providence, RI (1996). The interface of knots and physics (San Francisco, CA, 1995)
6. Bar-Natan, D.: *Non-Associative Tangles Geometric Topology (Athens, GA, 1993).* AMS/IP Studies in Advanced Mathematics, 2.1, pp. 139–183. American Mathematical Society, Providence, RI (1997)

7. Bar-Natan, D.: On associators and the Grothendieck-Teichmuller group I. Selecta Math. (N.S.) **4**(2), 183–212 (1998)
8. Bar-Natan, D., Le, T.T.Q., Thurston, D.: Two applications of elementary knot theory to Lie algebras and Vassiliev invariants. Geom. Topol. **7**, 1–31 (2003)
9. Belyĭ, G. V.: Galois extensions of a maximal cyclotomic field, Izv. Akad. Nauk SSSR Ser. Mat. **43**(2), 267–276, 479 (1979)
10. Brown, F.: Mixed tate motives over Spec(Z). Ann. Math. **175**(2), 949–976 (2012)
11. Brown, F.: Single-valued periods and multiple zeta values. Forum Math. Sigma **2**(e25), 37 (2014)
12. Cartier, P.: Construction combinatoire des invariants de Vassiliev-Kontsevich des nœuds. C. R. Acad. Sci. Paris Ser. I Math. **316**(11), 1205–1210 (1993)
13. Chmutov, S., Duzhin, S., Mostovoy, J.: *Introduction to Vassiliev Knot Invariants*, Cambridge University Press, 2012
14. Deligne, P.: Le groupe fondamental de la droite projective moins trois points. In: Galois Groups Over $\mathbb{Q}$, Berkeley, CA, 1987. Mathematical Sciences Research Institute Publications, vol. 16, pp. 79–297. Springer, New York-Berlin (1989)
15. Deligne, P., Goncharov, A.: Groupes fondamentaux motiviques de Tate mixte. In: Annales Scientifiques l'Ecole Normale. Supérieure, vol. 38, no. 1, pp. 1–56 (2005)
16. Drinfel'd, V.G.: On quasitriangular quasi-Hopf algebras and a group closely connected with Gal($\overline{\mathbb{Q}}/\mathbb{Q}$). Leningr. Math. J. **2**(4), 829–860 (1991)
17. Furusho, H.: The multiple zeta value algebra and the stable derivation algebra. Publ. Res. Inst. Math. Sci. **39**(4), 695–720 (2003)
18. Furusho, H.: p-adic multiple zeta values I-p-adic multiple polylogarithms and the p-adic KZ equation. Inv. Math. **155**(2), 253–286 (2004)
19. Furusho, H.: Multiple zeta values and Grothendieck-Teichmüller groups. AMS Contemp. Math. **416**, 49–82 (2006)
20. Furusho, H.: p-adic multiple zeta values II—tannakian interpretations. Am. J. Math. **129**(4), 1105–1144 (2007)
21. Furusho, H.: Pentagon and hexagon equations. Ann. Math. **171**(1), 545–556 (2010)
22. Furusho, H.: Galois action on knots I: action of the absolute Galois group. Quant. Topol. **8**(2), 295–360 (2017)
23. Hain, R.: The Hodge de Rham theory of relative Malcev completion. In: Annales scientifiques de I'Ecole normale supérieure, vol. 31, no. 1, pp. 47–92 (1998)
24. Ihara, Y.: Braids, Galois groups, and some arithmetic functions. In: Proceedings of the International Congress of Mathematicians, Vol. I, II (Kyoto: 99–120 p., 1991). Mathematical Society, Tokyo, Japan (1990)
25. Joyal, A., Street, R.: Braided tensor categories. Adv. Math. **102**(1), 20–78 (1993)
26. Kassel, C.: *Quantum Groups*. Graduate Texts in Mathematics, vol. 155. Springer-Verlag, New York (1995)
27. Kassel, C., Rosso, M., Turaev, V.: Quantum groups and knot invariants, Panoramas et Synthèses, 5. Société Mathématique de France, Paris (1997)
28. Kassel, C., Turaev, V.: Chord diagram invariants of tangles and graphs. Duke Math. J. **92**(3), 497–552 (1998)
29. Kontsevich, M.: Vassiliev's knot invariants. In: I. M. Gel'fand Seminar Part 2. Advances in Soviet Mathematics, vol. 16, pp. 137–150. American Mathematical Society, Providence, RI (1993)
30. Le, T.T.Q., Murakami, J.: The universal Vassiliev-Kontsevich invariant for framed oriented links. Compos. Math. **102**(1), 41–64 (1996)
31. Le, T.T.Q., Murakami, J.: Kontsevich's integral for the Kauffman polynomial. Nagoya Math. J. **142**, 39–65 (1996)
32. Lochak, P., Schneps, L.: Every acyclotomic element of the profinite Grothendieck-Teichmüller group is a twist. Rev. Roum. Math. Pures Appl. **60**(2), 117–128 (2015)
33. Majid, S.: A Quantum Groups Primer. London Mathematical Society Lecture Note Series, vol. 292. Cambridge University Press, Cambridge (2002)

34. Piunikhin, S.: Combinatorial expression for universal Vassiliev link invariant. Comm. Math. Phys. **168**(1), 1–22 (1995)
35. Rossi, C.A., Willwacher, T.P.: *Etingof's conjecture about Drinfeld associators*, Preprint, arXiv:1404.2047
36. Ševera, P., Willwacher, T.: Equivalence of formalities of the little discs operad. Duke Math. J. **160**(1), 175–206 (2011)
37. Turaev, V.G.: Operator invariants of tangles, and R-matrices. Izv. Akad. Nauk SSSR Ser. Mat. **53**(2), 1073–1107, 1135 (1989). Translation in Math. USSR-Izv. **35**(2), 411–444 (1990)
38. Ünver, S.: Drinfel'd-Ihara relations for p-adic maulti-zeta values. J. Number Theory **133**(5), 1435–1483 (2013)

On Distribution Formulas for Complex and l-adic Polylogarithms

Hiroaki Nakamura and Zdzisław Wojtkowiak

Dedicated to the memory of Professor Jean-Claude Douai

Abstract We study an l-adic Galois analogue of the distribution formulas for polylogarithms with special emphasis on path dependency and arithmetic behaviors. As a goal, we obtain a notion of certain universal Kummer–Heisenberg measures that enable interpolating the l-adic polylogarithmic distribution relations for all degrees.

Keywords Arithmetic fundamental group · Galois actions on étale paths · Functional equations of polylogarithms

1 Introduction

One of the most important and useful functional equations of classical complex polylogarithms is a series of distribution relations

$$Li_k(z^n) = n^{k-1}\left(\sum_{i=0}^{n-1} Li_k(\zeta_n^i z)\right) \qquad (\zeta_n = e^{2\pi i/n}). \tag{1}$$

H. Nakamura (✉)
Department of Mathematics, Osaka University, Osaka, Japan
e-mail: nakamura@math.sci.osaka-u.ac.jp

Z. Wojtkowiak
Départment of Mathematics, Laboratoire Jean Alexandre Dieudonné, U.R.A. au C.N.R.S., Université de Nice-Sophia Antipolis, No 168, Parc Valrose -B.P.N 71, 06108 Nice Cedex 2, France
e-mail: wojtkow@math.unice.fr

J. I. Burgos Gil et al. (eds.), *Periods in Quantum Field Theory and Arithmetic*, Springer Proceedings in Mathematics & Statistics 314,
https://doi.org/10.1007/978-3-030-37031-2_21

J. Milnor [10, (7), (32)] says that a function $\mathscr{L}_s(z)$ has (multiplicative) Kubert identities of degree $s \in \mathbb{C}$, if it satisfies

$$\mathscr{L}_s(z) = n^{s-1} \sum_{w^n = z} \mathscr{L}_s(w) \tag{2}$$

for every positive integer n. The aforementioned classical identity (1) for $Li_k(z)$ is, of course, a typical example of Kubert identity of degree k, assuming, however, correct choice of branches of the multivalued function Li_k in all terms of the identity. To avoid the ambiguity of branch choice, we would rather consider $\mathscr{L}_s(z)$ as a function $\mathscr{L}_s(z; \gamma)$ of paths γ on $\mathbf{P}^1 - \{0, 1, \infty\}$ from the unit vector $\overrightarrow{01}$ to z. The main aim of this paper is to study generalizations of the above distribution relation for multiple polylogarithms and their l-adic Galois analogs (l-adic iterated integrals) with special emphasis on path dependency.

Let K be a subfield of $\mathbb{C}$ with the algebraic closure $\overline{K} \subset \mathbb{C}$. The *$l$-adic polylogarithmic characters*

$$\tilde{\chi}_k^z : G_K \to \mathbb{Z}_\ell \qquad (k = 1, 2, \dots)$$

are introduced in [12] as $\mathbb{Z}_\ell$-valued 1-cochains on the absolute Galois group $G_K := \mathrm{Gal}(\overline{K}/K)$ for any given path γ from $\overrightarrow{01}$ to a K-rational point z on $\mathbf{P}^1 - \{0, 1, \infty\}$. Our study in [NW2] showed that $\tilde{\chi}_k^z : G_K \to \mathbb{Z}_\ell$ behave nicely as l-adic analogues of the classical polylogarithms $Li_k(z)$. The l-adic polylogarithms and l-adic iterated integrals are $\mathbb{Q}_\ell$-valued variants (and generalizations) of the above 1-cochains $\tilde{\chi}_k^z : G_K \to \mathbb{Z}_\ell$. (See Sects. 2 and 3 for their precise definitions.) We will give a geometrical proof of distribution relations for classical multiple polylogarithms and their l-adic analogues in considerable generality. In particular, we will obtain several versions of Kubert identities with explicit path systems for:

- classical multiple polylogarithms (Theorem 4),
- l-adic iterated integrals (Proposition 6, Theorem 17),
- l-adic polylogarithms and polylogarithmic characters (Theorem 19, Corollary 21).

The polylogarithm is interpreted as a certain coefficient of an extension of the Tate module by the logarithm sheaf arising from the fundamental group of $V_1 := \mathbf{P}^1 - \{0, 1, \infty\}$. The motivic construction dates back to the fundamental work of Beilinson-Deligne [1], Huber-Wildeshaus [7] (see also [5] Sect. 6 and references therein for more recent generalizations). In this article, we mainly work on the l-adic realization which forms a $\mathbb{Z}_\ell$- or $\mathbb{Q}_\ell$-valued 1-cochain on the Galois group G_K. In the collaboration [3] of the last author with J.-C. Douai, it was shown that certain linear combinations of l-adic polylogarithms at various points give rise to 1-cocycles on G_K, which lead to an l-adic version of Zagier's conjecture. See also Remark 9 and [11] Sect. 3.2

We will intensively make use of a system of simple cyclic covers $V_n := \mathbf{P}^1 - \{0, \mu_n, \infty\}$ over $V_1 = \mathbf{P}^1 - \{0, 1, \infty\}$, where μ_n is the group of n-th roots of unity $\{1, \zeta_n, \dots, \zeta_n^{n-1}\}$ ($\zeta_n := e^{2\pi i/n}$), and $\{0, \mu_n, \infty\}$ denotes $\{0, \infty\} \cup \mu_n$ by abuse of notation. We consider the family of cyclic coverings $V_n \to V_1$ and open immersions

$V_n \hookrightarrow V_1$ together with induced relations between their fundamental group(oid)s. Our basic idea is to understand the distribution relations of polylogarithms as the "trace property" of relevant coefficients ("iterated integrals") arising in those fundamental groups.

As observed in [13] and will be seen in Sect. 3 below, unlike in the classical complex case, there generally occur lower degree terms in l-adic case when a distribution relation is naively derived. This problem prevents artless approaches to l-adic Kubert identities i.e., distribution formulas of homogeneous form (with no lower degree terms). Our line of studies in Sects. 2–6 will lead us to understand why and how to make use of $\mathbb{Q}_\ell$-paths (l-adic paths with 'denominators') to eliminate such lower degree terms dramatically. Consequently in Sect. 7, as a primary goal of this paper, we arrive at introducing a generalization of the Kummer–Heisenberg measure of [12] so as to interpolate those l-adic distribution relations of polylogarithms for all degrees.

Remark 1 We have already studied in [19, 20] the distribution relations for those l-adic polylogarithms under certain restricted assumptions (see [20, Proposition 11.1.4] for l-adic dilogarithms, [20, Corollary 11.2.2, 11.2.4] for l-adic polylogarithms on restricted Galois groups, and [19, Theorem 2.1] for l-adic polylogarithmic characters with $\ell \nmid n$).

Basic setup, notations and convention

Below, we understand that all algebraic varieties are geometrically connected over a fixed field $K \subset \mathbb{C}$ and that all morphisms between them are K-morphisms. A path on a K-variety V is a topological path on $V(\mathbb{C})$ or an étale path on $V \otimes \overline{K}$ whose distinction will be obvious in contexts. The notation $\gamma : x \rightsquigarrow y$ means a path from x to y, and write $\gamma_1\gamma_2$ for the composed path tracing γ_1 first and then γ_2 afterwards. We write $\chi : G_K \to \mathbb{Z}_\ell^\times$ for the l-adic cyclotomic character (ℓ: a fixed prime). The Bernoulli polynomials $B_k(T)$ ($k = 0, 1, \ldots$) are defined by the generating function $\frac{ze^{Tz}}{e^z - 1} = \sum_{k=0}^{\infty} B_k(T)\frac{z^k}{k!}$, and the Bernoulli numbers are set as $B_k := B_k(0)$. For a vector space H, we write H^* for its dual vector space.

Assume $K \supset \mu_n$. We shall be concerned with two kinds of standard morphisms defined by

$$\begin{cases} \mathscr{J}_\zeta : V_n \hookrightarrow V_1 & \mathscr{J}_\zeta(z) = \zeta z \qquad (\zeta \in \mu_n); \\ \pi_n : V_n \to V_1 & \pi_n(z) = z^n. \end{cases}$$

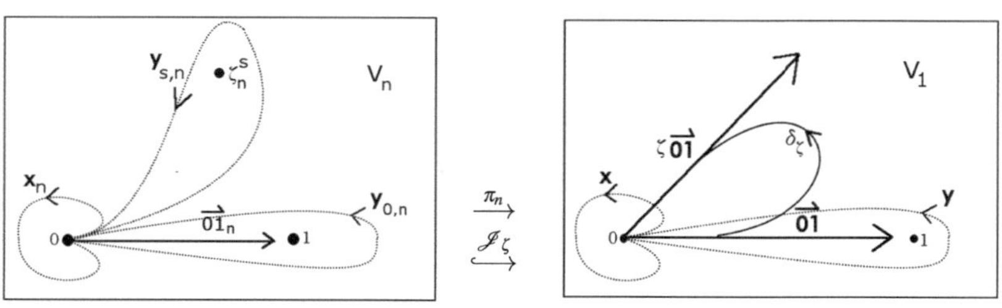

As easily seen, each $\mathscr{J}_\zeta$ is an open immersion, while π_n is an n-cyclic covering. Write $\overrightarrow{01}_n$ for the tangential base point represented by the unit tangent vector on V_n. Since $\mathscr{J}_1 : V_n \hookrightarrow V_1$ maps $\overrightarrow{01}_n$ to $\overrightarrow{01}_1$ (often written just $\overrightarrow{01}$), it induces the surjection homomorphism

$$\pi_1(V_n, \overrightarrow{01}_n) \twoheadrightarrow \pi_1(V_1, \overrightarrow{01}). \tag{3}$$

On the other hand, although the image $\pi_n(\overrightarrow{01}_n)$ is not exactly the same as $\overrightarrow{01}_1$ as a tangent vector, they give the same tangential base point on V_1 in the sense that they give equivalent fiber functors on the Galois category of finite étale covers of V_1. Henceforth, for simplicity, we shall regard $\pi_n(\overrightarrow{01}_n) = \overrightarrow{01}_1 = \overrightarrow{01}$, and regard $\pi_1(V_n, \overrightarrow{01}_n)$ as a subgroup of $\pi_1(V_1, \overrightarrow{01})$ by the homomorphism

$$\pi_1(V_n, \overrightarrow{01}_n) \hookrightarrow \pi_1(V_1, \overrightarrow{01}) \tag{4}$$

induced from π_n.

For each $\zeta \in \mu_n$, introduce a path $\delta_\zeta : \overrightarrow{01} \rightsquigarrow \zeta\overrightarrow{01} = \mathscr{J}_\zeta(\overrightarrow{01}_n)$ on V_1 to be the arc from $\overrightarrow{01}$ to $\zeta\overrightarrow{01}$ anti-clockwise oriented. Using the path δ_ζ, we obtain the identification $\pi_1(V_1, \overrightarrow{01}) \xrightarrow{\sim} \pi_1(V_1, \zeta\overrightarrow{01})$.

Let x, y be standard loops based at $\overrightarrow{01}_1$ on $V_1 = \mathbf{P}^1 - \{0, 1, \infty\}$ turning around the punctures 0, 1 once anticlockwise respectively. We introduce loops $x_n, y_{0,n} \dots, y_{n-1,n}$ based at $\overrightarrow{01}_n$ on V_n characterized by:

$$\begin{cases} x_n & := \pi_n^{-1}(x^n) = \mathscr{J}_1^{-1}(x), \\ y_{s,n} & := \mathscr{J}_1^{-1}(\delta_\zeta) \cdot \mathscr{J}_{\zeta^{-1}}^{-1}(y) \cdot \mathscr{J}_1^{-1}(\delta_\zeta)^{-1} \qquad (\zeta = e^{\frac{2\pi i s}{n}},\ s = 0, \dots, n-1) \end{cases}$$

so that $x_n, y_{0,n}, \dots, y_{n-1,n}$ freely generate $\pi_1(V(\mathbb{C}), \overrightarrow{01}_n)$.

Note that, in view of the above inclusion (4), we have the identifications:

$$x_n = x^n, \quad y_{s,n} = x^s y x^{-s}. \tag{5}$$

2 Complex Distribution Relations

For $n = 1, 2, \dots$, let

$$\omega(V_n) := \frac{dz}{z} \otimes \left(\frac{dz}{z}\right)^* + \sum_{i=0}^{n-1} \frac{dz}{z - \zeta_n^i} \otimes \left(\frac{dz}{z - \zeta_n^i}\right)^* \in \Omega^1_{\log}(V_n) \otimes \Omega^1_{\log}(V_n)^*$$

be the canonical one-form on V_n. Traditionally, we set

$$X_n := \left(\frac{dz}{z}\right)^* \quad \text{and} \quad Y_{i,n} := \left(\frac{dz}{z-\zeta_n^i}\right)^*.$$

Let $\mathscr{R}_n := \mathbb{C}\langle\langle X_n, Y_{i,n} \mid 0 \le i < n\rangle\rangle$ be the non-commutative algebra of formal power series over $\mathbb{C}$ generated by non-commuting variables X_n and $Y_{i,n}$ $(0 \le i < n)$. Consider the trivial bundle

$$\mathscr{R}_n \times V_n \to V_n$$

equipped with the (flat) connection $\nabla\colon \Phi \mapsto d\Phi - \Phi\,\omega(V_n)$ for smooth functions $\Phi : V_n \to \mathscr{R}_n$. For a piecewise smooth path $\gamma : [0,1] \to V_n$ from $\gamma(0) = a$ to $\gamma(1) = z$, let $\Phi : [0,1] \to \mathscr{R}_n$ be the solution to the differential equation $d\Phi = \Phi\,\omega(V_n)$ pulled back on γ with $\Phi(0) = 1$ and define $\Lambda(a \overset{\gamma}{\rightsquigarrow} z) \in \mathscr{R}_n$ to be $\Phi(1)$. (cf. [6] Sect. 2, [16] Sect 1; we here follow Hain's path convention in loc. cit.) In the case a being the tangential base point $\overrightarrow{01}$, we interpret $\Lambda(\overrightarrow{01} \overset{\gamma}{\rightsquigarrow} z)$ in a suitable manner introduced in [2], [17, Sect. 3.2].

Let M_n be the set of all monomials (words) in X_n and $Y_{i,n}$ $(0 \le i < n)$. Then, we can expand

$$\Lambda(a \overset{\gamma}{\rightsquigarrow} z) = 1 + \sum_{w \in \mathsf{M}_n} \mathsf{Li}_w(a \overset{\gamma}{\rightsquigarrow} z) \cdot w \tag{6}$$

in $\mathscr{R}_n$. If $w = X_n^{a_0} Y_{i_1,n} X_n^{a_1} \cdots Y_{i_k,n} X_n^{a_k}$, then

$$\mathsf{Li}_w(a \overset{\gamma}{\rightsquigarrow} z) = \int_{a,\gamma}^{z} \underbrace{\frac{dz}{z} \cdots \frac{dz}{z}}_{a_0} \cdot \frac{dz}{z-\zeta_n^{i_1}} \cdots \frac{dz}{z-\zeta_n^{i_k}} \cdot \underbrace{\frac{dz}{z} \cdots \frac{dz}{z}}_{a_k}, \tag{7}$$

the iterated integral along γ.

Definition 2 For a word $w = X_n^{a_0} Y_{i_1,n} X_n^{a_1} \cdots Y_{i_k,n} X_n^{a_k}$, we define its X-weight by

$$\mathsf{wt}_X(w) = a_0 + \cdots + a_k.$$

Let the cyclic cover

$$\pi_{rn,r} : V_{rn} \longrightarrow V_r \tag{8}$$

be given by $\pi_{rn,r}(z) = z^n$. Then, we have

$$\Big[\mathrm{Id} \otimes (\pi_{rn,r})_*\Big](\omega(V_{rn})) = \Big[(\pi_{rn,r})^* \otimes \mathrm{Id}\Big](\omega(V_r)).$$

This implies that the induced map from $(\pi_{rn,r})_*$ on complete tensor algebras (denoted by the same symbol):

$$\begin{array}{ccc} \hat{T}(\Omega^1_{\log}(V_{rn})^*) & \longrightarrow & \hat{T}(\Omega^1_{\log}(V_r)^*) \\ \| & & \| \\ \mathbb{C}\langle\langle X_{rn}, Y_{j,rn} \mid 0 \le j < rn\rangle\rangle & \longrightarrow & \mathbb{C}\langle\langle X_r, Y_{i,r} \mid 0 \le i < r\rangle\rangle \end{array}$$

preserves the associated power series:

$$(\pi_{rn,r})_* \left(\Lambda(\overrightarrow{01} \overset{\gamma}{\rightsquigarrow} z) \right) = \Lambda(\overrightarrow{01} \overset{\pi_{rn,r}(\gamma)}{\rightsquigarrow} z^n). \tag{9}$$

Note that

$$\begin{cases} (\pi_{rn,r})_*(X_{rn}) = nX_r, \\ (\pi_{rn,r})_*(Y_{j,rn}) = Y_{i,r} \quad (i \equiv j \bmod r). \end{cases} \tag{10}$$

Definition 3 For $w \in \mathsf{M}_{rn}$, we mean by (w mod r) the word in M_r obtained by replacing each letter $X_{rn}, Y_{j,rn}$ $(0 \le j < rn)$ appearing in w by $X_r, Y_{i,r}$ (where i is an integer with $0 \le i < r$, $i \equiv j \bmod r$) respectively. If r is a common divisor of m and n, $w \in \mathsf{M}_m$, $w' \in \mathsf{M}_n$ and (w mod r) = (w' mod r), then we shall write

$$w \equiv w' \mod r.$$

Theorem 4 *Notations being as above, let γ be a path on V_{rn} from $\overrightarrow{01}$ to a point z. Then, for any word $w \in \mathsf{M}_r$, we have the distribution relation*

$$\mathsf{Li}_w(\overrightarrow{01} \overset{\pi_{rn,r}(\gamma)}{\rightsquigarrow} z^n) = n^{\mathsf{wt}_X(w)} \sum_{\substack{u \in \mathsf{M}_{rn} \\ u \equiv w \bmod r}} \mathsf{Li}_u(\overrightarrow{01} \overset{\gamma}{\rightsquigarrow} z).$$

Proof The theorem follows immediately from the formula (9): Write $\Lambda(\overrightarrow{01} \overset{\gamma}{\rightsquigarrow} z) = 1 + \sum_{u \in \mathsf{M}_{rn}} \mathsf{Li}_u(\overrightarrow{01} \overset{\gamma}{\rightsquigarrow} z) \cdot u$ in $\mathscr{R}_{rn}$. Applying (9), we obtain

$$1 + \sum_{w \in \mathsf{M}_r} \mathsf{Li}_w(\overrightarrow{01} \overset{\pi_{rn,r}(\gamma)}{\rightsquigarrow} z^n) \cdot w = 1 + \sum_{u \in \mathsf{M}_{rn}} \mathsf{Li}_u(\overrightarrow{01} \overset{\gamma}{\rightsquigarrow} z) \cdot (\pi_{rn,r})_*(u).$$

Given any specific $w \in \mathsf{M}_r$ in LHS, collect from RHS all the coefficients of $(\pi_{rn,r})_*(u)$ for those u satisfying (u mod r) = w. Noting that $(\pi_{rn,r})_*(u) = n^{\mathsf{wt}_X(u)} w = n^{\mathsf{wt}_X(w)} w$ for them, we settle the assertion of the theorem. □

The above theorem generalizes the distribution relation (1) for the classical polylogarithm $Li_k(z)_\gamma$ along the path $\gamma : \overrightarrow{01} \rightsquigarrow z$. Indeed, in the notation above, since $\frac{dz}{1-z} = -\frac{dz}{z-1}$, we may identify

$$Li_k(z)_\gamma = -\mathsf{Li}_{YX^{k-1}}(\overrightarrow{01} \overset{\gamma}{\rightsquigarrow} z).$$

Applying the theorem to the special case $\pi_{n,1} : V_n \to V_1$, $w = YX^{k-1}$ where $Y = Y_{0,1}$, $X = X_1$, we obtain

$$\int_{\overrightarrow{01},\pi_{n,1}(\gamma)}^{z^n} \frac{dz}{z-1} \cdot \underbrace{\frac{dz}{z} \cdots \frac{dz}{z}}_{k-1} = n^{k-1} \sum_{\zeta \in \mu_n} \int_{\overrightarrow{01},\gamma}^{z} \frac{dz}{z-\zeta} \cdot \underbrace{\frac{dz}{z} \cdots \frac{dz}{z}}_{k-1}.$$

Each term of RHS turns out to be $Li_k(\zeta z)$ along the path $\delta_\zeta \cdot \mathscr{I}_\zeta(\gamma) : \overrightarrow{01} \rightsquigarrow \zeta\overrightarrow{01} \rightsquigarrow \zeta z$, after integrated by substitution $z \to \zeta z$. Noting that the integration here over $\delta_\zeta : \overrightarrow{01} \rightsquigarrow \zeta\overrightarrow{01}$ vanishes (cf. [17] Sect. 3), we obtain (1) with path system specified as follows:

$$Li_k(z^n)_{\pi_{n,1}(\gamma)} = n^{k-1} \sum_{\zeta \in \mu_n} Li_k(\zeta z)_{\delta_\zeta \cdot \mathscr{I}_\zeta(\gamma)}. \tag{11}$$

3 l-adic Case (general)

We shall look at the l-adic analogue of the previous section by recalling the following construction which essentially dates back to [18]. Let $K \subset \mathbb{C}$ and consider

$$\pi_1^\ell(V_n \otimes \overline{K}, \overrightarrow{01}),$$

the pro-ℓ (completion of the étale) fundamental group of $V_n \otimes \overline{K}$. It is easy to see the loops $x_n, y_{1,n}, \ldots, y_{n-1,n}$ introduced in Sect. 1 form a free generator system of the pro-ℓ fundamental group. Consider the (multiplicative) Magnus embedding into the ring of non-commutative power series

$$\iota_{\mathbb{Q}_\ell} : \pi_1^\ell(V_n \otimes \overline{K}, \overrightarrow{01}) \hookrightarrow \mathbb{Q}_\ell\langle\langle X_n, Y_{i,n} \mid 0 \le i \le n-1 \rangle\rangle \tag{12}$$

defined by $\iota_{\mathbb{Q}_\ell}(x_n) = \exp(X_n)$, $\iota_{\mathbb{Q}_\ell}(y_{i,n}) = \exp(Y_{i,n})$ (cf. [21, 15.1]). For simplicity, we often identify elements of $\pi_1^\ell(V_n \otimes \overline{K}, \overrightarrow{01})$ with their images by $\iota_{\mathbb{Q}_\ell}$. Let us write again M_n for the set of monomials in $X_n, Y_{i,n}$ $(i = 0, \ldots, n-1)$ (although variables have different senses from the previous section where they were duals of differential forms). We shall also employ the usage '$\mathsf{wt}_X(w)$' and '$w \equiv w' \bmod r$' by following the same manners as Definitions 2 and 3.

Recall that we have a canonical Galois action G_K on (étale) paths on $V_n \otimes \overline{K}$ with both ends at K-rational (tangential) points. Given a path γ from such a point a to a point $z \in V_n(K)$, we set, for any $\sigma \in G_K$,

$$\mathfrak{f}_\sigma^\gamma := \gamma \cdot \sigma(\gamma)^{-1} \in \pi_1^\ell(V_n \otimes \overline{K}, a), \tag{13}$$

where the RHS is understood to be the image in the pro-ℓ quotient. When $a = \overrightarrow{01}$, we expand $\mathfrak{f}_\sigma^\gamma \in \pi_1^\ell(V_n \otimes \overline{K}, \overrightarrow{01})$ in the form

$$\mathfrak{f}_\sigma^\gamma = 1 + \sum_{w \in \mathsf{M}_n} \mathsf{Li}_w(\overrightarrow{01} \overset{\gamma}{\rightsquigarrow} z)(\sigma) \cdot w \tag{14}$$

in $\mathbb{Q}_\ell\langle\langle X_n, Y_{i,n} \mid 0 \leq i \leq n-1 \rangle\rangle$, and associating the coefficient

$$\mathsf{Li}_w(\overrightarrow{01} \overset{\gamma}{\rightsquigarrow} z)(\sigma) := \mathsf{Coeff}_w(\mathfrak{f}_\sigma^\gamma)$$

of $w \in \mathsf{M}_n$ to $\sigma \in G_K$, we define the l-adic Galois 1-cochain

$$\mathsf{Li}_w(\overrightarrow{01} \overset{\gamma}{\rightsquigarrow} z) \left(= \mathsf{Li}_w^{(\ell)}(\overrightarrow{01} \overset{\gamma}{\rightsquigarrow} z)\right) : G_K \to \mathbb{Q}_\ell$$

for every monomial $w \in \mathsf{M}_n$. We call each $\mathsf{Li}_w(\overrightarrow{01} \overset{\gamma}{\rightsquigarrow} z)$ the l-adic iterated integral associated to $w \in \mathsf{M}_n$ and to the path γ on V_n.

Remark 5 The above naming 'l-adic iterated integral' is intended to be an analog of the iterated integral appearing in the complex case (6) (7). They represent general coefficients of the associator in the Magnus expansion. Conceptually, the associator lies in the pro-unipotent hull of the fundamental group and the monodromy information encoded in the total set of them is equivalent to that encoded in the general coefficients with respect to any fixed Hall basis of the corresponding Lie algebra. This line of formulation was, in fact, taken up, e.g., in [18] Sect. 5. However for the purpose of pursuing the distribution formulas in the present paper, the simple form of trace properties (9), (10) along the cyclic coverings $\pi_{rn,r} : V_{rn} \longrightarrow V_r$ is most essential. This is why we start with Magnus expansions $\mathfrak{f}_\sigma^\gamma$ in $\mathbb{Q}_\ell\langle\langle X_n, Y_{i,n}\rangle\rangle_i$ rather than with Lie expansions of $\log \mathfrak{f}_\sigma^\gamma$ with respect to a Hall basis in $\mathrm{Lie}\langle\langle X_n, Y_{i,n}\rangle\rangle_i$. But we shall discuss their relations in the polylogarithmic part of $n = 1$ in Sect. 4.

Now, as in Sect. 2, let us consider the morphism $\pi_{rn,r} : V_{rn} \to V_r$ given by $\pi_{rn,r}(z) = z^n$ for $n, r > 0$, and let γ be a path on V_{rn} from $\overrightarrow{01}$ to a K-rational point z. By our construction, the l-adic analogue of the equality (9) holds, i.e., $\pi_{rn,r}$ preserves the l-adic associators:

$$(\pi_{rn,r})_*(\mathfrak{f}_\sigma^\gamma) = \mathfrak{f}_\sigma^{\pi_{rn,r}(\gamma)} \quad (\sigma \in G_K). \tag{15}$$

However, unlike the complex case (10), $\pi_{rn,r}$ does not preserve the expansion coefficients homogeneously, i.e., it maps as

$$\begin{cases} (\pi_{rn,r})_*(X_{rn}) & = nX_r, \\ (\pi_{rn,r})_*(Y_{j,rn}) & = \exp(kX_r)Y_{i,r}\exp(-kX_r) \qquad (j = i + kr,\ 0 \leq i < r). \end{cases} \tag{16}$$

Proof of (16) Note that the cyclic projections $\pi_{rn,r}$ identify $\{\pi_1(V_n)\}_n$ as a sequence of subgroups of $\pi_1(V_1)$ as in (4), and regard $x_{rn} = x_r{}^n = x^{rn}$, $y_{j,rn} = x^j y x^{-j} = (x^r)^k x^i y x^{-i} (x^r)^{-k} = x_r^k y_{i,r} x_r^{-k}$. Although $\pi_{rn,r}$ does not keep injectivity on the complete envelops, it does induce a functorial homomorphism on them. The formula follows then from $x_n = \exp(X_n)$, $y_{s,n} = \exp(Y_{s,n})$. □

This causes generally (lower degree) error terms to appear in distribution relations for l-adic iterated integrals.

Still, if we restrict ourselves to the words whose X-weights are zero, we have the following

Proposition 6 *Notations being as above, if $w \in \mathsf{M}_r$ is a word with $\mathsf{wt}_X(w) = 0$, i.e., of the form $w = Y_{i_k,r} \cdots Y_{i_1,r}$, then it holds that*

$$\mathsf{Li}_w(\overrightarrow{01} \overset{\pi_{rn,r}(\gamma)}{\rightsquigarrow} z^n)(\sigma) = \sum_{\substack{u \in \mathsf{M}_{rn} \\ u \equiv w \bmod r}} \mathsf{Li}_u(\overrightarrow{01} \overset{\gamma}{\rightsquigarrow} z)(\sigma) \qquad (\sigma \in G_K).$$

Proof In the expansion of $(\pi_{rn,r})_*(\mathfrak{f}_\sigma^\gamma) = \mathfrak{f}_\sigma^{\pi_{rn,r}(\gamma)}$, the contributions to the coefficient of w come only from the first 'Y-only' term of each $u \in \mathsf{M}_{rn}$ with $u \equiv w \bmod r$. The proposition follows from this observation. □

Remark 7 In the l-adic Galois case, the distribution relations of Proposition 6 are used in [23] to construct measures on $\mathbb{Z}_\ell^r$ which generalize the measure on $\mathbb{Z}_\ell$ in [12]. The general distribution formula analogous to Theorem 4 for arbitrary words in M_r hold only up to lower degree terms in the l-adic Galois case. More generally, any covering maps between smooth algebraic varieties will give some kind of distribution relations.

4 l-adic Polylogarithms (Review)

Henceforth, we shall closely look at the case of l-adic polylogarithm where $r = 1$ and only those words $w \in \mathsf{M}_1$ involving $Y_{0,1}$ only once are concerned, in the setting of the previous section. For simplicity, we write $x := x_1$, $y := y_{0,1}$ and $X := \log(x)$, $Y := \log(y)$, and will be concerned with those coefficients of the words YX^{k-1} of $\mathfrak{f}_\sigma^\gamma$.

Let us recall some basic facts from [12, 13]. We introduced, for any path $\gamma : \overrightarrow{01} \rightsquigarrow z$ on $V_1 = \mathbf{P}^1 - \{0, 1, \infty\}$, the *$l$-adic polylogarithms*

$$\ell i_m(z, \gamma) : G_K \to \mathbb{Q}_\ell \tag{17}$$

(with regard to the fixed free generator system $\{x, y\}$ of $\pi_1^\ell(V_1 \otimes \overline{K}, \overrightarrow{01})$) to be the Lie expansion coefficients of the associator $\mathfrak{f}_\sigma^\gamma = \gamma \cdot \sigma(\gamma)^{-1}$ for $\sigma \in G_K$ modulo the ideal I_Y of Lie monomials including Y twice or more:

$$\log(\mathfrak{f}_\sigma^\gamma)^{-1} \equiv \rho_z(\sigma)X + \sum_{m=1}^{\infty} \ell i_m(z,\gamma)(\sigma)\mathrm{ad}(X)^{m-1}(Y) \mod I_Y. \tag{18}$$

Here, $\rho_z : G_K \to \mathbb{Z}_\ell(1)$ designates the Kummer 1-cocycle for power roots of z along γ. Note, however, that the other coefficients $\ell i_m(z,\gamma)(\sigma) \in \mathbb{Q}_\ell$ are generally not valued in $\mathbb{Z}_\ell$ due to applications of log respectively to x, y and $\mathfrak{f}_\sigma^\gamma \in \pi_1^\ell(V_1 \otimes \overline{K}, \overrightarrow{01})$. In fact, we can bound the denominators of $\ell i_m(z,\gamma)(\sigma)$ by relating them with more explicitly defined $\mathbb{Z}_\ell$-valued 1-cochains called the *l-adic polylogarithmic characters*

$$\tilde{\chi}_m^z (= \tilde{\chi}_m^{z,\gamma}) : G_K \to \mathbb{Z}_l \qquad (m \geq 1) \tag{19}$$

defined by the Kummer properties for $n \geq 1$:

$$\zeta_{\ell^n}^{\tilde{\chi}_m^z(\sigma)} = \sigma\left(\prod_{a=0}^{\ell^n-1}(1-\zeta_{\ell^n}^{\chi(\sigma)^{-1}a} z^{1/\ell^n})^{\frac{a^{m-1}}{\ell^n}}\right) \Big/ \prod_{a=0}^{\ell^n-1}(1-\zeta_{\ell^n}^{a+\rho_z(\sigma)} z^{1/\ell^n})^{\frac{a^{m-1}}{\ell^n}}, \tag{20}$$

where $(1-\zeta_{\ell^n}^{\alpha} z^{1/\ell^n})^{\frac{\beta}{\ell^n}}$ means the β-th power of a carefully chosen ℓ^n-th root of $(1-\zeta_{\ell^n}^{\alpha} z^{1/\ell^n})$ along γ. It is shown in [12, p.293 Corollary] that, for each $\sigma \in G_K$, the l-adic polylogarithm $\ell i_m(z,\gamma)(\sigma) \in \mathbb{Q}_\ell$ can be expressed by the Kummer- and l-adic polylogarithmic characters $\rho_z(\sigma)$, $\tilde{\chi}_m^z(\sigma) \in \mathbb{Z}_\ell$ as follows:

$$\ell i_m(z,\gamma)(\sigma) = (-1)^{m+1}\sum_{k=0}^{m-1}\frac{B_k}{k!}(-\rho_z(\sigma))^k \frac{\tilde{\chi}_{m-k}^z(\sigma)}{(m-k-1)!} \qquad (m \geq 1). \tag{21}$$

One has then the following relations among $\ell i_m(z,\gamma)(\sigma) \in \mathbb{Q}_\ell$ (17), $\tilde{\chi}_m^z(\sigma) \in \mathbb{Z}_\ell$ (19) and $\mathsf{Li}_{YX^{m-1}}(\overrightarrow{01} \overset{\gamma}{\rightsquigarrow} z)(\sigma) \in \mathbb{Q}_\ell$ (Sect. 3):

Proposition 8 *(i) Notations being as above, we have*

$$\tilde{\chi}_m^z(\sigma) = (-1)^{m+1}(m-1)!\sum_{k=1}^{m}\frac{\rho_z(\sigma)^{m-k}}{(m+1-k)!}\ell i_k(z,\gamma)(\sigma) \quad (m \geq 1).$$

(ii) Moreover, the expansion of $\mathfrak{f}_\sigma^\gamma$ *in* $\mathbb{Q}_\ell\langle\langle X, Y\rangle\rangle$ *partly looks like*

$$\mathfrak{f}_\sigma^\gamma = 1 + \sum_{i=1}^{\infty}\frac{(-\rho_z(\sigma))^i}{i!}X^i - \sum_{i=0}^{\infty}\frac{\tilde{\chi}_{i+1}^z(\sigma)}{i!}YX^i + \ldots(\textit{other terms}).$$

In particular, we have

$$\mathsf{Li}_{YX^{m-1}}(\overrightarrow{01} \overset{\gamma}{\rightsquigarrow} z)(\sigma) = -\frac{\tilde{\chi}_m^z(\sigma)}{(m-1)!} \quad (m \geq 1).$$

Proof (i) Follows immediately from inductively reversing the formula (21). (ii) Also follows easily from discussions in [13, p.284–285]: Suppose $\mathfrak{f}_\sigma^\gamma$ has monomial expansion as

$$\mathfrak{f}_\sigma^\gamma = 1 + \sum_{i=1}^{\infty} c_i \frac{X^i}{i!} - \sum_{i=0}^{\infty} d_{i+1} Y X^i + \text{...(other terms)}.$$

First, from (18), we see that $\mathfrak{f}_\sigma^\gamma \equiv e^{cX}$ modulo $Y = 0$ with a constant $c := -\rho_z(\sigma)$, hence that $c_i = c^i$. Next, to look at the coefficients of monomials of the forms X^i, YX^i $(i = 0, 1, 2, \dots)$ closely, we take reduction modulo the ideal $J_Y := \langle XY, Y^2 \rangle$ of $\mathbb{Q}_\ell\langle\langle X, Y\rangle\rangle$. Observe then the congruence:

$$\begin{aligned}\log(\mathfrak{f}_\sigma^\gamma) &\equiv (\mathfrak{f}_\sigma^\gamma - 1)\left\{1 - \frac{1}{2}(\mathfrak{f}_\sigma^\gamma - 1) + \frac{1}{3}(\mathfrak{f}_\sigma^\gamma - 1)^2 - + \cdots\right\} \\ &\equiv \left(-\sum_{i=0}^{\infty} d_{i+1} Y X^i\right)\left\{1 - \frac{1}{2}(e^{cX} - 1) + \frac{1}{3}(e^{cX} - 1)^2 - + \cdots\right\} \\ &\equiv \left(-\sum_{i=0}^{\infty} d_{i+1} Y X^i\right)\left\{\sum_{k=0}^{\infty} \frac{B_k}{k!} c^k X^k\right\} \qquad (\text{mod } J_Y)\end{aligned}$$

and find that the coefficient of YX^{m-1} in $\log(\mathfrak{f}_\sigma^\gamma)$ is

$$-\sum_{k=0}^{m-1} \frac{B_k}{k!} c^k d_{m-k} \qquad (*)$$

for $m \geq 1$.[1] On the other hand, the formula (18) combined with (21) calculates the same coefficient, which is $(-1)^{m-1}$-multiple of that of $\mathrm{ad}(X)^{m-1}(Y)$, as to be

$$(-1)^{m-1} \ell i_m(z, \gamma) = \sum_{k=0}^{m-1} \frac{B_k}{k!} (-\rho_z(\sigma))^k \frac{\tilde{\chi}_{m-k}^z(\sigma)}{(m-k-1)!} \qquad (**)$$

for $m \geq 1$. Comparing those $(*)$ and $(**)$ inductively on $m \geq 1$, we conclude our desired identities $d_{i+1} = -\tilde{\chi}_{i+1}(\sigma)/i!$ $(i \geq 0)$. □

Remark 9 The l-adic polylogarithm was constructed as a certain lisse $\mathbb{Q}_\ell$-sheaf on $V_1 = \mathbf{P}^1 - \{0, 1, \infty\}$ as in [1, 5, 7, 22]. The fiber over a point $z \in V_1(K)$ forms a polylogarithmic quotient torsor of l-adic path classes from $\overrightarrow{01}$ to z. We have the G_K-action on the path space whose specific coefficients are the l-adic (Galois) polylogarithms in our sense (17), viz., realized as $\mathbb{Q}_\ell$-valued 1-cochains on G_K. See also, e.g., [11] Sect. 3 for a concise account from the viewpoint of non-abelian cohomology in a mixed Tate category.

---

[1] Note that there are misprints in [13, p.284] where exponents $\circledast = 2, 3$ of $(e^{(\log z)X} - 1)^{\circledast}$ should read $\circledast = 1, 2$ respectively in the 2nd and 3rd terms in line -11.

5 Distribution Relations for $\tilde{\chi}_m^z$

Suppose now that $\mu_n \subset K \subset \mathbb{C}$ and that we are given a point $z \in V_n(K)$ together with a(n étale) path $\gamma : \overrightarrow{01}_n \rightsquigarrow z$ on $V_n \otimes \overline{K} = \mathbf{P}^1_{\overline{K}} - \{0, \mu_n, \infty\}$. We consider the l-adic polylogarithmic characters $\tilde{\chi}_m^{z^n}, \tilde{\chi}_m^{\zeta z} : G_K \to \mathbb{Z}_\ell$ ($\zeta \in \mu_n$) along the paths $\pi_n(\gamma) : \overrightarrow{01} \rightsquigarrow z^n$ and $\delta_\zeta \mathscr{J}_\zeta(\gamma) : \overrightarrow{01} \rightsquigarrow \zeta z$ respectively. In this section, we shall show the following l-adic analog of the distribution formula:

Theorem 10 *Notations being as above, we have*

$$\tilde{\chi}_k^{z^n}(\sigma) = \sum_{d=1}^{k} \binom{k-1}{d-1} n^{d-1} \sum_{s=0}^{n-1} (s\chi(\sigma))^{k-d} \tilde{\chi}_d^{\zeta_n^s z}(\sigma) \qquad (\sigma \in G_K,\ \zeta_n = e^{\frac{2\pi i}{n}},\ 0^0 = 1).$$

Consider now the l-adic Lie algebras $L_{\mathbb{Q}_\ell}(\overrightarrow{01}_n)$ and $L_{\mathbb{Q}_\ell}(\overrightarrow{01})$ associated to $\pi_1^\ell(V_n, \overrightarrow{01}_n)$ and $\pi_1^\ell(V_1, \overrightarrow{01}_1)$ respectively, and set specific elements of them by $X_n := \log x_n$, $Y_{i,n} := \log y_{i,n}$ ($i = 0, \ldots, n-1$), $X := \log x$ and $Y := \log y$.

In the following of this section, we shall fix $\sigma \in G_K$ and frequently omit mentioning σ that is potentially appearing in each term of our functional equation. In particular, the quantities χ, ρ_z designate the values $\chi(\sigma)$, $\rho_z(\sigma)$ at $\sigma \in G_K$ respectively. For our fixed $\sigma \in G_K$, let us determine the polylogarithmic part of the Galois transformation $\mathfrak{f}_\sigma^\gamma := \gamma \cdot \sigma(\gamma)^{-1}$ of the path $\gamma : \overrightarrow{01}_n \rightsquigarrow z$ in the form:

$$\begin{aligned} \log(\mathfrak{f}_\sigma^\gamma)^{-1} &\equiv CX_n + \sum_{s=0}^{n-1} \sum_{m=1}^{\infty} C_{s,m}\, \mathrm{ad}(X_n)^{m-1}(Y_{s,n}) \\ &\equiv CX_n + \sum_{s=0}^{n-1} \mathsf{C}_s(\mathrm{ad}X_n)(Y_{s,n}) \mod I_{Y_*}, \end{aligned} \tag{22}$$

where, I_{Y_*} represents the ideal generated by those terms including $\{Y_{0,n}, \ldots, Y_{n-1,n}\}$ twice or more, and $\mathsf{C}_s(t) = \sum_{m=1}^{\infty} C_{s,m} t^{m-1} \in \mathbb{Q}_\ell[[t]]$ ($s = 0, \ldots, n-1$).

We determine the above coefficients C, $C_{s,m}$ by applying the morphisms $\mathscr{J}_\zeta$ ($\zeta \in \mu_n$). Let us set

$$\mathsf{L}^{(\zeta)}(t) := \mathsf{L}_1(\zeta z) + \mathsf{L}_2(\zeta z)t + \mathsf{L}_3(\zeta z)t^2 + \cdots \quad (\zeta \in \mu_n);$$

$$\mathsf{L}^{(n)}(t) := \mathsf{L}_1(z^n) + \mathsf{L}_2(z^n)t + \mathsf{L}_3(z^n)t^2 + \cdots,$$

with

$$\begin{cases} \mathsf{L}_0(\zeta z) := \rho_{\zeta z} = \rho_z + \frac{s}{n}(\chi - 1), \\ \mathsf{L}_1(\zeta z) := \rho_{1-\zeta z}, \\ \mathsf{L}_k(\zeta z) := \frac{\tilde{\chi}_k^{\zeta z}(\sigma)}{(k-1)!} \quad (k \geq 2), \\ (\zeta = e^{2\pi i s/n},\ s = 0, 1, \ldots, n-1); \end{cases} \qquad \begin{cases} \mathsf{L}_0(z^n) := \rho_{z^n} = n\rho_z, \\ \mathsf{L}_1(z^n) := \rho_{1-z^n} = \sum_{\zeta \in \mu_n} \rho_{1-\zeta z}, \\ \mathsf{L}_k(z^n) := \frac{\tilde{\chi}_k^{z^n}(\sigma)}{(k-1)!} \quad (k \geq 2). \end{cases}$$

Then,

Lemma 11

(1) $C = \mathsf{L}_0(z) = \rho_z$.
(2) $\mathsf{C}_0(t) = \mathsf{L}^{(1)}(-t)\frac{\rho_z t}{e^{\rho_z t}-1}$.
(3) $\mathsf{C}_s(t) = \mathsf{L}^{(\zeta)}(-t)e^{((\frac{s}{n}-1)\chi - \frac{s}{n})t}\frac{\rho_z t}{e^{\rho_z t}-1} \qquad (s = 1, \ldots, n-1;\ \zeta = e^{-\frac{2\pi i s}{n}})$.

The proof of this lemma will be given later in this section.

Proof of Theorem 10 **assuming Lemma** 11
We apply the morphism $\pi_n : V_n \to V_1$ to $\log(\mathfrak{f}_\sigma^\gamma)^{-1}$. We first observe that $\pi_n(X_n) = nX,\ \pi_n(Y_{s,n}) = x^s Y x^{-s} = \sum_{k=0}^{\infty} \frac{s^k}{k!}(\mathrm{ad}X)^k(Y) = e^{s\cdot \mathrm{ad}X}(Y)$ for $s = 0, \ldots, n-1$. Hence,

$$\pi_n(\log(\mathfrak{f}_\sigma^\gamma)^{-1}) = CnX + \sum_{s=0}^{n-1} \mathsf{C}_s(n\,\mathrm{ad}X)\left(\sum_{k=0}^{\infty}\frac{s^k}{k!}(\mathrm{ad}X)^k\right)(Y). \tag{23}$$

The above LHS equals to

$$\log(\mathfrak{f}_\sigma^{\pi_n(\gamma)})^{-1} = \rho_{z^n}X + \sum_{k=1}^{\infty} \ell i_k(z^n, \pi_n(\gamma))(\mathrm{ad}X)^{k-1}(Y). \tag{24}$$

From the formula (21) we see that

$$\sum_{k=1}^{\infty} \ell i_k(z^n, \pi_n(\gamma))t^k = t\,\mathsf{L}^{(n)}(-t)\frac{\rho_z nt}{e^{\rho_z nt}-1}, \tag{25}$$

hence that the equality of RHSs of (23) and (24) results in:

$$\sum_{s=0}^{n-1} \mathsf{C}_s(nt)e^{st} = \mathsf{L}^{(n)}(-t)\frac{\rho_z nt}{e^{\rho_z nt}-1}. \tag{26}$$

Substituting $\mathsf{C}_s(t)$ $(s = 0, \ldots, n-1)$ by Lemma 11 (2), (3), the above left side equals

$$\left(\mathsf{L}^{(1)}(-nt) + \sum_{s=1}^{n-1} \mathsf{L}^{(\zeta)}(-nt)e^{((\frac{s}{n}-1)\chi - \frac{s}{n})nt}e^{st}\right)\frac{\rho_z nt}{e^{\rho_z nt}-1} \tag{27}$$

where, in the summation $\sum_s$, we understand $\zeta = e^{-\frac{2\pi is}{n}}$. As $((\frac{s}{n}-1)\chi - \frac{s}{n})nt + st = -(n-s)\chi t$, the replacement of ζ by $\zeta_n^s = e^{\frac{2\pi is}{n}}$ enables us to collect the sum as $\sum_{s=0}^{n-1} \mathsf{L}^{(\zeta)}(-nt)e^{-s\chi t}$. Finally, substituting t for $-t$, we obtain

$$\mathsf{L}^{(n)}(t) = \sum_{s=0}^{n-1} \mathsf{L}^{(\zeta)}(nt)e^{s\chi t} \qquad (\zeta = e^{\frac{2\pi is}{n}}). \tag{28}$$

Theorem 10 follows from comparing the coefficients of the above equation. □

We prepare the following combinatorial lemma concerning the Baker–Campbell–Hausdorff sum: $S \underset{\mathsf{CH}}{\oplus} T = \log(e^S e^T)$. Let

$$\beta(t) = \frac{t}{e^t - 1} = \sum_{n=0}^{\infty} B_n \frac{t^n}{n!}$$

be the generating function for Bernoulli numbers.

Lemma 12 *Let K be a field of characteristic 0 and let $\alpha, \ell_0, \ell_1, \dots \in K$. Let $\ell(X,Y) = \ell_0 X + \ell_+(\mathrm{ad}X)(Y) = \ell_0 X + \sum_{k=1}^{\infty} \ell_k (\mathrm{ad}X)^{k-1}(Y)$ be an arbitrary element of the formal Lie series ring $\mathrm{Lie}_K\langle\langle X, Y\rangle\rangle$ with $\ell_+(t) \in K[[t]]$. Then, we have the following congruence formulas modulo I_Y.*

(i)

$$\ell(X,Y) \underset{\mathsf{CH}}{\oplus} \alpha X \equiv (\alpha + \ell_0)X + \left(\frac{\beta((\alpha+\ell_0)\mathrm{ad}X)}{\beta(\alpha\,\mathrm{ad}X)}\ell_+(\mathrm{ad}X)\right)(Y);$$

(ii)

$$\alpha X \underset{\mathsf{CH}}{\oplus} \ell(X,Y) \equiv (\alpha + \ell_0)X + \left(\frac{\beta((\alpha+\ell_0)\mathrm{ad}X)}{\beta(\ell_0\,\mathrm{ad}X)}\ell_+(\mathrm{ad}X)e^{\alpha\,\mathrm{ad}X}\right)(Y).$$

Proof Both formulas follow from the polylogarithmic BCH formula and with a representation of the core generating function. See [13, Proposition 5.9 and (5.8)]. □

Proof of Lemma 11:

Apply the morphisms $\mathscr{I}_\zeta$ $(\zeta \in \mu_n)$ to determine the coefficients $C_{m,s}$ of the polylogarithmic terms of $\log \mathfrak{f}_\sigma^\gamma$ in (22).

Case $\zeta = 1$: Observe that $\mathscr{I}_1(X_n) = X$, $\mathscr{I}_1(Y_{0,n}) = Y$ and $\mathscr{I}_1(Y_{i,n}) = 0$ $(i \neq 0)$. Then, it follows from (22) that

$$\mathscr{I}_1(\log(\mathfrak{f}_\sigma^\gamma)^{-1}) = \log(\mathfrak{f}_\sigma^{\mathscr{I}_1(\gamma)})^{-1} \equiv CX + \big(\mathsf{C}_0(\mathrm{ad}X)\big)(Y) \mod I_Y.$$

We immediately see that the first coefficient C is given by

$$C = \rho_z = \mathsf{L}_0(z), \tag{29}$$

and that the other polylogarithmic coefficients are given by (21) as follows:

$$\mathsf{C}_0(t)=\sum_{k=1}^{\infty}\ell i_k(z,\mathscr{J}_1(\gamma))t^{k-1}=\mathsf{L}^{(1)}(-t)\frac{\rho_z t}{e^{\rho_z t}-1}. \tag{30}$$

Case $\zeta\neq 1$: Assume $\zeta=e^{-\frac{2\pi i s}{n}}$ $(s=1,\dots,n-1)$. We observe in this case that $\delta_\zeta\,\mathscr{J}_\zeta(X_n)\,\delta_\zeta^{-1}=X$, $\delta_\zeta\,\mathscr{J}_\zeta(Y_{s,n})\,\delta_\zeta^{-1}=xYx^{-1}=\sum_{k=0}^{\infty}\frac{(\mathrm{ad}X)^k(Y)}{k!}=e^{\mathrm{ad}X}(Y)$ and $\mathscr{J}_\zeta(Y_{i,n})=0$ $(i\neq 0)$. Therefore, it follows from (22) that

$$\delta_\zeta\cdot\mathscr{J}_\zeta(\log(\mathfrak{f}_\sigma^\gamma)^{-1})\cdot\delta_\zeta^{-1}\equiv CX+\big(\mathsf{C}_s(\mathrm{ad}X)e^{\mathrm{ad}X}\big)(Y)\mod I_Y. \tag{31}$$

On the other side, since $\mathfrak{f}_\sigma^{\delta_\zeta\mathscr{J}_\zeta(\gamma)}=\delta_\zeta\mathfrak{f}_\sigma^{\mathscr{J}_\zeta(\gamma)}\delta_\zeta^{-1}\mathfrak{f}_\sigma^{\delta_\zeta}$ by (13), we have

$$\begin{aligned}\delta_\zeta\cdot\mathscr{J}_\zeta(\log(\mathfrak{f}_\sigma^\gamma)^{-1})\cdot\delta_\zeta^{-1}&=\delta_\zeta\cdot\log(\mathfrak{f}_\sigma^{\mathscr{J}_\zeta(\gamma)})^{-1}\cdot\delta_\zeta^{-1}\\&=\Big(-\log(\mathfrak{f}_\sigma^{\delta_\zeta})^{-1}\Big)\underset{\mathsf{CH}}{\oplus}\Big(\log(\mathfrak{f}_\sigma^{\delta_\zeta\mathscr{J}_\zeta(\gamma)})^{-1}\Big)\\&\equiv\Big(-\frac{n-s}{n}(\chi-1)X\Big)\underset{\mathsf{CH}}{\oplus}\Big(\ell i_0(\zeta z)X+\sum_{k=1}^{\infty}\ell i_k(\zeta z)(\mathrm{ad}X)^{k-1}(Y)\Big)\end{aligned} \tag{32}$$

mod I_Y, where $\ell i_k(\zeta z)$ $(k\geq 0)$ are taken along the path $\delta_\zeta\,\mathscr{J}_\zeta(\gamma)$. Note here that $\ell i_0(\zeta z)=\mathsf{L}_0(\zeta z)=\rho_z+\frac{n-s}{n}(\chi-1)$ and that (21) implies

$$\sum_{k=1}^{\infty}\ell i_k(\zeta z,\delta_\zeta\mathscr{J}_\zeta(\gamma))t^{k-1}=\mathsf{L}^{(\zeta)}(-t)\,\beta(\mathsf{L}_0(\zeta z)\,t)=\mathsf{L}^{(\zeta)}(-t)\frac{\mathsf{L}_0(\zeta z)t}{e^{\mathsf{L}_0(\zeta z)t}-1}. \tag{33}$$

Putting this into (32) and using Lemma 12 (ii), we find

$$\delta_\zeta\cdot\mathscr{J}_\zeta(\log(\mathfrak{f}_\sigma^\gamma)^{-1})\cdot\delta_\zeta^{-1}\equiv\rho_z X+\left(\mathsf{L}^{(\zeta)}(-\mathrm{ad}X)e^{-\frac{n-s}{n}(\chi-1)\mathrm{ad}X}\frac{\rho_z\mathrm{ad}X}{e^{\rho_z\mathrm{ad}X}-1}\right)(Y) \tag{34}$$

mod I_Y. Comparing this with (31), we obtain

$$\mathsf{C}_s(t)=\mathsf{L}^{(\zeta)}(-t)e^{(-\frac{n-s}{n}(\chi-1)-1)t}\frac{\rho_z t}{e^{\rho_z t}-1}\qquad(s=1,\dots,n-1;\ \zeta=e^{-\frac{2\pi i s}{n}}). \tag{35}$$

Thus, the proof of Lemma 11 is completed. □

Remark 13 In [13, Theorem 5.7], we gave a general tensor criterion to have a functional equation of (complex and l-adic) polylogarithms from a collection of morphisms $\{f_i:X\to\mathbf{P}^1-\{0,1,\infty\}\}_{i\in I}$ and their formal sum $\sum_{i\in I}c_i[f_i]$. In our above case, it holds that the collection $\{\pi_n,\mathscr{J}_0,\dots,\mathscr{J}_{n-1}:V_n\to V_1\}$ satisfies the criterion with coefficients $1,-n^{k-1},\dots,-n^{k-1}$ (as observed already in [4, (1.9) (iii)]). Explicit evaluation of the error terms $\mathsf{E}_k:=\mathsf{E}_k(\sigma,\gamma)$ discussed in [13] (that explains part of lower degree inhomogeneous terms of our functional equation) can be obtained a posteriori from (25), (30), (33) and (28) as:

$$\sum_{k=1}^{\infty} \mathsf{E}_k t^k = \frac{\rho_z n t^2}{e^{\rho_z n t} - 1} \sum_{s=1}^{n-1} \mathsf{L}^{(\zeta_n^s)}(-nt)(e^{-s\chi t} - e^{-s(\chi-1)t}).$$

Note that the lower degree terms other than E_k are explained by the Roger type normalization (difference from ℓi_k and $\tilde{\chi}_k$) and the effects from compositions of paths $\overrightarrow{01} \rightsquigarrow \zeta\overrightarrow{01} \rightsquigarrow \zeta z$ of Baker–Campbell–Hausdorff type.

Remark 14 Replacing $\mathsf{L}^{(n)}(t)$, $\mathsf{L}^{(\zeta)}(nt)$ in (28) by those generating functions for $\ell i_k(z^n, \pi_n(\gamma))$, $\ell i_k(\zeta z, \delta_\zeta \mathscr{I}_\zeta(\gamma))$ by (25), (30) and (33), we obtain an equation

$$\sum_{k=1}^{\infty} \ell i_k(z^n, \pi_n(\gamma)) t^{k-1}$$
$$= \frac{\rho_z n t}{e^{\rho_z n t} - 1} \sum_{s=0}^{n-1} e^{s\chi t} \left(\frac{e^{-\mathsf{L}_0(\zeta_n^s z) n t} - 1}{-\mathsf{L}_0(\zeta_n^s z) n t} \right) \sum_{k=1}^{\infty} \ell i_k(\zeta_n^s z, \mathscr{I}_{\zeta_n^s}(\gamma))(-nt)^{k-1}$$

in $\mathbb{Q}_\ell[[t]]$. From this, for every fixed $k \geq 1$, one may express $\ell i_k(z^n, \pi_n(\gamma))$ as a linear combination of the $\ell i_d(\zeta_n^s z, \delta_{\zeta_n^s} \mathscr{I}_{\zeta_n^s}(\gamma))$ $(s = 0, \dots, n-1, d = 1, \dots, k)$. However, those coefficients are apparently more complicated than those in Theorem 10 where the polylogarithmic characters $\tilde{\chi}_k^{z^n}$, $\tilde{\chi}_d^{\zeta z}$ are treated.

6 Homogeneous Form

We keep the notations in Sect. 5 with assuming $\mu_n \subset K$. Let $\pi_{\mathbb{Q}_\ell}(\overrightarrow{01}_n)$ denote the *l-adic pro-unipotent fundamental group* of $V_n \otimes \overline{K}$ based at $\overrightarrow{01}_n$ which is by definition the pro-unipotent hull of the image of the Magnus embedding (12) consisting of all the group-like elements of the complete Hopf algebra $\mathbb{Q}_\ell\langle\langle X_n, Y_{i,n} \mid 0 \leq i \leq n-1\rangle\rangle$. We also define the *l-adic pro-unipotent path space* (or $\mathbb{Q}_\ell$*-path space* for short) $\pi_{\mathbb{Q}_\ell}(\overrightarrow{01}_n, v)$ for a K-(tangential) point v on V_n to be the $\mathbb{Q}_\ell$-rational extension of the path torsor $\pi_1^\ell(V_n \otimes \overline{K}, \overrightarrow{01}_n, v)$ via $\pi_1^\ell(V_n \otimes \overline{K}, \overrightarrow{01}_n) \subset \pi_{\mathbb{Q}_\ell}(\overrightarrow{01}_n)$. Note that both $\pi_{\mathbb{Q}_\ell}(\overrightarrow{01}_n)$ and $\pi_{\mathbb{Q}_\ell}(\overrightarrow{01}_n, v)$ have natural actions by G_K compatible with identification

$$\pi_1^\ell(V_n \otimes \overline{K}, \overrightarrow{01}_n) \subset \pi_{\mathbb{Q}_\ell}(\overrightarrow{01}_n), \quad \pi_1^\ell(V_n \otimes \overline{K}, \overrightarrow{01}_n, v) \subset \pi_{\mathbb{Q}_\ell}(\overrightarrow{01}_n, v).$$

Let us introduce rational modifications of the loops $y_{s,n}$ $(s = 0, \dots, n-1)$ and the paths δ_ζ $(\zeta \in \mu_n)$ respectively as follows. For $s = 0, \dots, n-1$ and $\zeta = e^{2\pi i s/n}$, set

$$\tilde{y}_{s,n} := x_n^{-\frac{s}{n}} y_{s,n} x_n^{\frac{s}{n}} \in \pi_{\mathbb{Q}_\ell}(\overrightarrow{01}_n),$$
$$\varepsilon_\zeta := x^{-\frac{s}{n}} \cdot \delta_\zeta \in \pi_{\mathbb{Q}_\ell}(\overrightarrow{01}, \zeta\overrightarrow{01}).$$

Note that, in the case $n = 1$, we have $x = x_1$, $y = \tilde{y}_{0,1}$ by definition.

The following lemma is the key to homogenize the l-adic distribution formula.

Lemma 15 *(i) For every $\sigma \in G_K$ and $\zeta \in \mu_n$, we have $\sigma(\varepsilon_\zeta) = \varepsilon_\zeta$. Moreover, for any path γ from $\zeta\overrightarrow{01}$ to a K-point w on V_1, we have $\varepsilon_\zeta\, \mathfrak{f}_\sigma^\gamma\, \varepsilon_\zeta^{-1} = \mathfrak{f}_\sigma^{\varepsilon_\zeta \gamma}$.*

(ii) The natural extensions of the homomorphisms on $\pi_{\mathbb{Q}_\ell}(\overrightarrow{01}_*)$ induced by $\mathscr{J}_\zeta : V_n \to V_1$, $\pi_{rn,r} : V_{rn} \to V_r$ (denoted by the same symbols) map the loops x_n, $\tilde{y}_{s,n}$ $(s = 0, \ldots, n-1)$ as follows.

(a) $\varepsilon_\zeta\, \mathscr{J}_\zeta(x_n)\, \varepsilon_\zeta^{-1} = x.$

(b) $\pi_{rn,r}(x_{rn}) = x_r{}^n.$

(c) $\varepsilon_\zeta\, \mathscr{J}_\zeta(\tilde{y}_{s,n})\, \varepsilon_\zeta^{-1} = \begin{cases} y & (\zeta = e^{-2\pi is/n}), \\ 1 & (\zeta \neq e^{-2\pi is/n}). \end{cases}$

(d) $\pi_{rn,r}(\tilde{y}_{j,rn}) = \tilde{y}_{i,r} \quad (0 \le i < r,\ 0 \le j < rn,\ i \equiv j \bmod r).$

Proof (i): Let $\zeta = e^{2\pi is/n}$ $(s = 0, \ldots, n-1)$. By the assumption $\mu_n \subset K$, we have $\chi(\sigma) \equiv 1 \bmod n$ for $\sigma \in G_K$. The first assertion follows immediately from the formula

$$\sigma(\delta_\zeta) = x^{\frac{s}{n}(\chi(\sigma)-1)}\delta_\zeta,$$

which can be easily seen from an argument similar to the proof of [12, Proposition 1] with replacement of $\bar{F}((t-z))$ by $\bar{F}\{\{\zeta t\}\}$. The second claim follows easily from the definition (13): $\mathfrak{f}_\sigma^p = p \cdot \sigma(p)^{-1}$ for any path $p : a \rightsquigarrow b$.

(ii): (a), (b) and the case $\zeta \neq e^{-2\pi is/n}$ of (c) are trivial. (d) follows from (b) and the fact $\pi_{rn,r}(y_{j,rn}) = x_r^k y_{i,r} x_r^{-k}$ with $j = i + kr$, $0 \le i < r$ (16). It remains to prove (c) in the case $\zeta = e^{-2\pi is/n}$. Suppose first that ζ is different from 1, i.e., $\zeta = e^{-2\pi is/n}$ for any fixed $s = 1 \ldots n-1$. Then $\varepsilon_\zeta = x^{-\frac{n-s}{n}} \cdot \delta_\zeta$. Since $\delta_\zeta\, \mathscr{J}_\zeta(y_{s,n})\, \delta_\zeta^{-1} = xyx^{-1}$, (a) implies $\delta_\zeta \mathscr{J}_\zeta(\tilde{y}_{s,n})\delta_\zeta^{-1} = x^{-\frac{s}{n}} xyx^{-1} x^{\frac{s}{n}} = x^{\frac{n-s}{n}} y x^{-\frac{n-s}{n}}$. It follows then that $\varepsilon_\zeta \mathscr{J}_\zeta(\tilde{y}_{s,n})\varepsilon_\zeta^{-1} = y$. Next, suppose $\zeta = 1$ (i.e., $s = 0$). Then, it is easy to settle this case by $\mathscr{J}_1(y_{0,n}) = y$. We thus complete the proof of (c). □

Now, we embed $\pi_{\mathbb{Q}_\ell}(\overrightarrow{01}_n)$ and its Lie algebra $L_{\mathbb{Q}_\ell}(\overrightarrow{01}_n)$ into the non-commutative power series ring $\mathbb{Q}_\ell\langle\langle \mathscr{X}_n, \mathscr{Y}_{s,n} \mid 0 \le s < n \rangle\rangle$ by setting $\mathscr{X}_n := X_n = \log x_n$, $\mathscr{Y}_{s,n} := \log \tilde{y}_{s,n}$, and denote by $\mathscr{M}_n$ the set of monomials in $\mathscr{X}_n, \mathscr{Y}_{s,n}$ $(s = 0, \ldots, n-1)$. For $w \in \mathscr{M}_n$, let $\mathsf{wt}_X(w)$ denote the number of $\mathscr{X}_n$ appearing in w. We shall also employ the monomial congruence '$w \equiv w' \bmod r$' by following the same manner as Definition 3 after replacing X_n, $Y_{i,n}$ by $\mathscr{X}_n$, $\mathscr{Y}_{i,n}$ $(n \in \mathbb{Z}_{>0}, 0 \le i < n)$ respectively. For the case $n = 1$, we will also simply write $\mathscr{X} = \mathscr{X}_1$, $\mathscr{Y} = \mathscr{Y}_{0,1}$.

Definition 16 Let z be a point in $V_n(K)$. Given a $\mathbb{Q}_\ell$-path $p \in \pi_{\mathbb{Q}_\ell}(\overrightarrow{01}, z)$ and any $\sigma \in G_K$, we set $\mathfrak{f}_\sigma^p := p \cdot \sigma(p)^{-1}$ and expand it in the form

$$\mathfrak{f}_\sigma^p = 1 + \sum_{w \in \mathscr{M}_n} \mathscr{L}i_w(\overrightarrow{01} \overset{p}{\rightsquigarrow} z)(\sigma) \cdot w$$

in $\mathbb{Q}_\ell\langle\langle \mathscr{X}_n, \mathscr{Y}_{i,n} \mid 0 \le i \le n-1 \rangle\rangle$. (Recall that, in (14), another (non-commutative) expansion of $\mathfrak{f}_\sigma^\gamma$ for $\gamma \in \pi_1^\ell(V_n \otimes \overline{K}, \overrightarrow{01}, z)$ was considered by using a different set of variables.) We call the above coefficient character

$$\mathscr{L}i_w(\overrightarrow{01} \overset{p}{\rightsquigarrow} z) \left(= \mathscr{L}i_w^{(\ell)}(\overrightarrow{01} \overset{p}{\rightsquigarrow} z)\right) : G_K \to \mathbb{Q}_\ell$$

the *l-adic iterated integral* associated to the word $w \in \mathscr{M}_n$ and to the $\mathbb{Q}_\ell$-path p on V_n.

Theorem 17 *Let p be a $\mathbb{Q}_\ell$-path on V_{rn} from $\overrightarrow{01}$ to a point $z \in V_{rn}(K)$. Then, for any word $w \in \mathscr{M}_r$, we have the distribution relation*

$$\mathscr{L}i_w(\overrightarrow{01} \overset{\pi_{rn,r}(p)}{\rightsquigarrow\rightsquigarrow} z^n)(\sigma) = n^{\mathrm{wt}_X(w)} \sum_{\substack{u \in \mathscr{M}_{rn} \\ u \equiv w \bmod r}} \mathscr{L}i_u(\overrightarrow{01} \overset{p}{\rightsquigarrow} z)(\sigma)$$

for $\sigma \in G_K$.

Proof The assertion follows in the same way as Theorem 4 after the above Lemma 15 (b), (d). □

Next, let us concentrate on the polylogarithmic part on V_1. Recall that both $\pi_{\mathbb{Q}_\ell}(\overrightarrow{01})$ and its Lie algebra $L_{\mathbb{Q}_\ell}(\overrightarrow{01})$ are embedded in $\mathbb{Q}_\ell\langle\langle X, Y \rangle\rangle$, where $X = \mathscr{X}_1$ and $Y = \mathscr{Y}_{0,1}$.

Definition 18 Let z be a point in $V_1(K) = \mathbf{P}^1(K) - \{0, 1, \infty\}$ and $p : \overrightarrow{01} \rightsquigarrow z$ a $\mathbb{Q}_\ell$-path. Consider the associator $\mathfrak{f}_\sigma^p := p \cdot \sigma(p)^{-1} \in \pi_{\mathbb{Q}_\ell}(\overrightarrow{01})$ for $\sigma \in G_K$, and define

$$\boldsymbol{\rho}_{z,p} : G_K \to \mathbb{Q}_\ell, \quad \ell i_m(z, p) : G_K \to \mathbb{Q}_\ell$$

by the non-commutative expansion corresponding to (18):

$$\log(\mathfrak{f}_\sigma^p)^{-1} \equiv \boldsymbol{\rho}_{z,p}(\sigma) X + \sum_{m=1}^{\infty} \ell i_m(z, p)(\sigma)(\mathrm{ad}X)^{m-1}(Y) \mod I_Y,$$

where I_Y represents the ideal generated by those terms including Y twice or more. Using these, we also define

$$\tilde{\boldsymbol{\chi}}_m^{z,p} : G_K \to \mathbb{Q}_\ell$$

for $m \geq 1$ by the equation extending Proposition 8 (i):

$$\tilde{\boldsymbol{\chi}}_m^{z,p}(\sigma) = (-1)^{m+1}(m-1)! \sum_{k=1}^{m} \frac{\boldsymbol{\rho}_{z,p}(\sigma)^{m-k}}{(m+1-k)!} \ell i_k(z,p)(\sigma). \tag{36}$$

Since $\mathbb{Q}_\ell$-paths generally do not give bijection systems between fibers of endpoints on finite étale covers, no simple interpretation is available for $\boldsymbol{\rho}_{z,p}$ or $\tilde{\boldsymbol{\chi}}_m^{z,p}$ by Kummer properties: For example, the above $\tilde{\boldsymbol{\chi}}_m^{z,p}(\sigma)$ $(\sigma \in G_K)$ generally has a denominator in $\mathbb{Q}_\ell$, i.e., may not be valued in $\mathbb{Z}_\ell$. This makes it difficult to understand $\tilde{\boldsymbol{\chi}}_m^{z,p}(\sigma)$ in terms of Kummer properties at finite levels of an arithmetic sequence like (20).

Once $\boldsymbol{\rho}_{z,p}$, $\ell i_m(z,p)$ and $\tilde{\boldsymbol{\chi}}_m^{z,p} : G_K \to \mathbb{Q}_\ell$ are defined as in the above Definition, the identities as in Proposition 8 (ii) and (21) can be extended in obvious ways for them by formal transformations of generating functions. In the same way, it holds that

$$-\frac{\tilde{\boldsymbol{\chi}}_m^{z,p}(\sigma)}{(m-1)!} = \mathscr{L}i_{YX^{m-1}}(\overrightarrow{01} \overset{p}{\rightsquigarrow} z)(\sigma) \tag{37}$$

for $p \in \pi_{\mathbb{Q}_\ell}(\overrightarrow{01}, z)$ and $\sigma \in G_K$.

Theorem 19 *Suppose $\mu_n \subset K \subset \mathbb{C}$ and let p be a $\mathbb{Q}_\ell$-path on V_n from $\overrightarrow{01}$ to a point $z \in V_n(K)$. Then,*

$$\ell i_k(z^n, \pi_n(p))(\sigma) = n^{k-1} \sum_{\zeta \in \mu_n} \ell i_k(\zeta z, \varepsilon_\zeta \mathscr{J}_\zeta(p))(\sigma)$$

holds for $\sigma \in G_K$.

Proof We first put the Lie expansion of $\log(\mathfrak{f}_\sigma^p)^{-1}$ in $\mathscr{X}_n = X_n = \log x_n$, $\mathscr{Y}_{s,n} = \log \tilde{y}_{s,n}$ $(s = 0, \ldots, n-1)$ in the Lie algebra $L_{\mathbb{Q}_\ell}(\overrightarrow{01}_n)$ as:

$$\begin{aligned} \log(\mathfrak{f}_\sigma^\gamma)^{-1} &\equiv DX_n + \sum_{s=0}^{n-1}\sum_{m=1}^{\infty} D_{s,m}\, (\mathrm{ad}\mathscr{X}_n)^{m-1}(\mathscr{Y}_{s,n}) \qquad (38)\\ &\equiv DX_n + \sum_{s=0}^{n-1} (\mathsf{D}_s(\mathrm{ad}\mathscr{X}_n))\, (\mathscr{Y}_{s,n}) \mod I_{\mathscr{Y}_*}, \end{aligned}$$

where, $I_{\mathscr{Y}_*}$ represents the ideal generated by those terms including $\{\mathscr{Y}_{0,n}, \ldots, \mathscr{Y}_{n-1,n}\}$ twice or more, and $\mathsf{D}_s(t) = \sum_{m=1}^{\infty} D_{s,m}\, t^{m-1} \in \mathbb{Q}_\ell[[t]]$ $(s = 0, \ldots, n-1)$. We shall determine those coefficients D and $\mathsf{D}_{s,m}$ by applying the morphisms $\mathscr{J}_\zeta$. For any fixed $\zeta = \zeta_n^{-s}$ $(s = 0, \ldots, n-1)$, by Lemma 15 (i), we obtain $\mathfrak{f}_\sigma^{\varepsilon_\zeta \mathscr{J}_\zeta(p)} = \varepsilon_\zeta \cdot \mathscr{J}_\zeta(p \cdot \sigma(p)^{-1}) \cdot \sigma(\varepsilon_\zeta)^{-1} = \varepsilon_\zeta \cdot \mathscr{J}_\zeta(\mathfrak{f}_\sigma^p) \cdot \varepsilon_\zeta^{-1}$, hence

$$\varepsilon_\zeta \cdot \mathscr{J}_\zeta(\log(\mathfrak{f}_\sigma^p)^{-1}) \cdot \varepsilon_\zeta^{-1} = \log(\mathfrak{f}_\sigma^{\varepsilon_\zeta \mathscr{J}_\zeta(p)})^{-1}.$$

As the right hand side comes from the associator for the path $\varepsilon_\zeta \mathscr{J}_\zeta(p) : \overrightarrow{01} \rightsquigarrow \zeta z$, it should coincide, by definition, with

$$\boldsymbol{\rho}_{\zeta z, \varepsilon_\zeta \mathscr{J}_\zeta(p)}(\sigma) X + \sum_{k=1}^{\infty} \ell i_k(\zeta z, \varepsilon_\zeta \mathscr{J}_\zeta(p))(\sigma)\, (\mathrm{ad} X)^{k-1}(Y),$$

while, the left hand side can be calculated after Lemma 15 (ii) (a), (c) to equal to

$$DX + \sum_{k=1}^{\infty} D_{s,k}\, (\mathrm{ad} X)^{k-1}(Y)$$

with s given by $\zeta = e^{-2\pi i s/n}$. Therefore, we conclude

$$D = \boldsymbol{\rho}_{\zeta z, \varepsilon_\zeta \mathscr{J}_\zeta(p)}(\sigma), \tag{39}$$

$$D_{s,k} = \ell i_k(\zeta z, \varepsilon_\zeta \mathscr{J}_\zeta(p))(\sigma) \tag{40}$$

for $\zeta = e^{-2\pi i s/n}$ $(s = 0, \ldots, n-1)$. Now, apply the projection morphism $\pi_n := \pi_{n,1} : V_n \to V_1$ and interpret the both sides of equality $\pi_n(\log(\mathfrak{f}_\sigma^p)^{-1}) = \log(\mathfrak{f}_\sigma^{\pi_n(p)})^{-1}$. Then, we obtain

$$DnX + \sum_{s=0}^{n-1} \sum_{k=1}^{\infty} D_{s,k}\, (n\, \mathrm{ad} X)^{k-1}(Y) = \boldsymbol{\rho}_{z^n, \pi_n(p)} X + \sum_{k=1}^{\infty} \ell i_k(z^n, \pi_n(p))\, \mathrm{ad}(X)^{k-1}(Y).$$

Comparing the coefficient of $(\mathrm{ad} X)^{k-1}(Y)$ in the above and (40), we conclude the proof of the theorem. □

In the above proof, for a given $\mathbb{Q}_\ell$-path $p : \overrightarrow{01} \rightsquigarrow z$ on V_n, we considered the collection of $\mathbb{Q}_\ell$-paths

$$\mathscr{P}_n := \{\varepsilon_\zeta \mathscr{J}_\zeta(p) : \overrightarrow{01} \rightsquigarrow \zeta z \mid \zeta = \zeta_n^s \in \mu_n\ (s = 0, 1, \ldots, n-1)\}$$

on $V_1 = \mathbf{P}^1 - \{0, 1, \infty\}$. Note that each $\varepsilon_\zeta \mathscr{J}_\zeta(p)$ can also be written as the composite of paths on V_1:

$$\varepsilon_\zeta \cdot [\zeta p] = x^{-\frac{s}{n}} \cdot \delta_\zeta \cdot [\zeta p] : \tag{41}$$

$$\overrightarrow{01} \overset{x^{-\frac{s}{n}}}{\rightsquigarrow} \overrightarrow{01} \overset{\delta_\zeta}{\rightsquigarrow} \zeta\overrightarrow{01} \overset{[\zeta p]}{\rightsquigarrow} \zeta z$$

where $[\zeta p] : \zeta\overrightarrow{01} \rightsquigarrow \zeta z$ means a path obtained by "rotating" $p : \overrightarrow{01} \rightsquigarrow z$ by the automorphism of $\mathbf{P}^1 - \{0, \infty\}$ with multiplication by ζ.

Corollary 20 *Notations being as above, the maps* $\boldsymbol{\rho}_{\zeta z, p}(\sigma) : G_K \to \mathbb{Q}_\ell$ *are all the same for the* $\mathbb{Q}_\ell$*-paths* $[p : \overrightarrow{01} \rightsquigarrow \zeta z] \in \bigcup_{n=1}^{\infty} \mathscr{P}_n$.

Proof As seen in (39), we have the common D upon applying $\mathscr{J}_\zeta$ to the first term of $\log(\mathfrak{f}_\sigma^p)^{-1}$. The assertion follows from this and the fact that $\mathscr{P}_n$ contains $\mathscr{P}_1 \neq \emptyset$. □

From this corollary, we immediately see that the above theorem also gives homogeneous functional equations for the rationally extended l-adic polylogarithmic characters.

Corollary 21 *Notations being as in Theorem 19, let $\tilde{\boldsymbol{\chi}}_k^{z^n,\pi_n(p)}$ and $\tilde{\boldsymbol{\chi}}_k^{\zeta z,\varepsilon_\zeta \mathscr{J}_\zeta(p)}$ ($\zeta \in \mu_n$) be the extended l-adic polylogarithmic characters. Then, we have*

$$\tilde{\boldsymbol{\chi}}_k^{z^n,\pi_n(p)}(\sigma) = n^{k-1} \sum_{\zeta\in\mu_n} \tilde{\boldsymbol{\chi}}_k^{\zeta z,\varepsilon_\zeta \mathscr{J}_\zeta(p)}(\sigma) \qquad (\sigma \in G_K).$$

Proof The assertion follows from Theorem 19 by applying Corollary 20 to the definition of l-adic polylogarithmic characters for $\mathbb{Q}_\ell$-paths (Definition 18). □

7 Translation in Kummer–Heisenberg Measure

Let $\gamma : \overrightarrow{01} \rightsquigarrow z$ be an l-adic path in $\pi_1^\ell(\mathbf{P}^1_{\overline{K}} - \{0, 1, \infty\}; \overrightarrow{01}, z)$ and $p := x^{-\frac{s}{n}}\gamma$ be the pro-unipotent path in $\pi_{\mathbb{Q}_\ell}(\overrightarrow{01}, z)$ produced by the composition with $x^{-\frac{s}{n}}$ for any fixed $s \in \mathbb{Z}_\ell$ and $n \in \mathbb{N}$. By definition we have $\mathfrak{f}_\sigma^p = x^{-\frac{s}{n}}\mathfrak{f}_\sigma^\gamma x^{\frac{s}{n}\chi(\sigma)}$ for $\sigma \in G_K$. Since $x^{-\frac{s}{n}} = \exp(\frac{-s}{n}X) \equiv 1$ modulo the right ideal $X\cdot\mathbb{Q}_\ell\langle\langle X, Y\rangle\rangle$, it follows from Proposition 8 (ii) that

$$\begin{aligned}
-\frac{\tilde{\boldsymbol{\chi}}_k^{z,p}(\sigma)}{(k-1)!} &= \mathsf{Coeff}_{YX^{k-1}}(\mathfrak{f}_\sigma^p) = \mathsf{Coeff}_{YX^{k-1}}\left(1\cdot \mathfrak{f}_\sigma^\gamma \cdot \exp\left(\frac{s\,\chi(\sigma)}{n}X\right)\right) \\
&= \sum_{i=0}^{k-1} \mathsf{Coeff}_{YX^i}(\mathfrak{f}_\sigma^\gamma)\cdot \frac{\left(\frac{s}{n}\chi(\sigma)\right)^{k-i-1}}{(k-i-1)!} = -\sum_{i=0}^{k-1} \frac{\tilde{\chi}_{i+1}^{z,\gamma}(\sigma)}{i\,!}\cdot\frac{\left(\frac{s}{n}\chi(\sigma)\right)^{k-i-1}}{(k-i-1)!}.
\end{aligned}$$

Thus we obtain

$$\tilde{\boldsymbol{\chi}}_k^{z,p}(\sigma) = \sum_{i=0}^{k-1}\binom{k-1}{i}\left(\frac{s}{n}\chi(\sigma)\right)^{k-i-1}\tilde{\chi}_{i+1}^{z,\gamma}(\sigma) \qquad (\sigma \in G_K). \tag{42}$$

Recall then that, in [12], introduced is a certain $\mathbb{Z}_\ell$-valued measure (called the Kummer–Heisenberg measure) $\boldsymbol{\kappa}_{z,\gamma}(\sigma)$ on $\mathbb{Z}_\ell$ for every path $\gamma : \overrightarrow{01} \rightsquigarrow z$ and $\sigma \in G_K$, which is characterized by the integration properties:

$$\tilde{\chi}_k^{z,\gamma}(\sigma) = \int_{\mathbb{Z}_\ell} a^{k-1} d\boldsymbol{\kappa}_{z,\gamma}(\sigma)(a) \qquad (k \geq 1). \tag{43}$$

Putting this into (42), we may rewrite the RHS to get

$$\tilde{\chi}_k^{z,p}(\sigma) = \int_{\mathbb{Z}_\ell} \left(a + \frac{s}{n}\chi(\sigma)\right)^{k-1} d\boldsymbol{\kappa}_{z,\gamma}(\sigma)(a). \tag{44}$$

Note that $\frac{s}{n} + \mathbb{Z}_\ell = \frac{s}{n}\chi(\sigma) + \mathbb{Z}_\ell$ as a subset of $\mathbb{Q}_\ell$ when $\mu_n \subset K$. Comparison of (43) and (44) leads us to introduce the following

Definition 22 Suppose $\mu_n \subset K$, and let $\sigma \in G_K$ and $p = x^{-\frac{s}{n}}\gamma \in \pi_{\mathbb{Q}_\ell}(\overrightarrow{01}, z)$ be as above. Define a $\mathbb{Z}_\ell$-valued measure $\boldsymbol{\kappa}_{z,p}(\sigma)$ on the coset $\frac{s}{n} + \mathbb{Z}_\ell(\subset \mathbb{Q}_\ell)$ by the property:

$$\tilde{\chi}_k^{z,p}(\sigma) = \int_{\frac{s}{n}+\mathbb{Z}_\ell} a^{k-1} d\boldsymbol{\kappa}_{z,p}(\sigma)(a) \qquad (k \geq 1).$$

A verification of this new notion of the extended measure $\boldsymbol{\kappa}_{z,p}(\sigma)$ is that our distribution relations in Corollary 21 can be summarized into a single relation of measures:

Theorem 23 *For $s \in \mathbb{Z}_\ell$, let $[n] : \frac{s}{n} + \mathbb{Z}_\ell \to \mathbb{Z}_\ell$ $(a \mapsto na)$ denote the continuous map of multiplication by $n \in \mathbb{N}$, and denote by $[n]_*\boldsymbol{\kappa}$ the push-forward measure on $\mathbb{Z}_\ell$ obtained from any measure $\boldsymbol{\kappa}$ on $\frac{s}{n} + \mathbb{Z}_\ell$ by $U \mapsto \boldsymbol{\kappa}([n]^{-1}(U))$ for the compact open subsets U of $\mathbb{Z}_\ell$. Then,*

$$\boldsymbol{\kappa}_{z^n,\pi_n(\gamma)}(\sigma) = \sum_{\zeta\in\mu_n} [n]_*\boldsymbol{\kappa}_{\zeta z,\varepsilon_\zeta \mathscr{I}_\zeta(\gamma)}(\sigma) \qquad (\sigma \in G_K).$$

Proof The formula follows immediately from Corollary 21 and the characteristic property (44) of the Kummer–Heisenberg measure. □

Question 24 In the above discussion, we defined $\boldsymbol{\kappa}_{z,p}(\sigma)$ only for $\mathbb{Q}_\ell$-paths $p : \overrightarrow{01} \rightsquigarrow z$ of the form $p = x^\alpha\gamma$ with $\alpha \in \mathbb{Q}_\ell$ and $\gamma : \overrightarrow{01} \rightsquigarrow z$ being l-adic (i.e., $\mathbb{Z}_\ell$-integral) paths. It is natural to conjecture existence of a suitable measure $\boldsymbol{\kappa}_{z,p}(\sigma)$ for a more general $\mathbb{Q}_\ell$-path $p : \overrightarrow{01} \rightsquigarrow z$ satisfying the property of Definition 22. The support of this measure should be a parallel transport $R(p, \sigma)$ of $\mathbb{Z}_\ell$ in $\mathbb{Q}_\ell$ such that $x^{R(p,\sigma)} \subset x^{\mathbb{Q}_\ell}$ is the image of $\pi_1^\ell(V_1 \otimes \overline{K}; \overrightarrow{01}, z) \cdot \sigma(p)^{-1}$ via the projection $\pi_{\mathbb{Q}_\ell}(\overrightarrow{01}) \twoheadrightarrow x^{\mathbb{Q}_\ell}$.

8 Inspection of Special Cases

In this section, we shall closely look at special cases of the l-adic distribution formula. Let us first consider dilogarithms, i.e., for the case of $k = 2$. By Theorem 10, we have

Corollary 25 *Let $\mu_n \subset K$ and $\gamma : \overrightarrow{01} \rightsquigarrow z \in V_n(K)$ be an l-adic path which induces paths $\pi_n(\gamma) : \overrightarrow{01} \rightsquigarrow z^n$ and $\delta_\zeta \mathscr{J}_\zeta(\gamma) : \overrightarrow{01} \rightsquigarrow \zeta z$ ($\zeta = \zeta_n^s \in \mu_n$) on $V_1 = \mathbf{P}^1 - \{0, 1, \infty\}$. Along these paths, we have the following $\mathbb{Z}_\ell$-valued functional equation*

$$\tilde{\chi}_2^{z^n}(\sigma) = n \sum_{s=0}^{n-1} \tilde{\chi}_2^{\zeta_n^s z}(\sigma) + \sum_{s=1}^{n-1} s\chi(\sigma)\rho_{1-\zeta_n^s z}(\sigma) \qquad (\sigma \in G_K),$$

where $\rho_{1-\zeta_n^s z}$ is the same as the 1st polylogarithmic character $\tilde{\chi}_1^{\zeta_n^s z} : G_K \to \mathbb{Z}_\ell$. □

In particular when $n = 2$, the above formula is specialized to the following.

Corollary 26 *For $\gamma : \overrightarrow{01} \rightsquigarrow z$ on $V_2 = \mathbf{P}^1 - \{0, \pm 1, \infty\}$, let $\pi_2(\gamma) : \overrightarrow{01} \rightsquigarrow z^2$, $\mathscr{J}_1(\gamma) : \overrightarrow{01} \rightsquigarrow z$ and $\delta_{-1}\mathscr{J}_{-1}(\gamma) : \overrightarrow{01} \rightsquigarrow -z$ be the induced paths on $\mathbf{P}^1 - \{0, 1, \infty\}$. Note here that $\delta_{-1} : \overrightarrow{01} \rightsquigarrow -\overrightarrow{01}$ is the positive half rotation. Along these paths, we have a functional equation of the l-adic polylogarithmic characters*

$$\tilde{\chi}_2^{z^2}(\sigma) = 2(\tilde{\chi}_2^{z}(\sigma) + \tilde{\chi}_2^{-z}(\sigma)) + \chi(\sigma)\rho_{1+z}(\sigma) \qquad (\sigma \in G_K).$$

□

Putting $z = \overrightarrow{10}$ in the above, and recalling $\tilde{\chi}_{2k}^{\overrightarrow{10}}(\sigma) = \frac{B_{2k}}{2(2k)}(\chi(\sigma)^{2k} - 1)$ ($\sigma \in G_{\mathbb{Q}}$) from [NW2] Proposition 5.13, we immediately obtain

Corollary 27 *Along the path $\gamma_{-1} : \overrightarrow{01} \rightsquigarrow (z=1) \rightsquigarrow (z=-1)$ induced by the positive half arc on the unit circle on $\mathbf{P}^1 - \{0, 1, \infty\}$, we have the following $\mathbb{Z}_\ell$-valued equation:*

$$\tilde{\chi}_2^{z=-1}(\sigma) = -\frac{\chi(\sigma)^2 - 1}{48} - \frac{1}{2}\chi(\sigma)\rho_2(\sigma) \qquad (\sigma \in G_{\mathbb{Q}}).$$

□

This result is an l-adic analog of the classical result $Li_2(-1) = -\frac{\pi^2}{12}$ ([Le]), and is compatible with [13, Remark 5.14 and Remark after (6.31)].

To confirm validity of our above narrow stream of geometrical arguments toward Corollary 27, we here present an alternative direct proof in a purely arithmetic way as below:

Arithmetic proof of Corollary 27. We (only) make use of the characterization of $\tilde{\chi}_m^z$ by the Kummer properties (20). Applying it to our case $m = 2$, $z = -1$ where $\rho_z(\sigma) = \frac{1}{2}(\chi(\sigma) - 1)$, we obtain

$$\zeta_{\ell^n}^{\tilde{\chi}_2^{z=-1}(\sigma)} = \sigma \left(\prod_{a=0}^{\ell^n-1} (1 - \zeta_{2\ell^n}^{2\chi(\sigma)^{-1}a+1})^{\frac{a}{\ell^n}} \right) \Big/ \prod_{a=0}^{\ell^n-1} (1 - \zeta_{2\ell^n}^{2a+\chi(\sigma)})^{\frac{a}{\ell^n}}. \qquad (*)$$

We evaluate both the denominator and numerator of the above right hand side, first by pairing two factors indexed by a and $a' = -\chi(\sigma) - a$ and by simplifying their product by

$$\left(1 - \zeta_{2\ell^n}^{-2a-\chi(\sigma)}\right)^{\frac{1}{\ell^n}} = \left(1 - \zeta_{2\ell^n}^{2a+\chi(\sigma)}\right)^{\frac{1}{\ell^n}} \cdot \zeta_{2\ell^n}^{l^n - \langle 2a+\chi(\sigma)\rangle}$$

with $0 \leq \langle 2a + \chi(\sigma)\rangle \leq 2\ell^n$ being the unique residue of $2a + \chi(\sigma)$ mod $2\ell^n$. Pick a disjoint decomposition of the index set $S := \{0 \leq a \leq \ell^n - 1\}$ into $S_+ \cup S_- \cup S_0$ so that, for all $a \in S$,

(i) $a \in S_+$ iff $\langle -\chi(\sigma) - a\rangle \in S_-$;

(ii) $a \in S_0$ iff $a \equiv -a - \chi(\sigma)$ mod ℓ^n.

Then, one finds:

$$\prod_{a \in S - S_0} (1 - \zeta_{2\ell^n}^{2\chi(\sigma)^{-1}a+1})^{\frac{a}{\ell^n}} = \prod_{a \in S_\pm} (1 - \zeta_{2\ell^n}^{2\chi(\sigma)^{-1}a+1})^{\frac{-\chi(\sigma)}{\ell^n}} \zeta_{2\ell^n}^{(\ell^n - \langle 1+2\chi(\sigma)^{-1}a\rangle)(-a-\chi(\sigma))},$$

$$\prod_{a \in S - S_0} (1 - \zeta_{2\ell^n}^{2a+\chi(\sigma)})^{\frac{a}{\ell^n}} = \prod_{a \in S_\pm} (1 - \zeta_{2\ell^n}^{2a+\chi(\sigma)})^{\frac{-\chi(\sigma)}{\ell^n}} \zeta_{2\ell^n}^{(\ell^n - \langle 2a+\chi(\sigma)\rangle)(-a-\chi(\sigma))}.$$

Noting that $\prod_{a \in S}(1 - \zeta_{2\ell^n}^{2a+1}) = 2$, we obtain the squared sides of $(*)$ as

$$\zeta_{\ell^n}^{2\tilde{\chi}_2^{z=-1}(\sigma)} = \frac{\sigma\left(2^{\frac{-\chi(\sigma)}{\ell^n}} \prod_{a \in S} \zeta_{2\ell^n}^{(\ell^n - \langle 1+2\chi(\sigma)^{-1}a\rangle)(-a-\chi(\sigma))}\right)}{2^{\frac{-\chi(\sigma)}{\ell^n}} \prod_{a \in S} \zeta_{2\ell^n}^{(\ell^n - \langle 2a+\chi(\sigma)\rangle)(-a-\chi(\sigma))}}.$$

Here, note that contribution from S_0 (which is empty when $\ell = 2$) is included into the factor $2^{\frac{-\chi(\sigma)}{\ell^n}}$ both in the numerator and the denominator. Now, choose integers $c, \bar{c} \in \mathbb{Z}$ so that $c \equiv \chi(\sigma)$, $c\bar{c} \equiv 1$ mod $2\ell^n$. Then, we obtain the following congruence equation mod ℓ^n:

$$\begin{aligned}
2\tilde{\chi}_2^{z=-1}(\sigma) &\equiv -\chi(\sigma)\rho_2(\sigma) \\
&\quad + \frac{1}{2}\sum_{a \in S} \chi(\sigma)(-a-c)(\ell^n - \langle 1 + \bar{c}a\rangle) - (-a-c)(\ell^n - \langle 1 + 2\bar{c}a\rangle) \\
&\equiv -\chi(\sigma)\rho_2(\sigma) \\
&\quad + \frac{1}{2}\sum_{a \in S}(-a-c)\left[\frac{\chi(\sigma)-1}{2} + \left\{\frac{2a+c}{2\ell^n}\right\} - c\left\{\frac{1+2\bar{c}a}{2\ell^n}\right\}\right] \\
&\equiv -\chi(\sigma)\rho_2(\sigma) + \frac{1}{2}\sum_{b \in S} b\left[c\left\{\frac{1+2\bar{c}b}{2\ell^n}\right\} - \left\{\frac{c+2b}{2\ell^n}\right\} + \frac{1-c}{2}\right].
\end{aligned}$$

By basic properties of the Bernoulli polynomial $B_2(X) = X^2 - X + \frac{1}{6}$ (cf. [8]), the last sum is congruent modulo $\frac{\ell^n}{48}\mathbb{Z}$ to

$$\sum_{b\in S}\frac{\ell^n}{2}\left[c^2B_2\left(\left\{\frac{1+2\bar{c}b}{2\ell^n}\right\}\right)-B_2\left(\left\{\frac{2b+c}{2\ell^n}\right\}\right)\right]$$
$$=\frac{1}{2}(\chi(\sigma)^2-1)B_2(\frac{1}{2})=-\frac{1}{24}\chi(\sigma)^2-1.$$

Summing up, we find the congruence relations

$$2\tilde{\chi}_2^{z=-1}(\sigma)\equiv-\chi(\sigma)\rho_2(\sigma)-\frac{1}{24}(\chi(\sigma)^2-1)\mod\frac{\ell^n}{48}\mathbb{Z}$$

for all n, hence the equality in $\mathbb{Z}_\ell$. This concludes the proof of the corollary. □

Turning to Theorem 10, by specialization to the case $n=2$ (but for general k), we obtain:

Corollary 28 *Along the paths from $\overrightarrow{01}$ to $\pm z$, z^2 on $\mathbf{P}^1-\{0,1,\infty\}$ used in Corollary 26, it holds that*

$$\tilde{\chi}_k^{z^2}(\sigma)=2^{k-1}\tilde{\chi}_k^{z}(\sigma)+\sum_{d=1}^{k}\binom{k-1}{d-1}2^{d-1}\chi(\sigma)^{k-d}\tilde{\chi}_d^{-z}(\sigma)\qquad(k\geq1,\ \sigma\in G_K).$$

Upon observing special cases of the above formula, we find that $\tilde{\chi}_4^{z=-1}$ does not factor through $\mathrm{Gal}(\mathbb{Q}(\mu_{\ell^\infty})/\mathbb{Q})$, because it involves a nontrivial term from $\tilde{\chi}_3^{\overrightarrow{10}}(\sigma)$ which does not vanish on $\mathrm{Gal}(\overline{\mathbb{Q}}/\mathbb{Q}(\mu_{\ell^\infty}))$ by Soulé [14].

With regard to the classical formula $Li_{2k}(-1)=(-1)^{k+1}(1-2^{2k-1})B_{2k}\frac{\pi^{2k}}{(2k)!}$ ([Le]), we should rather figure out its l-adic analog in terms of the "$\mathbb{Q}_\ell$-adic" polylogarithmic characters introduced in Definition 18. In fact,

Corollary 29 *Let $\gamma_{-1}:\overrightarrow{01}\rightsquigarrow(z=-1)$ be the path in Corollary 27. Then, along the $\mathbb{Q}_\ell$-path $x^{-\frac{1}{2}}\gamma_{-1}:\overrightarrow{01}\rightsquigarrow(z=-1)$, it holds that*

$$\tilde{\boldsymbol{\chi}}_{2k}^{z=-1}(\sigma)=\frac{(1-2^{2k-1})}{2^{2k}}\frac{B_{2k}}{2k}(\chi(\sigma)^{2k}-1)\qquad(\sigma\in G_{\mathbb{Q}}).$$

Proof Applying Corollary 21 to the case where $n=2$ and $p:\overrightarrow{01}\rightsquigarrow\overrightarrow{10}$ is the straight path on $V_2=\mathbf{P}^1-\{0,\pm1,\infty\}$, we obtain

$$\tilde{\boldsymbol{\chi}}_k^{\overrightarrow{10},\pi_2(p)}(\sigma)=2^{k-1}(\tilde{\boldsymbol{\chi}}_k^{\overrightarrow{10},\mathscr{I}_1(p)}(\sigma)+\tilde{\boldsymbol{\chi}}_k^{z=-1,\gamma_{-1}}(\sigma)).$$

Since $\pi_2(p)$ and $\mathscr{I}_1(p)$ are the same standard path $\overrightarrow{01}\rightsquigarrow\overrightarrow{10}$ on V_1, the values $\tilde{\boldsymbol{\chi}}_k^{\overrightarrow{10},\pi_2(p)}(\sigma)$ and $\tilde{\boldsymbol{\chi}}_k^{\overrightarrow{10},\mathscr{I}_1(p)}(\sigma)$ coincide with the (extended) Soulé value $\tilde{\chi}_k^{\overrightarrow{10}}(\sigma)$ (cf. [12, Remark 2]). The desired formula follows then from a basic formula from [13, Proposition 5.13]: $\tilde{\chi}_{2k}^{\overrightarrow{10}}(\sigma)=\frac{B_{2k}}{2(2k)}(\chi(\sigma)^{2k}-1)$ $(\sigma\in G_{\mathbb{Q}})$. □

Unlike the $\mathbb{Z}_\ell$-integral analog stated in Corollary 27, the above right hand side generally has denominators in $\mathbb{Q}_\ell$. This is due to the concern of $x^{-\frac{1}{2}} \in \pi_{\mathbb{Q}_l}(\overrightarrow{01})$ which does not lie in $\pi_1^\ell(V_1 \otimes \overline{K}, \overrightarrow{01})$ when $\ell = 2$.

Acknowledgements This work was partially supported by JSPS KAKENHI Grant Number JP26287006.

References

1. Beilinson, A., Deligne, P.: Interprétation motivique de la conjecture de Zagier reliant polylogarithms et régulateurs. Proc. Symp. in Pure Math. (AMS) 55–2, 97–121 (1994)
2. Deligne, P.: Le groupe fondamental de la droite projective moins trois points. In Ihara, Y., Ribet, K., Serre, J.-P. (eds.) Galois group over $\mathbb{Q}$, vol. 16, pp. 79–297. MSRI Publications (1989)
3. Douai, J.-C., Wojtkowiak, Z.: Descent for ℓ-adic polylogarithms. Nagoya Math. J. **192**, 59–88 (2008)
4. Gangl, H.: Families of functional equations for polylogarithms. Comtemp. Math. (AMS) **199**, 83–105 (1996)
5. Huber, A., Kings, G.: Polylogarithms for families of commutative group schemes. arXiv preprint arXiv:1505.04574
6. Hain, R.: On a generalization of Hilbert's 21st problem. Annales scientifiques de l'École Normale Supérieure, vol. 19, pp. 609–627 (1986)
7. Huber, A., Wildeshaus, J.: Classical motivic polylogarithm according to Beilinson and Deligne. Doc. Math. **3**, 27–133 (1998). Correction Doc. Math. **3**, 297–299 (1998)
8. Lang, S.: Cyclotomic fields I and II. In: Graduate Texts in Math. vol. 121. Springer (1990)
9. Lewin, L.: Polylogarithms and associated functions. North Holland (1981)
10. Milnor, J.: On polylogarithms Hurwitz zeta functions, and the Kubert identities. L'Enseignement Math. **29**, 281–322 (1983)
11. Nakamura, H., Sakugawa, K., Wojtkowiak, Z.: Polylogarithmic analogue of the Coleman-Ihara formula II. RIMS Kôkyûroku Bessatsu B **64**, 33–54 (2017)
12. Nakamura, H., Wojtkowiak, Z.: On explicit formulae for ℓ-adic polylogarithms. Proc. Symp. Pure Math. (AMS) **70**, 285–294 (2002)
13. Nakamura, H., Wojtkowiak, Z.: Tensor and homotopy criterions for functional equations of ℓ-adic and classical iterated integrals. In: Coates, J., et al. (eds.) Non-abelian Fundamental Groups and Iwasawa Theory. London Mathematical Society Lecture Note Series, vol. 393, pp. 258–310 (2012)
14. Soulé, C.: On higher p-adic regulators. Lecture Notes in Math, vol. 854, pp. 372–401. Springer (1981)
15. Wojtkowiak, Z.: A note on functional equations of the p-adic polylogarithms. Bull. Soc. Math. Fr. **119**, 343–370 (1991)
16. Wojtkowiak, Z.: Functional equations of iterated integrals with regular singularities. Nagoya Math. J. **142**, 145–159 (1996)
17. Wojtkowiak, Z.: Monodromy of iterated integrals and non-abelian unipotent periods, Geometric Galois Actions II. In: London Mathematical Society Lecture Note Series, vol. 243, pp. 219–289 (1997)
18. Wojtkowiak, Z.: On ℓ-adic iterated integrals, I Analog of Zagier Conjecture. Nagoya Math. J. **176**, 113–158 (2004)
19. Wojtkowiak, Z.: A note on functional equations of ℓ-adic polylogarithms. J. Inst. Math. Jussieu **3**, 461–471 (2004)
20. Wojtkowiak, Z.: On ℓ-adic iterated integrals, II—Functional equations and ℓ-adic polylogarithms. Nagoya Math. J. **177**, 117–153 (2005)

21. Wojtkowiak, Z.: On ℓ-adic iterated integrals, III—Galois actions on fundamental groups. Nagoya Math. J. **178**, 1–36 (2005)
22. Wojtkowiak, Z.: On ℓ-adic iterated integrals V, linear independence, properties of ℓ-adic polylogarithms, ℓ-adic sheave. In: Jacob, S. (ed.) The Arithmetic of Fundamental Groups, PIA 2010, pp. 339–374. Springer (2012)
23. Wojtkowiak, Z.: On ℓ-adic Galois L-functions. In: Algebraic Geometry and Number Theory, Summer School, Galatasaray University, Istanbul, 2014. Progress in Math. Birkhauser, vol. 321, pp. 161–209, Springer International Publishing (2017). arXiv:1403.2209v1 10 March (2014)

On a Family of Polynomials Related to $\zeta(2,1)=\zeta(3)$

Wadim Zudilin

Abstract We give a new proof of the identity $\zeta(\{2,1\}^l)=\zeta(\{3\}^l)$ of the multiple zeta values, where $l=1,2,\dots$, using generating functions of the underlying generalized polylogarithms. In the course of study we arrive at (hypergeometric) polynomials satisfying 3-term recurrence relations, whose properties we examine and compare with analogous ones of polynomials originated from an (ex-)conjectural identity of Borwein, Bradley and Broadhurst.

Keywords Multiple zeta values · Hypergeometric function · Hypergeometric polynomial · Generalized orthogonality

1 Introduction

The first thing one normally starts with, while learning about the multiple zeta values (MZVs)

$$\zeta(\boldsymbol{s})=\zeta(s_1,s_2,\dots,s_l)=\sum_{n_1>n_2>\dots>n_l\ge1}\frac{1}{n_1^{s_1}n_2^{s_2}\cdots n_l^{s_l}},$$

is Euler's identity $\zeta(2,1)=\zeta(3)$—see [3] for an account of proofs and generalizations of the remarkable equality. One such generalization reads

$$\zeta(\{2,1\}^l)=\zeta(\{3\}^l)\qquad\text{for}\quad l=1,2,\dots,\tag{1}$$

where the notation $\{\boldsymbol{s}\}^l$ denotes the multi-index with l consecutive repetitions of the same index $\boldsymbol{s}$. The only known proof of (1) available in the literature makes use of the duality relation of MZVs, originally conjectured in [6] and shortly after established in [12]. The latter relation is based on a simple iterated-integral representation of MZVs (see [12] but also [3, 4, 14] for details) but, unfortunately, it is not capable of

W. Zudilin (✉)
Department of Mathematics, IMAPP, Radboud University, PO Box 9010, 6500 Nijmegen, GL, Netherlands
e-mail: w.zudilin@math.ru.nl

J. I. Burgos Gil et al. (eds.), *Periods in Quantum Field Theory and Arithmetic*, Springer Proceedings in Mathematics & Statistics 314,
https://doi.org/10.1007/978-3-030-37031-2_22

establishing similar-looking identities

$$\zeta(\{3,1\}^l) = \frac{2\pi^{4l}}{(4l+2)!} \qquad \text{for} \quad l = 1, 2, \dots. \tag{2}$$

The equalities (2) were proven in [4] using a simple generating series argument.

The principal goal of this note is to give a proof of (1) via generating functions and to discuss, in this context, a related ex-conjecture of the alternating MZVs. An interesting outcome of this approach is a family of (hypergeometric) polynomials that satisfy a 3-term recurrence relation; a shape of the relation and (experimentally observed) structure of the zeroes of the polynomials suggest their bi-orthogonality origin [7, 8, 11].

2 Multiple Polylogarithms

For $l = 1, 2, \dots$, consider the generalized polylogarithms

$$\mathrm{Li}_{\{3\}^l}(z) = \sum_{n_1 > n_2 > \cdots > n_l \ge 1} \frac{z^{n_1}}{n_1^3 n_2^3 \cdots n_l^3},$$

$$\mathrm{Li}_{\{2,1\}^l}(z) = \sum_{n_1 > m_1 > n_2 > m_2 > \cdots > n_l > m_l \ge 1} \frac{z^{n_1}}{n_1^2 m_1 n_2^2 m_2 \cdots n_l^2 m_l},$$

$$\mathrm{Li}_{\{\overline{2},1\}^l}(z) = \sum_{n_1 > m_1 > n_2 > m_2 > \cdots > n_l > m_l \ge 1} \frac{z^{n_1}(-1)^{n_1+n_2+\cdots+n_l}}{n_1^2 m_1 n_2^2 m_2 \cdots n_l^2 m_l};$$

if $l = 0$ we set all these functions to be 1. Then at $z = 1$,

$$\zeta(\{3\}^l) = \mathrm{Li}_{\{3\}^l}(1) \quad \text{and} \quad \zeta(\{2,1\}^l) = \mathrm{Li}_{\{2,1\}^l}(1),$$

and we also get the related alternating MZVs

$$\zeta(\{\overline{2},1\}^l) = \mathrm{Li}_{\{\overline{2},1\}^l}(1)$$

from the specialization of the third polylogarithm.

Since

$$\begin{gathered}\left((1-z)\frac{\mathrm{d}}{\mathrm{d}z}\right)\left(z\frac{\mathrm{d}}{\mathrm{d}z}\right)^2 \mathrm{Li}_{\{3\}^l}(z)=\mathrm{Li}_{\{3\}^{l-1}}(z),\\ \left((1-z)\frac{\mathrm{d}}{\mathrm{d}z}\right)^2\left(z\frac{\mathrm{d}}{\mathrm{d}z}\right) \mathrm{Li}_{\{2,1\}^l}(z)=\mathrm{Li}_{\{2,1\}^{l-1}}(z),\\ \left((1+z)\frac{\mathrm{d}}{\mathrm{d}z}\right)^2\left(z\frac{\mathrm{d}}{\mathrm{d}z}\right) \mathrm{Li}_{\{\overline{2},1\}^l}(z)=\mathrm{Li}_{\{\overline{2},1\}^{l-1}}(-z)\end{gathered}$$

for $l=1,2,\dots$, the generating series

$$\begin{gathered}C(z;t)=\sum_{l=0}^{\infty}\mathrm{Li}_{\{3\}^l}(z)t^{3l},\\ B(z;t)=\sum_{l=0}^{\infty}\mathrm{Li}_{\{2,1\}^l}(z)t^{3l} \quad\text{and}\quad A(z;t)=\sum_{l=0}^{\infty}\mathrm{Li}_{\{\overline{2},1\}^l}(z)t^{3l}\end{gathered}$$

satisfy linear differential equations. Namely, we have

$$\begin{gathered}\left(\left((1-z)\frac{\mathrm{d}}{\mathrm{d}z}\right)\left(z\frac{\mathrm{d}}{\mathrm{d}z}\right)^2-t^3\right)C(z;t)=0,\\ \left(\left((1-z)\frac{\mathrm{d}}{\mathrm{d}z}\right)^2\left(z\frac{\mathrm{d}}{\mathrm{d}z}\right)-t^3\right)B(z;t)=0\end{gathered}$$

and

$$\left(\left((1-z)\frac{\mathrm{d}}{\mathrm{d}z}\right)^2\left(z\frac{\mathrm{d}}{\mathrm{d}z}\right)\left((1+z)\frac{\mathrm{d}}{\mathrm{d}z}\right)^2\left(z\frac{\mathrm{d}}{\mathrm{d}z}\right)-t^6\right)A(z;t)=0,$$

respectively. The identities (1) and identities

$$\frac{1}{8^l}\zeta(\{2,1\}^l)=\zeta(\{\overline{2},1\}^l) \qquad \text{for}\quad l=1,2,\dots,$$

conjectured in [4] and confirmed in [13] by means of a nice though sophisticated machinery of double shuffle relations and the 'distribution' relations (see also an outline in [2]), translate into

$$C(1;t)=B(1;t)=A(1;2t).$$

Note that

$$C(1;t)=\sum_{l=0}^{\infty}t^{3l}\sum_{n_1>n_2>\dots>n_l\ge 1}\frac{1}{n_1^3n_2^3\cdots n_l^3}=\prod_{j=1}^{\infty}\left(1+\frac{t^3}{j^3}\right). \tag{3}$$

At the same time, the differential equation for $C(z;t)=\sum_{n=0}^{\infty}C_n(t)z^n$ results in

$$-n^3C_n+(n+1)^3C_{n+1}=t^3C_n$$

implying

$$\frac{C_{n+1}}{C_n}=\frac{n^3+t^3}{(n+1)^3}=\frac{(n+t)(n+e^{2\pi i/3}t)(n+e^{4\pi i/3}t)}{(n+1)^3}$$

and leading to the hypergeometric form

$$C(z;t)={}_3F_2\biggl(\begin{matrix} t,\ \omega t,\ \omega^2 t \\ 1,\ 1\end{matrix}\biggm| z\biggr), \tag{4}$$

where $\omega=e^{2\pi i/3}$. We recall that

$${}_{m+1}F_m\biggl(\begin{matrix} a_0,\ a_1,\ \dots,\ a_m \\ b_1,\ \dots,\ b_m\end{matrix}\biggm| z\biggr)=\sum_{n=0}^{\infty}\frac{(a_0)_n(a_1)_n\cdots(a_m)_n}{n!\,(b_1)_n\cdots(b_m)_n}z^n,$$

where $(a)_n=\Gamma(a+n)/\Gamma(a)$ denotes the Pochhammer symbol (also known as the 'shifted factorial' because $(a)_n=a(a+1)\cdots(a+n-1)$ for $n=1,2,\dots$). It is not hard to see that the sequences $A_n(t)$ and $B_n(t)$ from $A(z;t)=\sum_{n=0}^{\infty}A_n(t)z^n$ and $B(z;t)=\sum_{n=0}^{\infty}B_n(t)z^n$ do not satisfy 2-term recurrence relations with polynomial coefficients. Thus, no hypergeometric representations of the type (4) are available for them.

3 Special Polynomials

The differential equation for $B(z;t)$ translates into the 3-term recurrence relation

$$n^3B_n-(n+1)^2(2n+1)B_{n+1}+(n+2)^2(n+1)B_{n+2}=t^3B_n \tag{5}$$

for the coefficients $B_n=B_n(t)$; the initial values are $B_0=1$ and $B_1=0$.

Lemma 1 *We have*

$$\begin{aligned} B_n(t)&=\frac{1}{n!}\sum_{k=0}^{n}\frac{(\omega t)_k(\omega^2 t)_k(t)_{n-k}(-t+k)_{n-k}}{k!\,(n-k)!} \\ &=\frac{(t)_n(-t)_n}{n!^2}\,{}_3F_2\biggl(\begin{matrix} -n,\ \omega t,\ \omega^2 t \\ -t,\ 1-n-t\end{matrix}\biggm| 1\biggr). \end{aligned} \tag{6}$$

Proof The recursion (5) for the sequence in (6) follows from application of the Gosper–Zeilberger algorithm of creative telescoping. The initial values for $n=0$ and 1 are straightforward. □

Remark 1 The hypergeometric form in (6) was originally prompted by [9, Theorem 3.4]: the change of variable $z\mapsto 1-z$ in the differential equation for $B(z;t)$ shows that the function $f(z)=B(1-z;t)$ satisfies the hypergeometric differential equation with upper parameters $-t,\ -\omega t,\ -\omega^2 t$ and lower parameters $0,\ 0$.

It is not transparent from the formula (6) (but immediate from the recursion (5)) that $B_n(t)\in t^3\mathbb{Q}[t^3]$ for $n=0,1,2,\dots$; the classical transformations of ${}_3F_2(1)$ and their representations as ${}_6F_5(-1)$ hypergeometric series (see [1]) do not shed a light on this belonging either.

Lemma 2 *We have*

$$B(1;t)=\prod_{j=1}^{\infty}\biggl(1+\frac{t^3}{j^3}\biggr). \tag{7}$$

Proof This follows from the derivation

$$\begin{aligned}
B(1;t)=\sum_{n=0}^{\infty}B_n(t)&=\sum_{n=0}^{\infty}\frac{1}{n!}\sum_{k=0}^{n}\frac{(\omega t)_k(\omega^2 t)_k(t)_{n-k}(-t+k)_{n-k}}{k!\,(n-k)!}\\
&=\sum_{k=0}^{\infty}\frac{(\omega t)_k(\omega^2 t)_k}{k!^2}\sum_{m=0}^{\infty}\frac{(t)_m(-t+k)_m}{m!\,(k+1)_m}\\
&=\sum_{k=0}^{\infty}\frac{(\omega t)_k(\omega^2 t)_k}{k!^2}\cdot{}_2F_1\biggl(\begin{matrix}t,\ -t+k\\ k+1\end{matrix}\biggm|1\biggr)\\
&=\frac{1}{\Gamma(1-t)\Gamma(1+t)}\sum_{k=0}^{\infty}\frac{(\omega t)_k(\omega^2 t)_k}{k!\,(1-t)_k}\\
&=\frac{1}{\Gamma(1-t)\Gamma(1+t)}\cdot{}_2F_1\biggl(\begin{matrix}\omega t,\ \omega^2 t\\ 1-t\end{matrix}\biggm|1\biggr)\\
&=\frac{1}{\Gamma(1-t)\Gamma(1+t)}\cdot\frac{\Gamma(1-t)}{\Gamma(1-(1+\omega)t)\Gamma(1-(1+\omega^2)t)}\\
&=\frac{1}{\Gamma(1+t)\Gamma(1+\omega t)\Gamma(1+\omega^2 t)}=\prod_{j=1}^{\infty}\biggl(1+\frac{t^3}{j^3}\biggr),
\end{aligned}$$

where we applied twice Gauss's summation [1, Sect. 1.3]

$${}_2F_1\biggl(\begin{matrix}a,\ b\\ c\end{matrix}\biggm|1\biggr)=\frac{\Gamma(c)\,\Gamma(c-a-b)}{\Gamma(c-a)\,\Gamma(c-b)}$$

valid when $\Re(c-a-b)>0$. □

Finally, we deduce from comparing (3) and (7),

Theorem 1 *The identity* $\zeta(\{3\}^l) = \zeta(\{2,1\}^l)$ *is valid for* $l = 1, 2, \dots$.

4 A General Family of Polynomials

It is not hard to extend Lemma 1 to the one-parameter family of polynomials

$$\begin{aligned} B_n^\alpha(t) &= \frac{1}{n!}\sum_{k=0}^n \frac{(\omega t)_k(\omega^2 t)_k(\alpha+t)_{n-k}(\alpha-t+k)_{n-k}}{k!\,(n-k)!} \\ &= \frac{1}{n!}\sum_{k=0}^n \frac{(\alpha+\omega t)_k(\alpha+\omega^2 t)_k(t)_{n-k}(\alpha-t+k)_{n-k}}{k!\,(n-k)!}. \end{aligned} \tag{8}$$

Lemma 3 *For each* $\alpha \in \mathbb{C}$*, the polynomials* (8) *satisfy the 3-term recurrence relation*

$$\begin{aligned} ((n+\alpha)^3 - t^3)B_n^\alpha - (n+1)(2n^2+3n(\alpha+1)+\alpha^2+3\alpha+1)B_{n+1}^\alpha \\ + (n+2)^2(n+1)B_{n+2}^\alpha = 0 \end{aligned}$$

and the initial conditions $B_0^\alpha = 1$, $B_1^\alpha = \alpha^2$. *In particular,* $B_n^\alpha(t) \in \mathbb{C}[t^3]$ *for* $n = 0, 1, 2, \dots$.

In addition, we have $B_n^\alpha \in t^3\mathbb{Q}[t^3]$ for $\alpha = 0, -1, \dots, -n+1$ (in other words, $B_n^\alpha(0) = 0$ for these values of α).

Lemma 4 $B_n^{1-n-\alpha}(t) = B_n^\alpha(t)$.

Proof This follows from the hypergeometric representation

$$B_n^\alpha(t) = \frac{(\alpha+t)_n(\alpha-t)_n}{n!^2}\, {}_3F_2\left(\begin{matrix} -n,\ \omega t,\ \omega^2 t \\ \alpha-t,\ 1-\alpha-n-t \end{matrix}\,\middle|\, 1\right).$$ □

Here is one more property of the polynomials that follows from Euler's transformation [1, Sect. 1.2].

Lemma 5 *We have*

$$\sum_{n=0}^\infty B_n^\alpha(t)z^n = (1-z)^{1-2\alpha}\sum_{n=0}^\infty B_n^{1-\alpha}(t)z^n.$$

Proof Indeed,

$$\begin{aligned}\sum_{n=0}^{\infty} B_n^{\alpha}(t)z^n &= \sum_{k=0}^{\infty} \frac{(\omega t)_k(\omega^2 t)_k}{k!^2}\, z^k \cdot {}_2F_1\left(\left.\begin{matrix}\alpha+t,\ \alpha-t+k\\ k+1\end{matrix}\right| z\right)\\ &= \sum_{k=0}^{\infty} \frac{(\omega t)_k(\omega^2 t)_k}{k!^2}\, z^k \cdot (1-z)^{1-2\alpha}\,{}_2F_1\left(\left.\begin{matrix}1-\alpha+t,\ 1-\alpha-t+k\\ k+1\end{matrix}\right| z\right)\\ &= (1-z)^{1-2\alpha}\sum_{n=0}^{\infty} B_n^{1-\alpha}(t)z^n.\end{aligned}$$

□

Alternative Proof of Lemma 2. It follows from Lemma 5 that

$$B_n^1(t) = \sum_{k=0}^{n} B_k(t),$$

hence $B(1;t) = \lim_{n\to\infty} B_n^1(t)$ and the latter limit is straightforward from (8). □

Note that, with the help of the standard transformations of ${}_3F_2(1)$ hypergeometric series, we can also write (8) as

$$B_n^{\alpha}(t) = \frac{(\alpha-\omega t)_n(\alpha-\omega^2 t)_n}{n!^2}\,{}_3F_2\left(\left.\begin{matrix}-n,\ \alpha+t,\ t\\ \alpha-\omega t,\ \alpha-\omega^2 t\end{matrix}\right| 1\right),$$

so that the generating functions of the continuous dual Hahn polynomials lead to the generating functions

$$\sum_{n=0}^{\infty} \frac{n!}{(\alpha-t)_n}\, B_n^{\alpha}(t)z^n = (1-z)^{-t}\,{}_2F_1\left(\left.\begin{matrix}\alpha+\omega t,\ \alpha+\omega^2 t\\ \alpha-t\end{matrix}\right| z\right)$$

and

$$\sum_{n=0}^{\infty} \frac{(\gamma)_n\, n!}{(\alpha-\omega t)_n(\alpha-\omega^2 t)_n}\, B_n^{\alpha}(t)z^n = (1-z)^{-\gamma}\,{}_3F_2\left(\left.\begin{matrix}\gamma,\ \alpha+t,\ t\\ \alpha-\omega t,\ \alpha-\omega^2 t\end{matrix}\right| \frac{z}{z-1}\right),$$

where γ is arbitrary.

Finally, numerical verification suggests that for real α the zeroes of B_n^{α} viewed as polynomials in $x=t^3$ lie on the real half-line $(-\infty, 0]$.

5 Polynomials Related to the Alternating MZV Identity

Writing

$$\begin{aligned}A(z;t)&=\sum_{n=0}^{\infty}A_n(t)z^n\\&=1+\tfrac{1}{4}t^3z^2-\tfrac{1}{6}t^3z^3+\big(\tfrac{1}{192}t^3+\tfrac{11}{96}\big)t^3z^4-\big(\tfrac{1}{240}t^3+\tfrac{1}{12}\big)t^3z^5\\&\quad+\big(\tfrac{1}{34560}t^6+\tfrac{23}{5760}t^3+\tfrac{137}{2160}\big)t^3z^6+O(z^7)\end{aligned}$$

and using the equation

$$\left((1+z)\frac{\mathrm{d}}{\mathrm{d}z}\right)^2\left(z\frac{\mathrm{d}}{\mathrm{d}z}\right)A(z;t)=t^3A(-z;t),$$

we deduce that

$$(n^3-T)A_n+(n+1)^2(2n+1)A_{n+1}+(n+2)^2(n+1)A_{n+2}=0, \tag{9}$$

where $T=(-1)^nt^3$. Producing two shifted copies of (9),

$$((n-1)^3+T)A_{n-1}+n^2(2n-1)A_n+(n+1)^2nA_{n+1}=0, \tag{10}$$

$$((n-2)^3-T)A_{n-2}+(n-1)^2(2n-3)A_{n-1}+n^2(n-1)A_n=0, \tag{11}$$

then multiplying recursion (9) by $n(n-1)^2(2n-3)$, recursion (10) by $-(n-1)^2(2n+1)(2n-3)$, recursion (11) by $(2n+1)((n-1)^3+T)$ and adding the three equations so obtained we arrive at

$$\begin{aligned}&(2n+1)((n-1)^3+T)((n-2)^3-T)A_{n-2}\\&\quad-n(n-1)(2n-1)(2n(n-1)(n^2-n-1)-3T)A_n\\&\quad+(n+2)^2(n+1)n(n-1)^2(2n-3)A_{n+2}=0.\end{aligned}$$

This final recursion restricted to the subsequence A_{2n}, namely

$$\begin{aligned}&(4n+5)((2n)^3-t^3)((2n+1)^3+t^3)A_{2n}\\&\quad-(4n+3)(2n+1)(2n+2)(2(2n+1)(2n+2)(4n^2+6n+1)-3t^3)A_{2n+2}\\&\quad+(4n+1)(2n+1)^2(2n+2)(2n+3)(2n+4)^2A_{2n+4}=0,\end{aligned} \tag{12}$$

and, similarly, to A_{2n+1} gives rise to two families of so-called Frobenius–Stickelberger–Thiele polynomials [10]. The latter connection, however, sheds no light on the asymptotics of $A_n(t)\in\mathbb{Q}[t^3]$. Unlike the case of $B(z;t)$ treated in Sect. 3 we cannot find closed form expressions for those subsequences. Here is the case most visually related to the recursion (12):

$$
\begin{aligned}
&(4n+5)\frac{(2n)^3-t^3}{t-2n}\,\frac{(2n+1)^3+t^3}{t+2n+1}\,A'_n \\
&\quad -(4n+3)(2n+1)(2n+2)((2n+1)(2n+2)+(8n^2+12n+1)t+3t^2)A'_{n+1} \\
&\quad +(4n+1)(2n+1)^2(2n+2)(2n+3)(2n+4)^2A'_{n+2}=0,
\end{aligned}
$$

where

$$
A'_n=\frac{1}{2^n\,(1/2)_n\,n!}\sum_{k=0}^{n}\frac{(\omega t/2)_k(\omega^2t/2)_k(t/2)_{n-k}(1/2)_{n-k}}{k!\,(n-k)!}(-1)^k.
$$

The latter polynomials are not from $\mathbb{Q}[t^3]$.

If we consider $\widetilde{A}_n(t)=\sum_{k=0}^{n}A_k(t)$ then (it is already known [5, 13] that)

$$
(n^3-(-1)^nt^3)\widetilde{A}_{n-1}+(2n+1)n\widetilde{A}_n-(n+1)^2n\widetilde{A}_{n+1}=0,\qquad n=1,2,\dots.
$$

As before, the standard elimination translates it into

$$
\begin{aligned}
&(2n+3)((n-1)^3+T)(n^3-T)\widetilde{A}_{n-2} \\
&\quad -(2n+1)n(n-1)(2(n^2+n+1)^2-6-T)\widetilde{A}_n \\
&\quad +(2n-1)(n+2)^2(n+1)^2n(n-1)\widetilde{A}_{n+2}=0,
\end{aligned}
$$

where $T=(-1)^nt^3$. One can easily verify that

$$
\widetilde{A}_n(t)=1+\dots+\frac{t^{3\lfloor n/2\rfloor}}{2^{\lfloor n/2\rfloor}\lfloor n/2\rfloor!\,n!}
$$

but we also lack an explicit representation for them.

We have checked numerically a fine behaviour (orthogonal-polynomial-like) of the zeroes of A_n and $\widetilde{A}_n$ viewed as polynomials in $x=t^3$ (both of degree $[n/2]$ in x). Namely, all the zeroes lie on the real half-line $(-\infty,0]$. This is in line with the property of the polynomials B_n^α (see the last paragraph in Sect. 4).

Acknowledgements The work originated from discussions during the research trimester on Multiple Zeta Values, Multiple Polylogarithms and Quantum Field Theory at ICMAT in Madrid (September–October 2014) and was completed during the author's visit in the Max Planck Institute for Mathematics in Bonn (March–April 2015); I thank the staff of the institutes for the wonderful working conditions experienced during these visits. I am grateful to Valent Galliano, Erik Koelink, Tom Koornwinder and Slava Spiridonov for valuable comments on earlier versions of the note. I am thankful as well to the two anonymous referees for the valuable feedback on the submitted version.

References

1. Bailey, W.N.: Generalized hypergeometric series. Cambridge Math. Tracts **32**, Cambridge Univ. Press, Cambridge (1935); 2nd reprinted edn.: Stechert-Hafner, New York–London (1964)
2. Borwein, J., Bailey, D.: Mathematics by Experiment. Plausible reasoning in the 21st century, 2nd edn. A K Peters, Ltd., Wellesley, MA (2008)
3. Borwein, J.M., Bradley, D.M.: Thirty-two Goldbach variations. Intern. J. Number Theory **2**(1), 65–103 (2006)
4. Borwein, J.M., Bradley, D.M., Broadhurst, D.J.: Evaluations of k-fold Euler/Zagier sums: a compendium of results for arbitrary k. Electron. J. Combin. **4**, # R5 (1997); Printed version: J. Combin. **4**(2), 31–49 (1997)
5. Borwein, J.M., Bradley, D.M., Broadhurst, D.J., Lisoněk, P.: Special values of multiple polylogarithms. Trans. Amer. Math. Soc. **353**(3), 907–941 (2001)
6. Hoffman, M.E.: Multiple harmonic series. Pacific J. Math. **152**(2), 275–290 (1992)
7. Iserles, A., Nørsett, S.P.: On the theory of biorthogonal polynomials. Trans. Amer. Math. Soc. **306**, 455–474 (1988)
8. Ismail, M.E.H., Masson, D.R.: Generalized orthogonality and continued fractions. J. Approx. Theory **83**, 1–40 (1995)
9. Mimachi, K.: Connection matrices associated with the generalized hypergeometric function ${}_3F_2$. Funkcial. Ekvac. **51**(1), 107–133 (2008)
10. Spiridonov, V.P., Tsujimoto, S., Zhedanov, A.: Integrable discrete time chains for the Frobenius-Stickelberger-Thiele polynomials. Comm. Math. Phys. **272**(1), 139–165 (2007)
11. Spiridonov, V., Zhedanov, A.: Spectral transformation chains and some new biorthogonal rational functions. Comm. Math. Phys. **210**(1), 49–83 (2000)
12. Zagier, D.: Values of zeta functions and their applications. In: Joseph, A., et al. (eds.), 1st European Congress of Mathematics (Paris, 1992), vol. II, Progr. Math. **120**, pp. 497–512, Birkhäuser, Boston (1994)
13. Zhao, J.: On a conjecture of Borwein, Bradley and Broadhurst. J. Reine Angew. Math. **639**, 223–233 (2010); Extended version: Double shuffle relations of Euler sums. Preprint http://arxiv.org/abs/math.AG/0412539, arXiv: 0705.2267 [math.NT] (2007)
14. Zudilin, W.: Algebraic relations for multiple zeta values. Uspekhi Mat. Nauk **58**(1), 3–32 (2003); English transl.: Russian Math. Surveys **58**:1, 1–29 (2003)

MIX
Papier aus verantwortungsvollen Quellen
Paper from responsible sources
FSC® C105338

If you have any concerns about our products,
you can contact us on
ProductSafety@springernature.com

In case Publisher is established outside the EU,
the EU authorized representative is:
Springer Nature Customer Service Center GmbH
Europaplatz 3, 69115 Heidelberg, Germany

Printed by Libri Plureos GmbH
in Hamburg, Germany